Biotics and the Gut–Brain Axis in Neurological Disorders

This book reviews the pivotal role of prebiotics, probiotics, and postbiotics in modulating the gut–brain axis, offering insights into their potential therapeutic applications and impact on neurological health. The initial chapters provide a foundational overview of prebiotics and introduce readers to the concept of the gut–brain axis. These chapters discuss bidirectional communication within the gut–brain axis and the impacts of gut dysbiosis on neurological health. Further chapters investigate the relationship between the gut–brain axis and autoimmune diseases, as well as the role of probiotics in major neurodegenerative diseases such as Alzheimer's and Parkinson's, amyotrophic lateral sclerosis (ALS), Huntington's disease, and conditions like multiple sclerosis (MS), epilepsy, and seizures. The book also examines the influence of prebiotics on mental health disorders such as anxiety and depression, and their impact on ischemic stroke via the gut–brain axis. It further covers the role of probiotics in reactive oxygen species (ROS) and inflammation-mediated neurodegenerative diseases and explores the correlation between gut microbiota, malnutrition, and neurological disorders. The book explores signaling pathways within the gut–brain axis and their relationships with neurological disorders, the application of prebiotics as functional foods, and the therapeutic potential of postbiotics and psychobiotics. This book is intended for scientists and researchers in the fields of microbiology, neuroscience, and gastroenterology. The book:

- Provides a thorough understanding of prebiotics, probiotics, and postbiotics, and their roles in the gut–brain axis;
- Explores bidirectional communication within the gut–brain axis and its impact on neurological health;
- Examines the relationship between the gut microbiota and various neurological disorders;
- Discusses the influence of prebiotics and probiotics on mental health disorders such as anxiety and depression;
- Investigates the connection between the gut–brain axis and autoimmune diseases;
- Covers advanced tools and bioinformatics approaches for studying gut microbiota and neurodegeneration.

Biotics and the Gut–Brain Axis in Neurological Disorders

Edited by
Neeraj Mishra, Sumel Ashique, and Mithun Bhowmick

CRC Press
Taylor & Francis Group
Boca Raton London New York

CRC Press is an imprint of the
Taylor & Francis Group, an **informa** business

Designed cover image: Shutterstock

First edition published 2026
by CRC Press
2385 NW Executive Center Drive, Suite 320, Boca Raton FL 33431

and by CRC Press
4 Park Square, Milton Park, Abingdon, Oxon, OX14 4RN

CRC Press is an imprint of Taylor & Francis Group, LLC

ISBN: 9781032882390 (hbk)
ISBN: 9781032882406 (pbk)
ISBN: 9781003536802 (ebk)

DOI: 10.1201/9781003536802

Typeset in Times
by Deanta Global Publishing Services, Chennai, India

Contents

Acknowledgment

We would like to express our sincere gratitude to Honorable Founder-President Dr. Ashok K. Chauhan, Ritnand Balved Education Foundation (RBEF), and Dr. Aseem Chauhan, Chancellor, Amity University, Madhya Pradesh. We are thankful to our Honorable Pro-Chancellor Lt. Gen. V. K. Sharma, AVSM (Retd.), Amity University Madhya Pradesh, Gwalior, India, Offg. Vice Chancellor Prof. (Dr.) R. S. Tomar, Amity University Madhya Pradesh, Gwalior, Pro-Vice Chancellor (Research) Prof. (Dr.) M. P. Kaushik, Amity University Madhya Pradesh, Gwalior, India, for their continuous encouragement. Finally, we would like to express our wholehearted appreciation to the faculty and staff members of Amity Institute of Pharmacy, Amity University, Madhya Pradesh, Gwalior, and the faculty members of Bengal College of Pharmaceutical Sciences and Research. Editors are also thankful to Bharat Technology, Uluberia for constant help via e-library during chapter preparation. We are also grateful to all the contributors for their hardwork in bringing out this wonderful book.

Preface

This book explores the role of gut microbiota in various neurological disorders and the potential application of prebiotics, probiotics, and postbiotics in modulating gut dysbiosis. Gut microbiota plays a significant role in the pathophysiology and physiology of the host, which ensures targeted therapeutic approaches by utilizing live healthy microorganisms and bacterial metabolites. The correlation between the gut–brain axis and the central nervous system (CNS) involves chemical, neural, immune, and endocrine pathways which ultimately influence gut dysbiosis. This dysbiosis affects the healthy bidirectional relationship (gut–brain axis) and can lead to neurodegeneration, neuroinflammation, and several other neurological disorders.

The book explores the intricate interplay between gut microbiota and the brain in the context of neurological disorders, focusing on the roles of prebiotics, probiotics, and postbiotics in modulating this gut–brain axis. The gut microbiota plays a crucial role in regulating brain function and behavior through the microbiota gut–brain axis, which encompasses bidirectional communication pathways between the gut and the brain.

The book begins with an overview of gut microbiota and its influence on neurological function, highlighting the role of dysbiosis in the pathogenesis of neurological disorders. It then delves into the mechanisms by which prebiotics, probiotics, and postbiotics exert their effects on the gut–brain axis, including modulation of neurotransmitter production, regulation of neuroinflammation, and modulation of gut barrier function. The book also discusses the potential therapeutic applications of prebiotics, probiotics, and postbiotics in managing these disorders, highlighting the latest research findings and clinical trials in the field.

This book will serve as a valuable resource for researchers, clinicians, and students interested in understanding the role of gut microbiota in neurological disorders and exploring novel therapeutic approaches targeting the gut–brain axis.

About the Editors

Neeraj Mishra has been serving as Professor and Head of the Department at the Amity Institute of Pharmacy, Amity University, Gwalior, India, since July 2019. With approximately 20 years of experience in teaching and research, Dr. Mishra has made significant contributions to the field. He has over 125 publications in both international and national journals, focusing on novel drug delivery systems, oral delivery of synbiotics, localized drug delivery, and targeted and controlled delivery of nanocarriers and microparticles for treating breast and colon cancer, as well as neurodegenerative disorders. He has authored more than 15 books and 25 book chapters and holds three international patents and three Indian patents. His achievements have been recognized with several awards, including the "Distinguished Professor Award" from DST-NSTMIS, SPAICS, Indore, in 2019, and the "Outstanding Scientist Award" at the 6th International Scientist Awards on Engineering, Science, and Medicine in 2020. He also received the "Distinguished Professor Award" from the Indian Pharmaceutical Association, MP State Branch, in 2021, and the "Best Academician Award" at the Indian Pharmacy Graduate Association, M.P. State International Conference in 2023.

Sumel Ashique is Assistant Professor in the Department of Pharmaceutical Technology, Bharat Technology, Howrah, India. With five years of teaching experience, he has made notable contributions to his field. He has published more than 100 papers in reputable international and national journals. His expertise includes drug delivery, nanotechnology, and targeted treatment strategies. He also serves as a reviewer for several prestigious international journals. He holds five granted patents from IP Australia, India, and Canada. He has contributed to 20 book chapters and has edited 18 books.

Mithun Bhowmick is Professor and Principal at the Bengal College of Pharmaceutical Sciences & Research in Durgapur, West Bengal, India. He has received numerous prestigious awards throughout his career. Dr. Bhowmick has actively participated in and presented research papers at various international and national conferences. He has published more than 50 research and review papers in esteemed national and international journals. His accomplishments include several patents from Australia, Germany, and India, as well as design patents and copyrights. Dr. Bhowmick is a reviewer for numerous pharmaceutical journals and is a lifetime member of several professional organizations, including the Association of Pharmaceutical Teachers of India (APTI), the Indian Pharmacy Graduate's Association (IPGA), and the Institute of Scholars (InSc.). He is also a registered fellow member of the Centre for Education Growth and Research (CEGR) in New Delhi.

1 Prebiotics

General Overview, Characteristics, and Application in Health Benefits

Ayush Madan[1,2], Rahul Kumar[3], Pranshul Sethi[4], and Debasis Mitra[3]

[1] Department of Biotechnology, School of Research & Technology, People's University, Bhanpur, Bhopal, Madhya Pradesh, India
[2] Institute of Tropical Aquaculture and Fisheries (AKUATROP), Universiti Malaysia Terengganu, Terengganu, Malaysia
[3] Department of Microbiology, Graphic Era (Deemed to be University), Clement town, Dehradun, India
[4] Department of Pharmacology, College of Pharmacy, Shri Venkateshwara University, Gajraula, Uttar Pradesh, India

1.1 INTRODUCTION TO PREBIOTICS

Gibson and Roberfroid conceived the concept of prebiotics in 1995 [1]. It was then defined as a nondigestible component of food which selectively accelerates the proliferation and activity of specific microbes in the colon and thereby enhances host well-being. This definition remained unchanged for more than 15 years. According to this definition, prebiotics are restricted to a few compounds from the carbohydrate class, such as lactulose, GOS, and short- and long-chain β-fructans (FOS and inulin). Dietary prebiotics are selectively fermented substances that cause positive alterations in the composition and activity of the intestinal microbiota, thereby improving host health. At the 6th ISAPP meeting (2008), experts proposed updated prebiotic criteria: resistance to gastric acid and mammalian enzymes, fermentability by gut microbiota, and selective stimulation of beneficial bacteria that confer host health benefits [2]. Although not all prebiotics are carbohydrate derived, one can distinguish them from others by applying the following two conditions: (1) fibers are carbohydrates that possess a degree of polymerization (DP) of three or more, and (2) they cannot be hydrolyzed by endogenous enzymes of the small intestine. One should keep in mind that solubility or fermentability of the fiber is not necessary [3, 4]. Moreover, new definitions of prebiotics are being published in scientific literature [4]. Nevertheless, the 2008 definition remains widely accepted in recent years. Despite the absence of a universally acknowledged definition, the key point in the original and extended definitions is that a prebiotic diet is associated with better human health. A further key aspect of the original definition is selectivity, which identifies a prebiotic's ability to stimulate an individual gut flora. However, this concept has recently been called into question [5]. This was demonstrated to improve the prebiotic effect by cross-feeding the ingestion of a source food by another species. The application of word selectivity in the definition of prebiotics has been questioned by this implication. A historical review of how the concept of prebiotics evolved can be found in an earlier article [5]. Many substances have been investigated to determine their potential as prebiotics. The most common forms of prebiotics include fructooligosaccharides (FOS), galactooligosaccharides (GOS), and *trans*-galactooligosaccharides (TOS). Short-chain fatty acids (SCFAs), including lactic acid, butyric acid, and propionic acid, are formed when gut flora ferment prebiotics. Prebiotics are food substances that are not digestible and help stimulate the growth of beneficial bacteria in the intestine. Common prebiotics include insulin, resistant starch, galactose, and fructose. SCFAs formed during fermentation enhance gut health. Prebiotics can improve digestive health, immune system, weight, and brain health. Food recommendations include prebiotic foods without recommended daily dietary intake. Prebiotic supplements are possible in cases of inability to obtain enough through diet, but consult a physician before using any supplement.

DOI: 10.1201/9781003536802-1

1.1.1 Distinction between Prebiotics, Probiotics, and Synbiotics

1.1.1.1 Probiotics

Probiotics are found in humans eating fermented foods early on [6]. Elie Metchnikoff theorized that applying yogurt-friendly bacteria to alter the gut microbiome would enhance well-being. This has spawned the concept of probiotics in medicine [7]. These microbes can reduce the pathogenic bacteria that secrete toxic compounds into the human gut and effectively manage the gut microbiota. Earlier, the term "probiotics" was coined based on the use of several microbes for the purpose of treating diseases [8]. In a paper titled "Anti-und Probiotika," Ferdinand Vergin came up with the term "probiotics" for the first time in the medical lexicon in 1954 [9]. Moreover, he emphasized the adverse effects that antibiotics have on intestinal microflora compared to advantageous bacteria known as "probiotics" that have a beneficial effect on human well-being [9]. Subsequently, for the first time, probiotic supplementation was developed as a drug against pig infections caused by *Escherichia coli*. The same results revealed that people consuming yoghurt fermented by *Lactobacillus* species had decreased blood cholesterol levels. Harrison et al. [10] reported that infants who ate *Lactobacillus acidophilus* had reduced serum cholesterol [10]. The World Health Organization (WHO) believes that probiotics are the second most significant immune defense system, if a few bacteria that widely recommend antibiotics fail. Many gut bacteria, such as *Lactobacillus*, *Bifidobacterium*, and *Enterococcus*, are beneficial for maintaining the stability of gut microbiota. Probiotics have been demonstrated to be beneficial to human health in several ways, such as antimicrobials, lactose intolerance, diarrhea disease treatment, ulcer treatment, immune stimulation, food preservation, and prevention of colon cancer [11].

1.1.1.2 Prebiotics

Prebiotics is a comparatively recent term defined as "nondigestible food ingredients that beneficially affect the host by selectively stimulating the growth and activity of one or a limited number of bacteria in the colon for improving the host's health" [12]. Taking a broader perspective, a panel of experts from the International Scientific Association for Probiotics and Prebiotics (ISAPP) redefined it as "a substrate selectively used by host microorganisms yielding a beneficial health effect." The definition captures substances that are non-carbohydrate in nature, uses locations other than the body, such as the gastrointestinal system, and much more than class in food [13]. Short-chain carbohydrates are typically used as substrates for the growth of probiotics in the upper gut, but are not absorbed. Cocoa flavanols are another class of compounds proposed as prebiotics. It has been demonstrated experimentally that flavanols can be used to stimulate LAB growth of lactic acid bacteria [14]. Prebiotics are present in numerous plant foods such as wheat, oats, chicory, onion, asparagus, garlic, and Jerusalem artichokes. These are bacteriogenic stimulants in the gut that induce metabolic activity in the colon.

1.1.1.3 Synbiotic

Synbiotics consist of a mixture of prebiotics and probiotics that benefit the health of humans and animals [15]. Probiotic microorganisms in synbiotic food products require prebiotics as unique substrates for growth [16]. A team of experts has recently updated the definition of synbiotics, based on the framework formulated by the International Scientific Association for Probiotics and Prebiotics (ISAPP). They differentiate between two types of synbiotics: complementary and synergistic. Probiotics and prebiotics interact to bring one or more health benefits, but co-dependent functions do not require them to exist in a complementary synbiotic. In a synergistic synbiotic, there is a substrate that co-administered microorganisms selectively [17]. These recommendations will be useful in the future to understand the relationship between prebiotics and probiotics and the development of symbiotic products with medicinal and health applications. Many studies have shown that the consumption of synbiotic-rich foods enhances host nutrition and well-being. Synbiotics with a concomitant reduction in coliforms were found to increase the concentration of probiotic bacteria, *Bifidobacterium* and *Lactobacillus*, in the fecal sample.

1.1.2 Importance of Prebiotics in Nutrition

Prebiotics are food ingredients that are not digested, but help stimulate the growth and activity of beneficial gut bacteria, which are important for maintaining and enhancing overall well-being (Figure 1.1). Dietary fibers in foods, such as garlic, onions, leeks, asparagus, bananas, and whole grains, are common sources of prebiotics. Prebiotics promote the growth of beneficial bacteria, including bifidobacteria and lactobacilli, which are important for maintaining a healthy balance of the gut microbiota. They also suppress harmful bacteria and decrease the risk of infections and digestive disorders. Prebiotics enhance digestion by stimulating normal bowel activity and preventing constipation. They also minimize symptoms of irritable bowel syndrome (IBS) and other gastrointestinal illnesses. Prebiotics also improve nutrient absorption and enhance the uptake of important minerals, such as magnesium and calcium. They assist the immune system by boosting the immune response and offering additional defense against infection. Prebiotics also have anti-inflammatory properties, assisting in the control of chronic inflammatory diseases and immune function. They assist with weight control through decreased hunger, improved feelings of fullness, and a healthy metabolism. They reduce the risk of type 2 diabetes by increasing insulin sensitivity and helping regulate blood sugar levels. Prebiotics also affect mental health by modulating the gut–brain axis, have a positive effect on mood and cognitive processes, and possibly lower the risk of anxiety and depression. Generally, prebiotics improve quality of

FIGURE 1.1 The importance of prebiotics in nutrition.

life by maintaining numerous aspects of wellness, such as digestion, immunity, metabolism, and mental health.

1.2 ROLE IN HUMAN DIET

The health and well-being of individuals are greatly supported through a healthy gut microbiome, which is what dietary supplements chiefly aim to deliver. Prominent gut colonizers lactobacilli can eliminate gut mucosal inflammation, metabolize lactose in individuals who are lactose-intolerant, relieve constipation, prevent traveler's diarrhea, and treat IBS [18]. In addition, bifidobacteria are often found in the gut of healthy humans and ferment certain oligosaccharides, making them reference markers of prebiotic potential. Commensal Clostridia, a member of the Firmicutes phylum, are important gut colonizers identified as critical controllers of immunological, physiological, and metabolic functions. Additionally, *Akkermansia muciniphila* and *Clostridium butyricum* have been reported to produce SCFAs and possess anti-inflammatory properties. Conversely, it has been shown that some strains of *Clostridium* and *Bacillus* enhance gut health by inhibiting the growth of pathogenic bacteria [19] (Figure 1.2).

1.3 TYPES OF PREBIOTICS

Carbohydrates that are not digested are generally considered prebiotics. However, prebiotics must meet the following requirements to be considered adequate:

a. They must be resistant to gastric acidity and mammalian enzymes.
b. They must be amenable to fermentation by gut bacteria.
c. They must increase the viability and activity of beneficial microbes.

Prebiotics exist in several forms, each with unique health advantages. For example, it has long been believed that the three main prebiotics are inulin, GOS, and FOS. Nonetheless, several other substances and dietary fibers

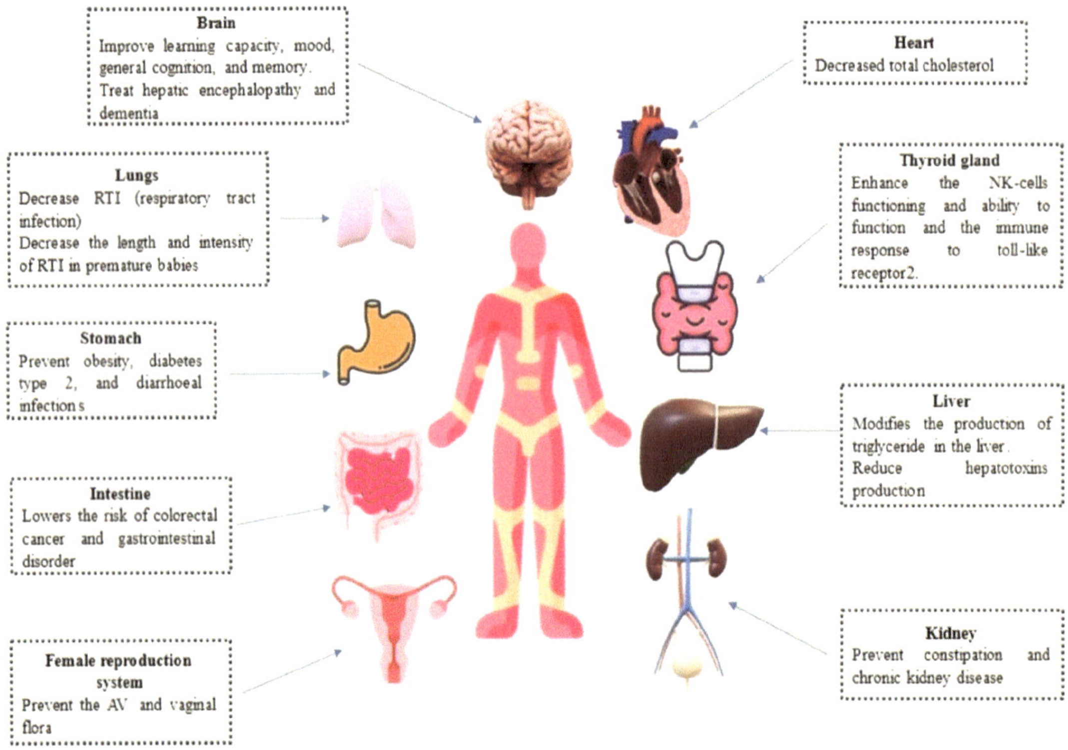

FIGURE 1.2 The comprehensive health benefits of prebiotics across different body systems.

have emerged as potential prebiotics with various health advantages. Literature has shown that the eight types of prebiotic nutritional fibers listed in Table 1.1 benefit the consumer's health. The most well-known prebiotics, together with information on their types, suppliers, production structure, and possible advantages are provided in Table 1.1.

1.4 BIOLOGICAL MECHANISMS

Intestinal microbiota, a sophisticated consortium of bacteria within the gastrointestinal tract, is essential for maintaining host well-being. It influences health through three primary biological processes: fermentation by the gut microbiota, production of SCFAs, and regulation of intestinal microbiome composition. Fermentation is an important metabolic process in which carbohydrates are broken down into alcohols or organic acids by microorganisms such as yeasts or microorganisms under anaerobic conditions. In the human gut, this process usually occurs in the large intestine with the help of different gut microbiota networks. These microorganisms are crucial for the breakdown of nondigestible carbohydrates, which slow down digestion in the upper gastrointestinal tract. The most prevalent and active bacteria in the fermentation process are anaerobic bacteria, of which Firmicutes, Bacteroidetes, Actinobacteria, and Proteobacteria are dominant. Each phylum contains multiple species with special metabolic functions and is responsible for the degradation of different substrates (Table 1.2) [31–38].

1.4.1 Mechanism of Fermentation

The gut fermentation process is initiated by the enzymatic hydrolysis of complex carbohydrates to break them down into simpler, easier-to-ferment sugars. This hydrolysis is facilitated by microbial enzymes, including glycoside hydrolases and polysaccharide lyases. Enzymes hydrolyze polysaccharides, such as cellulose, hemicellulose, and pectin, into monosaccharides and oligosaccharides. They are then subjected to microbial fermentation after hydrolysis via diverse metabolic pathways. The Embden–Meyerhof (glycolysis) and pentose phosphate pathways are the two main pathways for carbohydrate catabolism in gut bacteria. Fermentation of these sugars leads to the formation of a number of end products, including gases like carbon dioxide (CO_2), hydrogen (H_2), and methane (CH_4), and a variety of organic acids, most of which are SCFAs like acetate, propionate, and butyrate. Table 1.3 provides a summary of the fermentation end products and their effects on the host [31, 32, 39–44].

1.4.2 Production of Gases

Fermentation efficiency is affected by a number of factors, such as gut microbiota composition, microbial species diversity and abundance, and availability of important microbial players. The gut microbiota, a complex community of bacteria, archaea, viruses, and fungi, is responsible for the fermentation of prebiotics [45]. The relationship between these microbial communities and prebiotics regulates the

TABLE 1.1
Types of Prebiotics, Their Production Methods, Benefits, and Chemical Structures

Types of Prebiotics	Production Methods	Benefits	Chemical Structure	References
Pectin Oligosaccharides (POS)	Enzymatic breakdown mediated by pectinase	Anti-inflammatory impact	(1–4)-α-D-GalA (galacturonic acid)-(1,2)-α-L-Rha	[20]
Guar gum	Enzymatic breakdown employing cellulase	Boost glucose and cholesterol levels	β-D-mannopyranosyl (1–4) linked with α-D-galactopyranosyl (1–6) residues	[21]
Fructans	Hydrolysis of enzymes with Fructozyme L	Adjust the physiology of the gut to increase glucose levels and defend against infections	Fructose with β-(2→1) linkage	[22]
Xylooligosaccharides (XOS)	Plant xylans hydrolyzed by enzymes	Nondigestibility, beneficial impact on gut flora, and noncarcinogenic nature	Xylose units linked through β-(1→4) bonds	[23, 24]
Fructooligosaccharides (FOS)	Polymerization of monomeric fructose	Increased immunity, reduced triglycerides, improved mineral absorption, inhibition of harmful microbes, cancer prevention, and diabetic management	Glucose and fructose units linked by β(2→1) glycosidic linkages	[25–27]
Glactooligosaccharides	Lactose *trans*-galactosylation with β-galactosidase	Amplify the bifidogenic activity	Galactose and glucose bound by β-(1→3) and β-(1→4) linkages	[28]
Isomalto-oligosaccharides (IMO)	Conversion of liquid starch to glucose	Enhance the gut microbiota	Glucose bonds by α-(1→4) type	[29]
Soyabean oligosaccharides (SOS)	Not specified	Boost IgG levels, adjust body weight, and strengthen the immune system	Galactose α-(1-6) linked to glucose (Raffinose) galactose α-(1–6) linked to terminal galactose (Stachyose)	[30]

TABLE 1.2
Bacterial Phyla Involved in Gut Fermentation

Bacterial Phylum	Notable Species	Role in Fermentation	Reference
Firmicutes	*Clostridium*, *Lactobacillus*, *Eubacterium*	Production of butyrate, breakdown of complex carbohydrates	[31]
Bacteroidetes	*Bacteroides*, *Prevotella*	Production of propionate, degradation of polysaccharides	[32]
Actinobacteria	*Bifidobacterium*	Production of acetate and lactate, fermentation of oligosaccharides	[31]
Proteobacteria	*Escherichia*, *Enterobacter*	Various roles, including the production of gases and involvement in secondary fermentation processes	[31]
Verrucomicrobia	*Akkermansia*	Mucin degradation, production of acetate and propionate	[33]
Fusobacteria	*Fusobacterium*	Protein fermentation, production of butyrate	[34]
Cyanobacteria	*Cyanobacterium*	Nitrogen fixation, limited role in carbohydrate fermentation	[35]
Lentisphaerae	*Victivallis*	Fermentation of carbohydrates, production of propionate and acetate	[36]
Spirochaetes	*Treponema*	Fiber degradation, hydrogen production	[37]
Tenericutes	*Mycoplasma*	Various roles, some species involved in fermentation processes	[38]

TABLE 1.3
Fermentation End Products

Product	Type	Impact on Host	References
Gases	CO_2, H_2, CH_4	Influences gut motility and sensation	[31, 39]
Organic acids	Acetate	SCFA; energy source, lipid synthesis, cholesterol metabolism	[40]
	Propionate	SCFA; gluconeogenesis precursor, influences blood glucose	[41]
	Butyrate	SCFA; primary energy source for colonocytes, anti-inflammatory, enhances gut barrier integrity	[42, 43]
Alcohols	Ethanol	Energy source, impacts liver function	[44]
	Methanol	Toxic at high levels, can impact metabolic processes	[31]
Amines	Cadaverine, putrescine	Impact on cell growth and differentiation, potential toxicity at high levels	[31]
Phenols	Phenol, *p*-Cresol	Antimicrobial properties can be toxic to host cells	[31, 44]
Sulfides	Hydrogen sulfide	Impacts gut barrier function, potential toxin	[31]
Lactate	Lactic acid	Lowers pH, inhibits pathogen growth, source of energy for some bacteria	[31, 42]
Succinate	Succinic acid	Precursor for propionate production can be converted to SCFAs	[42, 44]
Formate	Formic acid	Involved in energy metabolism, precursor to methane in some bacteria	[31, 39]
Branched-chain fatty acids	Isobutyrate, isovalerate	Produced from amino acid fermentation, impacts gut health and odor	[32]

efficiency and results of fermentation [46]. Bifidobacteria are very effective in fermenting prebiotics such as inulin, fructooligosaccharides (FOS), and GOS, generating acetate and lactate, which decrease the gut pH and inhibit pathogen growth [47]. Lactobacilli, such as bifidobacteria, ferment prebiotics to generate lactic acid and other SCFAs, which are critical for ensuring a balanced gut environment and gut health. Firmicutes and Bacteroidetes prevail in the gut microbiota and perform unique functions in carbohydrate degradation [48]. Firmicutes degrade complex carbohydrates, whereas Bacteroidetes are generalist polysaccharide degraders [49]. Generally, the structure of the gut microbiota, microbial species diversity, and microbial species abundance are important determinants of fermentation efficiency and effectiveness (Table 1.4) [50].

1.5 IMPACT OF MICROBIOTA COMPOSITION ON SCFA PRODUCTION

The gut microbiota plays a primary role in determining food composition and fermentation, whereas dysbiosis contributes to harmful metabolites such as ammonia and hydrogen sulfide, which can initiate gastrointestinal disorders and drive diseases such as inflammatory bowel disease (IBD) [53]. Diet strongly influences the composition of gut microbiota and fermentation. A high intake of dietary fiber and prebiotics favors the growth of healthy bacteria, which leads to increased fermentation and SCFA production. However, an increase in fat and a decrease in fiber intake may favor the growth of harmful bacteria and inhibit fermentation. Substrates within the gut, including dietary fibers, prebiotics, and complex carbohydrates, have a significant impact on the fermentation process. Dietary fibers are resistant to the digestion of complex carbohydrates that arrive in the colon [54]. Prebiotics specifically stimulate the growth and activity of gut bacteria. Inulin, a fructan in chicory root, garlic, and onions, is selectively fermented by bifidobacteria and lactobacilli to produce acetate and butyrate, which reduces gut pH and suppresses pathogenic bacteria. Fructooligosaccharides are selectively fermented by bifidobacteria to enhance their growth and activity. GOS, present in human milk and certain legumes, stimulates bifidobacteria, producing acetate and lactate, enhancing gut health, and inhibiting pathogen colonization [55, 56] (Table 1.5).

1.6 PRODUCTION OF SHORT-CHAIN FATTY ACIDS

Short-chain fatty acids are important metabolites generated by the anaerobic fermentation of dietary fibers in the gut microbiota. The major SCFAs are acetate, propionate, and butyrate, which play pivotal roles in numerous physiological activities and exert profound effects on host health. Acetate is the most prevalent SCFA in the colon, and is generated by gut bacteria via different metabolic pathways. Propionate is synthesized via the succinate, acrylate, and propanediol pathways, primarily via the Bacteroidetes. It is transported to the liver, where it serves as a precursor for gluconeogenesis, affecting insulin sensitivity and blood glucose levels. Butyrate is primarily synthesized via the butyryl-CoA pathway and fermenting carbohydrates to pyruvate. SCFAs play a crucial role in promoting intestinal health, including gut barrier integrity, anti-inflammatory

TABLE 1.4
Various Species of Gut Microbiota and Their Fermentation Efficiency

Species/Group	Role in Fermentation	Primary SCFAs Produced	Health Benefits	References
Bifidobacteria	Efficiently ferment prebiotics like inulin, FOS, and GOS; produce acetate and lactate	Acetate, lactate	Lower gut pH, inhibit pathogens, support gut health	[46, 48]
Lactobacilli	Ferment prebiotics to produce lactic acid and other SCFAs; maintain gut health	Lactic acid, other SCFAs	Maintain balanced gut environment, promote gut health	[49]
Firmicutes	Associated with breakdown of complex carbohydrates; produce various SCFAs	Various SCFAs	Support carbohydrate metabolism, maintain gut health	[50]
Bacteroidetes	Versatile in degrading a wide range of polysaccharides; produce SCFAs and other metabolites	Various SCFAs	Support carbohydrate metabolism, maintain gut health	[50]
Proteobacteria	Less efficient in fermenting prebiotics; includes some opportunistic pathogens	Various SCFAs, some less beneficial	Can be pathogenic; less focus on fermentation benefits	[47]
Actinobacteria	Includes beneficial genera like *Bifidobacterium*; involved in carbohydrate fermentation	Acetate, lactate	Similar benefits to bifidobacteria; support gut health	[48]
Methanobrevibacter	Archaea that consume hydrogen and produce methane; influence overall fermentation efficiency	Methane (CH_4), not SCFAs	Reduce hydrogen levels, impacting other fermentation processes	[52]
Faecalibacterium prausnitzii	Major butyrate producer; crucial for anti-inflammatory effects in the gut	Butyrate	Anti-inflammatory, maintain gut barrier integrity	[52]

action, immune modulation, and systemic effects. Insight into the functions of SCFAs has major clinical implications, such as their potential value in patients with inflammatory bowel disease, metabolic disorders, and colorectal cancer (Table 1.6).

1.7 CLINICAL IMPLICATIONS OF MODULATING THE GUT MICROBIOME

Modulating the composition of the intestinal microbiome can help prevent and treat various diseases, including IBD, obesity, and metabolic syndrome. For instance, increasing the number of SCFA-producing bacteria through dietary modifications or by taking probiotics and prebiotics can reduce inflammation and improve metabolic health [62]. The gut–brain axis, a two-way verbal communication system between the gut and the brain, is promoted by the gut microbiome. Changes in gut microbiota composition were associated with intellectual fitness issues, despair, and tension. Probiotics and prebiotics can regulate the gut–mind axis and most likely enhance mental fitness [63]. Balanced and healthy intestinal microbiomes are vital for appropriate immune features. Beneficial microorganisms assist in the regulation of immune responses, enhance anti-inflammatory cytokine manufacturing, maintain the integrity of the intestinal barrier, and protect against pathogens [64].

1.8 HEALTH BENEFITS OF PREBIOTICS

Prebiotics are dietary fibers found in plants that stimulate the growth of healthy microorganisms in the gut with many health benefits, especially for the digestive system. Prebiotics such as inulin and FOS increase stool bulk and frequency by attracting water into the colon, making stool softer, and avoiding constipation. They also stimulate peristaltic activities and help with regular bowel movement. Fermentation of prebiotics by colonic bacteria releases SCFAs, which reduce the pH of the colon and increase the absorption of key minerals such as calcium and magnesium. The regular consumption of prebiotics is associated with a reduced risk of diverticulitis, colorectal cancer, and other gut conditions. SCFAs released during fermentation are anti-inflammatory, maintain the integrity of the intestinal lining, and inhibit the growth of pathogenic microorganisms. Prebiotics also support the gut microbiota by serving as a source of food for health-promoting bacteria, such as bifidobacteria and lactobacilli, which ferment prebiotics to yield SCFAs, promoting a healthy environment for their growth. This selective stimulation maintains a balanced intestinal microbiota that is essential for overall well-being. A healthy intestinal microbiota is responsible for preventing dysbiosis, an imbalance in gut bacteria that has been linked to several other diseases such as obesity, diabetes, and inflammatory bowel disease. Prebiotics tend to increase

TABLE 1.5
Various Types of Substrates and Their Fermentation Efficiency

Substrate Type	Examples	Role in Fermentation	Primary SCFAs Produced	Health Benefits	References
Soluble fibers	Pectins	Efficiently fermented by Bacteroidetes, producing SCFAs like acetate and propionate	Acetate, propionate	Lowers blood cholesterol, maintains gut health	[44]
Soluble fibers	Beta-glucans	Fermented by Firmicutes, leading to production of butyrate	Butyrate	Anti-inflammatory, supports colonocyte health	[57]
Insoluble fibers	Cellulose	Less efficiently fermented; adds bulk to stool	Minimal SCFAs	Improves bowel movements, adds bulk to stool	[57]
Prebiotics	Inulin	Selectively fermented by bifidobacteria and lactobacilli, producing acetate and butyrate	Acetate, butyrate	Lowers gut pH, inhibits pathogens, supports gut health	[58]
Prebiotics	Fructooligosaccharides	Selectively fermented by bifidobacteria, producing SCFAs, especially acetate and butyrate	Acetate, butyrate	Promotes growth and activity of bifidobacteria, enhances gut health	[51]
Prebiotics	Galactooligosaccharides	Effective in stimulating bifidobacteria, producing acetate and lactate	Acetate, lactate	Improves gut health, reduces pathogen colonization	[49]
Resistant starches	RS1	Fermented by various bacteria, producing butyrate	Butyrate	Supports colon health, has anti-inflammatory effects	[59]
Resistant starches	RS2	Fermented by various bacteria, producing butyrate	Butyrate	Supports colon health, has anti-inflammatory effects	[59]
Resistant starches	RS3	Fermented by various bacteria, producing butyrate	Butyrate	Supports colon health, has anti-inflammatory effects	[59]
Resistant starches	RS4	Fermented by various bacteria, producing butyrate	Butyrate	Supports colon health, has anti-inflammatory effects	[59]
Proteins	Various proteins	Fermentation produces SCFAs, BCFAs, ammonia, and phenolic compounds	Various SCFAs, BCFAs	May produce harmful metabolites in excess, affects gut health	[52]
Polyphenols	Flavonoids, tannins	Partially fermented, producing beneficial metabolites with antioxidant and anti-inflammatory properties	Various beneficial metabolites	Provides antioxidant and anti-inflammatory benefits	[60]
Sugars	Sucrose, glucose	Quickly fermented by a wide range of gut bacteria, producing a mix of SCFAs	Acetate, propionate, butyrate	Quick energy source, supports gut health	[57]
Sugar alcohols	Sorbitol, mannitol	Fermented by gut bacteria, producing SCFAs and gases; can cause bloating in excess	Acetate, propionate, butyrate	May cause bloating in excess, but supports gut health in moderation	[58]
Oligosaccharides	Raffinose	Fermented by bifidobacteria and lactobacilli, producing SCFAs	Acetate, butyrate	Supports gut health, reduces pathogen colonization	[51]
Oligosaccharides	Stachyose	Fermented by bifidobacteria and lactobacilli, producing SCFAs	Acetate, butyrate	Supports gut health, reduces pathogen colonization	[49]
Oligosaccharides	Verbascose	Fermented by bifidobacteria and lactobacilli, producing SCFAs	Acetate, butyrate	Supports gut health, reduces pathogen colonization	[51]
Polysaccharides	Starch	Fermented by various gut bacteria, producing SCFAs	Acetate, propionate, butyrate	Supports energy metabolism, maintains gut health	[59]
Polysaccharides	Glycogen	Fermented by gut bacteria, producing SCFAs	Acetate, propionate, butyrate	Supports energy metabolism, maintains gut health	[59]
Polysaccharides	Pectin	Efficiently fermented by Bacteroidetes, producing SCFAs	Acetate, propionate	Lowers blood cholesterol, maintains gut health	[44]
Polysaccharides	Hemicellulose	Partially fermented by gut bacteria, producing SCFAs	Acetate, propionate	Supports gut health, aids in digestion	[57]
Non-carbohydrate	Lignin	Not fermented; adds bulk to stool and aids in bowel movements	None	Adds bulk to stool, aids in bowel movements	[57]

TABLE 1.6
Different Pathways to Produce SCFAs, Their Functions, and Clinical Implications

SCFA	Production Pathways	Functions	Clinical Implications	References
Acetate	Acetyl-CoA pathway: Carbohydrates → Pyruvate → Acetyl-CoA → Acetate Wood–Ljungdahl pathway: $CO_2 + H_2$ → Acetate	Primary energy source for peripheral tissues, involved in cholesterol metabolism, enhances gut barrier function, influences appetite regulation, supports immune function	Reduced acetate levels linked to metabolic syndrome and obesity; essential for maintaining energy balance, potential role in regulating appetite and reducing fat accumulation	[42]
Propionate	Succinate pathway: Carbohydrates → Pyruvate → Oxaloacetate → Succinate → Propionate Acrylate pathway: Lactate → Acrylate → Propionate Propanediol pathway: Deoxy sugars → Propanediol → Propionate	Inhibits cholesterol synthesis, gluconeogenesis in the liver, modulates immune function, anti-inflammatory properties, potential to improve insulin sensitivity	Propionate can reduce blood cholesterol levels and improve glucose metabolism; potential therapeutic role in diabetes and hyperlipidemia, may help in managing inflammatory conditions	[61]
Butyrate	Butyryl-CoA pathway: Carbohydrates → Pyruvate → Acetyl-CoA → Butyryl-CoA → Butyrate	Primary energy source for colonocytes, anti-inflammatory effects, enhances gut barrier integrity, promotes mucus production in the gut, inhibits growth of pathogenic bacteria	Butyrate is crucial for maintaining colon health, anti-inflammatory properties beneficial in inflammatory bowel disease and colorectal cancer prevention, potential to enhance gut barrier function and prevent leaky gut syndrome	[52]

beneficial microorganism numbers while decreasing harmful numbers, thereby maintaining a balanced microbiota structure.

1.9 PREVENTION AND MANAGEMENT OF GASTROINTESTINAL DISORDERS

Irritable bowel syndrome is a widespread gastrointestinal disorder that presents as abdominal discomfort, bloating, and abnormal bowel habits. Symptoms are relieved by prebiotics, such as inulin and FOS, which enhance the composition of intestinal microbiota and reduce inflammation. Prebiotics may also relieve constipation by enhancing stool bulk and water content, softening stool, and easing bypassing. Prebiotics can also be used to treat diseases such as Crohn's disease and ulcerative colitis by creating SCFAs generated by prebiotic fermentation, which possess anti-inflammatory properties and can control IBD symptoms and enhance gut fitness. Prebiotics play a significant role in gut health and possess potential benefits in the prevention of numerous diseases. They selectively stimulate the activity and growth of beneficial bacteria, such as lactobacilli and bifidobacteria, which are crucial for normal gut physiology and may be useful in the treatment of IBS. Prebiotics have been suggested as possible therapeutic options for the prevention and treatment of IBS because they normalize disturbed gut microbiota, enhance intestinal barrier function, and reduce inflammation. They also promote the development of regulatory T cells (Tregs), which aid immunological tolerance and suppress intestinal inflammation. Prebiotics play a critical role in medicine because they have many effects on human health, such as maintaining and enhancing gut microbiota, aiding in digestive challenges, immunity strengthening, metabolic regulation, and mental and cardiac wellness.

1.10 EMERGING TRENDS IN PREBIOTIC RESEARCH

Prebiotics, which are nondigestible food components, have emerged as an important component of human health and disease prevention. Recent studies have shown novel applications of prebiotics and have shed light on their influence on the microbiome and health. Infants are of specific significance for the development of healthy gut microbiota and immune system, and infant formulas have been explored for the possibility of prebiotic supplementation. Future-generation prebiotics, such as inulin, FOS, and GOS, are being studied and manufactured, providing enhanced gut function, increased immunity, and lower inflammation. Microbiome profiling and personalized nutrition are new avenues for the promising growth of prebiotics. New technologies for defining the microbiome involve metagenomics and metabolomics, which can provide more information about an individual's gut microbiota. This study shows that by determining the individual's specific identity and functions of the gut microbiota, a prebiotic strategy can be created that has the greatest influence on them. Synbiotics, a combination of probiotics and prebiotics, promote the colonization and survival of beneficial bacteria in the gastrointestinal tract and have beneficial effects on health. Synbiotics are formulated to enhance the chances of survival and gut colonization of translocated probiotic bacteria and to promote the growth of native-friendly bacteria. New prebiotics such as seaweed-derived prebiotics from marine algae, arabinoxylooligosaccharides (AXOS) from plant sources, isomaltooligosaccharides (IMOs) from starch, and xylooligosaccharides (XOS) from plant cell walls represent a great leap in nutritional science for improving human well-being through food interventions.

1.11 CONCLUSION

In summary, probiotics represent a breakthrough in recognizing how dietary constituents can contribute to human well-being by selectively stimulating beneficial intestinal bacteria. Owing to their distinctive features, they can survive in the colon, where they are used as substrates for fermentation by resident microbiota, which results in a balanced microbial population. This, in turn, brings about numerous health advantages such as enhanced digestion, stronger immune response, and even protection against diseases such as obesity and cardiovascular conditions. Prebiotics can prove to be a therapeutic and preventive method for increasing gut health and overall physiological resilience, as research on their mechanisms and effects continues. Their relevance in present-day nutritional guidelines for long-term health promotion is placed in focus, emphasizing their capability to promote balanced gut flora supplemented with probiotics.

REFERENCES

1. Gibson, G. R., & Roberfroid, M. B. (1995). Dietary modulation of the human colonic microbiota: Introducing the concept of prebiotics. *The Journal of Nutrition*, *125*(6), 1401–1412.
2. Gibson, G. R., Scott, K. P., Rastall, R. A., Tuohy, K. M., Hotchkiss, A., Dubert-Ferrandon, A., ... & Buddington, R. (2010). Dietary prebiotics: Current status and new definition. *Food Science and Technology Bulletin: Functional Foods*, *7*(1), 1–19.
3. Slavin, J. (2013). Fiber and prebiotics: Mechanisms and health benefits. *Nutrients*, *5*(4), 1417–1435.
4. Roberfroid, M. B. (2008). Prebiotics: Concept, definition, criteria, methodologies, and products. In G. R. Gibson & M. B. Roberfroid (Eds.), Handbook of prebiotics (pp. 57–86). CRC Press.
5. Bindels, L. B., Delzenne, N. M., Cani, P. D., & Walter, J. (2015). Towards a more comprehensive concept for prebiotics. *Nature Reviews Gastroenterology & Hepatology*, *12*(5), 303–310.
6. Gogineni, V. K., Morrow, L. E., Gregory, P. J., & Malesker, M. A. (2013). Probiotics: History and evolution. *Journal of Ancient Diseases & Preventive Remedies*, *1*(2), 1–7.
7. Mackowiak, P. A. (2013). Recycling Metchnikoff: Probiotics, the intestinal microbiome and the quest for long life. *Frontiers in Public Health*, *1*, 56101.
8. Yk, L., Nomoto, K., Salminen, S., & Gorbach, S. L. (1999). *Handbook of probiotics*. John Wiley & Sons.
9. Sarowska, J., Choroszy-Król, I., Regulska-Ilow, B., Frej-Madrzak, M., & Jama-Kmiecik, A. (2013). The therapeutic effect of probiotic bacteria on gastrointestinal diseases. Advances in Clinical and Experimental Medicine, 22(5), 759–766.
10. Harrison, V. C., Peat, G., & Heese, H. (1974). Fetal growth in relation to histamine concentration in urine. *BJOG: An International Journal of Obstetrics & Gynaecology*, *81*(9), 686–690.
11. Masood, M. I., Qadir, M. I., Shirazi, J. H., & Khan, I. U. (2011). Beneficial effects of lactic acid bacteria on human beings. *Critical Reviews in Microbiology*, *37*(1), 91–98.
12. Gibson, G. R., & Roberfroid, M. B. (1995). Dietary modulation of the human colonic microbiota: Introducing the concept of prebiotics. *The Journal of Nutrition*, *125*(6), 1401–1412.
13. Gibson, G. R., Hutkins, R., Sanders, M. E., Prescott, S. L., Reimer, R. A., Salminen, S. J., ... & Reid, G. (2017). Expert consensus document: The International Scientific Association for Probiotics and Prebiotics (ISAPP) consensus statement on the definition and scope of prebiotics. *Nature Reviews Gastroenterology & Hepatology*, *14*(8), 491–502.
14. Tzounis, X., Rodriguez-Mateos, A., Vulevic, J., Gibson, G. R., Kwik-Uribe, C., & Spencer, J. P. (2011). Prebiotic evaluation of cocoa-derived flavanols in healthy humans by using a randomized, controlled, double-blind, crossover intervention study. *The American Journal of Clinical Nutrition*, *93*(1), 62–72.
15. Mackowiak, P. A. (2013). Recycling Metchnikoff: Probiotics, the intestinal microbiome and the quest for long life. *Frontiers in Public Health*, *1*, 56101.
16. Perrin, S., Warchol, M., Grill, J. P., & Schneider, F. (2001). Fermentations of fructo-oligosaccharides and their components by Bifidobacterium infantis ATCC 15697 on batch culture in semi-synthetic medium. *Journal of Applied Microbiology*, *90*(6), 859–865.
17. Swanson, K. S., Gibson, G. R., Hutkins, R., Reimer, R. A., Reid, G., Verbeke, K., ... & Sanders, M. E. (2020). The International Scientific Association for Probiotics and Prebiotics (ISAPP) consensus statement on the definition and scope of synbiotics. *Nature Reviews Gastroenterology & Hepatology*, *17*(11), 687–701.
18. Oak, S. J., & Jha, R. (2019). The effects of probiotics in lactose intolerance: A systematic review. *Critical Reviews in Food Science and Nutrition*, *59*(11), 1675–1683.
19. Lopetuso, L. R., Scaldaferri, F., Petito, V., & Gasbarrini, A. (2013). Commensal Clostridia: Leading players in the maintenance of gut homeostasis. *Gut Pathogens*, *5*, 1–8.

20. Chung, W. S. F., Meijerink, M., Zeuner, B., Holck, J., Louis, P., Meyer, A. S., ... Duncan, S. H. (2017). Prebiotic potential of pectin and pectic oligosaccharides to promote anti-inflammatory commensal bacteria in the human colon. *FEMS Microbiology Ecology*, *93*(11), fix127.
21. Mudgil, D., Barak, S., Patel, A., & Shah, N. (2018). Partially hydrolyzed guar gum as a potential prebiotic source. *International journal of biological macromolecules*, *112*, 207–210.
22. Franco-Robles, E., & López, M. G. (2015). Implication of fructans in health: Immunomodulatory and antioxidant mechanisms. *The Scientific World Journal*, *2015*(1), 289267.
23. Mohanty, D., Misra, S., Mohapatra, S., & Sahu, P. S. (2018). Prebiotics and synbiotics: Recent concepts in nutrition. *Food Bioscience*, *26*, 152–160.
24. Lecerf, J. M., Dépeint, F., Clerc, E., Dugenet, Y., Niamba, C. N., Rhazi, L., ... Pouillart, P. R. (2012). Xylo-oligosaccharide (XOS) in combination with inulin modulates both the intestinal environment and immune status in healthy subjects, while XOS alone only shows prebiotic properties. *British Journal of Nutrition*, *108*(10), 1847–1858.
25. Bali, V., Panesar, P. S., Bera, M. B., & Panesar, R. (2015). Fructo-oligosaccharides: Production, purification and potential applications. *Critical Reviews in Food Science and Nutrition*, *55*(11), 1475–1490.
26. Khanvilkar, S. S., & Arya, S. S. (2015). Fructooligosaccharides: Applications and health benefits: A review. *Agro Food Industry Hi Tech*, *26*(6), 8–12.
27. Aqil, F., Munagala, R., Agrawal, A. K., & Gupta, R. (2019). Anticancer phytocompounds: Experimental and clinical updates. In I. F. F. Shah, H. M. A. Khan, & M. A. Akhtar (Eds.), New look to phytomedicine (pp. 237–272). Academic Press.
28. Alander, M., Mättö, J., Kneifel, W., Johansson, M., Kögler, B., Crittenden, R., ... Saarela, M. (2001). Effect of galacto-oligosaccharide supplementation on human faecal microflora and on survival and persistence of Bifidobacterium lactis Bb-12 in the gastrointestinal tract. *International Dairy Journal*, *11*(10), 817–825.
29. Sorndech, W., Nakorn, K. N., Tongta, S., & Blennow, A. (2018). Isomalto-oligosaccharides: Recent insights in production technology and their use for food and medical applications. *LWT – Food Science and Technology*, *95*, 135–142.
30. Patel, S., & Goyal, A. (2011). Functional oligosaccharides: Production, properties and applications. *World Journal of Microbiology and Biotechnology*, *27*, 1119–1128.
31. Flint, H. J., Scott, K. P., Duncan, S. H., Louis, P., & Forano, E. (2012). Microbial degradation of complex carbohydrates in the gut. Gut Microbes, 3(4), 289–306.
32. Riviere, A., Selak, M., Lantin, D., Leroy, F., & De Vuyst, L. (2016). Bifidobacteria and butyrate-producing colon bacteria: Importance and strategies for their stimulation in the human gut. Frontiers in Microbiology, 7, 979.
33. Derrien, M., Vaughan, E. E., Plugge, C. M., & De Vos, W. M. (2004). Akkermansia muciniphila gen. nov., sp. nov., a human intestinal mucin-degrading bacterium. International Journal of Systematic and Evolutionary Microbiology, 54(5), 1469–1476.
34. Castellarin, M., Warren, R. L., Freeman, J. D., Dreolini, L., Krzywinski, M., Strauss, J., ... Holt, R. A. (2012). Fusobacterium nucleatum infection is prevalent in human colorectal carcinoma. Genome Research, 22(2), 299–306.
35. Prasanna, A. N., Gerna, G., Ng, K. P., & Venkatesh, M. (2020). Insights into the phylogeny and coding potential of bacterial and fungal dark matter. Fungal Diversity, 104, 313–333.
36. Cho, J. C., & Giovannoni, S. J. (2006). Lentisphaera araneosa gen. nov., sp. nov., a transparent exopolymer-producing marine bacterium, and the description of a novel bacterial phylum, Lentisphaerae. Environmental Microbiology, 8(3), 311–321.
37. Paster, B. J., et al. (1996). Phylogenetic analysis of the spirochetes. Journal of Bacteriology, 178(10), 2889–2898.
38. Razin, S., Yogev, D., & Naot, Y. (1998). Molecular biology and pathogenicity of mycoplasmas. Microbiology and Molecular Biology Reviews, 62(4), 1094–1156.
39. Samuel, B. S., & Gordon, J. I. (2006). A humanized gnotobiotic mouse model of host-archaeal-bacterial mutualism. Proceedings of the National Academy of Sciences, 103(26), 10011–10016.
40. Perry, R. J., et al. (2016). Acetate mediates a microbiome–brain–β-cell axis to promote metabolic syndrome. Nature, 534(7606), 213–217.
41. Weitkunat, K., et al. (2017). Importance of propionate for the repression of hepatic lipogenesis and improvement of insulin sensitivity in high-fat diet-induced obesity. Molecular Nutrition & Food Research, 60(12), 2611–2621.
42. Louis, P., & Flint, H. J. (2017). Formation of propionate and butyrate by the human colonic microbiota. Environmental Microbiology, 19(1), 29–41.
43. Peng, L., et al. (2009). Butyrate enhances the intestinal barrier by facilitating tight junction assembly via activation of AMP-activated protein kinase in Caco-2 cell monolayers. The Journal of Nutrition, 139(9), 1619–1625.
44. Roberfroid, M. (2007). Prebiotics: The concept revisited. Journal of Nutrition, 137(3 Suppl 2), 830S–837S.
45. Chang, P. V., et al. (2014). The microbial metabolite butyrate regulates intestinal macrophage function via histone deacetylase inhibition. Proceedings of the National Academy of Sciences, 111(6), 2247–2252.
46. Macfarlane, G. T., Steed, H., & Macfarlane, S. (2008). Bacterial metabolism and health-related effects of galactooligosaccharides and other prebiotics. *Journal of Applied Microbiology*, *104*(2), 305–344.
47. Nagpal, R., & Yamashiro, Y. (2018). Gut microbiota composition in healthy Japanese infants and young adults born by different delivery modes and fed different diets. *Journal of Bioscience and Bioengineering*, *126*(6), 715–725.
48. Gibson, G. R., & Roberfroid, M. B. (1995). Dietary modulation of the human colonic microbiota: Introducing the concept of prebiotics. Journal of Nutrition, 125(6), 1401–1412. https://doi.org/10.1093/jn/125.6.1401
49. Rastall, R. A., & Gibson, G. R. (2015). Recent developments in prebiotics to selectively impact beneficial microbes and promote intestinal health. *Current Opinion in Biotechnology*, *32*, 42–46.
50. Tap, J., Mondot, S., Levenez, F., Pelletier, E., Caron, C., Furet, J. P., Ugarte, E., Muñoz-Tamayo, R., Paslier, D. L., Nalin, R., Dore, J., & Leclerc, M. (2009). Towards the human intestinal microbiota phylogenetic core. *Environmental Microbiology*, *11*(10), 2574–2584.
51. Macfarlane, G. T., & Macfarlane, S. (2007). Models for intestinal fermentation: Association between food components, delivery systems, bioavailability, and functional interactions in the gut. *Current Opinion in Biotechnology*, *18*(2), 154–160.

52. Louis, P., & Flint, H. J. (2009). Diversity, metabolism and microbial ecology of butyrate-producing bacteria from the human large intestine. *FEMS Microbiology Letters*, *294*(1), 1–8.
53. Becattini, S., Taur, Y., & Pamer, E. G. (2016). Antibiotic-induced changes in the intestinal microbiota and disease. *Trends in Molecular Medicine*, *22*(6), 458–478.
54. Peterson, D. A., Frank, D. N., Pace, N. R., & Gordon, J. I. (2008). Metagenomic approaches for defining the pathogenesis of inflammatory bowel diseases. *Cell Host & Microbe*, *3*(6), 417–427.
55. Conlon, M. A., & Bird, A. R. (2015). The impact of diet and lifestyle on gut microbiota and human health. *Nutrients*, *7*(1), 17–44.
56. David, L. A., Maurice, C. F., Carmody, R. N., Gootenberg, D. B., Button, J. E., Wolfe, B. E., Ling, A. V., Devlin, A. S., Varma, Y., Fischbach, M. A., Biddinger, S. B., Dutton, R. J., & Turnbaugh, P. J. (2014). Diet rapidly and reproducibly alters the human gut microbiome. *Nature*, *505*(7484), 559–563.
57. Slavin, J. L. (2013). Fiber and prebiotics: Mechanisms and health benefits. *Nutrients*, *5*(4), 1417–1435.
58. Roberfroid, M. (2005). Inulin-type fructans: Functional food ingredients. *Journal of Nutrition*, *135*(5), 1202–1207.
59. Birt, D. F., Boylston, T., Hendrich, S., Jane, J. L., Hollis, J., Li, L., McClelland, J., Moore, S., Phillips, G. J., Rowling, M., Schalinske, K., Scott, M. P., & Whitley, E. M. (2013). Resistant starch: Promise for improving human health. *Advances in Nutrition*, *4*(6), 587–601.
60. Del Rio, D., Rodriguez-Mateos, A., Spencer, J. P. E., Tognolini, M., Borges, G., & Crozier, A. (2013). Dietary (poly)phenolics in human health: Structures, bioavailability, and evidence of protective effects against chronic diseases. *Antioxidants & Redox Signaling*, *18*(14), 1818–1892.
61. Reichardt, N., Duncan, P., Young, A., Walker, M. C., Parker, E. R., & Bowles, A. C. (2014). Phylogenetic distribution of three pathways for propionate production within the human gut microbiota. The ISME Journal, 8(7), 1323–1335. https://doi.org/10.1038/ismej.2014.14
62. Mills, S., et al. (2017). Precision nutrition and the microbiome, part I: Current state of the science. Nutrients, 9(8), 672.
63. Clarke, G., et al. (2014). The microbiome–gut–brain axis during early life regulates the hippocampal serotonergic system in a sex-dependent manner. Molecular Psychiatry, 18(6), 666–673.
64. Belkaid, Y., & Hand, T. W. (2014). Role of the microbiota in immunity and inflammation. Cell, 157(1), 121–141. https://doi.org/10.1016/j.cell.2014.03.011

2 Introduction to the Gut–Brain Axis

Ashish Garg[*], Abhavya Shukla, Sweta Garg, Prakash Pandey, Vishal Singh, and Gopal Rai

*Guru Ramdas Khalsa Institute of Science and Technology, Pharmacy, Jabalpur, Madhya Pradesh, India

2.1 INTRODUCTION

2.1.1 Overview of the Gut–Brain Axis

The gut–brain axis (GBA) is a bidirectional communication open in the central nervous system (CNS), including the brain and the other enteric nervous system (ENS) in the gastrointestinal tract. Within this complexity, there are various types of signaling by which neural, hormonal, immunological, and metabolic pathways with brain and with gut interact within this complex system. Recent scientific advances have shown the important role that the gut microbiota plays in maintaining the communication with mental health, cognition, and overall well-being, and the role of the microbial communities in the gut with regard to mental health, cognition, and well-being has been highlighted. The gut–brain axis includes many systems starting from the autonomic nervous system (ANS) that regulates involuntary physiological processes and the hypothalamic pituitary adrenal (HPA). Often described as a major parasympathetic nerve, the vagus nerve is thought to be a principal line of communication between the gut and the brain, carrying signals about the state of the gut's environment and perturbing brain function. At the same time, a similarly documented number of gut-derived molecules, such as short-chain fatty acids (SCFAs), neurotransmitters, and immune modulators, have also been shown to impact activities in the CNS, including mood, cognition, and behavior. One of the most intriguing aspects of the gut–brain axis is its ability to modulate mood and mental health. Many studies have linked gut microbiota and mental disorders, including depression, anxiety, neurodevelopmental and neurodegeneration disorders like autism spectrum disorder (ASD), Alzheimer's disease, and Parkinson's disease. Psychobiotics and gut microbiota-targeted therapies are emerging as promising approaches for improving mental health and emotional well-being (Cryan et al., 2019). The gut–brain axis research is interdisciplinary, covering neurology, gastroenterology, microbiology, and psychiatry. Recent advances in genomic sequencing have enabled more complete characterization of the diversity of the gut microbiome, and animal models have identified substantial effects of changes in gut microbiota composition on behavior and neurophysiology. Recent literature in general seems to reach the general consensus regarding the gut–brain axis as an essential part of human health because it affects not only the digestion, but also cognition, stress, and emotional regulation (Carabotti et al., 2015).

2.1.2 Gut–Brain Connection: Historical Context and Emerging Interest

Talking about the brain communication with the gut is not new. According to Hippocrates, one of the ancient Greek physician, often called 'father of medicine,' stated that 'all disease starts from the gut' (Khlevner et al., 2018). Early recognition of the importance of the gut to health reflects the long historical awareness of its vital role, without understanding the mechanisms. Several theories over the centuries have been suggested about the relationship between gut function and mental health, but recent technological improvements have permitted in-depth study into the communication between gut and brain. Auerbach, a British scientist of the 19th century, was the first to describe what he dubbed the 'enteric nervous system,' meaning the 'second brain,' a multitude of neurons in the GI tract walls that control digestion outside of the brain. As a result, the work led to an understanding that there is a complex and autonomous nervous system in the gut which is able to regulate many aspects of digestive function. Nevertheless, such a bidirectional interaction between the gut and the brain remained uninvestigated until the late 20th century (Carabotti et al., 2015). Around the 1970s, researchers found that the gut contained serotonin-producing cells, implying that the GI system may be involved in controlling our mood and behavior. In fact, 90% of the body's serotonin is made in the gut, which touts it as an essential organ in mood regulation. At about the same time, gastroenterology and psychiatry and neuroscience started to intersect to recognize how the gut can affect emotional states and mental health. Studies in this field, of which neurogastroenterology is an early part, have found such conditions as irritable bowel syndrome (IBS) a disorder in which gastrointestinal and psychological symptoms, such as anxiety and depression (Drossman, 2006), are often seen together.

DOI: 10.1201/9781003536802-2

However, the modern gut–brain axis came to light in the 1990s when scientists started unraveling how the two are actually interacting with each other, looking at the immune system, neurotransmitter signaling, and metabolic processes. Early studies used animal models, but in the 21st century, advances in microbiome research and in genetic sequencing technologies made it possible to investigate in more depth how gut microbiota modulate brain function and behavior.

The discovery of the human microbiome, along with the funding of the Human Microbiome Project (HMP) in 2007, kickstarted gut–brain axis research (Tumbaugh et al., 2007). As high-throughput sequencing technology emerged, researchers could, for the first time, characterize the diversity of microbial communities making their home in the human gut and link the composition of these communities and their health outcomes. The result was the identification of, for example, specific bacteria, such as *Lactobacillus* and *Bifidobacterium* genera, producing neuroactive compounds such as gama-aminobutyric acid (GABA) acting directly on the CNS.

As the 2010s gained way, the gut–brain axis became a growing research focus, with increasing interest in its therapeutic potential. Probiotics, prebiotics, and dietary interventions are investigated in clinical trials to find out how they affect mental health, while psychobiotics were and still are created. Another 2011 study was pivotal in demonstrating that mice that are germ-free and lack gut microbiota have aberrant stress responses and abnormal neurochemistry, which show that gut microbes are involved in emotional behavior (Sudo et al., 2004). Since then, many studies have demonstrated that the gut microbiota affect the stress response, influence the production of neurotransmitter, and change behavior.

Over the past years, studies of the gut–brain axis have been expanded to include a broad variety of neurological and psychiatric conditions. The idea that microbiota dysbiosis, that is, the imbalance of gut microbial communitie, might be associated with autism, multiple sclerosis, Parkinson's disease, and other conditions has been studied. For instance, last year, a study found that the Parkinson's gut microbiota profiles were distinct from healthy people, implying that microbial aberrations may play a part in the progression of these diseases (Sampson and Mazmanjan 2015). Likewise, this research has also shown that people who are suffering from depression tend to have smaller diversity of microbes, and lower amounts of beneficial microflora like *Bifidobacterium* (Jiang et al., 2015).

The discovery of the gut–brain axis already occurred with other diseases like cancer, diabetes, or heart disease, but the COVID-19 pandemic further accelerated the research with the interest in viral infections and gut health as well as on mental well-being. Gastrointestinal symptoms in many patients have led scientists to find the SARS-CoV-2 virus can infect cells in the GI tract. In addition, recent data suggest that gut microbiota changes related to post-viral syndrome are responsible for chronic fatigue, cognitive dysfunction, and depression (Plummer et al., 2023).

Today the gut–brain axis is considered to be one of the central fields of research in understanding how the gut impacts on the digestion of the gut as well as its influence on mental health, immune function, and disease. Today, scientists take a personalized medicine approach that targets the gut microbiota to improve a person's overall health and have the potential to treat psychiatric and neurological disorders. Future work will probably unveil yet another way to modulate the gut–brain axis, much like how most therapeutic strategies we now know about were unknown just a few short years ago.

2.2 ANATOMY AND PHYSIOLOGY OF THE GUT–BRAIN AXIS

The gut–brain axis is a two-way communication which links the CNS and the ENS for facilitating the communication between the brain and the GI (gastrointestinal) tract. The gut–brain axis therefore encompasses neural, hormonal, immune, and microbial communication necessary to maintain healthy GI tract function and psychological health. Figure 2.1 explored the connection between gut microbiota and neurological disorders (NDs).

2.2.1 Central Nervous System and Enteric Nervous System

The brain and the spinal cord integrate and coordinate body functions, and taking control of the sensory input arriving at the brain. It takes in inputs from external aspects of the environment, but also from internal bodily signals. The 'second brain' ENS is a giant network of neurons residing within the walls of the GI tract. The CNS and ENS work independently of each other to coordinate digestion and the peristalsis, alter the secretion of digestive enzymes by disaggregating and/or deforming luminal contents, and maintain the gut barrier. Despite this independence, the two systems themselves are constantly communicating with each other via the gut–brain axis. Of the 500 million neurons in the ENS, this equals approximately the number of neurons in the spinal cord (Furness, 2012). It is subdivided into two main plexuses: this myenteric plexus and submucosal plexus. Control of gut motility is controlled by the myenteric plexus and secretion of enzymes and blood flow in the gut by the submucosal plexus. Together these plexuses each help to keep gut homeostasis in check by responding to mechanical and chemical stimuli from the GI environment. The ENS contains neurons that secrete neurotransmitters (acetylcholine, serotonin, and dopamine) that can modulate gut motility, secretion, and pain perception (Guzel and Mirowska-Guzel, 2022).

Central of the interaction of CNS and the ENS is the ANS, made of the sympathetic and parasympathetic divisions. The sympathetic system inhibits gut motility and

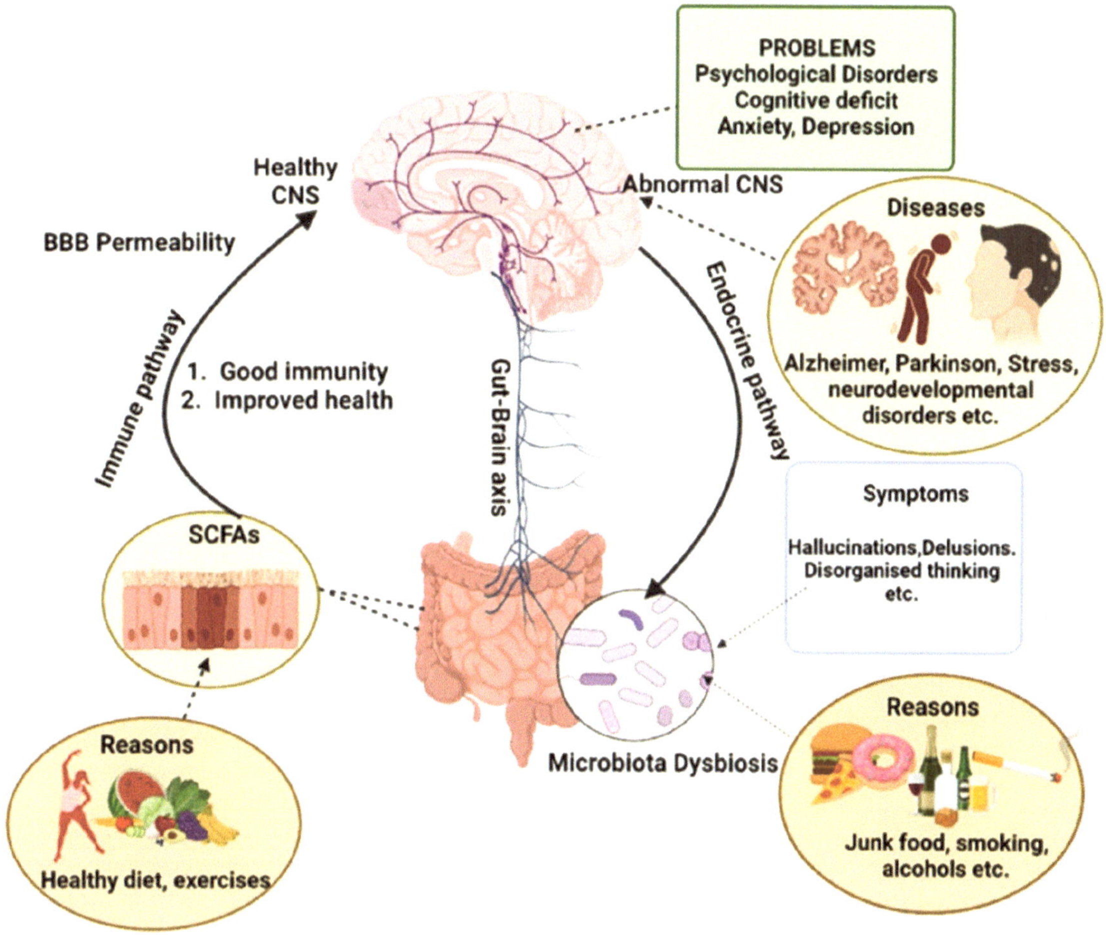

FIGURE 2.1 Exploring the connection between gut microbiota and neurological disorders (NDs). (Adopted under Creative Commons CC-BY license; Ashique et al., 2024.)

secretion, slowing down digestion during stress responses to support long-term fight or flight responses. Unlike this, the system parasympathetic, mostly controlled by the vagus nerve, activates digestive activity by helping restore normal gut functions once the stress is over. This bidirectional information exchange affects digestion and homeostasis of the system. Under stress, energy is steered toward vital survival functions, and under relaxed conditions the body focuses on digestion and nutrient absorption (Guzel and Mirowska-Guzel, 2022).

2.2.2 Role of the Vagus Nerve

The vagus nerve is central to the gut–brain axis as the main neural conductor of communication between the brain and gut. The vagus nerve (or cranial nerve X) is made from the brainstem and travels to a multitude of organs, including the heart, lungs, and the GI tract. About 80–90% of the nerve fibers in the vagus nerve are afferent, meaning they convey information from the gut to the brain (Berthoud and Neuhuber, 2000). The sensory input includes the mechanical stretching of the gut wall, the chemical composition of the gut contents, and the signals from gut hormones and neurotransmitters.

Several important functions of the gut–brain axis are mediated via the vagus nerve, specifically gut motility and appetite regulation as well as the immune responses. It controls peristalsis: the contractions of the muscle that push food through the digestive tract, and stimulates the release of digestive enzymes and bile acids. The vagus nerve has also an important role in the modulation of inflammation by its anti-inflammatory reflex. Vagal efferent responses to inflammatory signals in the gut release acetylcholine that inhibit production of pro-inflammatory cytokines, thereby helping regulate immune activity (Tracey, 2009).

In the past few years, the vagus nerve has become something of an object of interest for being involved in psychobiological processes. Vagal stimulation has experimentally been demonstrated to affect mood and behavior, yet is shown to mediate stress responses and emotional states. Vagus nerve stimulation of rodent models has been shown to increase brain-derived neurotrophic factor (BDNF), a key for neuronal plasticity and resilience to stress (Engineer et al., 2011). These findings have generated interest in vagus nerve stimulation as a therapeutic mode for the treatment of, for example, depression, anxiety, and inflammatory bowel disease.

The vagus nerve acts as a conduit from the gut microbes to brain function in the context of the microbiota–gut–brain axis. It has been shown that some gut bacteria can modulate vagal nerve activity. For instance, by taking *Lactobacillus rhamnosus*, a beneficial gut bacterium, mice given to the brain through their vagus nerve changed GABA receptor expression and exhibited reduced anxiety-like behaviors (Bravo et al., 2011). These effects were abolished by disruption of the vagus nerve (vagotomy), demonstrating that vagal signaling is required for gut microbiota to communicate with the brain.

2.2.3 Microbiota-Gut–Brain Communication Pathways

The gut microbiota, the trillions of microorganisms living within the GI tract, are essential to the workings of the gut–brain axis. Through neural, endocrine, immune, and metabolic mechanisms, these microbial communities communicate with the brain. We know from the research that the composition of our gut microbiota can impact a whole slew of physiological and psychological processes: cognition, mood, and behavior (Cryan et al., 2012).

2.2.3.1 Neural Pathways

The microbiota interacts with the brain via a key-neural route known as the vagus nerve. Again as mentioned above, some of the gut bacteria produce neuroactive substances such as GABA, serotonin, and dopamine, which help in modulating CNS activity. For example, modulation of vagal afferent signaling to alleviate anxiety has been shown for *Bifidobacterium longum* in rodents (Bercik and Collins, 2014). Chronic changes in vagal tone, which may be controlled by gut microbiota, have also been linked to mental illness such as depression (Tillisch et al., 2013).

2.2.3.2 Endocrine Pathways

In addition, the gut microbiota communicates with the brain through another essential pathway, the HPA (otherwise known as the hypothalamic pituitary axis), which regulates the body's response to stress. This triggers the hypothalamic release of corticotropin-releasing hormone (CRH), followed by stimulation of release of ACTH from the pituitary gland and cortisol from the adrenal glands. We have recently shown that gut microbiota can modulate the HPA axis, and some bacterial strains attenuate HPA activation in response to stress. *Lactobacillus helveticus and B. longum* have been demonstrated to lower cortisol in animal models and humans, and may therefore play a part in regulating stress and anxiety (O'Mahony et al., 2009).

2.2.3.3 Immune Pathways

But the gut microbiota also influence the brain through the immune system. A number of bacterial products have been identified that interact with gut-associated lymphoid tissue (GALT) and produce microbial-associated molecular patterns (MAMPs) and thus help regulate immune function. On these MAMPs, the cytokines, chemokines, and more immune modulators are produced that can cross the blood–brain barrier (BBB) or indirectly affect CNS by way of the vagus nerve. Impaired gut microbiota (i.e., dysbiosis) is associated with neuroinflammatory conditions, including multiple sclerosis, Alzheimer's disease, and Parkinson's disease (Mulak and Bonaz, 2015). Modulation of systemic and CNS inflammation by certain probiotics has been demonstrated and implies a role for the microbiota in immune-mediated communication between the gut and the brain (Dinan and Cryan, 2013).

2.2.3.4 Metabolic Pathways

A number of metabolic by-products produced by gut microbes, albeit SCFAs such as butyrate, propionate, and acetate, can impact brain function. SCFAs are generated through the fermentation of dietary fibers in gut bacteria and are absorbed into the blood stream and cause systemic effects. For example, butyrate, especially, has been found to stimulate neurogenesis, protect integrity of the BBB, and also to protect astrocytes (Bourassa et al., 2016). Moreover, gut microbiota can modulate the production of the neuropeptides, leptin and ghrelin, which regulate appetite and energy homeostasis, for an additional link between gut microbiota and CNS function.

Microbial metabolite studies have centered on the associations between microbial metabolites and neurodegenerative disease development. SCFAs have been extensively seen to interface with the activity of the brain's resident immune cells, also known as microglia, which are necessary to clear damaged neurons as well as pathogens. SCFA production in the gut has been linked to increased neuroinflammation and the progression of diseases like Alzheimer's or Parkinson's (Erny et al., 2015).

The aggregation of these communications into the term microbiota–gut–brain axis indicates that this network of communication is dynamic and multifaceted, integrating neural, hormonal, immune, and metabolic signals to shape brain and GI activity. These complex interactions have paved new ways in which the molecular basis of such neurological and psychiatric conditions can be tackled, and that modulation of the gut microbiota by diet, probiotics, and prebiotics has become a promising new avenue for therapeutic strategies for mental health.

2.3 KEY COMPONENTS OF THE GUT–BRAIN AXIS

A gut–brain axis is a network of the complex, bidirectional communication between the brain and the GI tract. They all work together; some components include gut microbiota, neurotransmitters, and hormonal signaling pathways, which together are important to support homeostasis and in turn impact mental health.

2.3.1 Gut Microbiota and Their Functions

Gut microbiota refers to the diverse community of microorganisms residing in the gastrointestinal tract. These microorganisms, which include bacteria, archaea, viruses, and fungi, outnumber human cells and play a crucial role in various physiological processes, including metabolism, immune function, and brain health. The gut microbiota is involved in the fermentation of dietary fibers, producing SCFAs, which have anti-inflammatory properties and can influence brain function (Aggarwal et al., 2023).

2.3.1.1 Composition of Gut Microbiota

The composition of gut microbiota differs among individuals significantly, and is influenced by diet, age, genetics, and environment (Belkaid and Hand, 2014). Typically, they function with a diverse array of bacterial species such as Firmicutes, Bacteroidetes, and Actinobacteria that are crucial for metabolic function and modulation of the immune system (Sender et al., 2016). Microbiota composition, or dysbiosis, has been linked to several disorders of health, including obesity, inflammatory bowel disease, and mental illness like depression and anxiety (Juarez et al., 2022).

2.3.1.2 Functions of Gut Microbiota

Gut microbiota serves several key functions that directly impact both gut and brain health.

Metabolism of Nutrients: Gut bacteria ferment non-digestible carbohydrates and produce SCFAs like acetate, propionate, and butyrate, which provide energy for colonocytes, influence appetite regulation, and contribute to the modulation of inflammation (Louis et al., 2017).

Immune System Modulation: The gut microbiome is crucial for the development and function of the immune system. It helps maintain the intestinal barrier, preventing pathogen invasion and modulating immune responses through the production of signaling molecules (Round and Mazmanian, 2009).

Synthesis of Neurotransmitters: Certain gut bacteria are capable of synthesizing neurotransmitters such as GABA, serotonin, and dopamine, which can influence mood and behavior. For example, approximately 90% of serotonin, a key neurotransmitter regulating mood, is produced in the gut (Yano et al., 2015).

Influencing Brain Function: The gut microbiota can communicate with the CNS via the vagus nerve and other pathways, potentially affecting cognitive functions and emotional behavior (Hsiao et al., 2013).

2.3.2 Neurotransmitters and the Gut

Chemical messengers, neurotransmitters, help neurons in the brain and throughout the nervous system communicate with one another. Synthesis and modulation of a number of neurotransmitters also depends upon the gut microbiota and thus can affect mood, cognition, and overall mental health.

2.3.2.1 Serotonin

One of the most well-known neurotransmitters produced in the gut is serotonin. This plays an important role in controlling mood, anxiety, and general mental environment. Somewhat surprisingly, the enterochromaffin cells of the gut (O'Mahony et al., 2009) approximate 90% of the body's serotonin. Serotonin levels can be influenced by gut microbiota, through the production of SCFAs, that impact the expression of tryptophan hydroxylase, the enzyme for serotonin synthesis (O'Mahony et al., 2009). Altered serotonin levels have been correlated with dysbiosis, and may be involved in the etiology of mood disorder (Foster et al., 2013).

2.3.2.2 GABA

GABA is an inhibitory neurotransmitter that inhibits neuronal activity in many parts of the nervous system. Some bacteria, such as *Lactobacillus* and *Bifidobacterium* species, have the ability to produce GABA (Xiong et al., 2023). Indeed, GABA synthesis in the gut might modulate the brain's GABAergic signaling to influence anxiety and stress responses (Dicks, 2022). In animal models, studies have shown that probiotics containing GABA-producing bacteria can reduce anxiety-like behavior (Merkouris et al., 2024).

2.3.2.3 Dopamine

Another important neurotransmitter to dopamine is involved with reward and enjoyment, and mood regulation. While most dopamine is made in the brain, gut microbiota may indirectly affect the level of dopamine. Cases in point include bacteria that can metabolize dietary compounds into precursors of dopamine, affecting availability for neurotransmission (Hamamah et al., 2022). Brain dopamine signaling is involved in neuropsychiatric disorders, including schizophrenia and attention-deficit hyperactivity disorder (ADHD) (Sittipo et al., 2022) and has been shown to be dysregulated.

2.3.3 Hormonal Regulation in Gut–Brain Communication

The gut to brain communication depends upon hormonal signaling. Several hormones produced in the gut and brain can influence the other's function as well as overall health.

2.3.3.1 Gut Hormones

Ghrelin: Ghrelin produces an effect on hunger known as the 'hunger hormone,' stimulating appetite, produced in the stomach. We have shown it to influence brain reward centers in ways and to extents that impact mood and cognitive function (Gil-Campos et al., 2006). Gut microbiota affect levels of *Ghrelin* and therefore a connection between appetite regulation and microbial composition (Fetissov, 2017).

Leptin: Derived from adipose tissue, leptin signals satiety and helps to balance energy. At the same time, it can influence mood and cognitive function. Shi et al. (2015) indicate that leptin may get its effects on the brain by inducing neuroinflammation and neuronal plasticity. Leptin signaling was also reported to be altered in obesity and related mood disorders (Kelesidis et al., 2010).

Cholecystokinin (CCK): Food intake triggers the CCK release, and CCK is important for digestion and satiety. Modulation of neurocircuitry in the brain is thought to influence anxiety and stress responses (Lam et al., 2005).

2.3.3.2 Brain Hormones

Cortisol: The adrenal glands release a hormone called cortisol when body is under stress. Chronic stress modulates gut permeability and dysbiosis and therewith gut hormone and neurotransmitter production (Madison and Kiecolt-Glase, 2019). A feedback loop between the gut and the brain has been suggested to associate elevated cortisol levels with anxiety and depression that occur in response to stress (Labanski et al., 2020).

Oxytocin: Oxytocin is referred to as the 'love hormone,' and plays a role in social bonding and emotional control. Gut microbiota may influence oxytocin signaling, influencing mood and social behaviors (Olff et al., 2013), and this was suggested by research. The importance of the gut–brain axis for emotional well-being becomes clear with this relationship.

The gut–brain axis is a complex system of communication system that links gut microbes, neurotransmitters, and hormonal regulation. The gut microbiota effect on mental health is huge, through the production of neurotransmitters, modulation of immune response, and hormonal signaling. Being able to understand these connections is fundamental for developing specific interventions for mental health disorders like depression and anxiety and overall wellness. Research on the gut–brain axis is ongoing with promise for finding new therapeutic ways to exploit the microbiome's brain healthy role.

2.4 RECENT ADVANCES IN RESEARCH ON THE GUT–BRAIN AXIS

The gut–brain axis is a complex and bidirectional communication network that links the GI tract and the brain. Recent advances in research have highlighted the significant role of gut microbiota in mental health, particularly in conditions such as anxiety, depression, and stress. Additionally, dysbiosis, or the imbalance of gut microbiota, has been implicated in various neurological disorders, including Parkinson's and Alzheimer's diseases. This section explores these aspects, emphasizing the gut microbiota's influence on mood and behavior.

2.4.1 Role of Gut Microbiota in Mental Health

The gut microbiota consists of trillions of microorganisms residing in the human gastrointestinal tract, playing crucial roles in digestion, metabolism, and immune function. Emerging evidence suggests that gut microbiota significantly impacts mental health by influencing brain function and behavior.

2.4.1.1 Gut Microbiota and Mood Disorders

Research has established a connection between gut microbiota and mood disorders, particularly anxiety and depression. Animal studies have shown that alterations in gut microbiota can lead to behavioral changes consistent with anxiety and depression. For instance, germ-free mice, which lack gut microbiota, exhibit increased anxiety-like behaviors when compared to their microbiota-replete counterparts (Cryan et al., 2019). The microbiota's influence on mood may be mediated through several mechanisms, including the production of neurotransmitters and modulation of inflammation. Some gut bacteria can produce neurotransmitters such as GABA, serotonin, and dopamine, all of which are vital for mood regulation (O'Mahony et al., 2009). For example, a study by Yano et al. demonstrated that gut bacteria are essential for serotonin production, with approximately 90% of the body's serotonin synthesized in the gut (Yano et al., 2015).

Moreover, inflammation is a key factor in the pathophysiology of mood disorders. Dysbiosis can lead to increased intestinal permeability, often referred to as 'leaky gut,' allowing the translocation of bacterial endotoxins into circulation. This results in systemic inflammation, which has been linked to depression and anxiety (Davis, 2016). A systematic review by Cryan et al. highlighted the correlation between inflammatory markers and the severity of depressive symptoms (Cryan et al., 2012).

2.4.1.2 Probiotics and Mental Health

Probiotics, defined as live microorganisms that confer health benefits to the host, have garnered attention for their potential in improving mental health. Numerous clinical trials have examined the effects of probiotics on anxiety and depression. For instance, a meta-analysis revealed that probiotic supplementation significantly reduces depression scores in individuals with major depressive disorder (Ribera et al., 2024). Specific strains, such as *L. rhamnosus and B. longum,* have shown promise in modulating mood and reducing anxiety. Animal studies have demonstrated that these probiotics can influence the HPA axis, a critical component in stress response regulation (Bravo et al., 2011). By balancing the gut microbiome, probiotics may help restore mental health in individuals suffering from anxiety and depression.

2.4.2 Microbiota Dysbiosis and Neurological Disorders

Gut microbiota composition is commonly referred to as dysbiosis, which can have major neurological effects. Dysbiosis shares a growing evidence link with neurodegenerative diseases like Parkinson's and Alzheimer's.

2.4.2.1 Gut Microbiota and Parkinson's Disease

It is a progressive neurodegenerative disorder characterized by motor and nonmotor symptoms, including cognitive decline and mood disorders. They have also found that gut microbiota composition is altered in humans with PD. Hill-Burns et al. showed significant difference in PD patients' gut microbiota from controls; they found decreased Firmicutes and increased pro-inflammatory bacteria (Hill-Burns et al., 2017).

The gut–brain connection in PD may be from the production of neurotoxic substances by some gut microbiota. So, for instance, some gut bacteria can produce alpha synuclein, a protein that is involved in the pathogenesis of PD. The further supporting hypothesis of a gut–brain axis in PD is the observation that gastrointestinal symptoms commonly precede motor ones in PD (Gao et al., 2024).

2.4.2.2 Gut Microbiota and Alzheimer's Disease

Another neurodegenerative disorder that involves cognitive decline and memory loss is Alzheimer's disease (AD). A role of dysbiosis in the pathogenesis of AD has been suggested by the emerging evidence. An altered microbiota composition in individuals with AD has been shown in studies: a reduction in microbial diversity and a rise in pathogenic microorganisms (Varesi et al., 2022).

It is possible that gut microbiota relationship to AD involves modulation of neuroinflammation and amyloid beta production. Increased systemic inflammation has been reported to worsen neuroinflammation in the brain, and dysbiosis has been related to this. For example, in a study, Zhan et al. have shown that gut microbiota can influence production of amyloid beta, an AD pathology hallmark, through inflammatory cytokine activity (Varesi et al., 2022).

Also, the gut microbiota may modulate the brain response to amyloid beta through the gut–brain axis. Hsiao and team found that gut bacteria could influence levels of amyloid precursor protein, affecting brain amyloid beta accumulation (Hsiao et al., 2013).

2.4.3 Gut Microbiota's Influence on Mood and Behavior

The influence of the gut microbiota on mood and behavior goes well beyond mood disorders and neurological diseases. In fact, the gut as part of the gut–brain axis has a huge impact on mindset and emotional responses, as well as cognitive function.

2.4.3.1 Gut Microbiota and Stress Response

Gut microbiota composition and function can be profoundly influenced by stress, and this seems to be a bidirectional relationship between gut and brain. We know that chronic stress changes the gut microbiota to a dysbiotic state, further exacerbating stress-related symptoms (Huang et al., 2019). Bailey et al. performed a study which revealed that the abundance of certain bacterial taxa changed in response to chronic stress and was strongly correlated with increased anxiety-like behavior (Galland, 2014). The gut–brain connection in stress response through modulation by gut microbiota is a critical mechanism. Dysbiosis can dysregulate the body's response to stress (HPA axis), resulting in excessive cortisol, and, therefore, lower moods and behavioral changes (de Almeida et al., 2017).

2.4.3.2 Gut Microbiota and Cognitive Function

Gut microbiota may affect cognitive function and behavior as well, as emerging evidence suggests. Alterations in gut microbiota are known to impact brain processes relating to learning and memory. Sudo et al., for example, showed that germ-free mice were unimpaired at performing the learning tasks, compared to mice with their normal gut microbiota (Sudo, 2004).

Gut bacteria may contribute in part to the production of SCFAs, and in turn cognitive function. But the neuroprotective effects and cognitive function-enhancing effects of SCFAs, including butyrate, have been demonstrated (Clarke et al., 2014). Additionally, the gut microbiota might modulate brain-derived neurotrophic factor (BDNF), a protein necessary for neuroplasticity and cognitive function (Duman and Aghajanian, 2012).

2.4.3.3 Gut Microbiota and Behavioral Changes

Both animal and human studies, including those on gut microbiota alterations, have documented behavioral changes. The research indicates that probiotics can affect how mood and behavior change. Tillisch et al. found in a randomized-controlled trial that participants that consumed probiotics reported lower negative emotions and higher positive emotions (Tillisch et al., 2013).

In addition, gut microbiota may contribute to the regulation of social behavior and stress resilience. Using a stress mouse model, Desbonnet et al. showed that stress decreased social behavior, and administration of probiotics was capable of improving social behavior (Foster et al., 2017).

New research on the gut–brain axis emphasizes the important contribution of gut microbiota to mental health, especially for anxiety, depression, and stress. Maintaining a healthy gut microbiome has been implicated in neurological diseases such as Parkinson's and Alzheimer's, providing evidence that an intact and healthy gut microbiome is important to healthy brain health. Making sense of how gut microbiota shape mood, behavior, and cognitive function means fresh approaches for therapeutic interventions that leverage changes to the microbiome to improve mental health.

2.5 MECHANISMS OF GUT–BRAIN INTERACTION

The Gut–brain axis is a complex and bidirectional communication between gastrointestinal tract and CNS. Elucidating gut microbiota impact on brain function, behavior, and mental health requires understanding of mechanisms underpinning gut–brain interactions. This section explores three primary mechanisms: this thesis will focus on inflammation and immune system involvement, metabolic pathways and gut–brain signaling, and the role of SCFAs.

2.5.1 Inflammation and Immune System Involvement

The gut and brain are connected by a critical mechanism for inflammation. Good gut microbe balance influences immune response that in turn can affect brain, behavior. Chronic inflammation caused by dysbiosis, irregular balance in gut microbiota, can also result in many neuropsychiatric conditions.

2.5.1.1 The Gut Microbiota and the Immune System

Immune response depends on the gut-associated lymphoid tissue (GALT). These gut microbiota play a critical role because they support homeostasis of the body through approximately 70% of the body's immune cells that live in the gut. Regulatory T cells (Tregs) are critical to maintaining tolerance to non-pathogenic antigens, and commensal bacteria stimulate production of these immune cells, including Tregs (Round and Mazmanian, 2009).

Disruption in the composition of the gut microbiota can perturb immune homeostasis and trigger either an overactive immune response or chronic inflammation. For instance, Round and Mazmanian showed that particular gut bacteria may drive Tregs differentiation (Round and Mazmanian, 2010). However, rather than promoting anti-inflammatory effects, dysbiosis has been shown to activate pro-inflammatory cytokines like tumor necrosis factor alpha (TNF-α) and interleukin 6 (IL-6), which are associated with the pathophysiology of depression and anxiety (Miller et al., 2016).

2.5.1.2 The Brain and Systemic Inflammation

Chronic systemic inflammation can also lead to a loss of brain function. Cerebrovascular and cellular mechanisms (CCM) can initiate upregulation of the expression of pro-inflammatory cytokines that pass over the BBB or can induce production of prostaglandins that affect neuronal activity and alter behavior. For example, studies following Dantzer et al. showed that inflammation plays a role in the development of depressive symptoms, with pro-inflammatory cytokine levels leading to an 'inflammatory response syndrome' in the brain (Dantzer et al., 2008).

The reciprocal relationship between gut dysbiosis and neuroinflammation is also further demonstrated. Dysbiosis can also promote gut inflammation, which on top of exacerbating neuroinflammatory responses can contribute to the incidence of neurodegenerative diseases like Alzheimer's and Parkinson's (Yang et al., 2019). It's a testament to the necessity of stabilizing the gut microbiome with proper balance preventing the bodily risks of chronic inflammation and its consequences on mental health.

2.5.2 Metabolic Pathways and Gut–Brain Signaling

Brain function is altered by the products of metabolisms and signaling molecules from the gut microbiota. These interactions could impact on neurotransmitter production, energy metabolism, and neuronal signaling, which can ultimately shape behavior and cognitive function.

2.5.2.1 Neurotransmitter Production

Neurotransmitters produced by the gut microbiota or the gut microbiolects alone are important for mood regulation and cognitive functions. For example, some gut bacteria are able to make the inhibitory neurotransmitter, GABA, a neurotransmitter linked to anxiety and stress (Bravo et al., 2011). Strandwitz et al. (2019) found specific gut bacteria like Bacteroides fragilis, which are capable of producing GABA, that can influence brain activity, thus indicating how gut microbiota may have impact on brain activity.

In addition, gut microbiota may affect serotonin synthesis, a neurotransmitter putatively involved in mood regulation. Most serotonin is produced in the gut, but the bioavailability to the brain by the gut microbiota composition can be affected. Studies have confirmed that certain strains of bacteria enhance serotonin synthesis and increase mood and reduce anxiety (Yano et al., 2015).

2.5.2.2 Gut–Brain Signaling Pathways

Multiple pathways from vagus nerve, endocrine, and immune-mediated pathways are involved in gut–brain signaling. The vagus nerve primarily acts as a key channel to convey information out of the gut and into the brain by transmitting messages from the gut on inflammation, microbial metabolisms, and food intake. The role of the vagus nerve on gut–brain interaction has been shown to reduce anxiety and depression-like behavior in animal models (Bonaz et al., 2017).

Gut–brain communication also requires endocrine signaling. We still don't know exactly what hormones the gut produces, but they include ghrelin, leptin, and glucagon-like peptide-1 (GLP-1), all of which affect appetite, metabolism, and mood. Neuroprotective effects and even improving cognition has been demonstrated for GLP1, in particular (Baggio and Drunken, 2014). Gut hormone brain signaling, moreover, shows that gut–brain communication is complex.

2.5.2.3 Brain Function and Energy Metabolism

Energy metabolism is also influenced by the gut microbiota, which may also influence brain health. SCFAs are formed by gut bacteria that ferment dietary fibers and may act as an energy source for cells in the intestinal tract while affecting other tissues, too. Previously, SCFAs, including butyrate, propionate, and acetate, were demonstrated to stimulate brain function by crossing the BBB and regulating neuroinflammatory responses (Nogal et al., 2021).

The work of Jiang et al. showed that butyrate treatment improved cognitive function in mice via increased synaptic plasticity and decreased neuroinflammation (Jiang et al., 2021). This suggests that gut microbiota can generate large effects on brain health and cognitive function, indicating that a healthy gut microbiome is important.

2.5.3 Role of Short-Chain Fatty Acids

Gut microbiota produces SCFAs through fermentation of dietary fibers, and they're important in gut–brain communication. It is known that these fatty acids, including acetate, propionate, and butyrate, have many effects on brain function and behavior.

2.5.3.1 SCFAs Production and Function

The major sources of SCFAs are the colon, where they drive gut health as an energy source for colonocytes. But their reach is far from the gut; they systemically circulate and can travel to the brain, for example. They can also cross BBB and modulate neuronal activity, neurotransmitter release, and inflammation (Moțățăianu et al., 2023).

Particular attention has been drawn to butyrate's neuroprotective effects. For example, I have shown it to promote neurogenesis, increase synaptic plasticity, and decrease neuroinflammation. According to a study by Cuenod et al., butyrate administration improved cognitive performance and reduced anxiety-like behaviors in mice, suggesting it could be a therapeutic option in mood disorders (Cavaliere et al., 2022).

2.5.3.2 SCFAs and Neuroinflammation

SCFA's are also known to modulate neuroinflammation. This family can inhibit the activation of microglia, the auxiliary immune cells located in the CNS, in the onset of the neuroinflammation (Altieri et al., 2023). SCFAs may protect against neurodegenerative diseases with a history of chronic inflammation, including Alzheimer's and Parkinson's (Zou et al., 2023), by reducing microglial activation.

Additionally, SCFAs can modulate the release of anti-inflammatory cytokines, helping to create an anti-inflammatory brain environment. Butyrate, for example, was found to improve IL-10 production, an anti-inflammatory cytokine that can quench the production of neuroinflammatory responses (Föh et al., 2022). The results are consistent with therapeutic potential for SCFAs in combatting neuroinflammation and associated cognitive decline.

2.5.3.3 Dietary Sources of SCFAs

Diet plays a role on SCFA production. But more SCFA is produced in the gut if you eat a high-fiber diet that is also rich in fruits, vegetables, and whole grains. On the other hand, too low of a fiber diet can cause dysbiosis and decrease SCFA production, and can have impacts on mental health (Slavin, 2013).

Dietary fiber interventions have been shown to improve composition of the gut microbiota and to enhance SCFA production. An example of this is when a randomized controlled trial (Fu et al., 2022) increased dietary fiber consumption, leading to increases in butyrate levels and improvements in cognitive function among older adults. The apparent potential to improve gut–brain communication through dietary interventions to promote mental well-being is highlighted.

Elucidating the role of gut microbiota in mental health requires the understanding of gut–brain interaction mechanisms. This complex communication network is characterized by inflammation, immune system involvement, metabolic pathways and gut–brain signaling, and the role of SCFAs. Dysbiosis can be a contributing factor for mood disorders and neurodegenerative diseases to chronic inflammation and dysbiosis. Improving mental health and cognitive function could be achieved through targeting these mechanisms via dietary interventions, probiotics, and lifestyle changes.

2.6 GUT–BRAIN AXIS AND THE ROLE OF DIET

The gut–brain axis is a complex communication network between the brain of the CNS and the gut through many pathways, including the vagus nerve, immune signaling, and microbial metabolites. Recent work emphasizes the importance of diet in modulating the gut microbiota and, hence, the mind through the gut microbiota axis. The impact of nutrition on gut microbiota, probiotics, prebiotics, psychobiotics, and dietary interventions aiming to modulate the gut–brain axis, and the role of probiotics, prebiotics, and psychobiotics on mental health are discussed in this section.

2.6.1 Nutrition on Gut Microbiota

The composition of the gut microbiota is especially dependent on nutrition. This type of diet can stitch increased Storey atomic acids (SCFAs), for example, butyrate, acetate, and propionate (Slavin, 2013). Gut residents crucially rely on a cocktail of SCFAs, which help to keep hair follicle ducts and gut flora intact, and modulate inflammation and perhaps brain function. These findings support a role for high-fiber diet in protecting against neurodegenerative diseases, via partial changes in gut microbiota composition (Portincasa et al., 2022).

On the other hand, diets high in saturated fats and sugars can contribute to dysbiosis, defined by a decrease in diversity of microbes and an increase in pro-inflammatory bacteria. It has been associated with numerous mental disorders (Föh et al., 2022). Dysbiosis thus can induce an inflammatory response leading to increase in blood–brain barrier permeability and increased release of inflammatory cytokines, to the brain and promote neuroinflammation and altered mood (Dutta et al., 2019).

Fiber is just one component of the diet to affect gut microbiota; other dietary components like fruit, vegetables, tea, and red wine contain polyphenols that have also been shown to modulate gut microbiota. These compounds were shown to promote the growth of beneficial bacteria while inhibiting growth of pathogens (including other bacteria associated with the gut microbiome) and promoting a healthy gut environment (Gao et al., 2022). Diet, gut microbiota, and mental health interaction is essential to nutritional strategies for improving overall well-being.

2.6.2 Translation of Probiotics, Prebiotics, and Psychobiotics in Mental Health

Probiotics are live microorganisms that, when consumed in adequate amounts and at the correct dose, provide health benefits to the host. And they're studied for their potential mental health effects, because they're known to positively influence gut microbiota. There is evidence from several clinical trials using specific probiotic strains, including *Lactobacillus* and *Bifidobacterium*, to lower anxiogenic and depressive symptoms (Yano et al., 2015; Bravo et al., 2011). They believe the SCFAs and neurotransmitters, such as serotonin, that are produced are responsible for these effects on mood and behavior.

Additionally, prebiotics, nondigested food components, that stimulate the growth of healthy gut bacteria contribute to our mental health. Prebiotics such as fructooligosaccharides (FOS) and inulin can improve gut health by stimulating good bacteria growth (Slavin, 2013). Prebiotics might indirectly improve mental health by promoting the production of neurotransmitters and the reduction of the inflammation (Bonaz et al., 2017).

They have paid considerable attention to 'psychobiotics,' a term used to refer specially to probiotics with a direct action on mental health. Their research shows that psychobiotics can affect gut–brain communication pathways, and might help in lowering stress, anxiety, and depressive symptoms (Strandwitz, et al., 2019). One example would be that the consumption of the probiotic strain *L. rhamnosus* was shown to reduce anxiety and depression-like behavior in animal models through modulation of the HPA axis and GABA, an important inhibitory neurotransmitter (Strandwitz et al., 2019).

2.6.3 Modulating the Gut–Brain Axis by Dietary Interventions

The gut and brain axis modulation through dietary interventions is becoming more recognized as a possible mental health intervention. Fruits, vegetables, whole grains, legumes, nuts, and healthy fats (like olive oil) are higher in foods in the Mediterranean diet that have been linked to better cognitive function and less depression. This diet encourages a diverse and balanced gut microbiota with production of beneficial metabolites SCFAs.

Additionally, a ketogenic diet consisting of low carbohydrates and high fat provided evidence of improved mood and cognitive function in select groups of people. It was shown that this diet can affect gut microbiota composition and boost the production of ketone bodies with neuroprotective effects (Silva et al., 2020; Ge et al., 2023). Although such diets are investigated here, further investigation will be needed to understand the sustainability and long-term effects.

Inclusion of some fermented foods like yogurt, kefir, sauerkraut, and kimchi in diet also helps with mental and gut health. These foods are high in probiotics and can help replenish the healthy gut microbiota while decreasing inflammation (Silva et al., 2020). The consumption of fermented foods has been linked with lower mood-related symptoms (i.e., anxiety and depression), indicating a potential as dietary intervention for mental health (Föh et al., 2022, Bravo et al., 2011).

In summary, diet and gut microbiota status are linked at the gut–brain axis that helps regulate our mental health. The knowledge of how nutrition affects gut microbiota composition and function is an important contribution to devise dietary strategies to promote mental health. Promising avenues for mental health outcomes might be through clinical practice integration of probiotics, prebiotics, and dietary interventions.

2.7 CLINICAL IMPLICATIONS OF THE GUT–BRAIN AXIS

More recently, the GBA has become more prominent as this important course operates in health and disease. An understanding of the nervous system's interaction with the gut microbiota reveals the ability to contribute to clinical conditions. In this section, we examine the clinical implications of the gut–brain axis for IBS and mental health disorders, the role of the gut–brain axis in ASD and cognitive function, as well as therapeutic approaches aimed at the gut–brain axis.

2.7.1 Irritable Bowel Syndrome and Mental Health Disorders Gut–Brain Axis

Irritable bowel syndrome is a functional gastrointestinal disorder that includes abdominal pain, bloating, and altered bowel habits. There is increasing evidence that IBS is related, or involved, to these psychological factors as anxiety, depression, and stress (Menees and Chey, 2018). This interplay relies on the gut–brain axis, and stress and emotional problems can make gut problems worse, which can worsen upon stress and emotional problems.

This has been shown in research with patients who have IBS, and which found that they commonly display changes in gut microbiota composition, marked by reduced diversity in their microbiota and overgrowth of particular bacterial species (Wang et al., 2020). By this dysbiosis increased intestinal permeability was hypothesized to increase intestinal permeability and may release pro-inflammatory cytokines which can affect brain function and behavior (Raskov et al., 2016). In addition, the gut–brain bidirectional communication is mediated by the gut major conduit for this communication, i.e., the vagus nerve, the signal which conveys gut health and emotional state (Bonaz et al., 2017)

Treatment strategies directed toward the gut–brain axis might represent promising directions for the management of IBS and the mental health disorders associated with IBS. In fact, and according to studies (Goodoory et al., 2023), probiotic supplementation can help improve gut health and alleviate symptoms of IBS, and it may also help with anxiety and depression of affected people. In addition, cognitive behavioral therapy (CBT) and mindfulness-based interventions have also shown to efficacy in treating both IBS and these associated psychological symptoms (Sperber et al., 2017), highlighting the importance of a combined approach for the treatment of both gut and mental health (Sperber et al., 2017).

2.7.2 Role in Autism Spectrum Disorder and Cognitive Function

Autism spectrum disorder is a neurodevelopmental disorder with elements of deficits in social communication and restricted, repetitive behaviors. Recent studies suggest that the gut–brain axis may play a role in ASD pathophysiology, as disturbances of gut microbiota may be involved in ASD (Zhang et al., 2023).

Gastrointestinal symptoms, including constipation, diarrhea, and abdominal pain, are common in children with ASD. Previous studies have shown dysbiosis, an imbalance in gut microbiota, in individuals with ASD and that they are related with severity of behavior symptoms. Neuronal activity and inflammatory responses can modulate neurotransmitters, and SCFAs produced by the gut microbiota can directly and indirectly influence brain development and function (Strandwitz, 2019).

In addition, the gut–brain axis may also contribute to cognitive function in adults with ASD. The fact that interventions that could balance gut microbiota (e.g., dietary changes or probiotic use) could have positive effects on social behavior and cognitive function is emerging (Fattorusso et al., 2019). Nevertheless, longer-term clinical studies would be needed to validate efficacy and the mechanisms underpinning these interventions.

2.7.3 Potential Therapeutic Approaches Targeting the Gut–Brain Axis

There have been numerous therapeutic approaches that target the gut–brain axis in order to improve health outcome given the intimate coupling between the gut microbiota and the brain. Dietary modifications, probiotics, prebiotics, and psychobiotics are included with these approaches.

Dietary Modifications: Dietary patterns that promote a healthy gut microbiota are readily adoptable and have dramatic effects on mental health. For example, the low fat diet, rich in fruits, vegetables, whole grains, nuts, and healthy fats (Nogal et al., 2021), has been linked with improved cognitive function and with a reduced risk of mood disorders. Additionally, diets rich in fiber, as well as increasing beneficial gut bacteria and SCFA production, may help influence mood and behavior (Wu et al., 2021).

Probiotics and Prebiotics: Live microorganisms, a type referred to as probiotics, have potential to improve mental health outcomes. Particular strains of *Lactobacillus* and *Bifidobacterium* have shown to reduce anxiety and depression (Le Morvan de Sequeira et al., 2022). Another potential way to improve gut health, and potentially baby's psychological symptoms, is by taking prebiotics, nondigested fibers that promote growth of helpful bacteria (Slavin, 2013). Probiotics and prebiotics may be used together to provide optimum gut–brain axis function and overall health.

Psychobiotics: Psychobiotics is the study of probiotic that affects mental health. Evidence supports psychobiotics modifying the gut–brain axis to reduce, for example, anxiety and depressive symptoms (Strandwitz et al., 2019). One example is *L. rhamnosus,* which has shown an effect as an anxiolytic in preclinical studies and may therefore have therapeutic potential for mood disorders (Kumar et al., 2023).

Mind-Body Therapies: Mindfulness meditation, yoga, and cognitive behavioral therapy (CBT) have demonstrated efficacy in the management of both gastrointestinal and psychological symptoms. These interventions decrease stress and increase coping as they improve overall quality of life (Sperber et al., 2017). Mind–body therapies integrated with dietary interventions may have an added benefit to managing those conditions that are related to the gut–brain axis.

The gut–brain axis is a focus of the never-ending research into novel therapeutic targets and strategies. Optimal gut health and mental well-being (Zeevi et al., 2015) promise personalized nutrition, and tailor dietary interventions according to people's microbiota profiles. The development of microbiome-based therapeutics, such as fecal microbiota transplantation (FMT), is a rapidly developing field and may be a means to treat conditions associated with dysbiosis (Vrieze et al., 2012).

Overall, human gut–brain axis is a dynamic yet complex interplay of the gastrointestinal system and the central nervous system. This understanding has the potential to offer rehabilitative therapeutic frameworks across the spectrum of clinical conditions ranging from IBS to ASD, and mental health disorders. With the discovery of the intricate internal workings of gut–brain axis, targeted interventions that restore gut health may become brain gain interventions to enrich mental well-being and general health.

2.8 CHALLENGES AND FUTURE DIRECTIONS IN GUT–BRAIN AXIS RESEARCH

Research into the GBA has also become an intriguing arena for the research of the interplay between the gastrointestinal and the CNS. However, various issues remain in the dissection of the complexities of gut–brain interactions. In this section, these challenges are discussed and future directions, including personalized medicine, microbiota therapies, and research priorities of novel treatment strategies, are explored.

2.8.1 Gut–Brain Interactions Are Complex

Intercourse between the gut microbiota, the ENS, vagus nerve, and numerous signagling molecules, including neurotransmitters and hormones, characterize the gut–brain axis. The complexity involved here poses challenges for research, in particular identifying which specific mechanisms and pathways govern gut–brain communication.

Microbiota Composition and Function: The human gut microbiota is home to several trillion microorganisms (Sender et al., 2016), including bacteria, archaea, viruses, and fungi, with a wide degree of inter-subject heterogeneity. The variability of some of these microbial taxa poses challenges to interpretation of how the activity of certain microbes affects brain function and behavior. Many factors (diet, age, genetics, and environmental exposures) affect the functions of gut microbiota, with implication for causal relationships (Levy et al., 2017). Recent technological advances in high-throughput sequencing have generated information about microbial diversity but have failed to elucidate underlying function.

Bidirectional Communication: Bilateral communication between the gut and the brain occurs through the gut–brain axis. This complexity, however, is gained in this bidirectional nature that makes research challenging in determining the directionality of the influence (Murray et al., 2022). For example, altering gut microbiota composition can affect stress, and in turn stress may alter gut microbiota composition, and changes in gut health comprise changes in mood and cognitive function. These dynamics need to be understood and the capabilities required for that understanding include sophisticated experimental designs and methodologies that are able to accurately capture these dynamics.

Individual Variability: Variability in gut–brain interactions occurs due to differences in genetics, microbiota composition, and lifestyle factors. This can complicate the development of one-size-fits-all type of treatment approaches (Chandrasekaran et al., 2024), as susceptibility to such disorders can depend on these differences, and responses to interventions can depend on differences. Personalized medicine, or tailoring interventions to match particular characteristics, could provide solutions to this challenge, provided that the research to produce effective strategies is conducted.

Ethical Considerations: Human intervention studies and the use of microbiota-based therapies are ethical considerations in research on the gut–brain axis that sometimes involve human subjects and animal models. For research, ethical frameworks have to guarantee that research is done responsibly and respect the rights and well-being of the participants (Lange et al., 2022).

2.8.2 Personalized Medicine and Microbiota Therapies

A promising approach to tackle gut–brain interactions is personal medicine. Healthcare providers can optimize treatment outcome by tailoring interventions to individual microbiota profiles and other personal factors.

Microbiota Profiling: Now, with advances in microbiome-sequencing technologies, individual gut microbiota composition is profiled in detail. Analysis of personal microbiota can also yield insight into how certain communities of microorganisms might affect health states. This means that, for instance, people who have more of this or that beneficial bacteria may perform better when given probiotic (Gibbons et al., 2022). This information can help clinicians decide which interventions may help restore balance in dysbiotic microbiota.

Dietary Interventions: Gut microbiota composition and function is a critically modifiable diet-dependent function. The profiles of the microbiota could be used to generate personalized dietary recommendations, which might improve treatment efficacy for conditions associated with the gut–brain axis. For example, a diet high in fiber and fermented foods may promote the growth of beneficial bacteria and good gut health (Roberfroid, 2007). Also, nutritionally directed at targeted populations of the microbes could help alleviate symptoms of anxiety, depression, and other psychiatric disorders.

Microbiota-Based Therapies: Fecal microbiota transplantation (FMT), a novel therapeutic approach for restoring healthy gut microbiota in dysbiosis, has emerged. Although the use of FMT has shown promise in the treatment of recurrent Clostridium difficile infection and is being explored for potential benefits in the treatment of IBS, ASD, and metabolic disorders (Wang et al., 2019), little is known about its safety and mechanisms of action for these uses. Despite that, FMT protocols are still standardized and optimal donor profiles are identified. Future studies can seek to understand the mechanisms by which FMT affects gut–brain communication and psychological outcomes.

Targeted Probiotics and Prebiotics: Another area where the personalized would be targeted probiotics and prebiotics is psychobiotics. Probiotics that confer mental health benefits through modulation of the gut–brain axis are called psychobiotics (Dinan et al., 2013). Psychobiotic formulation efficacy may be enhanced if tailored to a person's gut microbiota composition and mental health status. But these approaches still need to be rigorously tested in clinical trials to prove if the methods work and precisely what mechanisms are at play for those particular strains.

2.8.3 Future Research Priorities and Potential for New Treatments

The gut–brain axis will offer in future research the opportunity to significantly understand and make clinical applications. In order to uncover how gut microbiota influence brain function through metabolites (Olsson et al., 2022), immune signals, and neural circuits (Wang et al., 2019), mechanistic studies are needed. Diet, stress, and treatments can readily be established as causes by longitudinal tracking of changes in microbiota over time. To validate microbiota-targeted therapies, both symptom relief and long-term outcomes are assessed in clinical trials (Olsson et al., 2022). For novel solutions, it is important to rely on interdisciplinary collaborations in fields such as microbiology, neuroscience, and nutrition. Research is relevant to public health because mental health and gastrointestinal health interrelate. BioMark™ will accelerate biomarkers identification and personalized interventions via emerging technologies, including metabolomics and AI. While many challenges still exist, personalized medicine and microbiota therapies are paving the way for better mental health and well-being (Loh et al., 2024).

2.9 CONCLUSION

A complex and dynamic communication network, the gut–brain axis interlinks the gastrointestinal system to the CNS. It seems research in this field is continuing to be pushed even further as researchers start to realize we can and do have a direct correlation with mental well-being and gut health, and what that means as it pertains to mental illness. Such an understanding highlights the need to investigate how the gut–brain interactions work and whether they can be therapeutic targets. Only recently has the connection between dysbiosis—an imbalance in gut microbial communities—and various mental health conditions such as anxiety, depression, and stress-related disorders become widely recognized. This relationship indicates the possibility of microbiota-aimed therapies, probiotics, or prebiotics in gut health modulation and enhancement of psychological outcomes. Innovative research at this intersection, showing the surprising synergy between microbiome research and mental health treatment, is illustrated by the emerging field of psychobiotics–probiotics that promote mental health. Furthermore, the pathways by which gut microbiota influence the CNS vary: they include the production of neurotransmitters, SCFAs, and immune-signaling molecules. The results indicated that these metabolites, in fact, have direct influence on brain function and behavior, supporting a role for the gut in mental health. Challenges remain in gut–brain axis research, and much more research must be done to elucidate these mechanisms in greater detail to identify specific microbial targets of therapeutic intervention. In addition, our understanding of gut–brain interactions is confounded by the complexity of individual microbiota profiles and of host genetics, and by environmental factors. These challenges may be overcome by solutions involving personalized medicine approaches to tailor interventions based on individual characteristics. Gut health could be profiled as microbiota, and dietary interventions based on gut health could lead to more effective treatment strategies, but studies on humans and animals need to consider ethical issues surrounding such studies. The future holds much promise for interdisciplinary collaborations between microbiologists, neuroscientists, and mental health professionals to guide future work in understanding the gut–brain axis and, more broadly, to develop new modes of therapeutics. This research can totally revolutionize the treatment strategies of mental health disorders and weigh in the importance of good gut health to the overall well-being. Understanding this axis will continue to yield novel therapies for the gut microbiome and offer vast potential to develop therapies targeted at the gut microbiome that are more effective and personalized. MRI data that reveals the future of gut–brain axis research may offer an opportunity to fundamentally change mental healthcare and provide a germ of hope to those suffering from psychiatric disorders and underscore the potential for nutrition and gut health to support one's mental wellness.

REFERENCES

Aggarwal, N., Kitano, S., Puah, G. R. Y., Kittelmann, S., Hwang, I. Y., & Chang, M. W. (2023). Microbiome and human health: Current understanding, engineering, and enabling technologies. Chemical Reviews, 123(1), 31–72. https://doi.org/10.1021/acs.chemrev.2c00431

Altieri, C., Speranza, B., Corbo, M. R., Sinigaglia, M., & Bevilacqua, A. (2023). Gut-microbiota, and multiple sclerosis: Background, evidence, and perspectives. Nutrients, 15(4), 942. https://doi.org/10.3390/nu15040942

Ashique, S., Mohanto, S., Ahmed, M. G., Mishra, N., Garg, A., Chellappan, D. K., Omara, T., Iqbal, S., & Kahwa, I. (2024). Gut-brain axis: A cutting-edge approach to target neurological disorders and potential synbiotic application. Heliyon, 10(13), e34092. https://doi.org/10.1016/j.heliyon.2024.e34092

Baggio, L. L., & Drucker, D. J. (2014). Glucagon-like peptide-1 receptors in the brain: Controlling food intake and body weight. The Journal of Clinical Investigation, 124(10), 4223–4226. https://doi.org/10.1172/JCI78371

Belkaid, Y., & Hand, T. W. (2014). The role of the microbiota in immunity and inflammation. Cell, 157(1), 121–141.

Bercik, P., & Collins, S. M. (2014). The effects of inflammation, infection and antibiotics on the microbiota-gut-brain axis. Advances in Experimental Medicine and Biology, 817, 279–289.

Berthoud, H.-R., & Neuhuber, W. L. (2000). Functional and chemical anatomy of the afferent vagal system. Autonomic Neuroscience, 85(1–3), 1–17.

Bonaz, B., Sinniger, V., & Pellissier, S. (2017). The vagus nerve in the neuro-immune axis: Implications in the pathology of the gastrointestinal tract. Frontiers in Immunology, 8, 1452. https://doi.org/10.3389/fimmu.2017.01452

Bourassa, M. W., Alim, I., Bultman, S. J., & Ratan, R. R. (2016). Butyrate, neuroepigenetics and the gut microbiome: Can a high fiber diet improve brain health? Neuroscience Letters, 625, 56–63.

Bravo, J. A., Forsythe, P., Chew, M. V., Escaravage, E., Savignac, H. M., Dinan, T. G., Bienenstock, J., & Cryan, J. F. (2011). Ingestion of Lactobacillus strain regulates emotional behavior and central GABA receptor expression in a mouse via the vagus nerve. Proceedings of the National Academy of Sciences, 108(38), 16050–16055.

Carabotti, M., Scirocco, A., Maselli, M. A., & Severi, C. (2015). The gut-brain axis: Interactions between enteric microbiota, central and enteric nervous systems. Annals of Gastroenterology, 28(2), 203–209.

Cavaliere, G., Catapano, A., Trinchese, G., Cimmino, F., Penna, E., Pizzella, A., Cristiano, C., Lama, A., Crispino, M., & Mollica, M. P. (2022). Butyrate improves neuroinflammation and mitochondrial impairment in cerebral cortex and synaptic fraction in an animal model of diet-induced obesity. Antioxidants, 12(1), 4. https://doi.org/10.3390/antiox12010004

Chandrasekaran, P., Weiskirchen, S., & Weiskirchen, R. (2024). Effects of probiotics on gut microbiota: An overview. International Journal of Molecular Sciences, 25, 6022. https://doi.org/10.3390/ijms25116022

Clarke, G., Stilling, R. M., Kennedy, P. J., Stanton, C., Cryan, J. F., & Dinan, T. G. (2014). Minireview: Gut microbiota: The neglected endocrine organ. Molecular Endocrinology, 28(8), 1221–1238. https://doi.org/10.1210/me.2014-1108

Cryan, J. F., & Dinan, T. G. (2012). Mind-altering microorganisms: The impact of the gut microbiota on brain and behaviour. Nature Reviews Neuroscience, 13(10), 701–712.

Cryan, J. F., O'Riordan, K. J., Cowan, C. S. M., Sandhu, K. V., Bastiaanssen, T. F. S., Boehme, M., Codagnone, M. G., Cussotto, S., Fulling, C., Golubeva, A. V., Guzzetta, K. E., Jaggar, M., Long-Smith, C. M., Lyte, J. M., Martin, J. A., Molinero-Perez, A., Moloney, G., O'Connor, R., Cruz-Pereira, J. S., Peterson, V. L., Rea, K., Ritz, N. L., Sherwin, E., Spichak, S., Teichman, E. M., Vanhaecke, T., Mikkelsen, K., Spector, T. D., Stilling, R. M., Thompson, L. R., Touret, F., Ventura-Silva, A. P., Villanueva-Pareja, I., Wall, R., & Bastiaanssen, T. F. S. (2019). Psychobiotics and the gut-brain axis: The paradigm of microbiota-gut-brain signals in health and disease. Annual Review of Pharmacology and Toxicology, 59, 297–322.

Cryan, J. F., O'Riordan, K. J., Cowan, C. S. M., Sandhu, K. V., Bastiaanssen, T. F. S., Boehme, M., Codagnone, M. G., Cussotto, S., Fulling, C., Golubeva, A. V., Guzzetta, K. E., Jaggar, M., Long-Smith, C. M., Lyte, J. M., Martin, J. A., Molinero-Perez, A., Moloney, G., Morelli, E., Morillas, E., O'Connor, R., & Dinan, T. G. (2019). The microbiome-gut-brain axis. Nature Reviews Gastroenterology & Hepatology, 16(5), 328–339.

Dantzer, R., O'Connor, J. C., Freund, G. G., Johnson, R. W., & Kelley, K. W. (2008). From inflammation to sickness and depression: When the immune system subjugates the brain. Nature Reviews Neuroscience, 9(1), 46–56.

Davis, C. D. (2016). The gut microbiome and its role in obesity. Nutrition Today, 51(4), 167–174. https://doi.org/10.1097/NT.0000000000000167

de Almeida Magalhães, T., Correia, D., de Carvalho, L. M., Damasceno, S., & Brunialti Godard, A. L. (2017). Maternal separation affects expression of stress response genes and increases vulnerability to ethanol consumption. Brain and Behavior, 8(1), e00841. https://doi.org/10.1002/brb3.841

Dicks, L. M. T. (2022). Gut bacteria and neurotransmitters. Microorganisms, 10(9), 1838. https://doi.org/10.3390/microorganisms10091838

Dinan, T. G., Stanton, C., & Cryan, J. F. (2013). Psychobiotics: A novel class of psychotropic. Biological Psychiatry, 74(10), 720–726. https://doi.org/10.1016/j.biopsych.2013.05.001

Drossman, D. A. (2006). The functional gastrointestinal disorders and the Rome III process. Gastroenterology, 130(5), 1377–1390.

Duman, R. S., & Aghajanian, G. K. (2012). Synaptic dysfunction in depression: Potential therapeutic targets. Science, 338(6103), 68–72.

Dutta, S. K., Verma, S., Jain, V., Surapaneni, B. K., Vinayek, R., Phillips, L., & Nair, P. P. (2019). Parkinson's disease: The emerging role of gut dysbiosis, antibiotics, probiotics, and fecal microbiota transplantation. Journal of Neurogastroenterology and Motility, 25(3), 363–376. https://doi.org/10.5056/jnm19044

Engineer, N. D., Riley, J. R., Seale, J. D., Vrana, W. A., Shetake, J. A., Sudanagunta, S. P., Borland, M. S., & Kilgard, M. P. (2011). Reversing pathological neural activity using targeted plasticity. Nature, 470, 101–104.

Erny, D., Hrabě de Angelis, A. L., Jaitin, D., Wieghofer, P., Staszewski, O., David, E., Keren-Shaul, H., Mahlakoiv, T., Jakobshagen, K., Buch, T., Schwierzeck, V., Utermöhlen, O., Chun, E., Garrett, W. S., McCoy, K. D., Diefenbach, A., Staeheli, P., Stecher, B., Amit, I., & Prinz, M. (2015). Host microbiota constantly control maturation and function of microglia in the CNS. Nature Neuroscience, 18(7), 965–977.

Fattorusso, A., Di Genova, L., Dell'Isola, G. B., Mencaroni, E., & Esposito, S. (2019). Autism spectrum disorders and the gut microbiota. Nutrients, 11(3), 521. https://doi.org/10.3390/nu11030521

Fetissov, S. O. (2017). Role of the gut microbiota in host appetite control: Bacterial growth to animal feeding behaviour. Nature Reviews Endocrinology, 13(1), 11–25. https://doi.org/10.1038/nrendo.2016.150

Föh, B., Buhre, J. S., Lunding, H. B., Moreno-Fernandez, M. E., König, P., Sina, C., Divanovic, S., & Ehlers, M. (2022). Microbial metabolite butyrate promotes induction of IL-10+IgM+ plasma cells. PLoS One, 17(3), e0266071. https://doi.org/10.1371/journal.pone.0266071

Foster, J. A., Rinaman, L., & Cryan, J. F. (2017). Stress & the gut-brain axis: Regulation by the microbiome. Neurobiology of Stress, 7, 124–136. https://doi.org/10.1016/j.ynstr.2017.03.001

Foster, J. A., & McVey Neufeld, K.-A. N. (2013). Gut-brain axis: How the microbiome influences anxiety and depression. Trends in Neurosciences, 36(5), 267–272.

Fu, J., Zheng, Y., Gao, Y., & Xu, W. (2022). Dietary fiber intake and gut microbiota in human health. Microorganisms, 10, 2507. https://doi.org/10.3390/microorganisms10122507

Furness, J. B. (2012). The enteric nervous system and neurogastroenterology. Nature Reviews Gastroenterology & Hepatology, 9(5), 286–294.

Galland, L. (2014). The gut microbiome and the brain. Journal of Medicinal Food, 17(12), 1261–1272. https://doi.org/10.1089/jmf.2014.7000

Gao, L., Davies, D. L., & Asatryan, L. (2022). Sodium butyrate supplementation modulates neuroinflammatory response aggravated by antibiotic treatment in a mouse model of binge-like ethanol drinking. International Journal of Molecular Sciences, 23(24), 15688. https://doi.org/10.3390/ijms232415688

Gao, V., Crawford, C. V., & Burré, J. (2024). The gut-brain axis in Parkinson's disease. Cold Spring Harbor Perspectives in Medicine, a041618. Advance online publication. https://doi.org/10.1101/cshperspect.a041618

Ge, X., Zheng, M., Hu, M., Fang, X., Geng, D., Liu, S., Wang, L., Zhang, J., Guan, L., Zheng, P., Xie, Y., Pan, W., Zhou, M., Zhou, L., Tang, R., Zheng, K., Yu, Y., & Huang, X. F. (2023). Butyrate ameliorates quinolinic acid-induced cognitive decline in obesity models. The Journal of Clinical Investigation, 133(4), e154612. https://doi.org/10.1172/JCI154612

Gibbons, S. M., Gurry, T., Lampe, J. W., Chakrabarti, A., Dam, V., Everard, A., Goas, A., Gross, G., Kleerebezem, M., Lane, J., Maukonen, J., Penna, A. L. B., Pot, B., Valdes, A. M., Walton, G., Weiss, A., Zanzer, Y. C., Venlet, N. V., & Miani, M. (2022). Perspective: Leveraging the gut microbiota to predict personalized responses to dietary, prebiotic, and probiotic interventions. Advances in Nutrition, 13(5), 1450–1461. https://doi.org/10.1093/advances/nmac075

Gil-Campos, M., Aguilera, C. M., Cañete, R., & Gil, A. (2006). Ghrelin, a new hormone involved in energy homeostasis. Endocrine, 30(3), 353–359.

Goodoory, V. C., Khasawneh, M., Black, C. J., Quigley, E. M. M., Moayyedi, P., & Ford, A. C. (2023). Efficacy of probiotics in irritable bowel syndrome: Systematic review and meta-analysis. Gastroenterology, 165(5), 1206–1218. https://doi.org/10.1053/j.gastro.2023.07.018

Guzel, T., & Mirowska-Guzel, D. (2022). The role of serotonin neurotransmission in gastrointestinal tract and pharmacotherapy. Molecules, 27(5), 1680. https://doi.org/10.3390/molecules27051680

Hamamah, S., Aghazarian, A., Nazaryan, A., Hajnal, A., & Covasa, M. (2022). Role of microbiota-gut-brain axis in regulating dopaminergic signaling. Biomedicines, 10(2), 436. https://doi.org/10.3390/biomedicines10020436

Hill-Burns, E. M., Debelius, J. W., Morton, J. T., Wissemann, W. T., Lewis, M. R., Wallen, Z. D., Peddada, S. D., Factor, S. A., Molho, E., Zabetian, C. P., Knight, R., & Payami, H. (2017). Parkinson's disease and the gut microbiome: A case-control study. Movement Disorders, 32(5), 750–758.

Hsiao, E. Y., McBride, S. W., Hsien, S., Sharon, G., Hyde, E. R., McCue, T., Codelli, J. A., Chow, J., Reisman, S. E., Petrosino, J. F., Patterson, P. H., & Mazmanian, S. K. (2013). Microbiota modulate behavioral and physiological abnormalities associated with neurodevelopmental disorders. Cell, 155(7), 1451–1463.

Huang, T. T., Lai, J. B., Du, Y. L., Xu, Y., Ruan, L. M., & Hu, S. H. (2019). Current understanding of gut microbiota in mood disorders: An update of human studies. Frontiers in Genetics, 10, 98. https://doi.org/10.3389/fgene.2019.00098

Jiang, H., Ling, Z., Zhang, Y., Mao, H., Ma, Z., Yin, Y., Wang, W., Tang, W., Tan, Z., Shi, J., Li, L., & Ruan, B. (2015). Altered fecal microbiota composition in patients with major depressive disorder. Brain, Behavior, and Immunity, 48, 186–194.

Jiang, Y., Li, K., Li, X., Xu, L., & Yang, Z. (2021). Sodium butyrate ameliorates the impairment of synaptic plasticity by inhibiting the neuroinflammation in 5XFAD mice. Chemico-Biological Interactions, 341, 109452. https://doi.org/10.1016/j.cbi.2021.109452

Juarez, V. M., Montalbine, A. N., & Singh, A. (2022). Microbiome as an immune regulator in health, disease, and therapeutics. Advanced Drug Delivery Reviews, 188, 114400. https://doi.org/10.1016/j.addr.2022.114400

Kelesidis, T., Kelesidis, I., Chou, S., & Mantzoros, C. S. (2010). Narrative review: The role of leptin in human physiology: Emerging clinical applications. Annals of Internal Medicine, 152(2), 93–100. https://doi.org/10.7326/0003-4819-152-2-201001190-00008

Khlevner, J., Park, Y., & Margolis, K. G. (2018). Brain-gut axis: Clinical implications. Gastroenterology Clinics of North America, 47(4), 727–739. https://doi.org/10.1016/j.gtc.2018.07.002

Kumar, A., Pramanik, J., Goyal, N., Chauhan, D., Sivamaruthi, B. S., Prajapati, B. G., & Chaiyasut, C. (2023). Gut microbiota in anxiety and depression: Unveiling the relationships and management options. Pharmaceuticals, 16(4), 565. https://doi.org/10.3390/ph16040565

Labanski, A., Langhorst, J., Engler, H., & Elsenbruch, S. (2020). Stress and the brain-gut axis in functional and chronic-inflammatory gastrointestinal diseases: A transdisciplinary challenge. Psychoneuroendocrinology, 111, 104501. https://doi.org/10.1016/j.psyneuen.2019.104501

Lam, T. K., Schwartz, G. J., & Rossetti, L. (2005). Hypothalamic sensing of fatty acids. Nature Neuroscience, 8(5), 579–584. https://doi.org/10.1038/nn1456

Lange, L., Berg, G., Cernava, T., Champomier-Vergès, M. C., Charles, T., Cocolin, L., Cotter, P., D'Hondt, K., Kostic, T., Maguin, E., Makhalanyane, T., Meisner, A., Ryan, M., Kiran, G. S., de Souza, R. S., Sanz, Y., Schloter, M., Smidt, H., Wakelin, S., & Sessitsch, A. (2022). Microbiome ethics: Guiding principles for microbiome research, use and knowledge management. Environmental Microbiome, 17, 50. https://doi.org/10.1186/s40793-022-00444-y

Le Morvan de Sequeira, C., Hengstberger, C., Enck, P., & Mack, I. (2022). Effect of probiotics on psychiatric symptoms and central nervous system functions in human health and disease: A systematic review and meta-analysis. Nutrients, 14(3), 621. https://doi.org/10.3390/nu14030621

Levy, M., Kolodziejczyk, A. A., Thaiss, C. A., & Elinav, E. (2017). Dysbiosis and the immune system. Nature Reviews Immunology, 17(4), 219–232.

Loh, J. S., Mak, W. Q., Tan, L. K. S., et al. (2024). Microbiota–gut–brain axis and its therapeutic applications in neurodegenerative diseases. Signal Transduction and Targeted Therapy, 9, 37. https://doi.org/10.1038/s41392-024-01743-1

Louis, P., & Flint, H. J. (2017). Formation of propionate and butyrate by the human gut microbiota. Current Opinion in Clinical Nutrition & Metabolic Care, 20(2), 188–193.

Madison, A., & Kiecolt-Glaser, J. K. (2019). Stress, depression, diet, and the gut microbiota: Human-bacteria interactions at the core of psychoneuroimmunology and nutrition. Current Opinion in Behavioral Sciences, 28, 105–110. https://doi.org/10.1016/j.cobeha.2019.01.011

Menees, S., & Chey, W. (2018). The gut microbiome and irritable bowel syndrome. F1000Research, 7, F1000 Faculty Rev-1029. https://doi.org/10.12688/f1000research.14592.1

Merkouris, E., Mavroudi, T., Miliotas, D., Tsiptsios, D., Serdari, A., Christidi, F., Doskas, T. K., Mueller, C., & Tsamakis, K. (2024). Probiotics' effects in the treatment of anxiety and depression: A comprehensive review of 2014–2023 clinical trials. Microorganisms, 12(2), 411. https://doi.org/10.3390/microorganisms12020411

Miller, A. H., & Raison, C. L. (2016). The role of inflammation in depression: From evolutionary imperative to modern treatment target. Nature Reviews Immunology, 16(1), 22–34. https://doi.org/10.1038/nri.2015.5

Moţăţăianu, A., Şerban, G., & Andone, S. (2023). The role of short-chain fatty acids in microbiota–gut–brain cross-talk with a focus on amyotrophic lateral sclerosis: A systematic review. International Journal of Molecular Sciences, 24, 15094. https://doi.org/10.3390/ijms242015094

Mulak, A., & Bonaz, B. (2015). Brain-gut-microbiota axis in Parkinson's disease. World Journal of Gastroenterology, 21(37), 10609–10620.

Murray, E. R., Kemp, M., & Nguyen, T. T. (2022). The microbiota-gut-brain axis in Alzheimer's disease: A review of taxonomic alterations and potential avenues for interventions. Archives of Clinical Neuropsychology, 37(3), 595–607. https://doi.org/10.1093/arclin/acac008

Nogal, A., Valdes, A. M., & Menni, C. (2021). The role of short-chain fatty acids in the interplay between gut microbiota and diet in cardio-metabolic health. Gut Microbes, 13(1), 1–24. https://doi.org/10.1080/19490976.2021.1897212

Olff, M., Frijling, J. L., Kubzansky, L. D., Bradley, B., Ellenbogen, M. A., Cardoso, C., Bartz, J. A., Yee, J. R., & van Zuiden, M. (2013). The role of oxytocin in social bonding, stress regulation and mental health: An update on the moderating effects of context and interindividual differences. Psychoneuroendocrinology, 38(9), 1883–1894. https://doi.org/10.1016/j.psyneuen.2013.06.019

Olsson, L. M., Boulund, F., Nilsson, S., Khan, M. T., Gummesson, A., Fagerberg, L., Engstrand, L., Perkins, R., Uhlén, M., Bergström, G., Tremaroli, V., & Bäckhed, F. (2022). Dynamics of the normal gut microbiota: A longitudinal one-year population study in Sweden. Cell Host & Microbe, 30(5), 726–739.e3. https://doi.org/10.1016/j.chom.2022.03.002

O'Mahony, S. M., Marchesi, J. R., Scully, P., Codling, C., Ceolho, A. M., Quigley, E. M. M., Cryan, J. F., & Dinan, T. G. (2009). Early life stress alters behavior, immunity, and microbiota in rats: Implications for irritable bowel syndrome and psychiatric illnesses. Biological Psychiatry, 65(3), 263–267.

Plummer, A. M., Matos, Y. L., Lin, H. C., Ryman, S. G., Birg, A., Quinn, D. K., Parada, A. N., & Vakhtin, A. A. (2023). Gut-brain pathogenesis of post-acute COVID-19 neurocognitive symptoms. Frontiers in Neuroscience, 17, 1232480. https://doi.org/10.3389/fnins.2023.1232480

Portincasa, P., Bonfrate, L., Vacca, M., De Angelis, M., Farella, I., Lanza, E., Khalil, M., Wang, D. Q.-H., Sperandio, M., & Di Ciaula, A. (2022). Gut microbiota and short chain fatty acids: Implications in glucose homeostasis. International Journal of Molecular Sciences, 23, 1105. https://doi.org/10.3390/ijms23031105

Raskov, H., Burcharth, J., Pommergaard, H. C., & Rosenberg, J. (2016). Irritable bowel syndrome, the microbiota and the gut-brain axis. Gut Microbes, 7(5), 365–383. https://doi.org/10.1080/19490976.2016.1218585

Ribera, C., Sánchez-Ortí, J. V., Clarke, G., Marx, W., Mörkl, S., & Balanzá-Martínez, V. (2024). Probiotic, prebiotic, synbiotic and fermented food supplementation in psychiatric disorders: A systematic review of clinical trials. Neuroscience and Biobehavioral Reviews, 158, 105561. https://doi.org/10.1016/j.neubiorev.2024.105561

Roberfroid, M. (2007). Prebiotics: The concept revisited. The Journal of Nutrition, 137(3 Suppl 2), 830S–837S. https://doi.org/10.1093/jn/137.3.830S.

Round, J. L., & Mazmanian, S. K. (2009). The gut microbiota shapes intestinal immune responses during health and disease. Nature Reviews Immunology, 9(5), 313–323.

Round, J. L., & Mazmanian, S. K. (2010). Inducible Foxp3+ regulatory T-cell development by a commensal bacterium of the intestine. Proceedings of the National Academy of Sciences, 107(27), 12204–12209.

Sampson, T. R., & Mazmanian, S. K. (2015). Control of brain development, function, and behavior by the microbiome. Cell Host & Microbe, 17(5), 565–576.

Sender, R., Fuchs, S., & Milo, R. (2016). Revised estimates for the number of human and bacteria cells in the body. PLoS Biology, 14(8), e1002533.

Shi, Z., Li, B., & Brooks, V. L. (2015). Role of the paraventricular nucleus of the hypothalamus in the sympathoexcitatory effects of leptin. Hypertension, 66(5), 1034–1041. https://doi.org/10.1161/HYPERTENSIONAHA.115.06017

Silva, Y. P., Bernardi, A., & Frozza, R. L. (2020). The role of short-chain fatty acids from gut microbiota in gut-brain communication. Frontiers in Endocrinology, 11, 25. https://doi.org/10.3389/fendo.2020.00025

Sittipo, P., Choi, J., Lee, S., & Lee, Y. K. (2022). The function of gut microbiota in immune-related neurological disorders: A review. Journal of Neuroinflammation, 19(1), 154. https://doi.org/10.1186/s12974-022-02510-1

Slavin, J. L. (2013). Fiber and prebiotics: Mechanisms and health benefits. Nutrients, 5(4), 1417–1435.

Sperber, A. D., Dumitrascu, D., Fukudo, S., Gerson, C., Ghoshal, U. C., Gwee, K. A., Hungin, A. P. S., Kang, J. Y., Minhu, C., Schmulson, M., Bolotin, A., Friger, M., Freud, T., & Whitehead, W. (2017). The global prevalence of IBS in adults remains elusive due to the heterogeneity of studies: A Rome Foundation working team literature review. Gut, 66(6), 1075–1082. https://doi.org/10.1136/gutjnl-2015-311240

Strandwitz, P., Kim, K. H., Terekhova, D., Liu, J. K., Sharma, A., Levering, J., McDonald, D., Dietrich, D., Ramadhar, T. R., Lekbua, A., Mroue, N., Liston, C., Stewart, E. J., Dubin, M. J., Zengler, K., Knight, R., Gilbert, J. A., Clardy, J., & Lewis, K. (2019). GABA-modulating bacteria of the human gut microbiota. Nature Microbiology, 4(3), 396–403.

Sudo, N., Chida, Y., Aiba, Y., Sonoda, J., Oyama, N., Yu, X. N., Kubo, C., & Koga, Y. (2004). Postnatal microbial colonization programs the hypothalamic-pituitary-adrenal system for stress response in mice. The Journal of Physiology, 558(1), 263–275.

Tillisch, K., Labus, J., Kilpatrick, L., Jiang, Z., Stains, J., Ebrat, B., Guyuron, F., Kundu, R., Swan, B., Wu, V., Wade, S., & Mayer, E. A. (2013). Consumption of fermented milk product with probiotic modulates brain activity. Gastroenterology, 144(7), 1394–1401.

Tracey, K. J. (2009). Reflex control of immunity. Nature Reviews Immunology, 9(6), 418–428.

Turnbaugh, P. J., Ley, R. E., Hamady, M., Fraser-Liggett, C. M., Knight, R., & Gordon, J. I. (2007). The human microbiome project. Nature, 449, 804–810.

Varesi, A., Pierella, E., Romeo, M., Piccini, G. B., Alfano, C., Bjørklund, G., Oppong, A., Ricevuti, G., Esposito, C., Chirumbolo, S., & Pascale, A. (2022). The potential role of gut microbiota in Alzheimer's disease: From diagnosis to treatment. Nutrients, 14(3), 668. https://doi.org/10.3390/nu14030668

Vrieze, A., Van Nood, E., Holleman, F., Salojärvi, J., Kootte, R. S., Bartelsman, J. F., Dallinga-Thie, G. M., Ackermans, M. T., Serlie, M. J., Oozeer, R., Derrien, M., Druesne, A., Van Hylckama Vlieg, J. E., Bloks, V. W., Groen, A. K., Heilig, H. G., Zoetendal, E. G., Stroes, E. S., de Vos, W. M., ... Nieuwdorp, M. (2012). Transfer of intestinal microbiota from lean donors increases insulin sensitivity in individuals with metabolic syndrome. Gastroenterology, 143(4), 913–916.e7. https://doi.org/10.1053/j.gastro.2012.06.031

Wang, J. W., Kuo, C. H., Kuo, F. C., Wang, Y. K., Hsu, W. H., Yu, F. J., Hu, H. M., Hsu, P. I., Wang, J. Y., & Wu, D. C. (2019). Fecal microbiota transplantation: Review and update. Journal of the Formosan Medical Association, 118(Suppl 1), S23–S31. https://doi.org/10.1016/j.jfma.2018.08.011

Wang, L., Alammar, N., Singh, R., Nanavati, J., Song, Y., Chaudhary, R., & Mullin, G. E. (2020). Gut microbial dysbiosis in the irritable bowel syndrome: A systematic review and meta-analysis of case-control studies. Journal of the Academy of Nutrition and Dietetics, 120(4), 565–586. https://doi.org/10.1016/j.jand.2019.05.015

Wu, Y., Xu, H., Tu, X., & Gao, Z. (2021). The role of short-chain fatty acids of gut microbiota origin in hypertension. Frontiers in Microbiology, 12, 730809. https://doi.org/10.3389/fmicb.2021.730809

Xiong, R. G., Li, J., Cheng, J., Zhou, D. D., Wu, S. X., Huang, S. Y., Saimaiti, A., Yang, Z. J., Gan, R. Y., & Li, H. B. (2023). The role of gut microbiota in anxiety, depression, and other mental disorders as well as the protective effects of dietary components. Nutrients, 15(14), 3258. https://doi.org/10.3390/nu15143258

Yang, D., Zhao, D., Ali Shah, S. Z., Wu, W., Lai, M., Zhang, X., Li, J., Guan, Z., Zhao, H., Li, W., Gao, H., Zhou, X., & Yang, L. (2019). The role of the gut microbiota in the pathogenesis of Parkinson's disease. Frontiers in Neurology, 10, 1155. https://doi.org/10.3389/fneur.2019.01155

Yano, J. M., Yu, K., Donaldson, G. P., Shastri, G. G., Ann, P., Ma, L., Nagler, C. R., Ismagilov, R. F., Mazmanian, S. K., & Hsiao, E. (2015). Indigenous bacteria from the gut microbiota regulate host serotonin biosynthesis. Cell, 161(2), 264–276.

Zeevi, D., Korem, T., Zmora, N., Israeli, D., Rothschild, D., Weinberger, A., Ben-Yacov, O., Lador, D., Avnit-Sagi, T., Lotan-Pompan, M., Suez, J., Mahdi, J. A., Matot, E., Malka, G., Kosower, N., Rein, M., Zilberman-Schapira, G., Dohnalová, L., Pevsner-Fischer, M., ... Segal, E. (2015). Personalized nutrition by prediction of glycemic responses. Cell, 163(5), 1079–1094. https://doi.org/10.1016/j.cell.2015.11.001

Zhang, J., Zhu, G., Wan, L., Liang, Y., Liu, X., Yan, H., Zhang, B., & Yang, G. (2023). Effect of fecal microbiota transplantation in children with autism spectrum disorder: A systematic review. Frontiers in Psychiatry, 14, 1123658. https://doi.org/10.3389/fpsyt.2023.1123658

Zou, B., Li, J., Ma, R. X., Cheng, X. Y., Ma, R. Y., Zhou, T. Y., Wu, Z. Q., Yao, Y., & Li, J. (2023). Gut microbiota is an impact factor based on the brain-gut axis to Alzheimer's disease: A systematic review. Aging and Disease, 14(3), 964–1678. https://doi.org/10.14336/AD.2022.1127

3 Introduction and Overview of Prebiotics, Probiotics, and Postbiotics

Phool Chandra[1,*], *Nishat Fatma*[2], *and Neetu Sachan*[3]

[1]Teerthanker Mahaveer College of Pharmacy, Teerthanker Mahaveer University, Moradabad, Uttar Pradesh, India
[2]MIT College of Pharmacy Moradabad, Uttar Pradesh, India
[3]Maharana Pratap College of Pharmacy, Mandhana Kanpur, Uttar Pradesh, India

3.1 INTRODUCTION

Human health is greatly influenced by the homeostasis of the gut microbiota. Unfortunately, existing conventional treatments for certain gut microbiota-related illnesses only address the symptoms; it is challenging to address the underlying reasons, and they may even have unfavorable side effects like disruptions of the gut microbiota. Probiotics, prebiotics, and postbiotics can protect and treat a variety of diseases, according to an increasing number of clinical studies and evidences (1). The gastrointestinal (GI) microbiota, or group of microorganisms living in the human GI region, is made up of approximately a thousand of symbiotic and/or pathogenic microorganisms, which include viruses (such as bacteriophages), bacteria, or archaea, as well as unicellular eukaryotes, along with fungi and other nonbacterial or non-archaeal microbial species. The human gut microbiota, which is made up of billions of different microorganisms, has evolved to accommodate the intestine's distinctively varied conditions. The microbiota uses reciprocal biochemical communication for interaction with the host and quickly adapt to changes in the food that is ingested. This allows the host to maintain optimal immunological function and facilitate the exchange of resources (2). The host and microbiota interact as a unit to protect the equilibrium from possible pathogen invasion and to go through natural selection. Although there doesn't seem to be a single, ideal gut microbiota composition, dysbiosis, a disruption of the microbiota composition, is frequently linked to human illness, suggesting that human health depends on a balanced population of microbes. The host evolves together to the point where it not only tolerates but also requires commensals beneficial microbes to colonize it in order to support various elements of immunological development as well as function. On the other hand, dysbiosis, or abnormalities in the equilibrium of the microbiota in the gut, might be involved in developing certain illnesses. It is thought that human disorders such as inflammatory bowel disease (IBD)—which is described by Crohn's disease as well as ulcerative colitis—weight gain, type 2 diabetes, and more distant systemic diseases like cardiovascular diseases, specific cancers, and neurodegenerative disorders are caused by defects in mucosal tolerance (3–5). The use and source of probiotics and prebiotics continues to be influenced by personalized nutrition and precision medicine, as interest in modifying the microbial biomarkers of health and illness develops (Figures 3.1 and 3.2). Because probiotics, prebiotics, and postbiotics can alter the gut microbiota to enhance well-being and manage illness, their significance in medical research has also increased, as shown in Tables 3.1 and 3.2 (6).

3.2 PREBIOTICS

3.2.1 Definition and Characteristics

In 1995, Glenn Gibson and Marcel Roberfroid presented the prebiotics idea for the first time. It is now defined as "a substrate that is specifically consumed by host microorganisms providing a health benefit" instead of "a prebiotic." Prebiotics can be directly applied to various body areas that have been colonized by microbes, like the skin and vagina, even though the majority of them are now taken orally. A prebiotic is a type of indigestible food element that acts as a selective expansion and/or activity booster for one or a small number of beneficial organisms in the colon, hence enhancing the host's well-being (7). Prebiotics can be found in food ingredients such as some lipids (ethers and esters), various peptides and proteins, and nondigestible carbohydrates (oligo- and polysaccharides). These substances' chemical structure prevents them from being hydrolyzed by human enzymes used for digestion or absorbed in the upper gastrointestinal tract. One term for these gradients would be "colonie foods," meaning meals that enter the colon and act as intermediates for the endogenous colonies of microorganisms so indirectly supplying the host with energy, necessary micronutrients, and metabolic substrates. Many fermentable carbohydrates have been shown to have

DOI: 10.1201/9781003536802-3

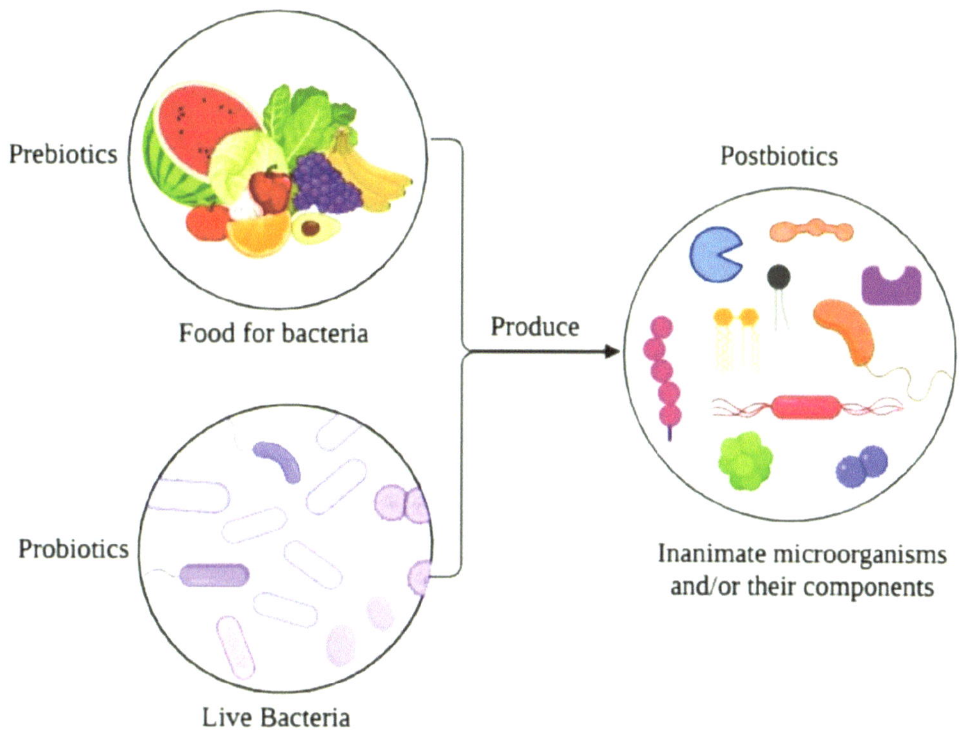

FIGURE 3.1 Source of prebiotics, probiotics, and postbiotics.

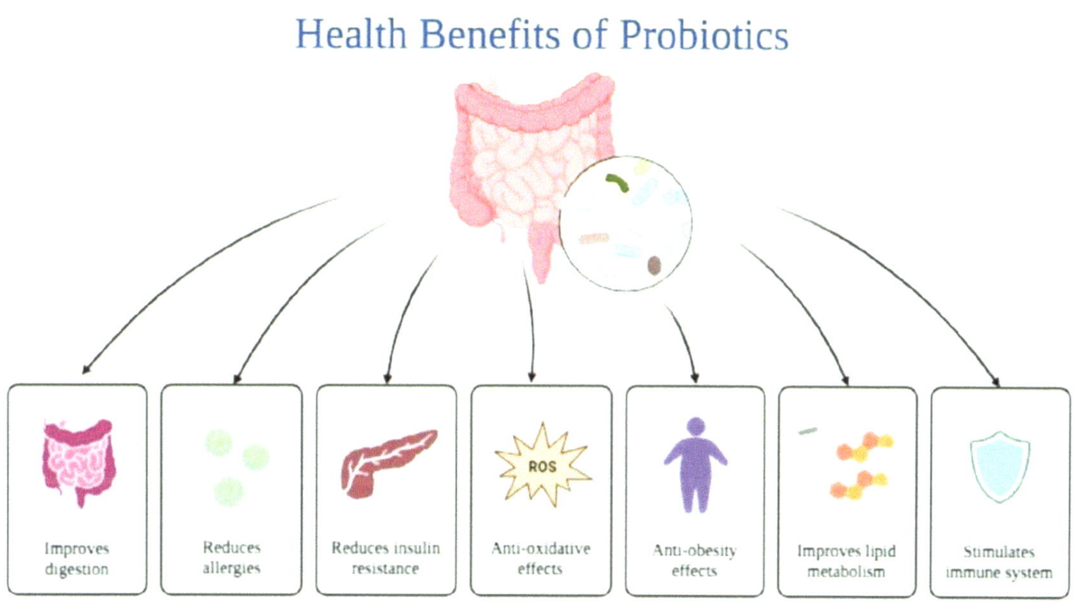

FIGURE 3.2 Health benefits of probiotics.

a prebiotic effect; however, the nondigestible oligosaccharides fructans and galactans are the dietary prebiotics with the most extensive collection of research supporting their health advantages for humans. Bifidobacteria preferentially digest certain oligosaccharides (8). This phenomenon can be described by structure-to-function correlations; the enzymes β-fructanosidase and β-galactosidase, which are widely present in bifidobacteria, can easily break down the linking connections in FOS and GOS, respectively. Additionally, it appears that this species preferentially metabolizes the degree of polymerization (DP) or length of chain size typical of oligosaccharides. A crucial prerequisite

TABLE 3.1
Source, Function, and Stability of Prebiotics, Probiotics, and Postbiotics

Category	Prebiotics	Probiotics	Postbiotics
Definition	Nondigestible food ingredients that stimulate the growth and activity of beneficial bacteria	Live microorganisms that provide health benefits when consumed in adequate amounts	Metabolic by-products or compounds produced by probiotics during fermentation
Sources	Dietary fibers, fruits, vegetables, and whole grains	Fermented foods, supplements	Fermented foods, supplements, metabolic extracts
Function	Promote the growth of beneficial gut bacteria, improve digestion, and enhance immune function	Enhance gut microbiota, support immune function, prevent and treat gastrointestinal issues, and improve digestion	Exert health benefits such as anti-inflammatory effects, modulation of the immune system, and gut health
Stability	Generally stable during food processing and digestion	Sensitive to heat, acid, and storage conditions; require proper formulation for stability	Generally, more stable than probiotics, as they are not live organisms
Examples	Inulin, fructooligosaccharides (FOS), galactooligosaccharides (GOS), and resistant starch	*Lactobacillus*, *Bifidobacterium*, *Saccharomyces boulardii*, yogurt, kefir, and sauerkraut	Short-chain fatty acids (SCFAs), enzymes, peptides, cell wall fragments, and butyrate

TABLE 3.2
Health Benefits and Challenges of Prebiotics, Probiotics, and Postbiotics

Category	Types	Health Benefits	Challenges
Prebiotics	Fructans Galactooligosaccharides Starch and glucose-derived oligosaccharides Resistant starch Hemicellulose-derived oligosaccharides	• Promote growth of beneficial gut bacteria • Improve digestion • Enhance immune function • Aid in calcium absorption	• Can cause gas and bloating in some individuals • Limited food sources • Variable individual responses
Probiotics	*Lactobacillus* species *Bifidobacterium* species *S. boulardii* *Streptococcus thermophilus*	• Improve gut health • Enhance immune function • Prevent and treat diarrhea • Reduce symptoms of irritable bowel syndrome (IBS) and inflammatory bowel disease (IBD)	• Stability during processing and storage • Survival through the gastrointestinal tract • Potential for overgrowth in immunocompromised individuals
Postbiotics	Metabolites produced by probiotics Short-chain fatty acids (SCFAs) Bacterial lysates	• Reduce inflammation • Enhance immune response • Support gut barrier function • Potential anticancer properties	• Standardization and identification of active components • Stability during storage and processing • Limited research on long-term effects

for prebiotic selection in the target regions and an indication of their capacity to perform so in a competitive developing mixed-cultural ecosystems like the human gut is the presence of the proper transport machinery to gather up and supply these substrates within the microbial cytoplasm. Prebiotics are not substances that modify the microbiota's composition by means other than those requiring the host microbes to actively utilize them (9).

3.2.2 Types of Prebiotics

Prebiotics are available in a wide variety of forms. Most of them are made up mostly of oligosaccharide carbohydrates (OSCs), which are a subset of the several types of carbohydrates. Most pertinent publications focus on OSCs; however, there are also certain findings indicating that prebiotics are more than just carbohydrates (10).

3.2.2.1 Fructans

A form of small-chain carbohydrates, or sugars, called fructans are polymers of the fructose molecule. It is made up of inulin and FOS, also known as oligofructose, which is a linear chain of fructose joined by a β-bond. Humans lack the digestive enzyme fructans, which prevents it from being broken down in the small intestine. Thus, the gut bacteria are capable of fermenting it. Numerous fruits and vegetables, including barley, wheat, rye, ripe bananas or onions, artichokes, and garlic, as well as cereal and grains, can contain it. Foods can also be supplemented with fructans to improve their fiber content. According to certain earlier research, fructans have the ability to especially enhance the growth of lactic acid bacteria (11).

3.2.2.2 Galactooligosaccharides

Galactooligosaccharides, which are lactose extension products, are separated into two subgroups: (i) GOS that include additional galactose at C3, C4, or C6, and (ii) GOS that are made by enzymatic *trans*-glycosylation of lactose. The main by-product of this process is a mixture of tri- to pentasaccharides with galactose in β (1→6), β (1→3), and β (1→4) linkages. TOS, standing for *trans*-galactooligosaccharides, is another name for this kind of GOS. Galactooligosaccharides are the primary plant sugar-based component of prebiotics. They are comprised of terminal glucose and galactosyl sequences connected by β-glycosidic linkages. It marginally promotes Firmicutes, Bacteroidetes or Enterobacteria, Bifidobacteria, and Lactobacilli. GOSs have a strong stimulating impact on *Bifidobacterium* and *Lactobacillus*. GOS also stimulates Firmicutes, Bacteroidetes, and Enterobacteria, but not as much as it does *Bifidobacteria*. It contributes in avoiding the onset of dermatitis, constipation, obesity, and colic (a condition that prevents newborns from screaming excessively). It helps those who are lactose-intolerant with symptoms like bloating, cramping, and stomach ache (12, 13).

3.2.2.3 Starch and Glucose-Derived Oligosaccharides

This particular form of starch is unable to be broken down in the small intestine. The combination of starch and starch breakdown products, or resistant starches, are made up of four main groups: RS1, RS2, RS3, and RS4. Resistant starch (RS) is a type of starch that can't be broken down in the upper gastrointestinal tract. Because RS produces a high amount of butyrate, which is known to benefit health, it has been proposed that RS belongs to the prebiotic category. Different Firmicute groups exhibit the maximum assimilation with a high concentration of RS. A study conducted in vitro suggests that *Bifidobacterium adolescentis*, *Eubacterium rectale*, and *Bacteroides thetaiotaomicron* may also break down RS, but to a lower extent. However, in the absence of *Ruminococcus bromii*, RS breakdown is not feasible in the combination of bacterial and fecal incubations. Because resistant starch produces a lot of butyrate, it is beneficial to health. It promotes the growth of lactobacilli and bifidobacteria, two types of good bacteria, in the digestive system (14).

3.2.2.4 Hemicellulose-Derived Oligosaccharides

Hemicellulose-derived oligosaccharides (HDOs), which include mannooligosaccharides (MOS) or Xylooligosaccharides (XOS), are the emerging prebiotics. Xylooligosaccharides are made up of xylose units arranged in short-chain sugar oligomers. It is made up of xylotetraose, xylotriose, and xylobiose. Enzymatic hydrolysis and xylan-degrading enzymes like *endo*-β-1,4-xylanase or xylan 1,4-β-xylosidase are two methods that can be used to obtain it. Fruits, vegetables, milk, or honey, bamboo shoots, and corn cobs, wheat straw, sugarcane bagasse, and hardwood xylan are among the foods that naturally contain it. Mannooligosaccharides are short chains of repeated mannose units joined by glycosidic linkages. They are a type of nonstarch hemicellulosic polysaccharide generated from mannan. The two types are α-MOS and β-MOS. The breakdown of *Saccharomyces cerevisiae* cell wall can provide α-MOS by physicochemical hydrolysis. A mixture of β-1,4-mannanase, β-mannosidase, or α-galactosidase, as well as β-glucosidase enzymes, can hydrolyze to produce β-MOS. Further, it is a prebiotic of the second generation. It is found in guar, konjac, and locust bean gums. By raising the proliferation of probiotics, it lowers pH and enhances the structural components of the gut. With their strong prebiotic activity, arabinoxylans can be generated. It improves one's overall health. It reduces the possibility of a number of intricate medical issues (15).

3.2.3 Mechanisms of Action

Prebiotics can generally withstand digestion in the smaller intestines by remaining in the gastrointestinal system due to the absence of enzymes in the human gut that break down their polymer bonds. After that, the body transports the intact prebiotics to the bigger intestine, where the gastrointestinal flora breaks them down and ferments them in a certain way to form certain secondary metabolites. These metabolites are then absorbed by the epithelium of the intestine or go to the liver via the portal vein. These metabolites can have positive effects on host physiological functions, including immunity regulation or pathogen resistance, improved intestinal barrier function, increased absorption of minerals, and decreased blood lipid levels (16). Beneficial bacteria, such as acetate, butyrate, and propionate, metabolize the most prevalent SCFAs in the colon and are helpful in preserving systemic and gastrointestinal wellness. Additionally, one particular benefit of prebiotics is that they support the developmental process of the target bacteria. By defending or encouraging the creation of advantageous fermentation products, they might encourage the proliferation of profitable microbes to compete with different species after consumption of particular prebiotics (for instance, inulin, FOS, along with GOS). Furthermore, the beneficial effects of prebiotics extend beyond the gastrointestinal tract to include the central nervous, immunological, and cardiovascular systems (17).

3.3 PROBIOTICS

3.3.1 Definition and Characteristics

The theory of conventional probiotics originated from the findings of Elie Metchnikoff in 1907, who discovered a connection between the longevity and well-being of elderly Bulgarians and their regular use of fermented dairy items high in lactobacilli (LAB), such as yogurt. According to Metchnikoff, the natural gut microbiota may benefit from the presence of bacteria. Since then, germs that improve the host's health have evolved to represent probiotics. Probiotics are termed after the Greek words "pro" and "bios," which mean "for life." Elie Metchnikoff made the initial discovery that consuming lactic acid bacteria can extend one's life. Probiotics are defined as chemicals and organisms that support the equilibrium of intestinal microbiota (18). These include hepatic encephalopathy, indigestion, gastroenteritis, stomach acidity, colitis, diarrhea, constipation, recolonization by microorganisms, immunostimulant, hypercholesterolemia, and carcinogenesis. Probiotics are defined as "living microorganisms that, once administered in adequate quantities, confer a health benefit on the host showed in" by the International Scientific Association for Probiotics and Prebiotics (19).

3.3.2 Types of Probiotics

Numerous types of bacteria can be found in probiotics. The most prevalent types of bacteria are those found in the *Lactobacillus* and *Bifidobacterium* genera. Probiotics can also be made from other bacteria or yeasts, for example, *Saccharomyces boulardii*. Considering its description, it is astounding how many different types of microorganisms can have probiotic effects. Probiotics come in a vast range of microorganisms and can be generically categorized into three primary groups: Bifidobacteria, Lactobacilli, and others. Since the LAB group is the most represented probiotic within the LAB group, the majority of current research on probiotic species focuses on this group (20, 21).

3.3.2.1 *Lactobacillus*

In the investigation of human gastrointestinal microorganisms, which is very closely linked to human health, *Lactobacillus* is a crucial probiotic. It not only permits the body synthesis of vital vitamins and amino acids and absorbs minerals, but it also has the potential to improve intestinal microecology by preventing the growth of pathogenic microorganisms. They are typical and generally safe. They are found in the human vagina and gastrointestinal system, where they coexist in symbiotic relationships and constitute a minor fraction of the gastrointestinal flora. Its surroundings become acidic because to lactic acid generation, which stops certain dangerous bacteria from growing. For the production of B-complex vitamins and the maintenance of immune quiescence, lactic acid bacteria are crucial to overall health. When *Lactobacillus* spores are taken orally, they travel via the stomach and enter the duodenum, where they quickly germinate and proliferate. *Lactobacillus* is active in metabolism in the intestine upon germination, which produces lactic acid (22).

3.3.2.2 *Bifidobacterium*

The name *Bifidobacterium* comes from the genus of specialized, Gram-positive anaerobic bacteria that frequently splits at the end. It is a type of physiological bacteria that is essential to human health and is found in large quantities in probiotic foods. In order to maintain intestinal health, *Bifidobacterium* may adapt to anaerobic gastrointestinal life, multiply and metabolize at the middle and ends of the small and large intestines, and exude bifidogenic substances that have probiotic benefits. At present, there are 32 species along with 9 subspecies of *Bifidobacterium*; 14 of these have been identified in human beings. They can be found as bifurcated 204 Y-shaped rods, club-shaped rods, and short curved rods. About 30 bifidobacteria species have been identified. The species that are utilized as probiotics include *Bifidobacterium breve*, *Bifidobacterium longum* or *Bifidobacterium infantis*, *Bifidobacterium adolescentis*, *Bifidobacterium bifidum* as well as *Bifidobacterium animalis*, *Bifidobacterium thermophilum*, and *Bifidobacterium lactis*. The organism's bifid Y-shaped morphology is the source of the name (23).

3.3.2.3 Other Bacteria Species

Gram-positive Parthenobacteria, such as *Enterococcus*, are widely employed in the food business nowadays, in addition to *Lactobacillus* and *Bifidobacterium*. One important characteristic of *Enterococcus strains* as probiotics is their capacity to coexist, compete, and adhere to host cells in the colon. A probiotic yeast strain known for its possible health advantages is *S. boulardii*. It is well-known for its antibiotic resistance and capacity to endure in the gastrointestinal system. *S. boulardii* is a probiotic supplement that has been researched for its ability to prevent and treat a range of gastrointestinal disorders. A popular lactic acid bacteria employed in the production of dairy goods like cheese and yogurt is *Streptococcus*. Because of its significant health benefits and capacity to maintain passage across the gastrointestinal tract, it is regarded as a probiotic strain. The species *thermophilus* aids in fermentation and could benefit intestinal health. The probiotic bacteria *Propionibacterium freudenreichii* is mostly utilized in the making of Swiss cheese. It is well renowned for its capacity to generate propionic acid, which enhances the flavors and smoothness of the cheese (24, 25).

3.3.3 Mechanisms of Action

Probiotics maintain the balance of gut microbiota, thereby preventing the production of inflammatory mediators: the body's continual reaction to potentially dangerous antigens that compromise intestinal integrity. IgA production is stimulated, and this is critical for mucosal immunity. Increased production of interferon-γ can activate macrophages, neutrophils, and natural killer (NK) cells, enhancing their capacity to eliminate pathogens and tumor cells. Additionally, it can promote the expression of major histocompatibility complex (MHC) class I and II proteins, thereby improving antigen presentation. By competing with other bacteria for adhesion sites and increasing mucus production, gastrointestinal microflora prohibits other bacteria from sticking around (26). The strong acids produced by lactobacilli and bifidobacteria, such as lactic and acetic acids, lower the pH of the intestines (5.8). Lower blood ammonia levels may arise from the conversion of poisonous ammonia (NH_3) to the nondiffusible ammonium ion (+NH), which is one impact of lowering the gastrointestinal tract's pH. Although the development of *Escherichia coli* and *Clostridium perfringens* was totally suppressed at pH 5.0 and 4.5, the rate of growth of bifidobacteria was mostly unaffected at low pH. Propionic or butyric acid, which probiotics generate, encourages the growth and differentiation of intestinal epithelial cells. Specific goblet cells release glycoproteins that condense into viscous gels that ensnare bacteria and prevent them from diffusing too far into the intestinal epithelium. Additionally, these bacteria create digestive enzymes like lysozyme, casein phosphatase, and lactase, which may increase dairy products' ability to digest and tolerate lactose (27, 28).

3.4 POSTBIOTICS

3.4.1 Definition and Characteristics

The word "postbiotic" originates from the Greek terms "bios," which means life, and "post," which means after. Furthermore, bacteria (or their substrates) are the center of the "biotic" group of terms, which include probiotics, prebiotics, symbiotics, and postbiotics. Postbiotics are described as "a preparation of inactive microorganisms &/or their components that provides a health benefit on the host" by the International Scientific Association for Probiotic and Prebiotics. With new technology, approaches, and applications being explored, it is conceivable to change scientific understanding of the benefits these postbiotic chemicals have on a person's diet and general health. They consist of many different substances, including vitamins, bacteriocins, or functional proteins, as well as peptides, SCFAs, polyamines, inactivated microbial cells, and various other bioactive metabolites. Studies have shown that dead cells, either complete or broken, may also have important health effects on humans (29). While postbiotics (metabolic products of probiotics), which are formed from probiotic cells, have health advantages for the host when taken in sufficient amounts, the health impacts of intestinal microbiomes rely on the vitality of the microbiome and on inactive microbiome products (30, 31).

3.4.2 Types of Postbiotics

3.4.2.1 Cell-Free Supernatants

Directly from cell cultures, one can collect cell-free supernatants that contain physiologically active compounds formed by yeast and bacteria through the surrounding liquid. Following the incubation time, the microorganisms are extracted using centrifugation. To guarantee sterility, the resultant mixture is lastly filtered. Various microorganisms' cultures yield supernatants, which exhibit varying activity. On gastrointestinal epithelial cells, macrophages, and neutrophils, *Lactobacillus acidophilus* and *Lactobacillus casei* supernatants have anti-inflammatory or antioxidant effects by decreasing the release of the pro-inflammatory cytokine tumor necrosis factor α (TNF-α) and improving the secretion of the anti-inflammatory cytokine interleukin-10 (IL-10). In the meantime, colon cancer cells can't invade when supernatants from *Lactobacillus rhamnosus* GG and *L. casei* cultures are used. Because cell-free supernatants have direct anticancer activity and can lower oxidative stress in vivo, they may be clinically helpful in the treatment of cancer (32). Conversely, it was discovered that the morphological composition and maturation of the gastrointestinal barrier were positively impacted by *Lactobacillus plantarum* supernatants. Early delivery of these supernatants to lambs was linked to a reduction in the quantity of intestinal pathogens and an increase in the intestine's absorption surface (33).

3.4.2.2 Exopolysaccharides

Microorganisms develop biopolymers with various chemical characteristics as they proliferate. These biopolymers have the ability to break apart from the bacterial cell wall and generate a diverse collection of materials known as exopolysaccharides (EPSs). Although their biological role is unclear, EPSs are presently used as water-binding, stabilizing, and emulsifying agents in the food business. However, there has been a new interest in the usage of EPSs in functional foods and pharmaceutical products. By connecting with dendritic cells (DCs) or macrophages and promoting T and NK cell proliferation, EPSs have the possibility of influencing the immunological response. Furthermore, in an in vitro model, an EPS isolated from tofu, a product of *L. plantarum*, increased the phagocytic capability of macrophages and triggered nitric oxide (NO) secretion. Meanwhile, the efficacy of the vaccination against foot-and-mouth disease was enhanced by the adjuvant use of an EPS produced from *L. casei*. Certain EPSs that are generated by *Lactobacillus* strains that are separated from fermented durian fruit have antioxidant and antibacterial characteristics. It has been demonstrated that the antioxidant potential of uronic acid, an EPS derived from *Lactobacillus helveticus* that is also remarkably responsible for the antioxidant qualities of green tea, can be explained by its capacity to bind iron ions (34, 35).

3.4.2.3 Enzymes

Reactive oxygen species (ROS) can harm lipids, proteins, or carbohydrates, including nucleic acids. To combat this, microorganisms have established defense mechanisms. Antioxidant enzymes, specifically glutathione peroxidase (GPx) and peroxide dismutase (SOD), catalase, or NADH-oxidase, are essential in battling against ROS. It was discovered that two strains of *Lactobacillus fermentum* had a high concentration of GPx, and they were subsequently shown to exhibit strong antioxidant capabilities in vitro. Postbiotics made from *L. plantarum* were shown to have antioxidant qualities. Furthermore, in a rat model of Crohn's disease, genetically altered *Lactobacillus* strains that produce catalase or SOD were superior to their unaltered counterparts in terms of symptom relief. Moreover, strains of *Lactobacillus* that exhibited elevated catalase activity were superior at mitigating inflammation (36).

3.4.2.4 Bacterial Lysates

The process of obtaining bacterial lysates (BLs) involves the chemical or mechanical breakdown of both Gram-positive and Gram-negative bacteria that are frequently seen in the environment. The concept of gut–lung axis or the functional link between the respiratory system and the intestine's immune system is the basis for their therapeutic application. Specifically, research has demonstrated that lyophilized B cells taken orally go to the Peyer's patches in the small intestinal tract, where they activate DCs and ultimately activate T and B cells. After that, mature lymphocytes move to the respiratory tract's mucous membrane, where they first activate the innate immune system and encourage the release of IgA. In fact, numerous clinical trials on a range of illnesses, including children's recurrent upper respiratory tracts infections, have validated the safety of using BLs. According to the hygiene hypothesis, there appears to be a systematic link between an increase in allergy illnesses and a decline in infection incidence in highly industrialized nations. Thus, in the event of inadequate access to microorganisms, employing BLs, which simulate the appearance of bacteria to activate the immune system, is a compelling strategy (37, 38).

3.4.3 Mechanisms of Action

The majority of human illnesses are linked to dysbiosis in the gastrointestinal microbiota, whose balance is essential for preserving health and postponing the onset of many illnesses. Diseases such as obesity, inflammation, and cancer may become more likely to occur when the natural gut microbiota is disturbed. Probiotics, prebiotics, or postbiotics can aid in the formation and alteration of the immune system, as well as the structure and functioning of the gut microbiota and the intestinal barrier. According to the most recent studies on the subject, postbiotics are crucial to the "addition" method through which probiotics treat illnesses, which primarily involve altering the gut flora. The "addition" procedure entails using postbiotics which are secreted by probiotics to modify the gut microbiota in order to treat or slow the progression of disease. The host's general health is enhanced by these postbiotics, which contain SCFAs, proteins, peptides or hormonal substances, and neuroactive substances. The metabolic capacities of viable probiotic bacteria are the main emphasis of the "subtraction" mechanism. These probiotic bacteria minimize the gastrointestinal absorption of these dangerous substances and reduce their influence on the body by breaking down or absorbing hazardous metabolites formed by metabolic disorders and dangerous external substances. This procedure eventually aids in the alleviation and management of the related disorders by concurrently modifying the gut flora and lowering shock (2, 39).

3.5 SYNERGIES AND INTERACTIONS

The combined effects of prebiotics, probiotics, and postbiotics enhance gut health and general well-being. Prebiotics refer to indigestible food parts, typically fibers, that function as substrates to encourage the expansion and activity of advantageous gut flora (2). Prebiotics are substances that the host is unable to digest, but which probiotics can utilize to ferment and aid in the multiplication and metabolism of gastrointestinal probiotics for the body's overall health. Probiotic microorganisms generate postbiotics, which have antibacterial, immune-modulating, and anti-inflammatory qualities. These abilities improve the defense against infection and contributes to health-promoting elements (40). Prebiotics have the ability to stimulate the growth of gut flora, whereas probiotics establish postbiotics by "adding" to control the gut microbiota and "removing" toxic metabolites and external substances to lessen their effects on the body and help prevent or treat disease (41).

3.6 CLINICAL APPLICATIONS AND THERAPEUTIC POTENTIAL

The study of the gut microbiota and their role in health and disease has progressively gained popularity in recent years. The healthy growth and functioning of the human immune response are significantly influenced by the gut flora. It is critical to support novel treatment approaches given the limitations of conventional medicine today, the need to treat diseases, the advancement of biomedical technology, and the management of gut microbiota-related disorders (42). Treatments for illnesses affecting the neurological system, immunological system, circulatory system, and digestive system have advanced recently. Gastrointestinal conditions such as IBS, IBD, antibiotic-associated diarrhea, and potentially colorectal cancer are successfully treated with localized therapy. Their effects on the gut–brain axis suggest possible uses in the treatment of neuropsychiatric disorders, such as depression and anxiety (43).

3.7 CONCLUSION

Probiotics, prebiotics, and postbiotics have been shown via significant research and clinical data to be effective in restoring gut microbial equilibrium and treating a range of disorders. Prebiotics have a substantial effect on human health, making them appealing and desirable agents to enhance human well-being against conditions including cancer, vascular disease, weight gain, and mental disorders. Numerous research has demonstrated the beneficial benefits of prebiotics on human health; however, to validate the health claims, carefully planned, prolonged clinical trials and genetic investigations are required. Through the identification of prebiotics' basic principles, scientists hope to develop improved food supplements that will improve human health. One promising approach to managing and treating certain major illnesses is the use of prebiotic dietary supplements to restore the natural balance of the gut microbiota. Current research indicates that these postbiotics have multifaceted effects, which include antioxidant, anti-inflammatory, immunomodulatory, and anticancer capabilities.

REFERENCES

1. Ji J, Jin W, Liu S, Jiao Z, Li X. Probiotics, prebiotics, and postbiotics in health and disease. *MedComm.* 2023;4(6):e420.
2. Li HY, Zhou DD, Gan RY, Huang SY, Zhao CN, Shang AO, et al. Effects and mechanisms of probiotics, prebiotics, synbiotics, and postbiotics on metabolic diseases targeting gut microbiota: a narrative review. *Nutrients.* 2021;13(9):3211.
3. Vallianou N, Stratigou T, Christodoulatos GS, Tsigalou C, Dalamaga M. Probiotics, prebiotics, synbiotics, postbiotics, and obesity: current evidence, controversies, and perspectives. *Curr Obes Rep.* 2020;9:179–92.
4. Wang B, Yao M, Lv L, Ling Z, Li L. The human microbiota in health and disease. *Engineering.* 2017;3(1):71–82.
5. Langdon A, Crook N, Dantas G. The effects of antibiotics on the microbiome throughout development and alternative approaches for therapeutic modulation. *Genome Med.* 2016;8:1–16.
6. Francino M. Antibiotics and the human gut microbiome: dysbioses and accumulation of resistances. *Front Microbiol.* 2016;6:164577.
7. Megur A, Daliri EBM, Baltriukienė D, Burokas A. Prebiotics as a tool for the prevention and treatment of obesity and diabetes: classification and ability to modulate the gut microbiota. *Int J Mol Sci.* 2022;23(11):6097.
8. Singh RD, Banerjee J, Arora A. Prebiotic potential of oligosaccharides: a focus on xylan derived oligosaccharides. *Bioact Carbohydrates Diet Fibre.* 2015;5(1):19–30.
9. La Fata G, Rastall RA, Lacroix C, Harmsen HJM, Mohajeri MH, Weber P, et al. Recent development of prebiotic research—statement from an expert workshop. *Nutrients.* 2017;9(12):1376.
10. Davani-Davari D, Negahdaripour M, Karimzadeh I, Seifan M, Mohkam M, Masoumi SJ, et al. Prebiotics: definition, types, sources, mechanisms, and clinical applications. *Foods.* 2019;8(3):92.
11. Singh SP, Jadaun JS, Narnoliya LK, Pandey A. Prebiotic oligosaccharides: special focus on fructooligosaccharides, its biosynthesis and bioactivity. *Appl Biochem Biotechnol.* 2017;183(2):613–35.
12. Ambrogi V, Bottacini F, Cao L, Kuipers B, Schoterman M, van Sinderen D. Galacto-oligosaccharides as infant prebiotics: production, application, bioactive activities and future perspectives. *Crit Rev Food Sci Nutr.* 2023;63(6):753–66.
13. Mei Z, Yuan J, Li D. Biological activity of galacto-oligosaccharides: a review. *Front Microbiol.* 2022;13:993052.
14. Li C, Hu Y. New definition of resistant starch types from the gut microbiota perspectives–a review. *Crit Rev Food Sci Nutr.* 2023;63(23):6412–22.
15. Jana UK, Kango N, Pletschke B. Hemicellulose-derived oligosaccharides: emerging prebiotics in disease alleviation. *Front Nutr.* 2021;8:670817.
16. Guarino MPL, Altomare A, Emerenziani S, Di Rosa C, Ribolsi M, Balestrieri P, et al. Mechanisms of action of prebiotics and their effects on gastro-intestinal disorders in adults. *Nutrients.* 2020;12(4):1037.
17. Markowiak-Kopeć P, Śliżewska K. The effect of probiotics on the production of short-chain fatty acids by human intestinal microbiome. *Nutrients.* 2020;12(4):1107.
18. Sampaio KB, Fusco V, de Brito Alves JL, de Souza EL. Probiotics: concepts, evolution, and applications. In: *Probiotics for human nutrition in health and disease.* Elsevier; 2022. pp. 3–24.
19. Salminen S, Collado MC, Endo A, Hill C, Lebeer S, Quigley EMM, et al. The International Scientific Association of Probiotics and Prebiotics (ISAPP) consensus statement on the definition and scope of postbiotics. *Nat Rev Gastroenterol Hepatol.* 2021;18(9):649–67.
20. Abou-Kassem DE, Elsadek MF, Abdel-Moneim AE, Mahgoub SA, Elaraby GM, Taha AE, et al. Growth, carcass characteristics, meat quality, and microbial aspects of growing quail fed diets enriched with two different types of probiotics (Bacillus toyonensis and Bifidobacterium bifidum). *Poult Sci.* 2021;100(1):84–93.
21. Zommiti M, Feuilloley MGJ, Connil N. Update of probiotics in human world: a nonstop source of benefactions till the end of time. *Microorganisms.* 2020;8(12):1907.
22. Jeong JJ, Park HJ, Cha MG, Park E, Won SM, Ganesan R, et al. The Lactobacillus as a probiotic: focusing on liver diseases. *Microorganisms.* 2022;10(2):288.
23. Wong CB, Odamaki T, Xiao J zhong. Insights into the reason of Human-Residential Bifidobacteria (HRB) being the natural inhabitants of the human gut and their potential health-promoting benefits. *FEMS Microbiol Rev.* 2020;44(3):369–85.
24. Mao Q, Sun X, Sun J, Zhang F, Lv A, Hu X, et al. A candidate probiotic strain of Enterococcus faecium from the intestine of the crucian carp Carassius auratus. *AMB Express.* 2020;10:1–9.
25. Zazharskyi VV, Davydenko P, Kulishenko O, Borovik IV, Zazharska NM, Brygadyrenko VV. Antibacterial and fungicidal activities of ethanol extracts of 38 species of plants. *Biosyst Divers.* 2020;28(3):281–9.

26. Latif A, Shehzad A, Niazi S, Zahid A, Ashraf W, Iqbal MW, et al. Probiotics: mechanism of action, health benefits and their application in food industries. *Front Microbiol.* 2023;14:1216674.
27. Tabasco R, de Palencia PF, Fontecha J, Peláez C, Requena T. Competition mechanisms of lactic acid bacteria and bifidobacteria: fermentative metabolism and colonization. *LWT-Food Sci Technol.* 2014;55(2):680–4.
28. Stanley RA. *Degradation and utilisation of mucins by anaerobic bacteria from the colon.* ResearchSpace@ Auckland; 1982.
29. Vinderola G, Sanders ME, Salminen S. The concept of postbiotics. *Foods.* 2022;11(8):1077.
30. Thorakkattu P, Khanashyam AC, Shah K, Babu KS, Mundanat AS, Deliephan A, et al. Postbiotics: current trends in food and pharmaceutical industry. *Foods.* 2022;11(19):3094.
31. Moradi M, Molaei R, Guimarães JT. A review on preparation and chemical analysis of postbiotics from lactic acid bacteria. *Enzyme Microb Technol.* 2021;143:109722.
32. Qadi WSM, Mediani A, Kasim ZM, Misnan NM, Sani NA, Jamar NH. Biological characterization and metabolic variations among cell-free supernatants produced by selected plant-based lactic acid bacteria. *Metabolites.* 2023;13(7):849.
33. Vincenzi A, Goettert MI, de Souza CFV. An evaluation of the effects of probiotics on tumoral necrosis factor (TNF-α) signaling and gene expression. *Cytokine Growth Factor Rev.* 2021;57:27–38.
34. Ahmed Z, Ahmad A. Biopolymer produced by the lactic acid bacteria: production and practical application. In: *Microbial production of food ingredients and additives.* Elsevier; 2017. p. 217–57.
35. Khalil ES, Abd Manap MY, Mustafa S, Alhelli AM, Shokryazdan P. Probiotic properties of exopolysaccharide-producing Lactobacillus strains isolated from tempoyak. *Molecules.* 2018;23(2):398.
36. Żółkiewicz J, Marzec A, Ruszczyński M, Feleszko W. Postbiotics—a step beyond pre-and probiotics. *Nutrients.* 2020;12(8):2189.
37. Radaic A, de Jesus MB, Kapila YL. Bacterial anti-microbial peptides and nano-sized drug delivery systems: the state of the art toward improved bacteriocins. *J Control Release.* 2020;321:100–18.
38. Jurkiewicz D, Zielnik-Jurkiewicz B. Bacterial lysates in the prevention of respiratory tract infections. *Polish J Otolaryngol.* 2018;72(5):1–8.
39. Magrone T, Jirillo E. The interplay between the gut immune system and microbiota in health and disease: nutraceutical intervention for restoring intestinal homeostasis. *Curr Pharm Des.* 2013;19(7):1329–42.
40. Markowiak P, Śliżewska K. Effects of probiotics, prebiotics, and synbiotics on human health. *Nutrients.* 2017;9(9):1021.
41. Ansari F, Neshat M, Pourjafar H, Jafari SM, Samakkhah SA, Mirzakhani E. The role of probiotics and prebiotics in modulating of the gut-brain axis. *Front Nutr.* 2023;10:1173660.
42. Purchiaroni F, Tortora A, Gabrielli M, Bertucci F, Gigante G, Ianiro G, et al. The role of intestinal microbiota and the immune system. *Eur Rev Med Pharmacol Sci.* 2013;17(3).
43. Kim CH. Control of lymphocyte functions by gut microbiota-derived short-chain fatty acids. *Cell Mol Immunol.* 2021;18(5):1161–71.

4 Bidirectional Communication in the Gut–Brain Axis

Priya Chaudhary[1], Divya Kumari[3], Apurva[4], Narotam Sharma[2], and Pracheta Janmeda[3]*

[1] Centre for Herbal Pharmacology and Environmental Sustainability, Chettinad Hospital and Research Institute, Chettinad Academy of Research and Education, Chennai, Tamil Nadu, India
[2] DNA Labs-A Centre for Applied Sciences, Dehrdaun, Uttarakhand, India
[3] Department of Bioscience and Biotechnology, Banasthali Vidyapith, Banasthali, Rajasthan, India
[4] Department of Biotechnology, School of Applied and Life Sciences, Uttaranchal University, Dehrdaun, Uttarakhand, India

4.1 INTRODUCTION

A complex integrated system not only properly maintains equilibrium of gastrointestine but also affects mood, enthusiasm, and some of the cognitive roles that have been revealed by investigations on the relationship of the brain and the gut. The phrase "brain–gut axis" (GBA) includes the linkage of these crosstalks (Carabotti et al., 2015). Its major role is to include the status of the gut with the affective and cognitive centers of the brain, along with the fusion of intestinal operations. Neuro-immuno-endocrine mediators are included in the mechanism of GBA interactions. It is a part of the limbic system, an important region of brain which is responsible for emotional and memory reactions. Cortisol is one of major stress hormones that influence the brain and associated animal organs. Therefore, the functions of intestine functional effector cells were affected by the brain, including immune cells, epithelial cells, enterochromaffin cells, interstitial cells, enteric neurons, and smooth muscle cells, by means of a combination of neurological and hormonal channels of interaction. On the other hand, the gut microbiota, whose function in brain–gut reciprocity interactions is currently being evaluated, influences the exact same cells (Kageyama et al., 2021). While every individual's microbiota composition is unique, the relative abundance and location of various microbial phylotypes along the intestine are comparable in individuals who are healthy. Firmicutes and Bacteroides are the two more common phyla, making up at least three-fourth of the microbiome (Sasso et al., 2023).

A growing array of immunological, neurological, hormonal, and metabolic processes involved in the microbiota-brain–gut axis-mediated regulation of host homeostasis are being identified. All of these procedures will be covered separately in later sections of this chapter, as it appears doubtful that any one of them is in charge of mediating every microbiota-gut–brain relationship (Ullah et al., 2023). The analysis of neuroactive chemicals, hormones, and metabolites which are produced by eukaryote and prokaryotes cells in the host that lead to an interkingdom communication system will receive special focus. Targeting particular neural, hormonal, and immunological pathways may direct to the development of novel therapeutic agents or probiotics that produce neuroactive molecules for the management of gastrointestinal and/or neurological problems.

4.2 MAST CELLS AND THE BRAIN–GUT AXIS IN THE NERVOUS SYSTEM

4.2.1 Mast Cells: Origin and Activation

Macrophages (MCs) play a physiological function that goes far beyond allergies. According to Dahlin and Hallgren (2015), they have a role in biological processes like tissue repair, wound healing, and angiogenesis. It is significant that patients with inflammatory bowel disease (IBD) have more MCs in their colon and terminal ileum. A bidirectional interaction system within the GI and ENS–CNS is created by the nearest morphological connection between MC and afferent nerve endings. Particularly, MC intermediaries activate nerves, which release neurotransmitters that can improve MC activity. According to reports, MCs coexist with GI mucosal sensory nerve cells that carry neuropeptides in the gut. These fibers include visceral afferents that express rapid receptor potential vanilloid 1 (TRPV1) receptors. One conceivable mechanism connecting vagal nerve pathway impairment with GI symptoms is a brain–MC interaction (Carabotti et al., 2015). Intestinal MC

DOI: 10.1201/9781003536802-4

degranulation is influenced by the CNS via vagal nerves, whereas degranulation is inhibited by sympathetic activity. The reverse is likewise true if MCs are directly impacted by the ENS. It's interesting to note that MCs function as effector cells by releasing physiologically activated mediators while also acting as sensory cells stimulated by immunological and nonimmune stimuli (Traina, 2017).

4.3 MAST CELL AND GUT–BRAIN AXIS

The brain–gut axis is an integrated system that facilitates bidirectional crosstalk within the brain and the gut through neuronal, endocrine, and circulatory signals. Behaviors and cognitive processes are influenced by the gut microbiome. The organism's first strong barrier against infections is the gut microbial ecology. The intestine's microbes can communicate with components of the host's endocrine system to alter the host's behavior. Mediators generated in the gut microbiota and the microbiota itself can impact the brain (Conte et al., 2020). The aggressiveness of pathogenic bacteria could have been disastrous in the absence of benign microbes. According to Rea et al. (2017), peripheral innate immune system activation-induced inflammation can affect both CNS behaviors and cognitive function. In the gastrointestinal system, there is proof of two-way communication between MCs and neurons. MCs are the traditional immune cell that is stimulated by neural stimuli and neurotransmitters (Figure 4.1).

4.4 ROLE OF MICROBIOTA IN GBA

Clinical and analytical findings determined that enteric microbiome plays a major role in GBA. Meanwhile, new research indicates that dysbiosis in autism may be related to microbiota, which may influence anxiety and depression-like behaviors. Dysbiosis is also present in functional gastrointestinal disorders (FGID) that are connected to a disturbance of GBA and strongly associated with mood disorders. Data show that there are both gut–brain and brain–gut deregulations, with the existing predominating situation such as irritable bowel syndrome (IBS) (Koloski et al., 2012). The disruption of the GBA causes changes in the production and motility of the intestines, as well as cellular abnormalities in the immunological and enteroendocrine systems and visceral hypersensitivity. Numerous lines of proof encourage the importance of the gut microbiota, which may interact with several of these distinct pathophysiological IBS goals. These include alterations in the constitution of the microbiome with flaws in its viability and variation, the emergence of post-infectious IBS, the combination of overgrowth of bacteria in the small intestine, and the efficacy of certain beneficial bacteria and nonsystemic antibiotics in treating IBS patients (Pimentel et al., 2011).

4.5 FROM GUT MICROBIOTA TO BRAIN

From past few years, there has been an increase in the number of investigations that have been done, with a motive on

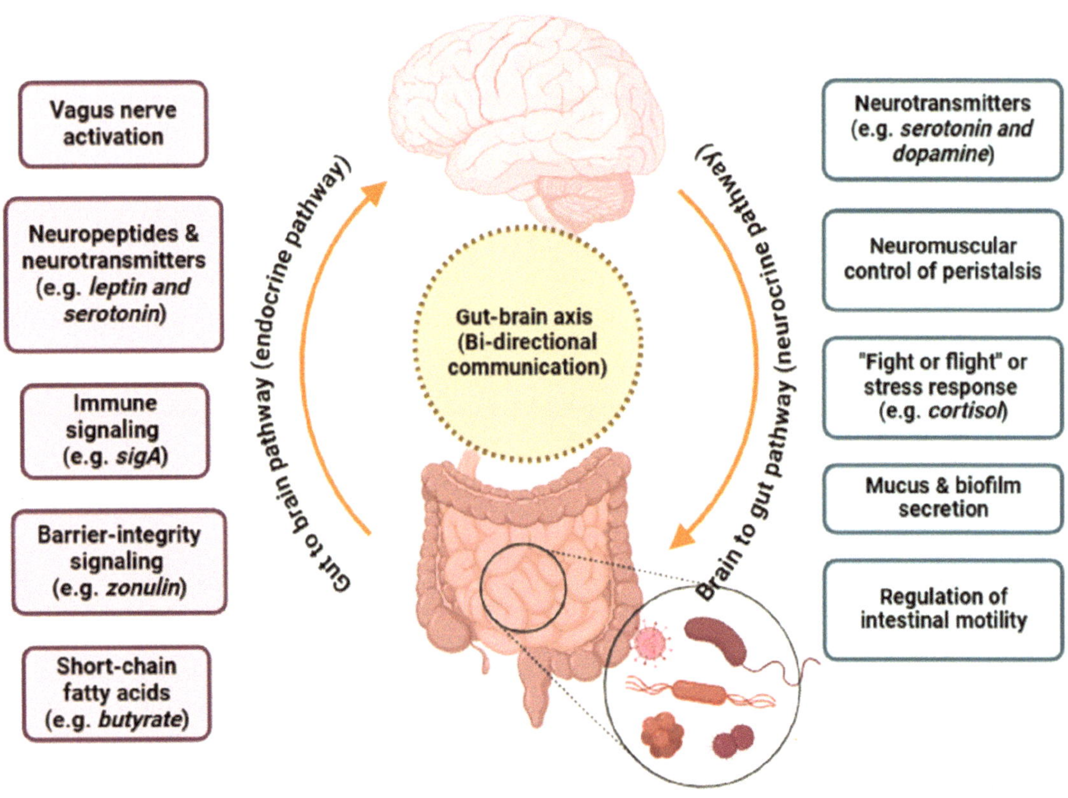

FIGURE 4.1 Bidirectional communication.

TABLE 4.1
An Overview of Current Preclinical and Clinical Research on the Gut Microbiota's Role in Alzheimer's Disease

Animal/Human	Treatment	Observation	Reference
Preclinical studies			
F1 3xTg mice	Modifications to the gut microbiome of bacteria	BGM may decrease neuroinflammatory reactions linked to cognitive decline and cerebral amyloidosis	Bello-Medina et al., 2022
3xTg-AD mice	One plasmid (pExu) containing a eukaryotic expression cassette encoding the human p62 protein is carried by the *Lactobacillus lactis* strain	We show that using these incredibly safe, noninvasive, nonpathogenic microorganisms to deliver the p62 protein is a novel and practical treatment strategy for AD	Cecarini et al., 2020
3xTg-AD and wild-type mice	SLAB51 probiotic formulation with several strains	We present data indicating that long-term SLAB51 supplementation boosted HIF-1α expression in the brain	Bonfili et al., 2022
5xFAD Alzheimer's disease mouse model	It has been documented that the prebiotic mannan oligosaccharide (MOS) changes the composition of the gut microbiome	The 5xFAD mice's cognitive and mental deficiencies are much reduced by MOS, which may be partially due to the gut's altered microbiome	Liu et al., 2021
Clinical studies			
AD patients	Modifications to the gut microbiome of bacteria	Changes in the gut microbiota may have an impact on the pathophysiology and cognitive disability of AD	Yu et al., 2023
AD patients	Ketogenic diet (KD)	By reducing symptoms and postponing the disease's course	Oliveira et al., 2024
PD patients	There was a strong correlation found between the prevalence of pro-inflammatory bacteria	PD patients had lower levels of fecal SCFAs but higher levels of plasma SCFAs, which were linked to particular alterations in the gut microbiota	Chen et al., 2022
PD patients	The gut microbiota of the transgenic PD mice underwent a significant alteration	It appears that patients who had higher baseline levels of inflammation may benefit more from rifaximin therapy	Hong et al., 2022

the effect of the microbiota on GBA, primarily including animal models. According to Diaz Heijtz et al. (2011), less colonization of bacteria is associated with the alteration of neurotransmitters in both the nervous and digestive systems. Reduced gene expression for enzymes that regulate neurotransmitter production and transport as well as that of muscle contractile proteins was linked to neuromuscular defects (Hooper et al., 2001). Simultaneously, memory impairment has also been documented in GF mice, which is likely due to a change in the level of expression in case of brain-derived neurotrophic factor (BDNF), one of the key indicators of memory. Lastly, because of reports of increased serotonin turnover and changed concentration of associated metabolites in the limbic region of GF animals, the presence of the microbiota also results in modification of the serotoninergic system (Diaz Heijtz et al., 2011).

4.6 FROM BRAIN TO GUT MICROBIOTA

Psychological stressors vary in their ability to modify the intestinal microbiota's makeup and total biomass, regardless of their length. In actuality, brief stressors also have an effect on the microbiota; for example, a 2-hour social stressor exposure can majorly alter the community composition and decreases the relative abundance of the valuable microbiome phyla (Galley et al., 2014). These effects could be mediated both directly through host-enteric microbiome signaling and indirectly through alterations in the intestinal circumstances via the parallel neuroendocrine efferent mechanism (Table 4.1).

According to several investigations, bacteria have binding regions for enteric neurotransmitters that the host produces. These regions can influence how some microbiota constituents maintain and raise susceptibility to infectious and inflammatory stimuli (Clarke et al., 2006). According to Macfarlane and Dillon (2007), these processes are essential for preserving the mucus layer and biofilm, which are sites where unique bacterial colonies multiply in a range of unique metabolic niches and microhabitats attached to the mucosa. The gut flora may then be impacted by the disturbance of the normal mucosal habitat brought on by an imbalance in GBA.

4.7 MAST CELL AND GASTROINTESTINAL AND INTESTINAL MICROBIOTA

The vagus nerve serves as a conduit for interaction between the gut microbiota and the brain, which is referred as the "gut–brain axis." Studies on the linkage existing between the mental health, the brain's cognitive centers, and gut microbiota are focusing more on the immune, endocrine, and neurological pathways (Sinagra et al., 2020). Greater concentration of substance P and lower concentration of proteases in the enteric nervous system (ENS) can impact the microbiota which is responsible for the immune system's regulation (Tognini, 2017). The complicated process of GBA's may affects synaptic plasticity, pain perception, hippocampal neurogenesis, and some of the neuroactive substances like neurotransmitters, glutamate, and tryptophan metabolites. It has been determined that any alteration to the gut microbiome can affect the host behavior. Thus, it may trigger the secretion of host immunological substance like inflammatory mediator (Chaudhry et al., 2023).

4.8 GUT MICROORGANISMS STIMULATE PERIPHERAL IMMUNE RESPONSES

Lu et al. (2018) revealed marked impairments in CNS development, function, mood, and behavior when compared to controls with a replete microbiota. These results suggest that gut bacteria plays a crucial function in brain function throughout life. Preliminary evidence supporting this notion came from investigations determining the effect of mice's native gut bacteria, or SFB, on susceptibility to investigational autoimmune encephalomyelitis (EAE), an animal model of multiple sclerosis (MS). GF animal models are highly resistant to EAE, in contrast to SPF animal model (Lee et al., 2011). According to Cosorich et al. (2017), intestinal Th17 cell responses in humans have a positive correlation with MS illness and a negative correlation with the relative frequency of *Prevotella* in the small intestine. This suggests that *Prevotella* is a key regulator of neurological inflammation in both MS and EAE. Together, these data show that gut microorganisms can influence T-cell responses during CNS inflammation.

4.9 LOW-LEVEL SYSTEMIC INFLAMMATION IN ILLNESSES RELATED TO THE NERVOUS SYSTEM

The human body uses severe inflammation as a natural defense mechanism against damage or infection, aiding in regeneration of tissues and healing of wounds. Many biomarkers, such as circulating and cellular indicators, were engaged in the low-grade inflammatory reaction brought on by diet and obesity. These biomarkers have been associated with an elevated danger of chronic illnesses (Crotti et al., 2018). Many investigations conducted lately have documented anatomical alterations in the brains of patients suffering from neuropsychological and heart disease. In the meantime, it has been documented that the anatomical brain changes driving neuropsychiatric illnesses are related to inflammation. This is because astrocytic activity and microglia influence synaptic pruning, which in turn influences gray matter volume. Acute illness-related increases in pro-inflammatory chemicals in peripheral blood are suggestive of poor-quality systemic inflammatory processes, which have been connected to serious mental health issues like stress, depression, and schizophrenia (Upthegrove et al., 2014). Nonsteroidal anti-inflammatory medicines (NSAIDs) and cytokine inhibitors benefit people with distress by lowering the signs and symptoms when used with antidepressant treatments.

Aspirin adjunctive medication may be helpful for psychosis, even though some studies did not discover an impact (Girgis et al., 2017).

4.10 LONG-TERM IMPACT OF GUT MICROORGANISMS ON CNS-RESIDENT IMMUNE CELL FUNCTION

4.10.1 Microglia

Tissue-resident myeloid cells called microglia are abundant across the CNS and serve a variety of vital roles in controlling and adjusting the circuitry of neurons throughout formation and into maturity. Microglia use phagocytosis, stimulation of complement, and release of cytokines to carry out their neuronal tasks, just like tissue-resident macrophages in other organs (Salter and Stevens, 2017). Abnormal microglia activity has been linked to neurological development, neuropsychiatric, and neurological disorders. While a number of genetic variables that control microglia activation have been identified, the role of external factors, including the gut microbiome, is only now becoming clear. According to Erny et al. (2015), dietary fiber's fermented product formed from microbes, SCFAs, has the ability to rectify microglia dysfunction in GF and ABX mice. This implies that microbial compounds could potentially serve as a universal pathway for gut–brain connections. In line with this, the gut microbiota's metabolite of tryptophan in food play a role in controlling the synthesis of VEGF-B and TGFα by microglia, which reduce CNS inflammation in human MS and EAE.

4.10.2 Astrocytes

Despite microglia, astrocytes are considered to be the most significant glial cells which participate in the maintenance of CNS. In addition to endothelial cells, which constitute of blood–brain barrier, astrocytes also play significant role in the growth of CNS by presenting antigens and resulting chemokines and cytokines secretion. Dietary tryptophan produced by the gut microbial species comprises the capacity that influences the activity of astrocyte (Rothhammer et al., 2016).

Type 1 interferon results in the activation of anti-inflammatory response in the astrocytes via the aryl hydrocarbon receptor (AHR). Indoles produced by ampicillin-sensitive gut bacterial species are considered to be important for AHR activity. To initiate microglial synapse engulfment, astrocytes release the cytokine IL-33, which belongs to the IL-1 family (Vainchtein et al., 2018). While the association of the microbiota in the gut in this coordination of brain circuit remodelling by astrocytes has not been investigated, this points to a cytokine-mediated route. Taken together, our data show that immune-related astrocyte functions are regulated by gut microbial metabolism.

4.11 INNATE AND ADAPTIVE IMMUNE CELLS FOUND IN THE BRAIN

Earlier studies have connected the synthesis of IL-22 at tissues of the mouth to AHR signaling, which is controlled by microbial by-products of tryptophan. According to more recent studies, certain roles in antitumor and antihelminthics resistance reactions are modulated by cell-intrinsic AHR stimulation in intestinal lining cells (ILCs), B cells, and intestinal epithelial cells (Metidji et al., 2018). The impact of AHR ligands on astrocytes suggests that other AHR-expressing innate (ILCs) and adaptive (T cells) immune cell types that exist in the central nervous system may be subject to comparable modulation of function. IgA with intestinal origin and plasma cells located in the CNS are also linked to the pathophysiology of EAE (Rojas et al., 2019).

4.12 MUTUAL INTERACTION BETWEEN MAST CELLS AND GUT MICROBIOTA

The gut microbiome maintains a bidirectional relation between the MCs and the GBA via immunological, neurological, and metabolic processes. According to Coretti et al. (2017), the brain effects the composition and function of gut microbiome by causing alterations in the epithelia's intestinal permeability, which ultimately caused mucosa to become inflamed.

According to Bonaz et al. (2018), the neurological route is determined to be the essential one for the gut response's adaptation as it regulates intestine and brain homeostasis through the vagal nerve. Furthermore, the serotonin (5-HT)-signaling system controls the gut motility and GI secretions in addition to mood and cognition. A growing body of research showed that the MC activation disrupts both GI permeability and the blood–brain barrier by releasing neuroimmune active substances comprising serotonin, 5-hydroxytryptamine (commonly known as 5-HT), histamine, melatonin, catecholamine, and gamma-aminobutyric acid (GABA). These chemicals are capable of penetrating the intestinal mucosa; entering into the circulation, crossing the blood–brain barrier, and ultimately impacting the brain function (Parashar and Udayabanu, 2017).

4.13 MAST CELLS ESSENTIAL FUNCTIONS WITHIN THE NEURO-IMMUNE RESPONSE NETWORK

MCs or potent pro-inflammatory effector cells are the significant immune cells that are located in the brain and the CNS. They are closely related with the microvessels in the neurovascular unit (NVU) and the abluminal surface of the blood–brain barrier. Through the suppression of various immunological responses that are linked to the generation of interleukin-10 by CD5 (+) B and FoxP3+ by Treg cells, MCs can regulate the immune cell homeostasis. Moreover, MCs are known to perform specific roles of phagocytes in the brain and mediates the production of pro-inflammatory cytokines, and regulators of the CNS. The control of neuroinflammatory mechanism and other related immune cell trafficking requires the activated MCs and the immune network for the communication. In autoimmune diseases, degranulated MCs release significant amount of histamine and tryptase, respectively. Moreover, MC-derived IL-4 has been considered to be crucial for the advancement of multiple sclerosis. MCs direct the release of cytokines, chemokines, and other components by activating microglia within the amyloid plaques of AD (Jones et al., 2019).

4.14 ENGAGING THE NEUROIMMUNE NETWORK IN GUT MICROBIOME

There is some growing evidence that dysbioses in the gut microbiome could disturb the balance of neuroimmune responses, which is responsible for the development of neurodegenerative disorders, carcinoma, and other related autoimmune diseases (Caballero-Villarraso et al., 2017). In the circumventricular organs, cytokines can stimulate neurons and glial cells. They can also go to other parts of the brain and influence the roles of gut microorganisms and mast cells in neural activity in neurodegenerative diseases (Hamadi et al., 2016).

Toll-like receptors (TLRs) are abundantly present over the MCs, epithelial cells, lymphoid cells, dendritic cells (DCs), and macrophages. The function of TLRs is to monitor the gut pathogens and stimulate the tolerance of immune system's to a normal microbiota. TLRs are responsible for the identification of pathogens that are related with bacterial antigens and molecular patterns (PAMPs) in order to direct the intestinal immunity (Lin et al., 2019).

When MCs and the gut bacterial species interact, the mucousal layer is degraded, which leads to MC activation. This phenomenon also directs the secretion of various immunomodulatory mediators, for instance matrix metallopeptidase, tryptase, chymase, and histamine that affects the excitation of nerve (De Zuani et al., 2018).

4.14.1 Alzheimer's Disease

In the AD individuals, brain inflammation may result in CNS glia to release mast cell chemoattractants like serum amyloid A and acute phase proteins, which direct mast cells to specific region where amyloid has been deposited (Maslinska et al., 2007). Jones et al. (2019) reported a decrease in the rate of cognitive deterioration in patients with mild-to-moderate AD who received masitinib as an add-on medication during a 24-week phase 2 randomized, placebo-controlled trial. The genesis and progression of

AD have been linked to stress-induced neuropathological processes that are facilitated by mast cell disruption of the blood–brain barrier. This was possibly accomplished by preventing mast cells that are close to the blood–brain barrier from releasing mediators that would reduce the permeability of the barrier, resulting in a drop in local pro-inflammatory molecules and additional mast cell migration into the brain. This is because masitinib is unlikely to cross the blood–brain barrier. There are presently phase 2 and phase 3 clinical trials ongoing for misitinib.

4.14.2 Parkinson's Disease

In an animal model, Hong et al. (2018) discovered that the CCL2 secreted by microglia and astrocytes directs the recruitment of mast cell into the substantia nigra. One of the cross-linking enzyme, such as tissue transglutaminase 2 (TG2), was observed to be expressed by recruited mast cells, followed by the generation of pro-inflammatory mediators which are associated with the death of dopaminergic neurons. Additionally, the studies discovered that the PD individual's serum had higher concentration of TG2 expression in comparison to control patients (Hong et al., 2018).

These observations point out the interesting function of mast cells in the pathophysiology of Parkinson's disease, which should be exercised when observing the condition in humans. Model-dependent systems and *in vitro* studies might not be suitable in replicating the intricate local microenvironments that significantly modify the phenotype of mast cells (Hurley et al., 2015).

4.14.3 Multiple Sclerosis

The clinical signs of multiple sclerosis include ataxia, poor coordination, hyperreflexia, stiffness, fatigue, decreased visual and sensory perception, and problems in cognition. Most patients have lesions in the brain or in the brain and spinal cord, whereas others solely have lesions in the spinal cord (Rutsch et al., 2020). Microbes, together with the substances they produce or release into the environment, play a significant part in the pathogenesis of multiple sclerosis (MS). GF mice have been used to create preclinical models of attenuated multiple sclerosis-like disease. Compared to mice receiving the microbiome of healthy individuals, those receiving the intestinal microbiota of MS patients had higher levels of experimental autoimmune encephalomyelitis and lower proportions of anti-inflammatory regulatory T cells (Cekanaviciute et al., 2017).

4.14.4 Autism Spectrum Disorders

In ASD, the amounts of chemical transmitters such as glutamate, oxytocin, GABA, and serotonin 5-HT complex are mediated by the gut flora (Almehmadi et al., 2019). As it becomes clear how much microbial metabolites such as taurine, bile acid metabolites, SCFAs, and 5-aminovaleric acid affect ASD symptoms, new technologies are being employed in this quickly developing field of research (Morais et al., 2021). Few and mostly contradictory ASD research have focused on the microbiome's role in pathogenesis. Some, however, show the distinctions between the bacteria found in ASD patients and controls, including Firmicutes, Clostridiales, Prevotella, Bifidobacterium, and Clostridium perfringens. This leads to alterations in the makeup of the gut microbiota, a decline in the quality of the diet, and a nutritional deficit. It is clear that these neurological illnesses are associated with lower concentrations of helpful bacteria and higher concentrations of harmful bacteria, yet it is impossible to conclude that any particular bacteria is compatible or related to the onset of ASD (Ho et al., 2020).

4.15 MC IS A POTENTIAL THERAPEUTIC TARGET FOR GBA-MEDIATED NEURODEGENERATIVE DISEASES

Given the connection among gut dysbiosis and neurodegenerative disorders (NDDs), dietary treatments aimed at addressing gut dysbiosis may hold promise for managing symptoms and reducing the neuroinflammatory and degenerative processes associated with NDDs. Complementary nutritional therapies targeted at modifying the arrangement of the gut microbiota and related metabolites could support current therapeutic techniques in establishing eubiosis and promoting overall gut health.

Consuming fruits, vegetables, legumes, and cereals is emphasized in the diet. Prebiotics are substances that the good bacteria in our stomachs specifically use to fuel their growth and the production of their by-products. When taken in sufficient proportions, probiotics—live, nonpathogenic microorganisms—promote health benefits. Synbiotics are specific concoctions of probiotics and prebiotics that work together to increase the survivability and therapeutic effects of both. FMT involves introducing carefully selected contributor fecal bacteria into the patient's gastrointestinal tract with the goal of reestablishing a healthy gut microbiome and improving gut microbiota functionality and diversity (De Caro et al., 2019).

4.16 THE CONNECTION BETWEEN MCS ACTIVATION IN INFLAMMATION AND TOLERANCE AND PRO-INFLAMMATORY CYTOKINES

Neurons, glia cells, and blood vessel epithelial cells are examples of brain cells. These cells are in charge of providing nourishment and defense for neuron (Conti et al., 2019). According to Winklewski et al. (2016), infectious pathogens can penetrate into the blood–brain barrier that shields the brain and make interaction with the microglia cells in order to create Th1-type inflammatory cytokines. Both acute and long-term CNS disorders get influenced by the release of the cytokines. A number of symptoms related with different types of brain illness involve the attribution to inflammation in the amygdala and hypothalamus regions, which is caused by the release of cytokines by microglia and mast cells. As a result, the release of pro-inflammatory substances exacerbates the pathological state of individual's. The neurotensin (NT) which is a neurotransmitter is observed to activate various types of immune cells in the brain, including human microglia and the human cell line SV40 (Theoharides et al., 2015).

4.17 GUT MICROBES ARE POTENT REGULATORS OF HOST IMMUNE RESPONSES

The host's immunological response to the gut microbiota is tightly limited to its surface in a healthy condition. The intestinal lumen and surrounding structures are separated by a single layer of epithelium. Microbiota division is accomplished by a variety of methods (Bevins and Salzman, 2011). Microbes living within the intestinal epithelium are kept apart by a thick layer of mucus. Organizing the mucus layer is the hyperglycosylated mucin MUC2. But MUC2 does more than just provide static protection; it also imprints enteric dendritic cells (DCs) with an anti-inflammatory state, thereby limiting the immune response of intestinal antigens. One feature that is essential for limiting trans-epithelial leakage is tight connections. The metabolite indole, for example, is one way that microbial signals fortify the epithelial barrier by upregulating tight junctions and related cytoskeletal proteins. Furthermore, the function of the mucosal barrier is preserved by antimicrobial peptides (AMPs) and secretory IgA antibodies. Through processes involving the sampling of gut bacteria for antigen presentation, intestinal DCs are thought to be essential in the compartmentalization of enteric microbiota (Zheng et al., 2020).

4.18 CONCLUSION

Trillions of indigenous bacteria inhabit the GI tract, shaping host immune activity both locally and extraintestinally. Irregular immune system responses are becoming recognized as causes of neuroinflammatory, neuropsychiatric, and neurodegenerative disorders. New discoveries from human and animal research indicate that the gut microbiota plays a key role in immunological dysfunction in a variety of CNS diseases. Gut microbes directly regulate systemic and intestinal immunological homeostasis, influencing neuroinflammation and potentially contributing to low-grade chronic inflammation in neuropsychiatric diseases. Chronic intestinal inflammation causes pathologic stimulation of the immune system against gut bacteria, which is a substantial precursor for dementia. In addition, the gut microbiota influences the activity of immune cells in the central nervous system, which may have implications for neuroinflammatory and neurodevelopmental diseases. As a result, addressing the microbiota-immune axis is a prospective therapeutic strategy for addressing CNS and associated illnesses.

REFERENCES

Almehmadi, K.A., Tsilioni, I., Theoharides, T.C. (2019). Increased expression of miR-155p5 in Amygdala of children with autism spectrum disorder. *Autism Research*, *13*(1), 18–23. https://doi.org/10.1002/aur.2205.

Bello-Medina, P.C., Corona-Cervantee, K., Torres, N.G.Z., González, A., Pérez-Morales, M., González-Franco, D.A., Gómez, A., García-Mena, J., Díaz-Cintra, S., Pacheco-López, G. (2022). Chronic-antibiotics induced gut microbiota dysbiosis rescues memory impairment and reduces β-Amyloid aggregation in a preclinical Alzheimer's disease model. *International Journal of Molecular Science*, *23*(15), 8209. https://doi.org/10.3390/ijms23158209.

Bevins, C.L., Salzman, N.H. (2011). Paneth cells, antimicrobial peptides and maintenance of intestinal homeostasis. *Nature Reviews Microbiology*, *9*(5), 356–368. https://doi.org/10.1038/nrmicro2546.

Bonaz, B., Bazin, T., Pellissier, S. (2018). The vagus nerve at the interface of the microbiota-gut-Brain Axis. *Frontier in Neuroscience*, *12*, 49. https://doi.org/10.3389/fnins.2018.00049.

Bonfili, L., Gong, C., Lombardi, F., Cifone, M.G., Eleuteri, A.M. (2022). Strategic modification of gut microbiota through oral bacteriotherapy influences hypoxia inducible factor-1α: Therapeutic implication in Alzheimer's disease. *International Journal of Molecular Science*, *23*(1), 357. https://doi.org/10.3390/ijms23010357.

Caballero-Villarraso, J., Galvan, A., Escribano, B.M., Tunez, I. (2017). Interrelationships among gut microbiota and host: Paradigms, role in neurodegenerative diseases and future prospects. *CNS and Neurological Disorders- Drug Targets*, *16*(8), 945–964. https://doi. org/10.2174/1871527316666170714120118.

Carabotti, M., Scirocco, A., Maselli, M.A., Severi, C. (2015). The gut-brain axis: Interactions between enteric microbiota, central and enteric nervous systems. *Annals of Gastroenterology*, *28*(2), 203–209.

De Caro, C., Leo, A., Nesci, V., Ghelardini, C., di Cesare Mannelli, L., Striano, P., Avagliano, C., Calignano, A., Mainardi, P., Constanti, A., Citraro, R., De Sarro, G., Russo, E. (2019). Intestinal inflammation increases convulsant activity and reduces antiepileptic drug efficacy in a mouse model of epilepsy. *Science Report*, *9*(1), 13983. https://doi.org/10.1038/s41598-019-50542-0.

Cecarini, V., Bonfili, L., Gogoi, O., Lawrence, S., Venanzi, F.M., Azevedo, V., Mancha-Agresti, P., Drumond, M.M., Rossi, G., Berardi, S., Galosi, L., Cuccioloni, M., Angeletti, M., Suchodolski, J.S., Pilla, R., Lidbury, J.A., Eleuteri, A.M. (2020). Neuroprotective effects of p62(SQSTM1)-engineered lactic acid bacteria in Alzheimer's disease: A pre-clinical study. *Aging (Albany NY)*, *12*(16), 15995–16020. https://doi.org/10.18632/aging.103900.

Cekanaviciute, E., Yoo, B.B., Runia, T.F., Debelius, J.W., Singh, S., Nelson, C.A., Kanner, R., Bencosme, Y., Lee, Y.K., Hauser, S.L., Crabtree-Hartman, E., Sand, I.K., Garcias, M., Zhu, Y., Casaccia, P., Cree, B.A., Knight, R., Mazmanian, S.K., Baranzini, S.B. (2017). Gut bacteria from multiple sclerosis patients modulate human T cells and exacerbate symptoms in mouse models. *Proceedings of the National Academy of Sciences, 114*, 10713–10718. https://doi.org/10.1073/pnas.1711235114

Chaudhry, T.S., Senapati, S.G., Gadam, S., Mannam, H.P.S.S., Voruganti, H.V., Abbasi, Z., Abhinav, T., Challa, A.B., Pallipamu, N., Bheemisetty, N., Arunachalam, S.P. (2023). The impact of microbiota on the gut-brain axis: Examining the complex interplay and implications. *Journal of Clinical Medicine, 12*(16), 5231. https://doi.org/10.3390/jcm12165231.

Chen, S.-J., Chen, C.-C., Liao, H.-Y., Lin, Y.-T., Wu, Y.-W., Liou, J.-M., Wu, M.-S., Kuo, C.-H., Lin, C.-H. (2022). Association of fecal and plasma levels of short-chain fatty acids with gut microbiota and clinical severity in patients with parkinson disease. *Neurology, 98*(8), e848–e858. https://doi.org/10.1212/WNL.0000000000013225.

Clarke, M.B., Hughes, D.T., Zhu, C., Boedeker, E.C., Sperandio, V. (2006). The QseC sensor kinase: A bacterial adrenergic receptor. *Proceedings of the National Academy of Sciences, 103*(27), 10420–10425. https://doi.org/10.1073/pnas.0604343103.

Conte, C., Sichetti, M., Traina, G. (2020). Gut-brain axis: Focus on neurodegeneration and mast cells. *Applied Sciences, 10*(5), 1828. https://doi.org/10.3390/app10051828.

Conti, P., D'Ovidio, C., Conti, C., Gallenga, C.E., Lauritano, D., Caraffa, A., Kritas, S.K., Ronconi, G. (2019). Progression in migraine: Role of mast cells and pro-inflammatory and anti-inflammatory cytokines. *European Journal of Pharmacology, 844*, 87–94. https://doi.org/10.1016/j.ejphar.2018.12.004.

Coretti, L., Cristiano, C., Florio, E., Scala, G., Lama, A., Keller, S., Cuomo, M., Russo, R., Pero, R., Paciello, O., Raso, G.M., Meli, R., Cocozza, S., Calignano, A., Chiariotti, L., Lembo, F. (2017). Sex-related alterations of gut microbiota composition in the BTBR mouse model of autism spectrum disorder. *Science Report, 7*, 45356. https://doi.org/10.1038/srep45356.

Cosorich, I., Dalla-Costa, G., Sorini, C., Ferrarese, R., Messina, M.J., Dolpady, J., Radice, E, Mariani, A., Testoni, P.A., Canducci, F., Comi, G., Martinelli, V., Falcone, M. (2017). High frequency of intestinal TH17 cells correlates with microbiota alterations and disease activity in multiple sclerosis. *Science Advances, 3*(2017), e1700492. https://doi.org/10.1126/sciadv.1700492.

Crotti, G., Gianfagna, F., Bonaccio, M., Di Castelnuovo, A., Costanzo, S., Persichillo, M., De Curtis, A., Cerletti, C., Donati, M.B., de Gaetano, G., Lacoviello, L. (2018). Body mass index and mortality in elderly subjects from the molisani study: A possible mediation by low-grade inflammation? *Immunological Investigations, 47*, 774–789. https://doi.org/10.1080/08820139.2018.1538237.

Dahlin, J.S., Hallgren, J. (2015). Mast cell progenitors: Origin, development and migration to tissues. *Molecular Immunology, 63*(1), 9–17. https://doi.org/10.1016/j.molimm.2014.01.018.

Diaz Heijtz, R., Wang, S., Anuar, F., Qian, Y., Björlholm, B., Samuelsson, A., Hibberd, M.L., Forssberg, H., Pettersson, S. (2011). Normal gut microbiota modulates brain development and Behavior. *Proceedings of the National Academy of Sciences, 108*(7), 3047–3052. https://doi.org/10.1073/pnas.1010529108.

Erny, D., de Angelis, A.L.H., Jaitin, D., Wieghofer, P., Statszewski, O., David, E., Keren-Shaul, H., Mahlakoiv, T., Jakobshagen, K., Buch, T., Schwierzeck, V., et al. (2015). Host microbiota constantly control maturation and function of microglia in the CNS. *Nature Neuroscience, 18*(7), 965–977. https://doi.org/10.1038/nn.4030.

Galley, J.D., Nelson, M.C., Yu, Z., Dowd, S.E., Walter, J., Kumar, P.S., Lyte, M., Bailey, M.T. (2014). Exposure to a social stressor disrupts the community structure of the colonic mucosa-associated microbiota. *BMC Microbiology, 14*, 189. https://doi.org/10.1186/1471-2180-14-189.

Girgis, R.R., Ciarleglio, A., Choo, T., Haynes, G., Bathon, J.M., Cremers, S., Kantrowitz, J.T., Lieberman, J.A., Brown, A.S. (2017). A randomized, double-blind, placebo-controlled clinical trial of tocilizumab, an interleukin-6 receptor antibody, for residual symptoms in schizophrenia. *Neuropsychopharmacology, 43*(6), 1317–1323. https://doi.org/10.1038/npp.2017.258.

Hamadi, N., Sheikh, A., Madjid, N., Lubbad, L., Amir, N., Al-Deen, S., Shebab, S., Khelifi-Touhami, F., Adem, A. (2016). Increased pro-inflammatory cytokines, glial activation and oxidative stress in the hippocampus after short-term bilateral adrenalectomy. *BMC Neuroscience, 17*(1), 61. https://doi.org/10.1186/s12868-016- 0296-1.

Ho, L.K.H., Tong, V.J.W., Syn, N., Nagarajan, N., Tham, E.H., Tay, S.K., Shorey, S., Tambyah, P.A., Law, E.C.N. (2020). Gut microbiota changes in children with autism spectrum disorder: A systematic review. *Gut Pathogens, 12*, 1–18. https://doi.org/10.1186/s13099-020-0346-1.

Hong, C.-T., Chan, L., Chen, K.-Y., Lee, H.-H., Huang, L.-K., Yang, Y.-C.S.H., Liu, Y.-R., Hu, C.-J. (2022). Rifaximin modifies gut microbiota and attenuates inflammation in Parkinson's disease: Preclinical and clinical studies. *Cells, 11*(21), 3468. https://doi.org/10.3390/cells11213468.

Hong, G.U., Cho, J.W., Kim, S.Y., Shin, J.H., Ro, J.Y. (2018). Inflammatory mediators resulting from transglutaminase 2 expressed in mast cells contribute to the development of Parkinson's disease in a mouse model. *Toxicology and Applied Pharmacology, 358*, 10–22. https://doi.org/10.1016/j.taap.2018.09.003

Hooper, L.V., Wong, M.H., Thelin, A., Hansson, L., Falk, P.G., Gordon, J.I. (2001). Molecular analysis of commensal host-microbial relationships in the intestine. *Science, 291*(5505), 881–884. https://doi.org/10.1126/science.291.5505.881.

Hurley M.J., Durrenberger P.F., Gentleman S.M., Walls A.F., Dexter D.T. (2015). Altered expression of brain proteinase-activated receptor-2, trypsin-2 and serpin proteinase inhibitors in Parkinson's disease. *Journal of Molecular Neuroscience, 57*, 48–62. https://doi.org/10.1007/s12031-015-0576-8

Jones, M.K., Nair, A., Gupta, M. (2019). Mast cells in neurodegenerative disease. *Frontiers in Cellular Neuroscience, 13*, 171. https://doi.org/10.3389/fncel.2019.00171.

Kageyama, K., Iwasaki, Y., Daimon, M. (2021). Hypothalamic regulation of corticotropin-releasing factor under stress and stress resilience. *International Journal of Molecular Science, 22*(22), 12242. https://doi.org/10.3390/ijms222212242.

Koloski, N.A., Jones, M., Kalantar, J., Weltman, M., Zaguirre, J., Talley, N.J. (2012). The brain-gut pathway in functional gastrointestinal disorders is bidirectional: A 12-year prospective population-based study. *Gut, 61*(9), 1284–1290. https://doi.org/10.1136/gutjnl-2011-300474.

Lee, Y.K., Menezes, J.S., Umesaki, Y., Mazmanian, S.K. (2011). Proinflammatory T-cell responses to gut microbiota promote experimental autoimmune encephalomyelitis. *Proceedings of the National Academy of Science, 108*(Suppl. 1), 4615–4622. https://doi.org/10.1073/pnas.1000082107.

Lin, C., Zhao, S., Zhu, Y., Fan, Z., Wang, J., Zhang, B., Chen, Y. (2019). Microbiota-gut-brain axis and toll-like receptors in Alzheimer's disease. *Computational and Structural Biotechnology Journal*, *17*, 1309–1317. https://doi.org/10.1016/j.csbj.2019.09.008.

Liu, Q., Xi, Y., Wang, Q., Liu, J., Li, P., Meng, X., Liu, K., Chen, W., Liu, X., Liu, Z. (2021). Mannan oligosaccharide attenuates cognitive and behavioral disorders in the 5xFAD Alzheimer's disease mouse model via regulating the gut microbiota-brain axis. *Brain, Behavior, and Immunity*, *95*, 330–343. https://doi.org/10.1016/j.bbi.2021.04.005.

Lu, J., Lu, L., Yu, Y., Cluette-Brown, J., Martin, C.R., Claud, E.C. (2018). Effects of intestinal microbiota on brain development in humanized Gnotobiotic mice. *Science Report*, *8*(1), 5443. https://doi.org/10.1038/s41598-018-23692-w.

Macfarlane, S., Dillon, J.F. (2007). Microbial biofilms in the human gastrointestinal tract. *Journal of Applied Microbiology*, *102*(5), 1187–1196. https://doi.org/10.1111/j.1365-2672.2007.03287.x.

Maslinska, D., Laure-Kamionowska, M., Maslinski, K.T., Gujski, M., Maslinski, S. (2007). Distribution of tryptase-containing mast cells and metallothionein reactive astrocytes in human brains with amyloid deposits. *Inflammatory Research*, *56*(Suppl. 1), S17–S18. https://doi.org/10.1007/s00011-006-0508-8.

Metidji, A., Omenetti, S., Crotta, S., Li, Y., Nye, E., Ross, E., Li, V., Maradhana, M.R., Schiering, C., Stockinger, B. (2018). The environmental sensor AHR protects from inflammatory damage by maintaining intestinal stem cell homeostasis and barrier integrity. *Immunity*, *49*(2018), 353–362. https://doi.org/10.1016/j.immuni.2018.07.010.

Morais L.H., Schreiber H.L., Mazmanian S.K. (2021). The gut microbiota-brain axis in behavior and brain disorders. *Nature Reviews Microbiology*, *19*, 241–255. https://doi.org/10.1038/s41579-020-00460-0.

Oliveira, T.P.D., Morais, A.L.B., dos Reis, P.L.B., Palotás, A., Vieira, L.B. (2024). A potential role for the ketogenic diet in Alzheimer's disease treatment: Exploring pre-clinical and clinical evidence. *Metabolites*, *14*(1), 25. https://doi.org/10.3390/metabo14010025.

Parashar, A., Udayabanu, M. (2017). Gut microbiota: Implications in Parkinson's disease. *Parkinsonism Related Disorder*, *38*, 1–7. https://doi.org/10.1016/j. parkreldis.2017.02.002.

Pimentel, M., Lembo, A., Chey, W.D., Zakko, S., Ringel, Y., Yu, J., Mareya, S.M., Shaw, A.L., Bortey, E., Forbes, W.P., TARGET Study Group. (2011) Rifaximin therapy for patients with irritable bowel syndrome without constipation. *The New England Journal of Medicine*, *364*, 22–32. https://doi.org/10.1056/nejmoa1004409.

Rea, K., Dinan, T.G., Cryan, J.F. (2017). The brain-gut axis contributes to neuroprogression in stress-related disorders. *Modern Trends in Pharmacopsychiatry*, *31*, 152–161. https://doi.org/10.1159/000470813.

Rojas, O.L., Pröbstel, A.K., Porfilio, E.A., Wang, A.A., Charabati, M., Sun, T., Lee, D.S.W., Galicia, G., Ramaglia, V., Ward, L.A., Leung, L.Y.T., Najafi, G., Khaleghi, K., Garcillán, B., Li, A., Besla, R., Naouar, I., Cao, E.Y., Chiaranunt, P., Burrows, K., Robinson, H.G., Allanach, J.R., Yam, J., Luck, H., Campbell, D.J., Allman, D., Brooks, D.G., Tomura, M., Baumann, R., Zamvil, S.S., Bar-Or, A., Horwitz, M.S., Winer, D.A., Mortha, A., Mackay, F., Prat, A., Osborne, L.C., Robbins, C., Baranzini, S.E., Gommerman, J.L. (2019). Recirculating intestinal IgA-producing cells regulate neuroinflammation via IL-10. *Cell*, *176*(3), 610–624.e18. https://doi.org/10.1016/j.cell.2018.11.035.

Rothhammer, V., Mascanfroni, I.D., Bunse, L., Takenaka, M.C., Kenison, J.E., Mayo, L., Chao, C.C., Patel, B., Yan, R., Blain, M., Alvarez, J.I., Kébir, H., Anandasabapathy, N., Izquierdo, G., Jung, S., Obholzer, N., Pochet, N., Clish, C.B., Prinz, M., Prat, A., Antel, J., Quintana, F.J. (2016). Type I interferons and microbial metabolites of tryptophan modulate astrocyte activity and central nervous system inflammation via the aryl hydrocarbon receptor. *Nature Medicine*, *22*(6), 586–597. https://doi.org/10.1038/nm.4106.

Rutsch, A., Kantsj, J.B., Ronchi, F. (2020). The gut-brain axis: How microbiota and host inflammasome influence brain physiology and pathology. *Frontier in Immunology*, *11*, 604179. https://doi.org/10.3389/fimmu.2020.604179

Salter, M.W., Stevens, B. (2017). Microglia emerge as central players in brain disease. *Nature Medicine*, *23*(9), 1018–1027. https://doi.org/10.1038/nm.4397.

Sasso, J.M., Ammar, R.M., Tenchov, R., Lemmel, S., Kelber, O., Grieswelle, M., Zhou, Q.A. (2023). Gut microbiome-brain alliance: A landscape view into mental and gastrointestinal health and disorders. *ACS Chemical Neuroscience Journal*, *14*(10), 1717–1763. https://doi.org/10.1021/acschemneuro.3c00127.

Sinagra, E., Utzeri, E., Morreale, G.C., Fabbri, C., Pace, F., Anderloni, A., (2020). Microbiota-gut-brain axis and its affect inflammatory bowel disease: Pathophysiological concepts and insights for clinicians. *World Journal of Clinical Cases*, *8*(6), 1013–1025. https://doi.org/10.12998/wjcc.v8.i6.1013.

Theoharides, T.C., Stewart, J.M., Hatziagelaki, E., Kolaitis, G. (2015). Brain "fog," inflammation and obesity: Key aspects of neuropsychiatric disorders improved by luteolin. *Frontier in Neuroscience*, *9*, 225. https://doi.org/10.3389/fnins.2015.00225.

Tognini, P. (2017). Gut microbiota: A potential regulator of neurodevelopment. *Frontiers in Cellular Neuroscience*, *11*, 25. https://doi.org/10.3389/fncel.2017.00025.

Traina, G. (2017). Mast cells in the brain—Old cells, new target. *Journal of Integrative Neuroscience*, *16*(s1), S69–S83. https://doi.org/10.3233/jin-170068.

Ullah, H., Arbab, S., Tian, Y., Liu, C.Q., Chen, Y., Qijie, L., Khan, M.I.U., Hassan, I.U., Li, K. (2023). The gut microbiota-brain axis in neurological disorder. *Frontier in Neuroscience*, *17*, 1225875. https://doi.org/10.3389/fnins.2023.1225875.

Upthegrove, R., Manzanares-Teson, N., Barnes, N.M. (2014). Cytokine function in medication-naive first episode psychosis: A systematic review and meta-analysis. *Schizophrenia Research*, *155*(1–3), 101–108. https://doi.org/10.1016/j.schres.2014.03.005.

Vainchtein, I.D., Chin, G., Cho, F.S., Kelley, K.W., Miller, J.G., Chien, E.C., Liddelow, S.A., Nguyen, P.T., Nakao-Inoue, H., Dorman, L.C., Akil, O., Joshita, S., Barres, B.A., Paz, J.T., Molofsky, A.B., Molofsky, A.V. (2018). Astrocyte-derived interleukin-33 promotes microglial synapse engulfment and neural circuit development. *Science*, *359*(6381), 1269–1273. https://doi.org/10.1126/science.aal3589.

Winklewski, P.J., Radkowski, M., Demkow, U. (2016). Neuroinflammatory mechanisms of hypertension: Potential therapeutic implications. *Current Opinion in Nephrology and Hypertension*, *25*(5), 410–416. https://doi.org/10.1097/mnh.0000000000000250.

Yu, B., Wan, G., Cheng, S., Wen, P., Yang, X., Li, J., Tian, H., Gao, Y., Zhong, Q., Liu, J., Li, J., Zhu, Y. (2023). Disruptions of gut microbiota are associated with cognitive deficit of preclinical Alzheimer's disease: A cross-sectional study. *Current Alzheimer Research*, *20*(11), 875–889. https://doi.org/10.2174/0115672050303878240319054149.

Zheng, D., Liwinski, T., Elinav, E. (2020). Interaction between microbiota and immunity in health and disease. *Cell Research*, *30*, 492–506. https://doi.org/10.1038/s41422-020-0332-7

De Zuani, M., Dal Secco, C., Frossi, B. (2018). Mast cells at the crossroads of microbiota and IBD. *European Journal of Immunology*, *48*(12), 1929–1937. https://doi.org/10.1002/eji.201847504.

5 Gut Dysbiosis and Its Impact on Neurological Health

Hari Om[1#], Kavita[2#], and Pramod Kumar Kushwaha[2]*

[1] Department of Pharmaceutical Sciences and Technology, Institute of Chemical Technology, Matunga (E), Mumbai, India
[2] Department of Microbiology, School of Basic Sciences, Central University of Punjab, VPO Ghudda, Bathinda, Punjab, India
*Corresponding Author. #Equally Contributed.

5.1 INTRODUCTION

In the face of unprecedented environmental and lifestyle changes in the modern era, human health confronts significant challenges, among which the escalating prevalence of diverse neurological health stands prominent. Over the past decade, compelling evidence has emerged highlighting the pivotal role of gut microbiota in mediating brain function through microbial metabolites [1, 2]. Groundbreaking studies at the intersection of neuroscience and microbiology have elucidated dynamic relationships between animals and their internal microbial populations, essential for neurological systems' growth and optimal performance [3]. These interactions, encompassing neural, immunological, and chemical signaling pathways, underscore their complexity and essentiality in individual health and our understanding of neurological health.

The gut microbiota significantly impacts the host's health by regulating cells in nearby and distant organs, including the brain. This effect is mediated via the two-way communication pathway known as the gut–brain axis (GBA), which enables the exchange of information through hormonal signals, brain networks, and the immune system [4, 5]. This impact is facilitated by the dual-directional communication route called the GBA, which allows for information sharing through brain networks, hormonal pathways, and the immune system [6]. Healthy gut microbiota benefits the body by generating neurotransmitters and microbial metabolites that interact with intestinal epithelial cells and immune cells. However, changes in the composition of gut microbiota and its metabolite production are associated with various immune-related neurological issues, including neurodegeneration, developmental disorders, and problems with emotional regulation.

The brain, the orchestrator of human behavior and control, comprises intricate populations of neuronal and nonneuronal cells interconnected through sophisticated structural networks [7]. In contrast, the digestive tract houses over 98% of the body's bacterial population, collectively termed the gut microbiota [8]. Advancements in omics technologies have profoundly enriched our understanding of the gut microbiota's role as a critical regulator in gut–brain interactions [9, 10]. Animal and human studies have revealed compelling evidence that the gut microbiota plays a pivotal role in influencing brain behavior and cognitive development. The gut microbiota's production of hormones and immunological factors suggests exciting potential for therapeutic interventions through gut microbiome modulation [11, 12].

Reciprocal communication between the brain and the gut influences sensory–motor and secretory functions through intricate neuro-humoral networks, emphasizing the bidirectional nature of the gut–brain axis [13]. Recently recognized as a critical regulator of normal brain function and a potential risk factor in neuropathological diseases, the gut–brain axis holds promise for therapeutic intervention in neurological disorders [8]. However, elucidating the precise mechanisms underlying the gut microbiota's influence on brain disorders remains an ongoing challenge [14]. Advancements in technology are poised to transcend correlative research, offering insights into biological mechanisms that could transform our approach to treating human disease.

In this engaging chapter, we delve into the fascinating interplay between the gut and brain, shedding light on their intricate signaling pathways and the pivotal role of the microbiota in neurological disorders. From neuropsychiatric disorders such as schizophrenia and autism spectrum disorders to mood disorders such as anxiety and depression and even neurodegenerative disorders such as Alzheimer's disease, Parkinson's disease, and multiple sclerosis, we uncover the profound impact of the gut–brain connection on our overall well-being.

DOI: 10.1201/9781003536802-5

5.2 GUT MICROBIOTA AND DYSBIOSIS

The human gut microbiota is an intricate and fascinating ecosystem within the gastrointestinal tract, teeming with thousands of bacteria, viruses, and even some eukaryotes. Its composition is dynamic, and influences host health, mainly through roles in immune system maturation, nutrient absorption, metabolism, and the production of essential vitamins and short-chain fatty acids. The microbiota is mainly composed of a small number of critical phyla, with Bacteroidetes and Firmicutes being the most prevalent, making up over 90% of its population. Most Firmicutes are Gram-positive bacteria, predominantly *Clostridium* (95%), along with *Bacillus, Lactobacillus, Enterococcus*, and *Ruminococcus*. The Bacteroidetes group comprises Gram-negative bacteria, primarily *Prevotella* and *Bacteroides* [15, 16]. It can be influenced during infancy by factors such as preterm or full-term birth, delivery method (vaginal or cesarean), and diet (human milk or formula). Additionally, it evolves with age, from childhood to adulthood, and even in later stages of life. Factors like antibiotic treatments, exercise, Body Mass Index (BMI), and dietary choices significantly shape your gut microbiome [17–19].

The gut microbiome is an incredibly diverse and rich ecosystem within the human body. Humans and the bacteria in our digestive systems have evolved to form mutually beneficial relationships. Our bodies provide a perfect home for these bacteria; the gut microbiota offers numerous physiological benefits. For example, it ferments indigestible carbohydrates, producing short-chain fatty acids (SCFAs) that play a crucial role in shaping the development and function of our immune and nervous systems [20, 21]. The exploration of the gut microbiome has emerged as a critical area of study owing to its ever-changing nature and pivotal impact on the progression and prevention of diseases. The advantages to the host bestowed by a balanced gut microbiome are undeniable. However, this equilibrium is only sometimes achieved, and an imbalanced relationship can take root. This imbalance, known as dysbiosis, results from disruptions in the microbial community, leading to decreased diversity and a harmful shift in microorganisms, often linked to pathological conditions [22]. Numerous epidemiological studies have compellingly demonstrated that alterations in diet, living environment, and lifestyle choices are closely linked to a decline in overall diversity and essential microorganisms in the gut microbiome [23]. The modern way of life has brought about significant changes that impact our gut microbiome. Factors such as improved sanitation, higher rates of cesarean sections, smaller family sizes, sedentary lifestyles, increased use of antibiotics, reduced *Helicobacter pylori* infections, and poorer dietary choices, all play a crucial role in shaping our gut microbiome. It is essential to recognize the influence of these factors on our overall health and well-being [24]. Early gut dysbiosis has been strongly linked to the onset of diseases later in life, underscoring the urgency of understanding its origins. During the first few years of life, the gut microbiome is highly susceptible to dysbiosis, making it crucial to investigate this critical period. The emergence of metagenomic sequencing has led to a surge in research on the gut microbiome's role in influencing human physiology, digestion, detoxification, and the development of the nervous and immune systems [22]. The stomach and small intestine harbor a lower species of microbes than the colon.

Conversely, the colon is densely inhabited with an impressive 1,012 cells per gram of intestinal material, housing a diverse range of 300–1,000 bacterial species. This highlights the significance of the gut microbiome in maintaining overall health and well-being [25, 26]. This incredible substance is crucial in supporting the body's immune system, warding off infections, regulating nutrient absorption and metabolism, and generating beneficial B vitamins and vitamin K through lactic acid fermentation [27].

5.3 CAUSES OF DYSBIOSIS

5.3.1 Dietary Factors

The impact of diet on gut dysbiosis is significant and complex, altering the composition of gut microbiota over both short and long durations. Numerous studies highlight how dietary habits can significantly alter the gut microbiome. An in-depth analysis of fecal samples from 98 individuals conclusively demonstrated that diets rich in protein and animal fats are closely linked to microbiome profiles dominated by *Bacteroides* (enterotype I). Conversely, diets high in carbohydrates are strongly associated with *Prevotella*-abundant profiles (enterotype II). Specifically, fat intake correlated positively with Bacteroidetes and Actinobacteria but negatively with fiber, while Firmicutes and Proteobacteria exhibited the opposite relationship [28]. Research has shown that adopting either a high-fat/low-fiber diet or a low-fat/high-fiber diet for ten days can change the microbiome's composition in as little as 24 hours. Despite these shifts, there was no impact on enterotypes or individual differences [29].

This groundbreaking discovery revealed that switching to a purely plant-based diet can swiftly alter the composition and gene expression of the human gut microbiome. Remarkable changes were observed within just two days of starting the diet, and interestingly, these changes reverted to their original state within two days of discontinuing the diet [17]. It has been observed that *Prevotella* is positively correlated with fiber intake. Additionally, a high-fat animal-based diet has been shown to increase bile-resistant microorganisms such as *Bilophila* and *Bacteroides*. These patterns reflect the gut profiles of herbivorous and carnivorous mammals [29, 30]. High-resistant starch or non-starch polysaccharides and reduced carbohydrate diets can lead to rapid and reversible changes in gut bacterial composition in overweight individuals, with detectable effects within three to four days [31].

Comparative studies of children from rural Africa and Europe have unveiled a striking difference in their gut microbiota. African children, whose fiber-rich diets stand in stark contrast to their European counterparts' sugar-laden, animal-based diets, exhibit a significantly higher prevalence of *Bacteroidetes* genera like *Prevotella* and *Xylanibacter*. On the other hand, European children display a higher prevalence of Firmicutes and Enterobacteriaceae. This stark contrast underscores the profound impact of diet on gut microbiota composition [23]. The African diet is rich in microbiota-accessible carbohydrates (MACs) in plant polysaccharides and dietary fibers. In contrast, the Western diet lacks these MACs, reducing microbial gene richness and diversity [25].

African children have higher levels of SCFAs in their fecal samples, which are produced from fermenting complex carbohydrates like resistant starch and inulin [32]. SCFAs—such as acetate, propionate, and butyrate—play a pivotal role in energizing colon cells, supporting intestinal movement, upholding a resilient epithelial barrier, showcasing potent anti-inflammatory effects, and impeding the growth of cancerous cells. These remarkable compounds greatly influence overall gut health and well-being [31, 32]. They also stimulate hormonal and neuronal signals that suppress appetite [33]. Butyrate, the most extensively researched SCFA, can be derived from everyday foods like butter and cheese. It can also be produced through bacterial fermentation of dietary fiber, acetate, lactate fermentation by specific microbial species, or mucin degradation by *Clostridium* species [34].

Specific dietary components are linked to changes in the composition of gut microbes. For instance, consuming full-fat yogurt is associated with increased *Bifidobacterium animalis* and *Streptococcus thermophilus*, while coffee consumption is linked to butyrate-producing *Lawsonibacter asaccharolyticus*. Healthy plant-based foods are associated with butyrate-producing bacteria like *Roseburia hominis* and *Faecalibacterium prausnitzii*, whereas less healthy plant-based and animal-based foods are linked to *Clostridium* species [35]. Furthermore, it is essential to note that nutritious animal-based diets stand apart from less healthy animal-based food diets in terms of their impact on gut microbiome profiles.

5.3.2 Antibiotics Use

Antibiotics are commonly prescribed and have many benefits, but using them can lead to short-term and long-term health issues, such as an increased risk of asthma and autoimmune diseases [36, 37]. Preterm infants are at higher risk for bacterial infections and sepsis. Administering antibiotics early in life reduces the diversity of gut bacteria and can make infants more susceptible to antibiotic-resistant infections [38]. Antibiotics may disrupt the natural assembly of gut bacteria and reduce diversity, potentially leading to adverse health outcomes.

Clostridium difficile is a common infection after taking antibiotics, responsible for 15–25% of antibiotic-associated diarrhea and most antibiotic-associated colitis cases across all age groups [39]. Zimmermann and Curtis's systematic review concluded that antibiotics significantly and frequently affect the gut microbiome, with effects often lasting for months after treatment and potentially leading to long-term alterations in microbial composition and function [40]. Hagan et al. discovered that changes in the gut microbiome caused by antibiotics negatively impacted adult response to the flu vaccine, significantly altering the transcriptional and metabolic processes of peripheral blood mononuclear cells [41].

The long-term effects of antibiotics on the gut microbiome still need to be fully understood. The impact depends on the type of antibiotic, how it is administered, the dose, duration, and spectrum of activity [40]. Antibiotics cause microorganisms in the gut to harbor and exchange antibiotic-resistant genes. Palleja et al. analyzed the adult gut microbiome using shotgun metagenomics after a last-resort antibiotic course. While the microbiome largely recovered over six months, some species did not return [42]. The evidence indicates a remarkable resilience in the gut microbiome of healthy adults against the long-term effects of antibiotics. Nevertheless, it raises concerns about potential lasting impacts, particularly prolonged exposure.

A study by Korpela et al. showed that a single course of antibiotics in infants led to different levels of *Bifidobacterium*, *Enterobacteria*, and *Clostridium* for several months, with significant variation in *Bifidobacterium* levels in their stool samples [43]. Sun et al. discovered that antibiotics increased the gene expression of different cytokines in the colon of treated mice, indicating that antibiotic-induced gut changes could lead to inflammatory responses [44].

Understanding how antibiotics affect the gut microbiome is crucial for tailoring antibiotic treatments and probiotic use. Current sequencing techniques can identify gut microbiome taxa, but advanced methods are needed to analyze antibiotic resistance genes. This is especially important for informing antibiotic therapies in early life when the gut microbiome is more vulnerable to antibiotics. Broad-spectrum antibiotics can impact 30% of gut bacteria, significantly decreasing taxonomic richness, diversity, and evenness [45, 46]. Antibiotic-induced alterations can persist for months or even years.

Host-targeted drugs are affected by gut bacteria, but the effects are not entirely understood. A study comparing six host-targeted drugs and eight antibiotics found that host-targeted drugs had little impact on gut bacteria. In contrast, antibiotics significantly altered the microbial community, especially within the Firmicutes group. Antibiotics can seriously affect health by killing or inhibiting bacterial growth in the gut [47]. Antibiotics such as streptomycin, ampicillin, vancomycin, and metronidazole can change gut microbial composition, reduce diversity, and alter the Gram-positive to Gram-negative bacteria ratio, increasing Gram-negative species. [48]. Antibiotic use can raise the risk of *Clostridioides difficile* infection, especially in elderly hospitalized patients, with mortality rates exceeding 10% in those aged 80 and above [49]. Rifaximin treatments have increased beneficial bacteria such as *Faecalibacterium prausnitzii*, *Bifidobacterium*, and *Lactobacillus* [50].

Antibiotics also affect gene expression, protein activity, and overall gut microbiota metabolism, potentially driving the microbiota toward states similar to those in disease conditions [51]. For example, individuals treated with β-lactams exhibit unbalanced sugar metabolism akin to obesity [52]. Antibiotic treatment damages intestinal barriers, weakens the epithelial cell barrier, and compromises colonization resistance, leading to increased infectivity and acute gut infections [53]. Antibiotic-resistant bacteria may proliferate, and the gut microbiota's diversity may not fully recover, highlighting the need for ongoing surveillance and public awareness to manage antibiotic resistance. Antibiotics can alter the gut microbiota, leading to changes in metabolism and bile acid function, affecting the body's physiology [54]. Understanding how antibiotics interact with gut microbiota and impact host health is essential for developing strategies to minimize adverse effects and manage antibiotic resistance.

5.3.3 Stress and Lifestyle Factors

Xenobiotics, including drugs, environmental pollutants, and food additives, can modify gut microbiota composition by interacting with gut bacteria and affecting their growth and functions. Bacteria can metabolize these xenobiotics into molecules with different roles, toxicity, and persistence than the original compounds [55, 56]. Recreational drugs, such as marijuana, have been linked to gut microbial changes and dysbiosis, but the interaction is complex and not fully understood. Chronic marijuana users were found to have a *Prevotella: Bacteroides* ratio 13 times lower than nonusers, correlated with reduced cognitive functions [57]. Other recreational drugs, like cocaine and methamphetamine use, have been associated with adverse changes in gut microbial composition, but further research is necessary to understand the implications fully.

Environmental pollutants like pesticides and heavy metals can change the balance of gut bacteria, which can affect health [58–60]. Pesticides, such as herbicides, insecticides, and fungicides, can affect the gut microbiota, causing changes in relative abundances in model organisms, although findings can be inconsistent at times [61].

Food additives, including artificial sweeteners, emulsifiers, preservatives, and colorants, are currently under research for their potential impact on gut microbial communities. Research has shown that artificial sweeteners such as aspartame, acesulfame, saccharin, sucralose, and cyclamate can disrupt the gut microbiota, reducing bacterial diversity and impair glucose metabolism [62, 63]. Emulsifiers have been associated with higher inflammation and alterations in the gut microbiota [64]. Research has shown that preservatives and colorants may affect gut microbial community structure, but further investigation is needed to understand their impact on gut microbiota and overall health [62].

Studies have shown that both cigarettes and E-cigarettes lead to lower diversity of gut bacteria. However, the specific differences in bacterial phyla or genera between smokers and nonsmokers vary across studies [65]. Certain chemicals like polycyclic aromatic hydrocarbons, aldehydes, nitrosamines, benzene, and heavy metals can enter the gut, act as antimicrobials, alter pH levels, impact organic acid production, and be transformed by gut bacteria into harmful compounds. These substances can promote the growth of certain bacteria, while inhibiting others, potentially leading to an imbalance in gut bacteria [65, 66]. Nicotine has been associated with alterations in the composition of gut microbes—increasing Proteobacteria and Bacteroidetes, while reducing Actinobacteria and Firmicutes [67].

Individuals with alcohol dependence show dysbiosis, and the gut microbiota is thought to contribute to the development of alcoholic liver disease. Patients with alcoholic liver disease exhibit altered levels of Bacteroidetes and Enterococcaceae [68]. Changes in gut microbiota have been associated with heightened acetate levels due to enzymes produced by the host [69].

Engaging in physical activity can change the makeup and activity of the gut microbiome. Studies have shown that exercise leads to a rise in butyrate levels in animal models and temporary alterations in the gut microbiome of humans [70, 71]. Athletes have a diverse gut microbiome with higher Firmicutes levels and lower Bacteroidetes levels than sedentary individuals. The dietary habits of athletes, such as increased protein consumption, may influence these differences [72]. Exercise intensity also affects gut microbiota composition: aerobic exercise improves diversity and gut health, and intense physical training in athletes is linked to decreased variety, intestinal distress, and inflammation. Mechanisms include autonomic nervous system communication, neuroendocrine signaling, serotonin regulation, intestinal integrity, permeability, raised body temperature, motility, and reduced intestinal blood flow. Intense exercise can make the intestinal wall more permeable and reduce gut mucus thickness. This may allow pathogens to enter the bloodstream, leading to increased inflammation and altering the balance of gut bacteria [71, 73]. Exercise-induced alterations in gut motility, including quicker transit times and enhanced movement of gas, can lead to physical modifications in the gastrointestinal tract. These changes may affect acidity, mucus production, the formation of biofilms, and the accessibility of nutrients, all of which can influence the composition of gut bacteria [55]. Active premenopausal women showed higher levels of butyrate producers compared to less active women, potentially due to their exercise habits. Dietary differences may also have influenced the results [74]. Controlled longitudinal studies have shown that exercise induces alterations in the gut microbiome and metabolic changes, observed in individuals with a lean physique experiencing increased SCFA production, *Faecalibacterium*, and decreased *Bacteroides* [70]. Additionally, variations between lean and obese individuals underscore the influence of weight and lifestyle choices on gut microbiota.

5.3.4 Environmental Factors

The composition of gut microbiota varies significantly between individuals from different countries due to their dietary habits. For example, comparisons between Korean and American adult twin pairs reveal that diet and environmental factors lead to distinct bacterial compositions [75]. Migration and changes in living environments also play a critical role in dysbiosis. Thai immigrants moving to the United States show rapid changes in gut microbiota, including loss of native species, decreased diversity, and a shift toward increased *Bacteroides* and decreased *Prevotella*. These changes correlate with a rise in metabolic diseases and obesity among immigrants, likely due to the adoption of a Western lifestyle and diet, along with a history of food insecurity [76, 77]. This evidence underscores the profound impact of environmental changes on gut health.

Living with pets can change the types of bacteria in the gut, especially Firmicutes, in adults [78]. In infants, exposure to furry pets is associated with changes in gut microbiota profiles, such as increased *Ruminococcus* and *Oscillospira*, which are linked to obesity and allergy predisposition [79]. However, some preliminary data suggest that pet exposure might offer protective effects against allergies by modulating the gut microbiota [80].

Ethnicity and genetic background also contribute to dysbiosis, although to a lesser extent than environmental factors like diet and living conditions. Host genetics had a limited role in determining gut microbial composition in healthy individuals with different genetic ancestries but a shared environment. This suggests that environmental influences are crucial in developing and maintaining gut health (Figure 5.1).

5.4 IMPACT OF GUT DYSBIOSIS AFFECTING NEUROLOGICAL HEALTH

Contemporary changes in environment and lifestyle are significantly impacting human health. A major global issue today is the significant increase in various neurological diseases. It has become clear since the beginning of this century that the microbiota in the gut can affect brain functioning by altering signaling pathways and generating microbial by-products [81].

5.4.1 Neurological Disorder

Diseases of the central and peripheral nervous systems that can affect the autonomous nervous system, spinal cord, brain, peripheral and cranial nerves, nerve roots, and neuromuscular plaque are referred to as neurological diseases. A shift in the gut microbiota's composition has been linked to neurological and neuropsychiatric illnesses [1].

5.4.1.1 Alzheimer's Disease

Alzheimer's disease (AD) is a prevalent and advancing neurological disorder affecting the CNS that has a significant negative effect on patients' psychological as well as social functioning and their general standard of life. In AD

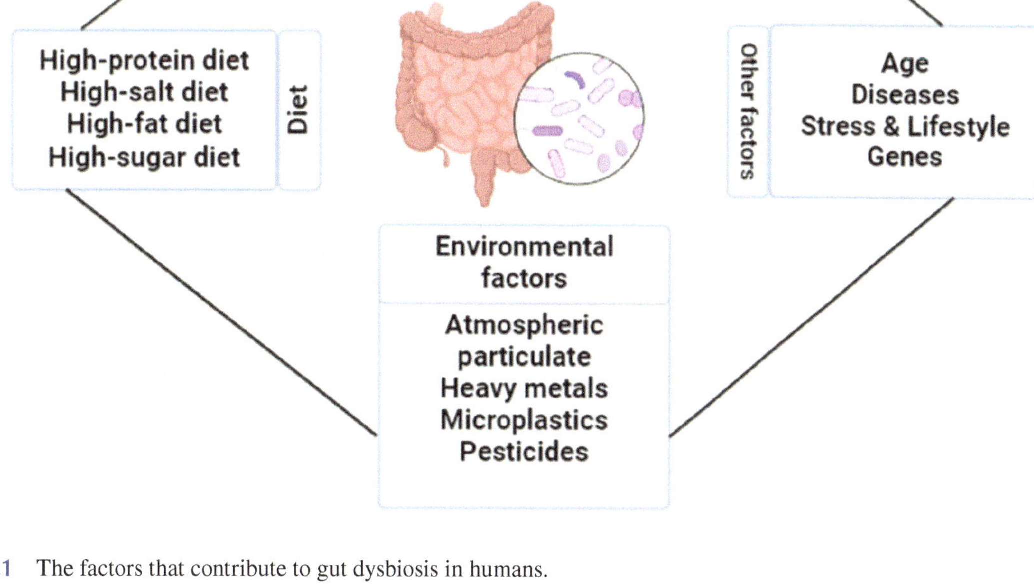

FIGURE 5.1 The factors that contribute to gut dysbiosis in humans.

patients, microtubule-associated protein Tau accumulates in cortical neuronal dendrites and axons, and there is often a progressive loss of cells in the brain and cognitive deterioration. Amyloid-β (Aβ) plaques are often abnormally accumulated around or beyond neurons [82].

In addition to being directly linked to behavioral, emotional, and cognitive regulation, the gut microbiota is critical for several brain functions, including myelinization, neurogenesis, and microglial activation. In AD mice models, such as 5xFAD and APP/PS1, several groups have identified changes in the gut flora [83]. An inverse link between the abundance of Firmicutes and Bacteroidetes species has been seen in both models of gut dysbiosis; that is, a rise in one corresponds to a reduction in the other. Interestingly, research has also shown that APP/PS1 mice raised under sterile settings have noticeably lower amounts of Aβ deposits than those raised in nonsterile conditions. Similarly, it has been noted that there is a significant rise in brain Aβ deposits when mice produced in sterile settings are transplanted with mice grown in normal conditions [84].

Comparing AD patients to healthy individuals, they exhibit gut dysbiosis and notable alterations in the intestinal microbiome's makeup. Mainly, bacteria from the phylum Firmicutes and the genus *Bifidobacterium* are less prevalent in AD patients [85].

In the same study, individuals with AD increased in the Bacteroidetes phylum. Actinobacteria levels were lower in AD patients. Changes in *Bifidobacterium* were primarily responsible for these differences. The beneficial effects of Actinobacteria, especially those belonging to the *Bifidobacterium* genus, have been well documented. These bacteria are a significant component of the human gut. Notably, several species of *Bifidobacterium* have been linked to reduced intestinal permeability and anti-inflammatory characteristics [86].

Furthermore, *Bifidobacterium* supplements have been demonstrated to enhance the characteristics that influence the gut mucosal barrier in mice and reduce intestinal LPS levels. Interestingly, in germ-free mice harboring human gut microbiota, higher *Bifidobacterium* levels are linked to decreased bacterial translocation to the systemic circulation. However, higher *Bacteroides* levels have been demonstrated to enhance bacterial translocation [82].

5.4.1.2 Parkinson's Disease

Parkinson's disease (PD) ranks as the second most common neurodegenerative condition that impacts older adults, with a typical survival span of 15 years from diagnosis to demise. The frequency of PD is estimated to be 0.4%, and by 2040, instances are expected to have doubled. The manifestations of PD encompass both motor and non-motor symptoms, including dementia, a potentially fatal condition. A typical age at which PD manifests itself is 60 years. When a sizable percentage of the neurons in the substantia nigra are already destroyed, neurological symptoms, including tremors, bradykinesia, and stiffness, are used to diagnose most PD patients [87].

CNS microglia's activity also depends on microbial peptides and metabolic products. Alterations in the microbiota may disrupt the inflammation homeostasis and indirectly impact the pathophysiology and course of PD [88]. Additionally, a few genetically modified organisms, such as *Clostridium* spp., *Lactobacillus salivarius, Parabacteroides distasonis*, and *Bacteroides vulgatus*, may alter neuro-inflammatory signaling, PD etiology, and brain function. The gut microbiome and its metabolites and environmental influences have sparked scientific interest. Lewy pathology in the enteric nervous system was reported in the 1980s, signaling a malfunctioning gut–brain axis in PD [89].

In a recent study, stool samples from 80 individuals diagnosed with Parkinson's disease and 72 people without the disease were examined. Confounding variables such as sex, age, and weight loss were considered. Compared to controls, PD patients showed a significant rise in the Lactobacillaceae, Enterobacteriaceae, and Enterococcaceae families and a significant decrease in the Lachnospiraceae family. Changes in Enterobacteriaceae and Lachnospiraceae levels were associated with a rise in disease severity and motor impairments. Significant changes in pathways supporting GI tract inflammation, as well as those involved in the metabolism of amino acids and SCFAs, were shown by functional analysis [90, 91].

5.4.1.3 Multiple Sclerosis

Multiple sclerosis (MS) is the most prevalent neurological condition affecting young individuals in North America and Europe. Axonal injury and demyelination result from $CD4^+$ and $CD25^+$ regulatory T-cells targeting brain and spinal cord cells in an autoimmune inflammatory process. Although none have been proven to be crucial, a mix of hereditary and environmental variables are probably involved in the pathophysiology of MS [92]. Growing evidence shows that modifications to the gut microbiome can significantly influence it. Although the precise processes are unknown, alterations in gut microbiota might trigger a pro-inflammatory state that leads to CNS injury and the eventual onset of MS [93].

Helicobacter, Clostridium, and *Enterococcus* are three bacteria likely to be implicated. Segmented filamentous bacteria can stimulate intestinal Th17 cells, increasing systemic autoimmunity and contributing to the immune response to gut infections. Some studies, however, suggest that particular bacteria can also give anti-inflammatory benefits. *Bacteroides fragilis*' polysaccharide product influences mucosal Treg cell responses and systemic Th1 in mice. MS patients had a lower percentage of *Fusobacteria, Eubacteriumrectale, Faecalibacterium*, and *Corynebacterium* than healthy controls, with an increase in Firmicutes, *Shigella, Escherichia*, and *Clostridium. Faecalibacterium* spp. populations have been reduced, leading to decreased levels of the metabolite butyrate. This has resulted in fewer pro-inflammatory cytokines, antigen-presenting cells, and Treg cells [94]. In MS patients, there is a decrease in *Bacteroides* spp., such as *Bacteroides coprocola* and *Bacteroides stercoris*, in the gut microbiota and a negative correlation between the amount of *Prevotella copri* and the risk of developing MS [95].

Certain compounds produced by bacteria play a role in the connection between the gut and the brain. For example, it was found that individuals with MS have significantly reduced levels of the *Bactroidetes* species metabolite, lipid 654, in their blood compared to healthy individuals, suggesting it could be a biomarker for MS. Additionally, toxins B and D from *Clostridium perfringens* have been linked to symptoms similar to those of MS, such as impaired coordination and vision problems. In patients with MS, these toxins may lead to retina inflammation by either breaking down barrier veins or binding to receptors in the vascular system. These *C. perfringens* toxins are not found in the systems of healthy individuals [96].

5.4.2 Cognitive Functions and Mental Health

Cognitive development refers to how individuals think, investigate, and solve problems. Developing information, skills, problem-solving abilities, and dispositions enhances one's understanding of their surroundings. Cognitive development includes information processing, intellect, reasoning, language development, and memory [97]. In 1970, researchers discovered that stress may influence the bacteria in mice's stomachs. In 2004, scientists at Kyusu University discovered that exposure to certain microorganisms significantly affected brain chemical levels. From birth until 36 months, research indicates that brain development and gut microbiota evolve in tandem. During this time, key brain developments include growth and myelination, synaptogenesis, structural processes, synaptic remodeling, and pruning of synapses and neurons [98].

Genetic variables and a milieu of environmental cues, such as nutrition, nursing, antibiotics, and learning experiences, influence cognitive development and gut microbiota [99]. Although these signals separately impact the development of the brain and the gut microbiota, they might also produce combined effects on the interactions between these two entities. For instance, the gut–brain axis represents a two-way communication pathway where signals are exchanged via the endocrine, neurological, circulatory, and immunological systems [100].

5.4.3 Mood Disorders: Depression and Anxiety

One of the UN's sustainable development goals is mental health. The most common mental illnesses include eating disorders, anxiety, depression, autism spectrum disorder (ASD), bipolar disorder, and schizophrenia [101]. The composition of the gut microbiota is complex; some microorganisms may protect against mental health issues, whereas others might contribute to the development and advancement of mental disorders [102].

Research has demonstrated that the gut microbiota can significantly impact brain function and mental health through pathways involving the vagus nerve, the microbial synthesis of neuroactive compounds, the microbial regulation of neuro-immune signaling, tryptophan metabolism mediated by the microbiota, and the microbial control of neuroendocrine function. Moreover, the microbiota in the stomach may create and control neurotransmitters like glutamate, dopamine, and serotonin, which are crucial for neurological and immunological processes in the brain [103].

5.4.3.1 Anxiety

Dysbiosis of the microbiota significantly contributes to the emergence and intensification of anxiety disorders. Research has demonstrated that disruptions in the diversity and makeup of intestinal bacteria can directly impact mental well-being, particularly anxiety. Individuals with anxiety-related conditions often exhibit distinct microbial profiles characterized by reduced diversity and altered abundances of specific bacterial groups. Often, there is a notable reduction in the presence of advantageous SCFA-producing bacteria, including *Faecalibacterium* and Firmicutes. This shift typically occurs alongside an increase in potentially detrimental bacteria such as *Fusobacterium* and Bacteroidetes [104].

This imbalance is not only associated with the intensity of anxiety but also affects the gut–brain axis, an intricate system of communication between the gut and the brain. The gut microbiome can affect brain activity and behavioral patterns via this axis, possibly exacerbating anxiety symptoms. Additionally, social factors like isolation can further exacerbate gut dysbiosis, leading to a vicious cycle of worsening anxiety. Therefore, understanding and addressing dysbiosis is crucial in managing anxiety disorders, offering potential avenues for therapeutic interventions, such as targeted probiotic therapies, to restore gut balance and alleviate anxiety symptoms.

5.4.3.2 Depression

Depression poses a substantial challenge to public health due to its strong correlation with increased suicide rates and various negative impacts. Research indicates a connection between imbalances in gut microbiota and the development and exacerbation of depression [105]. The research investigated how the fecal microbiota composition differs among four species and compared the presence of 16 bacterial families in healthy individuals versus those with major depressive disorder (MDD). Additionally, a separate study identified a decrease in alpha diversity within the gut microbiota in individuals undergoing depressive episodes. Further, insights from a comprehensive microbiome cohort study indicated that individuals suffer from depression [106].

Moreover, the findings showed that while individuals with MDD presented lower quantities of Bacteroidetes, there was a notable increase in levels of *Clostridium XI, Klebsiella, Prevotella*, and *Streptococcus*. In addition, it was observed that, when compared to their healthy counterparts, premenopausal women experiencing depression exhibited elevated levels of *Klebsiella aerogenes*, an organism capable of breaking down estradiol. Additionally, studies revealed that when feces from depressed individuals were transferred to germ-free mice, the mice began to display behaviors akin to depression [107]. Various genera of gut microbiota, such as *Klebsiella, Prevotella*, and *Clostridium*, are positively correlated with depression. Additionally, an imbalance in gut microbiota appears to play a crucial role in the underlying mechanisms of depression through its impact on the expression of particular proteins in tissues related to the GBA [103, 108].

5.4.4 Neurodevelopment and Neurodegenerative Diseases

Development of the brain is a complex process that usually starts in the third week of pregnancy and lasts until early adulthood. The function of GI microflora has recently been identified as one of the elements that might influence brain development. Recently, it has been shown that the gut microbiota influences behavioral regulation in humans and animals [109].

Dysregulation of the gut microbiome in early life can impact neurodevelopment. The microbiome is thought to significantly influence development, particularly neurological development, via the MGBA. It is acquired at birth and develops during infancy and adolescence. Numerous elements of neurodevelopment have been demonstrated to be impaired in germ-free animal models, including neurogenesis, myelination, dendritic spine development, and aberrant ENS formation. These abnormalities can be corrected by transplanting a normal microbiome into the recipient after birth [110].

It also seems that the gut microbiome is necessary for the successful growth of the HPA axis, as maturing microbiomes stimulate inflammatory markers, including IL-6, IL-1β, and TNF-α, which aid in the maturation of the HPA axis. The development of thinking ability and functional communication has been linked to the variety of bacteria in the gut, even though human studies on the subject are far more limited. This is particularly the case in studies on ASD, where factors such as the mother's microbiome, the establishment of the microbiota in pregnancy, the procedure of delivery, and the use of antibiotics by both the mother and the child have been linked to microbiome growth and diminished neurodevelopment [111, 112]. One new theory about MGBA connectivity is that it functions through the microbiome's possible impact on brain microRNA expression, which might impact the expression of neurodevelopmental genes (Table 5.1) [86, 113].

5.5 MECHANISMS OF GUT MICROBIOTA DYSBIOSIS IN NEUROLOGICAL DISEASES

5.5.1 Neurotransmitter

Gut dysbiosis has profound implications for neurological health, primarily through its effects on neurotransmitter production and regulation. The GBA is a complex communication network that links the gastrointestinal tract to the central nervous system, which is central to this process [138]. Microorganisms in the gut can directly produce neuroactive substances and encourage the host to generate metabolites and neurotransmitters. These elements are vital for normal nervous system operations and influence microglia development, the brain's primary immune cells [139].

Gut microbes play a role in producing or regulating critical neurotransmitters like serotonin, dopamine, norepinephrine, acetylcholine, and GABA. For example, *Bifidobacterium infantis* increases blood plasma tryptophan levels, affecting central serotonin transmission, which is crucial for mood regulation. *Bifidobacterium* and *Lactobacillus* species are involved in GABA production, which decreases neuronal excitability and encourages relaxation. Other microbes such as *Escherichia, Bacillus*, and *Saccharomyces* contribute to the production of norepinephrine, while *Streptococcus, Candida, Escherichia*, and *Enterococcus* are involved in serotonin production. Also, *Lactobacillus* species can produce acetylcholine, which is necessary for memory, learning, and muscle activation [140].

An imbalance caused by dysbiosis can affect the equilibrium of these neurotransmitters, potentially resulting in a range of neurological conditions. Low levels of serotonin, dopamine, and GABA are linked to depression and anxiety. Imbalances in acetylcholine can lead to cognitive decline and diseases like Alzheimer's [138].

Beyond their role in producing neurotransmitters, the gut microbiota also impacts immune responses and neuroinflammation by creating SCFAs like butyrate, propionate, and acetate. These compounds, which result from the fermentation of dietary fiber, are crucial for the communication between the gut and the brain. They influence the immune system, epigenetic mechanisms, and neuroplasticity by acting on vagal afferents to modulate inflammation and hormone balance and serve as signaling molecules for G protein-coupled receptors (GPCRs). These receptors play a crucial role in regulating gene expression by blocking the activity of histone deacetylases. An imbalance in SCFA production due to gut dysbiosis can increase neuroinflammation and disrupt neurotransmitter pathways, potentially leading to various neurological conditions [141].

TABLE 5.1
Increased and Decreased Microorganisms of Gut Dysbiosis Linked with Different Neurological Conditions

Disease	Decreased Microorganisms	Increased Microorganisms	References
Alzheimer's disease (AD)	*Bifidobacterium, Clostridium, Dialister, Turicibacter*	*Bacteroides, Blautia, Bilophila, Allistipes, Gamella*	[114–117]
Anxiety	*Bifidobacterium longum, Lactobacillus helveticus*	*Citrobacter rodentium, Campylobacter jejuni, Trichuris muris*	[118–122]
Autism spectrum disorder (ASD)	*Enterococcus, Coprococcus, Blautia, Bifidobacterium, Prevotella, Veillonella, Dialister*	*Clostridium, Bacillus* spp., *Bacteroides, Klebsiella oxytoca, Ruminococcus,* Proteobacteria, *Lactobacillus*	[118, 123, 124]
Depression	*Subdoligranulum and Coprococcus, Ruminococcaceae, Dialister,* Firmicutes, *Lachnoclostridium, Faecalibacterium,* Actinobacteria, Bacteroidetes, *Bifidobacterium, and Lactobacillus*	*Bacteroidales, Alistipes, Clostridium, Oscillibacter*	[125–130]
Huntington's disease	*Clostridium, Bacteroides, Megamonas, Allistipes*	*Bifidobacterium, Faecalibacterium, Prevotella Gemmiger*	[117]
Multiple sclerosis (MS)	*Lactobacillus, Butyricimonas, Collinsella, Slackia, Prevotella, Parabacteroides distasonis, Eubacterium rectale*	*Clostridium, Blautia, Akkermansia, Methanobrevibacter, Acinetobacter calcoaceticus, C. perfringens*	[117, 123, 124, 131–135]
Parkinson's disease (PD)	*Clostridium, Blautia, Prevotella, Bacteroides coprocola, Bacteroides massiliensis, Faecalibacterium, Dorea longicatena, Bacteroides dorei, Ruminococcus callidus Bacteroides plebeius, Stoquefichus massiliensis, Prevotella copri, Coprococcus eutactus*	*Akkermansia, Catabacter hongkongensis,Christensenella, Catabacter, Christensenella minuta, Bifidobacterium, Ruminococcus bromii, Lactobacillus, Oscillospira, and Papillibacter cinnamivorans*	[117, 118, 136, 137]
Schizophrenia	*Coprococcus, Roseburia,* and *Blautia*	*Clostridium sporogenes,* Veillonellaceae, Lactobacillaceae, Halothiobacillaceae, Brucellaceae, Micrococcineae, *Succinivibrio, Megasphaera, Collinsella, Klebsiella, and Methanobrevibacter*	[118, 123, 124, 131, 132, 134]

Microbial-associated molecular patterns (MAMPs) produced by gut microbes, such as lipopolysaccharides, also contribute to the interaction between the gut and brain by engaging with receptors such as Toll-like receptors, which are involved in immune responses. This process can result in peripheral immune cells moving to the brain, which triggers neuroinflammation and affects the functioning of the central nervous system [142].

5.5.2 Influencing of Endocrine Pathways

The mechanism of gut dysbiosis affecting neurological health involves alterations in the endocrine signaling pathways. The gut microbiota produces SCFAs, which regulate the production of gut hormones. [143]. Dysbiosis can disrupt the balance of SCFAs, leading to altered activation of GPCRs in the colon. This disruption significantly impacts the release of essential hormones such as peptide YY (PYY) and glucagon-like peptide 1 (GLP1) from enteroendocrine L-cells. These hormones are crucial for maintaining normal neurological functions and influencing mood, memory, and learning [144].

Furthermore, SCFAs communicate with the brain through free fatty acid receptors (FFARs) by activating vagal afferents. Dysbiosis, therefore, can lead to impaired signaling to the brain, affecting the CNS and peripheral nervous system (PNS). GLP1, which has receptors in various body parts, including the brain, heart, lungs, intestines, and pancreas, is involved in neuroprotection, neuroplasticity, and the reduction of neurodegenerative markers. Disruption in GLP1 signaling due to dysbiosis may contribute to neurological disorders such as Alzheimer's disease (Figure 5.2) [145, 146].

PYY, another hormone affected by dysbiosis, regulates appetite and gastric motility. It is also found in the brain, specifically within the hypothalamus and pituitary gland. Dysbiosis-induced changes in PYY levels can impact appetite control and brain activity, potentially leading to neuropsychiatric conditions [147].

Moreover, dysbiosis might affect other metabolic hormones like ghrelin, leptin, and insulin, which affect brain function. Leptin, for instance, plays a role in energy balance and appetite regulation through its action on hypothalamic receptors. Alterations in leptin signaling due to gut

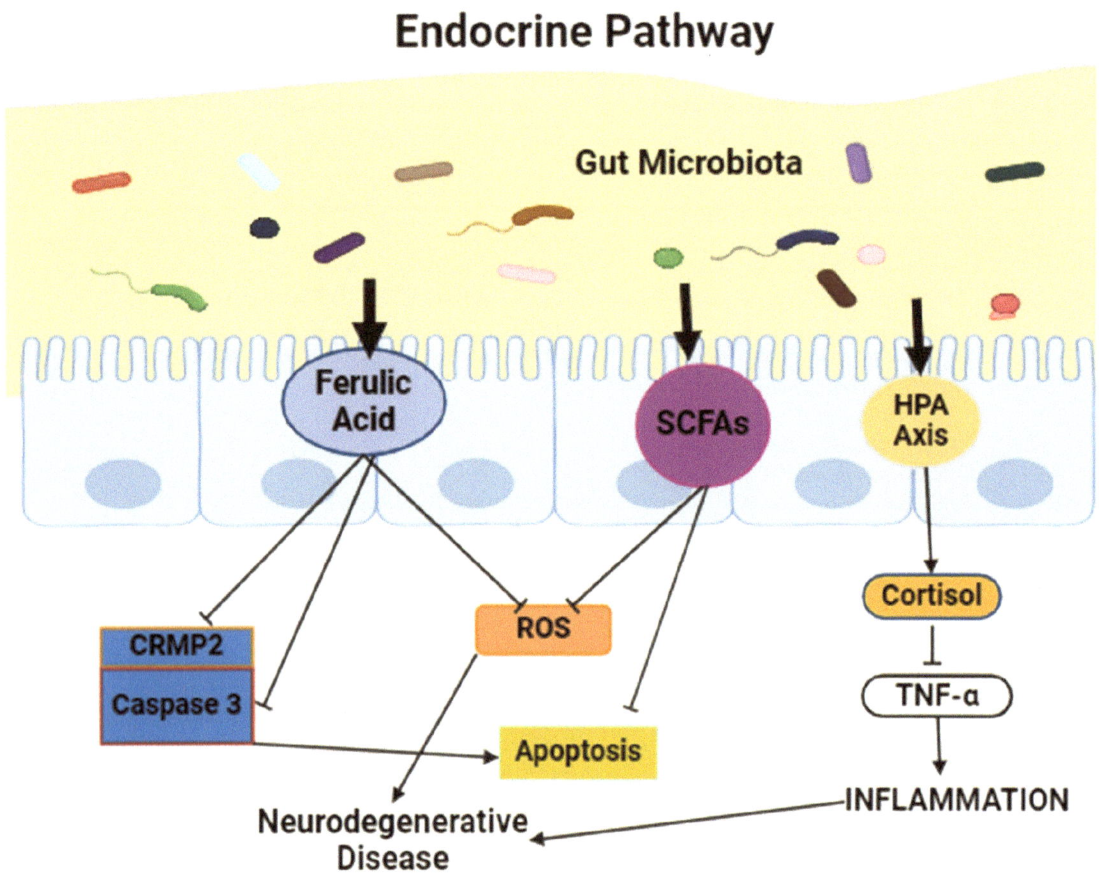

FIGURE 5.2 This figure illustrates the endocrine pathways through which the gut microbiota influences neurological signaling. The SCFAs stimulate neurotransmitter synthesis, inhibit ROS production by upregulating FoxP+ transcription, and prevent apoptosis by inhibiting caspase cascades. One of the primary mechanisms involves activating the hypothalamic-pituitary-adrenal (HPA) axis through the gut microbiota, which releases cortisol. Cortisol plays a vital role in dampening inflammatory responses and affecting neurological functions.

microbiota imbalances could further exacerbate neurological health issues [148].

5.5.3 Influencing Immune Signaling Pathways

An imbalance in gut microbial populations can lead to neurological diseases by affecting immune signaling pathways [149]. The gut microbiota profoundly impacts the mucosal immune system's formation, maturation, and function; maintaining homeostasis and defending against pathogens is crucial, but dysbiosis throws off this equilibrium, triggering the immune response in the gut and brain. This activation can trigger neuroinflammation and contribute to the onset of neurological disorders [150].

Dysbiosis affects the brain through the GBA, a two-way communication pathway between the gut and the CNS. The immune system serves as a vital mediator of this communication, with cytokines originating in the gut entering the bloodstream and being transmitted to the brain. Immune cells detect MAMPs emanating from dysbiotic microbiota through toll-like receptors, triggering the release of pro-inflammatory cytokines, including IL-1β, IL-6, TNF-α, and IL-17. Crossing these cytokines into the brain through the BBB can provoke inflammation within the brain, which might contribute to developing neurological disorders (Figure 5.3) [151].

Moreover, dysbiosis can modulate inflammatory metabolism by altering the release of several cytokines, such as IL-4, IL-10, and IFN-γ, which play essential roles in sustaining the equilibrium of the immune system. This altered cytokine profile can influence the immune responses in the gut, leading to conditions such as irritable bowel syndrome (IBS), where the enteric nervous system (ENS) becomes dysregulated. This condition involves heightened permeability of the gut lining, triggering of the mucosal immune system, and disturbances in the gut's sensory and epithelial functions. These changes can subsequently impact the gastrointestinal tract and the GBA activities [152].

Furthermore, specific microbial populations in the gut have been linked to particular neurological conditions [153]. For instance, studies have shown that in PD, an imbalance in the gut microbiota is linked to higher levels of TLR4 and CD3 in the colon's lining, along with a decrease in proteins that maintain cellular barriers. This contributes

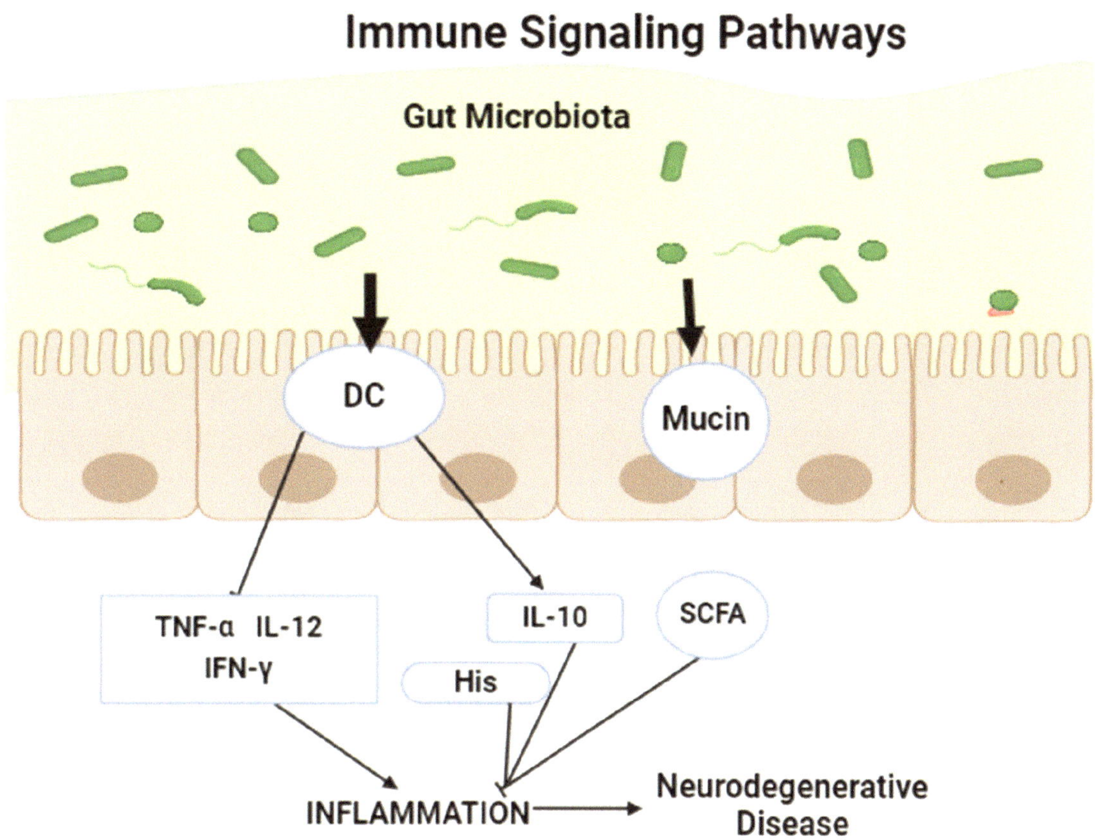

FIGURE 5.3 The figure illustrates the immunological pathways through which the gut microbiota influences neurological signaling by modulating the immune system. Molecular patterns associated with microbes (MAMPs) found on the surface of intestinal bacteria engage with immune cell receptors, including toll-like receptors (TLRs) and receptors for intercellular adhesion molecules (ICAMs) on dendritic cells (DCs). These engagements initiate a response that combats inflammation, marked by an increase in the levels of anti-inflammatory cytokines such as interleukin-10 (IL-10) and IL-4, along with a decrease in the generation of pro-inflammatory cytokines, namely tumor necrosis factor-alpha (TNFα), interleukin-1 beta (IL-1β), and interleukin-6 (IL-6).

to inflammation and exacerbates issues with movement. Similarly, in AD, certain bacteria such as *Shigella* and *Escherichia coli* correlate with systemic expression of pro-inflammatory cytokines, contributing to neuroinflammation. In MS, dysbiosis leads to an imbalance in Th17 and regulatory T-cells, exacerbating the disease's progression [152].

5.5.4 Influencing Neuronal Pathways

Gut dysbiosis can profoundly affect neurological health through its impact on neuronal pathways. One of the fundamental mechanisms by which dysbiosis influence neurological diseases involves the vagus nerve, a critical component of the ENS that connects the gut to the CNS [145]. This nerve is crucial for the two-way communication between the gut and the brain, and any disturbances in its functioning can result in significant neurological effects [154].

The vagus nerve is the longest cranial nerve and includes sensory and motor fibers. Sensory neurons account for approximately 80% of their communication capacity, transmitting microbial-derived metabolites and other signals from the gut to the brain. Motor neurons, on the other hand, handle about 20% of the communication, relaying signals from the brain to the gut. When gut microbiota become dysbiotic, this balance is disrupted, leading to altered signaling through the vagus nerve. Dysbiosis can cause inflammatory changes in the gut, which are then communicated to the brain via vagal afferent pathways, potentially influencing neurological functions and behavior (Figure 5.4) [155].

Studies show that gut dysbiosis can directly impact the vagus nerve's signaling to the brain. For example, research involving pathogen-infected animals has shown that disruption of the vagus nerve impairs the brain's response to gut-derived signals, highlighting the importance of this nerve in maintaining healthy gut–brain communication. Treating beneficial bacteria like *Bifidobacterium longum* in chronic colitis models can alleviate anxiety-like behavior. However, these impacts are diminished when the vagus nerve is impaired, highlighting the vagus nerve's essential function in facilitating the microbiota's effect on the brain. [143].

Dysbiosis can also impact the brain through mechanisms independent of direct immune responses. For instance,

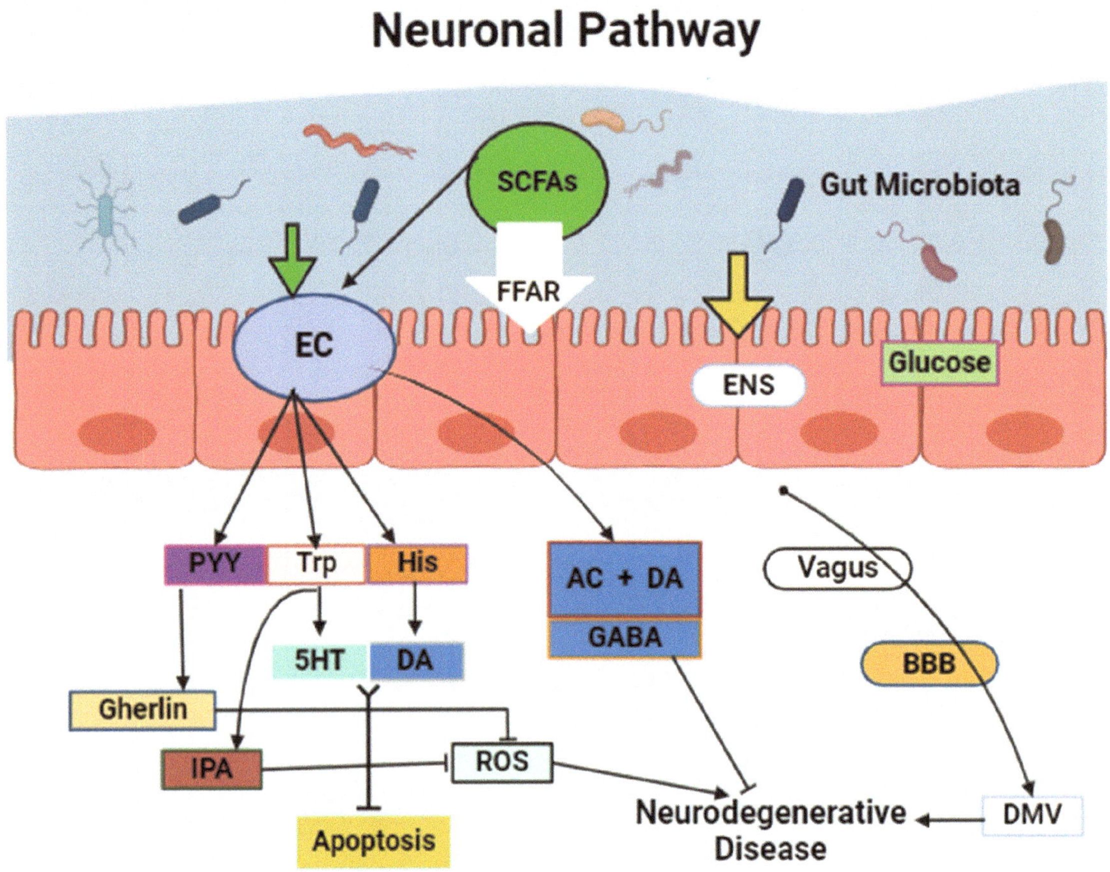

FIGURE 5.4 This illustration demonstrates the main routes through which gut microbiota impact nerve signaling within the central nervous system (CNS). Some gut bacteria can directly synthesize neurotransmitters like dopamine (DA) and acetylcholine (ACh). These substances then move into the bloodstream and pass through the blood–brain barrier (BBB), influencing activity within the CNS. Additionally, other bacteria stimulate enterochromaffin cells (ECs) in the gut lining to release neurotransmitters like serotonin (5HT) and gamma-aminobutyric acid (GABA). These ECs also produce neuroactive molecules, including peptide YY (PYY), tryptophan (Trp), and histamine (His), which further influence CNS signaling.

specific bacterial strains have been shown to interact with vagal nerve fibers, influencing brain function and behavior. An example is the effect of *Campylobacter jejuni*, a bacterium associated with gastrointestinal distress, which has been observed to increase anxiety-related behaviors and activate brain regions connected to vagal afferents. This suggests that microbial imbalances can alter neuronal pathways, disrupt normal brain function and behavior, and contribute to neurological diseases [156].

5.6 THERAPEUTIC APPROACHES TO RESTORING GUT HEALTH

5.6.1 Diet

Diet plays a pivotal role in restoring gut microbiota balance, which can significantly affect neurological health. Dietary interventions offer a promising avenue for modulating the gut microbiota and, in turn, influencing brain health and function. Eating lots of fiber from fruits, vegetables, and whole grains helps good gut bacteria grow [157, 158]. These microorganisms generate SCFAs known for their anti-inflammatory effects, which are essential in preserving the health of the blood–brain barrier. Specifically, butyrate has demonstrated its ability to boost brain-derived neurotrophic factors, a crucial protein that aids neuron survival. It encourages the formation of new neurons and plays significant roles in cognitive processes such as learning and memory. Reducing the intake of saturated fats and refined sugars, detrimental dietary components, while increasing the consumption of foods rich in anti-inflammatory compounds, such as fruits and vegetables packed with polyphenols and omega-3 fatty acids, can help restore gut microbiota balance and reduce neuroinflammation [159].

The Mediterranean diet, which includes high fiber, antioxidants, unsaturated fats, and low red meat intake, has been associated with increased beneficial lactic acid bacteria in the gut. In comparison, diets predominant in the West, which are high in animal fats and proteins, tend to promote higher populations of *Enterobacteria* and *Bacteroides*. This dietary pattern is linked to elevated inflammation and a greater likelihood of health problems [160]. Research indicates that adhering to a vegetarian diet can enhance the immune system and influence the composition of the gut microbiome. Additionally, a vegan diet, rich in dietary fiber and antioxidants, has been linked with various health benefits, especially concerning conditions such as diabetes and hypertension [161, 162]. However, poorly planned strict diets may lead to nutritional deficiencies, posing health risks.

Polyphenols, found in red wine, coffee, berries, tea, and vegetables are known to boost the presence of advantageous gut microbiota like butyrate-producing bacteria, *Bifidobacteria, Bacteroides vulgatus, Lactobacillus*, and *Akkermansia muciniphila*, and at the same time diminish the numbers of bacteria that produce lipopolysaccharides [163, 164].

Probiotics and prebiotics are also crucial in dietary strategies to restore gut microbiota and enhance neurological health. Probiotics, which are live beneficial bacteria, can be introduced through fermented foods such as yogurt, sauerkraut, kefir, and supplements. These microorganisms can help re-establish a healthy microbiome, outcompeting harmful bacteria and reducing gut inflammation [165, 166]. Foods like onions, garlic, bananas, and asparagus contain prebiotics, which are fibers that nourish the good bacteria in our gut, promoting their growth and activity.

5.6.2 Fecal Microbiota Transplantation

Fecal microbiota transplantation (FMT) is a modern treatment that involves transferring healthy gut microbes from a donor to a recipient's gastrointestinal tract to alter the recipient's gut microbiome [167, 168]. This technique has become a broad and largely nontargeted strategy to combat gut dysbiosis, leveraging the donor's microbial communities to outcompete and replace the recipient's. Over the years, FMT has evolved from a crude procedure involving the transfer of niche use of fresh donor feces to a widely accepted treatment method with standardized FMT products. These advancements emphasize pharmaceutical formulation, pharmacokinetics, pharmacodynamics, and toxicity, enhancing the safety and efficacy of FMT [169]. FMT has the advantage of transferring beneficial microbes and their supporting ecosystem, helping to manage conditions like *C. difficile* infection and inflammatory bowel disease [170]. The recent FDA approval of RBX2660 (Rebyota) and SER-109 (Vowst) for preventing recurrent *C. difficile* infection highlights the therapeutic potential of FMT [171–173]. SER-109 is a capsule-based oral microbiome therapeutic that eliminates the need for invasive endoscopy procedures [172].

In neurological health, the gut–brain axis plays a crucial role, and FMT has shown potential for modulating this axis to alleviate neurodegenerative disease symptoms. There is an association between imbalances in gut microbiota and changes in microbial metabolites with the onset of neurodegenerative conditions and activation of glial cells. FMT from healthy or wild-type mice has been shown to improve gut dysbiosis and reduce neurodegeneration in mouse models of neurodegenerative diseases in vivo. [174–178]. Promoting beneficial microbes and elevating valuable metabolites helps restore the integrity of the intestinal and blood–brain barrier. This reduces systemic inflammation, glial activation, and neuroinflammation in recipient mice, showing the potential of FMT as a therapeutic strategy for neurodegenerative conditions.

The effectiveness of FMT in neurological health depends on factors such as the rearing environment of FMT donors and the gender of the recipients. A study showed that FMT from *C9orf72*-mutant mice in a pro-survival environment improved autoimmune inflammatory phenotypes in recipient C9orf72-mutant mice [179]. FMT from mice in a pro-inflammatory environment did not lead to improvements. FMT engraftment and efficacy may vary based on the gender of the recipients. In a Huntington's disease mouse model, male mice showed more excellent resistance to FMT engraftment than female mice, resulting in no improvement in cognitive function for males [180]. Additionally, this difference in FMT effectiveness based on sex has been seen in clinical trials aimed at treating irritable bowel syndrome, with women patients showing a markedly better response to FMT compared to men [181].

These findings suggest that future studies should consider the donor's environment and the recipient's gender as crucial factors influencing the effectiveness and reaction to FMT. By accounting for these variables, researchers can optimize FMT protocols and enhance therapeutic outcomes for various diseases associated with gut dysbiosis, including neurodegenerative disorders. The potential of FMT in modulating the GBA presents a promising avenue for therapeutic strategies in neurological health, warranting further exploration and standardized methodologies to ensure reliable and comprehensive outcomes.

5.6.3 Probiotics, Prebiotics, and Postbiotics

Probiotics, prebiotics, and the emerging concept of postbiotics play significant roles in microbiota modulation, particularly concerning neurological health. Probiotics, defined by the International Scientific Association for Probiotics and Prebiotics (ISAPP) as "live microorganisms that, when administered in adequate amounts, confer a health benefit on the host," have been extensively researched for their potential to restore microbial balance and support various aspects of health [182]. Probiotics improve gut and BBB health, reducing inflammation and glial activation, which are critical in neurodegenerative diseases [183]. Certain probiotic strains like *Lactobacillus* and *Bifidobacterium* have been shown to improve cognitive functions in animal models of neurodegenerative diseases by reducing intestinal inflammation and restoring healthy gut microbiota [184].

Prebiotics are nondigestible substances that promote the growth of beneficial gut bacteria and increase the production of health-promoting metabolites [182]. ISAPP defines prebiotics as "a substrate that is selectively utilized by host microorganisms, conferring a health benefit" [185]. Substances like fructo-oligosaccharides, inulin, and galacto-oligosaccharides, categorized as prebiotics, help to enhance the population of good bacteria like *Lactobacillus* and *Bifidobacterium* [186]. Recent developments have broadened the scope of prebiotics to encompass newly discovered beneficial gut bacteria, such as *Eubacterium, Roseburia*, and *Faecalibacterium* [185, 186].

Postbiotics, a relatively new concept, refers to the bioactive compounds produced by probiotics during fermentation. These compounds include SCFAs, enzymes, peptides, polysaccharides, cell surface proteins, and other metabolites [187, 188]. Postbiotics can exert health benefits similar to live probiotics without the need for viable organisms. They are particularly beneficial in cases where the use of live bacteria may pose risks, such as in individuals with compromised immune systems. Postbiotics like SCFAs, especially butyrate, have neuroprotective effects on neurological health.

The combined effects of probiotics, prebiotics, and postbiotics have shown promise in neurodegenerative disease models. These combinations can help restore healthy gut bacteria and improve metabolic functions, especially those related to producing SCFAs. By improving the gut microbiota composition and function, synbiotics can alleviate impairments in biological barriers, reduce systemic lipopolysaccharide levels, and attenuate neuroinflammation, improving neurological outcomes. However, the clinical applicability of probiotics, prebiotics, and postbiotics in neurological health remains debatable. While animal studies provide substantial evidence of their benefits, translating these findings to humans requires further investigation. Variations in study populations, probiotic strains, dosages, and follow-up periods contribute to inconsistent results in clinical trials. Moreover, concerns regarding the safety of these interventions, especially in individuals with compromised intestinal barriers or immune systems, highlight the need for more rigorous safety data. Despite these challenges, the potential of probiotics, prebiotics, and postbiotics as modulators of the GBA presents a promising avenue for therapeutic strategies in neurological health, warranting further exploration and standardized methodologies to ensure reliable and comprehensive outcomes.

5.7 CLINICAL IMPLICATIONS AND FUTURE DIRECTIONS

Despite the promising nature of these interventions, several clinical challenges need to be addressed. The safety and efficacy of microbial therapies must be thoroughly evaluated, particularly in vulnerable populations such as children, the elderly, and the immunocompromised individuals. Theoretical risks are associated with microbial interventions, including sepsis, adverse metabolic activities, gene transfer, and immune response activation [189]. Careful patient selection, stringent donor screening, and robust clinical protocols are essential to minimize these risks and ensure the safe application of microbiota-targeted therapies.

The future of gut microbiota research in neurological health will advance significantly by integrating advanced scientific techniques and comprehensive study designs. Utilizing high-resolution metagenomic analysis and multi-omics strategies will be essential for decoding the intricate relationships between gut microbiota and brain function. These techniques will enable scientists to pinpoint particular microbial strains and their metabolic products that positively or negatively affect neurological health. Understanding these interactions at a molecular level will pave the way for precision medicine approaches tailored to individual microbiome profiles. Expanding research beyond the bacteriome to include the virome and bacteriophages will provide a more holistic understanding of the microbiota-GBA [190]. Developing engineered bacteria and other innovative microbial therapies holds promise for conditions where traditional probiotics and prebiotics are ineffective. Further research should aim to identify the best formulations, dosages, and treatment lengths for microbial therapies in different neurological conditions. Identifying the appropriate patient populations and disease stages for these therapies will also be critical for maximizing their therapeutic potential.

As research progresses, the goal is establishing microbiota-targeted therapies as routine clinical strategies for neurological health. This will require extensive clinical trials to confirm their efficacy and safety and develop standardized protocols for their use. Integrating microbiota-targeted therapies into clinical practice can transform the management of neurological disorders, offer new hope for patients, and advance the neurogastroenterology field.

5.8 CONCLUSION

The complex connection between an imbalance in gut bacteria and brain health highlights the intricate nature of the GBA. Through mechanisms involving neurotransmitter regulation, endocrine and immune signaling pathways, and neuronal pathways, dysbiosis significantly impacts neurological function and contributes to the development and progression of various neurological disorders. The gut microbiota significantly affects essential neurotransmitters like serotonin, GABA, dopamine, and acetylcholine, crucial for regulating mood, cognitive function, and neuroprotection. Disruptions in their production and balance due to dysbiosis can lead to mood disorders, cognitive decline, and neurodegenerative diseases. The endocrine and immune pathways also highlight the gut's role in regulating hormones and cytokines that affect brain function and inflammation, further linking gut health to neurological conditions.

Therapeutic approaches aimed at restoring gut health, including dietary interventions, FMT, probiotics, prebiotics, and postbiotics, present promising strategies for mitigating the adverse effects of gut dysbiosis on the brain. High-fiber diets and specific microbial interventions can enhance the production of beneficial metabolites like SCFAs, reduce neuroinflammation, and support the integrity of the BBB. While these therapeutic strategies show potential, particularly in preclinical studies, their clinical applicability requires further investigation to address variability in individual responses, safety concerns, and standardized methodologies. Understanding the donor's environment and the recipient's characteristics, such as gender, is crucial for optimizing treatments like FMT. In conclusion, the GBA offers a promising frontier for understanding and treating neurological diseases. By advancing our knowledge of gut microbiota's role in brain health and developing targeted therapies to restore microbial balance, we can pave the way for innovative approaches to prevent and manage neurological disorders.

REFERENCES

1. M. Grochowska, T. Laskus, M. Radkowski, Gut microbiota in neurological disorders, Arch. Immunol. Ther. Exp. 67 (2019) 375–383. https://doi.org/10.1007/S00005-019-00561-6.
2. L.F. Iannone, A. Preda, H.M. Blottière, G. Clarke, D. Albani, V. Belcastro, M. Carotenuto, A. Cattaneo, R. Citraro, C. Ferraris, F. Ronchi, G. Luongo, E. Santocchi, L. Guiducci, P. Baldelli, P. Iannetti, S. Pedersen, A. Petretto, S. Provasi, K. Selmer, A. Spalice, A. Tagliabue, A. Verrotti, N. Segata, J. Zimmermann, C. Minetti, P. Mainardi, C. Giordano, S. Sisodiya, F. Zara, E. Russo, P. Striano, Microbiota–gut–brain axis involvement in neuropsychiatric disorders, Expert Rev. Neurother. 19 (2019) 1037–1050. https://doi.org/10.1080/14737175.2019.1638763.
3. L.H. Morais, The gut microbiota–brain axis in behaviour and brain disorders, *Nat. Rev. Microbiol.* 19 (2021) 241–255. https://doi.org/10.1038/s41579-020-00460-0.
4. S.M. Collins, M. Surette, P. Bercik, The interplay between the intestinal microbiota and the brain, *Nat. Rev. Microbiol.* 10 (2012) 735–742. https://doi.org/10.1038/NRMICRO2876.
5. M.H. Arbab, Interaction between gut microbiota and central and enteric nervous systems: the gut-brain axis concept, in Gut Microbiota: Health and Disease, 2023, pp. 169–178. https://doi.org/10.1002/9781119904786.ch15.
6. A. Burberry, M.F. Wells, F. Limone, A. Couto, K.S. Smith, J. Keaney, G. Gillet, N. van Gastel, J.Y. Wang, O. Pietilainen, M. Qian, P. Eggan, C. Cantrell, J. Mok, I. Kadiu, D.T. Scadden, K. Eggan, C9orf72 suppresses systemic and neural inflammation induced by gut bacteria, *Nature* 582 (2020) 89–94. https://doi.org/10.1038/S41586-020-2288-7.
7. G. Deidda, M. Biazzo, Gut and brain: investigating physiological and pathological interactions between microbiota and brain to gain new therapeutic avenues for brain diseases, *Front Neurosci* 15 (2021). https://doi.org/10.3389/FNINS.2021.753915/FULL.
8. Q. Ma, C. Xing, W. Long, H.Y. Wang, Q. Liu, R.F. Wang, Impact of microbiota on central nervous system and neurological diseases: the gut-brain axis, *J. Neuroinflammation* 16 (2019). https://doi.org/10.1186/S12974-019-1434-3.
9. Y. Bhattarai, J. Si, M. Pu, O.A. Ross, P.J. Mclean, L. Till, W. Moor, M. Grover, K.K. Kandimalla, K.G. Margolis, G. Farrugia, P.C. Kashyap, Role of gut microbiota in regulating gastrointestinal dysfunction and motor symptoms in a mouse model of Parkinson's disease, Gut Microbes 13 (2021). https://doi.org/10.1080/19490976.2020.1866974.
10. M. Zhu, X. Liu, Y. Ye, X. Yan, Y. Cheng, L. Zhao, F. Chen, Z. Ling, Gut microbiota: a novel therapeutic target for Parkinson's disease, Front Immunol. 13 (2022). https://doi.org/10.3389/FIMMU.2022.937555/FULL.
11. H. Wang, Y. Wang, Gut microbiota–brain axis, Chin. Med. J. 129 (2016) 2373–2380. https://doi.org/10.4103/0366-6999.190667.
12. J.M. Yano, K. Yu, G.P. Donaldson, G.G. Shastri, P. Ann, L. Ma, C.R. Nagler, R.F. Ismagilov, S.K. Mazmanian, E.Y. Hsiao, Indigenous bacteria from the gut microbiota regulate host serotonin biosynthesis, *Cell* 161 (2015) 264–276. https://doi.org/10.1016/J.CELL.2015.02.047.
13. J. Cryan, T.G. Dinan, Mind-altering microorganisms: the impact of the gut microbiota on brain and behaviour, *Nat. Rev. Neurosci.* 13 (2012). https://doi.org/10.1038/nrn3346.
14. C.R. Martin, V. Osadchiy, A. Kalani, E.A. Mayer, The brain–gut–microbiome axis, Cell. Mol. Gastroenterol. Hepatol. 6 (2018) 133–148. https://doi.org/10.1016/J.JCMGH.2018.04.003.
15. E. Rinninella, P. Raoul, M. Cintoni, F. Franceschi, G.A.D. Miggiano, A. Gasbarrini, M.C. Mele, What is the healthy gut microbiota composition? A changing ecosystem across age, environment, diet, and diseases, *Microorganisms* 7 (2019). https://doi.org/10.3390/MICROORGANISMS7010014.
16. M. Arumugam, J. Raes, E. Pelletier, D. Le Paslier, T. Yamada, D.R. Mende, G.R. Fernandes, J. Tap, T. Bruls, J.M. Batto, M. Bertalan, N. Borruel, F. Casellas, L. Fernandez, L. Gautier, T. Hansen, M. Hattori, T. Hayashi, M. Kleerebezem, K. Kurokawa, M. Leclerc, F. Levenez, C. Manichanh, H.B. Nielsen, T. Nielsen, N. Pons, J. Poulain, J. Qin, T. Sicheritz-Ponten, S. Tims, D. Torrents, E. Ugarte, E.G. Zoetendal, J. Wang, F. Guarner, O. Pedersen, W.M. de Vos, S. Brunak, J. Doré, J. Weissenbach, S.D. Ehrlich, P. Bork, M. Antolín, F. Artiguenave, H.M. Blottiere, M. Almeida, C. Brechot, C. Cara, C. Chervaux, A. Cultrone, C. Delorme, G. Denariaz, R. Dervyn, K.U. Foerstner, C. Friss, M. van de Guchte, E. Guedon, F. Haimet, W. Huber, J. van Hylckama-Vlieg, A. Jamet, C. Juste, G. Kaci, J. Knol, K. Kristiansen, O. Lakhdari, S. Layec, K. Le Roux, E. Maguin, A. Mérieux, R.M. Minardi, C. M'rini, J. Muller, R. Oozeer, J. Parkhill, P. Renault, M. Rescigno, N. Sanchez, S. Sunagawa, A. Torrejon, K. Turner, G. Vandemeulebrouck, E. Varela, Y. Winogradsky, G. Zeller, Enterotypes of the human gut microbiome, *Nature* 473 (2011) 174–180. https://doi.org/10.1038/NATURE09944.
17. L.A. David, C.F. Maurice, R.N. Carmody, D.B. Gootenberg, J.E. Button, B.E. Wolfe, A. V. Ling, A.S. Devlin, Y. Varma, M.A. Fischbach, S.B. Biddinger, R.J. Dutton, P.J. Turnbaugh, Diet rapidly and reproducibly alters the human gut microbiome, *Nature* 505 (2014) 559–563. https://doi.org/10.1038/NATURE12820.
18. T. Yatsunenko, F.E. Rey, M.J. Manary, I. Trehan, M.G. Dominguez-Bello, M. Contreras, M. Magris, G. Hidalgo, R.N. Baldassano, A.P. Anokhin, A.C. Heath, B. Warner, J. Reeder, J. Kuczynski, J.G. Caporaso, C.A. Lozupone, C. Lauber, J.C. Clemente, D. Knights, R. Knight, J.I. Gordon, Human gut microbiome viewed across age and geography, *Nature* 486 (2012) 222–227. https://doi.org/10.1038/NATURE11053.

19. L. Putignani, F. Del Chierico, A. Petrucca, P. Vernocchi, B. Dallapiccola, The human gut microbiota: a dynamic interplay with the host from birth to senescence settled during childhood, *Pediatr. Res.* 76 (2014) 2–10. https://doi.org/10.1038/PR.2014.49.
20. I. Laforest-Lapointe, M.C. Arrieta, Patterns of early-life gut microbial colonization during human immune development: an ecological perspective, *Front. Immunol.* 8 (2017). https://doi.org/10.3389/FIMMU.2017.00788/FULL.
21. L. V. Hooper, J.I. Gordon, Commensal host-bacterial relationships in the gut, *Science* 292 (2001) 1115–1118. https://doi.org/10.1126/SCIENCE.1058709.
22. K. Parkin, C.T. Christophersen, V. Verhasselt, M.N. Cooper, D. Martino, Risk factors for gut dysbiosis in early life, *Microorganisms* 9 (2021). https://doi.org/10.3390/MICROORGANISMS9102066.
23. C. De Filippo, D. Cavalieri, M. Di Paola, M. Ramazzotti, J.B. Poullet, S. Massart, S. Collini, G. Pieraccini, P. Lionetti, Impact of diet in shaping gut microbiota revealed by a comparative study in children from Europe and rural Africa, *Proc. Natl. Acad. Sci. U.S.A.* 107 (2010) 14691–14696. https://doi.org/10.1073/PNAS.1005963107.
24. A. Spor, O. Koren, R. Ley, Unravelling the effects of the environment and host genotype on the gut microbiome, *Nat. Rev. Microbiol.* 9 (2011) 279–290. https://doi.org/10.1038/NRMICRO2540.
25. P.J. Turnbaugh, R.E. Ley, M. Hamady, C.M. Fraser-Liggett, R. Knight, J.I. Gordon, The human microbiome project, *Nature* 449 (2007) 804–810. https://doi.org/10.1038/NATURE06244.
26. C.L. Sears, A dynamic partnership: celebrating our gut flora, *Anaerobe* 11 (2005) 247–251. https://doi.org/10.1016/J.ANAEROBE.2005.05.001.
27. H. Wu, The role of gut microbiota in immune homeostasis and autoimmunity, Gut Microbes 3 (2012) 4–14. https://doi.org/10.4161/gmic.19320.
28. S. Pedroza Matute, S. Iyavoo, Exploring the gut microbiota: lifestyle choices, disease associations, and personal genomics, *Front. Nutr.* 10 (2023). https://doi.org/10.3389/FNUT.2023.1225120.
29. G.D. Wu, J. Chen, C. Hoffmann, K. Bittinger, Y.Y. Chen, S.A. Keilbaugh, M. Bewtra, D. Knights, W.A. Walters, R. Knight, R. Sinha, E. Gilroy, K. Gupta, R. Baldassano, L. Nessel, H. Li, F.D. Bushman, J.D. Lewis, Linking long-term dietary patterns with gut microbial enterotypes, *Science* 334 (2011) 105–108. https://doi.org/10.1126/SCIENCE.1208344.
30. B.D. Muegge, J. Kuczynski, D. Knights, J.C. Clemente, A. González, L. Fontana, B. Henrissat, R. Knight, J.I. Gordon, Diet drives convergence in gut microbiome functions across mammalian phylogeny and within humans, *Science* 332 (2011) 970–974. https://doi.org/10.1126/SCIENCE.1198719.
31. R.J. Klement, V. Pazienza, Impact of different types of diet on gut microbiota profiles and cancer prevention and treatment, *Medicina* 55 (2019). https://doi.org/10.3390/MEDICINA55040084.
32. P. Barko, M. McMichael, K.S. Swanson, D.A. Williams, The gastrointestinal microbiome: a review, *J. Vet. Intern. Med.* 32 (2017) 9–25. https://doi.org/10.1111/jvim.14875.
33. E. Chambers, D. Morrison, G. Frost, Control of appetite and energy intake by SCFA: what are the potential underlying mechanisms? *Proc. Nutr. Soc.* 74 (2015) 328–336. https://doi.org/10.1017/S0029665114001657.
34. R.M. Stilling, M. Van De Wouw, G. Clarke, R.M. Stilling, C. Stanton, T.G. Dinan, J.F. Cryan, The neuropharmacology of butyrate: the bread and butter of the microbiota-gut-brain axis? *Neurochem. Int.* 99 (2016) 110–132. https://doi.org/10.1016/j.neuint.2016.06.011.
35. F. Asnicar, S.E. Berry, A.M. Valdes, L.H. Nguyen, G. Piccinno, D.A. Drew, E. Leeming, R. Gibson, C. Le Roy, H. Al Khatib, L. Francis, M. Mazidi, O. Mompeo, M. Valles-Colomer, A. Tett, F. Beghini, L. Dubois, D. Bazzani, A.M. Thomas, C. Mirzayi, A. Khleborodova, S. Oh, R. Hine, C. Bonnett, J. Capdevila, S. Danzanvilliers, F. Giordano, L. Geistlinger, L. Waldron, R. Davies, G. Hadjigeorgiou, J. Wolf, J.M. Ordovás, C. Gardner, P.W. Franks, A.T. Chan, C. Huttenhower, T.D. Spector, N. Segata, Microbiome connections with host metabolism and habitual diet from 1,098 deeply phenotyped individuals, *Nat. Med* 27 (2021) 321–332. https://doi.org/10.1038/S41591-020-01183-8.
36. A. Strzępa, F.M. Lobo, M. Majewska-Szczepanik, M. Szczepanik, Antibiotics and autoimmune and allergy diseases: causative factor or treatment?, *Int. Immunopharmacol.* 65 (2018) 328–341. https://doi.org/10.1016/J.INTIMP.2018.10.021.
37. J. Metsälä, A. Lundqvist, L.J. Virta, M. Kaila, M. Gissler, S.M. Virtanen, Prenatal and post-natal exposure to antibiotics and risk of asthma in childhood, *Clin. Exp. Allergy.* 45 (2015) 137–145. https://doi.org/10.1111/CEA.12356.
38. D. Van Der Waaij, C.E. Nord, Development and persistence of multi-resistance to antibiotics in bacteria; an analysis and a new approach to this urgent problem, *Int. J. Antimicrob. Agents.* 16 (2000) 191–197. https://doi.org/10.1016/S0924-8579(00)00227-2.
39. J.E. Belizário, J. Faintuch, Microbiome and gut dysbiosis, *Exp. Suppl.* 109 (2018) 459–476. https://doi.org/10.1007/978-3-319-74932-7_13.
40. P. Zimmermann, N. Curtis, The effect of antibiotics on the composition of the intestinal microbiota - a systematic review, *J. Infect.* 79 (2019) 471–489. https://doi.org/10.1016/J.JINF.2019.10.008.
41. T. Hagan, M. Cortese, N. Rouphael, C. Boudreau, C. Linde, M.S. Maddur, J. Das, H. Wang, J. Guthmiller, N.Y. Zheng, M. Huang, A.A. Uphadhyay, L. Gardinassi, C. Petitdemange, M.P. McCullough, S.J. Johnson, K. Gill, B. Cervasi, J. Zou, A. Bretin, M. Hahn, A.T. Gewirtz, S.E. Bosinger, P.C. Wilson, S. Li, G. Alter, S. Khurana, H. Golding, B. Pulendran, Antibiotics-driven gut microbiome perturbation alters immunity to vaccines in humans, *Cell.* 178 (2019) 1313-1328.e13. https://doi.org/10.1016/J.CELL.2019.08.010.
42. A. Palleja, K.H. Mikkelsen, S.K. Forslund, A. Kashani, K.H. Allin, T. Nielsen, T.H. Hansen, S. Liang, Q. Feng, C. Zhang, P.T. Pyl, L.P. Coelho, H. Yang, J. Wang, A. Typas, M.F. Nielsen, H.B. Nielsen, P. Bork, J. Wang, T. Vilsbøll, T. Hansen, F.K. Knop, M. Arumugam, O. Pedersen, Recovery of gut microbiota of healthy adults following antibiotic exposure, *Nat. Microbiol.* 3 (2018) 1255–1265. https://doi.org/10.1038/S41564-018-0257-9.
43. K. Korpela, A. Salonen, H. Saxen, A. Nikkonen, V. Peltola, T. Jaakkola, W. de Vos, K.L. Kolho, Antibiotics in early life associate with specific gut microbiota signatures in a prospective longitudinal infant cohort, *Pediatr. Res.* 88 (2020) 438–443. https://doi.org/10.1038/S41390-020-0761-5.

44. L. Sun, X. Zhang, Y. Zhang, K. Zheng, Q. Xiang, N. Chen, Z. Chen, N. Zhang, J. Zhu, Q. He, Antibiotic-induced disruption of gut microbiota alters local metabolomes and immune responses, *Front. Cell. Infect. Microbiol.* 9 (2019). https://doi.org/10.3389/fcimb.2019.00099.
45. L. Dethlefsen, D.A. Relman, Incomplete recovery and individualized responses of the human distal gut microbiota to repeated antibiotic perturbation, *Proc. Natl. Acad. Sci. USA.* 108 (2011) 4554–4561. https://doi.org/10.1073/PNAS.1000087107.
46. D.B. DiGiulio, R. Romero, H.P. Amogan, J.P. Kusanovic, E.M. Bik, F. Gotsch, C.J. Kim, O. Erez, S. Edwin, D.A. Relman, Microbial prevalence, diversity and abundance in amniotic fluid during preterm labor: a molecular and culture-based investigation, *PLoS One.* 3 (2008). https://doi.org/10.1371/JOURNAL.PONE.0003056.
47. J.H. Yang, P. Bhargava, D. McCloskey, N. Mao, B.O. Palsson, J.J. Collins, Antibiotic-induced changes to the host metabolic environment inhibit drug efficacy and alter immune function, *Cell Host Microbe.* 22 (2017) 757-765.e3. https://doi.org/10.1016/J.CHOM.2017.10.020.
48. C. Huang, S. Feng, F. Huo, H. Liu, Effects of four antibiotics on the diversity of the intestinal microbiota, *Microbiol. Spectr.* 10 (2022). https://doi.org/10.1128/SPECTRUM.01904-21.
49. R.L.P. Jump, Clostridium difficile infection in older adults, *Aging Health.* 9 (2013) 403. https://doi.org/10.2217/AHE.13.37.
50. F.R. Ponziani, M.A. Zocco, F. D'Aversa, M. Pompili, A. Gasbarrini, Eubiotic properties of rifaximin: disruption of the traditional concepts in gut microbiota modulation, *World J. Gastroenterol.* 23 (2017) 4491–4499. https://doi.org/10.3748/WJG.V23.I25.4491.
51. M. Ferrer, V.A.P. Martins dos Santos, S.J. Ott, A. Moya, Gut microbiota disturbance during antibiotic therapy: a multi-omic approach, *Gut Microbes.* 5 (2014) 64–70. https://doi.org/10.4161/gmic.27128.
52. E. Hernández, R. Bargiela, M.S. Diez, A. Friedrichs, A.E. Pérez-Cobas, M.J. Gosalbes, H. Knecht, M. Martínez-Martínez, J. Seifert, M. Von Bergen, A. Artacho, A. Ruiz, C. Campoy, A. Latorre, S.J. Ott, A. Moya, A. Suárez, V.A.P. Martins dos Santos, M. Ferrer, Functional consequences of microbial shifts in the human gastrointestinal tract linked to antibiotic treatment and obesity, *Gut Microbes.* 4 (2013) 306–315. https://doi.org/10.4161/GMIC.25321.
53. H. Duan, L. Yu, F. Tian, Q. Zhai, L. Fan, W. Chen, Antibiotic-induced gut dysbiosis and barrier disruption and the potential protective strategies, *Crit. Rev. Food Sci. Nutr.* 62 (2020) 1427–1452. https://doi.org/10.1080/10408398.2020.1843396.
54. R.H. Lathakumari, L.K. Vajravelu, A. Satheesan, S. Ravi, J. Thulukanam, Antibiotics and the gut microbiome: understanding the impact on human health, *Med. Microecol.* 20 (2024) 100106. https://doi.org/10.1016/J.MEDMIC.2024.100106.
55. B. Kil Song, K. Ok Cho, Y. Jo, J. Woo Oh, Y. Soo Kim, Colon transit time according to physical activity level in adults, *J. Neurogastroenterol. Motil.* 18 (2012) 64–69. https://doi.org/10.5056/jnm.2012.18.1.64.
56. N. Koppel, V.M. Rekdal, E.P. Balskus, Chemical transformation of xenobiotics by the human gut microbiota, *Science.* 356 (2017) 1246–1257. https://doi.org/10.1126/SCIENCE.AAG2770.
57. B.S. Alves De Oliveira, D.S. Milanezi, P. Do, V. Gonzaga, F.R. Detoni, R.N. Soriano, The gut microbiota in neurodegenerative diseases: revisiting possible therapeutic targets for cannabidiol, Heliyon. 3 (2017) e12172. https://doi.org/10.1016/j.heliyon.2022.e12172.
58. L. Chi, P. Tu, H. Ru, K. Lu, Studies of xenobiotic-induced gut microbiota dysbiosis: from correlation to mechanisms, Gut Microbes 13 (2021). https://doi.org/10.1080/19490976.2021.1921912.
59. Y. Jin, S. Wu, Z. Zeng, Z. Fu, Effects of environmental pollutants on gut microbiota, *Environ. Pollut.* 222 (2017) 1–9. https://doi.org/10.1016/j.envpol.2016.11.045.
60. P. Bist, S. Choudhary, Impact of heavy metal toxicity on the gut microbiota and its relationship with metabolites and future probiotics strategy: a review, Biol. Trace Elem. Res. 200 (2011) 5328–5350. https://doi.org/10.1007/s12011-021-03092-4.
61. F. Giambò, M. Teodoro, C. Costa, C. Fenga, M. Gregory Robson, Toxicology and microbiota: how do pesticides influence gut microbiota? A review, Int. J. Environ. Res. Public Health 18 (2021) 5510. https://doi.org/10.3390/ijerph18115510.
62. Y. Cao, H. Liu, N. Qin, X. Ren, B. Zhu, X. Xia, Impact of food additives on the composition and function of gut microbiota: a review, *Trends Food Sci. Technol.* 99 (2020) 295–310. https://doi.org/10.1016/j.tifs.2020.03.006.
63. J. Suez, T. Korem, D. Zeevi, G. Zilberman-Schapira, C.A. Thaiss, O. Maza, D. Israeli, N. Zmora, S. Gilad, A. Weinberger, Y. Kuperman, A. Harmelin, I. Kolodkin-Gal, H. Shapiro, Z. Halpern, E. Segal, E. Elinav, Artificial sweeteners induce glucose intolerance by altering the gut microbiota, Nat. 514 (2014). https://doi.org/10.1038/nature13793.
64. B. Chassaing, T. Van De Wiele, J. De Bodt, M. Marzorati, A.T. Gewirtz, Dietary emulsifiers directly alter human microbiota composition and gene expression ex vivo potentiating intestinal inflammation, *Gut.* 66 (2017) 1414–1427. https://doi.org/10.1136/gutjnl-2016-313099.
65. M. Antinozzi, M. Giffi, N. Sini, F. Gallè, F. Valeriani, C. De Vito, G. Liguori, V.R. Spica, M.S. Cattaruzza, Cigarette smoking and human gut microbiota in healthy adults: a systematic review, Biomedicines. 10 (2022). https://doi.org/10.3390/biomedicines10020510.
66. X. Gui, Z. Yang, M.D. Li, Effect of cigarette smoke on gut microbiota: state of knowledge, *Front. Physiol.* 12 (2021). https://doi.org/10.3389/FPHYS.2021.673341/FULL.
67. R. Wang, S. Li, L. Jin, W. Zhang, N. Liu, H. Wang, Z. Wang, P. Wei, F. Li, J. Yu, S. Lu, Y. Chen, Z. Li, C. Wu, Four-week administration of nicotinemoderately impacts blood metabolic profile and gut microbiota in a diet-dependent manner, *Biomed. Pharmacother.* 115 (2019). https://doi.org/10.1016/j.biopha.2019.108945.
68. C. MCC, Lisa A. Noel, Fernando C. G. L. Johnson, Paula S. Hill, Michael F. Smith, Thomas M. Miller, Gabriel A. Alba, Comparing the effects of acute alcohol consumption in germ-free and conventional mice: the role of the gut microbiota, BMC Microbiol. 14 (2014). https://doi.org/10.1186/s12866-014-0240-4.
69. S.-C. Wang, Y.-C. Chen, S.-J. Chen, C.-H. Lee, C.-M. Cheng, Alcohol addiction, gut microbiota, and alcoholism treatment: a review, Int. J. Mol. Sci. 21 (2020) 6413. https://doi.org/10.3390/ijms21176413.
70. J. Allen, Lindsey Mailing, Grace Niemiro, et al., Exercise alters gut microbiota composition and function in lean and obese humans, Med. Sci. Sports Exerc. 50 (2018). https://doi.org/10.1249/JES.0000000000000183.

71. L. Mailing, Jacob M. Allen, Timothy W. Buford, Christopher J. Fields, Jeffrey A. Woods, Exercise and the gut microbiome: a review of the evidence, potential mechanisms, and implications for human health, Exerc. Sport Sci. Rev. 47 (2019). https://doi.org/10.1249/JES.0000000000000183.
72. S.F. Clarke, E.F. Murphy, O. O'Sullivan, A.J. Lucey, M. Humphreys, A. Hogan, P. Hayes, M. O'Reilly, I.B. Jeffery, R. Wood-Martin, D.M. Kerins, E. Quigley, R.P. Ross, P.W. O'Toole, M.G. Molloy, E. Falvey, F. Shanahan, P.D. Cotter, Exercise and associated dietary extremes impact on gut microbial diversity, *Gut.* 63 (2014) 1913–1920. https://doi.org/10.1136/gutjnl-2013-306541.
73. B.H. Keirns, N.A. Koemel, C.M. Sciarrillo, K.L. Anderson, S.R. Emerson, Exercise and intestinal permeability: another form of exercise-induced hormesis?, *Am. J. Physiol. Gastrointest. Liver Physiol.* 319 (2020) G512–G518. https://doi.org/10.1152/AJPGI.00232.2020.
74. C. Bressa, M. Bailén-Andrino, J. Pé rez-Santiago, R. Gonzá lez-Soltero, M. Pérez, M. Gregoria Montalvo-Lominchar, J. Luis Maté-Muñoz, R. Domínguez, D. Moreno, M. Larrosa, Differences in gut microbiota profile between women with active lifestyle and sedentary women, *PLoS One.* 12 (2017). https://doi.org/10.1371/journal.pone.0171352.
75. S. Lee, J. Sung, J.E. Lee, G.P. Ko, Comparison of the gut microbiotas of healthy adult twins living in South Korea and the United States, *Appl. Environ. Microbiol.* 77 (2011) 7433–7437. https://doi.org/10.1128/AEM.05490-11.
76. L. Franzen, C. Smith, Acculturation and environmental change impacts dietary habits among adult Hmong, *Appetite.* 52 (2009) 173–183. https://doi.org/10.1016/j.appet.2008.09.012.
77. U. Mulasi-Pokhriyal, C. Smith, L. Franzen-Castle, Investigating dietary acculturation and intake among US-born and Thailand/Laos-born Hmong-American children aged 9–18 years, *Public Health Nutr.* (2011). https://doi.org/10.1017/S1368980011001649.
78. A.E. Kates, O. Jarrett, J.H. Skarlupka, A. Sethi, M. Duster, L. Watson, G. Suen, K. Poulsen, N. Safdar, Household Pet Ownership and the Microbial Diversity of the Human Gut Microbiota, *Front. Cell. Infect. Microbiol.* 10 (2020). https://doi.org/10.3389/FCIMB.2020.00073/FULL.
79. H.M. Tun, T. Konya, T.K. Takaro, J.R. Brook, R. Chari, C.J. Field, D.S. Guttman, A.B. Becker, P.J. Mandhane, S.E. Turvey, P. Subbarao, M.R. Sears, J.A. Scott, A.L. Kozyrskyj, S.S. Anand, M. Cyr, J.A. Denburg, J. Macri, T. Eiwegger, M. Azad, K. HayGlass, C. Ramsey, E. Simons, A.D. Befus, M. Brauer, D. Daley, M.S. Kobor, T.R. Kollmann, P.D. Paré, A. Sandford, S. Tebbutt, D.L. Holness, W.Y.W. Lou, J. Scott, F. Silverman, E. Chen, G. Miller, S. Dell, H. Grasemann, R. Hegele, T. Moraes, F. Ratjen, T. To, Q. Duan, P. Hystad, C. Laprise, T. Takaro, Exposure to household furry pets influences the gut microbiota of infants at 3-4 months following various birth scenarios, *Microbiome.* 5 (2017) 1–14. https://doi.org/10.1186/S40168-017-0254-X/TABLES/3.
80. M. Nermes, A. Endo, J. Aarnio, S. Salminen, E. Isolauri, Furry pets modulate gut microbiota composition in infants at risk for allergic disease, *J. Allergy Clin. Immunol.* 136 (2015) 1688–1690.e1. https://doi.org/10.1016/j.jaci.2015.07.029.
81. H. Ullah, S. Arbab, Y. Tian, C.Q. Liu, Y. Chen, L. Qijie, M.I.U. Khan, I.U. Hassan, K. Li, The gut microbiota–brain axis in neurological disorder, *Front. Neurosci.* 17 (2023) 1–16. https://doi.org/10.3389/fnins.2023.1225875.
82. T. Zhang, G. Gao, L. Kwok, Z. Sun, Gut microbiome-targeted therapies for Alzheimer ' s disease, *Gut Microbes.* 15 (2023). https://doi.org/10.1080/19490976.2023.2271613.
83. L. Bello-Corral, L. Alves-Gomes, J.A. Fernández-Fernández, D. Fernández-García, I. Casado-Verdejo, L. Sánchez-Valdeón, Implications of gut and oral microbiota in neuroinflammatory responses in Alzheimer's disease, *Life Sci.* 333 (2023). https://doi.org/10.1016/j.lfs.2023.122132.
84. E. Area-Gomez, E.A. Schon, Alzheimer disease, in *Advances in Experimental Medicine and Biology*, 2017, 149–156. https://doi.org/10.1007/978-981-10-4567-7_11.
85. S. Chandra, S.S. Sisodia, R.J. Vassar, The gut microbiome in Alzheimer ' s disease: what we know and what remains to be explored, *Mol. Neurodegener.* 7 (2023) 1–21. https://doi.org/10.1186/s13024-023-00595-7.
86. S. Khan, K.H. Barve, M.S. Kumar, Recent advancements in pathogenesis, diagnostics and treatment of Alzheimer's disease, *Curr. Neuropharmacol.* 18 (2020) 1106–1125. https://doi.org/10.2174/1570159x18666200528142429.
87. A. Keshavarzian, S.J. Green, P.A. Engen, R.M. Voigt, A. Naqib, C.B. Forsyth, E. Mutlu, K.M. Shannon, Colonic bacterial composition in Parkinson's disease, *Mov. Disord.* 30 (2015) 1351–1360. https://doi.org/10.1002/mds.26307.
88. S. Carding, K. Verbeke, D.T. Vipond, B.M. Corfe, L.J. Owen, Dysbiosis of the gut microbiota in disease, *Microb. Ecol. Heal. Dis.* 26 (2015) 1–9. https://doi.org/10.3402/mehd.v26.26191.
89. R. Medzhitov, C. Janeway, Innate immune recognition: mechanisms and pathways, *Immunol. Rev.* 173 (2000) 89–97. https://doi.org/10.1034/j.1600-065X.2000.917309.x.
90. T.R. Sampson, J.W. Debelius, T. Thron, S. Janssen, G.G. Shastri, Z.E. Ilhan, C. Challis, C.E. Schretter, S. Rocha, V. Gradinaru, M.F. Chesselet, A. Keshavarzian, K.M. Shannon, R. Krajmalnik-Brown, P. Wittung-Stafshede, R. Knight, S.K. Mazmanian, Gut microbiota regulate motor deficits and neuroinflammation in a model of Parkinson's disease, *Cell.* 167 (2016) 1469-1480.e12. https://doi.org/10.1016/j.cell.2016.11.018.
91. M. Romo-Vaquero, E. Fernández-Villalba, A.L. Gil-Martinez, L. Cuenca-Bermejo, J.C. Espín, M.T. Herrero, M.V. Selma, Urolithins: potential biomarkers of gut dysbiosis and disease stage in Parkinson's patients, *Food Funct.* 13 (2022) 6306–6316. https://doi.org/10.1039/d2fo00552b.
92. M. Sharifa, T. Ghosh, O.A. Daher, P. Bhusal, Y.A. Alaameri, J. Naz, C. Ekhator, S.B. Bellegarde, P. Bisharat, V. Vaghani, A. Hussain, Unraveling the gut-brain axis in multiple sclerosis: exploring dysbiosis, oxidative stress, and therapeutic insights, *Cureus.* 15 (2023) 4–11. https://doi.org/10.7759/cureus.47058.
93. P. Prajjwal, P. Inban, B. Natarajan, S. Gadam, M.D.M. Marsool, H. Tariq, P. Paras, N. Vora, S.T. Al-Aish, A.D.M. Marsool, O. Amir Hussin, Remyelination in multiple sclerosis, along with its immunology and association with gut dysbiosis, lifestyle, and environmental factors, *Ann. Med. Surg.* 85 (2023) 4417–4424. https://doi.org/10.1097/ms9.0000000000001127.
94. U. Adabanya, A. Awosika, A. Khan, E. Oluka, M. Adeniyi, Pediatric multiple sclerosis: an integrated outlook at the interplay between genetics, environment and brain-gut dysbiosis, *AIMS Neurosci.* 10 (2023) 232–251. https://doi.org/10.3934/Neuroscience.2023018.

95. D. Kujawa, L. Laczmanski, S. Budrewicz, A. Pokryszko-dragan, M. Podbielska, Targeting gut microbiota: new therapeutic opportunities in multiple sclerosis, *Gut Microbes.* 15 (2023). https://doi.org/10.1080/19490976.2023.2274126.
96. S.K. Yadav, K. Ito, S. Dhib-Jalbut, Interaction of the gut microbiome and immunity in multiple sclerosis: impact of diet and immune therapy, *Int. J. Mol. Sci.* 24 (2023). https://doi.org/10.3390/ijms241914756.
97. Q. Zhang, X. Dong, W. Jin, J. Fan, Early brain cognitive development in late preterm infants: an event-related potential and resting EEG study, *Ital. J. Pediatr.* 50 (2024) 1–9. https://doi.org/10.1186/s13052-023-01567-9.
98. F. Chang, Y. Huo, S. Zhang, H. Zeng, B. Tang, The impact of boarding schools on the development of cognitive and non - cognitive abilities in adolescents, *BMC Public Health.* (2023) 1–17. https://doi.org/10.1186/s12889-023-16748-8.
99. I. Shennon, B.C. Wilson, A.H. Behling, T. Portlock, R. Haque, T. Forrester, C.A. Nelson, J.M.O. Sullivan, The infant gut microbiome and cognitive development in malnutrition, *Clin. Nutr.* 43 (2024) 1181–1189. https://doi.org/10.1016/j.clnu.2024.03.029.
100. M. Angela, F. La, F. Granziera, D. Panetta, M. Pardo-tendero, M. Barone, S. Turroni, F. Faita, C. Kusmic, P. Brigidi, D. Monleon, P. Iozzo, Brain Behavior and Immunity Gut-derived metabolites mediating cognitive development in 5-year-old children: early-life transplant in mice has lasting effects throughout adulthood, *Brain Behav. Immun.* 114 (2023) 94–110. https://doi.org/10.1016/j.bbi.2023.08.009.
101. A.J. Mcguinness, A. Loughman, J.A. Foster, F. Jacka, Mood disorders: the gut bacteriome and beyond, *Biol. Psychiatry.* 95 (2024) 319–328. https://doi.org/10.1016/j.biopsych.2023.08.020.
102. T.T. Huang, J.B. Lai, Y.L. Du, Y. Xu, L.M. Ruan, S.H. Hu, Current understanding of gut microbiota in mood disorders: an update of human studies, *Front. Genet.* 10 (2019) 1–12. https://doi.org/10.3389/fgene.2019.00098.
103. R. Xiong, J. Li, J. Cheng, D. Zhou, S. Wu, S. Huang, A. Saimaiti, Z. Yang, R. Gan, H. Li, The role of gut microbiota in anxiety, depression, and other, *Nutrient.* 15 (2023).
104. A. Capuco, I. Urits, Current perspectives on gut microbiome dysbiosis and depression, *Adv. Ther.* 37 (2020) 1328–1346. https://doi.org/10.1007/s12325-020-01272-7.
105. B. Bicknell, A. Liebert, T. Borody, G. Herkes, C. McLachlan, H. Kiat, Neurodegenerative and neurodevelopmental diseases and the gut-brain axis: the potential of therapeutic targeting of the microbiome, *Int. J. Mol. Sci.* 24 (2023). https://doi.org/10.3390/ijms24119577.
106. P.M. Barbosa, E.R. Barbosa, The gut brain-axis in neurological diseases, *Int. J. Cardiovasc. Sci.* 33 (2020) 528–536. https://doi.org/10.36660/ijcs.20200039.
107. P.B. De Souza, L. de Araujo Borba, L. Castro de Jesus, A.P. Valverde, J. Gil-Mohapel, A.L.S. Rodrigues, Major depressive disorder and gut microbiota: role of physical exercise, *Int. J. Mol. Sci.* 24 (2023). https://doi.org/10.3390/ijms242316870.
108. L. Liu, H. Wang, X. Chen, Y. Zhang, H. Zhang, P. Xie, Review Gut microbiota and its metabolites in depression: from pathogenesis to treatment, *EBioMedicine.* 90 (2023) 104527. https://doi.org/10.1016/j.ebiom.2023.104527.
109. L. Ghezzi, C. Cantoni, E. Rotondo, D. Galimberti, The gut microbiome–brain crosstalk in neurodegenerative diseases, *Biomedicines.* 10 (2022) 1–17. https://doi.org/10.3390/biomedicines10071486.
110. C.J. Stewart, N.J. Ajami, J.L.O. Brien, D.S. Hutchinson, P. Daniel, M.C. Wong, M.C. Ross, R.E. Lloyd, H. Doddapaneni, G.A. Metcalf, D. Muzny, R.A. Gibbs, T. Vatanen, C. Huttenhower, R.J. Xavier, M. Rewers, W. Hagopian, early childhood from the TEDDY study, *Nature.* (2018). https://doi.org/10.1038/s41586-018-0617-x.
111. P.B. de Souza, L. de Araujo Borba, L. Castro de Jesus, A.P. Valverde, J. Gil-Mohapel, A.L.S. Rodrigues, Major depressive disorder and gut microbiota: role of physical exercise, *Int. J. Mol. Sci.* 24 (2023). https://doi.org/10.3390/ijms242316870.
112. K.A. Jellinger, Basic mechanisms of neurodegeneration: a critical update, *J. Cell. Mol. Med.* 14 (2010) 457–487. https://doi.org/10.1111/j.1582-4934.2010.01010.x.
113. F.S. Heravi, K. Naseri, H. Hu, Gut microbiota composition in patients with neurodegenerative disorders (Parkinson's and Alzheimer's) and healthy controls: a systematic review, *Nutrients.* 15 (2023). https://doi.org/10.3390/nu15204365.
114. X. Hu, T. Wang, F. Jin, Alzheimer's disease and gut microbiota, *Sci. China Life Sci.* 59 (2016) 1006–1023. https://doi.org/10.1007/s11427-016-5083-9.
115. N.M. Vogt, R.L. Kerby, K.A. Dill-McFarland, S.J. Harding, A.P. Merluzzi, S.C. Johnson, C.M. Carlsson, S. Asthana, H. Zetterberg, K. Blennow, B.B. Bendlin, F.E. Rey, Gut microbiome alterations in Alzheimer's disease, *Sci. Rep.* 7 (2017). https://doi.org/10.1038/s41598-017-13601-y.
116. C. Jiang, G. Li, P. Huang, Z. Liu, B. Zhao, The gut microbiota and Alzheimer's disease, *Adv. Alzheimer's Dis.* 9 (2022) 187–201. https://doi.org/10.3233/AIAD220013.
117. M. Kandpal, O. Indari, B. Baral, S. Jakhmola, D. Tiwari, V. Bhandari, R.K. Pandey, K. Bala, A. Sonawane, H.C. Jha, Dysbiosis of Gut microbiota from the perspective of the gut–brain axis: role in the provocation of neurological disorders, *Metab.* 12 (2022) 1064. https://doi.org/10.3390/METABO12111064.
118. N.A. Faruqui, D.H. Prium, S.A. Mowna, M.A. Ullah, Y. Araf, B. Sarkar, U.S. Zohora, M.S. Rahman, Gut microorganisms and neurological disease perspectives, *Future Neurol.* 16 (2021) 53. https://doi.org/10.2217/FNL-2020-0026.
119. P. Bercik, E.F. Verdu, J.A. Foster, J. MacRi, M. Potter, X. Huang, P. Malinowski, W. Jackson, P. Blennerhassett, K.A. Neufeld, J. Lu, W.I. Khan, I. Corthesytheulaz, C. Cherbut, G.E. Bergonzelli, S.M. Collins, Chronic gastrointestinal inflammation induces anxiety-like behavior and alters central nervous system biochemistry in mice, *Gastroenterology.* 139 (2010) 2102-2112.e1. https://doi.org/10.1053/J.GASTRO.2010.06.063.
120. P. Bercik, A.J. Park, D. Sinclair, A. Khoshdel, J. Lu, X. Huang, Y. Deng, P.A. Blennerhassett, M. Fahnestock, D. Moine, B. Berger, J.D. Huizinga, W. Kunze, P.G. Mclean, G.E. Bergonzelli, S.M. Collins, E.F. Verdu, The anxiolytic effect of Bifidobacterium longum NCC3001 involves vagal pathways for gut-brain communication, *Neurogastroenterol. Motil.* 23 (2011) 1132–1139. https://doi.org/10.1111/J.1365-2982.2011.01796.X.
121. M. Lyte, W. Li, N. Opitz, R.P.A. Gaykema, L.E. Goehler, Induction of anxiety-like behavior in mice during the initial stages of infection with the agent of murine colonic hyperplasia Citrobacter rodentium, *Physiol. Behav.* 89 (2006) 350–357. https://doi.org/10.1016/j.physbeh.2006.06.019.
122. L.E. Goehler, S.M. Park, N. Opitz, M. Lyte, R.P.A. Gaykema, Campylobacter jejuni infection increases anxiety-like behavior in the holeboard: possible anatomical

substrates for viscerosensory modulation of exploratory behavior, *Brain. Behav. Immun.* 22 (2008) 354–366. https://doi.org/10.1016/j.bbi.2007.08.009.

123. F. Liu, J. Li, F. Wu, H. Zheng, Q. Peng, H. Zhou, Altered composition and function of intestinal microbiota in autism spectrum disorders: a systematic review, *Transl. Psychiatry* 91. 9 (2019) 1–13. https://doi.org/10.1038/s41398-019-0389-6.
124. A. Mehra, G. Arora, G. Sahni, M. Kaur, H. Singh, B. Singh, S. Kaur, Gut microbiota and autism spectrum disorder: from pathogenesis to potential therapeutic perspectives, *J. Tradit. Complement. Med.* 13 (2023) 135–149. https://doi.org/10.1016/J.JTCME.2022.03.001.
125. A. Kumar, J. Pramanik, N. Goyal, D. Chauhan, B.S. Sivamaruthi, B.G. Prajapati, C. Chaiyasut, Gut microbiota in anxiety and depression: unveiling the relationships and management options, *Pharmaceuticals.* 16 (2023). https://doi.org/10.3390/PH16040565.
126. M. Valles-Colomer, G. Falony, Y. Darzi, E.F. Tigchelaar, J. Wang, R.Y. Tito, C. Schiweck, A. Kurilshikov, M. Joossens, C. Wijmenga, S. Claes, L. Van Oudenhove, A. Zhernakova, S. Vieira-Silva, J. Raes, The neuroactive potential of the human gut microbiota in quality of life and depression, *Nat. Microbiol.* 44. 4 (2019) 623–632. https://doi.org/10.1038/s41564-018-0337-x.
127. E. Aizawa, H. Tsuji, T. Asahara, T. Takahashi, T. Teraishi, S. Yoshida, M. Ota, N. Koga, K. Hattori, H. Kunugi, Possible association of bifidobacterium and lactobacillus in the gut microbiota of patients with major depressive disorder, *J. Affect. Disord.* 202 (2016) 254–257. https://doi.org/10.1016/j.jad.2016.05.038.
128. H. Jiang, Z. Ling, Y. Zhang, H. Mao, Z. Ma, Y. Yin, W. Wang, W. Tang, Z. Tan, J. Shi, L. Li, B. Ruan, Altered fecal microbiota composition in patients with major depressive disorder, *Brain. Behav. Immun.* 48 (2015) 186–194. https://doi.org/10.1016/j.bbi.2015.03.016.
129. A. Naseribafrouei, K. Hestad, E. Avershina, M. Sekelja, A. Linløkken, R. Wilson, K. Rudi, Correlation between the human fecal microbiota and depression, Neurogastroenterol. Motil. 26 (2014) 1155–1162. https://doi.org/10.1111/nmo.12378.
130. H. Kunugi, Gut microbiota and pathophysiology of depressive disorder, *Ann. Nutr. Metab.* 77 (2021) 11–20. https://doi.org/10.1159/000518274.
131. S. Vukusic, C. Confavreux, The natural history of multiple sclerosis, in *H*andbook of Multiple Sclerosis, 3rd ed., 2001, 433–447. https://doi.org/10.1007/s100720070018.
132. L. Calvo-Barreiro, H. Eixarch, X. Montalban, C. Espejo, Combined therapies to treat complex diseases: the role of the gut microbiota in multiple sclerosis, *Autoimmun. Rev.* 17 (2018) 165–174. https://doi.org/10.1016/j.autrev.2017.11.019.
133. David Mielcarz and Lloyd H. Kasper, The gut microbiome in multiple sclerosis, Curr. Treat. Options Neurol. 17 (2015) 18. https://doi.org/10.1007/s11940-015-0344-7.
134. Pavan Bhargava, Ellen M. Mowry, Gut microbiome and multiple sclerosis, Curr. Neurol. Neurosci. Rep. 14 (2014). https://doi.org/10.1007/s11910-014-0492-2.
135. H. Tremlett, D.W. Fadrosh, A.A. Faruqi, F. Zhu, J. Hart, S. Roalstad, J. Graves, S. Lynch, E. Waubant, G. Aaen, A. Belman, L. Benson, T. Charles Casper, T. Chitnis, M. Gorman, Y. Harris, L. Krupp, T.E. Lotze, J. Ness, C. Olsen, M. Rodriguez, J. Rose, T.C. Simmons, J.M. Tillema, B. Weinstock-Guttman, Gut microbiota in early pediatric multiple sclerosis: a case–control study, *Eur. J. Neurol.* 23 (2016) 1308–1321. https://doi.org/10.1111/ENE.13026.
136. V.A. Petrov, I. V. Saltykova, I.A. Zhukova, V.M. Alifirova, N.G. Zhukova, Y.B. Dorofeeva, A. V. Tyakht, B.A. Kovarsky, D.G. Alekseev, E.S. Kostryukova, Y.S. Mironova, O.P. Izhboldina, M.A. Nikitina, T. V. Perevozchikova, E.A. Fait, V. V. Babenko, M.T. Vakhitova, V.M. Govorun, A.E. Sazonov, Analysis of gut microbiota in patients with parkinson's disease, *Bull. Exp. Biol. Med.* 162 (2017) 734–737. https://doi.org/10.1007/S10517-017-3700-7/METRICS.
137. G.Ö. Uyar, H. Yildiran, A nutritional approach to microbiota in Parkinson's disease, *Biosci. Microbiota, Food Heal.* 38 (2019) 115. https://doi.org/10.12938/BMFH.19-002.
138. L.M.T. Dicks, Gut Bacteria and neurotransmitters, *Microorganisms.* 10 (2022) 1–24. https://doi.org/10.3390/microorganisms10091838.
139. X. Yang, J. Lou, W. Shan, J. Ding, Z. Jin, Y. Hu, Q. Du, Q. Liao, R. Xie, J. Xu, Pathophysiologic role of neurotransmitters in digestive diseases, *Front. Physiol.* 12 (2021) 1–18. https://doi.org/10.3389/fphys.2021.567650.
140. Anna M. Trier, Madison R. Mack, Brian S. Kim, The Neuroimmune Axis in Skin Sensation, Inflammation, and Immunity," J. Immunol. 202 (2019 May) 2829–2835. https://doi.org/10.4049/jimmunol.1801473.The.
141. M. van de Wouw, M. Boehme, J.M. Lyte, N. Wiley, C. Strain, O. O'Sullivan, G. Clarke, C. Stanton, T.G. Dinan, J.F. Cryan, Short-chain fatty acids: microbial metabolites that alleviate stress-induced brain–gut axis alterations, *J. Physiol.* 596 (2018) 4923–4944. https://doi.org/10.1113/JP276431.
142. G. Sharon, R. Sampson, Daniel H. Geschwind, Sarkis K. Mazmanian, The central nervous system and the gut microbiome. Cell 176 (2017) 139–148. https://doi.org/10.1016/j.cell.2016.10.027.The.
143. Q. Wang, Q. Yang, X. Liu, The microbiota-gut-brain axis and neurodevelopmental disorders, *Protein Cell.* 14 (2023) 762–775. https://doi.org/10.1093/procel/pwad026.
144. S. Cussotto, K. V Sandhu, T.G. Dinan, J.F. Cryan, The neuroendocrinology of the microbiota-gut-brain axis : a behavioural perspective, *Front. Neuroendocrinol.* (2018). https://doi.org/10.1016/j.yfrne.2018.04.002.
145. E.Y. Hsiao, S.W. Mcbride, S. Hsien, G. Sharon, E.R. Hyde, T. Mccue, J.A. Codelli, J. Chow, S.E. Reisman, J.F. Petrosino, P.H. Patterson, S.K. Mazmanian, Abnormalities associated with autism, *Cell.* 155 (2013) 1451–1463. https://doi.org/10.1016/j.cell.2013.11.024.The.
146. L.H. Morais, The gut microbiota–brain axis in behaviour and brain disorders, *Nat. Rev. Microbiol.* (n.d.). https://doi.org/10.1038/s41579-020-00460-0.
147. F.M. Gribble, F. Reimann, Enteroendocrine cells: chemosensors in the intestinal epithelium, *Annu. Rev. Physiol.* 78 (2016) 277–299. https://doi.org/10.1146/annurev-physiol-021115-105439.
148. C. Boktor, J. Haney, W. Wu, C. Rabut, M.S. Ladinsky, *Behaviour in Mice.* 602 (2022) 647–653. https://doi.org/10.1038/s41586-022-04396-8.A.
149. J.F. Cryan, K.J. O'riordan, C.S.M. Cowan, K. V Sandhu, T.F.S. Bastiaanssen, M. Boehme, M.G. Codagnone, S. Cussotto, C. Fulling, A. V Golubeva, K.E. Guzzetta, M. Jaggar, C.M. Long-Smith, J.M. Lyte, J.A. Martin, A. Molinero-Perez, G. Moloney, E. Morelli, E. Morillas, R. O'connor, J.S. Cruz-Pereira, V.L. Peterson, K. Rea, N.L. Ritz, E. Sherwin, S. Spichak, E.M. Teichman, M. van de

Wouw, A.P. Ventura-Silva, S.E. Wallace-Fitzsimons, N. Hyland, G. Clarke, T.G. Dinan, The microbiota-gut-brain axis, *Physiol. Rev.* 99 (2019) 1877–2013. https://doi.org/10.1152/physrev.00018.2018.
150. T.C. Fung, C.A. Olson, E.Y. Hsiao, systems in health and disease, *Nat. Neurosci.* 20 (2020) 145–155. https://doi.org/10.1038/nn.4476.Interactions.
151. G. Chen, X. Ran, B. Li, Y. Li, D. He, B. Huang, S. Fu, J. Liu, W. Wang, EBioMedicine sodium butyrate inhibits in fl ammation and maintains epithelium barrier integrity in a TNBS-induced In fl ammatory bowel disease mice model, *EBioMedicine.* 30 (2018) 317–325. https://doi.org/10.1016/j.ebiom.2018.03.030.
152. A. Schnell, L. Huang, M. Singer, A. Singaraju, R.M. Barilla, B.M.L. Regan, A. Bollhagen, P.I. Thakore, D. Dionne, T.M. Delorey, M. Pawlak, G. Meyer zu Horste, O. Rozenblatt-Rosen, R.A. Irizarry, A. Regev, V.K. Kuchroo, Stem-like intestinal Th17 cells give rise to pathogenic effector T cells during autoimmunity, *Cell.* 184 (2021) 6281–6298.e23. https://doi.org/10.1016/j.cell.2021.11.018.
153. C.C. Hoogenraad, L. Riol-blanco, Interleukin-17 : a social cytokine, *Cell.* 181 (2020) 517–519. https://doi.org/10.1016/j.cell.2020.03.060.
154. A. Agus, J. Planchais, H. Sokol, Review gut microbiota regulation of tryptophan metabolism in health and disease, *Cell Host Microbe.* 23 (2018) 716–724. https://doi.org/10.1016/j.chom.2018.05.003.
155. B. Bonaz, T. Bazin, S. Pellissier, The vagus nerve at the interface of the microbiota-gut-brain axis, *Front. Neurosci.* 12 (2018) 1–9. https://doi.org/10.3389/fnins.2018.00049.
156. S. Carloni, M. Rescigno, Seminars in Immunology The gut-brain vascular axis in neuroinflammation, *Semin. Immunol.* 69 (2023) 101802. https://doi.org/10.1016/j.smim.2023.101802.
157. P.D. Schley, C.J. Field, The immune-enhancing effects of dietary fibres and prebiotics, *Br J Nutr.* 87 (2002) S221–S230. https://doi.org/10.1079/bjn/2002541.
158. R.K. Singh, H.W. Chang, D. Yan, K.M. Lee, D. Ucmak, K. Wong, M. Abrouk, B. Farahnik, M. Nakamura, T.H. Zhu, T. Bhutani, W. Liao, Influence of diet on the gut microbiome and implications for human health, *J. Transl. Med.* 2017 151. 15 (2017) 1–17. https://doi.org/10.1186/S12967-017-1175-Y.
159. A.P. Simopoulos, Omega-3 fatty acids in inflammation and autoimmune diseases, *J Am Coll Nutr.* 21 (2002) 495–505. https://doi.org/10.1080/07315724.2002.10719248.
160. P. Illiano, R. Brambilla, C. Parolini., The mutual interplay of gut microbiota, diet and human disease, *FEBS J.* 287 (2020) 833–855. https://doi.org/10.1111/febs.15217.
161. A.S. Wilson, K.R. Koller, M.C. Ramaboli, L.T. Nesengani, S. Ocvirk, C. Chen, C.A. Flanagan, F.R. Sapp, Z.T. Merritt, F. Bhatti, T.K. Thomas, S.J.D. O'Keefe, Diet and the human gut microbiome: an international review, *Dig. Dis. Sci.* 65 (2020) 723–740. https://doi.org/10.1007/S10620-020-06112-W.
162. H. Sakkas, P. Bozidis, C. Touzios, D. Kolios, G. Athanasiou, E. Athanasopoulou, I. Gerou, C. Gartzonika, Nutritional status and the influence of the vegan diet on the gut microbiota and human health, *Med.* 56 (2020). https://doi.org/10.3390/medicina56020088.
163. U. Etxeberria, N. Arias, N. Boqué, M.T. Macarulla, M.P. Portillo, J.A. Martínez, F.I. Milagro, Reshaping faecal gut microbiota composition by the intake of trans-resveratrol and quercetin in high-fat sucrose diet-fed rats, *J. Nutr. Biochem.* 26 (2015) 651–660. https://doi.org/10.1016/j.jnutbio.2015.01.002.
164. A.M. Valdes, J. Walter, E. Segal, T.D. Spector, Role of the gut microbiota in nutrition and health, *BMJ.* 361 (2018) 36–44. https://doi.org/10.1136/bmj.k2179.
165. E. Dimidi, S.R. Cox, M. Rossi, K. Whelan, Fermented foods: definitions and characteristics, impact on the gut microbiota and effects on gastrointestinal health and disease, *Nutrients.* 11 (2019). https://doi.org/10.3390/nu11081806.
166. W. Cheng, C. Wu, J. Yu, The role of gut microbiota in cancer treatment: friend or foe?, *Gut.* 69 (2020) 1867–1876. https://doi.org/10.1136/gutjnl-2020-321153.
167. L.R. Lopetuso, S. Deleu, L. Godny, V. Petito, P. Puca, F. Facciotti, H. Sokol, G. Ianiro, L. Masucci, M. Abreu, I. Dotan, S.P. Costello, A. Hart, T.H. Iqbal, S. Paramsothy, M. Sanguinetti, S. Danese, H. Tilg, F. Cominelli, T.T. Pizarro, A. Armuzzi, G. Cammarota, A. Gasbarrini, S. Vermeire, F. Scaldaferri, The first international Rome consensus conference on gut microbiota and faecal microbiota transplantation in inflammatory bowel disease, *Gut.* 72 (2023) 1642–1650. https://doi.org/10.1136/gutjnl-2023-329948.
168. G. Cammarota, G. Ianiro, H. Tilg, M. Rajilić-Stojanović, P. Kump, R. Satokari, H. Sokol, P. Arkkila, C. Pintus, A. Hart, J. Segal, M. Aloi, L. Masucci, A. Molinaro, F. Scaldaferri, G. Gasbarrini, A. Lopez-Sanroman, A. Link, P. De Groot, W.M. De Vos, C. Högenauer, P. Malfertheiner, E. Mattila, T. Milosavljević, M. Nieuwdorp, M. Sanguinetti, M. Simren, A. Gasbarrini, European consensus conference on faecal microbiota transplantation in clinical practice, *Gut* (2017) 569–580. https://doi.org/10.1136/gutjnl-2016-313017.
169. A. Khoruts, C. Staley, M.J. Sadowsky, Faecal microbiota transplantation for Clostridioides difficile: mechanisms and pharmacology, *Nat. Rev. Gastroenterol. Hepatol.* 18 (2021) 67–80. https://doi.org/10.1038/s41575-020-0350-4.
170. M. Sorbara, E.P.-N.R. Microbiology, U. 2022, Microbiome-based therapeutics, Nat. Rev. Microbiol. 20 (2022): 365–380.
171. S. Khanna, M. Assi, C. Lee, D. Yoho, T. Louie, W. Knapple, H. Aguilar, J. Garcia-Diaz, G.P. Wang, S.M. Berry, J. Marion, X. Su, T. Braun, L. Bancke, P. Feuerstadt, Efficacy and safety of RBX2660 in PUNCH CD3, a Phase III, randomized, double-blind, placebo-controlled trial with a bayesian primary analysis for the prevention of recurrent clostridioides difficile infection, *Drugs.* 82 (2022) 1527–1538. https://doi.org/10.1007/S40265-022-01797-X.
172. M.D. Sims, S. Khanna, P. Feuerstadt, T.J. Louie, C.R. Kelly, E.S. Huang, E.L. Hohmann, E.E.L. Wang, C. Oneto, S.H. Cohen, C.S. Berenson, L. Korman, C. Lee, B. Lashner, C.S. Kraft, M. Ramesh, M. Silverman, D.S. Pardi, A. De, A. Memisoglu, D.A. Lombardi, B.R. Hasson, B.H. McGovern, L. von Moltke, Safety and tolerability of SER-109 as an investigational microbiome therapeutic in adults with recurrent clostridioides difficile infection: a phase 3, open-label, single-arm trial, *JAMA Netw. Open.* 6 (2023) e2255758. https://doi.org/10.1001/jamanetworkopen.2022.55758.
173. P. Feuerstadt, T.J. Louie, B. Lashner, E.E.L. Wang, L. Diao, J.A. Bryant, M. Sims, C.S. Kraft, S.H. Cohen, C.S. Berenson, L.Y. Korman, C.B. Ford, K.D. Litcofsky, M.-J. Lombardo, J.R. Wortman, H. Wu, J.G. Auniņš, C.W.J. McChalicher, J.A. Winkler, B.H. McGovern, M. Trucksis, M.R. Henn, L. von Moltke, SER-109, an oral microbiome

therapy for recurrent clostridioides difficile infection, *N. Engl. J. Med.* 386 (2022) 220–229. https://doi.org/10.1056/NEJMOA2106516.

174. Z. Xie, M. Zhang, Y. Luo, D. Jin, X. Guo, W. Yang, J. Zheng, H. Zhang, L. Zhang, C. Deng, W. Zheng, E.K. Tan, K. Jin, S. Zhu, Q. Wang, Healthy human fecal microbiota transplantation into mice attenuates MPTP-induced neurotoxicity via AMPK/SOD2 pathway, *Aging Dis.* 14 (2023) 2193–2214. https://doi.org/10.14336/AD.2023.0309.
175. J. Sun, J. Xu, Y. Ling, F. Wang, T. Gong, C. Yang, S. Ye, K. Ye, D. Wei, Z. Song, D. Chen, J. Liu, Fecal microbiota transplantation alleviated Alzheimer's disease-like pathogenesis in APP/PS1 transgenic mice, *Transl. Psychiatry.* 9 (2019). https://doi.org/10.1038/s41398-019-0525-3.
176. M.S. Kim, Y. Kim, H. Choi, W. Kim, S. Park, D. Lee, D.K. Kim, H.J. Kim, H. Choi, D.W. Hyun, J.Y. Lee, E.Y. Choi, D.S. Lee, J.W. Bae, I. Mook-Jung, Transfer of a healthy microbiota reduces amyloid and tau pathology in an Alzheimer's disease animal model, *Gut.* 69 (2020) 283–294. https://doi.org/10.1136/gutjnl-2018-317431.
177. M. Sun, Y. Zhu, Z. Zhou, X. Jia, Y. Xu, Q. Yang, C. Cui, Y. Shen, Neuroprotective effects of fecal microbiota transplantation on MPTP-induced Parkinson's disease mice: gut microbiota, glial reaction and TLR4/TNF-α signaling, *Brain Behav. Immun.* 70 (2018) 48–60. https://doi.org/10.1016/j.bbi.2018.02.005.
178. Z. Zhong, W. Chen, H. Gao, N. Che, M. Xu, L. Yang, Y. Zhang, M. Ye, Fecal microbiota transplantation exerts a protective role in MPTP-Induced Parkinson's disease via the TLR4/PI3K/AKT/NF-κB pathway stimulated by α-Synuclein, *Neurochem. Res.* 46 (2021) 3050–3058. https://doi.org/10.1007/S11064-021-03411-0.
179. A. Burberry, M.F. Wells, F. Limone, A. Couto, K.S. Smith, J. Keaney, G. Gillet, N. van Gastel, J.Y. Wang, O. Pietilainen, M. Qian, P. Eggan, C. Cantrell, J. Mok, I. Kadiu, D.T. Scadden, K. Eggan, C9orf72 suppresses systemic and neural inflammation induced by gut bacteria, *Nature.* 582 (2020) 89–94. https://doi.org/10.1038/s41586-020-2288-7.
180. C. Gubert, J.M. Choo, C.J. Love, S. Kodikara, B.A. Masson, J.J.M. Liew, Y. Wang, G. Kong, V.K. Narayana, T. Renoir, K.A. Lê Cao, G.B. Rogers, A.J. Hannan, Faecal microbiota transplant ameliorates gut dysbiosis and cognitive deficits in Huntington's disease mice, *Brain Commun.* 4 (2022). https://doi.org/10.1093/braincomms/fcac205.
181. T. Holvoet, M. Joossens, J.F. Vázquez-Castellanos, E. Christiaens, L. Heyerick, J. Boelens, B. Verhasselt, H. van Vlierberghe, M. De Vos, J. Raes, D. De Looze, Fecal microbiota transplantation reduces symptoms in some patients with irritable bowel syndrome with predominant abdominal bloating: short- and long-term results from a placebo-controlled randomized trial, *Gastroenterology.* 160 (2021) 145–157.e8. https://doi.org/10.1053/j.gastro.2020.07.013.
182. C. Hill, F. Guarner, G. Reid, G.R. Gibson, D.J. Merenstein, B. Pot, L. Morelli, R.B. Canani, H.J. Flint, S. Salminen, P.C. Calder, M.E. Sanders, Expert consensus document: the international scientific association for probiotics and prebiotics consensus statement on the scope and appropriate use of the term probiotic, *Nat. Rev. Gastroenterol. Hepatol.* 11 (2014) 506–514. https://doi.org/10.1038/nrgastro.2014.66.
183. J.S. Loh, W.Q. Mak, L.K.S. Tan, C.X. Ng, H.H. Chan, S.H. Yeow, J.B. Foo, Y.S. Ong, C.W. How, K.Y. Khaw, Microbiota–gut–brain axis and its therapeutic applications in neurodegenerative diseases, *Signal Transduct. Target. Ther.* 919 (2024) 1–53. https://doi.org/10.1038/s41392-024-01743-1.
184. L. Bonfili, V. Cecarini, O. Gogoi, C. Gong, M. Cuccioloni, M. Angeletti, G. Rossi, A. Maria Eleuteri, C.A.M. Eleuteri, Microbiota modulation as preventative and therapeutic approach in Alzheimer's disease, *FEBS J.* 288 (2021) 2836–2855. https://doi.org/10.1111/febs.15571.
185. S. Salminen, M.C. Collado, A. Endo, C. Hill, S. Lebeer, E.M.M. Quigley, M.E. Sanders, R. Shamir, J.R. Swann, H. Szajewska, G. Vinderola, The International Scientific Association of Probiotics and Prebiotics (ISAPP) consensus statement on the definition and scope of postbiotics, *Nat. Rev. Gastroenterol. Hepatol.* 18 (2021) 649–667. https://doi.org/10.1038/s41575-021-00440-6.
186. C. Lordan, D. Thapa, R.P. Ross, P.D. Cotter, Potential for enriching next-generation health-promoting gut bacteria through prebiotics and other dietary components, *Gut Microbes.* 11 (2019) 1–20. https://doi.org/10.1080/19490976.2019.1613124.
187. S. Salminen, M.C. Collado, A. Endo, C. Hill, S. Lebeer, E.M.M. Quigley, M.E. Sanders, R. Shamir, J.R. Swann, H. Szajewska, G. Vinderola, The International Scientific Association of Probiotics and Prebiotics (ISAPP) consensus statement on the definition and scope of postbiotics, *Nat. Rev. Gastroenterol. Hepatol.* 18 (2021) 649–667. https://doi.org/10.1038/S41575-021-00440-6.
188. B. Liang, D. Xing, The current and future perspectives of postbiotics, *Probiotics Antimicrob. Proteins.* (2023) 1626–1643. https://doi.org/10.1007/s12602-023-10045-x.
189. D. Kothari, S. Patel, S.K. Kim, Probiotic supplements might not be universally-effective and safe: a review, *Biomed. Pharmacother.* 111 (2019) 537–547. https://doi.org/10.1016/j.biopha.2018.12.104.
190. S.G. Sorboni, H.S. Moghaddam, R. Jafarzadeh-Esfehani, S. Soleimanpour, A comprehensive review on the role of the gut microbiome in human neurological disorders, *Clin. Microbiol. Rev.* 35 (2022). https://doi.org/10.1128/CMR.00338-20/ASSET/CA6CAEAC-BD20-42F0-B7CA-667E2075DCCA/ASSETS/IMAGES/LARGE/CMR.00338-20-F005.JPG.

6 Connection between Gut Microbiota and Neurodegenerative Diseases

Aniruddha Sen[1], Vijay Singh[1], Bidisha Halder[2], Sapna Saini[1], Sumel Ashique[3], Radheshyam Pal[3], Sabina Yasmin[4], and Mohammad Yousuf Ansari[5]*

[1] Department of Biochemistry, All India Institute of Medical Sciences, Gorakhpur, Uttar Pradesh, India
[2] Indian Institute of Science, Bengaluru, India
[3] Department of Pharmaceutical Sciences, Bengal College of Pharmaceutical Sciences and Research, Durgapur, West Bengal, India
[4] Department of Pharmaceutical Chemistry, College of Pharmacy, King Khalid University, Abha, Saudi Arabia
[5] Ibne Seena Pharmacy College, AMES Group of Institutions, Shahabad, Hardoi, Uttar Pradesh, India

6.1 INTRODUCTION

The gut microbiota are the complex and varied collection of microorganisms found in our digestive tract. Archaea, bacteria, viruses, and fungi together constitute a complex ecosystem. Through several body mechanisms, these living entities help their hosts to remain healthy. Comprising a network of nerves linking the gut to the central nervous system (CNS), the gut–brain axis has drawn a lot of research as it might affect brain health and disease. Recent research reveals how gut bacteria helps to regulate brain activity and function. This implies that this kind of bacterium might influence the beginning and course of neurodegenerative disorders. Hormonal, immunological, and neurological systems among other channels allow the gut's microorganisms to communicate with the CNS. A dysbiosis, or imbalance in the gut bacteria, can make brain diseases like Alzheimer's, Parkinson's, multiple sclerosis (MS), and amyotrophic lateral sclerosis (ALS) worse (1, 2). Living things called bacteria help the body eat, absorb nutrients, keep its metabolism running smoothly, and keep the defense system in check. The chemicals that these bacteria make might have an effect on brains and other faraway places. Microbes ferment dietary fibers to make short-chain fatty acids (SCFAs). These acids protect neurons and may help control brain function. Gut bacteria also plays a big role in controlling immune reactions, which could have a big effect on brain health. Dysbiosis has been linked to a number of allergic and inflammatory diseases. Findings like these show that a healthy intestinal flora is needed for the immune system to keep homeostasis. More and more studies are showing a link between gut bacteria and neurodegenerative diseases. This further shows how important all three types of health are. Understanding the influence of gut bacteria on the CNS could generate innovative treatment strategies that prioritize the gut microbiome. The objective of these methods is to prevent or mitigate neurodegenerative disorders (3 –5). Neurodegenerative diseases are characterized by the progressive degradation of neurons, which impedes the ability to think and move. The main cause of Alzheimer's disease, the most often occurring neurodegenerative illness, is the formation of amyloid-beta plaques and neurofibrillary tangles in the brain. Cognitive ability declines and memory loss follows from this. Parkinson's disease causes dopaminergic neurons in the substata nigra to die. Motor symptoms, including stiffness, tremors, and slow motion, follow from this. MS is an autoimmune condition whereby the immune system attacks the myelin membrane of the nerve cells, causing a range of neurological symptoms. In patients with ALS, delayed death of motor neurons results in muscle weakening and atrophy. Despite much study, the specific causes of many diseases remain unknown. Most treatments concentrate more on controlling symptoms than on curing disease. The likely function of gut flora in the development of these diseases offers a fresh perspective on knowledge and treatment of neurodegenerative diseases (6–9). Climate change's health impacts include dietary changes, changes in air and water quality, increased heat stress, and insect-borne diseases. These alterations also make gut flora more susceptible to inflammatory agents and gastrointestinal diseases. This emphasizes the importance of minimizing the consequences of climate change in order to protect overall health and intestinal conditions (10). The study aims to fully understand gut microbiota, how gut–brain communication affects brain function and behavior, the possible links between gut microbiota dysbiosis and the

DOI: 10.1201/9781003536802-6

onset and progression of neurodegenerative diseases, and the role of gut microbiota in specific neurodegenerative diseases like Alzheimer's, Parkinson's, multiple sclerosis, and amyotrophic later in life. The focus of this chapter is on gut microbiota as a potential treatment for neurological disorders. This will improve our understanding and treatment of these complex disorders (11, 12).

6.2 GUT MICROBIOTA AND THE GUT–BRAIN AXIS

6.2.1 Composition and Function of Gut Microbiota

The gut microbiome is a group of bacteria that live in the human digestive system. It is complicated and always changing. There are bacteria, archaea, viruses, and fungi in this environment. All of them play an important role in keeping the host healthy. The makeup of the gut microbiota varies a lot from person to person and is affected by things like diet, genetics, environment, and age. Most of the microbiota in the gut is made up of bacteria. Bacteroidetes and Firmicutes are the most common groups. These tiny living things are very important for many things, like breaking down and absorbing nutrients, making vitamins like vitamin K and vitamin B, and protecting us from dangerous bacteria by outcompeting them and making chemicals that kill them (13). Furthermore, controlling the host's metabolism and defenses are gut bacteria. Fiber-fermenting gut bacteria produce short-chain fatty acids (SCFAs) such as propionate, butyrate, and acetate, which serve as key energy sources for intestinal cells. These metabolites regulate intestinal motility and modulate inflammatory responses. Moreover, many bioactive compounds present in the brain are synthesized or influenced by gut-resident bacteria. Some of them are GABA, serotonin, and dopamine (14). These hormones affect both our actions and feelings. By eliminating infections, delivering medications, and altering the immune system, VLPs, which resemble viruses, impact our gut flora. They increase general health, regulate intestinal flora, and ward against disease. All of these products are meant to support a more consistent and healthy intestinal environment (15). Numerous medical disorders, including diabetes, obesity, inflammatory bowel disease, and neurodegenerative diseases, have been linked to dysbiosis, that is, changes in the gut flora (16).

6.2.2 The Gut–Brain Axis: Mechanisms of Communication

The gut–brain axis is a two-way communication network that connects the gastrointestinal tract with the CNS. The intricate interplay encompasses the neurological, immunological, and hormonal pathways that enable the transmission of signals between the gut bacteria and the brain (4).

6.2.2.1 Neural Pathways

Neural paths are very important for the stomach and brain to be able to interact with each other. The vagus nerve, which is the longest nerve in the head, sends messages from the stomach to the brain and is a very important part of this system. Vagal-afferent fibers can pick up on metabolites from microbes and other cues coming from the gut. They send this information to the brainstem, which changes many things that happen in the brain. Researchers have found that the gut bacteria can control how the vagus nerve works, which in turn impacts how the brain works and how people behave (17). The enteric nervous system (ENS), which is sometimes called the "second brain," is made up of a large network of neurons that control the functions of the digestive system and communicate with the brain and spinal cord (18, 19).

6.2.2.2 Immune Pathways

A lot of what happens in the gut and the brain is controlled by the immune system. Immune cells and cytokines make contact easier. In order to work with the host's immune system, the microbiota in the gut increases the production of cytokines and other signaling chemicals that can change how the brain works. Microbial metabolites, such as SCFAs, can regulate blood–brain barrier (BBB) permeability and alter neuroinflammatory responses (20). Dysbiosis can cause an inflammatory response that might contribute to neuroinflammation and perhaps worsen neurodegenerative diseases (21).

6.2.2.3 Hormonal Pathways

Hormonal pathways also allow signals to go between the stomach and the brain. The microbes in the gut may have an effect on the HPA axis, a critical mechanism in the body's stress response. Stress can alter the composition of gut microbiota, which can therefore influence the HPA axis' reaction. Specific microbial metabolites can influence cortisol release, a stress-related hormone. This, in turn, influences brain function and behavior (22). In addition, gut microbes help to synthesize and regulate other hormones like serotonin. The stomach produces the majority of serotonin, which plays an important role in mood regulation (23).

6.2.3 Role of Microbiota in Brain Function and Behavior

The gut–brain link allows the bacteria in the gut to significantly influence the brain function and behavior of humans. Recent studies point to the gut microbiota, the collection of microorganisms in the digestive tract—as influencing our perceptions, emotions, and behavior. Studies on animals have shown that germ-free mice, which don't have any intestinal microbiota, have different brain chemicals and behavior than animals that were raised normally. Some of these differences are increased behavior that seems like anxiety and changed amounts of neurotransmitters like serotonin and dopamine (24). Changes in human gut flora have been linked to several neuropsychiatric diseases, including anxiety, depression, and autism spectrum disorders, as well as other ailments. Treatments using probiotics and prebiotics aid a lot of various diseases by restoring the balance of gut flora to a healthy level. Some forms of probiotics, for instance, have been connected to a drop in anxiety and depression symptoms. Changing the flora in your gut could improve your therapy (25). The gut bacteria also aggravate neurological disorders such as Parkinson's and Alzheimer's. Dysbiosis can make the intestines more permeable, which makes it easier for bacterial endotoxins to get into the bloodstream and cause inflammation throughout the body and in the brain. Amyloid-beta plaques and alpha-synuclein clusters are two different signs of Alzheimer's disease and Parkinson's disease, respectively, that can form when there is inflammation (26).

6.3 MECHANISMS LINKING GUT MICROBIOTA AND NEURODEGENERATIVE DISEASES

6.3.1 Immune Modulation and Neuroinflammation

Mostly present in the colon, the gut flora greatly influences the development of immune cells, including B cells, T cells, macrophages, and dendritic cells (27, 28). Interactions among the gut bacteria and the host produce chemokines, cytokines, and antibodies that induce inflammation. For example, the gut microbiome helps keep the BBB in good shape and controls neurophysiological processes like neuronal development and neurotransmission (29, 30). Even though the brain is physically separate from the rest of the body, bacteria in the gut still affect the central nervous system (31). When there are no germs in the gut, the balance of microglial activity is thrown off. The gut microbiota affects how activated microglia are. Your gut flora controls the activity of hematopoietic stem cells and myeloid precursors in the bone marrow. This keeps the neutrophil count in your blood at a healthy level (32). From now on, the immune system and the bacteria in the gut are connected in a complicated way that controls the immune system. Zinc is important for preventing cancer and maintaining the health of gut bacteria because it helps repair DNA, boosts immunity, protects against free radicals, and maintains the integrity of the gut barrier. Getting enough zinc is important, and it can be done through food or supplements (33). The immune system and the bacteria in the gut work together. Neuroinflammation happens a lot in people with nerve diseases. A group of cells called microglial cells watch over the area around the brain. If viruses or bacteria get into the brain, neuroinflammation can happen. When this happens, microglial cells surround the pathogens and kill them, which keeps the brain stable. Neurodegenerative diseases, on the other hand, make it hard to treat neuroinflammation (34, 35).

6.3.1.1 Chronic Inflammation

One of the main signs of neurodegenerative diseases is often long-lasting inflammation. When the microbiome in the gut is out of balance, it can cause dysbiosis. This can make the intestines more permeable, which is also known as "leaky gut." This disease makes it easier for lipopolysaccharides (LPS), which are bacterial endotoxins, to get into the bloodstream. Endotoxins can get through the BBB. This triggers an immune reaction and leads to inflammation in the brain. When you keep stimulating your immune system, it can make pro-inflammatory cytokines like tumor necrosis factor-alpha (TNF-α) and interleukins (IL-1β, IL-6). These can make damage to neurons worse and may play a part in the development of neurodegenerative diseases like Parkinson's and Alzheimer's (36).

6.3.1.2 Impact on Neurodegenerative Diseases

An imbalance in the gut flora is linked to exacerbation of several neurodegenerative diseases and can cause neuroinflammation. Under the effect of chronic neuroinflammation, important symptoms of Alzheimer's disease, amyloid-beta plaques, and tau tangles develop and gather in those affected by the condition. Similarly, neuroinflammation may make the loss of dopaminergic neurons in the substantia nigra worse in people with Parkinson's disease. Researchers have found that people who have been labeled with Parkinson's disease often have changes in the types of bacteria that live in their gut. This is linked to the intestines becoming more permeable and having higher amounts of cytokines that cause inflammation (Table 6.1) (37).

TABLE 6.1
Microorganisms and Their Possible Roles in Neurodegenerative Diseases

Microorganisms	Possible Reason	Neurodegenerative Disease
E. coli, *Staphylococcus*, *Streptococcus*, *Mycobacteria*, *Salmonella*, *Klebsiella*, *Lactobacillus*, *Bifidobacterium*	Bacterial amyloids, GABA	Alzheimer's disease
Roseburia	Produces SCFAs	Parkinson's disease
Hydrogen-product bacteria	Reduced dopaminergic loss	Parkinson's disease
Coriobacteriales	Unknown	Huntington's disease
Clostridium	Decreased SCFAs	Multiple sclerosis

6.3.2 Neuroactive Metabolites and Brain Health

Gut microbiota makes different chemicals that can affect brain health and performance. Short-chain fatty acids, neurotransmitters, and other signaling molecules are some of these neuroactive metabolites. They can change the activity and defense of neurons.

6.3.2.1 Short-Chain Fatty Acids

Some examples of SCFAs are acetate, propionate, and butyrate. They are made when bacteria in the gut consume fibers in food. Numerous studies have shown that these molecules have many positive effects on the brain. In particular, butyrate can reduce inflammation and strengthen the barrier between the blood and the brain. It is also used as fuel by colonocytes and has been linked to controlling gene expression by blocking histone deacetylase. Short-chain fatty acids can help keep the immune system in check, lower toxic stress, and boost the production of neurotrophic factors. These things are very important for keeping neurons healthy and able to change (38).

6.3.2.2 Other Metabolites

The gut microbiome makes more than just SCFAs. These bacteria make many other chemicals that can change how the brain works. Neurotransmitters like serotonin, dopamine, and GABA are some of these chemicals. It's important to know that the gut makes about 90% of the body's serotonin. The serotonin that the gut makes can affect mood and behavior. Also, the bacteria in the gut can make dopamine and GABA precursors, which are important for controlling mood and neuronal activity (23). Neuropsychiatric illnesses have been linked to irregular production of these substances, and they may also play a role in neurodegenerative diseases (3).

6.3.3 Protein Misfolding and Clearance

Proteins frequently misfold and aggregate in numerous neurological disorders, such as Alzheimer's, Parkinson's, and amyotrophic lateral sclerosis. These mechanisms can be significantly influenced by the intestinal flora. Dysbiosis can result in more frequent protein misfolding due to the increased synthesis of microbial compounds. For instance, certain gastrointestinal microorganisms possess the capacity to produce amyloid-like proteins that may interact with human amyloid proteins and accelerate the binding process. Long-term inflammation induced by gastrointestinal dysbiosis can also compromise the immune system's ability to eradicate improperly packed proteins. This can cause these proteins to build up in the brain. Proteins must be folded correctly in order for them to work right. Neurodegenerative diseases can happen when proteins fold wrong and stick together in ways that are similar to how prions do (39). Chaperones are molecules that help proteins fold in the right way. A group of helpers called CLIPs, which are involved in protein production, keep an eye on the translated proteins. It is clear that the CLIPs are down-regulated with age, which causes proteotoxic stress. This stress is especially clear in people with Alzheimer's, Huntington's, and Parkinson's (40). In a way similar to how prion diseases are passed on, misfolded protein disorders change normal proteins into dangerous shapes. Pathogenic amyloids are able to move from one neuron to another, which is how diseases spread in neurons (41, 42). Over time, the misfolded proteins make it harder for neurons to talk to each other, which leads to the death of nerve cells (43).

6.3.4 Dysbiosis and Disease Pathogenesis

Dysbiosis, or imbalance in the gut microbiota, is becoming more and more recognized as a factor in the development of neurodegenerative diseases. It's possible for changes in the gut bacteria to stop the production of good chemicals, make the gut more permeable, and cause inflammation all over the body and in the nervous system. These alterations cause the environment to become less conducive, which contributes to the progression of neurodegenerative diseases. The researchers discovered that individuals with Parkinson's disease had lower levels of beneficial bacteria in their intestine, particularly Prevotella, and higher levels of harmful bacteria, particularly Enterobacteriaceae. Similar dysbiosis trends have been observed in other neurodegenerative diseases, indicating that the intestinal microbiota and neurodegeneration are interconnected (39, 44).

6.4 GUT MICROBIOTA IN SPECIFIC NEURODEGENERATIVE DISEASES

6.4.1 Alzheimer's Disease

Alzheimer's disease (AD) is the most prevalent form of dementia. Individuals with Alzheimer's disease experience a decline in their cognitive abilities and memory over time. Gut flora is a critical factor in the evolution of AD, as indicated by a recent study. The gut flora may influence brain activity in a variety of ways, including regulating the immune system, producing neuroactive compounds, and ensuring the gut barrier remains intact. Between 60% and 80% of older people have Alzheimer's disease, a common neurological disorder that causes dementia or cognitive decline (45), This suggests a link between gut bacteria and cognitive behavior. The bacteria in the gut affect how the brain works (46). The gut microbiota influences brain functioning, indicating a connection between gut microbiota and cognitive behavior (47). Because of this, gut bacteria play a part in how AD happens. More illnesses like *Klebsiella pneumoniae*, *Bacteroides fragilis*, and *Eggerthella lenta* were found in people with Alzheimer's disease (48). It has been found that people with Alzheimer's disease have less *Eubacterium rectale* and *Bacillus subtilis* in their stool than healthy people, but more *Escherichia* or *Shigella* (49). This shows how important it is for people with cognitive decline to have buildups of amyloid and bacteria. Microorganisms that make butyrate are less common in people with AD. Butyrate is a chemical that helps reduce inflammation and keep the immune system in check. To lower inflammation and boost anti-inflammatory benefits, it is important to keep the balance of the intestines (48).

6.4.1.1 Microbiota Dysbiosis and Amyloid-Beta Deposition

Amyloid-beta (Aβ) plaques, a hallmark of Alzheimer's disease, have been linked to an imbalance in the gut bacterial population, microbiota dysbiosis. Dysbiosis-induced increased permeability of the intestines helps bacterial metabolites and endotoxins flow into the circulation and then into the brain. When these chemicals get into the brain, they can cause inflammation and help Aβ clumps form. Researchers have found that some bacteria in the gut can make amyloid-like proteins that interact with human Aβ and speed up the process of it clumping together and building up in the brain (44). Also, studies on mice that don't have any germs have shown that the lack of gut microbiota leads to a drop in Aβ pathology. This clearly shows a link between gut bacteria and the formation of amyloid plaques (49).

6.4.1.2 Neuroinflammation and Cognitive Function

Neuroinflammation is a key part of how Alzheimer's disease gets worse. Making metabolites and SCFAs helps the gut's bacteria to regulate neuroinflammatory responses. One form of SCFA that can help lower inflammation and strengthen the blood–brain barrier, therefore preventing the entry of pro-inflammatory hormones into the brain. Dysbiosis reduces the abundance of helpful bacteria generating SCFAs, therefore exacerbating neuroinflammation and slowing down brain activity. Researchers have proven in models of Alzheimer's disease that changing the gut flora with probiotics or dietary change lowers neuroinflammation and improves brain function (50).

6.4.2 Parkinson's Disease

Parkinson's disease (PD) is a neurodegenerative condition marked by the deterioration of dopaminergic neurons in the substantia nigra, resulting in both motor and non-motor symptoms. The gut microbiota has been linked to the development of PD through different processes, such as communication between the gut and the brain, modulation of the immune system, and changes in metabolism.

6.4.2.1 Gut Dysbiosis as Biomarkers

Gut dysbiosis has some people wondering whether it could indicate Parkinson's disease. Persons with Parkinson's disease have quite different gut flora than that of healthy persons, according to studies. For example, there is typically a reduction in beneficial bacteria such as Prevotella and an increase in pro-inflammatory bacteria like Enterobacteriaceae. Such alterations in the gut microbiota can influence the permeability of both the intestinal barrier and the blood–brain barrier. This makes it easier for neurotoxic chemicals to get in and cause brain cells to die. Some types of bacteria found in the gut flora of people with Parkinson's disease could be used as biomarkers to help find the disease early and track its progress (51).

6.4.2.2 Influence on Motor and Non-Motor Symptoms

Both the physical and non-physical signs of Parkinson's disease can be changed by the gut microbiota. Dopaminergic neurons are dying, which is the main cause of motor signs like tremors, stiffness, and slow movement. The microbiota in the gut can change how neurotransmitters, like dopamine, are made and broken down, which is important for motor performance (52). When there is dysbiosis, the production of substances used to make dopamine can go down and signs of inflammation in the brain can go up. This can make motor symptoms worse. Non-motor signs of Parkinson's disease, like digestive problems, depression, and memory loss, are also affected by the gut microbiota. For example, people with Parkinson's disease often have digestive problems like constipation, which can happen years before they start having movement problems. Changing the types of microbes in the gut have been linked to these stomach problems, showing a strong link between gut health and PD (53).

6.4.3 Multiple Sclerosis

Multiple sclerosis is an autoimmune condition where the protective covering of nerve cells in the CNS is damaged, resulting in problems with movement and thinking abilities. The gut microbiota is becoming more widely acknowledged as a crucial element in the regulation of immune responses and the advancement of MS.

6.4.3.1 Immune Regulation and Microbial Metabolites

The gut microbiota has a substantial impact on the immune system of the host by producing a range of metabolites. Short-chain fatty acids, including butyrate, acetate, and propionate, are extensively researched microbial metabolites that contribute to immune modulation. The SCFAs are generated through the process of fermenting dietary fibers by gut bacteria. Research has demonstrated that these SCFAs exhibit anti-inflammatory characteristics. Butyrate has the ability to improve the strength of the gut barrier, decrease inflammation throughout the body, and regulate the activity of regulatory T cells (Tregs), which play a vital role in maintaining immunological tolerance and preventing autoimmune reactions (54). Furthermore, compounds produced from gut microbiota have the ability to impact the generation and function of cytokines and other immune mediators. Specific bacterial strains have the ability to generate indole derivatives, which have been proven to regulate the synthesis of interleukin-10 (IL-10). An anti-inflammatory cytokine, IL-10 is absolutely essential for controlling immunological reactions in MS (55). An imbalance in the bacteria in the gut, known as dysbiosis, can lead to too high levels of pro-inflammatory cytokines, including tumors necrosis factor-alpha (TNF-α) and interleukin-6 (IL-6). This can aggravate the inflammatory reaction and help to produce MS (56).

6.4.3.2 Therapeutic Interventions

At the time, scientists are looking into how the bacteria in the gut might be used to treat MS. Probiotics, prebiotics, and dietary therapies, all try to balance the gut bacteria and make more of the chemicals that are good for you. Probiotics are live microorganisms that are good for your health. They have been shown to have the ability to change the gut microbiota and reduce inflammation in people with MS. Some types of probiotics, like *Lactobacillus* and *Bifidobacterium*, have been shown to improve the function of the gut barrier and control immune reactions (MS). Studies have shown that certain probiotic strains, such as *Lactobacillus* and *Bifidobacterium*, can enhance the function of the gut barrier and regulate immunological responses (57). If you have MS, prebiotics can also help you. These are parts of food that the body can't break down but that help good bugs grow in the gut. Bacteria in the gut break down proteins in food, which makes SCFAs. Inflammation goes down and immune tolerance goes up when these SCFAs are present (58). Also, fecal microbiota transplantation (FMT) is being looked into as a possible way to help MS. FMT involves giving feces from a healthy donor to a person with MS in order to restore a healthy gut flora and make the disease symptoms better (59).

6.4.4 Amyotrophic Lateral Sclerosis

Amyotrophic lateral sclerosis is a degenerative neurological disorder that gradually causes the deterioration of motor neurons, resulting in muscle weakness and wasting. Recent studies have started to reveal the function of gut bacteria in the development and advancement of ALS.

6.4.4.1 Neuroinflammation and Neurodegeneration

ALS is marked in part by neuroinflammation, a phenomenon suspected to be involved in motor neuron destruction. The gut flora affects neuroinflammation since it generates various molecules and regulates immune responses. ALS sufferers with dysbiosis have been linked by researchers to more porous stomachs. This lets bacteria products like LPS into the bloodstream, which starts systemic inflammation (60). Neuroinflammation can get worse because of this systemic inflammation, which can hurt nerves in the central nervous system. It is also possible for the gut bacteria to change the production of neurotrophic factors and other substances that are good for neurons. Important for neurons to survive and function, brain-derived neurotrophic factor (BDNF) can be produced by gut-based bacterial SCFAs crossing the blood–brain barrier (61). Lower levels of BDNF and SCFAs brought on by dysbiosis can exacerbate ALS.

6.4.4.2 Microbiota-Based Therapies

Gut flora–oriented medicinal approaches are being investigated as potential treatments for ALS. Prebiotics and antibiotics have shown promise in preclinical studies by modulating the composition of gut bacteria and enhancing the production of beneficial metabolites. For example, FMT is under research as a possible ALS therapy as the administration of particular probiotic strains has showed a decrease in neuroinflammation and improvement of motor performance in animal models with ALS. Fecal microbiota transplantation is the process of transferring fecal matter from a donor with a healthy gut microbiota to a patient with ALS in order to restore a healthy gut microbial community. Initial research has indicated that FMT has the ability to alter the makeup of gut microbiota, decrease inflammation throughout the body, and enhance motor function in individuals with ALS (62). Furthermore, researchers are investigating dietary strategies that can stimulate the proliferation of advantageous gut bacteria and boost the synthesis of neuroprotective compounds as potential treatments for ALS.

6.5 THERAPEUTIC IMPLICATIONS

6.5.1 Dietary Interventions

Changing your diet alters the gut microbiota and how it functions, which affects the host's health and disease in turn. Many foods you eat can modify the gut microorganisms. These alterations produce disparities in the microbiota. They can either help good bacteria grow or turn them into dysbiosis. It has been shown that fiber, polyphenols, and other beneficial substances found in foods can help good bacteria like *Bifidobacterium* and *Lactobacillus* grow (63). These bacteria make SCFAs, which are similar to butyrate. These acids can lower swelling and make the barrier in the gut stronger (64). On the other hand, diets that are high in saturated fats and carbohydrates can help dangerous bacteria grow, which can make the intestines more permeable and cause inflammation all over the body (16). There have been many studies that show that making changes to your food can help treat neurodegenerative diseases. The Mediterranean diet is one example of a food plan that has been linked to a lower risk of getting Alzheimer's disease. This diet mostly consists in fruits, vegetables, whole grains, and good fats (64). Important for maintaining brain function is the growth of a varied and healthy population of bacteria in the gut made possible by this food plan. By altering the gut bacteria and enhancing mitochondrial activity, ketogenic diets high in fat and low in carbohydrates have also been found to alleviate Parkinson's disease symptoms (65).

6.5.2 Probiotics and Prebiotics

Taken in enough numbers, probiotics are live, health-conscious live microbes. Many times, they are used to restore a normal gut bacterial balance and increase gut barrier integrity. Many studies have been done on the possible utility of probiotics, especially species of *Lactobacillus* and *Bifidobacterium* on gut health and their possible therapeutic value for neurological diseases (66). These probiotics lower inflammation, control immune system response, and boost molecular synthesis, protecting the neurological system. Food components known as prebiotics cannot be broken down but nonetheless help beneficial bacteria in the gut to flourish and serve their purposes. Popular kind of prebiotics are inulin, fructooligosaccharides (FOS), and galactooligosaccharides (GOS). Prebiotics have been demonstrated to assist good bacteria in proliferating, therefore boosting the synthesis of SCFAs and strengthening the gut barrier (67). The combination of probiotics and prebiotics, referred to as synbiotics, can produce synergistic benefits in regulating the gut microbiota and enhancing the health of the host (68).

6.5.3 Fecal Microbiota Transplantation

Fecal microbiota transplanting aims to move feces from a healthy donor to a sick individual, thereby restoring a normal gut flora. Now under investigation as a potential treatment for many additional diseases, including neurological issues (69), FMT has been demonstrated to help cure recurrent *Clostridium difficile* infections. FMT seeks to restore a normal gut flora, therefore enhancing the intestinal barrier's performance, reducing inflammation in the body overall, and regulating immunological responses. First studies indicate that FMT can assist those suffering from neurodegenerative illnesses. In those with Alzheimer's disease, FMT has been shown to produce changes in beneficial gut flora, decreased inflammation, and improved brain function (70). Some studies suggest that FMT may also help those with Parkinson's disease, relieve their digestive issues, and enhance their movement ability (71). Although these early findings are encouraging, additional study is required to completely grasp the long-term consequences and safety of FMT in the treatment of neurodegenerative diseases.

6.5.4 Future Directions in Microbiota-Based Therapies

Scientists are constantly looking for novel ways to change the gut flora to complement therapies; bacterial-based medications are fast advancing. Next-generation probiotics represent one area of tremendous future research promise. The used bacteria of these probiotics have undergone genetic alteration to allow their synthesis of specific drugs or immune system regulation (72). Particularly influencing specific biological processes linked to neurodegenerative illnesses, these genetically engineered probiotics can offer customized therapy. Moreover, growing in frequency is the application of postbiotics. Made by gut bacteria, postbiotics are bioactive molecules with maybe medical advantages. Postbiotics including vitamins, bacteriocins, and SCFAs can lower inflammation, boost immunity, and shield nerve cells (73). Using these chemicals could lead to new ways of treating neurodegenerative diseases that don't involve live bacteria. At the moment, researchers are looking into precision medicine methods, which involve making medicines fit each person's unique gut microbiota and genetic profile. This personalized approach might help microbiota-based medicines work better while reducing bad effects. Microbiota-based treatments hold a lot of promise for making new and useful drugs as we learn more about the gut–brain axis and how it affects neurodegenerative illnesses. Table 6.2 lists the main medical benefits of changing the germs that live in the gut.

6.6 CONCLUSION

An increasing amount of study has shown that there is a complex link between gut microbiota and neurodegenerative diseases. The gut bacteria plays a clear role in the start and progression of neurodegenerative diseases such as AD, PD, MS, and ALS). Among the numerous ways gut bacteria affect brain function include immune system regulation, synthesis of neuroactive chemicals, and preservation of a robust gut barrier. Dysbiosis, an imbalance in the resident bacteria of the gut, has been linked to long-term inflammation, protein misfolding, and metabolic problems. Together, these factors increase the likelihood of neurological illnesses occurring. Treatments emphasizing the gut flora are changing your diet, using probiotics and prebiotics, and FMT. For managing and maybe altering some diseases, these have shown great promise.

Future studies should concentrate on pinpointing the precise mechanisms by which gut bacteria influence neurodegenerative disorders. Longitudinal studies are quite crucial if one wants to completely grasp the cause-and-effect correlations and evolving interactions between gut bacteria and neurodegeneration as they occur over time. Clinical studies are necessary to ascertain whether microbiota-based treatments including probiotics, prebiotics, and FMT, are safe and beneficial for persons with neurodegenerative disorders. Personalized medical tactics should be able to make treatments more successful as long as one considers gut microbiota and genetic profile. Bioinformatics, metagenomics, and metabolomics have come a long way. These fields will be very helpful in finding specific microbial signatures and molecules that are connected to neurodegenerative diseases. This will make it easier to come up with methods that are accurate and effective. Furthermore, knowing how nutrition, lifestyle, and surroundings influence the gut flora and how that influences brain function would provide us with all the knowledge required to devise fresh approaches to prevent and cure disorders.

Patients may benefit much from investigating the gut flora in neurodegenerative disorders. The Mediterranean diet and the ketogenic diet have shown potential in changing the gut microbiota and improve cognitive and motor functioning in people with AD and PD, respectively. Patients can be given probiotics and prebiotics to improve

TABLE 6.2
Therapeutic Implications of Gut Microbiota Interventions

Therapeutic Strategy	Description	Health Impacts	Examples
Dietary interventions	Dietary changes may impact gut microbiota composition	Promotes good bacteria, SCFA synthesis, lowers inflammation, strengthens the intestinal barrier, and has promise in neurological illnesses	High fiber, polyphenols, Mediterranean diet, and ketogenic diet
Probiotics and prebiotics	Probiotics are live helpful bacteria, whereas prebiotics are indigestible components that promote beneficial bacteria development	It regulates immunological response, reduces inflammation, strengthens the intestinal barrier, and has promise in neurological illnesses	*Lactobacillus*, *Bifidobacterium*, inulin, FOS, and GOS
Fecal microbiota transplantation	Transferring feces from healthy donors to patients to rebuild gut flora	Treats recurrent *C. difficile* infections, has potential advantages in neurological illnesses, decreases inflammation, and enhances cognitive and motor abilities	Fecal transplants
Future directions in microbiota-based therapies	Precision medical techniques, postbiotics (bioactive substances from bacteria), and next-generation probiotics (genetically designed)	Specific treatment approaches, immunomodulatory, anti-inflammatory, neuroprotective, and maybe helpful in neurodegenerative illnesses	Probiotics that have been genetically altered, SCFAs, bacteriocins, and vitamins

their gut health, lower inflammation, and help protect their nerve cells. Fecal microbiota transplantation is still under testing for neurological illnesses, but it presents a fresh and hopeful approach to treat them by restoring a normal gut bacterial count and so enhancing communication between the stomach and the brain. Innovative applications of new therapeutic techniques, including postbiotics and next-generation probiotics, abound. While postbiotics are made of beneficial molecules produced by gut bacteria, next-generation probiotics use genetically engineered organisms. By closely analyzing every person's gut microbe, precision medicine could be able to design customized plans that make treatments more effective while reducing side effects. We are learning more about the link between the gut and the brain. Using treatments that focus on the bacteria in neurodegenerative diseases could completely change the way we do therapy now and make a huge difference in how well patients do.

REFERENCES

1. Cryan JF, O'Riordan KJ, Cowan CSM, Sandhu KV, Bastiaanssen TFS, Boehme M, et al. The microbiota-gut-brain axis. *Physiol Rev.* 2019 Oct 1;99(4):1877–2013.
2. Dinan TG, Cryan JF. Gut-brain axis in 2016: brain-gut-microbiota axis - mood, metabolism and behaviour. *Nat Rev Gastroenterol Hepatol.* 2017 Feb;14(2):69–70.
3. Sampson TR, Mazmanian SK. Control of brain development, function, and behavior by the microbiome. *Cell Host Microbe.* 2015 May 13;17(5):565–76.
4. Silva YP, Bernardi A, Frozza RL. The role of short-chain fatty acids from gut microbiota in gut-brain communication. *Front Endocrinol (Lausanne).* 2020;11:25.
5. Blennow K, Hampel H. CSF markers for incipient Alzheimer's disease. *Lancet Neurol.* 2003 Oct;2(10):605–13.
6. Balestrino R, Schapira A h. v. Parkinson disease. *Eur J Neurol.* 2020;27(1):27–42.
7. Hemmer B, Kerschensteiner M, Korn T. Role of the innate and adaptive immune responses in the course of multiple sclerosis. Lancet Neurol. 2015;14(4):406–19. https://pubmed.ncbi.nlm.nih.gov/25792099/
8. Brown RH, Al-Chalabi A. Amyotrophic lateral sclerosis. *N Engl J Med.* 2017 Jul 13;377(2):162–72.
9. Quigley EMM. Gut microbiota and the role of probiotics in therapy. *Curr Opin Pharmacol.* 2011 Dec;11(6):593–603.
10. Sen A, Chatterjee A. Confronting the health impacts of climate change: a comprehensive exploration. In: Prabhakar PK, Leal Filho W, editors. *Preserving health, preserving earth: the path to sustainable healthcare.* Springer Nature Switzerland; 2024. p. 1–9. [cited 2024 Jun 21]. https://doi.org/10.1007/978-3-031-60545-1_1
11. Thursby E, Juge N. Introduction to the human gut microbiota. *Biochem J.* 2017 May 16;474(11):1823–36.
12. Mayer EA, Tillisch K, Gupta A. Gut/brain axis and the microbiota. *J Clin Invest.* 2015 Mar 2;125(3):926–38.
13. Turnbaugh PJ, Ley RE, Mahowald MA, Magrini V, Mardis ER, Gordon JI. An obesity-associated gut microbiome with increased capacity for energy harvest. *Nature.* 2006 Dec 21;444(7122):1027–31.
14. Bonaz B, Bazin T, Pellissier S. The vagus nerve at the interface of the microbiota-gut-brain axis. *Front Neurosci.* 2018;12:49.
15. Venishaa S, Kumar P, Sen A, Anand A, Kaushik M, Bhowmick M, et al. Unlocking the Role of Virus-like Particles (VLPs) for Cancer Treatment. Curr Cancer Ther Rev. 2024;20:1–27.
16. Gershon MD. The enteric nervous system: a second brain. *Hosp Pract.* 1999 Jul 15;34(7):31–32, 35–38, 41–42 passim.

17. Braniste V, Al-Asmakh M, Kowal C, Anuar F, Abbaspour A, Tóth M, et al. The gut microbiota influences blood-brain barrier permeability in mice. *Sci Transl Med.* 2014 Nov 19;6(263):263ra158.
18. Foster JA, McVey Neufeld KA. Gut-brain axis: how the microbiome influences anxiety and depression. *Trends Neurosci.* 2013 May;36(5):305–12.
19. Sen A, Mohanraj PS, Ranjan A, Rajendran V, ArulVijayaVani S, Balan Y, et al. Unraveling the role of tumor necrosis factor-alpha in diabetic peripheral neuropathy: a systematic review and meta-analysis. *Cureus.* 2023 Dec;15(12):e49926.
20. Sudo N, Chida Y, Aiba Y, Sonoda J, Oyama N, Yu XN, et al. Postnatal microbial colonization programs the hypothalamic-pituitary-adrenal system for stress response in mice. *J Physiol.* 2004 Jul 1;558(Pt 1):263–75.
21. Yano JM, Yu K, Donaldson GP, Shastri GG, Ann P, Ma L, et al. Indigenous bacteria from the gut microbiota regulate host serotonin biosynthesis. *Cell.* 2015 Apr 9;161(2):264–76.
22. Cryan JF, O'Mahony SM. The microbiome-gut-brain axis: from bowel to behavior. *Neurogastroenterol Motil.* 2011 Mar;23(3):187–92.
23. Steenbergen L, Sellaro R, van Hemert S, Bosch JA, Colzato LS. A randomized controlled trial to test the effect of multispecies probiotics on cognitive reactivity to sad mood. *Brain Behav Immun.* 2015 Aug;48:258–64.
24. Hooper LV, Littman DR, Macpherson AJ. Interactions between the microbiota and the immune system. *Science.* 2012 Jun 8;336(6086):1268–73.
25. Olszak T, An D, Zeissig S, Vera MP, Richter J, Franke A, et al. Microbial exposure during early life has persistent effects on natural killer T cell function. *Science.* 2012 Apr 27;336(6080):489–93.
26. Belkaid Y, Hand TW. Role of the microbiota in immunity and inflammation. *Cell.* 2014 Mar 27;157(1):121–41.
27. Erny D, Hrabě de Angelis AL, Jaitin D, Wieghofer P, Staszewski O, David E, et al. Host microbiota constantly control maturation and function of microglia in the CNS. *Nat Neurosci.* 2015 Jul;18(7):965–77.
28. Fung TC, Olson CA, Hsiao EY. Interactions between the microbiota, immune and nervous systems in health and disease. *Nat Neurosci.* 2017 Feb;20(2):145–55.
29. Grenham S, Clarke G, Cryan JF, Dinan TG. Brain-gut-microbe communication in health and disease. *Front Physiol.* 2011;2:94.
30. Deshmukh HS, Liu Y, Menkiti OR, Mei J, Dai N, O'Leary CE, et al. The microbiota regulates neutrophil homeostasis and host resistance to Escherichia coli K1 sepsis in neonatal mice. *Nat Med.* 2014 May;20(5):524–30.
31. Heneka MT, Carson MJ, El Khoury J, Landreth GE, Brosseron F, Feinstein DL, et al. Neuroinflammation in Alzheimer's disease. *Lancet Neurol.* 2015 Apr;14(4):388–405.
32. Qin J, Li R, Raes J, Arumugam M, Burgdorf KS, Manichanh C, et al. A human gut microbial gene catalogue established by metagenomic sequencing. *Nature.* 2010 Mar 4;464(7285):59–65.
33. Singh V, Dwivedi S, Singh BP, Goel A. Unveiling the zinc paradox: balancing act in reproductive health and prostate pathology. *Ind J Clin Biochem.* 2024 Feb 5 [cited 2024 Jun 19]. https://doi.org/10.1007/s12291-024-01183-2
34. Forsyth CB, Shannon KM, Kordower JH, Voigt RM, Shaikh M, Jaglin JA, et al. Increased intestinal permeability correlates with sigmoid mucosa alpha-synuclein staining and endotoxin exposure markers in early Parkinson's disease. *PLoS One.* 2011;6(12):e28032.
35. Dwivedi S, Singh V, Agrawal R, Misra R, Sadashiv, Fatima G, et al. Human monkeypox virus and host immunity: new challenges in diagnostics and treatment strategies. *Adv Exp Med Biol.* 2024 Jan 1;1451:219–37.
36. Houser MC, Tansey MG. The gut-brain axis: is intestinal inflammation a silent driver of Parkinson's disease pathogenesis? *NPJ Parkinsons Dis.* 2017;3:3.
37. Scheperjans F, Aho V, Pereira PAB, Koskinen K, Paulin L, Pekkonen E, et al. Gut microbiota are related to Parkinson's disease and clinical phenotype. *Mov Disord.* 2015 Mar;30(3):350–8.
38. Quigley EMM. Microbiota-brain-gut axis and neurodegenerative diseases. *Curr Neurol Neurosci Rep.* 2017 Oct 17;17(12):94.
39. Brehme M, Voisine C, Rolland T, Wachi S, Soper JH, Zhu Y, et al. A chaperome subnetwork safeguards proteostasis in aging and neurodegenerative disease. *Cell Rep.* 2014 Nov 6;9(3):1135–50.
40. Pearce MMP, Spartz EJ, Hong W, Luo L, Kopito RR. Prion-like transmission of neuronal huntingtin aggregates to phagocytic glia in the Drosophila brain. *Nat Commun.* 2015 Apr 13;6:6768.
41. Brettschneider J, Del Tredici K, Lee VMY, Trojanowski JQ. Spreading of pathology in neurodegenerative diseases: a focus on human studies. *Nat Rev Neurosci.* 2015 Feb;16(2):109–20.
42. Shankar GM, Li S, Mehta TH, Garcia-Munoz A, Shepardson NE, Smith I, et al. Amyloid-beta protein dimers isolated directly from Alzheimer's brains impair synaptic plasticity and memory. *Nat Med.* 2008 Aug;14(8):837–42.
43. Vogt NM, Kerby RL, Dill-McFarland KA, Harding SJ, Merluzzi AP, Johnson SC, et al. Gut microbiome alterations in Alzheimer's disease. *Sci Rep.* 2017 Oct 19;7(1):13537.
44. Hu X, Wang T, Jin F. Alzheimer's disease and gut microbiota. *Sci China Life Sci.* 2016 Oct;59(10):1006–23.
45. Jouanne M, Rault S, Voisin-Chiret AS. Tau protein aggregation in Alzheimer's disease: an attractive target for the development of novel therapeutic agents. *Eur J Med Chem.* 2017 Oct 20;139:153–67.
46. Diaz Heijtz R, Wang S, Anuar F, Qian Y, Björkholm B, Samuelsson A, et al. Normal gut microbiota modulates brain development and behavior. *Proc Natl Acad Sci USA.* 2011 Feb 15;108(7):3047–52.
47. Haran JP, Bhattarai SK, Foley SE, Dutta P, Ward DV, Bucci V, et al. Alzheimer's disease microbiome is associated with dysregulation of the anti-inflammatory P-glycoprotein pathway. *mBio.* 2019 May 7;10(3):e00632-19.
48. Cattaneo A, Cattane N, Galluzzi S, Provasi S, Lopizzo N, Festari C, et al. Association of brain amyloidosis with pro-inflammatory gut bacterial taxa and peripheral inflammation markers in cognitively impaired elderly. *Neurobiol Aging.* 2017 Jan;49:60–8.
49. Harach T, Marungruang N, Duthilleul N, Cheatham V, Mc Coy KD, Frisoni G, et al. Reduction of Abeta amyloid pathology in APPPS1 transgenic mice in the absence of gut microbiota. *Sci Rep.* 2017 Feb 8;7:41802.

50. Hill-Burns EM, Debelius JW, Morton JT, Wissemann WT, Lewis MR, Wallen ZD, et al. Parkinson's disease and Parkinson's disease medications have distinct signatures of the gut microbiome. *Mov Disord.* 2017 May;32(5):739–49.
51. Yang D, Zhao D, Ali Shah SZ, Wu W, Lai M, Zhang X, et al. The role of the gut microbiota in the pathogenesis of parkinson's disease. *Front Neurol.* 2019;10:1155.
52. Sampson TR, Debelius JW, Thron T, Janssen S, Shastri GG, Ilhan ZE, et al. Gut microbiota regulate motor deficits and neuroinflammation in a model of parkinson's disease. *Cell.* 2016 Dec 1;167(6):1469-1480.e12.
53. Arpaia N, Campbell C, Fan X, Dikiy S, van der Veeken J, deRoos P, et al. Metabolites produced by commensal bacteria promote peripheral regulatory T-cell generation. *Nature.* 2013 Dec 19;504(7480):451–5.
54. Zelante T, Iannitti RG, Cunha C, De Luca A, Giovannini G, Pieraccini G, et al. Tryptophan catabolites from microbiota engage aryl hydrocarbon receptor and balance mucosal reactivity via interleukin-22. *Immunity.* 2013 Aug 22;39(2):372–85.
55. Jangi S, Gandhi R, Cox LM, Li N, von Glehn F, Yan R, et al. Alterations of the human gut microbiome in multiple sclerosis. *Nat Commun.* 2016 Jun 28;7:12015.
56. Tankou SK, Regev K, Healy BC, Tjon E, Laghi L, Cox LM, et al. A probiotic modulates the microbiome and immunity in multiple sclerosis. *Ann Neurol.* 2018 Jun;83(6):1147–61.
57. Watzl B, Girrbach S, Roller M. Inulin, oligofructose and immunomodulation. *Br J Nutr.* 2005 Apr;93 Suppl 1:S49-55.
58. Salathé EP, Venkataraman R. Interaction of fluid movement and particle diffusion across capillary walls. *J Biomech Eng.* 1982 Feb;104(1):57–62.
59. Zhao W, Beers DR, Henkel JS, Zhang W, Urushitani M, Julien JP, et al. Extracellular mutant SOD1 induces microglial-mediated motoneuron injury. *Glia.* 2010 Jan 15;58(2):231–43.
60. Bercik P, Collins SM. The effects of inflammation, infection and antibiotics on the microbiota-gut-brain axis. *Adv Exp Med Biol.* 2014;817:279–89.
61. Hemarajata P, Versalovic J. Effects of probiotics on gut microbiota: mechanisms of intestinal immunomodulation and neuromodulation. *Therap Adv Gastroenterol.* 2013 Jan;6(1):39–51.
62. Gupta S, Allen-Vercoe E, Petrof EO. Fecal microbiota transplantation: in perspective. *Therap Adv Gastroenterol.* 2016 Mar;9(2):229–39.
63. Macfarlane S, Macfarlane GT. Regulation of short-chain fatty acid production. *Proc Nutr Soc.* 2003 Feb;62(1):67–72.
64. Scarmeas N, Stern Y, Tang MX, Mayeux R, Luchsinger JA. Mediterranean diet and risk for Alzheimer's disease. *Ann Neurol.* 2006 Jun;59(6):912–21.
65. Khalesi S, Sun J, Buys N, Jayasinghe R. Effect of probiotics on blood pressure: a systematic review and meta-analysis of randomized, controlled trials. *Hypertension.* 2014 Oct;64(4):897–903.
66. Miqdady M, Mistarihi JA, Azaz A, Rawat D. Prebiotics in the infant microbiome: the past, present, and future. Pediatr Gastroenterol Hepatol Nutr. 2020 Jan;23(1):1.
67. Kolida S, Gibson GR. Synbiotics in health and disease. *Annu Rev Food Sci Technol.* 2011;2:373–93.
68. Kelly CR, Kahn S, Kashyap P, Laine L, Rubin D, Atreja A, et al. Update on fecal microbiota transplantation 2015: indications, methodologies, mechanisms, and outlook. *Gastroenterology.* 2015 Jul;149(1):223–37.
69. Hazan S. Rapid improvement in Alzheimer's disease symptoms following fecal microbiota transplantation: a case report. *J Int Med Res.* 2020 Jun;48(6):300060520925930.
70. Xue LJ, Yang XZ, Tong Q, Shen P, Ma SJ, Wu SN, et al. Fecal microbiota transplantation therapy for Parkinson's disease: a preliminary study. *Medicine* (Baltimore). 2020 Aug 28;99(35):e22035.
71. Duan FF, Liu JH, March JC. Engineered commensal bacteria reprogram intestinal cells into glucose-responsive insulin-secreting cells for the treatment of diabetes. *Diabetes.* 2015 May;64(5):1794–803.
72. Aguilar-Toalá JE, Garcia-Varela R, Garcia HS, Mata-Haro V, González-Córdova AF, Vallejo-Cordoba B, et al. Postbiotics: an evolving term within the functional foods field. *Trends Food Sci Technol.* 2018 May 1;75:105–14.
73. Zmora N, Zeevi D, Korem T, Segal E, Elinav E. Taking it personally: personalized utilization of the human microbiome in health and disease. *Cell Host Microbe.* 2016 Jan 13;19(1):12–20.

7 Gut Microbiota in Mental Health Disorders

Saikat Santra[1], Rajdip Goswami[2], Bhaskar Pal[3], Subhajit Das[4], and Ayan Chatterjee[4]

*[1] JRSET College of Pharmacy, Panchpota, Chakdah, West Bengal, India
[2] Department of Pharmaceutical Technology, School of Health and Medical Sciences, Adamas University, West Bengal, India
[3] Department of Pharmacology, Charaktala College of Pharmacy, Charaktala, Mothabari, Malda, West Bengal, India
[4] School of Health Science and Technology, Medhavi Skills University, Sikkim, India

7.1 INTRODUCTION

7.1.1 Background

The gut–brain axis expresses a complex and bidirectional interactive network that links the GIT with the CNS (Arneth, B. M. et al., 2018). This connection has gained potential interest in recent years due to its principal role in influencing neurological health and disease (Iadecola, C. et al., 2017). The human gut microbiota, comprising trillions of microorganisms, is a key player in this communication network (Illiano, P. et al., 2020).

7.2 THE GUT–BRAIN AXIS–

7.2.1 Overview of Gut–Brain Communication

The GBA is a multifaceted, bidirectional interactive network that links the gastrointestinal tract and the central nervous system, playing a crucial role in maintaining physiological homeostasis (Chaudhry, T. S. et al., 2023). This axis involves a complex interplay between several systems: the gut microbiota, the ENS, the ANS, and the CNS (Fried, S. et al., 2021). Each of these components contributes to the dynamic interactions that regulate essential bodily functions, including digestion, metabolism, immune response, and even emotional and cognitive processes (Figure 7.1) (González Olmo, B. M. et al., 2021).

7.2.2 Neural Pathways

Neural pathways form a fundamental component of the gut–brain communication system, with the vagus nerve serving as a primary conduit (Metz and Pavlov, 2018). The vagus nerve, the longest of the cranial nerves, directly connects the gut to the brain, facilitating rapid communication. It transmits sensory information from the gut to the brainstem and mediates efferent signals that regulate gastrointestinal functions such as motility, secretion, and blood flow.

7.2.3 Hormonal and Immunological Pathways

Hormonal and immunological pathways are also critical in mediating interrelation between the GIT and the CNS (Khan, W. I. et al., 2010). The gut is an endocrine organ that releases a variety of hormones and neuropeptides, including serotonin, ghrelin, and GLP-1, which can influence brain function and behavior (Diz-Chaves et al., 2020; Moretti, S. et al., 2023).

7.3 GUT MICROBIOTA: COMPOSITION AND FUNCTIONS

All individuals have unique profile of gut microbiota, which plays the potential functions in nutrient absorption, structural framework of GI mucosal barrier, and immune activity in the host. Gut microbiota consists of various bacterial types taxonomically divided by genus, sequence, phyla, and family (Rinninella, E. et al., 2019).

7.3.1 Overview of Gut Microbiota

The gut microbiota containing greater than 3 million genetic materials makes a range of metabolomics, where the human genome contains nearly about 23,000 genes (Valdes, A. M. et al., 2018). The researchers phylogenetically recognized the constituents of gut microbiota by assessing directly after extracting DNA and RNA from feces. Most of these processes are premised on the extraction and amplification of DNA of 16S rRNA gene (Mizrahi-Man et al., 2013).

DOI: 10.1201/9781003536802-7

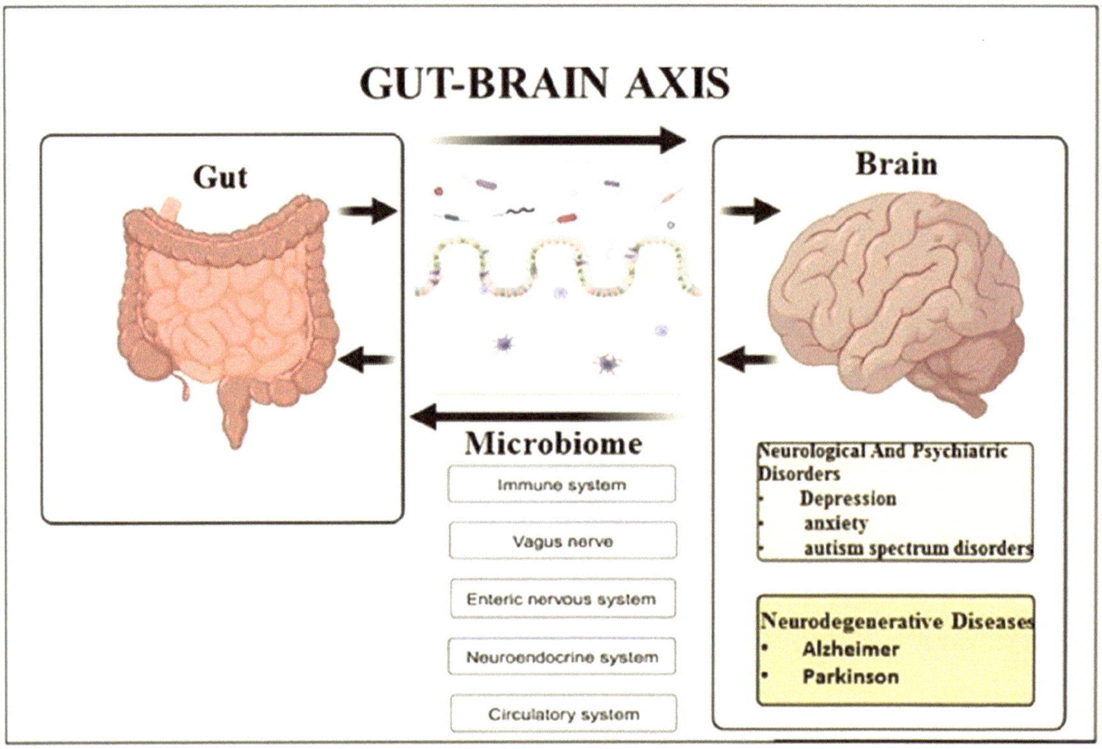

FIGURE 7.1 The above diagram shows the relation of the gut–brain axis with arrows depicting both the ways that are involved in the connection between the two. The microbiome resides in the gut that has connection with the brain through immunogenic system, vagus nerve, ENS, neuroendocrinology, and circulation. There is evidence that these interactions can affect neurological and psychiatric disorders like depression, anxiety, ASD, Alzheimer's, Parkinson's, and so on. Thus, the diagram focuses on the endogenous microbiome, its importance for brain functions, and effects of potential for mental and neurological diseases (Singh, R. et al., 2021).

7.3.2 Functions of Gut Microbiota

The human gut microbiome is the principal modulator of digestion along the GIT; symbiotic bacteria are contributing a potential role in synthesis, absorption, metabolism and extraction. The production and absorption of nutritional components such as bile acids (BA), amino acids (AA), vitamins, short-chain fatty acids (SCFAs), and lipids, and their subsequent metabolomic pathways.

7.3.3 Factors Influencing Gut Microbiota

There are different internal and external factors that can contribute to the constitution of the GI-microbiota and also affect the health functions. The gut bacterial constituents can differ in neonates with cesarean delivery and normal delivery (Dominguez-Bello et al., 2016; Tamburini et al., 2016; Portela et al., 2015; Kuhle, et al., 2015; Rutayisire et al., 2016).

7.4 GUT MICROBIOTA AND MENTAL HEALTH DISORDERS

The bidirectional connection between CNS and GI-microbiome is referred to as gut–brain axis, which has attracted potential interest in recent days. Increasing evidences show the impact of gut microbiota in gastrointestinal and extra gastrointestinal disorders. Dysbiosis and inflammation of GIT may lead to mental illnesses like anxiety and depression, which are prevalent in today's society. Probiotics has the potential role to restore the normal microbial composition, and therefore shows beneficial effects in the prevention and management of anxiety and depression.

7.4.1 Depression and Anxiety

Inflammation in GIT produces stress response in the microbiome by the formation of cytokines and different neuromodulators. Combined with the greater intestinal permeability, these materials are then progressed systemically. Higher levels of cytokines TNF-α and MCP in the blood induce the permeability of BBB, increasing the activity of rogue materials from the permeable GIT (Biesmans et al., 2015). Their production triggers brain function, and leads to anxiety, depression, and memory impairment (Muscatello et al., 2014; Clapp et al., 2017).

7.4.2 Schizophrenia

The GI-microbiome is a dynamic community of microbes which are diverse, personalized, and influenced by different factors like diet, sleep, and stress, and are dominated by phyla Bacteroidetes and Firmicutes. The inflammatory stimulations have long been exposed to schizophrenia

(Müller et al., 2015), but the origins are still unclear. Schizophrenic symptoms are associated with the formation of IL-6, IL-8, and TNF-α, and minimization of anti-inflammatory IL-10 (Balu, 2016), indicating that the genetic material driving inflammation is also necessary. Dysbiosis can ameliorate the bacterial infections and trigger inflammation; this hypothesis suggests that the bacterial infections are the major risk factors for schizophrenia.

7.4.3 Autism Spectrum Disorders

Mounting data found that changes in the concentration of gut fauna and their products are relevant not only to digestive disorders but also to autism. So far the local microbial blend that is located in the human gut influences human health through vast bacterial mammalian co-metabolism and exerts a significant impact on human health through gut–brain microbial symbiosis. Actually, the healthy composition of microbiota is able to reduce the severity of the symptoms of autism since microbial colonization plays a significant role in the formation of the CNS through neuroendocrine, neuroimmune, and autosomic systems.

Probably, the most interesting and hot topic in the field of microbiota is its connection with autism (Figure 7.2).

7.5 MECHANISMS LINKING GUT MICROBIOTA TO MENTAL HEALTH

The human gut microbiota, comprising trillions of microorganisms, plays a pivotal role in maintaining various physiological processes. Emerging research underscores its profound influence on mental health, implicating several mechanisms such as immune system modulation, production of neuroactive compounds, neuroinflammatory pathways, and the regulation of the hypothalamic pituitary adrenal (HPA) axis.

7.5.1 Immune System Modulation

7.5.1.1 Microbial Influences on Immune Cells

Studies have shown that gut microbiota modulate the activity of various immune cells, including macrophages, dendritic cells, and T cells.

7.5.2 Production of Neuro-Active Compounds

7.5.2.1 Microbial Synthesis of Neurotransmitters

The gut microbiota can synthesize and metabolize various neurotransmitters, including serotonin, dopamine, and GABA, which are critical for brain function and emotional regulation (Wiley N. C. et al., 2017).

7.5.2.2 Serotonin Production

Approximately 90% of the body's serotonin is produced in the gut. Specific gut bacteria, such as *Enterococcus* and *Streptococcus*, can produce serotonin. Serotonin plays a key role in mood regulation, and alterations in its levels are associated with mood disorders, such as depression and anxiety (Yano et al., 2015).

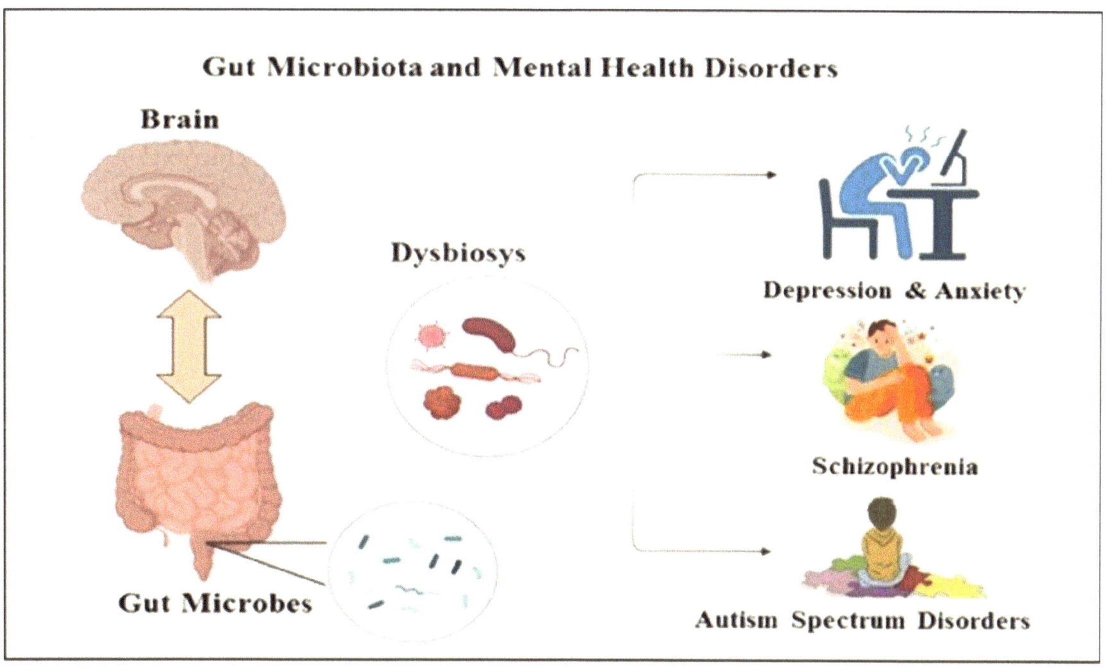

FIGURE 7.2 It depicts the connection between the gut microbial composition with neuropsychiatric disorders and diseases. Here it demonstrates how the gut has a bidirectional relationship with the brain and how it illustrates dysbiosis-enabling of various mental disorders affecting human beings such as depression, anxiety, schizophrenia, and ASD. The diagram also highlights the gut–brain axis as the main pathway for these disorders implying that the existence of some types of gut bacteria can have profound effects of brain and mental health (Oroojzadeh et al., 2022)..

7.5.3 Neuroinflammatory Pathways

7.5.3.1 Inflammation and Mental Health

Chronic inflammation has been increasingly recognized as a contributing factor to various mental health disorders. The gut microbiota can influence neuroinflammation through multiple pathways (Silva, Y. P. et al., 2020).

7.6 EVIDENCE FROM ANIMAL AND HUMAN STUDIES OF GUT MICROBIOTA

The gut microbiota has garnered significant interest for its profound influence on human health, particularly in relation to mental health.

7.6.1 Animal Model Insights

7.6.1.1 Germ-Free Mice Studies

GF mice, which are bred in sterile conditions and lack gut microbiota, have been pivotal in gut–brain axis research. These mice exhibit numerous behavioral and physiological abnormalities compared to conventionally raised mice (Sudo et al., 2004).

7.6.1.2 Specific Bacterial Strains and Mental Health

Research has identified specific bacterial strains that can modulate behavior and brain function. *Lactobacillus rhamnosus*, for instance, has been shown to reduce anxiety and depression-like behaviors in mice through its effects on the GABAergic system, increasing the expression of GABA receptors in the brain (Bravo et al., 2011).

7.6.1.3 Neuro Developmental Studies

Animal models have also provided insights into the role of gut microbiota in neurodevelopment. Early life microbial exposure is crucial for the normal development of the brain and the immune system. Studies have shown that disruption of gut microbiota during critical developmental windows, such as through antibiotic treatment, can lead to long-lasting effects on behavior and brain function (Cox et al., 2014).

7.6.2 Human Clinical Studies

7.6.2.1 Microbiota Diversity and Mental Health

Human studies have consistently shown that individuals with mental health disorders often exhibit altered gut microbiota composition compared to healthy controls. Reduced microbial diversity and imbalances in specific bacterial taxa have been linked to conditions such as depressive disease, anxiety, and ASD (Silva, Y. P. et al., 2020; Jiang et al., 2015).

7.7 MICROBIOTA-BASED THERAPIES

Some of the microbiota-based therapies encompass the concept of leveraging on the gut microbiome in order to enhance human well-being with reference to mental health. Probiotics, which are helpful living organisms, are used to maintain the balance of the gut and can also affect an individual's mood and thinking (Table 7.1, Figure 7.3) (Góralczyk-Bińkowska et al., 2022).

7.8 GUT MICROBIOTA AND STRESS RESPONSE

This relationship between glycobiome and stress response in the body is tight and critical, which is through the GB axis, the neural, hormonal, and immunological links. Stress influences changes in the component and count of the beneficial microbiota and thus impacts the body's response in both physical and mental health staking. Stress is controlled by hypothalamic pituitary adrenal axis, releasing cortisol hormone into the blood stream.

7.8.1 Gut–Brain Axis Modulation in Stress-Related Disorders

The GBA is a primary factor that governs the operation of the body, especially as it relates to stress. This two-way communication network consists of signaling and communication between CNS, ENS, and the microbiota present in the GIT. It is noteworthy that stress leads to shifts in the composition of the microbiota that in turn impacts the brain and behavior (Table 7.2) (Sandhu, K. et al., 2017).

7.9 CONCLUSION

These researches of gut microbiota in mental health disorders are conducted with a view to bring about a social change in understanding the bidirectional relations between gut and brain. New data reproduced new evidence of the relationship between the GI-microbiome and neurogenic disorders as both the source of the development of

TABLE 7.1
Microbiota-Based Therapies and Mental Health Outcomes

Therapy	Mental Health Outcomes	Reference
Probiotics	Reduced anxiety, improved mood	Bravo et al., 2011
Prebiotics	Enhanced cognitive function	Basso et al., 2022
Fecal microbiota transplantation (FMT)	Potential improvements in depression symptoms	Kurokawa et al., 2018
Diet and lifestyle interventions	Overall mental well-being improvement	Nelson et al., 2008

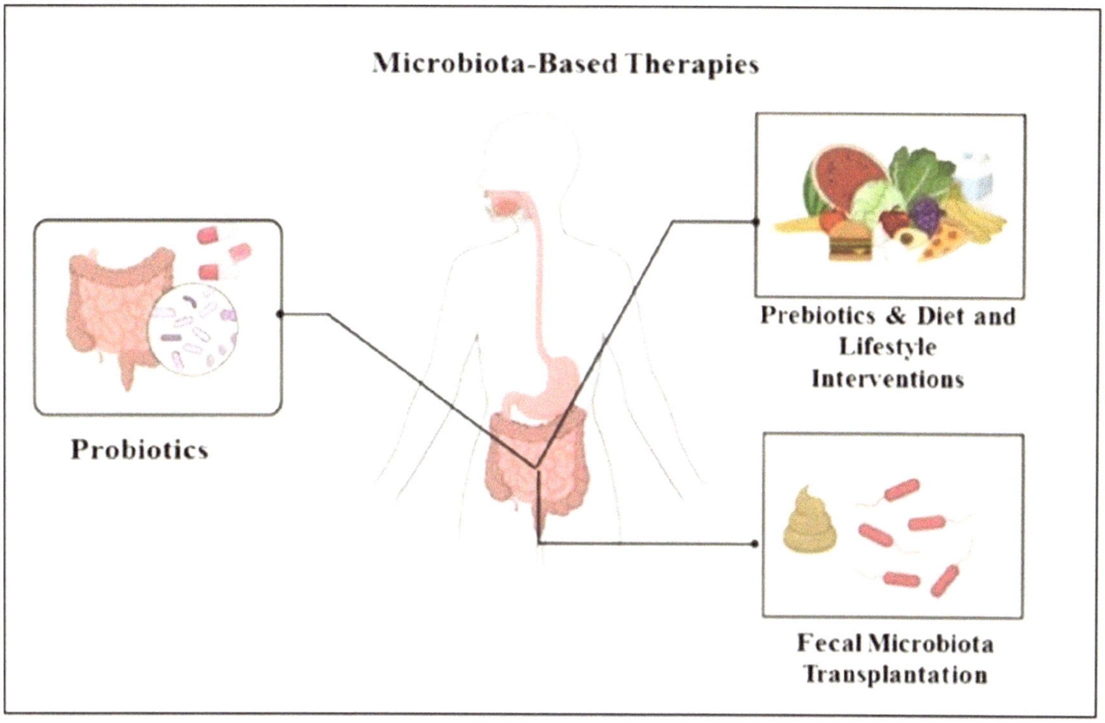

FIGURE 7.3 This image seeks to illustrate several microbiota-targeted interventions that may be employed in solving problems associated with bowel dysfunction to promote body health in a general sense. It highlights three primary interventions: probiotics, prebiotics in concert with dietary and lifestyle modifications, and fecal microbiota transplantation (FMT). Probiotics involve the consumption of live bacteria that is positively intended to enhance the functioning of the bowels. Thus, prebiotics, in conjunction with diet and lifestyle changes, target the overall management of healthy gut microbiome. FMT is therefore the process of administering a stool sample from a healthy person to a patient with the aim of rehabilitating the person's flora. These therapies are meant to address the gut–brain axis, with the goal to treat or prevent as far as possible different diseases, mainly focusing on the mental health (Hou et al., 2022).

pathologies, including anxiety, depression, schizophrenia, and autism spectrum disorder, and proven to be an actual regulator of these processes. The common link between the stomach and the brain, often described as the mind–stomach complex, is comprised of neuroendocrine and immune machinery. Several studies have noted that gut microbiota can become imbalanced, which has been associated with the progression of mental illness. This imbalance affects brain function through several mechanisms: involvement in the regulation of the immune response, synthesis of neuroactive steroids, and neuroinflammation in addition to influencing the HPA axis. The following chapter provides some support to this conceptual discussion in relation to the data from animal experiments and human research. Animal studies demonstrate stimulus–response reactions, where the makeup of microbial entities in the gut differ, and proportional human data map a connection between the variety of microbiomes and people's mental health. As a result, the present literature brings into a new potential of employing microbiota modification therapies, including probiotics, prebiotics, and FMT in managing some of the signs of mental disorders.

Then again, for further propositions of better treatments, larger research should be conducted concerning the connection of gut microbiota to diseases. Further development of this study on this connection should be investigations regarding definite microbial strains and metabolites influencing the mental state, microbiota-based products' lasting impact, and the possibility of administering the products with the traditional techniques of intervention.

TABLE 7.2
Clinical Implications and Future Directions

Area of Focus	Key Considerations	Reference
Personalized medicine	Tailoring microbiota-based interventions to individual microbiome profiles	Singh, R. et al., 2021
Long-term effects	Understanding the sustainability of therapeutic outcomes	Thursby and Juge, 2017
Integration with traditional treatments	Combining microbiota therapies with existing treatment	Kennedy et al., 2017
Regulatory and safety aspects	Ensuring safety and efficacy of microbiota-based therapies	Annahazi, A. et al., 2020

Thus, progressing the knowledge in the gastrointestinal and the cerebral correlations has been marked as a milestone in the mental health response. It leads to the creation of new possibilities of acquiring new forms of therapies that likely hold the potential of increasing the effectiveness of treating such mental health disorders.

ABBREVIATIONS

AA	amino acids
AD	Alzheimer's disease
ANS	autonomic nervous system
APP	antimicrobial peptide production
ASD	autism spectrum disorders
BA	bile acids
BBB	blood–brain barrier
BDNF	brain-derived neurotrophic factor
CNS	central nervous system
CRH	corticotrophin-releasing hormone
ENS	enteric nervous system
GABA	gamma-aminobutyric acid
GALT	gut-associated lymphoid tissue
GBA	gut–brain axis
GIT	gastointestinal tract
GLP-1	glucagon-like peptide-1
GF	germ free
GMD	gut microbiota dysbiosis
GRA	glutamate receptor antagonists
HPA	hypothalamic pituitary adrenal
IDO	indoleamine-2,3-dioxygenase
K3M	kynurenine-3-monooxygenase
NMDAR	*N*-methyl-D-aspartate receptor
PD	Parkinson's disease
rRNA	ribosomal RNA
RCTs	randomized controlled trials
SCFAs	short-chain fatty acids
Vit	vitamins

BIBLIOGRAPHY

Ailioaie, L. M., & Litscher, G. (2021). Probiotics, photobiomodulation, and disease management: Controversies and challenges. *International Journal of Molecular Sciences*, *22*(9), 4942. https://doi.org/10.3390/ijms22094942.

Al-Sheraji, S. H., Ismail, A., Manap, M. Y., Mustafa, S., Yusof, R. M., & Hassan, F. A. (2013). Prebiotics as functional foods: A review. *Journal of Functional Foods*, *5*(4), 1542–1553. https://doi.org/10.1016/j.jff.2013.08.009.

Annahazi, A., & Schemann, M. (2020). The enteric nervous system: "A little brain in the gut". *Neuroforum*, *26*(1), 31–42. https://doi.org/10.1515/nf-2019-0027

Ansari, F., Neshat, M., Pourjafar, H., Jafari, S. M., Samakkhah, S. A., & Mirzakhani, E. (2023). The role of probiotics and prebiotics in modulating of the gut-brain axis. *Frontiers in Nutrition*, *10*, 1173660. https://doi.org/10.3389/fnut.2023.1173660

Arneth, B. M. (2018). Gut–brain axis biochemical signalling from the gastrointestinal tract to the central nervous system: gut dysbiosis and altered brain function. *Postgraduate Medical Journal*, *94*(1114), 446–452. https://doi.org/10.1136/postgradmedj-2017-135424

Balu, D. T. (2016). The NMDA receptor and schizophrenia: from pathophysiology to treatment. *Advances in Pharmacology*, *76*, 351–382. https://doi.org/10.1016/bs.apha.2016.01.006

Barros, C. P., Guimarães, J. T., Esmerino, E. A., Duarte, M. C. K., Silva, M. C., Silva, R., ... Cruz, A. G. (2020). Paraprobiotics and postbiotics: Concepts and potential applications in dairy products. *Current Opinion in Food Science*, *32*, 1–8. https://doi.org/10.1016/j.cofs.2019.12.003

Bashan, A., Gibson, T. E., Friedman, J., Carey, V. J., Weiss, S. T., Hohmann, E. L., & Liu, Y. Y. (2016). Universality of human microbial dynamics. *Nature*, *534*(7606), 259–262. https://doi.org/10.1038/nature18301.

Basso, M., Johnstone, N., Knytl, P., Nauta, A., Groeneveld, A., & Kadosh, K. C. (2022). A systematic review of psychobiotic interventions in children and adolescents to enhance cognitive functioning and emotional behavior. *Nutrients*, *14*(3), 614. https://doi.org/10.3390/nu14030614

Berding, K., Vlckova, K., Marx, W., Schellekens, H., Stanton, C., Clarke, G., ... Cryan, J. F. (2021). Diet and the microbiota–gut–brain axis: Sowing the seeds of good mental health. *Advances in Nutrition*, *12*(4), 1239–1285. https://doi.org/10.1093/advances/nmaa181.

Berk, M., Williams, L. J., Jacka, F. N., O'Neil, A., Pasco, J. A., Moylan, S., ... Maes, M. (2013). So depression is an inflammatory disease, but where does the inflammation come from?. *BMC Medicine*, *11*(1), 200. https://doi.org/10.1186/1741-7015-11-200

Bermúdez-Humarán, L. G., Salinas, E., Ortiz, G. G., Ramirez-Jirano, L. J., Morales, J. A., & Bitzer-Quintero, O. K. (2019). From probiotics to psychobiotics: Live beneficial bacteria which act on the brain-gut axis. *Nutrients*, *11*(4), 890. https://doi.org/10.3390/nu11040890.

Biesmans, S., Bouwknecht, J. A., Ver Donck, L., Langlois, X., Acton, P. D., De Haes, P., ... Nuydens, R. (2015). Peripheral administration of tumor necrosis factor-alpha induces neuroinflammation and sickness but not depressive-like behavior in mice. *BioMed Research International*, *2015*(1), 716920. https://doi.org/10.1155/2015/716920

Bistoletti, M., Bosi, A., Banfi, D., Giaroni, C., & Baj, A. (2020). The microbiota-gut-brain axis: Focus on the fundamental communication pathways. *Progress in Molecular Biology and Translational Science*, *176*, 43–110. https://doi.org/10.1016/bs.pmbts.2020.08.012

Blekhman, R., Goodrich, J. K., Huang, K., Sun, Q., Bukowski, R., Bell, J. T., ... Clark, A. G. (2015). Host genetic variation impacts microbiome composition across human body sites. *Genome Biology*, *16*, 1–12. https://doi.org/10.1186/s13059-015-0759-1.

Boix-Amorós, A., Collado, M. C., & Mira, A. (2016). Relationship between milk microbiota, bacterial load, macronutrients, and human cells during lactation. *Frontiers in Microbiology*, *7*, 492.https://doi.org/10.3389/fmicb.2016.00492.

Bravo, J. A., Forsythe, P., Chew, M. V., Escaravage, E., Savignac, H. M., Dinan, T. G., Bienenstock, J., & Cryan, J. F. (2011). Ingestion of Lactobacillus strain regulates emotional behavior and central GABA receptor expression in a mouse via the vagus nerve. *Proceedings of the National Academy of Sciences*, *108*(38), 16050–16055. https://doi.org/10.1073/pnas.1102999108

Brestoff, J. R., & Artis, D. (2013). Commensal bacteria at the interface of host metabolism and the immune system. *Nature Immunology*, *14*(7), 676–684. https://doi.org/10.1038/ni.2640

Carlessi, A. S., Borba, L. A., Zugno, A. I., Quevedo, J., & Réus, G. Z. (2021). Gut microbiota–brain axis in depression: The role of neuroinflammation. *European Journal of Neuroscience*, *53*(1), 222–235. https://doi.org/10.1111/ejn.14631.

Casertano, M., Fogliano, V., & Ercolini, D. (2022). Psychobiotics, gut microbiota and fermented foods can help preserving mental health. *Food Research International*, *152*, 110892. https://doi.org/10.1016/j.foodres.2021.110892

Cerdó, T., Diéguez, E., & Campoy, C. (2019). Early nutrition and gut microbiome: Interrelationship between bacterial metabolism, immune system, brain structure, and neurodevelopment. *American Journal of Physiology-Endocrinology and Metabolism*, *317*(4), E617–E630. https://doi.org/10.1152/ajpendo.00188.2019

Chaudhry, T. S., Senapati, S. G., Gadam, S., Mannam, H. P. S. S., Voruganti, H. V., Abbasi, Z., ... Arunachalam, S. P. (2023). The impact of microbiota on the gut–brain axis: Examining the complex interplay and implications. *Journal of Clinical Medicine*, *12*(16), 5231. https://doi.org/10.3390/jcm12165231.

Clapp, M., Aurora, N., Herrera, L., Bhatia, M., Wilen, E., & Wakefield, S. (2017). Gut microbiota's effect on mental health: The gut-brain axis. *Clinics and Practice*, *7*(4), 987. https://doi.org/10.4081/cp.2017.987

Cox, L. M., Yamanishi, S., Sohn, J., Alekseyenko, A. V., Leung, J. M., Cho, I., Kim, S. G., Li, H., Gao, Z., Mahana, D., Zárate Rodriguez, J. G., Rogers, A. B., Robine, N., Loke, P., & Blaser, M. J. (2014). Altering the intestinal microbiota during a critical developmental window has lasting metabolic consequences. *Cell*, *158*(4), 705–721. https://doi.org/10.1016/j.cell.2014.05.052

de Jonge, W. J. (2013). The gut's little brain in control of intestinal immunity. *International Scholarly Research Notices*, *2013*(1), 630159. https://doi.org/10.1155/2013/630159

Devkota, S. (2016). Prescription drugs obscure microbiome analyses. *Science*, *351*(6272), 452–453. https://doi.org/10.1126/science.aaf1353.

Dicks, L. M. (2022). Gut bacteria and neurotransmitters. *Microorganisms*, *10*(9), 1838. https://doi.org/10.3390/microorganisms10091838

Diz-Chaves, Y., Herrera-Pérez, S., González-Matías, L. C., Lamas, J. A., & Mallo, F. (2020). Glucagon-like peptide-1 (GLP-1) in the integration of neural and endocrine responses to stress. *Nutrients*, *12*(11), 3304. https://doi.org/10.3390/nu12113304

Dominguez-Bello, M. G., De Jesus-Laboy, K. M., Shen, N., Cox, L. M., Amir, A., Gonzalez, A., ... Clemente, J. C. (2016). Partial restoration of the microbiota of cesarean-born infants via vaginal microbial transfer. *Nature Medicine*, *22*(3), 250–253. https://doi.org/10.1038/nm.4039

Doney, E., Cadoret, A., Dion-Albert, L., Lebel, M., & Menard, C. (2022). Inflammation-driven brain and gut barrier dysfunction in stress and mood disorders. *European Journal of Neuroscience*, *55*(9–10), 2851–2894. https://doi.org/10.1111/ejn.15239

Farzi, A., Fröhlich, E. E., & Holzer, P. (2018). Gut microbiota and the neuroendocrine system. *Neurotherapeutics*, *15*(1), 5–22. https://doi.org/10.1007/s13311-017-0600-5.

Flint, H. J., Scott, K. P., Louis, P., & Duncan, S. H. (2012). The role of the gut microbiota in nutrition and health. *Nature Reviews Gastroenterology & Hepatology*, *9*(10), 577–589. https://doi.org/10.1038/nrgastro.2012.156

Fried, S., Wemelle, E., Cani, P. D., & Knauf, C. (2021). Interactions between the microbiota and enteric nervous system during gut-brain disorders. *Neuropharmacology*, *197*, 108721. https://doi.org/10.1016/j.neuropharm.2021.108721.

Ganci, M., Suleyman, E., Butt, H., & Ball, M. (2019). The role of the brain–gut–microbiota axis in psychology: The importance of considering gut microbiota in the development, perpetuation, and treatment of psychological disorders. *Brain and Behavior*, *9*(11), e01408. https://doi.org/10.1002/brb3.1408

González Olmo, B. M., Butler, M. J., & Barrientos, R. M. (2021). Evolution of the human diet and its impact on gut microbiota, immune responses, and brain health. *Nutrients*, *13*(1), 196. https://doi.org/10.3390/nu13010196.

Goodrich, J. K., Waters, J. L., Poole, A. C., Sutter, J. L., Koren, O., Blekhman, R., ... Ley, R. E. (2014). Human genetics shape the gut microbiome. *Cell*, *159*(4), 789–799. https://doi.org/10.1016/j.cell.2014.09.053.

Góralczyk-Bińkowska, A., Szmajda-Krygier, D., & Kozłowska, E. (2022). The microbiota–gut–brain Axis in psychiatric disorders. *International Journal of Molecular Sciences*, *23*(19), 11245. https://doi.org/10.3390/ijms231911245

Graf, D., Di Cagno, R., Fåk, F., Flint, H. J., Nyman, M., Saarela, M., & Watzl, B. (2015). Contribution of diet to the composition of the human gut microbiota. *Microbial Ecology in Health and Disease*, *26*(1), 26164.https://doi.org/10.3402/mehd.v26.26164@zmeh20.2015.26.issue-s2

He, Q., Si, C., Sun, Z., Chen, Y., & Zhang, X. (2022). The intervention of prebiotics on depression via the Gut-Brain axis. *Molecules*, *27*(12), 3671. https://doi.org/10.3390/molecules27123671.

Hou, K., Wu, Z., Chen, X., Wang, J., Zhang, D., Xiao, C., Zhu, D., Koya, J. B., Wei, L., Li, J., & Chen, Z. (2022). Microbiota in health and diseases. *Signal Transduction and Targeted Therapy*, *7*(1). https://doi.org/10.1038/s41392-022-00974-4

Iadecola, C. (2017). The neurovascular unit coming of age: A journey through neurovascular coupling in health and disease. *Neuron*, *96*(1), 17–42. https://doi.org/10.1016/j.neuron.2017.07.030

Illiano, P., Brambilla, R., & Parolini, C. (2020). The mutual interplay of gut microbiota, diet and human disease. *The FEBS Journal*, *287*(5), 833–855. https://doi.org/10.1111/febs.15217

Jiang, H., Ling, Z., Zhang, Y., Mao, H., Ma, Z., Yin, Y., Wang, W., Tang, W., Tan, Z., Shi, J., Li, L., & Ruan, B. (2015). Altered fecal microbiota composition in patients with major depressive disorder. *Brain, Behavior, and Immunity*, *48*, 186–194. https://doi.org/10.1016/j.bbi.2015.03.016

Johnson, D., Thurairajasingam, S., Letchumanan, V., Chan, K. G., & Lee, L. H. (2021). Exploring the role and potential of probiotics in the field of mental health: Major depressive disorder. *Nutrients*, *13*(5), 1728. https://doi.org/10.3390/nu13051728

Ju, S., Shin, Y., Han, S., Kwon, J., Choi, T. G., Kang, I., & Kim, S. S. (2023). The gut–brain axis in schizophrenia: The implications of the gut microbiome and SCFA production. *Nutrients*, *15*(20), 4391.https://doi.org/10.3390/nu15204391

Kechagia, M., Basoulis, D., Konstantopoulou, S., Dimitriadi, D., Gyftopoulou, K., Skarmoutsou, N., & Fakiri, E. M. (2013). Health benefits of probiotics: A review. *International Scholarly Research Notices*, *2013*(1), 481651. https://doi.org/10.5402/2013/481651

Kennedy, B. M., Rehman, M., Johnson, W. D., Magee, M. B., Leonard, R., & Katzmarzyk, P. T. (2017). Healthcare providers versus patients' understanding of health beliefs and values. *Patient Experience Journal*, *4*(3), 29–37.

Khan, W. I., & Ghia, J. E. (2010). Gut hormones: Emerging role in immune activation and inflammation. *Clinical & Experimental Immunology*, *161*(1), 19–27. https://doi.org/10.1111/j.1365-2249.2010.04150.x

Kuhle, S., Tong, O. S., & Woolcott, C. G. (2015). Association between caesarean section and childhood obesity: A systematic review and meta-analysis. *Obesity Reviews*, *16*(4), 295–303. https://doi.org/10.1111/obr.12267

Kurokawa, S., Kishimoto, T., Mizuno, S., Masaoka, T., Naganuma, M., Liang, K., Kitazawa, M., Nakashima, M., Shindo, C., Suda, W., Hattori, M., Kanai, T., & Mimura, M. (2018). The effect of fecal microbiota transplantation on psychiatric symptoms among patients with irritable bowel syndrome, functional diarrhea and functional constipation: An open-label observational study. *Journal of Affective Disorders*, *235*, 506–512. https://doi.org/10.1016/j.jad.2018.04.038

Laterza, L., Rizzatti, G., Gaetani, E., Chiusolo, P., & Gasbarrini, A. (2016). The gut microbiota and immune system relationship in human graft-versus-host disease. *Mediterranean Journal of Hematology and Infectious Diseases*, *8*(1). https://doi.org/10.4084/mjhid.2016.025

Leclercq, S., Forsythe, P., & Bienenstock, J. (2016). Posttraumatic stress disorder: does the gut microbiome hold the key?. *The Canadian Journal of Psychiatry*, *61*(4), 204–213. https://doi.org/10.1177/0706743716635535.

Li, H. Y., Zhou, D. D., Gan, R. Y., Huang, S. Y., Zhao, C. N., Shang, A. O., ... & Li, H. B. (2021). Effects and mechanisms of probiotics, prebiotics, synbiotics, and postbiotics on metabolic diseases targeting gut microbiota: A narrative review. *Nutrients*, *13*(9), 3211. https://doi.org/10.3390/nu13093211

Li, J., Zhao, F., Wang, Y., Chen, J., Tao, J., Tian, G., & Cai, J. (2017). Gut microbiota dysbiosis contributes to the development of hypertension. *Microbiome*, *5*, 1–19.https://doi.org/10.1186/s40168-016-0222-x

Lordan, C., Thapa, D., Ross, R. P., & Cotter, P. D. (2020). Potential for enriching next-generation health-promoting gut bacteria through prebiotics and other dietary components. *Gut Microbes*, *11*(1), 1–20.https://doi.org/10.1080/19490976.2019.1613124

Marchesi, J. R., Adams, D. H., Fava, F., Hermes, G. D., Hirschfield, G. M., Hold, G., ... Hart, A. (2016). The gut microbiota and host health: A new clinical frontier. *Gut*, *65*(2), 330–339. https://doi.org/10.1136/gutjnl-2015-309990

Metz, C. N., & Pavlov, V. A. (2018). Vagus nerve cholinergic circuitry to the liver and the gastrointestinal tract in the neuroimmune communicatome. *American Journal of Physiology-Gastrointestinal and Liver Physiology*, *315*(5), G651-G658. https://doi.org/10.1152/ajpgi.00195.2018

Mikulska, J., Juszczyk, G., Gawrońska-Grzywacz, M., & Herbet, M. (2021). HPA axis in the pathomechanism of depression and schizophrenia: new therapeutic strategies based on its participation. *Brain Sciences*, *11*(10), 1298. https://doi.org/10.3390/brainsci11101298.

Mittal, R., Debs, L. H., Patel, A. P., Nguyen, D., Patel, K., O'Connor, G., ... Liu, X. Z. (2017). Neurotransmitters: The critical modulators regulating gut–brain axis. *Journal of Cellular Physiology*, *232*(9), 2359–2372. https://doi.org/10.1002/jcp.25518

Mizrahi-Man, O., Davenport, E. R., & Gilad, Y. (2013). Taxonomic classification of bacterial 16S rRNA genes using short sequencing reads: Evaluation of effective study designs. *PLoS One*, *8*(1), e53608.https://doi.org/10.1371/journal.pone.0053608

Modi, S. R., Collins, J. J., & Relman, D. A. (2014). Antibiotics and the gut microbiota. *The Journal of Clinical Investigation*, *124*(10), 4212–4218. http://doi.org/10.1172/JCI72333

Moretti, S., Schietroma, I., Sberna, G., Maggiorella, M. T., Sernicola, L., Farcomeni, S., ... Borsetti, A. (2023). HIV-1–host interaction in gut-associated lymphoid tissue (GALT): Effects on local environment and comorbidities. *International Journal of Molecular Sciences*, *24*(15), 12193.https://doi.org/10.3390/ijms241512193

Morris, G., Fernandes, B. S., Puri, B. K., Walker, A. J., Carvalho, A. F., & Berk, M. (2018). Leaky brain in neurological and psychiatric disorders: Drivers and consequences. *Australian & New Zealand Journal of Psychiatry*, *52*(10), 924–948.

Müller, N., Weidinger, E., Leitner, B., & Schwarz, M. J. (2015). The role of inflammation in schizophrenia. *Frontiers in Neuroscience*, *9*, 372.https://doi.org/10.3389/fnins.2015.00372

Muscatello, M. R. A., Bruno, A., Scimeca, G., Pandolfo, G., & Zoccali, R. A. (2014). Role of negative affects in pathophysiology and clinical expression of irritable bowel syndrome. *World Journal of Gastroenterology: WJG*, *20*(24), 7570. http://doi.org/10.3748/wjg.v20.i24.7570

Nelson, M. C., Story, M., Larson, N. I., Neumark-Sztainer, D., & Lytle, L. A. (2008). Emerging adulthood and college-aged youth: An overlooked age for weight-related behavior change. *Obesity*, *16*(10), 2205–2211. https://doi.org/10.1038/oby.2008.365

O'Riordan, K. J., Collins, M. K., Moloney, G. M., Knox, E. G., Aburto, M. R., Fülling, C., & Cryan, J. F. (2022). Short chain fatty acids: microbial metabolites for gut-brain axis signalling. *Molecular and Cellular Endocrinology*, *546*, 111572. https://doi.org/10.1016/j.mce.2022.111572

Obermajer, T., & Pogačić, T. (2016). Commentary: Relationship between milk microbiota, bacterial load, macronutrients, and human cells during lactation. *Frontiers in Microbiology*, *7*, 1281. https://doi.org/10.3389/fmicb.2016.01281

Oroojzadeh, P., Bostanabad, S. Y., & Lotfi, H. (2022). Psychobiotics: The influence of gut microbiota on the gut-brain axis in neurological disorders. *Journal of Molecular Neuroscience*, *72*(9), 1952–1964. https://doi.org/10.1007/s12031-022-02053-3.

Panagos, P. G., Vishwanathan, R., Penfield-Cyr, A., Matthan, N. R., Shivappa, N., Wirth, M. D., ... Sen, S. (2016). Breastmilk from obese mothers has pro-inflammatory properties and decreased neuroprotective factors. *Journal of Perinatology*, *36*(4), 284–290. https://doi.org/10.1038/jp.2015.199.

Pereira, T. M., Côco, L. Z., Ton, A. M., Meyrelles, S. S., Campos-Toimil, M., Campagnaro, B. P., & Vasquez, E. C. (2021). The emerging scenario of the gut-brain axis: The therapeutic actions of the new actor kefir against neurodegenerative diseases. *Antioxidants*, *10*(11), 1845. https://doi.org/10.3390/antiox10111845.

Peterson, C. T. (2020). Dysfunction of the microbiota-gut-brain axis in neurodegenerative disease: The promise of therapeutic modulation with prebiotics, medicinal herbs, probiotics, and symbiotic. *Journal of Evidence-based Integrative Medicine*, *25*, 2515690X20957225. https://doi.org/10.1177/2515690X20957225

Portela, D. S., Vieira, T. O., Matos, S. M., de Oliveira, N. F., & Vieira, G. O. (2015). Maternal obesity, environmental factors, cesarean delivery and breastfeeding as determinants of overweight and obesity in children: Results from a cohort. *BMC Pregnancy and Childbirth*, *15*, 1–10.https://doi.org/10.1186/s12884-015-0518-z

Ramani, R. (2008). Vagus nerve stimulation therapy for seizures. *Journal of Neurosurgical Anesthesiology*, *20*(1), 29–35. https://doi.org/10.1097/ANA.0b013e31815b7df1

Rea, K., Dinan, T. G., & Cryan, J. F. (2020). Gut microbiota: A perspective for psychiatrists. *Neuropsychobiology*, *79*(1), 50–62. https://doi.org/10.1159/000504495

Rehman, A., Sina, C., Gavrilova, O., Häsler, R., Ott, S., Baines, J. F., ... Rosenstiel, P. (2011). Nod2 is essential for temporal development of intestinal microbial communities. *Gut*, *60*(10), 1354–1362. https://doi.org/10.1136/gut.2010.216259

Rinninella, E., Raoul, P., Cintoni, M., Franceschi, F., Miggiano, G. A. D., Gasbarrini, A., & Mele, M. C. (2019). What is the healthy gut microbiota composition? A changing ecosystem across age, environment, diet, and diseases. *Microorganisms*, *7*(1), 14 https://doi.org/10.3390/microorganisms7010014

Ripke, S., Neale, B. M., Corvin, A., Walters, J. T., Farh, K. H., Holmans, P. A., ... Milanova, V. (2014). Biological insights from 108 schizophrenia-associated genetic loci. *Nature*, *511*(7510), 421. http://doi.org/10.1038/nature13595

Rogers, G. B., Keating, D. J., Young, R. L., Wong, M. L., Licinio, J., & Wesselingh, S. (2016). From gut dysbiosis to altered brain function and mental illness: mechanisms and pathways. *Molecular Psychiatry*, *21*(6), 738–748. https://doi.org/10.3390/ijms21207551

Rowland, I., Gibson, G., Heinken, A., Scott, K., Swann, J., Thiele, I., & Tuohy, K. (2018). Gut microbiota functions: Metabolism of nutrients and other food components. *European Journal of Nutrition*, *57*, 1–24. https://doi.org/10.1007/s00394-017-1445-8

Rutayisire, E., Wu, X., Huang, K., Tao, S., Chen, Y., & Tao, F. (2016). Cesarean section may increase the risk of both overweight and obesity in preschool children. *BMC Pregnancy and Childbirth*, *16*, 1–8.https://doi.org/10.1186/s12884-016-1131-5

Sandhu, K. V., Sherwin, E., Schellekens, H., Stanton, C., Dinan, T. G., & Cryan, J. F. (2017). Feeding the microbiota-gut-brain axis: Diet, microbiome, and neuropsychiatry. *Translational Research*, *179*, 223–244. https://doi.org/10.1016/j.trsl.2016.10.002

Severance, E. G., Prandovszky, E., Castiglione, J., & Yolken, R. H. (2015, May). Gastroenterology issues in schizophrenia: Why the gut matters. *Current Psychiatry Reports*, *17*, 1–10. https://doi.org/10.1007/s11920-015-0574-0

Shen, Y., Xu, J., Li, Z., Huang, Y., Yuan, Y., Wang, J., & Liang, Y. (2018). Analysis of gut microbiota diversity and auxiliary diagnosis as a biomarker in patients with schizophrenia: A cross-sectional study. *Schizophrenia Research*, *197*, 470–477. https://doi.org/10.1016/j.schres.2018.01.002

Shine, J. M., O'Callaghan, C., Walpola, I. C., Wainstein, G., Taylor, N., Aru, J., ... John, Y. J. (2022). Understanding the effects of serotonin in the brain through its role in the gastrointestinal tract. *Brain*, *145*(9), 2967–2981. https://doi.org/10.1093/brain/awac256

Silva, Y. P., Bernardi, A., & Frozza, R. L. (2020). The role of short-chain fatty acids from gut microbiota in gut-brain communication. *Frontiers in endocrinology*, *11*, 508738. https://doi.org/10.3389/fendo.2020.00025

Singh, R., Zogg, H., & Ro, S. (2021). Role of microRnas in disorders of gut–brain interactions: Clinical insights and therapeutic alternatives. *Journal of Personalized Medicine*, *11*(10), 1021. https://doi.org/10.3390/jpm11101021

Suda, K., & Matsuda, K. (2022). How microbes affect depression: Underlying mechanisms via the Gut–brain axis and the modulating role of probiotics. *International Journal of Molecular Sciences*, *23*(3), 1172. https://doi.org/10.3390/ijms23031172

Sudo, N., Chida, Y., Aiba, Y., Sonoda, J., Oyama, N., Yu, X. N., Kubo, C., & Koga, Y. (2004). Postnatal microbial colonization programs the hypothalamic-pituitary-adrenal system for stress response in mice. *The Journal of Physiology*, *558*(Pt 1), 263–275. https://doi.org/10.1113/jphysiol.2004.063388

Szeligowski, T., Yun, A. L., Lennox, B. R., & Burnet, P. W. (2020). The gut microbiome and schizophrenia: The current state of the field and clinical applications. *Frontiers in Psychiatry*, *11*, 156. https://doi.org/10.3389/fpsyt.2020.00156

Tamburini, S., Shen, N., Wu, H. C., & Clemente, J. C. (2016). The microbiome in early life: Implications for health outcomes. *Nature Medicine*, *22*(7), 713–722. https://doi.org/10.1038/nm.4142

Thursby, E., & Juge, N. (2017). Introduction to the human gut microbiota. *Biochemical Journal*, *474*(11), 1823–1836. https://doi.org/10.1042/BCJ20160510

Tognini, P. (2017). Gut microbiota: A potential regulator of neurodevelopment. *Frontiers in Cellular Neuroscience*, 11, 25. https://doi.org/10.3389/fncel.2017.00025

Tyagi, P., Tasleem, M., Prakash, S., & Chouhan, G. (2020). Intermingling of gut microbiota with brain: Exploring the role of probiotics in battle against depressive disorders. *Food Research International*, *137*, 109489. https://doi.org/10.1016/j.foodres.2020.109489

Valdes, A. M., Walter, J., Segal, E., & Spector, T. D. (2018). Role of the gut microbiota in nutrition and health. *BMJ*, *361*. https://doi.org/10.1136/bmj.k2179

van Son, J., Koekkoek, L. L., La Fleur, S. E., Serlie, M. J., & Nieuwdorp, M. (2021). The role of the gut microbiota in the gut–brain axis in obesity: mechanisms and future implications. *International Journal of Molecular Sciences*, *22*(6), 2993. https://doi.org/10.3390/ijms22062993

Vinderola, G., Sanders, M. E., & Salminen, S. (2022). The concept of postbiotics. *Foods*, *11*(8), 1077. https://doi.org/10.3390/foods11081077

Wen, L., & Duffy, A. (2017). Factors influencing the gut microbiota, inflammation, and type 2 diabetes. *The Journal of Nutrition*, *147*(7), 1468S–1475S. https://doi.org/10.3945/jn.116.240754

Wiley, N. C., Dinan, T. G., Ross, R. P., Stanton, C., Clarke, G., & Cryan, J. F. (2017). The microbiota-gut-brain axis as a key regulator of neural function and the stress response: Implications for human and animal health. *Journal of Animal Science*, *95*(7), 3225–3246. https://doi.org/10.2527/jas.2016.1256

Xiong, R. G., Li, J., Cheng, J., Zhou, D. D., Wu, S. X., Huang, S. Y., ... Li, H. B. (2023). The role of gut microbiota in anxiety, depression, and other mental disorders as well as the protective effects of dietary components. *Nutrients*, *15*(14), 3258. https://doi.org/10.3390/nu15143258

Yano, J. M., Yu, K., Donaldson, G. P., Shastri, G. G., Ann, P., Ma, L., Nagler, C. R., Ismagilov, R. F., Mazmanian, S. K., & Hsiao, E. Y. (2015). Indigenous bacteria from the gut microbiota regulate host serotonin biosynthesis. *Cell*, *161*(2), 264–276. https://doi.org/10.1016/j.cell.2015.02.047

Yoo, D. H., Kim, I. S., Van Le, T. K., Jung, I. H., Yoo, H. H., & Kim, D. H. (2014). Gut microbiota-mediated drug interactions between lovastatin and antibiotics. *Drug Metabolism and Disposition*, *42*(9), 1508–1513. https://doi.org/10.1124/dmd.114.058354

Zhou, L., & Foster, J. A. (2015). Psychobiotics and the gut–brain axis: In the pursuit of happiness. *Neuropsychiatric Disease and Treatment*, 715–723. https://doi.org/10.2147/NDT.S61997

Zielińska, D., Karbowiak, M., & Brzezicka, A. (2022). The role of psychobiotics to ensure mental health during the COVID-19 pandemic—a current state of knowledge. *International Journal of Environmental Research and Public Health*, *19*(17), 11022. https://doi.org/10.3390/ijerph191711022

8 Gut–Brain Axis and Autoimmune Diseases

Correlation between Both

Devid Chutia[1], Jadab Rajkonwar[2], and Srijita Chakrabarti[3]

[1] Institute of Pharmacy, Assam Don Bosco University, Tapesia, Assam, India
[2] Regional Medical Research Centre, Northeast Region (RMRC-NE)-ICMR, Dibrugarh, Assam, India
[3] Faculty of Pharmaceutical Science, Assam down town University, Guwahati, Assam, India,

8.1 INTRODUCTION

The gut–brain axis (GBA) is a two-way communication system between the brain and gut microbiota, influencing gastrointestinal and neurological disorders. It involves hormones, metabolism, microbes, and the immune system (Khawar et al., 2023). Gut microbiota, including Bacteroidetes, Firmicutes, Actinobacteria, Proteobacteria, Fusobacteria, and Verrucomicrobiota, helps maintain immune balance. However, the mechanism behind its effect on systemic immunity in autoimmune diseases (AID) remains unclear (Sai et al., 2024). Dysbiosis, an imbalance in gut microbiota, may contribute to AID by disrupting this complex interaction, leading to abnormal immune responses, including increased production of B cells, T cells, and pro-inflammatory cytokines. Factors like infections, genetics, environmental changes, and altered microbiota may trigger autoimmune disorders, with dietary changes and antibiotic use potentially driving the rise in these diseases (Xu et al., 2019).

8.2 ROLE OF MICROBIOTA IN GUT–BRAIN COMMUNICATION

The gut microbiota plays a crucial role in health by influencing organs, including the brain, through the microbiota-GBA. This bidirectional communication involves pathways such as the vagus nerve, neurotransmitters, metabolites, and immune responses. The vagus nerve links the gut and brain, allowing the brain to sense the gut environment (Sandhu et al., 2017; Zhao et al., 2018). The gut microbiota also affects the hypothalamic–pituitary–adrenal (HPA) axis, regulating stress and immune responses (Chakrabarti et al., 2022). Microbiota-produced neurotransmitters like GABA, serotonin, dopamine, and noradrenaline influence mood, behavior, and cognition. Disruptions in these neurotransmitters are linked to disorders such as depression, schizophrenia, and autism (Kasarello et al., 2023). Studies suggest that gut bacteria, like *Bifidobacterium* and *Lactobacilli*, generate GABA (Duranti et al., 2020; Zheng et al., 2020). The microbiota also impacts serotonin production by affecting tryptophan absorption (Yano et al., 2015). Dopamine, linked to mood and behavior, is affected by gut bacteria and is associated with disorders like Parkinson's and addiction (Maini et al., 2020). Understanding the microbiota's role in neurological diseases offers new therapeutic opportunities, including probiotics, prebiotics, and microbiotic products.

8.3 AUTOIMMUNE DISEASES ASSOCIATED WITH GUT–BRAIN AXIS

GBA, a revolutionary bidirectional communication system between the central nervous system (CNS) and the gastrointestinal (GI) tract, is increasingly recognized for its pivotal role in health. This intricate connection, facilitated through neural, hormonal, and immunological pathways, is increasingly recognized for its pivotal role in the onset and progression of AID. GBA profoundly influences these conditions, characterized by the immune system's misguided attack on the body's tissues. The pivotal role of GBA in AID underscores the importance of this connection, revolutionizing our comprehension of these conditions and shedding new light on this crucial aspect of health. By understanding and appreciating the role of GBA (Figure 8.1), we can better manage and treat AID, opening up a new frontier in therapeutic approaches (Richards et al., 2021). Pathogenesis and management of different AID are enlisted in Table 8.1.

DOI: 10.1201/9781003536802-8

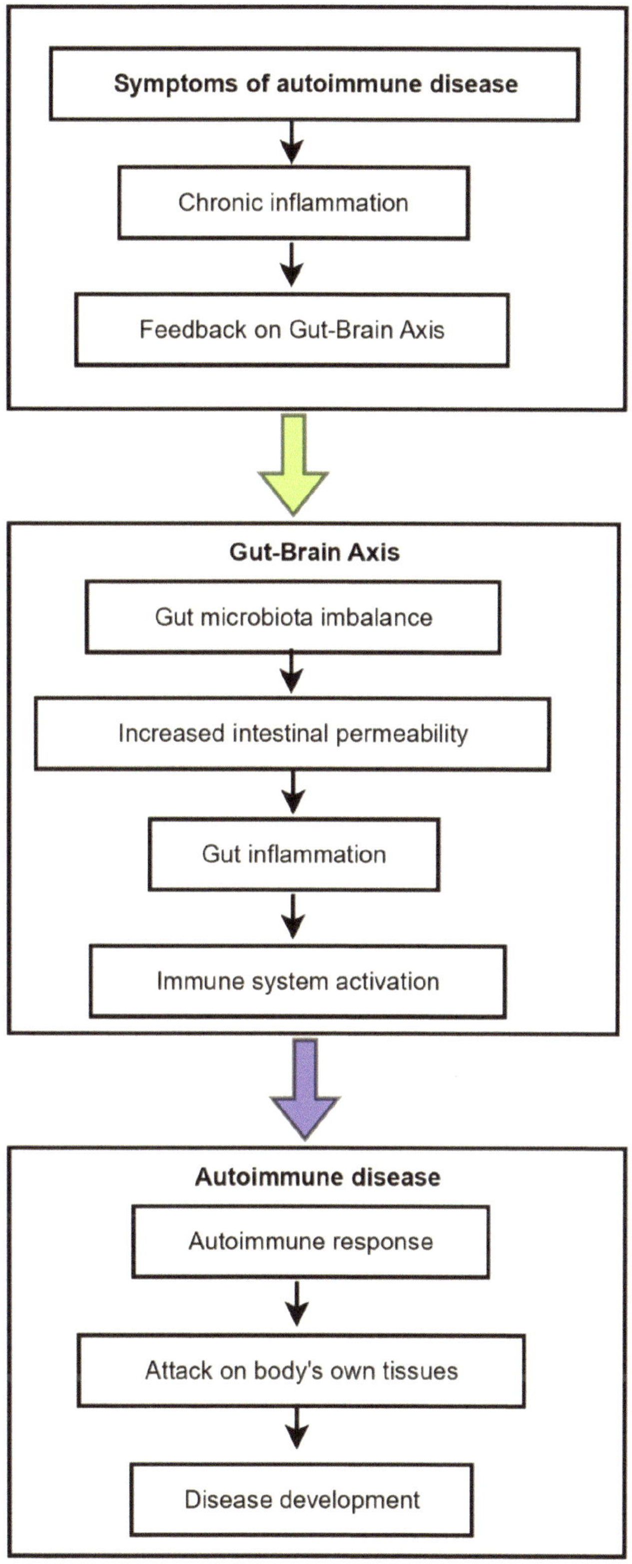

FIGURE 8.1 The pivotal role of GBA in autoimmune diseases underscores the importance of this connection, revolutionizing our comprehension of these conditions and shedding new light on this crucial aspect of health.

TABLE 8.1
Role of Gut Microbiota in Pathogenesis and Management of Autoimmune Diseases

AID	Cellular Pathogenesis	Management	Limitation	References
Multiple sclerosis (MS)	The gut microbiome regulates the immune system, and changes in its composition can influence MS development. A leaky gut may trigger or worsen the autoimmune response, leading to myelin damage and disrupted brain–body communication, with the GBA playing a key role	A fiber-rich diet, regular exercise, and medications help manage MS symptoms like fatigue, pain, and muscle spasms. Physical therapy, mental health support, and regular checkups improve quality of life and well-being	The variability in gut microbiota among MS patients, along with genetic, environmental, and gender factors, makes it challenging to identify consistent microbial patterns or metabolites linked to the disease	Dunalska et al., 2023
T1DM	The gut microbiota impacts T1DM progression by influencing immune responses and inflammation. The GBA, through neural, hormonal, and immune pathways, may contribute to autoimmune diseases like T1DM, affecting brain function, behavior, and blood sugar control	Fecal microbial transplantation (FMT) aims to modify the gut microbiota in T1DM patients, potentially improving immune responses. Dietary changes, probiotics, and GLP-1-based therapies can also help regulate immune function and improve glucose control	Gastrointestinal symptoms like nausea, vomiting, diarrhoea, and constipation can occur due to disruptions in the gut microbiota, which is often affected in diabetes, leading to dysbiosis	Bessac et al., 2018; Drewes et al., 2019
Rheumatoid arthritis (RA)	In RA, an imbalanced gut microbiota can lead to systemic inflammation, potentially contributing to depression-like behaviors. Studies showed that mice with collagen-induced arthritis (CIA), a model for RA, display depression-like symptoms, increased inflammatory markers like IL-6, and changes in gut microbiota composition	FMT from RA patients to antibiotic-treated mice has been found to cause depression-like behaviors, changes in gut microbiota, and increased levels of inflammatory cytokines such as IL-6 and TNF-α	Animal models like CIA mice for RA may not fully replicate human RA and its depression-like behaviors. Variability in human microbiota and short-term studies limit the applicability and understanding of long-term GBA impacts and recovery	Romero-Figueroa et al., 2023
Inflammatory bowel disease (IBD)	The GBA can contribute to anxiety and depression by causing neurobiological imbalances. IBDs trigger systemic inflammation that affects brain inflammation, hormonal balance, and blood–brain barrier integrity, potentially leading to neurodegenerative diseases. IBD involves immune imbalances, and altered microbial metabolites like short-chain fatty acids (SCFAs) can worsen inflammation and symptoms	IBD treatment includes anti-inflammatory drugs, immune suppressors, and biologics targeting specific immune pathways. A proper diet and lifestyle changes like stress management, exercise, and sleep are important. In some cases, surgery may be needed to remove damaged intestine sections or create a new waste pathway	Current research on serum biomarkers and the link between IBD and neurodegenerative diseases is inconclusive. The complex relationship is not fully understood, and limited long-term studies hinder the understanding of IBD's chronic progression and its impact on the GBA and neurodegenerative diseases	Bakshi et al., 2024

(*Continued*)

TABLE 23.3 (CONTINUED)

Psoriasis	Dysregulation of the gut microbiota is linked to inflammatory skin conditions like psoriasis, with IL-23/IL-17 pathways involved in both gut and skin inflammation. Regulating the microbiome can reduce skin inflammation and mental health issues. Gut microbes also produce neurotransmitters that influence immune responses and psoriasis severity	Modulating the gut microbiota with diet, prebiotics, and probiotics can reduce inflammation and improve skin health in psoriasis. Probiotics, fermented foods, and anti-inflammatory nutrients support gut health, while regulating neurotransmitters and inflammatory pathways like IL-23/IL-17 may ease skin symptoms and related mental health issues	Many studies on the gut microbiome and psoriasis involve small and diverse sample sizes, which may not represent the broader population. The long-term efficacy and safety of probiotics and other microbiome-modulating therapies remain unclear	Wang et al., 2021
Sjögren's syndrome (SS)	The GBA links the gut and brain, affecting mental health in SS. A study on NOD mice with SS found depression-like behavior and altered gut microbiota, connecting SS to mental health. It also noted reduced synaptic protein expression and highlighted the role of inflammation and autoantibodies in depression	Diet changes, probiotics, and anti-inflammatory treatments can help manage gut microbiota and reduce depression in SS. Monitoring autoantibodies aids treatment, while mental health support and antidepressants may be necessary, under medical supervision	The study used NOD mice to model SS, which may not fully reflect the human condition. It focused on depression-like behaviors, but SS patients may experience a broader range of psychiatric symptoms	Pu et al., 2022
Schizophrenia	Disruptions in gut microbiota, caused by diet, medication, and stress, can impact brain function and contribute to mental disorders like schizophrenia. Gut bacteria produce SCFAs that influence brain activity, affecting microglia, cytokines, and neurotransmitters like serotonin and dopamine, which may play a role in CNS disorders	A fiber-rich diet, along with stress-reducing activities like mindfulness, yoga, and exercise supports a healthy GBA and mental health in schizophrenia. Dietary interventions, omega-3s, B vitamins, and probiotics may help manage symptoms by restoring gut–brain communication and improving the microbiome	Gut microbiota composition varies due to diet, genetics, and environment, complicating the study of its impact on schizophrenia. External factors like diet, medication, and stress further challenge research. Integrating findings from microbiology, neuroscience, and psychiatry is key to developing effective treatments	Zhu et al., 2020
Parkinson's disease (PD)	PD affects both the CNS and enteric nervous system (ENS), with α-synuclein clumps found in both the brain and gut, suggesting the disease may begin in the gut. Gut microbiome changes may influence the GBA and contribute to PD, with gastrointestinal issues like constipation commonly observed in PD patients	Laxatives, stool softeners, probiotics, and prebiotics can help relieve constipation in PD patients. Regular exercise supports bowel health, while severe cases may require surgery. Anticholinergic drugs and dopaminergic agonists are also commonly prescribed	The complexity of gut–brain interactions in Parkinson's disease complicates targeted treatments and early biomarkers. Experimental models may not fully reflect human disease, and varying gastrointestinal symptoms make a universal treatment challenging	Zhang et al., 2023

(Continued)

TABLE 23.3 (CONTINUED)

Alzheimer's disease (AD)	The GBA links the gut and brain through neural, hormonal, and immune pathways. In Alzheimer's disease, changes in gut microbiota affect brain health by altering neurotransmitter production and inflammation, with gut metabolites crossing the blood–brain barrier and contributing to neurodegeneration	Regular exercise, stress management, and supplements like omega-3s can support brain and gut health in Alzheimer's. FMT, probiotics, prebiotics, and dietary changes may also improve cognitive function and alleviate symptoms	Lack of high-quality research on the GBA's role in mental disorders. Most studies used correlational analysis, unable to establish causality	Dhanawat et al., 2024
Celiac	It is an autoimmune disorder triggered by gluten, affecting the small intestine and potentially the brain. The GBA links the gut and brain, and in celiac disease, gluten fragments activate immune cells that can cause neuroinflammation, leading to symptoms like anxiety, depression, and cognitive issues	The main treatment is a strict gluten-free diet to prevent immune activation. Monitoring psychological symptoms like anxiety and depression is crucial, as they may be linked to the disease or inflammation. Managing immune responses is key to preventing neuroinflammation and behavioral symptoms	A major challenge in celiac disease is determining whether psychological symptoms are due to the disease or stress. Inconsistent responses to a gluten-free diet and unclear mechanisms linking gluten-induced immune activation to neuroinflammation complicate treatment	Levin & Malohat, 2023
Depression	The GBA links the gut and brain through immune, hormonal, and neural pathways, influencing mental health. Dysbiosis can trigger neuroinflammation, activating microglial cells and affecting mood. The gut microbiota also impacts immune signals, neurotransmitters, and stress hormones, contributing to depression and cognitive changes	Probiotics and synbiotics can restore gut balance and improve depression. Stress management, mindfulness, exercise, and anti-inflammatory diets also support mental health. Depression treatments include medications like selective serotonin reuptake inhibitors, serotonin-norepinephrine reuptake inhibitors, tricyclics, and other antidepressants such as bupropion and agomelatine	Each person's gut microbiota is unique, making treatments like probiotics and synbiotics inconsistent for neurological disorders. There's no standardized approach for using dietary changes or synbiotics, and more clinical research is needed to understand the GBA's potential in treating neurological health	Yixin Xu, 2024
Autism	Dysbiosis can cause neuroinflammation and activate microglial cells, affecting brain health and potentially contributing to autism. The gut microbiota influences the immune system, neurotransmitter production, and stress regulation, which may be disrupted in autism, impacting brain function and development	Dietary changes and synbiotics may help manage neurological disorders by reducing inflammation and supporting brain health. In autism, treatments like NMDA-receptor antagonists, cytokine regulation, and galantamine analogues show potential for improving symptoms	The GBA is complex, and limited understanding of its mechanisms hinders the development of targeted treatments for disorders like autism. Variations in gut microbiota due to diet, genetics, and environment affect the effectiveness of GBA-based therapies	Ashique et al., 2024

8.3.1 Multiple Sclerosis

Multiple sclerosis (MS) is an inflammatory condition characterized by progressive demyelination in the brain and spinal cord, leading to neuro-axonal injury and disability. It allows immune cells to enter the CNS, activating glial cells and causing inflammation and tissue damage (Pabian-Jewuła & Rylski, 2023). The GBA) plays a key role in MS pathogenesis, with gut microbiome imbalances (dysbiosis) linked to oxidative stress, systemic inflammation, and leaky gut syndrome, all contributing to MS. Understanding the GBA's role is vital for developing effective treatments. Therapies like fecal microbiota transplantation, probiotics, dietary changes, and gut-derived metabolites show promise in modulating the gut microbiome and immune responses for MS management (Alkinani & Phia, 2024).

8.3.2 Type 1 Diabetes

Type 1 diabetes (T1DM) is a condition where the pancreas fail to produce insulin, leading to high blood sugar levels and requiring lifelong insulin therapy. It often develops in childhood or adolescence and can cause complications, including white matter abnormalities in the brain, affecting cognitive and emotional functions (Eizirik et al., 2020). The GBA plays a crucial role in T1DM by regulating metabolism, appetite, and gastrointestinal function. In T1DM, chronic hyperglycemia can lead to enteric neuroinflammation, which may contribute to motility issues, but insulin treatment can mitigate this (Longo et al., 2023). The gut microbiota and its metabolites influence the GBA and offer potential therapeutic targets for diabetes management. The vagus nerve, a key part of the GBA, helps control appetite and metabolism, with interactions between the gut microbiota and vagus nerve potentially affecting feeding behavior and metabolic health in T1DM.

8.3.3 Rheumatoid Arthritis

Rheumatoid arthritis is a chronic inflammatory disorder that primarily affects the joints, causing pain, swelling, and stiffness, and can lead to complications in other organs like the skin, eyes, lungs, and heart. The GBA, a complex communication system between the gut and brain, plays a significant role in RA (Yu et al., 2021). Gut dysbiosis, or an imbalance in gut microbiota, can influence the immune system, triggering inflammation and altering intestinal permeability, which may precede arthritis (Piva et al., 2023). Specific gut bacteria, like *Bifidobacterium longum*, can alleviate RA symptoms by modulating the microbiome and immune responses. The GBA also affects brain function, potentially contributing to neuroinflammatory conditions and influencing psychological aspects such as depression and anxiety in RA patients. Personalized psychological support is crucial to improve the quality of life for these individuals.

8.3.4 Inflammatory Bowel Disease

Inflammatory bowel disease, including Crohn's disease and ulcerative colitis, involves chronic inflammation of the digestive tract, causing symptoms like diarrhoea, abdominal pain, weight loss, and fatigue. Risk factors include family history, smoking, high-fat diets, and urban living (Liu et al., 2022). The gut microbiota, which affects brain function and behavior through metabolites and neurotransmitters, plays a key role in IBD. Psychological stress can worsen gut inflammation, suggesting stress management as a potential therapy. Dysfunction in the gut–brain axis in IBD is linked to inflammation, nutritional deficiencies, and cognitive issues. Psychosomatic treatments are needed to address both mental health and physical health. Therapies like drug treatments, fecal microbiota transplantation, and probiotics show promise in regulating the gut microbiota and easing IBD symptoms (Zhu et al., 2022).

8.3.5 Rectal and Colon Ulcer

Rectal and colon ulcer (RCU) is a progressive inflammatory bowel disease affecting the rectum and colon, with a global prevalence of 5 million cases as of 2024. Its causes include genetic predisposition, environmental factors, gut dysbiosis, and immune dysregulation, leading to gut permeability and inflammation. Symptoms typically include bloody diarrhoea, with diagnosis confirmed through clinical, biological, endoscopic, and histological assessments (Peppas et al., 2021). Disturbances in the GBA are observed in RCPU patients, particularly involving mast cells and the Mrgprb2 receptor, which play a role in bacterial defence and neurogenic inflammation. Studies on Mrgprb2 knockout mice show worsened symptoms, including weight loss, colon damage, microbiota changes, and compromised intestinal barrier function. Additionally, neuropsychiatric symptoms like anxiety and sadness are common, with prolonged inflammation potentially causing systemic and neuroinflammatory changes, affecting macrophages and microglia in the brain (Ge et al., 2022).

8.3.6 Psoriasis

Psoriasis is a chronic immunomodulatory disease linked to obesity, diabetes, and cardiovascular issues (Ferraretto et al., 2023). The GBA plays a key role in its pathogenesis and management. The gut microbiota influences psoriasis by modulating neurotransmitters like dopamine, serotonin, and GABA. In psoriasis patients, an imbalance in gut microbiota is observed, with increased Firmicutes, Actinobacteria, and Verrucomicrobia, and decreased Bacteroidetes, Euryarchaeota, and Proteobacteria (Khawar et al., 2023). This imbalance affects pathways related to immune responses, inflammation, and vascular growth. The gut–brain–skin axis demonstrates the gut's role in skin health, with probiotics showing potential in restoring microbial balance and alleviating psoriasis symptoms.

8.3.7 Sjögren's Syndrome

Sjögren's syndrome is a chronic autoimmune disease characterized by lymphocytic infiltration of exocrine glands, causing dry mouth and dry eyes, and often associated with other autoimmune disorders like rheumatoid arthritis and lupus (Mayer & Tillisch, 2011). It can affect multiple organs, including the joints, skin, lungs, and nervous system. SS patients have a higher prevalence of depression, potentially linked to the GBA (Tsianos & Moutsopoulos, 1989). A study in non-obese diabetic mice, an SS model, showed depression-like behaviors, reduced synaptic protein expression in the prefrontal cortex, and altered gut microbiota composition. Similar GBA-related changes are seen in conditions like schizophrenia and short bowel syndrome, where gut microbiota and GI barrier dysfunction are linked to psychiatric symptoms.

8.3.8 Schizophrenia

Schizophrenia is a mental disorder associated with autoimmune dysfunction, with specific brain regions showing elevated autoimmune antibodies (Wakefield, 2002). The gut microbiota, through the GBA, may influence brain function, and dysbiosis (microbial imbalance) is linked to schizophrenia. The gut microbiota plays a role in immune system development, vitamin synthesis, and pathogen defence, with disruptions potentially leading to CNS dysfunction (Yang & Zhang, 2024). Specific bacterial families, such as *Lachnospiraceae* and its genera *Coprococcus* and *Roseburia,* are crucial for gut health and are reduced in schizophrenia. SCFA produced by gut microbes can cross the blood–brain barrier, affecting microglia and cytokine regulation, which are linked to schizophrenia. Environmental factors like diet, stress, and infections can also influence the gut microbiota and brain function, potentially accelerating brain aging in schizophrenia patients.

8.3.9 Parkinson's Disease

Parkinson's disease involves neuroinflammation and immune responses triggered by gut microorganisms, leading to misfolded proteins like α-synuclein, which contribute to disease progression (Gazmer et al., 2023). The GBA plays a key role in PD, with gut microbiota disruptions and dysbiosis increasing intestinal permeability, systemic inflammation, and neuroinflammation, all of which exacerbate neurodegeneration (Zheng et al., 2023). The enteric nervous system and gut α-synuclein are critical in PD development. Gastrointestinal dysfunction may appear early in PD, potentially before motor symptoms. Research suggests that a pathogen in the gut may initiate α-synuclein production, which then spreads to the CNS, highlighting the importance of understanding the microbiome–GBA interaction for personalized interventions.

8.3.10 Alzheimer's Disease

Alzheimer's disease is marked by cognitive decline and linked to changes in gut microbiota, which can influence brain inflammation and immune responses, potentially affecting disease progression (Iyen et al., 2019). Research suggests that gut dysbiosis contributes to neuroinflammation, amyloid-β accumulation, and tau hyperphosphorylation—key features of AD pathology (Deardorff, 2023). The gut microbiome also exacerbates systemic inflammation, worsening neuroinflammation in AD. Additionally, early gut dysbiosis and psychiatric symptoms are common in AD, indicating that changes in the gut environment may influence both psychiatric symptoms and neurodegeneration.

8.3.11 Celiac Disease

Celiac disease is an immune-mediated disorder caused by gluten intolerance, primarily affecting the gastrointestinal system. Consuming gluten triggers an immune response in the small intestine, leading to inflammation and malabsorption. The GBA plays a key role in both digestive and mental health, with inflammation in the small intestine potentially impacting brain function. Pro-inflammatory cytokines and immune system activation can affect the CNS, leading to symptoms like anxiety, depression, and cognitive difficulties (Andrew J. Wakefield, 2002). Individuals with celiac disease often show changes in gut microbiota, including a decrease in beneficial bacteria and an increase in pathogenic ones.

8.3.12 Depression

Depression is a hormone-dependent mental disorder marked by persistent sadness, loss of interest, and physical issues that affect thoughts, feelings, and behaviors. It is a global health concern, impacting over 300 million people worldwide, with higher prevalence in females (Alkinani & Phia, 2024). Dysbiosis in the gut microbiome may disrupt gut-to-brain communication, contributing to depressive symptoms. SCFA and neurotrophins play a role in this process (Ferraretto et al., 2023). Specific gut bacteria, like *Lactobacillus* and *Bifidobacterium*, are involved in the secretion of neurotransmitters linked to depression, suggesting that probiotics targeting these microbes could help manage depressive disorders.

8.3.13 Autism

Autism is a neurodevelopmental disorder affecting around 1% of the population, characterized by social communication difficulties, repetitive behaviors, and cognitive challenges (Blagonravova et al., 2023). People with autism often experience sensory sensitivities and eating issues. The GBA has been linked to autism, with gut microbiome dysbiosis and impaired gut barrier integrity potentially contributing to autistic behaviors (Ashique et al., 2024). Disruptions

TABLE 8.2
Clinical Trials on the Role of Microbiota in Treating Autoimmune Diseases

Name of Disease	Study Design with Patient's Details	Purpose of Clinical Trial	Adverse Effects/or Limitation of the Study	Outcomes	References
Autism spectrum disorder (ASD)	A pilot double-blind randomized placebo-controlled trial	To investigate the effect of *L. reuteri* (a product containing a combination of strains ATCC-PTA-6475 and DSM-17938) in children with ASD	The treatment does not alter overall autism severity, repetitive behaviors, microbiome composition, or immune profile	The trial showed that *L. reuteri* enhances social behavior in children with ASD and ATCC-PTA-6475 reverses the social deficits in a preclinical mouse model for ASD	Mazzone et al., 2024
Multiple sclerosis (MS)	A prospective study with 515 patients	To assess whether 25-hydroxyvitamin D3 levels are associated with MS risk	Intake of vitamin D several-fold higher than 2,000 U/d, which is considered to be the safe dose	515 potential MS cases were identified, among which 237 reported definite MS and 78 probable MS. Higher levels of 25-hydroxyvitamin D3 were found in the examined white population, resulting in a significant reduction in the risk of MS	Munger et al., 2006; Schepici et al., 2019
MS	A randomized study with 43 MS patients aged between 22 and 52 years	To evaluate how a semi-vegetarian diet, associated with vitamin D and other supplements can improve physical and inflammatory status in patients with Relapsing–Remitting (RR) MS and Progressive Primary (PR) MS	None	An increase in the ratio of eicosapentanoic acid + docosaexaenoic acid/arachidonic acid (n-3/n-6 polyunsaturated fatty acids) was observed following fish oil supplementation after three months of treatment	Riccio et al., 2016
MS	A randomized pilot study with 16 patients with MS and aged 18–60 years old	To demonstrate how intermittent fasting (IF) can improve the course of MS by causing less inflammation, demyelination and axonal damage. The effect of diet on the gut microbiota was also evaluated	Though experimental autoimmune encephalomyelitis animal model is approved by USFDA, however, the model does not perfectly mirror the human disease	IF increases intestinal bacterial diversity, especially in *Lactobacillaceae*, *Bacteroidaceae*, and *Prevotellaceae,* in both an experimental autoimmune encephalomyelitis (EAE) animal model and patients with RR-MS. In the EAE model, IF reduced WBC, inflammatory cytokines, and leptin, while increased cortisol and adiponectin levels. It also decreased IL-17-producing T cells and boosted regulatory T cells in the intestinal lamina propria, modulating systemic immune responses	Cignarella et al., 2018

(*Continued*)

TABLE 8.2 (CONTINUED)

Name of Disease	Study Design with Patient's Details	Purpose of Clinical Trial	Adverse Effects/or Limitation of the Study	Outcomes	References
Type 1 diabetes and celiac disease	Two-sample Mendelian randomization Study, a statistical approach that implies causal association from an exposure to an outcome	To reveal the causal association between gut microbiota and AID	The study did not analyze the two genders separately, so it might have influenced the results as most AID are more prevalent in women than in men	One causal bacterial genus, *Bifidobacterium,* was identified. The results of reverse Mendelian randomization analysis showed no evidence of reverse causality from type 1 diabetes and celiac disease to the *Bifidobacterium* genus	Xu et al., 2022
Type 1 diabetes	A randomised, double-blind placebo control trial with 23 children in the placebo group and 27 in the probiotic group	To investigate the mechanism of glycemic control in children with T1D followed by high-strength probiotics supplementation by assessing immune-regulatory markers	None	There was a significant increase in the percentage of induced T regulatory cells and median C-peptide level, also significant decrease in median glycated hemoglobin levels. The high-strength probiotics improved the immunoregulatory milieu, thus preserved the beta-cell function and better glycemic control	Lokesh et al., 2025
Parkinson's disease	A double-blind, placebo-controlled, randomized clinical trial with 35–75 years aged patients and with a mild-to-moderate symptom burden and dysbiosis of fecal microbiota	To assess the safety and symptomatic efficacy of colonic single-dose anaerobically prepared FMT	Gastrointestinal adverse events in FMT group were mentioned	As per the findings of the trial, FMT was safe; however, it did not offer clinically significant improvements	Scheperjans et al., 2024
Stress and depression	A cross-sectional study with 160 healthy Danish women aged 18-40 years	To investigate if gut microbiome and mental health connections observed in populations with a clinical diagnosis of healthy women experiencing stress and depressive symptoms	Small sample size	Results of the study indicated that the gut microbiota may impact neuroactive pathways, including Kynurenine pathway, and GABAergic, glutamatergic, and dopaminergic neurotransmission, even in healthy individuals, thus targeting gut microbiome could be a promising approach to treat psychiatric disorder	Bashir et al., 2024

(*Continued*)

TABLE 8.2 (CONTINUED)

Name of Disease	Study Design with Patient's Details	Purpose of Clinical Trial	Adverse Effects/or Limitation of the Study	Outcomes	References
Postpartum depression	Two-sample Mendelian randomization design	To investigate the causal relationship between four bacterial features in the gut microbiota genome-wide association study and postpartum depression	The investigation includes male and female participants so, it introduced bias though genetic variants located on the sex chromosomes were excluded from the analysis	The study stated that *Actinobacteria* and *Holdemanella* genera may have a potential beneficial impact in reducing the risk of postpartum depression	Zhang et al., 2024
Schizophrenia	A Mendelian randomization study	To evaluate the relation between gut microbiome and schizophrenia with 28 gut microbiome-associated genetic instrumental variants	-	The study suggested that genetically increased Lachnospiraceae, Lactobacillaceae, and Verrucomicrobiaceae reduced schizophrenia risk, thereby regulating disturbed intestinal microbiota, which might be a new therapeutic approach for schizophrenia patients	Q. Zeng et al., 2024
Alzheimer's disease	A two-sample Mendelian randomization study	To evaluate the causal impact of gut microbiota on AD	Limited gut microbiota data sources led to a potential selection bias	The study revealed potential causal effects of the gut microbiota species, including Order Selenomonadales, Family Pasteurellaceae, and Genus *Methanobrevibacter,* with elevated likelihood of developing AD, whereas Class Mollicutes, Genus *Ruminiclostridium 9*, Genus *Clostridium innocuum* group, and Genus *Eggerthella* exert a protective influence against the disease	Chen et al., 2024

in neuroinflammation and microglial activity suggest that altering gut composition and focusing on the GBA could offer new therapeutic approaches for autism spectrum disorders (ASD).

8.4 CLINICAL TRIALS ON THE ROLE OF MICROBIOTA IN TREATING AUTOIMMUNE DISEASES

Although recent researches are focused on gut microbiota and its role in human health, however, a number of case studies and clinical trials are still not satisfactory. To design and develop innovative therapeutic approaches, gut microbiota should be considered as an important biological parameter, and it is advisable to be investigated through randomized and unbiased clinical studies (Davoutis et al., 2024; L. Zeng et al., 2024). Some of the clinical trials on the role of microbiota in treating AID are enlisted in Table 8.2.

8.5 CONCLUSION AND FUTURE PROSPECTS

Dysbiosis, an imbalance in the microbiota, is linked to various diseases, characterized by a reduction in beneficial microbes and an increase in harmful bacteria. While microbiota-produced proteins and metabolites are essential for immune development, digestion, and energy balance, the host–microbiota interaction is complex. Recent advancements in meta-transcriptomics, metagenomics, and metabolomics suggest that microbial metabolites and genes offer deeper insights than microbial taxa alone. These developments focus on the entire gut ecosystem to better understand microbiota functions and their impact on human health. Additionally, non-bacterial communities in the gut also play key roles in shaping microbiota and influencing host physiology. Future research may lead to novel therapies and more personalized treatments.

REFERENCES

Alkinani, A. Y., & Phia, W. C. W. J. (2024). Investigating gut-brain interaction: Connection between gut microbes and neurological health. In *International journal of innovative science and research technology* (pp. 947–954). https://doi.org/10.38124/ijisrt/ijisrt24may1166.

Ashique, S., Mohanto, S., Ahmed, M. G., Mishra, N., Garg, A., Chellappan, D. K., Omara, T., Iqbal, S., & Kahwa, I. (2024). Gut-brain axis: A cutting-edge approach to target neurological disorders and potential synbiotic application. *Heliyon*, e34092–e34092. https://doi.org/10.1016/j.heliyon.2024.e34092.

Bakshi, I., Dey, S., Raut, A. J., Katta, S., & Sharma, P. (2024). Exploring the gut-brain axis: A comprehensive review of interactions between the gut microbiota and the central nervous system. *International Journal For Multidisciplinary Research.* https://doi.org/10.36948/ijfmr.2024.v06i03.19563.

Bashir, Z., Hugerth, L.W., Krog, M.C., Prast-Nielsen, S., Edfeldt, G., Boulund, F., Schacht, S.R., Tetens, I., Engstrand, L., Schuppe-Koistinen, I., & Fransson, E. (2024). Investigations of microbiota composition and neuroactive pathways in association with symptoms of stress and depression in a cohort of healthy women. *Frontiers in Cellular and Infection Microbiology*, *14*, 1324794.

Bessac, A., Cani, P. D., Meunier, E., Dietrich, G., Dietrich, G., Knauf, C., & Knauf, C. (2018). Inflammation and gut-brain axis during type 2 diabetes: Focus on the crosstalk between intestinal immune cells and enteric nervous system. *Frontiers in Neuroscience*, *12*, 725. https://doi.org/10.3389/FNINS.2018.00725.

Blagonravova, A. S., Г а л о в а, E. A., Shirokova, I. Y., & Galova, D. A. (2023). The gut-brain axis — clinical study results. *Experimental & Clinical Gastroenterology*, *6*, 5–13. https://doi.org/10.31146/1682-8658-ecg-214-6-5-13.

Chakrabarti, A., Geurts, L., Hoyles, L., Iozzo, P., Kraneveld, A.D., La Fata, G., Miani, M., Patterson, E., Pot, B., Shortt, C., & Vauzour, D. (2022). The microbiota–gut–brain axis: Pathways to better brain health. Perspectives on what we know, what we need to investigate and how to put knowledge into practice. *Cellular and Molecular Life Sciences*, *79*(2), 80. https://doi.org/10.1007/s00018-021-04060-w

Chen, A., Wang, Y., & Hu, Y.Q. (2024). Exploring causal relationships between gut microbiota and Alzheimer's disease: A bidirectional mendelian randomization study. *Journal of Alzheimer's Disease Reports*, *8*(1), 1031–1040.

Cignarella, F., Cantoni, C., Ghezzi, L., Salter, A., Dorsett, Y., Chen, L., Phillips, D., Weinstock, G.M., Fontana, L., Cross, A.H., & Zhou, Y. (2018). Intermittent fasting confers protection in CNS autoimmunity by altering the gut microbiota. *Cell Metabolism*, *27*(6), 1222–1235.

Davoutis, E., Gkiafi, Z., & Lykoudis, P.M. (2024). Bringing gut microbiota into the spotlight of clinical research and medical practice. *World Journal of Clinical Cases*, *12*(14), 2293.

Deardorff, D. K. (2023). The importance of gut-brain axis and use of probiotics as a treatment strategy for multiple sclerosis. *Multiple Sclerosis and Related Disorders*, *71*, 104547. https://doi.org/10.1016/j.msard.2023.104547.

Dhanawat, M., Malik, G., Wilson, K., Gupta, S., Gupta, N., & Sardana, S. (2024). The gut microbiota-brain axis: A new frontier in Alzheimer's disease pathology. *Cns & Neurological Disorders-Drug Targets*, *23*. https://doi.org/10.2174/0118715273302508240613114103.

Drewes, A. M., Brock, C., & Drewes, A. M. (2019). *Autonomic visceral neuropathy and gastrointestinal disorders* (pp. 851–861). https://doi.org/10.1007/978-3-030-11815-0_54.

Dunalska, A., Saramak, K., & Szejko, N. (2023). The role of gut microbiome in the pathogenesis of multiple sclerosis and related disorders. *Cells*, *12*(13), 1760. https://doi.org/10.3390/cells12131760.

Duranti, S., Ruiz, L., Lugli, G.A., Tames, H., Milani, C., Mancabelli, L., Mancino, W., Longhi, G., Carnevali, L., Sgoifo, A., & Margolles, A. (2020). Bifidobacterium adolescentis as a key member of the human gut microbiota in the production of GABA. *Scientific Reports*, *10*(1), 14112. https://doi.org/10.1038/s41598-020-70986-z

Eizirik, D. L., Pasquali, L., & Cnop, M. (2020). Pancreatic β-cells in type 1 and type 2 diabetes mellitus: Different pathways to failure. *Nature Reviews Endocrinology*, *16*(7), 349–362. https://doi.org/10.1038/s41574-020-0355-7.

Ferraretto, A., Donetti, E., García-Mena, J., & Pacheco-López, G. (2023). Editorial: The gut-skin-brain axis in human health and disease. *Frontiers in Nutrition*, *10*. https://doi.org/10.3389/fnut.2023.1155614.

Gazmer, A., Chakrabarty, M., Chutia, D., Bhattacharjee, A., & Bhuyan, N. R. (2023). Antiparkinsonian effects of Indian spice amomum subulatum fruit extract by modulating the behavioural, inflammatory markers, antioxidant, and histopathology parameters in Rats. *Brain Behavior and Immunity Integrative*, 100015. https://doi.org/https://doi.org/10.1016/j.bbii.2023.100015.

Ge, L., Liu, S., Li, S., Yang, J., Hu, G., Xu, C., & Song, W. (2022). Psychological stress in inflammatory bowel disease: Psychoneuroimmunological insights into bidirectional gut–brain communications. *Frontiers in Immunology*, *13*(October), 1–20. https://doi.org/10.3389/fimmu.2022.1016578.

Iyen, I. S., Ogofotha, Oghenekeno, G., & Oko, J. O. (2019). *Advances in Nutraceuticals*, *5*(2), 8–22. https://www.arcjournals.org/pdfs/ijmpnp/v5-i2/2.pdf.

Kasarello, K., Cudnoch-Jedrzejewska, A., & Czarzasta, K. (2023). Communication of gut microbiota and brain via immune and neuroendocrine signaling. *Frontiers in Microbiology*, *14*, 1118529. https://doi.org/10.3389/fmicb.2023.1118529.

Khawar, M. M., Ijaz, S., Goyal, P., Kandambige, D., Sharifa, M., Maslamani, A. N. J., Kutabi, S. Al, Saleh, I., Albshir, M. M., Almadhoun, M. K. I. K. H., Soomro, S. N., & Kumari, N. (2023). The gut-brain axis in autoimmune diseases: Emerging insights and therapeutic implications. *Cureus*, *15*. https://doi.org/10.7759/cureus.48655.

Levin, E. R., & Malohat, O. (2023). *The brain-gut axis and psychogastroenterology* (pp. 281–289). https://doi.org/10.1016/b978-0-323-91497-0.00097-7.

Liu, P., Li, Y., Wang, R., Ren, F., & Wang, X. (2022). Oxidative stress and antioxidant nanotherapeutic approaches for inflammatory bowel disease. In *Biomedicines* (Vol. 10, Issue 1). MDPI. https://doi.org/10.3390/biomedicines10010085.

Lokesh, M. N., Kumar, R., Jacob, N., Sachdeva, N., Rawat, A., Yadav, J., & Dayal, D. (2025). Supplementation of high-strength oral probiotics improves immune regulation and preserves beta cells among children with new-onset type 1 diabetes mellitus: A randomised, double-blind placebo control trial. *Indian Journal of Pediatrics*, *92*(3), 277–283

Longo, S., Rizzi, S., & Federici, M. (2023). Microbiota-gut-brain axis: relationships among the vagus nerve, gut microbiota, obesity, and diabetes. *Acta Diabetologica*, *60*(8), 1007–1017. https://doi.org/10.1007/s00592-023-02088-x.

Maini Rekdal, V., Nol Bernadino, P., Luescher, M. U., Kiamehr, S., Le, C., Bisanz, J. E., Turnbaugh, P. J., Bess, E. N., & Balskus, E. P. (2020). A widely distributed metalloenzyme class enables gut microbial metabolism of host-and diet-derived catechols. *Elife*, *9*, e50845. https://doi.org/10.7554/eLife.50845.

Mayer, E. A., & Tillisch, K. (2011). The brain-gut axis in abdominal pain syndromes. *Annual Review of Medicine*, *62*(1), 381–396. https://doi.org/10.1146/ANNUREV-MED-012309-103958.

Mazzone, L., Dooling, S.W., Volpe, E., Uljarević, M., Waters, J.L., Sabatini, A., Arturi, L., Abate, R., Riccioni, A., Siracusano, M., & Pereira, M. (2024). Precision microbial intervention improves social behavior but not autism severity: A pilot double-blind randomized placebo-controlled trial. *Cell Host & Microbe*, *32*(1), 106–116.

Munger, K. L., Levin, L. I., Hollis, B. W., Howard, N. S., & Ascherio, A. (2006). Serum 25-hydroxyvitamin D levels and risk of multiple sclerosis. *Jama*, *296*(23), 2832–2838.

Pabian-Jewuła, S., & Rylski, M. (2023). Does the functional polymorphism-1562C/T of MMP-9 gene influence brain disorders? *Frontiers in Cellular Neuroscience*, *17*. https://doi.org/10.3389/fncel.2023.1110967.

Peppas, S., Pansieri, C., Piovani, D., Danese, S., Peyrin-Biroulet, L., Tsantes, A. G., Brunetta, E., Tsantes, A. E., & Bonovas, S. (2021). The brain-gut axis: Psychological functioning and inflammatory bowel diseases. *Journal of Clinical Medicine*, *10*(3), 377. https://doi.org/10.3390/JCM1003037.

Piva, F., Gervois, P., Karrout, Y., Sané, F., & Romond, M. (2023). Gut–joint axis: Impact of bifidobacterial cell wall lipoproteins on arthritis development. *Nutrients*, *15*(23), 4861. https://doi.org/10.3390/nu15234861.

Pu, Y., He, Y., Zhao, X. F., Zhang, Q., Wen, J.-R., Hashimoto, K., & Liu, Y. (2022). Depression-like behaviors in mouse model of Sjögren's syndrome: A role of gut–microbiota–brain axis. *Pharmacology, Biochemistry and Behavior*, *219*, 173448. https://doi.org/10.1016/j.pbb.2022.173448.

Riccio, P., Rossano, R., Larocca, M., Trotta, V., Mennella, I., Vitaglione, P., Ettorre, M., Graverini, A., De Santis, A., Di Monte, E., & Coniglio, M. G. (2016). Anti-inflammatory nutritional intervention in patients with relapsing-remitting and primary-progressive multiple sclerosis: A pilot study. *Experimental Biology and Medicine*, *241*(6), 620-635.

Richards, P., Thornberry, N. A., & Pinto, S. (2021). The gut-brain axis: Identifying new therapeutic approaches for type 2 diabetes, obesity, and related disorders. *Molecular Metabolism*, *46*, 101175. https://doi.org/10.1016/J.MOLMET.2021.101175.

Romero-Figueroa, M. del S., Ramírez-Durán, N., Montiel-Jarquín, Á. J., & Horta-Baas, G. (2023). Gut-joint axis: Gut dysbiosis can contribute to the onset of rheumatoid arthritis via multiple pathways. *Frontiers in Cellular and Infection Microbiology*, *13*. https://doi.org/10.3389/fcimb.2023.1092118.

Sai, A., Shetty, G. B., Shetty, P., & Nanjeshgowda, H. L. (2024). Influence of gut microbiota on autoimmunity: A narrative review. *Brain Behavior and Immunity Integrative*, *5*, 100046.

Sandhu, K. V., Sherwin, E., Schellekens, H., Stanton, C., Dinan, T. G., & Cryan, J. F. (2017). Feeding the microbiota-gut-brain axis: Diet, microbiome, and neuropsychiatry. *Translational Research*, *179*, 223-244. https://doi.org/10.1016/j.trsl.2016.10.002

Scheperjans, F., Levo, R., Bosch, B., Lääperi, M., Pereira, P. A. B., Smolander, O. P., Aho, V. T. E., Vetkas, N., Toivio, L., Kainulainen, V., Fedorova, T. D., Lahtinen, P., Ortiz, R., Kaasinen, V., Satokari, R., & Arkkila, P. (2024). Fecal microbiota transplantation for treatment of Parkinson disease: A randomized clinical trial. *JAMA Neurology*, *81*(9), 925–938.

Schepici, G., Silvestro, S., Bramanti, P., & Mazzon, E. (2019). The gut microbiota in multiple sclerosis: An overview of clinical trials. *Cell Transplantation*, *28*(12), 1507–1527.

Tsianos, E. V., & Moutsopoulos, H. M. (1989). Sjögren's syndrome and the gut. *Baillière's Clinical Rheumatology*, *3*(2), 357–370. https://doi.org/10.1016/S0950-3579(89)80026-3.

Wakefield, A. J. (2002). The gut–brain axis in childhood developmental disorders. *Journal of Pediatric Gastroenterology and Nutrition*, *34*(S1). https://doi.org/10.1002/j.1536-4801.2002.tb07699.x.

Wang, X., Li, Y., Li, Y., Wu, L., Xiao, S., Ji, Y., Tan, Y., Jiang, C., & Zhang, G. (2021). Dysregulation of the gut-brain-skin axis and key overlapping inflammatory and immune mechanisms of psoriasis and depression. *Biomedicine & Pharmacotherapy*, *137*, 111065. https://doi.org/10.1016/J.BIOPHA.2020.111065.

Xu, H., Liu, M., Cao, J., Li, X., Fan, D., Xia, Y., Lu, X., Li, J., Ju, D., & Zhao, H. (2019). The dynamic interplay between the gut microbiota and autoimmune diseases. *Journal of Immunology Research*, *2019*(1), 7546047.

Xu, Q., Ni, J. J., Han, B. X., Yan, S. S., Wei, X. T., Feng, G. J., Zhang, H., Zhang, L., Li, B., & Pei, Y. F. (2022). Causal relationship between gut microbiota and autoimmune diseases: A two-sample Mendelian randomization study. *Frontiers in Immunology*, *12*, 746998.

Xu, Y. (2024). *Gut-to-brain axis provides novel therapeutic targets for depression*. https://doi.org/10.54254/2753-8818/29/20240754.

Yano, J. M., Yu, K., Donaldson, G. P., Shastri, G. G., Ann, P., Ma, L., Nagler, C. R., Ismagilov, R. F., Mazmanian, S. K., & Hsiao, E. Y. (2015). Indigenous bacteria from the gut microbiota regulate host serotonin biosynthesis. *Cell*, *161*(2), 264–276. https://doi.org/10.1016/j.cell.2015.02.047.

Yang, H., & Zhang, S. (2024). Meta-analysis of the gut-brain axis and mental disorder. *Theoretical and Natural Science*, *44*(1), 178–186. https://doi.org/10.54254/2753-8818/44/20240773.

Yu, X., Zhou, J., Zhao, F., Liu, X., Mao, Y., Diao, L., Wen, C., & Liu, M. (2021). Tomatidine suppresses the destructive behaviors of fibroblast-like synoviocytes and ameliorates type II collagen-induced arthritis in rats. *Frontiers in Pharmacology*, *12*, 670707. https://doi.org/10.3389/FPHAR.2021.670707.

Zeng, L., Yang, K., He, Q., Zhu, X., Long, Z., Wu, Y., Chen, J., Li, Y., Zeng, J., Cui, G., & Xiang, W. (2024). Efficacy and safety of gut microbiota-based therapies in autoimmune and rheumatic diseases: A systematic review and meta-analysis of 80 randomized controlled trials. *BMC Medicine*, *22*(1), 110.

Zeng, Q., Zhang, M., & Wang, R. (2024). Causal link between gut microbiome and schizophrenia: A Mendelian randomization study. *Psychiatric Genetics*, *34*(2), 43–53.

Zhang, J., Wei, L., Tan, H., & Pang, W. (2024). Gut microbiota and postpartum depression: A Mendelian randomization study. *Frontiers in Psychiatry*, *15*, 1282742.

Zhang, W., Ye, Y., Song, J., Sang, T., Xia, T., Xie, L., Qiu, X., Zeng, Q., & Luo, X. (2023). Research progress of microbiota-gut-brain axis in Parkinson's disease. *Journal of Integrative Neuroscience*. https://doi.org/10.31083/j.jin2206157.

Zhao, L., Xiong, Q., Stary, C.M., Mahgoub, O.K., Ye, Y., Gu, L., Xiong, X., & Zhu, S. (2018). Bidirectional gut-brain-microbiota axis as a potential link between inflammatory bowel disease and ischemic stroke. *Journal of Neuroinflammation*, *15*, 1–11.

Zheng, J., Wittouck, S., Salvetti, E., Franz, C. M. A. P., Harris, H. M. B., Mattarelli, P., O'Toole, P. W., ... Lebeer, S. (2020). A taxonomic note on the genus Lactobacillus: Description of 23 novel genera, emended description of the genus Lactobacillus Beijerinck 1901, and union of Lactobacillaceae and Leuconostocaceae. *International Journal of Systematic and Evolutionary Microbiology*, *70*(4), 2782–2858. https://doi.org/10.1099/ijsem.0.004107

Zheng, Y., Bonfili, L., Wei, T., & Eleuteri, A. M. (2023). Understanding the gut–brain axis and its therapeutic implications for neurodegenerative disorders. *Nutrients*, *15*. https://doi.org/10.3390/nu15214631.

Zhu, F., Wang, W., Ma, Q., Yang, Z., Fan, Y., Ju, Y., Guo, R., Wang, Q., Mu, X., Mu, X., Zhao, B., Gao, Y., He, X., Gao, F., Qian, L., Chen, C., Chen, Y., Gao, C., & Ma, X. (2020). Role of short-chain fatty acids in the gut-brain axis in schizophrenia: Contribution to immune activation and pathophysiology in humans and mice. In *bioRxiv*. https://doi.org/10.1101/2020.04.11.021915.

Zhu, Y., Zhao, Q., Huang, Q., Li, Y., Yu, J. J., Zhang, R., Liu, J., Yan, P., Xia, J., Guo, L., Liu, G., Yang, X., & Zeng, J. (2022). Nuciferine regulates immune function and gut microbiota in DSS-induced ulcerative colitis. *Frontiers in Veterinary Science*, *9*. https://doi.org/10.3389/fvets.2022.939377.

9 Signaling Events from the Gut *Regulate Brain Function and Its Link to Autoimmune Disorders*

Gagandeep Kaur[1], *Parul Sood*[1], *Shabnam*[2], *Rupesh K. Gautam*[3*], *and Nitin Jangra*[1]

[1] Chitkara University School of Pharmacy, Chitkara University, Himachal Pradesh, India
[2] College of Pharmacy, Guru Nanak Institutions, Mullana, Haryana, India
[3] Department of Pharmacology, Indore Institute of Pharmacy, Indore, India

9.1 INTRODUCTION

The human body is a perfect example where it is owing trillions of microorganisms which are believed to have influence over and regulate the host's physiology. The gastrointestinal tract of humans is home to the majority of bacteria, which are commonly referred to as gut microbiota (GM). Various factors exert influence over both composition and inhabitants of the gut microbiota (Chandel et al., 2024). Therefore, it can vary based on the attributes of the host. The four main phyla of microorganisms—Proteobacteria, Bacteroidetes, Firmicutes, and Actinobacteria—along with two additional phyla—Fusobacteria and Verrucomicrobia—comprise the whole gut microbiota (Eckburg et al., 2005). The commensal bacteria participate in intercellular communication, both within their own community and with the epithelial cells present in the gut of the host. This communication is essential for maintaining gut homeostasis and strengthening the host's immune response. GM provides several advantageous impacts on the host in a healthy condition. However, in the presence of sickness or disturbance, it has been documented to contribute to the advancement of various conditions, such as neurological disorders (Gubert et al., *2020*). Increasing data indicates a robust bidirectional communication process between the gut and the brain, crucial to preserve gut balance and brain function (Sharma et al., 2021; Post et al., 2023). Moreover, changes in the makeup of gut bacteria and disturbances in the balance of the gut environment, often associated with dietary habits, antibiotic use, and bacterial and viral infections, have been recognized as variables that contribute to onset of gut–brain disorders (Post et al., 2023). However, there is still considerable dispute over the exact chemical mechanism that is involved in the fluctuation of the two-way connection between the gut and the brain. Early enteric nervous system development in gut relies heavily on commensal gut bacteria (Cryan et al., 2012). Similarly, McVey et al. found in GF mice that gut microbiota restores normal extrinsic as well as intrinsic neuronal function and facilitates communication between intestine and CNS. Such findings emphasize the important role of the GM in early postnatal development of the CNS. This chapter offers a comprehensive review of the available data that reveals the role of GM with the cellular part as well as metabolites in the development of different neurological disorders. Furthermore, many therapy strategies employed to improve these illnesses are emphasized (O'Mahony et al., 2020).

9.2 OVERVIEW OF GUT–BRAIN AXIS

Understanding the interaction between gut and brain has uncovered an intricate communicating arrangement that not only maintains the balance of the gastrointestinal system, but also has various impacts on emotions, motivation, and advanced cognitive abilities. The intricate nature of these connections is encompassed by the term gut–brain axis (GBA). The gut–brain axis is an intricate system that involves numerous endocrine components that play a role in a wide range of physiological functions, including digestion, osmoregulation, and waste disposal. Due to the significant amount of hormones released from this axis and their intricate interactions inside and beyond it, it is challenging to attribute a precise and individual function to each hormone. The term "microbiota–gut–brain axis refers the biochemical communication GM role in biological as well as chemical signaling among gut and brain through GI tract (Mayer et al., 2022). The intestine-CNS axis encompasses multiple components, such as CNS, neuroendocrine and immune system, ANS branches (sympathetic, parasympathetic), hypothalamic-pituitary-adrenal axis (HPA axis), vagus nerve, as well as the GM (Martin et al., 2017).

DOI: 10.1201/9781003536802-9

9.3 MECHANISMS OF GUT–BRAIN AXIS COMMUNICATION

The gastrointestinal tract and central nervous system connect via an intricate network that encompasses multiple channels. Examples of pathways include neural pathways, hormonal pathways, and immune pathways (Keightley et al., 2015) as shown in Figure 9.1 Comprehending the mechanisms that govern the connections between the stomach and the brain is crucial for elucidating the underlying causes of different illnesses and creating specific treatment strategies that try to restore balance within this intricate system.

9.3.1 Neural Pathways

The gut–brain axis relies on neural pathways, particularly the vagus nerve, to facilitate bidirectional communication between the gut and the brain. Recent study has demonstrated that it has a vital function in regulating peripheral regulatory T cells (pTregs) in the gastrointestinal tract. The regulation of this process is governed by the vago-vagal reflex, which is facilitated by signals originating from the gut microbiota. Understanding the gut–brain axis mechanism may facilitate to develop neurostimulatory medications that especially focus on the complex gut–brain connection (Kanai et al., 2022). This bidirectional pathway carries sensory information throughout the gastrointestinal canal to the nucleus tractus solitarius (NTS) in the the brainstem, where it is consolidated and assessed. The NTS is a crucial center in the complex neuronal networks that control eating behavior and the management of energy balance. Researchers have used modern methods like optogenetics and brain activity monitoring to discover specific groups of NTS neurons, specifically those that release prodynorphin. These neurons have a significant impact on suppressing appetite and increasing energy expenditure (Chen et al., 2020). Afterward, efferent signals are transmitted back to the stomach, affecting several gastrointestinal activities such as movement, secretion, and perception of internal

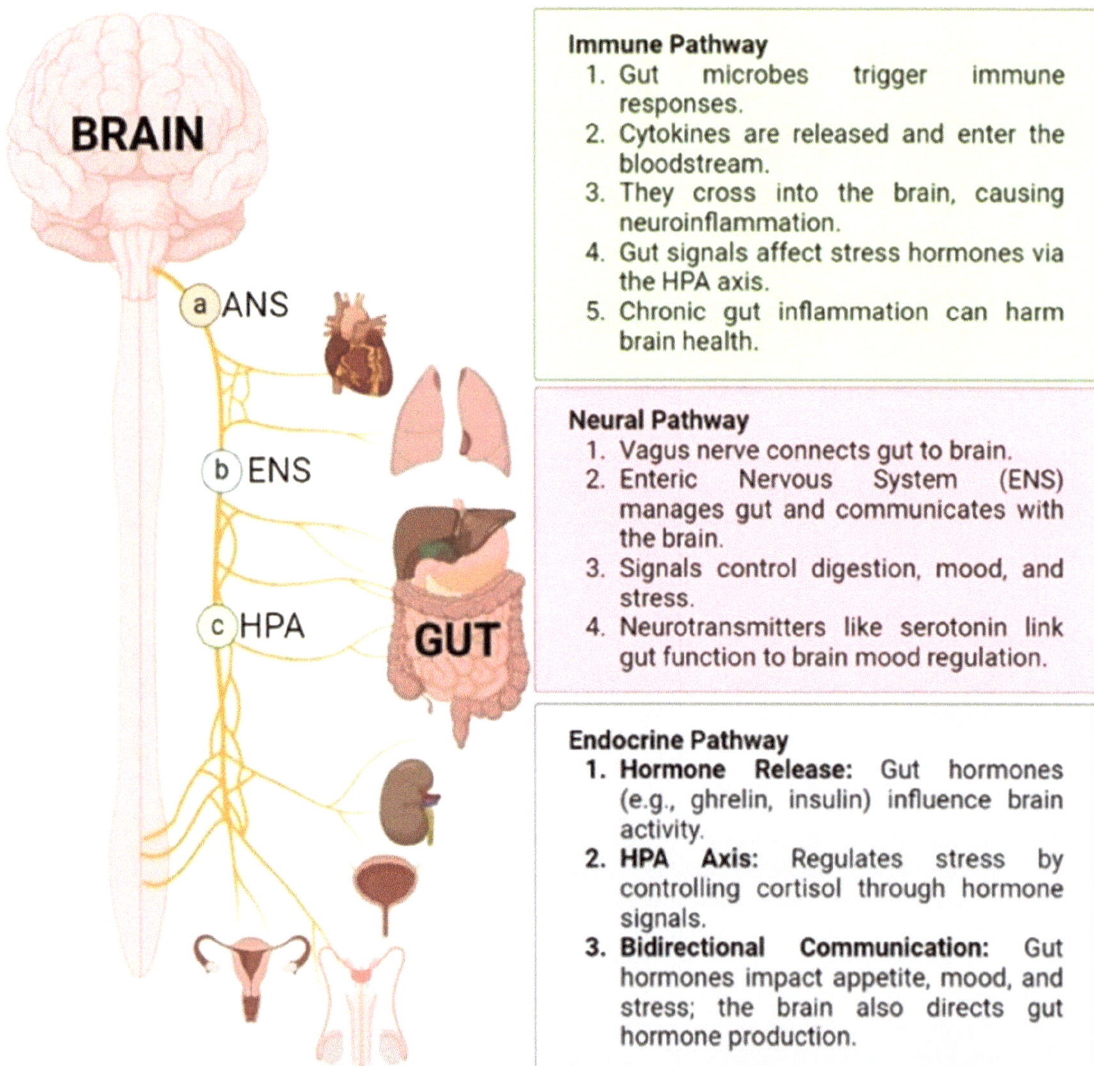

FIGURE 9.1 Mechanism of gut–brain axis communication.

organs. The neuronal communication network mentioned here has a crucial role in controlling hunger, fullness, and digestive processes (Ahima et al., 2008). This complex procedure guarantees the maintenance of metabolic balance and the effective breakdown of food (Wachsmuth et al., 2022).

9.3.2 Hormonal Pathways

The preservation of balance in the gut and the brain axis has been displayed to have diverse effects on both physical health and mental health. Enteroendocrine cells, found throughout the gastrointestinal system, release gut hormones that serve as crucial signaling molecules in this axis. The association between GM and hormones is generally acknowledged in the communication between the gastrointestinal tract and the brain. The microbiota has a vital role in regulating a range of diseases related to the gut–brain axis, including gastrointestinal issues and psychiatric disorders. Furthermore, gut hormones play various crucial roles in maintaining health and act as vital signals in this axis. GM significantly impacts the secretion and functions of these gastrointestinal hormones (Sun et al., 2020).

9.4 IMPACT OF GUT SIGNALING ON BRAIN FUNCTION

The presence of age-related irregularities in the gut microbiota has been coupled to cognitive decline, depression, and anxiety. However, the specific processes behind this association have not been investigated. Our work showed that transferring the gut microbiota from older mice to younger mice caused inflammation in both stomach and brain, leading to cognitive deterioration, depression, and anxiety. We discovered a decrease in the production of mucin and an increase in gut permeability, commonly stated to as "leaky gut," due to a loss in the presence of beneficial metabolites such as butyrate. This decline is attributed to a decrease in the population of butyrate-producing bacteria in the gut microbiota of older individuals. Consequently, the expression of butyrate receptors, specifically FFAR2 and FFAR3, were suppressed. Administering butyrate efficiently decreased inflammation, restored mucin expression and intestinal hurdles, and corrected brain dysfunction. Additionally, juvenile mice who had particularly experienced the loss of FFAR2/FFAR3 in their intestines exhibited comparable issues in both their gastrointestinal tract and brain, comparable to those observed in older animals. The findings of our study indicate that a decrease in the population of bacteria that produce butyrate in the GM of older individuals leads to lower levels of butyrate and reduced signaling through the FFAR2/FFAR3 receptors. This, in turn, results in reduced mucin production contributing to increase gut permeability, inflammation, and brain abnormalities. The results emphasize the importance of activating butyrate-FFAR2/FFAR3 receptors as a viable approach to reduce the harmful effects on gut and brain health in older persons caused by an imbalanced gut microbiota due to aging. The health of the stomach has a direct impact on CNS (Cong et al., 2015).

9.5 LINK BETWEEN GUT HEALTH AND AUTOIMMUNE DISORDERS

Autoimmune disorders are characterized by the overproduction of autoantibodies. Immune system is influenced by genetic and environmental factors, which result in the development of B cells that help to produce autoantibodies and T cells that react to self-components. Additionally, there is an abnormal generation of pro-inflammatory cytokines. The hypothesis suggests that the rise in autoimmune illnesses can be attributed to significant changes in the gut microbiota resulting from various factors, such as dietary alterations and the widespread usage of antibiotics. Autoimmune illnesses are influenced by a combination of genetic and environmental factors. AhR has the ability to contribute to autoimmune diseases by attaching itself to various substances that come from the diet, cells, and microbes. Subsequently, it converts both internal and external impulses into cellular responses (Kawajiri & Fujii-Kuriyama, 2017]. Qiu et al. discovered that rats lacking AhR had reduced levels of innate IL-22, resulting in an elevation of commensal SFB known as immune activator and subsequent rise in Th17 cells. This research highlighted AhR expression, which plays a protective effect in T-cell-induced experimental colitis attributed to its ability to reduce harmful Th17 cells (Qiu et al., 2013). Furthermore, mice exposed to 2,3,7,8-tetrachloro dibenzo-p-dioxin (TCDD), a ligand for the aryl hydrocarbon receptor (AhR), exhibited alterations in quantities of *Bacteroides fragilis* (an immunological suppressor) and SFB and in their gut microbiota. In addition, the existence of SFB in the gut microbiome had a notable impact on the host's reaction caused by TCDD, indicating the potential for therapeutic interactions between AhR ligands and important commensal bacteria (Stedtfeld et al., 2017). Dysbiosis, an abnormality in the gut microbiota, has been recognized as a potential contributing factor to autoimmune disorders. The gut microbiota contributes to autoimmunity by altering autoantigens at the molecular level and inducing cross-reactivity with autoantigens. When gut bacteria that are alive cross through a gut barrier that is not working properly at the level of cells, they directly interact with immune and tissue cells, which leads to the development of systemic autoimmunity (Dehner et al., 2019). The correlation between an alteration in GM and autoimmune illnesses may arise from diverse pathways that could influence the functioning and response of the human immune system. Through the modulation of the host immune response and the activation of antigen-presenting cells (APCs), such as dendritic cells (DCs), it is possible to induce antigen presentation and the synthesis of cytokines. As a result, this impacts the development and functioning of T cells. In addition, this impact disturbs the balance of Th17 and Treg cells responsible for maintaining body stability. Antigenic mimicry can lead to the resemblance of foreign antigens to self-antigens, hence initiating the proliferation of autoreactive T- and B-lymphocytes that are produced in response to infections. This stimulation has the potential to contribute to the development of autoimmunity. Nevertheless, the permeability of the intestinal mucosa is modified due to the regulation of expression of tight junction proteins. There is ample data indicating that the GM may play significant role in initiating and exacerbating disease progression in individuals with autoimmune disorders. Possible processes encompass molecular mimicry, changes in the permeability of the intestinal mucosa, the immunological response elicited by the microbiota, and antigenic mimicry. GM has the ability to impact the immune system's ability to distinguish between the body's own cells and foreign substances, which could potentially contribute to the development of autoimmune diseases. People with autoimmune illnesses often show signs of weakened intestinal barriers, which can expose their immune system to helpful gut bacteria. Moreover, when the mucosal immune system is unable to tolerate the presence of GM, it leads to aberrant and pathogenic immunological reactions, hence exacerbating the severity of the disease [Routy et al., 2018].

9.6 GUT–BRAIN AXIS AND SPECIFIC AUTOIMMUNE DISEASES

9.6.1 Type 1 Diabetes

Type 1 diabetes (T1D) is a condition where the immune system attacks the body's own cells and is associated with an imbalance in the microbial community. The development of T1D in both preclinical and clinical stages is mostly linked to GIT pathogenesis as shown in Figure 9.2. This includes conditions like celiac disease or an increase in intestinal leakage, which may be caused by microbial dysbiosis [Theyab et al., 2020].

A study conducted on mice found that Cytophaga–Flavobacter–Bacteroidetes (CFB) has an impact on the proportion of Th17 and Treg cells, which is essential for maintaining immunological homeostasis. This imbalance can result in the development of autoimmune disorders. A further study observed a rise in Bacteriodetes dorei levels among persons who have preclinical or clinical type 1 diabetes. *Bacteroides dorei* releases lipopolysaccharide (LPS) which stimulates the innate immune response. Consequently, this impedes the progress of autoimmunity. A study found that patients with T1D exhibit an increase in Proteobacteria, Actinobacteria, and Bacteroidetes, while lacking bacteria that produce butyrate (Brown et al., 2011). The observation that butyrate stimulates the production of mucin and the development of tight junctions provides an explanation for the heightened intestinal permeability observed in people with T1D. The microbiota in individuals with T1D is lacking in *Provetella* species, which are responsible for breaking down mucin. On the other hand, it is enriched with bacteria that create .propionates and acetates, which are known to negatively affect the activities of neutrophils, as shown in T1D patients (Huang et al., 2016). Additional data indicates that an imbalance in microbial composition disrupts the immune system in the gut, causing an overactive inflammatory response that can lead to TID, even in the absence of any noticeable changes in gut permeability. Another study has shown that dendritic cells (DCs) are incapable of promoting the development of FoxP3+Treg cells in patients with T1D. Consequently, when Treg cells are deficient in the gut, the immune system's ability to tolerate and distinguish self-antigens in the pancreatic beta-cells is diminished (Badami et al., 2011; Kahalehili et al., 2021). AhR is a transcriptional component that impedes the advancement of T1D. Curiously, the activation was predominantly confined to the small intestine. However, there were no changes identified in the differentiation of T cells in the spleen or peripheral lymph nodes (PLNs) (Kahalehili et al., 2021). A study revealed that rats exhibiting prediabetes exhibit a higher abundance of *Bifidobacterium*, *Lactobacillus*, and *Bacteroides*. Additional studies indicate that *Lactobacillus bifidus* has the ability to trigger the autoimmune response in mice that lack the IL-1 receptor antagonist (Graves et al., 2017). Mice lacking the MyD88 molecule, involved in the signaling pathways of Toll-like receptors (TLRs), are shielded against T1D when exposed to a microbiome signature marked by a substantial presence of *Lactobacillus* and a low Firmicutes to Bacteroidetes ratio. *Lactobacillus* exhibits a defensive impact against T1D, whereas *Bacillus cereus* postpones the initiation of T1D in NOD mice (Zheng et al., 2022).

Individuals diagnosed with T1D also exhibit a deficiency in bacteria that produce butyrate and bacteria that break down mucin (84). Preclinical T1D is linked to a decrease in the prevalence of bacteria that produce butyrate, such as *Roseburia* (Al Theyab et al., 2020). Gut microbes increase the generation of butyrate by breaking down mucin when the intake of dietary fibers is limited. The degradation of mucin in the stomach leads to an elevation in both the permeability and immunogenicity, potentially leading to the onset of T1D. Butyrate can potentially impact various processes, including the upregulation of tight junction proteins which in turn affects the susceptibility to T1D. Furthermore, it triggers programmed cell death in pro-inflammatory T-lymphocytes in a mouse cell line (Li et al., 2010). Butyrate promotes Th1 cell development by promoting the synthesis of IFN-gamma and T-bet through the inhibition of histone deacetylase, as well as by suppressing the expression of Rora, Rorgt, and IL-17 (Chen et al., 2019). The literature contains contradictory results about the effects of orally administering butyrate on the advancement of T1D

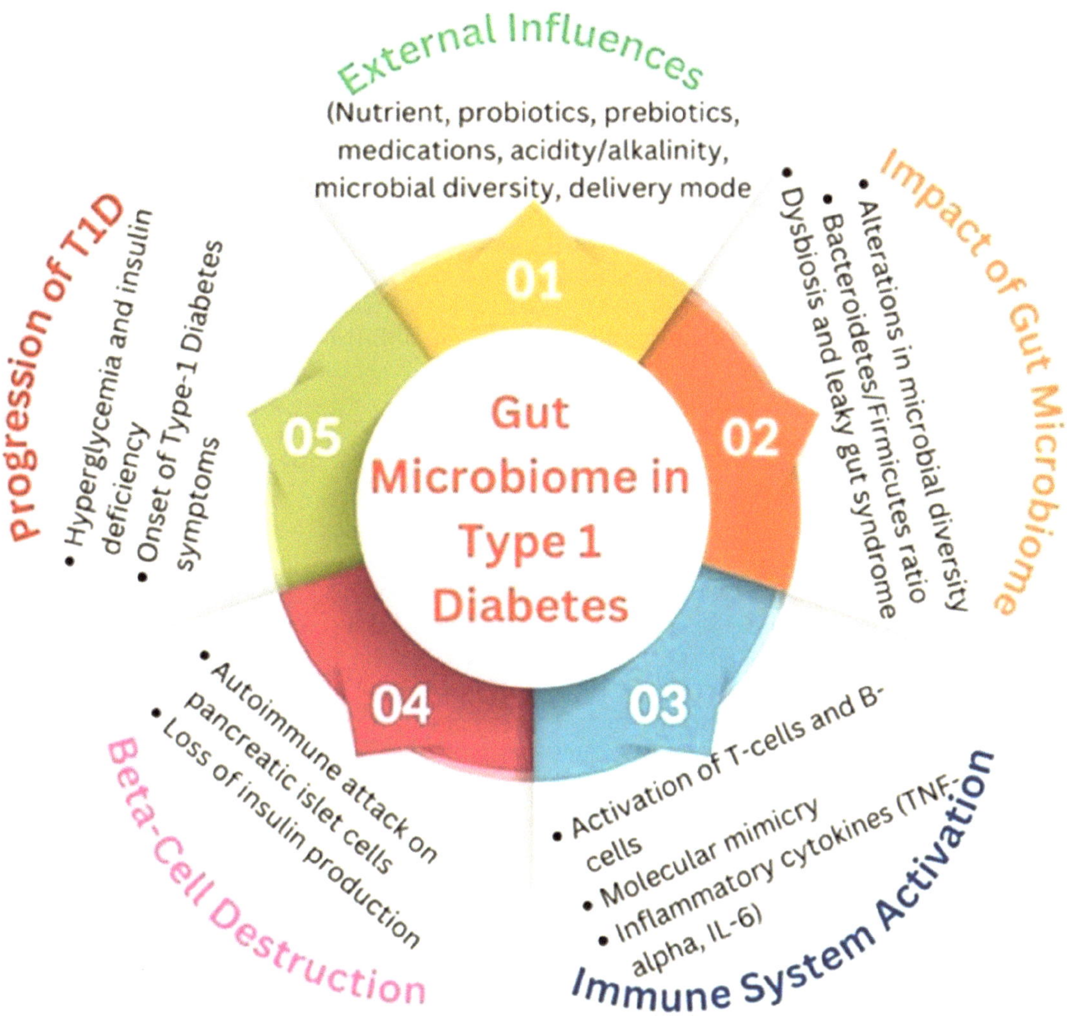

FIGURE 9.2 Role of gut microbiome in type 1 diabetes.

in humans as opposed to experimental animals (de Groot et al., 2020).

Various pieces of evidence suggest that viral infections, including rotavirus, cytomegalovirus (CMV), and enterovirus, have the potential to trigger or exacerbate the development of T1D. Nevertheless, the precise mechanisms behind this phenomenon remain incompletely understood (Vehik et al., 2019). Coxsackievirus B4 (CVB4) induces T1D by imitating host-specific enzymes, specifically the beta-cell glutamic acid decarboxylase enzyme. Interferon-gamma greatly decreased the occurrence of type 1 diabetes produced by CVB4 by affecting the response of pancreatic B cells. Another concept suggests that the virome triggers inflammation in the pancreas, especially hindering the ability of CD8+ cytotoxic T-lymphocytes to identify self-antigens of the pancreatic beta cells. Several studies have suggested a potential association between certain viral infections, and a lower occurrence of T1D. The NOD mouse model demonstrates that infection with mouse Norovirus-4 (MNV4) leads to an increase in the variety of gut bacteria and an expansion of Treg cells (Hassan & Baldridge, 2019).

To investigate whether the impact of MNV4 on immunity is a result of indirect manipulation of the gut microbiota, it is noteworthy that germ-free mice infected with MNV4 exhibit changes in their immune response. These changes include an elevation in Treg cells and modifications in different elements of the immune system, which include B cells, T cells, macrophages, as well as cytokine biomarkers. The presence of an imbalanced virome is linked to several autoimmune illnesses, such as T1D (Santiago-Rodriguez & Hollister, 2019). Most gut viruses come from lysogens, which are bacteria that have dormant phages (prophages) incorporated into their genomes. When certain signals, such a change in food, are received, these phages are released from the lysogens and then affect the composition of the microbiome and its susceptibility to autoimmune illnesses (Oh et al., 2019). Prophages may include immunomodulatory chemicals that have a direct impact on the immune system's reaction. For instance, the recently identified prophage FHKU.vir has a toxin that possesses superantigens. Superantigens facilitate the colonization of *Streptococcus pyogenes* by triggering non-selective differentiation of T cells. The research on the virome and its connections to host disease is significantly less extensive in comparison to the study of the gut microbiome. Furthermore, the literature presents ambiguous findings and an unverified correlation between virome shift and T1D. The primary obstacle in comprehending virome interaction lies in its direct impact on the dynamics and variety of bacterial communities. This results in a complex interplay of factors that makes it difficult to reach a conclusive outcome (Hsu et al., 2019). By inoculating gnotobiotic mice, which were already colonized with specific human bacteria, with phages, there was a clear decrease in the vulnerable species and an indirect impact on the remaining species, potentially due to inter-species interactions. A different investigation has found a correlation among the initial concentration of amyloid-producing *Escherichia coli* with the ratio of *E. coli* bacteriophage to *E. coli*. This hyperlink leads to a reduction in *E. coli* and plays a role in the progression of T1D (Tetz et al., 2019). To summarize, numerous research have examined the impact of microbial dysbiosis on the development of T1D through various mechanisms, such as the disturbance of the Th17/Treg balance, modifications in interleukin production, and changes in gut permeability. The consequences of these actions are influenced by certain microbial metabolites, including butyrate, LPS, and arachidonoyl-GPC. Reactivation of latent viral infections has the potential to disrupt the microbial ecology, leading to the development of T1D. However, our understanding of microbial dysbiosis in late-onset T1D is currently limited, as is our knowledge of whether nutritional therapy can impact the severity or progression of the disease. The research discovered an elevation in the abundance of *Veillonella* and *Clostridium*, but *Bifidobacterium*, *Lactobacillus*, and *Prevotella* were diminished in children with paediatric T1D in comparison to healthy children (Soyucen et al., 2014). Nevertheless, the existing research examining the correlation between late-onset T1D and microbiota shift do not establish a clear cause-and-effect relationship, as it is believed that T1D itself leads to changes in the microbiota as the disease advances.

9.6.2 Multiple Sclerosis

Multiple sclerosis (MS) is a common neurological illness that affects the CNS and causes an autoimmune response against the myelin layer. The condition is distinguished by impaired vision, motor impairment, and alterations in sensory perception (Dendrou et al., 2015). Research indicates that individuals with MS exhibit alteration in gut microbiota composition compared to healthy individuals. Patients with MS have a unique microbiome profile characterized by decreased abundance of *Bacteroides*, *Prevotella* and *Faecalibacterium prausnitzii*, and an increased presence of *Akkermansia muciniphila* (Jangi et al., 2016). Children with pediatric MS have a distinct decrease in the presence of Ruminococcaceae bacteria that produce short-chain fatty acids (SCFAs) as compared to healthy children (Schoeps et al., 2024). Nevertheless, the causality of this relationship remains uncertain. The alterations in gut microbiota lead to an elevation in the population of regulatory T cells (Treg), which suppress the activation of pro-inflammatory T cells (Berer et al., 2017). Elevated levels of Th1 and Th17 cells in the bloodstream resulted in an increase in the permeability of the BBB that trigger CNS inflammation (Dendrou et al., 2015; Zhao et al., 2020). Transplanting fecal microbiota from persons with MS into transgenic mice led to a notably higher incidence of autoimmune encephalomyelitis (EAE) compared to transplant from healthy individuals (Zhao et al., 2020). In a study, mice that were administered a wide range of antibiotics showed a prevention of motor dysfunction and damage to the axons in a mouse model of EAE. However, bacterial recolonization led to decreased motor function and compromised axonal integrity. The protective mechanism of antibiotics may involve the participation of $CD4^+$, $CD5^+$, and $CD39^+$ T cells, and $CD1d^+$ B cells in the CNS of mice (Mestre et al., 2019). Antibiotic-treated mice also experienced delay in the onset of EAE symptoms. In addition, the concentration of IFN-γ and IL-17A decreased, with increased IL-10 level concentration in mice bloodstream (Zeraati et al., 2019). Additionally, it was discovered that the reduction of microbiota caused by antibiotics could lead to a drop in hippocampus BDNF levels and an increase in learning and memory levels in mice induced with EAE. Furthermore, the researchers observed elevated levels of anxiety-like behavior and increased levels of hippocampal IL-1β and TNF-α in mice induced with EAE. These investigations have confirmed link between the gut microbiota and the severity of neurodegenerative illness in progressive MS model (Zeraati et al., 2019).

9.6.3 Systemic Lupus Erythematosus

Systemic lupus erythematosus (SLE) is an autoimmune illness that impacts several organs, including joints, blood, and kidneys. Its exact cause is still unknown. The primary feature of SLE is the formation and buildup of immune complexes as a result of the production of autoantibodies that specifically attack nuclear antigens. These complexes can be identified several years prior to the development of the disease (Lossius et al., 2012) This autoimmune assault leads to inflammation and the failure of organs. The precise process underlying the activation of autoreactive T cells and the formation of autoantibodies is not yet fully understood. However, several possibilities have been proposed, including genetic predisposition and exposure to environmental stimuli that exhibit molecular mimicry (Hagberg et al., 2020). The current understanding is that this inflammation is caused by an imbalance in Th17 and Treg cells, resulting in an elevated production of pro-inflammatory cytokines (Guo et al., 2020), which in turn contribute to systemic inflammation. It is evident that SLE patients have a significant infiltration of IL-17-producing cells in their kidney, lung, and skin tissues, which leads to damage in these organs (Koga et al., 2019). An exclusive gut microbiota pattern is shown in association with the occurrence of inflammatory flares and the advancement of the disease. This signature is distinguished by a decrease in the variety of microbiota, accompanied by an increase in certain species like *Campylobacter, Streptococcus*, and *Veillonella,* and a decrease in others like *Bifidobacterium* (Li et al., 2020). The presence of *Lactobacillus* is dramatically reduced, whereas the quantity of Lachnospiraceae is increased in mice with SLE (Kim et al., 2019). The prevalence of Lachnospiraceae is gender-dependent, with a higher quantity in females, hence exacerbating the severity of the condition. Curiously, mice that are prone to lupus exhibited an increase in the metabolic pathways associated with the genes responsible for movement and reproduction which could potentially be connected to the ability of microorganisms to induce inflammation throughout the body. Administering retinoic acid to increase the population of lactobacilli led to the amelioration of lupus symptoms, indicating that lactobacilli may have a beneficial effect in avoiding or mitigating inflammation. Research has documented a change in the ratio of Firmicutes to Bacteroidetes, along with alterations in SCFAs and Th17 levels in the blood serum of individuals with SEL. The modified microbiota in SEL is characterized by a disproportionate ratio of Treg/Th17/Th1. Specifically, two bacteria from the *Clostridium* genus disrupt the Th17/Th1 balance, resulting in the conversion of CD4+ cells into Th17 and consequent inflammation. *Bifidobacterium bifidum* inhibits the excessive activation of CD4+ cells (López et al., 2016). A study conducted by Hooper et al. (2012) revealed that the migration of *Enterococcus gallinarum* from the intestines to the liver resulted in the overproduction of autoimmune antibodies, inflammation, and mortality in mice with a genetic predisposition to this condition. The use of antibiotics not only specifically targeted the elimination of *E. gallinarum*, but also led to the complete eradication of autoantibodies in mice. The DNA of *E. gallinarum* extracted from the livers of individuals with autoimmune disorders elicited pro-inflammatory reactions in human liver cells, similar to the pattern found in mice (Manfredo et al., 2018). Latent EBV infection is also linked to the initiation of SLE. EBV's oncogenic potential may be attributed to its ability to evade the immune system. The antigens of Epstein–Barr virus (EBV) display molecular resemblance with antigens linked to systemic lupus erythematosus (SLE), leading to an autoimmune reaction upon EBV activation. In addition, EBV inhibits the production of anti-inflammatory interleukins, leading to increased systemic inflammation (Lossius et al., 2012). An experimental EBV peptide vaccine administered to animals during a study resulted in the production of antibodies that reacted with many targets and induced symptoms resembling SLE (Sundar et al., 2004).

9.6.4 Rheumatid Arithritis

Rheumatoid arthritis (RA) is a systemic autoimmune disorder that primarily impacts the joints, and in some cases can also affect other internal organs. It leads to inflammation and swelling. The initial evidence linking microbial dysbiosis and RA was reported in 1979, when it was discovered that germ-free rats have a 100% chance of developing RA when injected with an intradermal adjuvant. In contrast, conventional rats have a susceptibility ranging from 0% to 20% and exhibit weaker or delayed inflammation. Curiously, the inflammation caused by this is healed through the introduction of *E. coli* and slightly exacerbated by the introduction of lactobacilli (Waldenlind, 2022). The authors suggested that the lipopolysaccharide (LPS) of *E. coli* may play a role in resolving RA (Lorenz et al., 2013). Research indicates that the microbiome of individuals with RA contains a higher abundance of certain types of microorganisms, including *Provetella*, *Lactobacillus* sabotage, and segmented filamentous bacteria (Mousa et al., 2022). Elevated quantities of these bacteria result in heightened amounts of their pro-inflammatory by-products, such as serum amyloid protein A. Consequently, this process stimulates the production of Th1 and Th17 cells, which then disperse throughout the body and have a role in the emergence of many illnesses. Multiple species of the Provetella genus have been linked to various inflammatory and autoimmune conditions, such as low-grade inflammation, periodontitis, bacterial vaginosis, and rheumatoid arthritis (RA). Emerging data indicates that *Provetella copri* may play a role in the progression of RA. *P. copri* has the ability to cause inflammation in the mucosal lining by activating TLR-2. This activation leads to the differentiation of Th17 cells and the excessive production of pro-inflammatory interleukins such as IL-23 and IL-1. Additionally, it also attracts neutrophils to the site of inflammation (Mousa et al., 2022). This inflammatory response causes an increase in the permeability of the gut, resulting in the release of bacteria and pro-inflammatory chemicals. This, in turn, leads to a greater level of inflammation throughout the body and triggers an immunological response. An increased prevalence of *P. copri* is also linked to rheumatoid arthritis in mice lacking the NLRP6 gene, which is a pro-inflammatory gene involved in the inflammasome. These mice have been genetically altered to produce reduced amounts of pro-inflammatory cytokines. The cohabitation of mice with normal genetic traits and animals without specific genes led to the development of inflammatory symptoms in the normal mice. This indicates that the microorganisms in the gut have the ability to trigger autoimmune disorders, even in the absence of a hereditary predisposition. This discovery also indicates a potential involvement of *Provetella* in the development of arthritis as a harmful microorganism, and it could possibly function as a diagnostic indicator for RA. *Prevotella intestinalis*, a member of the same family, has been linked to intestinal inflammation, specifically colitis, in mice. This phenomenon is thought to happen because of decreased short-chain fatty acids and anti-inflammatory cytokine. Additional research indicates that *Prevotella histicola* exhibits anti-inflammatory properties and confers protection against arthritis in mice. Curiously, the process of suppressing inflammation appears to be the opposite of that caused by other pathobionts within the same species. Researchers and coworkers demonstrate that *P. histicola* controls the activity of dendritic cells (CD103+), leading to the production of regulatory T cells. These T cells inhibit the development of Th17 cells, thereby reducing the levels of pro-inflammatory interleukins and boosting the levels of anti-inflammatory interleukins, such as IL-10. In addition, it promotes tight junction protein production, resulting in a decrease in the permeability of the intestines. The findings highlight the potential of *P. histicola* as a probiotic for managing RA and perhaps other autoimmune disorders.

9.6.5 Spondyloarthritis

The GM plays an essential role in modulating host's immune system. Disruption of this connection can result in the development of autoimmune and inflammatory diseases, including spondyloarthritis (SpA). The study investigated the microbial dysbiosis and disturbed metabolic function observed in several forms of SpA, such as acute anterior uveitis, psoriatic arthritis, ankylosing spondylitis, and inflammation associated with SpA in the gut.

9.6.6 Huntington's Disease

Huntington's disease (HD) represents a neurodegenerative disorder marked by three primary symptoms: gradual decline in motor function, cognitive impairment, and behavioral abnormalities, along with involuntary weight loss (Wasser et al., 2020). The primary cause of this condition is mostly attributed to the growth and instability of a trinucleotide (cytosine–adenine–guanine, CAG) repeat within the huntingtin gene, which is widely expressed throughout the brain (MacDonald et al., 1993). Early signs of HD in both individuals and animals include neuronal loss in the basal ganglia, white matter atrophy, and impaired myelination (Teo et al., 2019; Shaffer et al., 2017). Recent investigations have revealed the possible existence of gut dysbiosis in persons with HD. In Huntington's disease (HD), unintentional weight loss is a significant clinical feature that has been attributed to gastrointestinal dysfunction in a mouse model of the disease (Van der Burg et al., 2011). An animal model study found a notable disparity in the microbiota makeup of HD mice when they reached 12 weeks of age. HD mice exhibited a notable rise in the abundance of Bacteroidetes and a decline in the abundance of Firmicutes (Kong et al., 2020). Furthermore, HD mice exhibited intestinal dysbiosis and motor impairments at 12 weeks of age. The condition of germ-free (GF) mice was modified, resulting in changes in the myelination of the corpus callosum and the flexibility of white matter in a mouse model of HD. This was achieved by using a bacterial artificial chromosome (Radulescu et al., 2019). Wasser and colleagues examined the makeup of gut bacteria in individuals with HD. HD patients displayed lower alpha-diversity microbial communities and significant variation in beta-microbial diversity (Wasser et al., 2020). Male groups had notable variations in the phylum-level composition of Euryarchaeota, Firmicutes, and Verrucomicrobia. Moreover, the particular association between gut microbiota, cognitive function, and clinical outcomes was found exclusively among patients who possess the HD gene expansion.

9.7 THERAPEUTIC IMPLICATION

The diet has a substantial impact on regulating inflammation related to neurological disorders, highlighting the important interactions between the brain and the gut. A study by Kurowska et al. explored how nutrition significantly influences inflammation related to neurological illnesses, specifically emphasizing the connection between the gut and the brain. The relationship between diets and neurological disorders can be influenced by different diets because of their impact on GM and inflammatory processes, which play a vital part in the onset and advancement of these disorders. The Mediterranean diet is characterized by its abundant quantities of whole grains, seeds, nuts, fruits, vegetables, and olive oil. Moreover, it promotes the moderate intake of poultry and fish rich in omega-3 fatty acids. The substance anti-inflammatory and antioxidant effects stem from its abundant presence of polyphenols, and other bioactive components. Studies have found a connection between following this specific diet and a decreased likelihood of experiencing cognitive decline and neurological disorders such as Alzheimer's and Parkinson's diseases. This is because it can regulate the GM and decrease systemic inflammation in the human body. The ketogenic diet is a dietary strategy that involves consuming a high amount of fat, a moderate amount of carbohydrates, and an adequate amount of protein. It has shown significant effectiveness in treating several neurological diseases.

9.7.1 Probiotics

The human gastrointestinal tract hosts a wide range of microbial communities that exert substantial effects on immune responses, digestive processes, medication breakdown, intestinal health, and protection against infections. A recent study has shown that GM interacts with brain through a two-way communication network called the gut–brain axis. This interaction involves multiple pathways, and encompasses humoral, immunological, endocrine, and neurological channels. A recent study has shown that GM plays a function in interacting with the brain. The transmission of information occurs through a two-way network called the gut–brain axis. An alteration in the GM disrupts many avenues of communication, resulting in neurological disorders and impairments in cognitive function. Both preclinical and clinical research studies have demonstrated that probiotics can successfully restore a balanced gut microbiota, reduce intestinal acidity, and alleviate inflammation and the growth of dangerous microbes in the gastrointestinal tract. Probiotics are live microorganisms, specifically bacteria and yeast, and contribute positively to overall growth without causing illness. When ingested, they enhance and strengthen the balance of bacteria, especially in the gastrointestinal system (Sánchez et al., 2017). The probiotic strains contribute to multiple physiological functions, including reducing intestinal acidity, facilitating intercellular communication, and suppressing the growth of dangerous bacteria, thus altering the host's immune response (Iacob et al., 2019). Moreover, it improves psychological and mental well-being, influencing mood, anxiety, concentration, memory, and cognitive capabilities (Thangaleela et al., 2022). The GM, which is linked to chronic inflammation and immune system functions, contributes to maintain mucosal barrier structural integrity. An imbalance in the gut microbiome can adversely affect mental well-being, suggesting that restoring a balanced GM might improve mental well-being. Probiotics also play a vital role in mitigating oxidative stress through the production of various antioxidant enzymes, antioxidants, and the ability to bind metal ions. Probiotics can impede the activation of Toll-like receptors (TLRs), hence limiting immune responses such as inflammatory reactions (Dobielska et al., 2022). Wang et al. (2023) demonstrated that reducing inflammation can strengthen blood–brain barrier and improve neurological functions. Probiotics known for their antioxidant and anti-inflammatory effects can positively influence cognitive performance in individuals with depression by addressing hypothalamic-pituitary-adrenal (HPA) axis dysfunctioning and boosting monoamines levels and neuroplasticity (Dobielska et al., 2022). Probiotics have the ability to alter the levels of several neurotransmitters and neuromodulators, such as glutamate, serotonin, and γ-aminobutyric acid. These substances have a crucial role in regulating the brain's function through metabolic pathways (Dicks, 2022). Clinical research investigating the use of probiotics to alleviate autoimmune conditions are summarized in Table 9.1.

9.7.2 Prebiotics

Prebiotics refer to indigestible dietary fibers that support the growth of beneficial microorganisms in the digestive system. By fostering an ideal environment of these microbial communities, prebiotics positively influence the composition of gut microbes (Sánchez et al., 2017). Recent study has demonstrated the capacity of prebiotics to influence immune responses and decrease the severity of diseases. Research on rheumatoid arthritis (RA) has demonstrated that prebiotics stimulate the growth of helpful bacteria while reducing pro-inflammatory bacteria linked to disease progression. For instance, supplementation with *Bacillus coagulans* and prebiotic inulin, either independently or in mixture, led to significant reduction in pro-inflammatory cytokines, serum amyloid Aβ, and fibronectin. This approach helped to slow down RA disease progression and improved its clinical characteristics (Mazziotta et al., 2023). Moreover, the synergistic impact of prebiotics, a particular dietary regimen referred to as synbiotics, and beneficial probiotic bacteria amplify favorable results. An instance of this occurred when *L. casei* was employed in conjunction with prebiotic oligofructose-enriched inulin. The combination of these factors led to the stimulation of anti-inflammatory reactions, a decrease in damage to the colon, boosted lactobacilli presence in feces, and an enhancement of myeloperoxidase action in rat colitis models (Plaza-Díaz et al., 2017). Synbiotics optimize the conditions for probiotics to flourish, hence reinforcing the beneficial relationship between probiotic strains and GM, leading to improved health advantages (Dahiya et al., 2022).

9.7.3 Postbiotics

Postbiotics including bacterial lysates enzymes, fragments of cell walls, SCFAs, antimicrobial peptides, exopolysaccharides, and cell-free supernatants have emerged as safer and more stable alternatives as compare to living microorganisms for therapeutic purposes (Rad et al., 2020). Studies have shown that postbiotics can influence the immune response in many autoimmune diseases. Specifically, LC-DG and its postbiotic have shown promise to reduce inflammation, joint degeneration, and bone erosion in rheumatoid arthritis, as well as reduce inflammatory mucosal response in patients suffering from postinfectious irritable bowel syndrome suggested as to decrease the concentrations of pro-inflammatory cytokines and TLR-4 expression, while elevating IL-10 levels simultaneously. Recent studies on postbiotics offer a hopeful treatment approach for managing autoimmune diseases (ADs). Because of their consistent and clearly defined properties, as well as their ability to modify the immune system and reduce inflammation, these compounds are well-suited for therapeutic usage.

9.7.4 Fecal Microbiota Transplantation

Fecal microbiota transplantation (FMT) is the process of introducing healthy waste products into the recipient's gastrointestinal system in order to re-establish a harmonious composition of gut bacteria. While the effectiveness of FMT in treating *Clostridium difficile* infections (CDIs) has been recognized, its impact on the management of ADs is now being investigated.

There has been a notable surge in interest in this topic in recent years, mostly due to the worldwide outbreak of CDI and the remarkable progress in metagenomic sequencing of the GM. These improvements have led to improved comprehension of the composition and functioning of GM. FMT is currently recommended as the most effective treatment for recurrent CDI. With further enhancements, FMT could be regarded as a potential therapeutic option for first cases of CDI. Furthermore, there is a keen interest in exploring

TABLE 9.1
Clinical Trials of Probiotics for Autoimmune Disorder Treatment

Subjects	Study	Result	Type of Study	Reference
5 patients	*Trichiuris suis ova* 2,500 ova orally, 1 time every two weeks, for three months	Decreased new MSI lesions, first two months number of eosinophils and hs-CRP increased than decreased	Pilot study	Fleming et al., 2011
65 patients	Two multi-strain probiotic capsules/ day for 12 weeks	Reduction in hs-CRP, TNF-α, and IFN-γ, as well ↑ed FOXP3 and TGB-β	Double-blind placebo- randomized controlled clinical trial	Rahimlou, Nematollahi et al., 2022
70 patients	Two multi-strain probiotic capsules/ day for six months	Significant increase in BDNF and a significant reduction in the IL-6 levels	Randomized double blind placebo-controlled clinical trial	Rahimlou, Hosseini et al., 2022
40 MS patients	*Saccharomyces boulardii* supplement for four months	Reduction in hs-CRP and ↑ed antioxidant capacity	Randomized double-blind clinical trial	Asghari et al., 2023
388 children and adolescents with TID	*Probiotics*	Significant -ve impact on the glycemic control of those individuals, *i.*	Randomized-controlled trial	Stefanaki et al., 2024
60 children with T1D	Oral powder of *Vivomixx* for six months	Median C-peptide and plasma IL-10 levels were significantly increased	Single-center randomized double-blind placebo-controlled study	Lokesh et al., 2024
33 children	Probiotic enterol	Significant correction in dysbiosis	Clinical trial	Akhmedova & Shamansurova, 2024
50 T1D patients	Capsule of multi-probiotic strains (*Streptococcus thermophilus, Bifidobacterium longum, Lactobacterium bulagricumi*) for 12 weeks	Significant reduction in FBG, post prandial glucose and LDL	Double-blind placebo-controlled clinical pilot study	Zhang et al., 2023
54 patients with OA	Prebiotic (oligofructose-enriched inulin for six months	Significant reduction in knee pain and trunk fat mass	Randomized controlled clinical trial	Fortuna et al., 2024
54 patients with OA	*Saccharomyces boulardii* for 12 weeks	Significant improvement in pain, oxidative stress, and inflammation	Randomized triple-blind placebo-controlled trial	Dolatkhah et al., 2024
42 patients with RA	Probiotic with multiple strains for 60 days	Significant reduction in WBCs, TNF-α and interleukin	Randomized and double-blind placebo-controlled study	Cannarella et al., 2021
88 RA patients	*P. copri*, a gut commensal	Elevated levels of IgA anti-*Pc*-p27 antibodies was reduced after treatment	Randomized controlled clinical trial	Seifert et al., 2023
46 RA patients	*Lactobacillus casei* 01 supplementation for eight weeks	Reduction in serum hs-CRP levels, tender and swollen joint counts, GH score and DAS28	A randomized double-blind clinical trial	Alipour et al., 2014

additional medical conditions that are linked to an imbalance in the GM, including bowel inflammation, irritable bowel syndrome (IBS), and diabetes mellitus. Nevertheless, there remains a scarcity of rigorous and reliable studies. Multiple ongoing clinical trials are currently exploring the effectiveness of FMT as a therapeutic intervention for ulcerative colitis. Several unresolved inquiries persist, such as the FMT methodology, including the most effective way of delivery, the characteristics of an ideal donor, concerns over safety, and the potential long-term consequences of FMT. A recent study discovered that FMT from mice with SLE can induce autoimmune reactions in mice that are free of any microorganisms (germ-free) and belong to the C57BL/6J strain. In a similar experiment, researchers transferred fecal microbiota from patients with SLE to mice that were free of any microorganisms. As a consequence, lupus-like traits and heightened inflammatory reactions were observed. The mice exhibited increased levels of autoantibodies linked to systemic lupus erythematosus (SLE), along with heightened levels of certain cytokines (IL-6, IL-8, INF-γ, and TNF-α). Additionally, there was an increase in B-lymphocyte subsets in the intestinal lamina propria, and an expansion of Th-17 and CD4+ CXCR3+ cells in the peripheral blood, while the number of immunomodulatory Treg cells was reduced. These effects were not observed in animals that received gut microbiota from healthy individuals.

Moreover, when RA fecal samples were introduced into SKG mice, a strain known to be susceptible to arthritis, there was a notable rise in Th17 cells in the intestines. Consequently, when these mice were treated with zymosan, they developed severe arthritis. The colon and regional lymph nodes displayed increased IL-17 responses to RPL23A, while the spleen did not. Furthermore, the co-cultivation of naive SKG mice T cells with *P. copri*–stimulated DCs led to the production of IL-17 and the fast development of arthritis (Romero-Figueroa et al., 2023).

Mice that were colonized with microbiota obtained from donors with inflammatory bowel disease (IBD) exhibited significant increase in mucosal Th17 cells, a reduction in tolerogenic RORγt+ Treg cells, and hightened vulnerability to colitis. On the other hand, when microbiota from a healthy donor was transplanted, it resulted in the production of RORγt+ Treg cells and led to better outcomes for the condition. An independent study investigated the influence of bacteria derived from individuals with MS on human T cells and the worsening of MS indications in pharmacological mice models. The study found that fecal bacteria linked to MS might trigger inflammatory reactions and the development of T cells, ultimately exacerbating MS-like symptoms in mice.

In the field of metabolic diseases, the application ofFMT in persons who have just been diagnosed with T1D has the potential to prevent the decline in the body's natural insulin production during one year. Studies have found that patients who underwent autologous FMT demonstrated notable preservation of stimulated C-peptide levels, in contrast to those patients who received allogenic FMT from a healthy donor. The maintenance of beta cell activity was discovered to be linked to particular bacterial taxa and plasma metabolites.

Other approaches such as use of oral capsules and transferring sterile fecal filtrate have also demonstrated effectiveness in treating disorders related to microbiota Nevertheless, it is imperative to conduct a thorough screening of these intervention strategies that aim to modify the makeup of gut microbiota to avoid any possible negative consequences in the recipients. Furthermore, there are other obstacles regarding the choice of donors, the establishment of standardized FMT techniques, and the examination of potential long-term consequences that must be tackled throughout preclinical and clinical trials.

In addition, recent advancement in developing gut microbiota–linked medicines, a new therapeutic method, has been offered using synthetic or genetically modified bacteria, commonly referred to as engineered bacteria. Manipulation of microbial genetic properties allows for the development of specialized capabilities that go beyond the natural characteristics of microorganisms. These modified microorganisms have been applied in a wide range of fields, including agriculture, healthcare, and industrial applications. This practice shows significant potential for advancing biotechnology and medicine in the future.

9.8 CONCLUSION

The human gastrointestinal system hosts a wide range of microbial communities that exert substantial influences on immunity, digestion, drug processing, intestinal health, and defence against infections. The study has demonstrated that the assemblage of microorganisms in the digestive system, referred to as GM, play significant role in communicating information with the brain. This communication occurs via a bidirectional network called the gut–brain axis, which encompasses humoral, immune-mediated, endocrine, and neurological channels. An alteration in the GM composition has a detrimental effect on the pathways responsible for communication, resulting in neurological problems and impairments in cognitive function. Research from both preclinical and clinical studies indicates that probiotics significantly help in restoring healthy balance of microorganisms in the gut, lower intestinal acidity, and reduce inflammation and harmful bacteria in the gastrointestinal tract. Moreover, probiotics enhance cellular communication and elevate the levels of brain-derived neurotrophic factors in the bloodstream. Probiotics are becoming increasingly recognized as a possible method to prevent and control neurological diseases and cognitive deficiencies. Although these initial findings are promising, it is imperative to continually monitor and address the safety concerns and potential risks associated with the use of probiotics.

REFERENCES

Ahima, R. S., & Antwi, D. A. (2008). Brain regulation of appetite and satiety. *Endocrinology and Metabolism Clinics of North America*, *37*(4), 811–823.

Akhmedova, S., & Shamansurova, Z. (2024). The usage of probiotics enterol in combined therapy of colon microbiocenosis in children with type 1 diabetes mellitus. *Science and Innovation*, *3*(D5), 220–224.

Al Theyab, A., Almutairi, T., Al-Suwaidi, A. M., Bendriss, G., McVeigh, C., & Chaari, A. (2020). Epigenetic effects of gut metabolites: Exploring the path of dietary prevention of type 1 diabetes. *Frontiers in Nutrition*, *7*, 563605.

Alipour, B., Homayouni-Rad, A., Vaghef-Mehrabany, E., Sharif, S. K., Vaghef-Mehrabany, L., Asghari-Jafarabadi, M., ... Mohtadi-Nia, J. (2014). Effects of L actobacillus casei supplementation on disease activity and inflammatory cytokines in rheumatoid arthritis patients: A randomized double-blind clinical trial. *International Journal of Rheumatic Diseases*, *17*(5), 519–527. https://doi.org/10.1111/1756-185X.12333.

Asghari, K. M., Dolatkhah, N., Ayromlou, H., Mirnasiri, F., Dadfar, T., & Hashemian, M. (2023). The effect of probiotic supplementation on the clinical and Para-clinical findings of multiple sclerosis: A randomized clinical trial. *Scientific Reports*, *13*(1), 18577.

Badami, E., Sorini, C., Coccia, M., Usuelli, V., Molteni, L., Bolla, A. M., ... Falcone, M. (2011). Defective differentiation of regulatory FoxP3+ T cells by small-intestinal dendritic cells in patients with type 1 diabetes. *Diabetes*, *60*(8), 2120–2124.

Berer, K., Gerdes, L. A., Cekanaviciute, E., Jia, X., Xiao, L., Xia, Z., ... Wekerle, H. (2017). Gut microbiota from multiple sclerosis patients enables spontaneous autoimmune encephalomyelitis in mice. *Proceedings of the National Academy of Sciences*, *114*(40), 10719–10724.

Brown, C. T., Davis-Richardson, A. G., Giongo, A., Gano, K. A., Crabb, D. B., Mukherjee, N., ... Triplett, E. W. (2011). Gut microbiome metagenomics analysis suggests a functional model for the development of autoimmunity for type 1 diabetes. *PLoS One*, *6*(10), e25792. https://doi.org/10.1371/journal.pone.0025792

Cannarella, L. A. T., Mari, N. L., Alcântara, C. C., Iryioda, T. M. V., Costa, N. T., Oliveira, S. R., ... Simão, A. N. C. (2021). Mixture of probiotics reduces inflammatory biomarkers and improves the oxidative/nitrosative profile in people with rheumatoid arthritis. *Nutrition*, *89*, 111282.

Chandel, P., Thapa, K., Kanojia, N., Rani, L., Singh, T. G., & Rohilla, P. (2024). Exploring therapeutic potential of phytoconstituents as a gut microbiota modulator in the management of neurological and psychological disorders. *Neuroscience*, *551*, 69–78.

Chen, J., Cheng, M., Wang, L., Zhang, L., Xu, D., Cao, P., ... Zhan, C. (2020). A vagal-NTS neural pathway that stimulates feeding. *Current Biology*, *30*(20), 3986–3998.

Chen, L., Sun, M., Wu, W., Yang, W., Huang, X., Xiao, Y., ... Cong, Y. (2019). Microbiota metabolite butyrate differentially regulates Th1 and Th17 cells' differentiation and function in induction of colitis. *Inflammatory Bowel Diseases*, *25*(9), 1450–1461. https://doi.org/10.1093/ibd/izz046

Cong, X., Henderson, W. A., Graf, J., & McGrath, J. M. (2015). Early life experience and gut microbiome: The brain–gut–microbiota signaling system. Advances in Neonatal Care, 15(5), 314–323.

Cryan, J. F., & Dinan, T. G. (2012). Mind-altering microorganisms: The impact of the gut microbiota on brain and behaviour. *Nature Reviews Neuroscience*, *13*(10), 701–712.

Dahiya, D., & Nigam, P. S. (2022). Probiotics, prebiotics, synbiotics, and fermented foods as potential biotics in nutrition improving health via microbiome-gut-brain axis. *Fermentation*, *8*(7), 303.

de Groot, P. F., Nikolic, T., Imangaliyev, S., Bekkering, S., Duinkerken, G., Keij, F. M., ... Nieuwdorp, M. (2020). Oral butyrate does not affect innate immunity and islet autoimmunity in individuals with longstanding type 1 diabetes: A randomised controlled trial. *Diabetologia*, *63*, 597–610.

Dehner, C., Fine, R., & Kriegel, M. A. (2019). The microbiome in systemic autoimmune disease: Mechanistic insights from recent studies. *Current Opinion in Rheumatology*, *31*(2), 201–207.

Dendrou, C. A., Fugger, L., & Friese, M. A. (2015). Immunopathology of multiple sclerosis. *Nature Reviews Immunology*, 15(9), 545–558.

Dicks, L. M., & Vermeulen, W. (2022). Do bacteria provide an alternative to cancer treatment and what role does lactic acid bacteria play? *Microorganisms*, *10*(9), 1733.

Dobielska, M., Bartosik, N. K., Zyzik, K. A., Kowalczyk, E., & Karbownik, M. S. (2022). Mechanisms of cognitive impairment in depression. May probiotics help? *Frontiers in Psychiatry*, *13*, 904426.

Dolatkhah, N., Jafari, A., Eslamian, F., Toopchizadeh, V., Saleh, P., & Hashemian, M. (2024). Saccharomyces boulardii improves clinical and paraclinical indices in overweight/obese knee osteoarthritis patients: A randomized triple-blind placebo-controlled trial. *European Journal of Nutrition*, *6*, 1–15.

Eckburg, P. B., Bik, E. M., Bernstein, C. N., Purdom, E., Dethlefsen, L., Sargent, M., ... Relman, D. A. (2005). Diversity of the human intestinal microbial flora. *Science*, *308*(5728), 1635–1638.

Fleming, J. O., Isaak, A., Lee, J. E., Luzzio, C. C., Carrithers, M. D., Cook, T. D., Field, A. S., Boland, J., & Fabry, Z. (2011). Probiotic helminth administration in relapsing–remitting multiple sclerosis: A phase 1 study. *Multiple Sclerosis Journal*, *17*(6), 743–754.

Fortuna, R., Wang, W., Mayengbam, S., Tuplin, E. W. N., Sampsell, K., Sharkey, K. A., ... Reimer, R. A. (2024). Effect of prebiotic fiber on physical function and gut microbiota in adults, mostly women, with knee osteoarthritis and obesity: A randomized controlled trial. *European Journal of Nutrition*, *6*, 1–13.

Graves, C. L., Li, J., LaPato, M., Shapiro, M. R., Glover, S. C., Wallet, M. A., & Wallet, S. M. (2017). Intestinal epithelial cell regulation of adaptive immune dysfunction in human type 1 diabetes. *Frontiers in Immunology*, *7*, 679.

Gubert, C., Kong, G., Renoir, T., & Hannan, A. J. (2020). Exercise, diet and stress as modulators of gut microbiota: Implications for neurodegenerative diseases. *Neurobiology of Disease*, *134*, 104621.

Guo, M., Wang, H., Xu, S., Zhuang, Y., An, J., Su, C., ... Liu, X. (2020). Alteration in gut microbiota is associated with dysregulation of cytokines and glucocorticoid therapy in systemic lupus erythematosus. *Gut Microbes*, *11*(6), 1758–1773.

Hagberg, N., Lundtoft, C., & Rönnblom, L. (2020). Immunogenetics in systemic lupus erythematosus: Transitioning from genetic associations to cellular effects. Scandinavian Journal of Immunology, 92(4), e12894. https://doi.org/10.1111/sji.12894

Hassan, E., & Baldridge, M. T. (2019). Norovirus encounters in the gut: Multifaceted interactions and disease outcomes. *Mucosal Immunology*, *12*(6), 1259–1267.

Hooper, L. V., Littman, D. R., & Macpherson, A. J. (2012). Interactions between the microbiota and the immune system. *Science*, *336*(6086), 1268–1273.

Hsu, B. B., Gibson, T. E., Yeliseyev, V., Liu, Q., Lyon, L., Bry, L., et al. (2019). Dynamic modulation of the gut microbiota and metabolome by bacteriophages in a mouse model. Cell Host & Microbe, 25(6), 803–814.e5. https://doi.org/10.1016/j.chom.2019.05.001

Huang, J., Xiao, Y., Xu, A., & Zhou, Z. (2016). Neutrophils in type 1 diabetes. *Journal of Diabetes Investigation*, *7*(5), 652–663.

Iacob, S., Iacob, D. G., & Luminos, L. M. (2019). Intestinal microbiota as a host defense mechanism to infectious threats. *Frontiers in Microbiology*, *9*, 3328.

Jangi, S., Gandhi, R., Cox, L. M., Li, N., Von Glehn, F., Yan, R., ... Weiner, H. L. (2016). Alterations of the human gut microbiome in multiple sclerosis. *Nature Communications*, *7*(1), 12015.

Kahalehili, H. M., Newman, N. K., Pennington, J. M., Kolluri, S. K., Kerkvliet, N. I., Shulzhenko, N., ... Ehrlich, A. K. (2021). Dietary indole-3-carbinol activates AhR in the gut, alters Th17-microbe interactions, and exacerbates insulitis in NOD mice. *Frontiers in Immunology*, *11*, 606441.

Kanai, T., & Teratani, T. (2022). Role of the vagus nerve in the gut-brain axis: Development and maintenance of gut regulatory t cells via the liver-brain-gut vago-vagal reflex. *Brain and Nerve= Shinkei Kenkyu no Shinpo*, *74*(8), 971–977.

Kawajiri, K., & Fujii-Kuriyama, Y. (2017). The aryl hydrocarbon receptor: A multifunctional chemical sensor for host defense and homeostatic maintenance. *Experimental Animals*, *66*(2), 75–89.

Keightley, P. C., Koloski, N. A., & Talley, N. J. (2015). Pathways in gut-brain communication: Evidence for distinct gut-to-brain and brain-to-gut syndromes. *Australian & New Zealand Journal of Psychiatry*, *49*(3), 207–214.

Kim, J. W., Kwok, S. K., Choe, J. Y., & Park, S. H. (2019). Recent advances in our understanding of the link between the intestinal microbiota and systemic lupus erythematosus. *International Journal of Molecular Sciences*, *20*(19), 4871.

Koga, T., Ichinose, K., Kawakami, A., & Tsokos, G. C. (2019). The role of IL-17 in systemic lupus erythematosus and its potential as a therapeutic target. *Expert Review of Clinical Immunology*, *15*(6), 629–637.

Kong, G., Lê Cao, K. A., Judd, L. M., Li, S., Renoir, T., & Hannan, A. J. (2020). Microbiome profiling reveals gut dysbiosis in a transgenic mouse model of Huntington's disease. *Neurobiology of Disease*, *135*, 104268.

Li, K., Wei, S., Hu, L., Yin, X., Mai, Y., Jiang, C., ... Zhao, B. (2020). Protection of fecal microbiota transplantation in a mouse model of multiple sclerosis. *Mediators of Inflammation*, *2020*(1), 2058272.

Li, N., Hatch, M., Wasserfall, C. H., Douglas-Escobar, M., Atkinson, M. A., Schatz, D. A., & Neu, J. (2010). Butyrate and type 1 diabetes mellitus: Can we fix the intestinal leak?. *Journal of Pediatric Gastroenterology and Nutrition*, *51*(4), 414–417.

Lokesh, M. N., Kumar, R., Jacob, N., Sachdeva, N., Rawat, A., Yadav, J., & Dayal, D. (2024). Supplementation of high-strength oral probiotics improves immune regulation and preserves beta cells among children with new-onset type 1 diabetes mellitus: A randomised, double-blind placebo control trial. *Indian Journal of Pediatrics*, *3*, 1–7.

López, P., de Paz, B., Rodríguez-Carrio, J., Hevia, A., Sánchez, B., Margolles, A., & Suárez, A. (2016). Th17 responses and natural IgM antibodies are related to gut microbiota composition in systemic lupus erythematosus patients. *Scientific Reports*, *6*(1), 24072.

Lorenz, W., Buhrmann, C., Mobasheri, A., Lueders, C., & Shakibaei, M. (2013). Bacterial lipopolysaccharides form procollagen-endotoxin complexes that trigger cartilage inflammation and degeneration: Implications for the development of rheumatoid arthritis. *Arthritis Research & Therapy*, *15*, 1–17.

Lossius, A., Johansen, J. N., Torkildsen, Ø., Vartdal, F., & Holmøy, T. (2012). Epstein-Barr virus in systemic lupus erythematosus, rheumatoid arthritis and multiple sclerosis—association and causation. *Viruses*, *4*(12), 3701–3730.

MacDonald, M. E., Ambrose, C. M., Duyao, M. P., Myers, R. H., Lin, C., Srinidhi, L., ... Harper, P. S. (1993). A novel gene containing a trinucleotide repeat that is expanded and unstable on Huntington's disease chromosomes. *Cell*, *72*(6), 971–983.

Manfredo Vieira, S., Hiltensperger, M., Kumar, V., Zegarra-Ruiz, D., Dehner, C., Khan, N., ... Kriegel, M. A. (2018). Translocation of a gut pathobiont drives autoimmunity in mice and humans. *Science*, *359*(6380), 1156–1161.

Martin, C. R., & Mayer, E. A. (2017). Gut-brain axis and behavior. *Intestinal Microbiome: Functional Aspects in Health and Disease*, *88*, 45–54.

Mayer, E. A., Nance, K., & Chen, S. (2022). The gut–brain axis. *Annual Review of Medicine*, *73*(1), 439–453.

Mazziotta, C., Tognon, M., Martini, F., Torreggiani, E., & Rotondo, J. C. (2023). Probiotics mechanism of action on immune cells and beneficial effects on human health. *Cells*, *12*(1), 184.

Mestre, L., Carrillo-Salinas, F. J., Mecha, M., Feliú, A., Espejo, C., Álvarez-Cermeño, J. C., ... Guaza, C. (2019). Manipulation of gut microbiota influences immune responses, axon preservation, and motor disability in a model of progressive multiple sclerosis. *Frontiers in Immunology*, *10*, 1374.

Mousa, W. K., Chehadeh, F., & Husband, S. (2022). Recent advances in understanding the structure and function of the human microbiome. *Frontiers in Microbiology*, *13*, 825338.

Oh, J. H., Alexander, L. M., Pan, M., Schueler, K. L., Keller, M. P., Attie, A. D., ... van Pijkeren, J. P. (2019). Dietary fructose and microbiota-derived short-chain fatty acids promote bacteriophage production in the gut symbiont Lactobacillus reuteri. *Cell Host & Microbe*, *25*(2), 273–284. https://doi.org/10.1016/j.chom.2018.11.016

O'Mahony, S. M., McVey Neufeld, K. A., Waworuntu, R. V., Pusceddu, M. M., Manurung, S., Murphy, K., ... Cryan, J. F. (2020). The enduring effects of early-life stress on the microbiota–gut–brain axis are buffered by dietary supplementation with milk fat globule membrane and a prebiotic blend. *European Journal of Neuroscience*, *51*(4), 1042–1058.

Plaza-Díaz, J., Ruiz-Ojeda, F. J., Vilchez-Padial, L. M., & Gil, A. (2017). Evidence of the anti-inflammatory effects of probiotics and synbiotics in intestinal chronic diseases. *Nutrients*, *9*(6), 555.

Post, Z., Manfready, R. A., & Keshavarzian, A. (2023, August). Overview of the gut–brain axis: From gut to brain and back again. In *Seminars in neurology*. Thieme Medical Publishers, Inc..

Qiu, J., Guo, X., Zong-ming, E. C., He, L., Sonnenberg, G. F., Artis, D., ... Zhou, L. (2013). Group 3 innate lymphoid cells inhibit T-cell-mediated intestinal inflammation through aryl hydrocarbon receptor signaling and regulation of microflora. *Immunity*, *39*(2), 386–399.

Radulescu, C. I., Garcia-Miralles, M., Sidik, H., Bardile, C. F., Yusof, N. A. B. M., Lee, H. U., ... Pouladi, M. A. (2019). Manipulation of microbiota reveals altered callosal myelination and white matter plasticity in a model of Huntington disease. *Neurobiology of Disease*, *127*, 65–75.

Rahimlou, M., Hosseini, S. A., Majdinasab, N., Haghighizadeh, M. H., & Husain, D. (2022). Effects of long-term administration of Multi-Strain Probiotic on circulating levels of BDNF, NGF, IL-6 and mental health in patients with multiple sclerosis: A randomized, double-blind, placebo-controlled trial. *Nutritional Neuroscience*, *25*(2), 411–422.

Rahimlou, M., Nematollahi, S., Husain, D., Banaei-Jahromi, N., Majdinasab, N., & Hosseini, S. A. (2022). Probiotic supplementation and systemic inflammation in relapsing-remitting multiple sclerosis: A randomized, double-blind, placebo-controlled trial. *Frontiers in Neuroscience*, *16*, 901846.

Romero-Figueroa, M. D. S., Ramírez-Durán, N., Montiel-Jarquín, A. J., & Horta-Baas, G. (2023). Gut-joint axis: Gut dysbiosis can contribute to the onset of rheumatoid arthritis via multiple pathways. *Frontiers in Cellular and Infection Microbiology*, *13*, 1092118.

Routy, B., Le Chatelier, E., Derosa, L., Duong, C. P., Alou, M. T., Daillère, R., ... Zitvogel, L. (2018). Gut microbiome influences efficacy of PD-1–based immunotherapy against epithelial tumors. *Science*, *359*(6371), 91–97.

Sánchez, B., Delgado, S., Blanco-Míguez, A., Lourenço, A., Gueimonde, M., & Margolles, A. (2017). Probiotics, gut microbiota, and their influence on host health and disease. *Molecular Nutrition & Food Research*, *61*(1), 1600240.

Santiago-Rodriguez, T. M., & Hollister, E. B. (2019). Human virome and disease: High-throughput sequencing for virus discovery, identification of phage-bacteria dysbiosis and development of therapeutic approaches with emphasis on the human gut. *Viruses*, *11*(7), 656. https://doi.org/10.3390/v11070656

Schoeps, V. A., Cortese, M., Munger, K. L., Mancuso, J. D., Niebuhr, D. W., Peng, X., ... Bjornevik, K. (2024). Smoking and multiple sclerosis risk in black people: A nested case-control study. *Multiple Sclerosis and Related Disorders*, *81*, 105375.

Seifert, J. A., Bemis, E. A., Ramsden, K., Lowell, C., Polinski, K., Feser, M., ... Holers, V. M. (2023). Association of antibodies to prevotella copri in anti–cyclic citrullinated peptide-positive individuals at risk of developing rheumatoid arthritis and in patients with early or established rheumatoid arthritis. *Arthritis & Rheumatology*, *75*(4), 507–516.

Shaffer, J. J., Ghayoor, A., Long, J. D., Kim, R. E. Y., Lourens, S., O'Donnell, L. J., ... Johnson, H. J. (2017). Longitudinal diffusion changes in prodromal and early HD: Evidence of white-matter tract deterioration. *Human Brain Mapping*, *38*(3), 1460–1477.

Sharma, V. K., Singh, T. G., Garg, N., Dhiman, S., Gupta, S., Rahman, M. H., Najda, A., Walasek-Janusz, M., Kamel, M., Albadrani, G. M., Akhtar, M. F., Saleem, A., Altyar, A. E., & Abdel-Daim, M. M. (2021). Dysbiosis and Alzheimer's disease: A role for chronic stress?. *Biomolecules*, *11*(5), 678.

Soyucen, E., Gulcan, A., Aktuglu-Zeybek, A. C., Onal, H., Kiykim, E., & Aydin, A. (2014). Differences in the gut microbiota of healthy children and those with type 1 diabetes. *Pediatrics International*, *56*(3), 336–343.

Stedtfeld, R. D., Chai, B., Crawford, R. B., Stedtfeld, T. M., Williams, M. R., Xiangwen, S., ... Hashsham, S. A. (2017). Modulatory influence of segmented filamentous bacteria on transcriptomic response of gnotobiotic mice exposed to TCDD. *Frontiers in Microbiology*, *8*, 1708.

Stefanaki, C., Rozou, P., Efthymiou, V., Xinias, I., Mastorakos, G., Bacopoulou, F., & Papagianni, M. (2024). Impact of probiotics on the glycemic control of pediatric and adolescent individuals with type 1 diabetes: A systematic review and meta-analysis. *Nutrients*, *16*(16), 2629.

Sun, L. J., Li, J. N., & Nie, Y. Z. (2020). Gut hormones in microbiota-gut-brain cross-talk. *Chinese Medical Journal*, *133*(7), 826–833.

Sundar, K., Jacques, S., Gottlieb, P., Villars, R., Benito, M. E., Taylor, D. K., & Spatz, L. A. (2004). Expression of the Epstein-Barr virus nuclear antigen-1 (EBNA-1) in the mouse can elicit the production of anti-dsDNA and anti-Sm antibodies. *Journal of Autoimmunity*, *23*(2), 127–140. https://doi.org/10.1016/j.jaut.2004.06.001

Teo, R. T. Y., Ferrari Bardile, C., Tay, Y. L., Yusof, N. A. B. M., Kreidy, C. A., Tan, L. J., & Pouladi, M. A. (2019). Impaireremyelination in a mouse model of Huntington disease. *Molecular Neurobiology*, *56*, 6873–6882.

Tetz, G., Brown, S. M., Hao, Y., & Tetz, V. (2019). Type 1 diabetes: An association between autoimmunity, the dynamics of gut amyloid-producing E. coli and their phages. *Scientific Reports*, *9*(1), 9685.

Thangaleela, S., Sivamaruthi, B. S., Kesika, P., & Chaiyasut, C. (2022). Role of probiotics and diet in the management of neurological diseases and mood states: A review. *Microorganisms*, *10*(11), 2268.

van der Burg, J. M., Winqvist, A., Aziz, N. A., Maat-Schieman, M. L., Roos, R. A., Bates, G. P., ... Wierup, N. (2011). Gastrointestinal dysfunction contributes to weight loss in Huntington's disease mice. *Neurobiology of Disease*, *44*(1), 1–8.

Vehik, K., Lynch, K. F., Wong, M. C., Tian, X., Ross, M. C., Gibbs, R. A., ... TEDDY Study Group. (2019). Prospective virome analyses in young children at increased genetic risk for type 1 diabetes. *Nature Medicine*, *25*(12), 1865–1872. https://doi.org/10.1038/s41591-019-0667-0

Wachsmuth, H. R., Weninger, S. N., & Duca, F. A. (2022). Role of the gut–brain axis in energy and glucose metabolism. *Experimental & Molecular Medicine*, *54*(4), 377–392.

Waldenlind, K. (2022). *The impact of autoimmune thyroid disease (AITD) on Rheumatoid Arthritis (Ra) and the impact of RA on Aitd.* Karolinska Institutet (Sweden).

Wang, H., Meng, X., Xing, S., Guo, B., Chen, Y., & Pan, Y. Q. (2023). Probiotics to prevent necrotizing enterocolitis and reduce mortality in neonates: A meta-analysis. *Medicine*, *102*(8), e32932

Wasser, C. I., Mercieca, E. C., Kong, G., Hannan, A. J., McKeown, S. J., Glikmann-Johnston, Y., & Stout, J. C. (2020). Gut dysbiosis in Huntington's disease: Associations among gut microbiota, cognitive performance and clinical outcomes. *Brain Communications*, *2*(2), fcaa110.

Zeraati, M., Enayati, M., Kafami, L., Shahidi, S. H., & Salari, A. A. (2019). Gut microbiota depletion from early adolescence alters adult immunological and neurobehavioral responses in a mouse model of multiple sclerosis. *Neuropharmacology*, *157*, 107685.

Zhang, X., Zhang, Y., Luo, L., Le, Y., Li, Y., Yuan, F., ... Xu, P. (2023). The beneficial effects of a multispecies probiotic supplement on glycaemic control and metabolic profile in adults with type 1 diabetes: A randomised, double-blinded, placebo-controlled pilot-study. *Diabetes, Metabolic Syndrome and Obesity*, *16*, 829–840.

Zheng, S. J., Luo, Y., & Xiao, J. H. (2022). The impact of intestinal microorganisms and their metabolites on type 1 diabetes mellitus. *Diabetes, Metabolic Syndrome and Obesity: Targets and Therapy*, *15*, 1123–1139.

10 Role of Probiotics in Alzheimer's Disease and Parkinson's Disease

Mohsin Vahid Khan[1], Mohd Rumman[1], Anas Islam[2], Mohd Kamil Hussain[3], and Gulam Rabbani[4]

[1] Department of Biosciences, Integral University, Lucknow, India
[2] School of Pharmaceutical Sciences, Integral University, Lucknow, India
[3] Department of Chemistry, Govt. Raza PG College, Rampur, India
[4] IT-medical Fusion Center, Gyeongbuk, Republic of Korea

10.1 INTRODUCTION

The notion of probiotics has been given by father of natural immunity, Elie Metchnikoff, a Nobel laureate who believed in achieving good health followed by deferred caducity on introducing edibles full of useful bacteria in routine diet (Kumar et al., 2020). The term probiotics comes from Greek origin, which signifies "for life," and is suggested to be coined and used by Ferdinand Vergin in his article "Anti- und Probiotika" published in 1954. This article revealed the advantageous consequence of beneficial bacteria on equating several antibacterial compounds, including antibiotics on intestinal microbiota (Markowiak & Slizewska, 2017). The World Health organization (WHO) and the Food and Agricultural Organization of the United Nations (FAO)-associated group of internationally acclaimed scientists working on probiotics defined it in most accepted description across the globe in year 2001. They refer "live microorganisms which when administered in adequate amounts confer a health benefit on host" as probiotics. Probiotics rich in beneficial bacteria along with yeast restores homoeostasis via maintaining gut biome of the overwhelming number of bacterial species residing in the human gut against further rise of other harmful bacteria even in experiencing the misbalancing conditions. This condition includes surgical procedure, unsuitable diet, immunosuppressive treatment, and gastrointestinal ailments (Alvarez-Olmos & Oberhelman, 2001; Doron & Gorbach, 2006; Slizewska, Markowiak-Kopec, & Slizewska, 2021; Snyder, 1997).

10.2 PROBIOTICS AND ITS BENEFICIAL EFFECTS ON HUMAN

Humans are exposed to various microorganisms, both pathogenic and nonpathogenic, which can pose health risks. Henry Tissier and Theodor Escherich's research on human intestine flora revealed that formula-fed babies lack fiber-digesting bifidobacteria, while breastfed infants have substantial bifidobacteria (Kechagia et al., 2013; Maduell, 2019). In today's scenario, probiotics has gained much popularity and is currently used for the treatment of several health complications. It has been proposed that obesity, a metabolic disorder affecting people of all ages, especially youngsters, across the world and referred to as pandemic of this century, can also be treated by using probiotics as an alternative therapy (Cai et al., 2023). Furthermore, probiotics may augment immunity in case of viral infection such as Covid-19 and human immunodeficiency (Blazquez-Bondia et al., 2022; Das, 2022). Table 10.1 and Figure 10.1 depict how probiotics produces beneficial consequences for several diseases and complications.

10.3 NEURODEGENERATIVE DISEASES (NDDS) AND PROTEIN AGGREGATION

The onset of dementia by usual manifestation of neurodegenerative disease has reached at catastrophic level. The loss of neurons is considered as the culprit of neurodegenerative disorders, a group of neurological ailment, resulting in the disruption of the peripheral nervous system (PNS) and the central nervous system (CNS) functioning (Wilson et al., 2023). Aggregation of proteins leads to alterations in proteostasis, causing neurodegenerative diseases. Several factors can trigger protein aggregation and cause neurodegenerative and non-neurodegenerative diseases. Proteins aggregates exhibit two morphological variants, amorphous and amyloids, characterized by ordered and unordered conformation of aggregates (Kaushik & Cuervo, 2015; Khan et al., 2015, 2017). Neurodegenerative infirmities include Alzheimer's disease, Parkinson's disease, Huntington's disease, Creutzfeldt disease, cerebral supra nuclear palsy and

DOI: 10.1201/9781003536802-10

TABLE 10.1
Beneficial Effects of Probiotics with Specific Species/Strains of Microorganisms

Microorganism sp./ Strain	Heath Benefits	Reference
Lactobacillus rhamnosus GG (LGG), *Saccharomyces boulardi*	Diarrhea prevention	Guarino, Guandalini, & Lo Vecchio, 2015
Lactobacillus brevis RK03	Antidiabetic, antihypertensive	Wu, Hsueh, Kuo, & Liu, 2018
Clostridium butyricum	Treatment of irreversible bowel syndrome	Sun et al., 2018
Akkermansia muciniphila	Augment insulin sensitivity, diminish glucose level (fasting) and serum triglyceride	Zhai, Feng, Arjan, & Chen, 2019
Lactobacillus fermentum, *Lactobacillus acidophilu*, *Bifidobacterium lactis*	Coronary heart disease by reducing plasma total cholesterol and LDL	Guo et al., 2011
Faecalibacterium prausnitzii	Anti-inflammatory effect	Martin et al., 2017
Bifidobacterium lactis Bb12 (BB12) and *Lactobacillus rhamnosus* GG (ATCC 53103)	Reduction in eczema sternness	Isolauri et al., 2000
Lactobacillus casei and *Lactobacillus acidophilus*	Antimicrobial activity against resistant strain	Karska-Wysocki, Bazo, & Smoragiewicz, 2010
Lactobacillus bulgaricus	Expressing antimicrobial peptide	Moro-Garcia et al., 2013
Lactobacillus casei Shirota	Aflatoxins, mycotoxins removal	Nikbakht Nasrabadi et al., 2013
Clostridium butyricum	Neuroprotective effect	Liu et al., 2015
Bifidobacterium lactis Bb12 (BB12) and *Lactobacillus rhamnosus* GG (LGG)	Colon cancer suppression	Rafter et al., 2007
Pediococcus acidilactici Kp10	Wide range of antimicrobial activity	Abbasiliasi et al., 2017
Brevibacillus laterosporus	Antibacterial and antitumor potential	Jiang et al., 2017
Lactobacillus gasseri SBT2055	Lowering adipose tissue mass and reduction of body weight	Kadooka et al., 2010

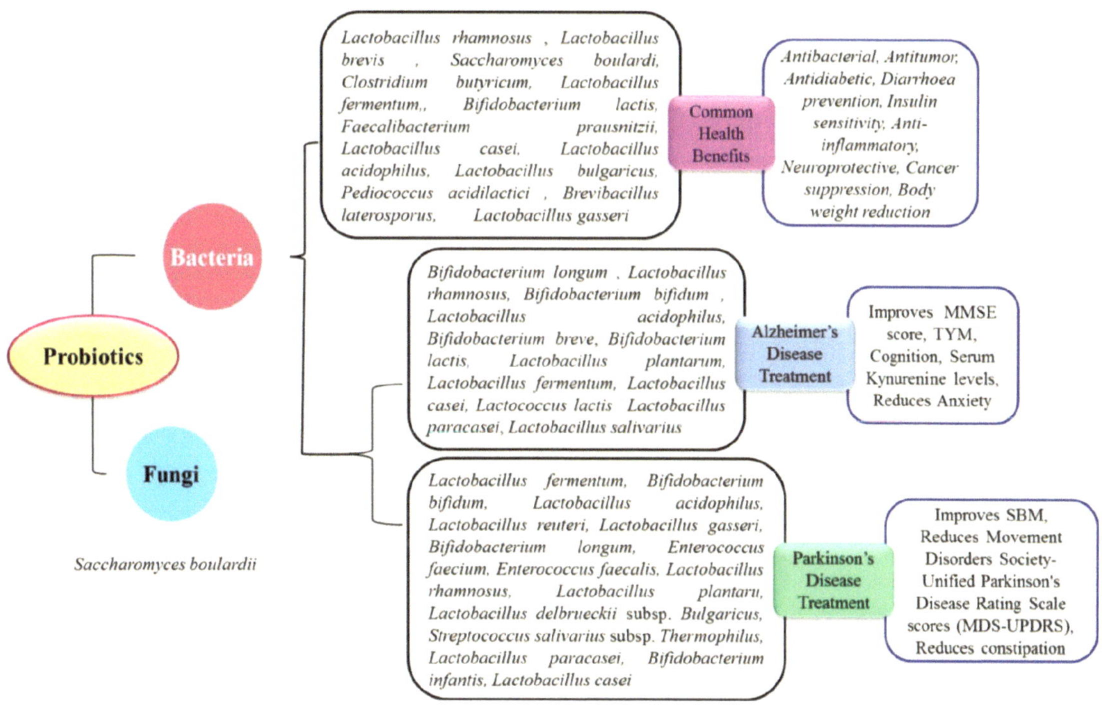

FIGURE 10.1 Role of probiotics with different bacterial strain providing common health benefits, and treatment of neurodegenerative diseases like Alzheimer's disease and Parkinson's disease.

dementia with levy bodies, etc., but the most common neurodegenerative disease encompasses Alzheimer's disease (AD) and Parkinson's disease (PD) (Davenport et al., 2023).

10.4 ALZHEIMER'S DISEASE

Alzheimer's disease is a progressive neurodegenerative disease affecting large populations worldwide. The eponym was given by Emil Kraepelin to this disease in regard to German psychiatric Alois Alzheimer for elucidating clinical and pathological characteristics of the disease for the first time. It is influenced by vascular impairment, aging, inflammation, environmental factors, and lifestyle. Dementia, a reduction in daily routines and cognitive abilities, is linked to Alzheimer's disease. Depression and anxiety are also neuropsychiatric symptoms. The disease is caused by misfolding and protein aggregation, with evidence suggesting that these factors contribute to its progression (Breijyeh & Karaman, 2020; Cipriani, Dolciotti, Picchi, & Bonuccelli, 2011; Mendez, 2021). Alzheimer's disease is characterized by the presence of plaques and neurofibrillary tangles, caused by Aβ peptide. Onset of Alzheimer's occurs when the Aβ-42 peptide variant forms due to proteolysis of amyloid beta precursor protein (APP) and undergoes oligomerization. The tau protein's neurofibrillary entanglement occurs through paired helical filaments (PHF) (DeTure & Dickson, 2019; O'Brien & Wong, 2011).

10.4.1 Gut Microbiota and Alzheimer's Disease

The gut microbiome significantly influences nuance of gut and brain function, with differences in gut microbiome populations influenced by factors like environmental, geographical, medication, and birth mother's gut microbiome. Alzheimer's disease-related pathogenesis shows link between dysbiosis of intestinal microbiota by resulting in the intensification of not only gut but also blood–brain barrier (BBB). Furthermore, increased inflammation, proinflammatory cytokine production, deformation in signaling pathway, lipopolysaccharide and excessive amyloid secretion are also mediated by gut microbiome disproportion (Denman, Park, & Jo, 2023; Jiang et al., 2017). Multitechnique investigation of gut microbiome evidences that a patient of Alzheimer's disease exhibits loss of microbiota assortment. Similarly, at level of phylum, in case of these patients, amplified Bacteroidetes and lessened Firmicutes have been observed. Furthermore, diminished abundance of *Eubacterium rectale* (anti-inflammatory bacteria) and augmented abundance of *Escherichial shigella* (proinflammatory bacteria) are found in Alzheimer's patients (Cattaneo et al., 2017; Vogt et al., 2017).

10.5 PARKINSON'S DISEASE

Parkinson's disease is a progressive neurodegenerative disorder similar to AD characterized by movement impairment and synucleinopathy. Neuropsychiatric symptoms exhibited by patients of Parkinson's disease are lessened sensation, passiveness, reduced cognition, psychosis, and discomforted sleep due to altered sleep–wake cycle. Nevertheless, bradykinesia, limb stiffness, hypomimia, tremble, postural instability, gait abnormalities, and tremor at rest are considered as recent criteria for physical symptoms to define Parkinson's disease (Bloem, Okun, & Klein, 2021; Dauer & Przedborski, 2003). Several environmental factors, including air pollution, chemicals like organochlorines in the form of pesticides, lifestyle, and even molecules of pharmaceutical importance (calcium channel blocker, ibuprofen and statins) are associated with the onset of Parkinson's disease (Meade, Fairlie, & Mason, 2019).

10.5.1 Gut Microbiota and Parkinson's Disease

It has been evidenced that there is direct correlation between gut microbiota and either early Parkinson's disease stage or with onset. They can potentially affect the central nervous system's pathophysiology by altering endocrine and neurological pathways associated with the gut–brain axis (GBA) (Zhang, Tang, & Guo, 2023). The dysbiosis-Parkinson's disease study suggests intestinal microbiota alteration in patient suffering from Parkinson's disease on comparing control which is assumed to be responsible for aggregation of α-synuclein, formation of Lewy body (LB), and inflammation. The gut microbiota dysbiosis is characterized by augmentation of Lactobacillaceae, *Ruminococcus bromii*, *Lactobacillus mucosae*, *Papillibacter cinnamivorans*, *Prevotella copri*, *Catabacter hongkongensis*, *Christensenella minuta*, *Oscillospira*, *Catabacter*, Bifidobacteriaceae, *Christensenella*, Verrucomicrobiaceae, and Ruminococcaceae abundance while decrease of *Bacteroidesdorei*, *Prevotella*, *Faecalibacterium*, *Stoqueichus massiliensis*, *Bacteroides massiliensis*, *Coprococcus eutactus*, *Dorea longicatena*, *Blautia glucerasea*, *Bacteroides coprocola*, Lachnospiraceae, *Bacteroides plebeius*, *Ruminococcus callidus*, *Bacteroides*, *Dorea*, and Prevotellaceae abundance (Berthouzoz et al., 2023; Petrov et al., 2017).

10.6 CURRENT MANAGEMENT STRATEGIES FOR NEURODEGENERATIVE DISEASES

Neurodegenerative diseases are disorders causing the degeneration of the nervous system, presenting significant clinical challenges. Current management strategies focus on the symptom relief and supportive care, while emerging therapies include pharmacological interventions, lifestyle modifications, rehabilitation, and emerging therapies.

10.6.1 Pharmacological Management

10.6.1.1 Alzheimer's Disease

Currently, the pharmacological treatment of Alzheimer's disease includes cholinesterase inhibitors (donepezil, rivastigmine, and galantamine) and *N*-methyl-D-aspartate (NMDA) receptor antagonists (memantine) (Grossberg, 2003; Tang, Wang, & Ren, 2023). Cholinesterase inhibitors enhance cholinergic neurotransmission by inhibiting acetylcholinesterase, while memantine regulates glutamate activity, proposing neuroprotective effects. Recent advancements also include anti-amyloid beta and tau-targeting therapies, such as aducanumab, which aim to address the underlying pathology of AD (J. Cummings et al., 2024).

10.6.1.2 Parkinson's Disease

Management strategies for Parkinson's disease primarily involve dopaminergic therapies, particularly levodopa combined with carbidopa (Murakami et al., 2023). Emerging therapies like gene therapy and deep brain stimulation are increasingly being explored as alternative management options for advanced stages of PD (Axelsen & Woldbye, 2018).

10.6.2 Nonpharmacological Management

10.6.2.1 Lifestyle Modifications

Adopting a healthy lifestyle can significantly influence the management of neurodegenerative diseases. Regular physical activity, cognitive stimulation, and proper nutrition are shown to play vital role in maintaining overall health and potentially delaying progression (Alanazi, 2024). Diets rich in antioxidants, omega-3 fatty acids, and vitamins are promoted in managing Alzheimer's and other forms of neurodegeneration (Morris, 2009).

10.6.2.2 Rehabilitation

Rehabilitation remains crucial in the management of neurodegenerative diseases. Physical therapy, occupational therapy, and speech therapy are tailored to individual patient needs, improving mobility, daily functioning, and communication (Santiago & Potashkin, 2023). Cognitive rehabilitation strategies may help individuals with mild cognitive impairment, and early stages of Alzheimer's disease retain cognitive function and improve quality of life (Bahar-Fuchs, Clare, & Woods, 2013).

10.6.2.3 Emerging Therapies

Gene therapy has shown positive impact in preclinical models for several neurodegenerative disorders (W. Chen, Hu, & Ju, 2020). Anti-inflammatory approaches, such as the use of non-steroidal anti-inflammatory drugs (NSAIDs) and monoclonal antibodies targeting neuroinflammation, are being studied for their potential to modify disease progression (Deardorff & Grossberg, 2017). Future therapies targeting mitochondrial dysfunction, proteostasis, and neuroprotection are anticipated to revolutionize the management of neurodegenerative diseases.

10.7 PROBIOTICS AND ALZHEIMER'S DISEASE TREATMENT

Certain strains of probiotics can produce neuroactive compounds, including gamma-aminobutyric acid (GABA) and serotonin, which play a vital role in mood regulation and cognitive function (Y. Chen & Xu, 2021; Tette, Kwofie, & Wilson, 2022). Preliminary studies suggest that certain probiotics may interfere with the aggregation of Aβ peptides, potentially slowing disease progression (Seira Curto et al., 2022). The advantageous effects on probiotics have been examined on various animal models as well as Alzheimer's patients with distinct age and disease status such as mild, moderate, and severe Alzheimer's disease. Some specific methods as primary and secondary outcomes are frequently measured for grading of mental status of the Alzheimer's patients prior to and after the probiotics treatment. The primary outcomes include Mini Mental State Examination (MMSE), Verbal Fluency Test (CFT), and Clock Drawing Test (CDT) scores (Hilgert, 2023; Tamtaji, Heidari-Soureshjani et al., 2019). Similarly, another cognitive screening method, Test Your Memory (TYM), is also used to access different cognitive status of Alzheimer's patients. Furthermore, Activities of Daily Living (ADL), Generalized Anxiety Disorder (GAD-7), and Lawton Instrumental Activities of Daily Living (IADL) scale, oxidative stress, metabolic profile, and inflammation biomarkers are considered as secondary outcomes (Agahi et al., 2018; Akhgarjand et al., 2022). Table 10.2 depicts beneficial effects of probiotics on Alzheimer's patients of varied anthropometric characteristics, containing different bacterial strains.

10.8 PROBIOTICS AND PARKINSON'S DISEASE TREATMENT

Recent research indicates that probiotics can effectively alleviate the negative effects of Parkinson's disease, particularly in addressing common gastrointestinal issues like constipation and dysbiosis. Probiotic strains can improve bowel regularity and overall gut health, enhancing patients' quality of life. Additionally, probiotics may have direct neuroprotective effects, modulating neuroinflammation, and promoting neurogenesis (Sarubbo, Cavallucci, & Pani, 2022). Probiotics may help with sleep disturbances, anxiety, and depressive symptoms in PD. Current research is still evolving, but clinical trials show that probiotic supplementation can improve bowel habits, relieve constipation, and improve daily activity. The Movement Disorders Society Unified Parkinson's Disease Rating Scale (MDS-UPDRS) and spontaneous bowel movements (SBM) scales indicate the potential benefits of probiotics in Parkinson's patients. However, further research is needed to fully understand

TABLE 10.2

Beneficial Effect of Probiotics with Specific Species/Strains of Microorganisms on Alzheimer's Disease Patients

Probiotics Strain	No. of Patients in Clinical Trial	Treatment Duration	Cognitive and Other Clinical Impact	Reference
Bifidobacterium longum R0175 and *Lactobacillus rhamnosus* HA-114	90	12 weeks	Reduced anxiety, improved daily activity, improved MMSE score	Akhgarjand et al., 2022
Lactobacillus acidophilus, Bifidobacterium longum, and *Bifidobacterium bifidum*	79	12 weeks	Augmented cognition, improved MMSE score	Tamtaji, Heidari-Soureshjani et al., 2019
Bifidobacterium breve A1	121	12 weeks	Improved MMSE score	Kobayashi, Kuhara, Oki, & Xiao, 2019
Bifidobacterium lactis, Lactobacillus plantarum, Lactobacillus acidophilus, Lactobacillus fermentum, Bifidobacterium longum, and *Bifidobacterium bifidum*	48	12 weeks	Improved Test Your Memory (TYM) score, Augmented cognition	Agahi et al., 2018
Lactobacillus plantarum C29	133	12 weeks	Augmented cognition	Hwang et al., 2019
Lactobacillus fermentum, Lactobacillus acidophilus, Bifidobacterium bifidum, and *Lactobacillus casei*	60	12 weeks	Improved MMSE score	Akbari et al., 2016
Lactococcus lactis W19, *Lactobacillus paracasei* W20, *Lactobacillus acidophilus* W22, *Bifidobacterium lactis* W51, *Bifidobacterium lactis* W52, *Lactobacillus casei* W56, *Lactobacillus plantarum* W62, *Bifidobacterium bifidum* W23, and *Lactobacillus salivarius* W24	20	4 weeks	Elevated serum kynurenine levels	Leblhuber et al., 2018

their potential benefits (Barichella et al., 2016; Georgescu et al., 2016; Tamtaji, Taghizadeh et al., 2019; Tan et al., 2021), as illustrated in Table 10.3.

10.9 MECHANISM OF PROBIOTICS ACTION

Probiotics are live microorganisms that improve health by modulating gut microbiota and enhancing the immune system. They have various mechanisms, including competitive inhibition of pathogens, gut microbiome modulation, bioactive metabolite production, and immunomodulation as shown in Figure 10.2. Their popularity stems from their potential health benefits in gastrointestinal health, immune function, and metabolic disease risk reduction. Understanding these mechanisms is crucial for clinical and therapeutic use.

10.9.1 Competitive Inhibition of Pathogens

One of the primary mechanisms by which probiotics exert their beneficial effects is through competitive inhibition. Probiotics can adhere to the intestinal epithelium, occupying binding sites that pathogenic bacteria might otherwise use (Raheem, Liang, Zhang, & Cui, 2021; van Zyl, Deane, & Dicks, 2020). Probiotics can outcompete pathogens for essential nutrients such as carbohydrates, amino acids, and vitamins (Chandrasekaran, Weiskirchen, & Weiskirchen, 2024; Zhou, Chen, Patil, & Dong, 2024). By consuming these resources more rapidly, probiotics can limit the availability of critical nutrients to harmful pathogens, thus inhibiting their growth. In the gastrointestinal tract, *Lactobacillus*, *Bifidobacterium*, and *Saccharomyces* are commonly studied probiotics known for their competitive inhibition capabilities. For instance, *Lactobacillus rhamnosus* has demonstrated the ability to inhibit the growth of pathogens such as *Escherichia coli* and *Clostridium difficile* (Dempsey & Corr, 2022; Fijan, 2014). By maintaining a balanced microbiota, probiotics can prevent conditions such as bacterial vaginosis and urinary tract infections. The competitive inhibition of pathogens extends beyond the gastrointestinal and urogenital tracts. Probiotics may also impact respiratory health by enhancing innate immunity and competing with pathogens such as *Streptococcus pneumoniae* and *Haemophilus influenzae* (Darbandi et al., 2021).

10.9.2 Modulation of Gut Microbiota

Probiotics are known to modulate the gut microbiota composition, promoting the growth of beneficial bacteria while inhibiting the proliferation of harmful ones. Through the fermentation of dietary fibers (prebiotics), probiotics

TABLE 10.3
Beneficial Effect of Probiotics with Specific Species/Strains of Microorganisms on Parkinson's Disease

Probiotics Strain	No. of Patients in Clinical Trial	Treatment Duration	Clinical Impact	Reference
Lactobacillus fermentum, Bifidobacterium bifidum, Lactobacillus acidophilus, and *Lactobacillus reuteri*	60	12 weeks	Reduced insulin resistance, diminished insulin level, and reduced MDS-UPDRS	Tamtaji, Taghizadeh et al., 2019
Bifidobacterium bifidum, Lactobacillus gasseri, Lactobacillus acidophilus, Bifidobacterium longum, Enterococcus faecium, Lactobacillus reuteri, Enterococcus faecalis, and *Lactobacillus rhamnosus*	280	4 weeks	Improved SBM	Tan et al., 2021
Bifidobacterium, Lactobacillus plantaru, Enterococcus faecium, Lactobacillus rhamnosus GG, *Lactobacillus delbrueckii* spp. *Bulgaricus, Streptococcus salivarius* spsp. *Thermophilus, Lactobacillus paracasei,* and *Lactobacillus acidophilus*	120	4 weeks	Prevents constipation	Barichella et al., 2016
Lactobacillus sp. and *Bifidobacterium* sp.	55	8 weeks	Improved bowel habits	Ibrahim et al., 2020
Bifidobacterium infantis and *Lactobacillus acidophilus*	40	12 weeks	Relieves abdominal pain and bloating symptoms	Georgescu et al., 2016
Lactobacillus casei Shirota	40	5 weeks	Improved bowel habits	Cassani et al., 2011

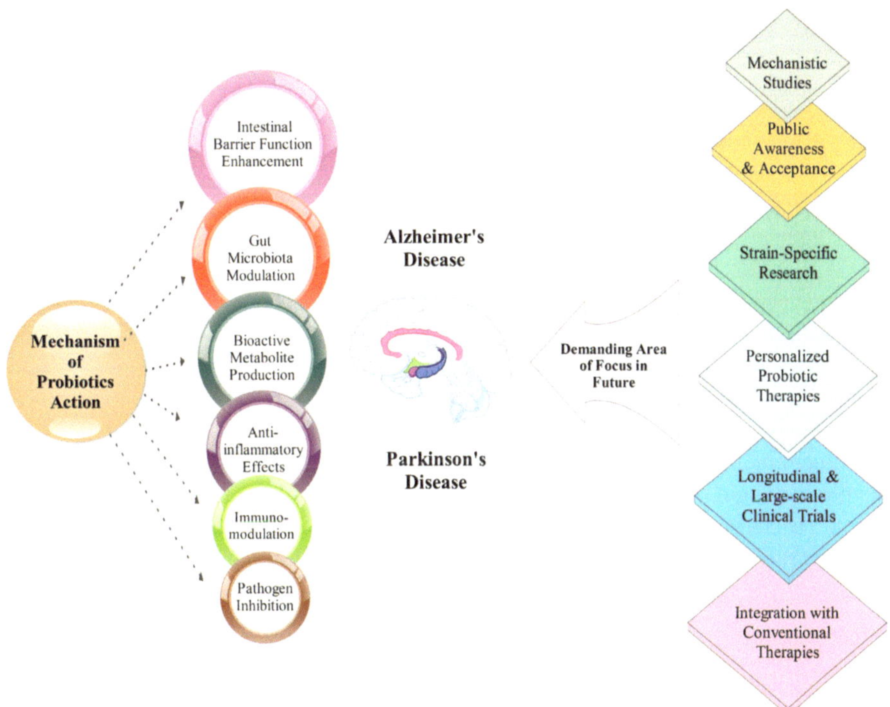

FIGURE 10.2 Distinct mechanism of action exhibited by probiotics and further aiming field for treating the neurodegenerative diseases.

stimulate the production of short-chain fatty acids (SCFAs) like butyrate, propionate, and acetate. SCFAs play a vital role in colonic health, provide energy for colonocytes, and possess anti-inflammatory properties, contributing to gut barrier integrity and overall well-being (Cook & Sellin, 1998; J. H. Cummings et al., 1987; den Besten et al., 2013). Antibiotic use, poor diet, and illness can disrupt gut microbiota, leading to dysbiosis. Probiotics can help restore microbial diversity and enhance resilience against pathogenic invasion (Valdes, Walter, Segal, & Spector, 2018; Zhou et al., 2024).

10.9.3 Enhancement of Intestinal Barrier Function

The intestinal barrier is crucial for gut health and well-being, facilitating nutrient absorption, immune function, and protection against pathogens. However, factors like diet, stress, medications, and infections can compromise this barrier, leading to "leaky gut." Recent research suggests probiotics can enhance intestinal barrier function by upregulating tight junction proteins and promoting mucin secretion, providing potential therapeutic strategies for gut-related disorders (Gou et al., 2022; Rose, Odle, Blikslager, & Ziegler, 2021). Studies have shown that certain probiotics can promote the proliferation and differentiation of intestinal epithelial cells (Xie et al., 2024; Yan & Polk, 2020). This leads to improved mucosal barrier function and resistance to pathogens. A randomized controlled trial involving patients with IBD demonstrated that a specific probiotic strain significantly improved intestinal permeability and clinical symptoms, highlighting its potential in managing inflammatory conditions (Mishra et al., 2022). Animal studies have shown that administration of probiotics after antibiotic treatment can prevent disruption of the intestinal barrier and restore gut microbiota balance (Kesavelu & Jog, 2023). Research on individuals with IBS has indicated that certain probiotics can reduce symptoms and improve gut barrier integrity, partly through modulation of immune responses.

10.9.4 Production of Bioactive Metabolites

Probiotics can also produce bioactive metabolites that contribute to their health benefits. These include SCFAs, enzymes, vitamins (such as B vitamins), and other metabolites that may exert systemic effects (Cook & Sellin, 1998; D. Wang et al., 2022). For instance, SCFAs have been shown to influence host metabolism and immune responses. Additionally, the metabolic by-products of probiotics can modulate gut–brain axis signaling, potentially impacting mood and cognitive function (Carabotti, Scirocco, Maselli, & Severi, 2015; Parker, Fonseca, & Carding, 2020). Lactic acid, acetic acid, and other organic acids produced by probiotics can lower the pH of the gut environment, inhibiting the growth of pathogenic bacteria and promoting a healthy microflora balance (Indira et al., 2019; Y. Wang et al., 2019).

10.9.5 Immunomodulation

Probiotics play a significant role in modulating the immune system. Probiotics enhance the activity of antigen-presenting cells (APCs), such as dendritic cells and macrophages (Saito et al., 2024; Zheng, Liwinski, & Elinav, 2020). By facilitating better antigen presentation, probiotics help prime T cells, leading to an improved adaptive immune response. There is compelling evidence that certain probiotics can reduce the risk of allergic diseases, such as eczema and asthma, particularly in infants and children. This effect is thought to be mediated through the modulation of Th1/Th2 balance and enhanced regulatory T cell function (N. Guo & Lv, 2023; Kim et al., 2018).

10.9.6 Anti-Inflammatory Effects

In addition to their role in modulating the immune response, probiotics exhibit anti-inflammatory properties (Cristofori et al., 2021). They can reduce the production of pro-inflammatory cytokines (such as TNF-alpha and IL-6) while promoting anti-inflammatory cytokines (like IL-10). This balance is crucial in managing various inflammatory conditions, including inflammatory bowel disease (IBD) and irritable bowel syndrome (IBS) (Aghamohammad et al., 2023; Katz, 2006; Satish Kumar et al., 2022; Smits et al., 2005).

10.10 CONCLUSION

Research shows that the gut microbiome influences the central nervous system through neurochemical pathways, immune modulation, and metabolite production. Probiotics, such as *Lactobacillus* and *Bifidobacterium* strains, can reduce anxiety and depression symptoms, enhance cognitive function, and potentially protect against neurodegenerative diseases. However, strain specificity is crucial, as not all probiotics are created equal. Current studies are in preliminary stages, with small sample sizes and variable methodologies. The effects of probiotics may not be universally applicable across individuals, necessitating more robust clinical trials. Standardization in probiotic formulation and administration is also needed, along with a thorough examination of their interaction with the CNS.

10.11 FUTURE PERSPECTIVES

Future research should focus on strain-specific research to identify bacteria with therapeutic potential for neurological disorders. Personalized probiotic therapies could be developed to address individual microbiome profiles, enhancing treatment effectiveness and minimizing side effects. Understanding the mechanisms of probiotics' effects on the CNS is crucial for new therapeutic strategies. Longitudinal and large-scale clinical trials are needed to establish a robust clinical framework. Probiotics could also be integrated with conventional treatments, enhancing treatment outcomes. Public awareness and acceptance of probiotics are also essential.

ABBREVIATIONS

AD Alzheimer's disease
ASD autism spectrum disorder
PD Parkinson's disease

REFERENCES

Abbasiliasi, S., Tan, J. S., Bashokouh, F., Ibrahim, T. A. T., Mustafa, S., Vakhshiteh, F., et al. (2017). In vitro assessment of Pediococcus acidilactici Kp10 for its potential use in the food industry. *BMC Microbiol*, *17*(1), 121.

Agahi, A., Hamidi, G. A., Daneshvar, R., Hamdieh, M., Soheili, M., Alinaghipour, A., et al. (2018). Does severity of Alzheimer's disease contribute to its responsiveness to modifying gut microbiota? A double blind clinical trial. *Front Neurol, 9*, 662.

Aghamohammad, S., Sepehr, A., Miri, S. T., Najafi, S., Pourshafie, M. R., & Rohani, M. (2023). Ameliorating inflammation in an in vitro model by screening the anti-inflammatory and immunomodulatory roles of putative probiotics in inflammatory bowel disease. *Heliyon, 9*(9), e19475.

Akbari, E., Asemi, Z., Daneshvar Kakhaki, R., Bahmani, F., Kouchaki, E., Tamtaji, O. R., et al. (2016). Effect of probiotic supplementation on cognitive function and metabolic status in Alzheimer's disease: A randomized, double-blind and controlled trial. *Front Aging Neurosci, 8*, 256.

Akhgarjand, C., Vahabi, Z., Shab-Bidar, S., Etesam, F., & Djafarian, K. (2022). Effects of probiotic supplements on cognition, anxiety, and physical activity in subjects with mild and moderate Alzheimer's disease: A randomized, double-blind, and placebo-controlled study. *Front Aging Neurosci, 14*, 1032494.

Alanazi, M. A. (2024). The role of physical activity in adjunctive nursing management of neuro-degenerative diseases among older adults: A systematic review of interventional studies. *Life (Basel), 14*(5), 597.

Alvarez-Olmos, M. I., & Oberhelman, R. A. (2001). Probiotic agents and infectious diseases: A modern perspective on a traditional therapy. *Clin Infect Dis, 32*(11), 1567–1576.

Axelsen, T. M., & Woldbye, D. P. D. (2018). Gene therapy for Parkinson's disease, an update. *J Parkinsons Dis, 8*(2), 195–215.

Bahar-Fuchs, A., Clare, L., & Woods, B. (2013). Cognitive training and cognitive rehabilitation for persons with mild to moderate dementia of the Alzheimer's or vascular type: A review. *Alzheimers Res Ther, 5*(4), 35.

Barichella, M., Pacchetti, C., Bolliri, C., Cassani, E., Iorio, L., Pusani, C., et al. (2016). Probiotics and prebiotic fiber for constipation associated with Parkinson disease: An RCT. *Neurology, 87*(12), 1274–1280.

Berthouzoz, E., Lazarevic, V., Zekeridou, A., Castro, M., Debove, I., Aybek, S., et al. (2023). Oral and intestinal dysbiosis in Parkinson's disease. *Rev Neurol (Paris), 179*(9), 937–946.

Blazquez-Bondia, C., Parera, M., Catala-Moll, F., Casadella, M., Elizalde-Torrent, A., Aguilo, M., et al. (2022). Probiotic effects on immunity and microbiome in HIV-1 discordant patients. *Front Immunol, 13*, 1066036.

Bloem, B. R., Okun, M. S., & Klein, C. (2021). Parkinson's disease. *Lancet, 397*(10291), 2284–2303.

Breijyeh, Z., & Karaman, R. (2020). Comprehensive review on Alzheimer's disease: Causes and treatment. *Molecules, 25*(24), 5789.

Cai, Y., Liu, P., Zhou, X., Yuan, J., & Chen, Q. (2023). Probiotics therapy show significant improvement in obesity and neurobehavioral disorders symptoms. *Front Cell Infect Microbiol, 13*, 1178399.

Carabotti, M., Scirocco, A., Maselli, M. A., & Severi, C. (2015). The gut-brain axis: Interactions between enteric microbiota, central and enteric nervous systems. *Ann Gastroenterol, 28*(2), 203–209.

Cassani, E., Privitera, G., Pezzoli, G., Pusani, C., Madio, C., Iorio, L., et al. (2011). Use of probiotics for the treatment of constipation in Parkinson's disease patients. *Minerva Gastroenterol Dietol, 57*(2), 117–121.

Cattaneo, A., Cattane, N., Galluzzi, S., Provasi, S., Lopizzo, N., Festari, C., et al. (2017). Association of brain amyloidosis with pro-inflammatory gut bacterial taxa and peripheral inflammation markers in cognitively impaired elderly. *Neurobiol Aging, 49*, 60–68.

Chandrasekaran, P., Weiskirchen, S., & Weiskirchen, R. (2024). Effects of probiotics on gut microbiota: An overview. *Int J Mol Sci, 25*(11), 6022.

Chen, W., Hu, Y., & Ju, D. (2020). Gene therapy for neurodegenerative disorders: Advances, insights and prospects. *Acta Pharm Sin B, 10*(8), 1347–1359.

Chen, Y., & Xu, J. (2021). Regulation of neurotransmitters by the gut microbiota and effects on cognition in neurological disorders. *Nutrients, 13*(6), 2099.

Cipriani, G., Dolciotti, C., Picchi, L., & Bonuccelli, U. (2011). Alzheimer and his disease: A brief history. *Neurol Sci, 32*(2), 275–279.

Cook, S. I., & Sellin, J. H. (1998). Review article: Short chain fatty acids in health and disease. *Aliment Pharmacol Ther, 12*(6), 499–507.

Cristofori, F., Dargenio, V. N., Dargenio, C., Miniello, V. L., Barone, M., & Francavilla, R. (2021). Anti-inflammatory and immunomodulatory effects of probiotics in gut inflammation: A door to the body. *Front Immunol, 12*, 578386.

Cummings, J., Osse, A. M. L., Cammann, D., Powell, J., & Chen, J. (2024). Anti-amyloid monoclonal antibodies for the treatment of Alzheimer's disease. *BioDrugs, 38*(1), 5–22.

Cummings, J. H., Pomare, E. W., Branch, W. J., Naylor, C. P., & Macfarlane, G. T. (1987). Short chain fatty acids in human large intestine, portal, hepatic and venous blood. *Gut, 28*(10), 1221–1227.

Darbandi, A., Asadi, A., Ghanavati, R., Afifirad, R., Darb Emamie, A., Kakanj, M., et al. (2021). The effect of probiotics on respiratory tract infection with special emphasis on COVID-19: Systemic review 2010–20. *Int J Infect Dis, 105*, 91–104.

Das, K. T., Pradhan, S., Chakrabarti, S., Mondal, K. C., & Gosh, K. (2022). Current status of probiotic and related health benefits. *Applied Food Research, 2*(2), 100185.

Dauer, W., & Przedborski, S. (2003). Parkinson's disease: Mechanisms and models. *Neuron, 39*(6), 889–909.

Davenport, F., Gallacher, J., Kourtzi, Z., Koychev, I., Matthews, P. M., Oxtoby, N. P., et al. (2023). Neurodegenerative disease of the brain: A survey of interdisciplinary approaches. *J R Soc Interface, 20*(198), 20220406.

Deardorff, W. J., & Grossberg, G. T. (2017). Targeting neuroinflammation in Alzheimer's disease: Evidence for NSAIDs and novel therapeutics. *Expert Rev Neurother, 17*(1), 17–32.

Dempsey, E., & Corr, S. C. (2022). Lactobacillus spp. for Gastrointestinal health: Current and future perspectives. *Front Immunol, 13*, 840245.

den Besten, G., van Eunen, K., Groen, A. K., Venema, K., Reijngoud, D. J., & Bakker, B. M. (2013). The role of short-chain fatty acids in the interplay between diet, gut microbiota, and host energy metabolism. *J Lipid Res, 54*(9), 2325–2340.

Denman, C. R., Park, S. M., & Jo, J. (2023). Gut-brain axis: gut dysbiosis and psychiatric disorders in Alzheimer's and Parkinson's disease. *Front Neurosci, 17*, 1268419.

DeTure, M. A., & Dickson, D. W. (2019). The neuropathological diagnosis of Alzheimer's disease. *Mol Neurodegener, 14*(1), 32.

Doron, S., & Gorbach, S. L. (2006). Probiotics: Their role in the treatment and prevention of disease. *Expert Rev Anti Infect Ther, 4*(2), 261–275.

Fijan, S. (2014). Microorganisms with claimed probiotic properties: An overview of recent literature. *Int J Environ Res Public Health, 11*(5), 4745–4767.

Georgescu, D., Ancusa, O. E., Georgescu, L. A., Ionita, I., & Reisz, D. (2016). Nonmotor gastrointestinal disorders in older patients with Parkinson's disease: Is there hope? *Clin Interv Aging, 11*, 1601–1608.

Gou, H. Z., Zhang, Y. L., Ren, L. F., Li, Z. J., & Zhang, L. (2022). How do intestinal probiotics restore the intestinal barrier? *Front Microbiol, 13*, 929346.

Grossberg, G. T. (2003). Cholinesterase inhibitors for the treatment of Alzheimer's disease: Getting on and staying on. *Curr Ther Res Clin Exp, 64*(4), 216–235.

Guarino, A., Guandalini, S., & Lo Vecchio, A. (2015). Probiotics for prevention and treatment of Diarrhea. *J Clin Gastroenterol, 49*(Suppl 1), S37–S45.

Guo, N., & Lv, L. L. (2023). Mechanistic insights into the role of probiotics in modulating immune cells in ulcerative colitis. *Immun Inflamm Dis, 11*(10), e1045.

Guo, Z., Liu, X. M., Zhang, Q. X., Shen, Z., Tian, F. W., Zhang, H., et al. (2011). Influence of consumption of probiotics on the plasma lipid profile: a meta-analysis of randomised controlled trials. *Nutr Metab Cardiovasc Dis, 21*(11), 844–850.

Hilgert, S. S., & Dias, D. P. M. (2023). The intestinal microbiota as an ally in the treatment of Alzheimer's disease. *Gut Microbiome 4*, 1–8.

Hwang, Y. H., Park, S., Paik, J. W., Chae, S. W., Kim, D. H., Jeong, D. G., et al. (2019). Efficacy and safety of lactobacillus plantarum C29-fermented soybean (DW2009) in individuals with mild cognitive impairment: A 12-Week, multi-center, randomized, double-blind, placebo-controlled clinical trial. *Nutrients, 11*(2), 305.

Ibrahim, A., Ali, R. A. R., Manaf, M. R. A., Ahmad, N., Tajurruddin, F. W., Qin, W. Z., et al. (2020). Multi-strain probiotics (Hexbio) containing MCP BCMC strains improved constipation and gut motility in Parkinson's disease: A randomised controlled trial. *PLoS One, 15*(12), e0244680.

Indira, M., Venkateswarulu, T. C., Abraham Peele, K., Nazneen Bobby, M., & Krupanidhi, S. (2019). Bioactive molecules of probiotic bacteria and their mechanism of action: A review. *3 Biotech, 9*(8), 306.

Isolauri, E., Arvola, T., Sutas, Y., Moilanen, E., & Salminen, S. (2000). Probiotics in the management of atopic eczema. *Clin Exp Allergy, 30*(11), 1604–1610.

Jiang, C., Li, G., Huang, P., Liu, Z., & Zhao, B. (2017). The gut microbiota and Alzheimer's disease. *J Alzheimers Dis, 58*(1), 1–15.

Jiang, H., Ji, C., Sui, J., Sa, R., Wang, X., Liu, X., et al. (2017). Antibacterial and antitumor activity of Bogorol B-JX isolated from Brevibacillus laterosporus JX-5. *World J Microbiol Biotechnol, 33*(10), 177.

Kadooka, Y., Sato, M., Imaizumi, K., Ogawa, A., Ikuyama, K., Akai, Y., et al. (2010). Regulation of abdominal adiposity by probiotics (Lactobacillus gasseri SBT2055) in adults with obese tendencies in a randomized controlled trial. *Eur J Clin Nutr, 64*(6), 636–643.

Karska-Wysocki, B., Bazo, M., & Smoragiewicz, W. (2010). Antibacterial activity of Lactobacillus acidophilus and Lactobacillus casei against methicillin-resistant Staphylococcus aureus (MRSA). *Microbiol Res, 165*(8), 674–686.

Katz, J. (2006). The role of probiotics in IBD. *Gastroenterol Hepatol (N Y), 2*(1), 16–18.

Kaushik, S., & Cuervo, A. M. (2015). Proteostasis and aging. *Nat Med, 21*(12), 1406–1415.

Kechagia, M., Basoulis, D., Konstantopoulou, S., Dimitriadi, D., Gyftopoulou, K., Skarmoutsou, N., et al. (2013). Health benefits of probiotics: A review. *ISRN Nutr, 2013*, 481651.

Kesavelu, D., & Jog, P. (2023). Current understanding of antibiotic-associated dysbiosis and approaches for its management. *Ther Adv Infect Dis, 10*, 20499361231154443.

Khan, M. V., Ishtikhar, M., Rabbani, G., Zaman, M., Abdelhameed, A. S., & Khan, R. H. (2017). Polyols (Glycerol and Ethylene glycol) mediated amorphous aggregate inhibition and secondary structure restoration of metalloproteinase-conalbumin (ovotransferrin). *Int J Biol Macromol, 94*(Pt A), 290–300.

Khan, M. V., Rabbani, G., Ishtikhar, M., Khan, S., Saini, G., & Khan, R. H. (2015). Non-fluorinated cosolvents: A potent amorphous aggregate inducer of metalloproteinase-conalbumin (ovotransferrin). *Int J Biol Macromol, 78*, 417–428.

Kim, H. W., Hong, R., Choi, E. Y., Yu, K., Kim, N., Hyeon, J. Y., et al. (2018). A probiotic mixture regulates T Cell balance and reduces atopic dermatitis symptoms in mice. *Front Microbiol, 9*, 2414.

Kobayashi, Y., Kuhara, T., Oki, M., & Xiao, J. Z. (2019). Effects of Bifidobacterium breve A1 on the cognitive function of older adults with memory complaints: A randomised, double-blind, placebo-controlled trial. *Benef Microbes, 10*(5), 511–520.

Kumar, R., Sood, U., Gupta, V., Singh, M., Scaria, J., & Lal, R. (2020). Recent advancements in the development of modern probiotics for restoring human gut microbiome dysbiosis. *Indian J Microbiol, 60*(1), 12–25.

Leblhuber, F., Steiner, K., Schuetz, B., Fuchs, D., & Gostner, J. M. (2018). Probiotic supplementation in patients with Alzheimer's dementia - an explorative intervention study. *Curr Alzheimer Res, 15*(12), 1106–1113.

Liu, J., Sun, J., Wang, F., Yu, X., Ling, Z., Li, H., et al. (2015). Neuroprotective effects of clostridium butyricum against vascular dementia in mice via metabolic butyrate. *Biomed Res Int, 2015*, 412946.

Maduell, E. F., . Pascual, C.C. (2019). The origins of gut microbiome research in Europe: From escherich to nissle. *Human Microbiome Journal, 14*, 100065.

Markowiak, P., & Slizewska, K. (2017). Effects of probiotics, prebiotics, and synbiotics on human health. *Nutrients, 9*(9), 1021.

Martin, R., Miquel, S., Benevides, L., Bridonneau, C., Robert, V., Hudault, S., et al. (2017). Functional characterization of novel faecalibacterium prausnitzii strains isolated from healthy volunteers: A step forward in the use of F. prausnitzii as a next-generation probiotic. *Front Microbiol, 8*, 1226.

Meade, R. M., Fairlie, D. P., & Mason, J. M. (2019). Alpha-synuclein structure and Parkinson's disease - lessons and emerging principles. *Mol Neurodegener, 14*(1), 29.

Mendez, M. F. (2021). The relationship between anxiety and Alzheimer's disease. *J Alzheimers Dis Rep, 5*(1), 171–177.

Mishra, J., Stubbs, M., Kuang, L., Vara, N., Kumar, P., & Kumar, N. (2022). Inflammatory bowel disease therapeutics: A focus on probiotic engineering. *Mediators Inflamm, 2022*, 9621668.

Moro-Garcia, M. A., Alonso-Arias, R., Baltadjieva, M., Fernandez Benitez, C., Fernandez Barrial, M. A., Diaz Ruisanchez, E., et al. (2013). Oral supplementation with Lactobacillus delbrueckii subsp. bulgaricus 8481 enhances systemic immunity in elderly subjects. *Age (Dordr), 35*(4), 1311–1326.

Morris, M. C. (2009). The role of nutrition in Alzheimer's disease: Epidemiological evidence. *Eur J Neurol, 16*(Suppl 1), 1–7.

Murakami, H., Shiraishi, T., Umehara, T., Omoto, S., & Iguchi, Y. (2023). Recent advances in drug therapy for Parkinson's disease. *Intern Med, 62*(1), 33–42.

Nikbakht Nasrabadi, E., Jamaluddin, R., Abdul Mutalib, M. S., Khaza'ai, H., Khalesi, S., & Mohd Redzwan, S. (2013). Reduction of aflatoxin level in aflatoxin-induced rats by the activity of probiotic Lactobacillus casei strain Shirota. *J Appl Microbiol, 114*(5), 1507–1515.

O'Brien, R. J., & Wong, P. C. (2011). Amyloid precursor protein processing and Alzheimer's disease. *Annu Rev Neurosci, 34*, 185–204.

Parker, A., Fonseca, S., & Carding, S. R. (2020). Gut microbes and metabolites as modulators of blood-brain barrier integrity and brain health. *Gut Microbes, 11*(2), 135–157.

Petrov, V. A., Saltykova, I. V., Zhukova, I. A., Alifirova, V. M., Zhukova, N. G., Dorofeeva, Y. B., et al. (2017). Analysis of gut microbiota in patients with Parkinson's disease. *Bull Exp Biol Med, 162*(6), 734–737.

Rafter, J., Bennett, M., Caderni, G., Clune, Y., Hughes, R., Karlsson, P. C., et al. (2007). Dietary synbiotics reduce cancer risk factors in polypectomized and colon cancer patients. *Am J Clin Nutr, 85*(2), 488–496.

Raheem, A., Liang, L., Zhang, G., & Cui, S. (2021). Modulatory effects of probiotics during pathogenic infections with emphasis on immune regulation. *Front Immunol, 12*, 616713.

Rose, E. C., Odle, J., Blikslager, A. T., & Ziegler, A. L. (2021). Probiotics, prebiotics and epithelial tight junctions: A promising approach to modulate intestinal barrier function. *Int J Mol Sci, 22*(13), 6729.

Saito, S., Okuno, A., Peng, Z., Cao, D. Y., & Tsuji, N. M. (2024). Probiotic lactic acid bacteria promote anti-tumor immunity through enhanced major histocompatibility complex class I-restricted antigen presentation machinery in dendritic cells. *Front Immunol, 15*, 1335975.

Santiago, J. A., & Potashkin, J. A. (2023). Physical activity and lifestyle modifications in the treatment of neurodegenerative diseases. *Front Aging Neurosci, 15*, 1185671.

Sarubbo, F., Cavallucci, V., & Pani, G. (2022). The influence of gut microbiota on neurogenesis: Evidence and hopes. *Cells, 11*(3).

Satish Kumar, L., Pugalenthi, L. S., Ahmad, M., Reddy, S., Barkhane, Z., & Elmadi, J. (2022). Probiotics in irritable bowel syndrome: A review of their therapeutic role. *Cureus, 14*(4), e24240.

Seira Curto, J., Surroca Lopez, A., Casals Sanchez, M., Tic, I., Fernandez Gallegos, M. R., & Sanchez de Groot, N. (2022). Microbiome impact on amyloidogenesis. *Front Mol Biosci, 9*, 926702.

Slizewska, K., Markowiak-Kopec, P., & Slizewska, W. (2021). The role of probiotics in cancer prevention. *Cancers (Basel), 13*(1).

Smits, H. H., Engering, A., van der Kleij, D., de Jong, E. C., Schipper, K., van Capel, T. M., et al. (2005). Selective probiotic bacteria induce IL-10-producing regulatory T cells in vitro by modulating dendritic cell function through dendritic cell-specific intercellular adhesion molecule 3-grabbing nonintegrin. *J Allergy Clin Immunol, 115*(6), 1260–1267.

Snyder, L., Peter, J. E., Henkin, T. M., & Champress, W. (1997). *Molecular genetics of bacteria*. ASM Press, 53–66.

Sun, Y. Y., Li, M., Li, Y. Y., Li, L. X., Zhai, W. Z., Wang, P., et al. (2018). The effect of Clostridium butyricum on symptoms and fecal microbiota in diarrhea-dominant irritable bowel syndrome: A randomized, double-blind, placebo-controlled trial. *Sci Rep, 8*(1), 2964.

Tamtaji, O. R., Heidari-Soureshjani, R., Mirhosseini, N., Kouchaki, E., Bahmani, F., Aghadavod, E., et al. (2019). Probiotic and selenium co-supplementation, and the effects on clinical, metabolic and genetic status in Alzheimer's disease: A randomized, double-blind, controlled trial. *Clin Nutr, 38*(6), 2569–2575.

Tamtaji, O. R., Taghizadeh, M., Daneshvar Kakhaki, R., Kouchaki, E., Bahmani, F., Borzabadi, S., et al. (2019). Clinical and metabolic response to probiotic administration in people with Parkinson's disease: A randomized, double-blind, placebo-controlled trial. *Clin Nutr, 38*(3), 1031–1035.

Tan, A. H., Lim, S. Y., Chong, K. K., MAA, A. M., Hor, J. W., Lim, J. L., et al. (2021). Probiotics for constipation in parkinson disease: A randomized placebo-controlled study. *Neurology, 96*(5), e772–e782.

Tang, B. C., Wang, Y. T., & Ren, J. (2023). Basic information about memantine and its treatment of Alzheimer's disease and other clinical applications. *Ibrain, 9*(3), 340–348.

Tette, F. M., Kwofie, S. K., & Wilson, M. D. (2022). Therapeutic anti-depressant potential of microbial GABA produced by lactobacillus rhamnosus strains for GABAergic signaling restoration and inhibition of addiction-induced HPA axis hyperactivity. *Curr Issues Mol Biol, 44*(4), 1434–1451.

Valdes, A. M., Walter, J., Segal, E., & Spector, T. D. (2018). Role of the gut microbiota in nutrition and health. *BMJ, 361*, k2179.

van Zyl, W. F., Deane, S. M., & Dicks, L. M. T. (2020). Molecular insights into probiotic mechanisms of action employed against intestinal pathogenic bacteria. *Gut Microbes, 12*(1), 1831339.

Vogt, N. M., Kerby, R. L., Dill-McFarland, K. A., Harding, S. J., Merluzzi, A. P., Johnson, S. C., et al. (2017). Gut microbiome alterations in Alzheimer's disease. *Sci Rep, 7*(1), 13537.

Wang, D., Pham, V. T., Steinert, R. E., Zhernakova, A., & Fu, J. (2022). Microbial vitamin production mediates dietary effects on diabetic risk. *Gut Microbes, 14*(1), 2154550.

Wang, Y., Wu, J., Lv, M., Shao, Z., Hungwe, M., Wang, J., et al. (2019). Metabolism characteristics of lactic acid bacteria and the expanding applications in food industry. *Front Bioeng Biotechnol, 9*, 612285.

Wilson, D. M., 3rd, Cookson, M. R., Van Den Bosch, L., Zetterberg, H., Holtzman, D. M., & Dewachter, I. (2023). Hallmarks of neurodegenerative diseases. *Cell, 186*(4), 693–714.

Wu, C. H., Hsueh, Y. H., Kuo, J. M., & Liu, S. J. (2018). Characterization of a potential probiotic lactobacillus brevis RK03 and efficient production of gamma-aminobutyric acid in batch fermentation. *Int J Mol Sci, 19*(1).

Xie, L. W., Lu, H. Y., Tang, L. F., Tang, F. L., Zhu, R. Q., Wang, D. F., et al. (2024). Probiotic consortia protect the intestine against radiation injury by improving intestinal epithelial homeostasis. *Int J Radiat Oncol Biol Phys, 120*(1), 189–204.

Yan, F., & Polk, D. B. (2020). Probiotics and probiotic-derived functional factors-mechanistic insights into applications for intestinal homeostasis. *Front Immunol, 11*, 1428.

Zhai, Q., Feng, S., Arjan, N., & Chen, W. (2019). A next generation probiotic, Akkermansia muciniphila. *Crit Rev Food Sci Nutr, 59*(19), 3227–3236.

Zhang, X., Tang, B., & Guo, J. (2023). Parkinson's disease and gut microbiota: from clinical to mechanistic and therapeutic studies. *Transl Neurodegener, 12*(1), 59.

Zheng, D., Liwinski, T., & Elinav, E. (2020). Interaction between microbiota and immunity in health and disease. *Cell Res, 30*(6), 492–506.

Zhou, P., Chen, C., Patil, S., & Dong, S. (2024). Unveiling the therapeutic symphony of probiotics, prebiotics, and postbiotics in gut-immune harmony. *Front Nutr, 11*, 1355542.

11 Role of Probiotics in Amyotrophic Lateral Sclerosis (ALS) and Huntington's Disease

Mirunalini Gobinath[1], *L Priyanka Dwarampudi*[1], *Shanmugam R*[2], *Subhajit Dutta*[2], *Parikshit Roychowdhury*[3], *and Swati Swagatika Swain*[3]

[1] Department of Pharmacognosy, JSS College of Pharmacy, JSS Academy of Higher Education & Research, Ooty, Tamil Nadu, India
[2] TIFAC CORE HD, JSS College of Pharmacy, JSS Academy of Higher Education & Research, Ooty, Tamil Nadu, India
[3] Department of Pharmaceutics, JSS College of Pharmacy, JSS Academy of Higher Education & Research, Ooty, Tamil Nadu, India

11.1 INTRODUCTION

Neurodegenerative diseases (NDD), as the name implies, cause gradual loss of neurons due to degeneration, which are irreversible. In other words, glial cell activation and neuroinflammation are some of the key features of NDDs. These changes can deteriorate the morphological and physiological activities of the nervous system, leading to disorders such as Parkinson's disease (PD), Huntington's disease (HD), amyotrophic lateral sclerosis (ALS), and Alzheimer's disease (AD). The onset and development of NDDs involves any variations in genetic and environmental factors. Some of the pathways that influence the disease pathophysiology are misfolded abnormal proteins that aggregate and cause neurotoxicity, cellular changes like mitochondrial dysfunction, oxidative stress, excitation of neurotransmitters, autophagy, lysosomal dysfunction, age-related degeneration, and neuroinflammation (Loh et al., 2024).

Since these diseases have no cure, many researchers have put forward their discoveries for management and to halt the complications of the disease. Among those, microbiome-based interventions are more popular. These include supplementation of probiotics and psychobiotics, fecal microbial transplantation from a healthy individual, and several techniques that maintain the homeostasis of gut microbiota (Dhyani et al., 2024).

The gastrointestinal tract of the human hosts around billions and trillions of microbes, including bacteria, viruses, fungi, and archaea, which are known as gut microorganisms. Approximately 2,000–4,000 species of gut microbes have been identified so far. Current evidence has stated that these commensal bacteria play a significant role in regulating the central nervous system and immune system. Gut bacteria are healthy microorganisms that constantly maintain their composition through the help of diet. Thus, any alterations in dietary modifications, aging, genetic predisposition, and environmental factors would directly influence the healthy composition of gut bacteria. Throughout the history of evolution, there is a shared mutualistic symbiosis between the human host and bacteria. This enhances the supply of essential nutrients and creates a pleasant environment between the host and microbes and thus exerts a substantial impact on normal physiology as well as pathological conditions (Ma et al., 2024).

Gut microbiomes are also called "the second brain" because they can modulate mental health via regulating cognitive and behavioral activities of brain directly or indirectly. Subsequently, many fellow scientists and researchers have proved that there is a dual communication between intestinal dysbiosis and the progress of NDDs. Thus, implementing therapeutic strategies that balance the altered gut microbiome is crucial (J. Wang et al., 2017).

11.2 ROLE OF PROBIOTICS AND PSYCHOBIOTICS IN NEURODEGENERATIVE DISEASES

In recent days, the value of gut microbiomes in NDDs has been increasingly acknowledged. Furthermore, there is plethora of preclinical evidence which states the altered composition of gut microbes might cause the pathogenesis of NDDs primarily by the activation of microglial functions. The collaboration within microbiota-gut–brain axis and immune system facilitates the regulation of glial cells, which is a focused target in the management of NDDs. In some cases, any defect in the gastrointestinal system is often correlated with some neurodegenerative disorders,

DOI: 10.1201/9781003536802-11

particularly when the gut is inflamed "leaky gut", there is a drastic change in the composition of intestinal microbiota, which in turn provokes modulation of vagus nerves. In some perspectives, long-term gut dysbiosis plays a leading role in the etiology of neurodegeneration. Dysbiosis significantly causes an interplay between commensal bacteria and harmful bacteria, which increases their composition and neurotoxic substances. These further precipitate the condition and cause the hallmarks of NDD (Wilson et al., 2023).

Some metabolites present in the gut, such as short-chain fatty acids (SCFA), have proven to decrease neuroinflammation and neurodegeneration. These SCFAs, including butyrate (C4), propionates (C3), and acetates (C2), are produced by the fermentation of complex carbohydrates, which are not easy to digest. Moreover, several studies state the link between SCFAs with physiology, such as intestinal homeostasis, immune system, glucose and cholesterol metabolic pathways, and many more (R. Wang et al., 2023).

The pathophysiology of NDDs is complex, which represents dysfunction of numerous neurotransmitters such as cholinergic, dopaminergic, GABAergic, glutamatergic, and serotonergic systems. On the other hand, commensal microbes present in the gut regulate these neurotransmitter pathways. One of the main neurotransmitters that acts as a connector between the microbial-gut–brain axis is serotonin. It is also called 5-hydroxytryptamine (5-HT). The reason why they are related to the gut is because they are secreted by the enterochromaffin cells of the gut. Even though 5-HT formed in the gut cannot bypass the BBB, its precursor 5-hydroxytryptophan and derivatives of serotonin, including melatonin and *N*-acetyl serotonin, have the tendency to cross the BBB and regulate the CNS. Intestinal bacteria *Turicibacter sanguinis* is involved in the biosynthesis of serotonin in the gut (Yano et al., 2015).

In case of AD, the serotonergic mechanism is involved in the pathology like amyloid plaques deposition. Numerous studies have stated that the lower the levels of serotonin, the higher the chance of accumulation of plaques. Interestingly, some genes that are related to the serotonin pathway is comparatively low in case of gut dysbiosis. Thus, this proves that probiotics and other supplements have direct or indirect involvement in regulation of serotonergic mechanisms. In some preclinical studies, the administration of prebiotics such as oligosaccharides, including fructose and galactose, have shown an increase in the serotonin concentrations in the brain. However, this isn't demonstrated in all cases of prebiotics (W. Zhang et al., 2023).

Likewise in Parkinson's, any dysfunction in serotonergic mechanism leads to the development of neuropsychiatric symptoms. These are primarily dependent upon the gut microbiome, i.e., abnormal synthesis and metabolism of various neurotransmitters is predominantly more when compared with healthy mice. In some preclinical studies, using FMT of probiotics and postbiotics from diseased PD mice in healthy mice demonstrated to have more gut dysbiosis than normal. This can cause major breakdown in the motor functions, decreased striatal dopaminergic and serotonergic mechanisms, and potential synergistic additive effects that precipitate the condition (Duranti et al., 2020).

Another major neurotransmitter γ-amino butyric acid (GABA), which is mostly present in astrocytes of the neurons, shows motor symptoms in both PD and HD. It is also proved that GABA can be directly synthesized and produced by gut microbiomes such as *Lactobacillus, Bacteroides, Bifidobacterium, Escherichia* sp., and *Parabacteroides.* Thus, in perspectives of AD, when dysbiosis occurs they might directly cause GABAergic hypoactivation, which in return potentiates neurodegenerative illness and cognitive impairment. This can be a challenging therapeutic implication in the discovery of GABAergic system targets and treat AD via modulating the gut microbiome (Tang et al., 2023).

Dopamine is a multidimensional neurotransmitter which has its crucial role in controlling various cognitive activities of the brain like motivation, reward, learning, and motor actions. Any abnormalities in their transmission initiated a wide range of neuropsychiatric illness, including PD. Gut microbiota has an indispensable role in balance of dopamine. Thus, dysbiosis leads to increased monoamine oxidase levels and stimulates degradation of dopamine, whereas in properly maintained gut microbial composition, dopamine level alleviates in mice model. These preclinical studies support the prospectives of microbiome-based therapeutics which ameliorate the dopaminergic system damage that have caused in PD (Zhao et al., 2021).

Cholinergic neurons release acetylcholine, which is responsible in several cognitive and motor functions such as learning, anticipating, thinking, planning, attention, memory, and other brain functions. Recent insights in gut microbial research have demonstrated that there is an interconnection between gut microbiota and cholinergic signaling. In certain in vivo studies, it is stated that when mice undergo antibiotic-induced gut dysbiosis, there is a drastic change in synaptic transmission of cholinergic neurons. In AD, cholinergic signaling indicates impairment in cognitive functions of the forebrain. The root cause of disruption in cholinergic signaling is due to the amyloid plaques and tau phosphorylation. In some preclinical studies, the administration of probiotics like *Lactobacillus plantarum* and prebiotics such as *Morinda officinalis* have refilled the levels of acetylcholine in AD-induced mice models. This indicates there is a tremendous potential for microbiome-based treatment for AD in the future (Chen et al., 2017).

In most of the NDDs, glutamate is an important neurotransmitter which engages in the pathogenesis of the disease. This excitatory neurotransmitter causes excitation and is responsible for neuronal plasticity. Much recent evidence proves that it is released by astrocytes for synaptic transmission and if glutamate becomes abnormally excessive, it leads to neuronal death and probably neurodegeneration. Similarly, like the above all neurotransmitters, glutamatergic signaling is also controlled over gut microbiota. This influence of glutamate in various NDDs, like AD, PD, and depression, have been well-studied. It is essential to perform additional investigation to fully understand the possibility of intrinsic mechanisms involved in microbiome-based therapeutics that target glutamatergic signals in PD (de Ceglia et al., 2023).

11.3 ROLE OF PROBIOTICS IN AMYOTROPHIC LATERAL SCLEROSIS

Of all the neurodegenerative illnesses, ALS is fatal, which involves the progressive disruption of motor neurons. Thus, it is commonly referred to as neuromuscular or motor neuron disease of NDDs. Distinctive features such as degeneration of upper and lower motor neurons that arises from the motor cortex, brain stem, and spinal cord are observed in clinical ALS. This loss affects voluntary muscles and primarily causes muscle weakness of the limbs and even respiratory muscles. This is the reason for severe ALS, at the end-stage people often die due to neuromuscular respiratory failure.

Mutations in TAR DNA-binding protein 43 (TDP43) get accumulated in excess and they often have altered conformation, which enhances their chances for aggregation and is termed as proteinopathy. Primary symptoms that are seen in ALS are gastrointestinal disturbances. This indicates that there is a strong interconnection between ALS symptoms and the gut. The interlink between homeostasis of ALS disease progression and intestinal symptoms is demonstrated through the $SOD1^{G93A}$ mice model. Diminished butyrate-producing bacteria *Butyrivibrio fibrisolvens* is related to the onset of leaky gut. Subsequently, restoring dysbiosis with bacterial metabolites improves the intestinal permeability and inflammation in the same mice model (Y. Zhang et al., 2024).

ALS, a.k.a. Lou Gehrig's disease, which is identified by stiffness, muscle weakness, and paralysis. It is also associated with gastrointestinal symptoms, thus proving there is a connection between microbial gut–brain axis and the disease. These imbalances are mentioned in many preclinical and clinical trials in animals and humans. Consequently, clinical studies report that the proportion and variation of gut microbiota in ALS individuals is remarkably varied when compared with any healthy population or other NDD groups. These variations in diversity or the composition are studied through high-throughput sequencing studies and other bioinformatic analysis or metagenomics to determine their taxonomy and microbial colonies. Levels of acylcarnitine, fatty acids, and bile acid metabolism might be the causative factors of ALS according to Mendelian randomization study. Additionally, scores showed on the updated ALS functional rating scale demonstrated a negative correlation with these metabolites (Goutman et al., 2022).

In the pathogenesis of ALS, gut microbiota progresses the onset of the disease via direct or indirect reasons. For instance, when gene mutations occur, there are primary symptoms of ALS, which are nothing but gastrointestinal manifestations. This successively causes gut microbiota to directly stimulate the CNS either by central inflammation or systemic inflammation. This indeed, increases the production of LPS from the intestinal flora, which penetrates through the intestinal membrane "leaky gut", and facilitates the ignition of pro-inflammatory cytokines. Furthermore, it induces the two-way link of the immune system, ENS, CNS, and neural pathways to increase the disease progression. This is one of the mechanisms where the microbial-gut–brain axis has an essential part in onset of ALS (Noor Eddin et al., 2024). Table 11.1 represents all the genes, proteins, locus, and their associated disease mechanisms with ALS.

Although there is a strong correlation between the pathogenesis of ALS and gut dysbiosis, molecular mechanisms which are involved in disease progression are still in debate. The key factors such as dysfunctional mitochondria, oxidative stress, impaired neurotransmitter flow, hyperactivation of glial cells like astrocytes and microglia, impaired immune system, and neuroinflammation are the major contributors to the molecular pathophysiology of ALS (Rooks & Garrett, 2016).

In addition, transgenic SOD1-G93A mouse model have demonstrated the interconnection between severity of the disease with gut microbial dysbiosis. In this study, they have shown that the administration of probiotic supplements of certain gut microbial species in germ-free mice treated with broad-spectrum antibiotic have proved changes in electron transport chain function, oxidative stress pathways, and mitochondrial dysfunction in them. This supports the concept of gut dysbiosis and progression of ALS (Blacher et al., 2019).

In both the central nervous system and the peripheral nervous system, nonneuronal cells such as oligodendrocytes, astrocytes, and microglia have a prominent role in regulating neuronal growth and development. The largest glial cells are the astrocytes, which play a main function of providing adequate nutrition and differentiation of neurons. They also maintain the ionic homeostasis between each neuron. Subsequently, hyperactivation of astrocytes might induce dysregulation of ionic homeostasis like calcium influx, stress to endoplasmic reticulum, and dysfunctional mitochondria, which leads to impaired energy metabolism. This triggers various neuroinflammatory factors and neurotoxins that cause motor neuron damage and disease progression of ALS. Factors that trigger astrocyte are short-chain fatty acids like butyrate, gut-microbial-induced histone deacetylase inhibitors, melatonin, lipopolysaccharides, damage associated molecular patterns (DAMPs), high-mobility group box 1 (HMGB1), TLR-4, and heat shock proteins and ALS susceptible genes like TDP43, mutant superoxide dismutase type1, NFκB, BACE1, and *FUS* (Anderson, 2023).

Other glial cells that have a potential role in ALS are microglia, which are also referred to as immune monitoring systems in CNS. This eliminates any damaged nerve cell and neuroinflammatory factors in order to maintain homeostasis, to promote neural development and tissue regeneration. When there is an injury or trauma or infection in the CNS, glial cells such as astrocytes and microglia release

TABLE 11.1
Genes that Are Associated with ALS and Their Disease Mechanism (Hardiman et al., 2017)

S. No.	Locus	Inheritance	Gene (Protein)	Pathophysiological Mechanism
1	ALS1	Autosomal dominant, or autosomal recessive	Superoxide dismutase1 (SOD1)	Changes in oxidative stress pathways
2	ALS2	Autosomal recessive	*ALS2*	Trafficking of endosomal-mediated transport
3	ALS3	Autosomal dominant	Unknown	Unknown
4	ALS4	Autosomal dominant	Senataxin (*SETX*)	RNA metabolism
5	ALS5	Autosomal recessive	Unknown	DNA damage and repair, axon growth in neurons
6	ALS6	Autosomal dominant or autosomal recessive	RNA-binding protein *FUS*	RNA metabolism
7	ALS7	Autosomal dominant	Unknown	Unknown
8	ALS8	Autosomal dominant	Vesicle-associated membrane protein-associated protein B/C (*VAPB*)	Stress in endoplasmic reticulum
9	ALS9	Autosomal dominant	Angiogenin (*ANG*)	RNA metabolism
10	ALS10	Autosomal dominant	TAR DNA-binding protein 43 (*TARDBP*)	RNA metabolism
11	ALS11	Autosomal dominant	Polyphosphoinositide phosphatase (*FIG4*)	Trafficking of endosomal-mediated transport
12	ALS12	Autosomal dominant or autosomal recessive	Optineurin (*OPTN*)	Autophagy
13	ALS13	Autosomal dominant	Ataxin 2 (*ATXN2*)	RNA metabolism
14	ALS14	Autosomal dominant	Valosin-containing protein (*VCP*)	Autophagy
15	ALS15	X-linked dominant	Ubiquilin-2 (*UBQLN2*)	Autophagy and ubiquitin-proteasome system
16	ALS16	Autosomal dominant	Sigma non-opioid intracellular receptor 1 (*SIGMAR1*)	Autophagy and ubiquitin-proteasome system
17	ALS17	Autosomal dominant	Charged multivesicular body protein 2B (*CHMP2B*)	Trafficking of endosomal-mediated transport
18	ALS18	Autosomal dominant	Profilin 1 (*PFN1*)	Cytoskeleton
19	ALS19	Autosomal dominant	Receptor tyrosine protein kinase erB 4 (*ERBB4*)	Neuronal development
20	ALS20	Autosomal dominant	Heterogeneous nuclear ribonucleoprotein A1 (*HNRNPA1*)	RNA metabolism
21	ALS21	Autosomal dominant	Matrin 3 (*MATR3*)	RNA metabolism
22	ALS22	Autosomal dominant	Tubulin α4A (*TUBA4A*)	Cytoskeleton

cytotoxic elements that induce neuronal precursor cells to the injured area. Then, this causes proliferation and differentiation of those precursor cells into astrocytes, which in turn gets activated and triggers inflammatory responses, including infiltration of macrophages and T cells, inflammatory vesicle proteins, and the release of pro-inflammatory cytokines, which leads to damaged motor neurons and pathogenesis of ALS (Liu et al., 2020).

In preliminary stages of the disease, M2 phenotypes of microglia promote the survival, repair, and regeneration of neural tissues by increasing protective signals of CD200, fractalkine, CD206, and Ym1 expression. As the disease progresses into late stages, microglial phenotypes like M1 get activated, which increase damaged motor neurons to produce threatening mutative SOD1 signals. This further enhances secretion of NADPH oxidase, NOX2, reactive oxygen species, reactive nitrogen species, ronulic acid, neurotoxins, and inflammatory cytokines such as TNF-α, Il-6, IL-1, and, IL-1β, which lead to motor nerve damage (Kent & Miron, 2024).

Many researchers state that there is an interconnection between intestinal permeability, neuroinflammation, and gut dysbiosis. When the intestinal flora faces dysbiosis, it produces virulent protein, which then enters the brain through vagus nerves and causes neurodegeneration of motor neurons and glial cells like microglia and astrocytes and further causes motor nerve damage. In preclinical studies of SOD1G93A mouse, models of ALS have proved that microbiota plays an indispensable role in their pathogenesis (Figure 11.1). On the other hand, gut dysbiosis can directly or indirectly trigger immune reactions, which cause the release of T cells, monocytes, macrophages, and other inflammatory factors in the CNS and lead to neuroinflammation. Sometimes these immune responses release

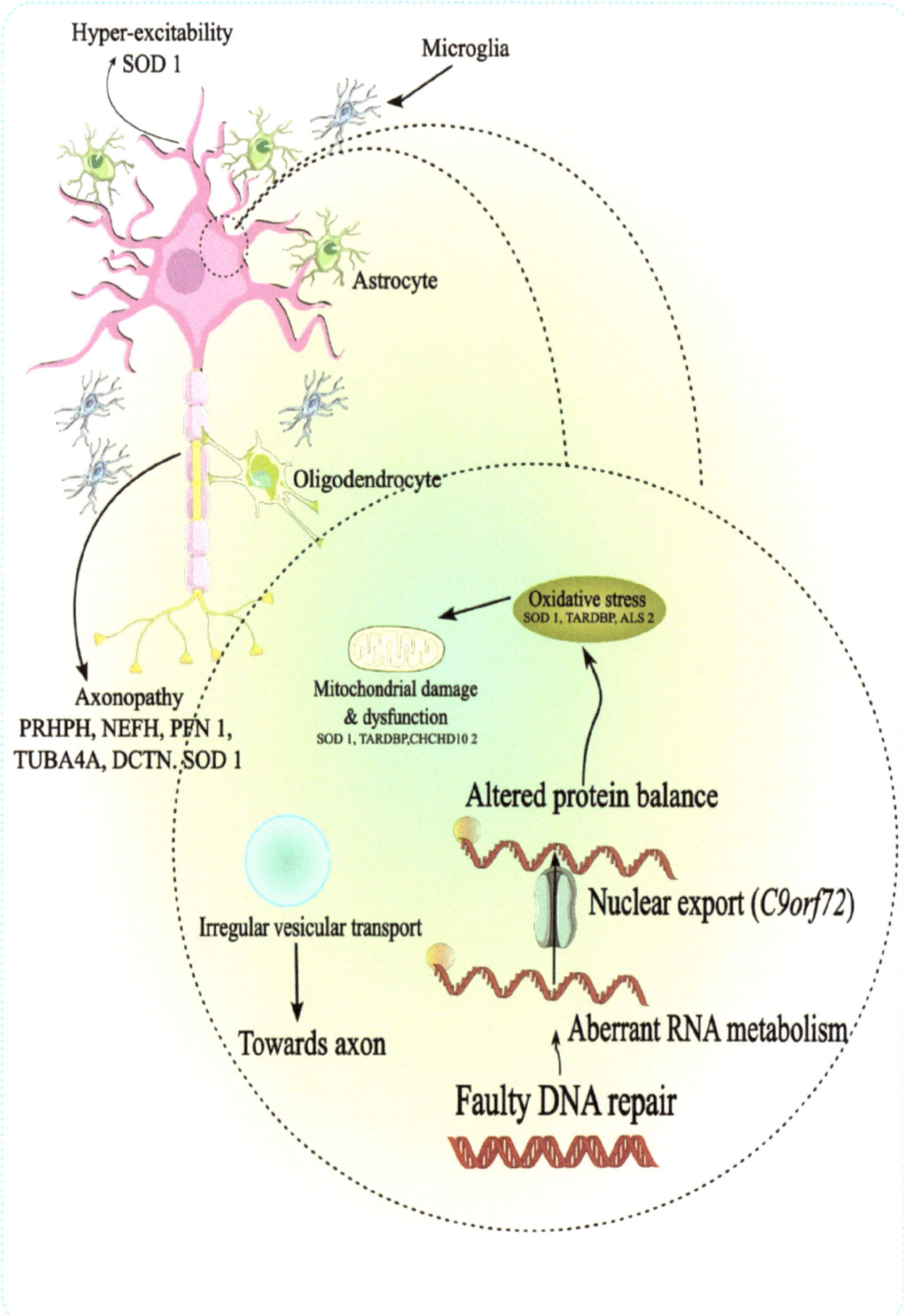

FIGURE 11.1 Molecular pathogenesis of ALS.

lipopolysaccharides, pro-inflammatory cytokines, and other toxins that cause increased intestinal damage and loss of permeability and BBB permeability, thus igniting the activation of glial cells and peripheral neuroinflammation. Therefore, changes in the composition and proportion of gut microbiota can influence the pathogenesis of ALS, and administering psychobiotics and synbiotics supplements might balance the altered gut microbiota (Castelli et al., 2021).

11.4 ROLE OF PROBIOTICS IN HUNTINGTON'S DISEASE

Huntington's disease (HD) is a rare NDD, which is characterized by continuous repeated sequences of DNA in huntingtin gene that leads to abnormal protein misfolding and aggregates in the brain structures. It is a multifaceted disease that shows neurological, psychiatric, motor, and GI-related symptoms. Abnormal choreiform movements such as repetitive quick and jerky involuntary motion problems and athetosis are the clinical features of the disease. These symptoms have no cure, and they usually stop only during sleep. Apart from this irregular eye movements, poor coordination and gastrointestinal symptoms like diarrhea, unexpected weight loss, and nutrition absorption deficiency are commonly associated with HD. Emerging research has shown that there is an interconnection between dysbiosis of the intestinal flora and the pathophysiology of HD (Tabrizi et al., 2020).

In addition, HD is an autosomal dominant disorder, which means there is almost 50% chance of inheriting to offsprings. Huntingtin gene encodes the chromosome 4p16.3, which contains triplet repeat of CAG for about 10–35 times in normal conditions. While on the contrary of HD, it gets repeated abnormally more than 36 times, which is often referred to as polyglutamine disease, as CAG codes for glutamine. This mutated huntingtin causes proteinopathy and aggregates mutated proteins within neuronal cells in putamen and caudate regions of basal ganglia, which further leads to excitation and excessive signaling of calcium channels, often causing neuronal cell death. This is how the HD symptoms related to motor and cognitive problems arise. CNS involvement of misfolded protein aggregates might lead to mood disorders like depression, dementia, and personality changes. Moreover, extended and tandem repeats of CAG often affect DNA replication. These pathophysiological mechanisms usually occur in dorsal striatum of caudate and putamen of basal ganglia and cause tissue atrophy (Krause et al., 2024).

In some aspects, HD shows similar characteristics to AD, which leads to β-amyloid plaques accumulation in microglia and astrocytes. This has been proved in various genome-wide association studies (GWAS) and have shown that connection to glial cells involvement is primarily due to AD genes. In both R6/1 and R6/2 transgenic mouse models, continuous expansion and repeats of mutated huntingtin and CAG sequences cause misfolded proteins to aggregate to microglia, leading to their activation through multiple processes (Loh et al., 2024).

Involvement of gut microbiota and gut–brain axis plays an indispensable role in HD pathogenesis. In many preclinical trials, gut dysbiosis has increased pro-inflammatory cytokines and increased production of hydrogen sulfide (H2S), which negatively affects the intestinal health. Also, certain microbial communities, including Bacteroidetes and Firmicutes, are imbalanced. These in order cause irregular mucosal secretions in the intestinal mucosal layer and decrease neuropeptides and irregular secretions of endocrinal hormones. These diminish the butyrate-forming bacterial species such as *Blautia producta* and *Eubacterium hallii*, which counterpart the reduction of acetic acid and butyric acid which have been linked to the pathophysiological mechanisms of various metabolic diseases. Besides this, *E. hallii* interacts with bile acid metabolism, promoting the gastrointestinal symptoms in HD (Tabrizi et al., 2020).

Gut microbiota also plays a role in the metabolism of tryptophan, which is converted to indole derivatives that aid as proteins or ligands of aryl hydrocarbon receptors (AHR) in glial cells. These receptors inhibit the signaling expression of NFκB and reduce neuroinflammation in the CNS. On the other hand, increased proliferation and functional expression of astrocytes and microglia, 3-indolepropionic acid, diminishes neurotoxic reactions of A1 astrocytes in certain ischemic mice models. In addition, the activation of microglia and neuroinflammation are the hallmarks of majority of neurodegeneration in case of gut dysfunction. These lead to misfolded protein aggregation such as tau and amyloids in AD, α-synuclein in PD, TDP43 in ALS, mutant huntingtin in HD and SOD1, and FUS in common NDDs. Deficiency in microglial autophagy is due to lysosomal dysfunction, which activates further production of microglial cells and astrocytes to the site of infection/injury and secretes pro-inflammatory cytokines, chemokines, NLRP3, and mTOR factors, which attenuate senescence. Despite these, the gut bacteria also influence the secretion of anti-inflammatory TRAIL+ (tumor necrosis factor-related apoptosis inducing ligand positive) in glial cells that alleviate neuroinflammation by T cell apoptosis TRAIL death receptor signals (DR5) (Sanmarco et al., 2021).

When the proportion of intestinal flora shifts into pathological state, which is referred to as dysbiosis, the intestinal barrier is compromised. The major physiology involved in epithelial layer of the intestine is to maintain the barrier integrity and control the movement of elements such as ions, water, solutes, and minerals. Tight junctions made up of multiple proteins, including zonula occludens, tricellulin, occludin, adhesion molecules, and cingulin, play a crucial role in barrier integration. Tight junctions and their proteins aid in barrier integrity and tissue homeostasis. However, in case of NDDs, like HD and PD, altered tight junctions lead to increased intestinal permeability. For instance, α-synuclein accumulates in ENS, which causes leaky gut in PD. It also reduces the tight junction proteins such as occludin (Yin et al., 2024).

Gene carriers of HD have impaired GI and several disorders like IBD. They have devastated symptoms like unintentional weight loss, diarrhea, malnutrition, malabsorption, imbalance in gut movements, and morphological abnormalities like shortened intestinal villi and decreased thickness of mucosa in the intestine. The gut barrier also prevents the entry of bacteria from the lumen into the systemic circulation. Subsequently, in case of leaky gut, the bacteria enter systemic circulation and to BBB and cause loss of BBB integrity and infections. Certain metabolic secretory products cause loss of integrity in the blood–brain barrier, which indicates a biomarker for neurodegeneration. Hence, increased BBB leakage is seen in NDDs like AD, PD, and HD, also in R6/2 HD mice model (Bates et al., 2015).

Enteric neuropeptides include vasoactive intestinal polypeptides, cocaine–amphetamine-regulated transcript (CART), and other vesicular acetylcholine transporters play a bidirectional role in the communication between gut–brain axis. These neuropeptides interconnect immune responses and CNS events. The mHTT expressed ubiquitously gets misfolded protein aggregates in ENS and causes morphological abnormality in the intestine in R6/2 HD mice models. When lipopolysaccharides increases due to dysbiosis, it exerts immune response such as increased TNF-α, IL-4, IL-6, IL-8, and anti-inflammatory cytokines like IL-4 and IL-10, which further increases mRNA expression of NFκB signaling. These can be inhibited by certain strains of clostridium and *Faecalibacterium prausnitzii*. Intestinal bacteria are associated with the fabrication of metabolites and neuroactive molecules, including acetylcholine, dopamine, GABA, and serotonin, which stimulate afferent neurons of the ENS and connect vagus nerve with CNS. Cholecystokinin (CCK) involves in gastric emptying, food intake, satiety, and hunger responses of the GIT by acting on CCK-1 receptor and vagal afferent neuronal fibers. These are controlled by short-chain fatty acids like butyrate and long-chain fatty acids like oleates, which also interfere with lipid, bile, protein metabolism, and CCK-mediated pathways. Thus, whereas many studies are still done to clearly explain the pathogenesis of dysbiosis in HD, current evidence suggests that incorporating probiotics, synbiotics, or psychobiotics will prevent the onset of various NDDs (Figure 11.2) (Chakraborty et al., 2024).

11.5 FUTURE PERSPECTIVES AND RESEARCH OUTLOOK

The incidence of probiotic therapeutics has been increasing tremendously; hence, it is one of the promising solutions for treating several diseases related to microbial dysbiosis. Enhancing probiotic strains with modern techniques such as nanotechnology and nanomedicine, gene-editing techniques such as CRISPR technology, and multi-omics approaches provide advanced engineered next-generation probiotics. They also address to the safety and efficacy of treatment, and ameliorates precise dosing and in-crafting personalized treatment strategies (Al-Fakhrany & Elekhnawy, 2024).

Gene-editing technologies (Table 11.2) include methods such as CRISPR Cas systems, suicide plasmids, transposons, lambda (λ) red and RecE/T systems, group 2 introns, and Cre/lox systems. These methods provide precise gene editing and improve the quality and viability, desired characteristics to the probiotics. Gene-editing technologies address the safety and efficacy of probiotic therapeutics in various diseases. These technologies improve mechanisms and develop advanced engineered probiotics. CRISPR Cas9 produces genetically modified organisms, which are easier and faster to edit and correct genetic mutations. They also modify genes with high precision (L. Wang et al., 2024).

So far, traditional strains such as *Bacteroides*, *Bacillus subtilis*, *Clostridium*, *Eubacterium*, *Bifidobacterium*, *Lactobacillus*, and *Saccharomyces* have undergone gene-editing technology and have improved stability and functionality that play a crucial role in biotherapeutics of engineered probiotics (Rothstein et al., 2020).

Additionally, nanomedicine and nanotechnology have been increasingly developed and explored in the field of probiotics because they provide stability, enhanced viability, and bioavailability to the encapsulated formulations. In recent years, nanotechnology has their potential in alternative to traditional formulation methods. The collaboration of nanomedicine with probiotics is often referred to as "nanoprobiotics," which is employed to overcome various limitations that have been faced in food and pharmaceutical industries till date. The exact mechanism involves incorporating probiotics into protective shells, i.e., nanoparticles which act as a physical barrier to the encapsulated elements. This mainly aims to ameliorate the viability, bioavailability, and promote targeted drug delivery and controlled drug release (Baral et al., 2021).

Moreover, encapsulation techniques of probiotics and postbiotics improve the survival of the commensal microbes that were incorporated and protect them from gastrointestinal environment or high temperature, oxygen, and light, i.e., storage and handling issues. They form a shield around probiotics and facilitate precise and controlled release of encapsulated elements at an appropriate concentration. This also improves functionality and effectiveness. Some examples of nanotechnology that are used in probiotics are nanoparticles, nanoemulsion, nanofibers, and nanoliposomes (Pandey et al., 2024).

Multi-omics approaches include merging of genomics and metagenomics in research with high-throughput genome-sequencing data and transcriptomics. These techniques help to enhance and produce advanced probiotics, which are more potent and viable than the original. Sometimes omics help in anticipating any mutational abnormalities and designing desirable characteristics in the microbial strains of probiotics. Transcriptomics and meta-transcriptomics include mRNA, miRNAs, lncRNAs, and non-coding RNAs that regulate post-transcriptional modifications. Proteomics gives the knowledge regarding proteins and their interactions with the environment. For instance, any specific growth phase of an organism or their physiological conditions during adequate environmental conditions. This is also used to identify any changes in acetylation, methylation, phosphorylation or glycosylation, and post-translational changes in sequences of DNA or RNA. Proteomics are collaborated with metabolomics and other omics approach to process potential probiotic strains. Integrating these unique future perspectives of research in probiotic strains helps us to craft more precise and personalized medicines for the future (Wallen et al., 2022).

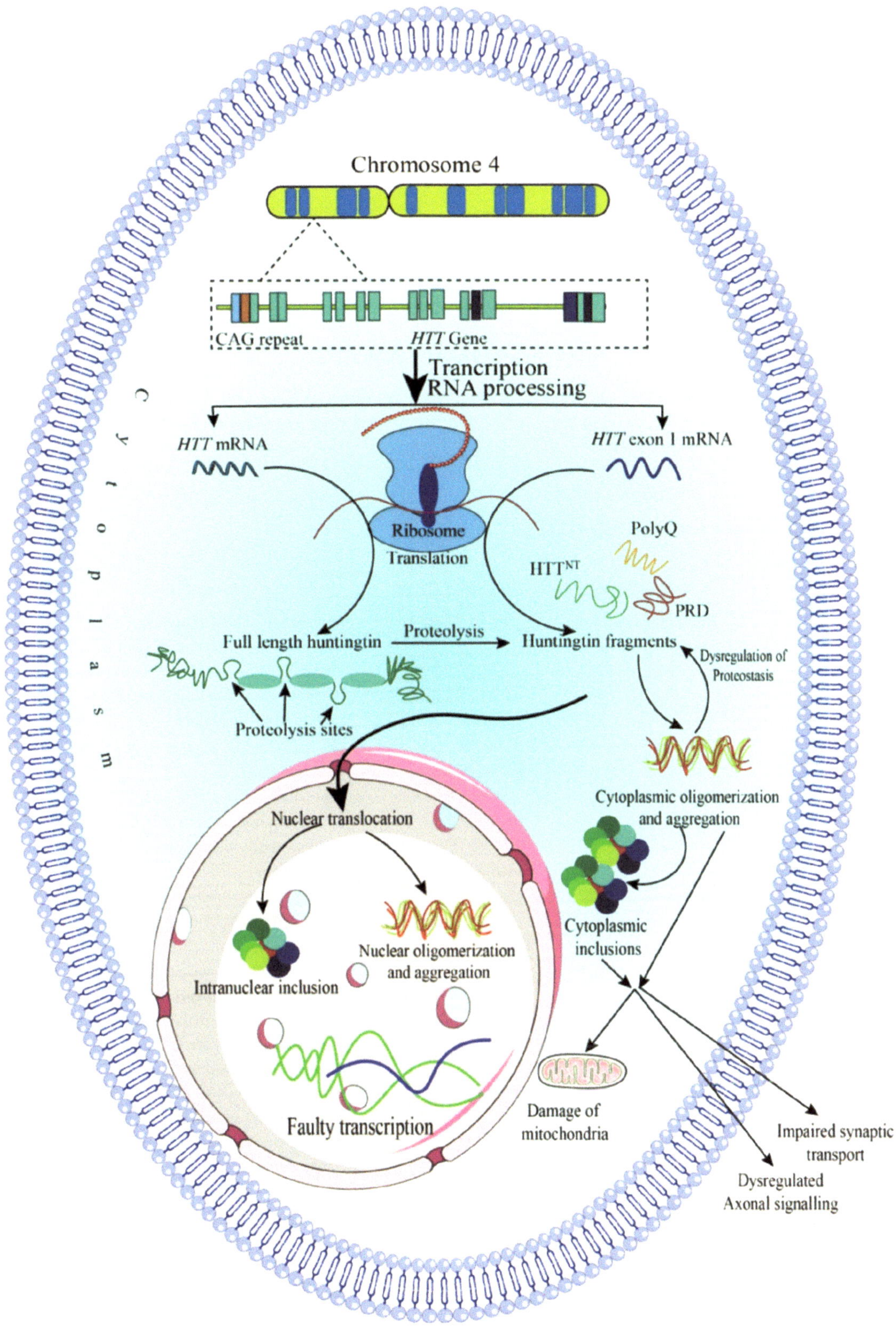

FIGURE 11.2 Molecular pathogenesis of HD.

TABLE 11.2
Basic Techniques of Gene-Editing and Applications in Probiotics Engineering (Zhong et al., 2023)

Methods	Applications
CRISPR Cas systems	For insertion, deletion, and point mutations of genes
Cre/lox recombination systems	For inversion, deletion, and translocation of genes
Group II introns	For insertion and deletion of genes
DsDNA recombination initiated by λ-red and RecE/T systems	For deletion, insertion, point mutations, and DNA repair
Suicide plasmids	For insertion, deletion, and replacement of genes
Transposons	For inactivation, insertion, and mutagenesis of genes

11.6 CLINICAL EVIDENCE IN PROBIOTICS FOR ALS AND HD

The clinical evidence of probiotics use in ALS and HD were mentioned in Tables 11.3 and 11.4 which explains both the preclinical data and animal models that were used to evaluate probiotics (Table 11.5).

TABLE 11.3
Preclinical Data of Probiotics in ALS and HD

S. No.	Animal Type/ Model	Disease Stage	Involved Organ	Symptoms	Reference
1	mSOD1^{G93A}	Presymptomatic and symptomatic	Liver	Hepatocyte and liver parenchymal cell atrophy, higher number of NKT cells, lowering in the number of T cells and a smaller size of the liver compared to that of the control mice	Finkelstein et al., 2011
2	mSOD1^{G93A} with and without the treatment with butyrate	Presymptomatic and symptomatic stages	Interorgan communication systems and inflammation	The treated group showed significantly lowered IL-17, IBA, and lipopolysaccharides in serum. Changes in lipid, amino acid, and animo acid synthesis were observed	Ogbu et al., 2022
3	mSOD1^{G93A}	Presymptomatic and symptomatic stages	Small intestine, colon, and enteric nervous system	Reduction in peristalsis and bowel movement, dysfunction and abnormalities in ENS	Y. Zhang et al., 2021
4	TDP43A315T	Presymptomatic and symptomatic stages	ENS, cecum, and small intestine	High prevalence of mutant TDP-43 of myenteric plexus neurons. Reduction in the colon circumference, abdominal rigidity, reduction in the numbers of neurons, and increase in the size of the alive surviving neurons. Necrosis of the neurons responsible for nitric acid synthase enzyme and presence of leaking small and large intestines and presence of inflammatory cytokines and LPS	Herdewyn et al., 2014
5	TDP43A315T	Pre-disease stage	Intestine	Moderate inflammation/dilation in the cecum and lower parts of ileum, empty stomach but dry fecal matter in ileum and watery in the jejunum and duodenum	Esmaeili et al., 2013

TABLE 11.4
Animal Models for Probiotics and Neurodegenerative Diseases

Model	Disease Manifestation/ Target	Intervention/ Outcomes	Techniques	Reference
SOD1^{G93A}	Intestinal leakage, dysregulated guy microbiota	Reduction of *Butyrivibrio fibriosolvens,* in turn reducing the butyrate production, reducing *E. coli* and *Fermicus* when the mice are 8 weeks old	The assessment was done using western blot, ELISA, RT-qPCR, and microbial sequencing of the fecal matter	Wu et al., 2015
SOD1^{G93A}	The study intervention focused on intestinal microflora to slow down the progression of the disease	Significant enhancement in the colonies of *Odoribacter*, *Eubacterium*, and *Bacteroide*, while reducing the number of *Tannerelia*. Mice that were treated with butyrate, intestinal bacterial balance was restored. They showed gut integrity enhancement, and their lifespan also improved	Live cellular imaging techniques, western blot analysis, immunofluorescence assay, RT-qPCR, cell transfection	Y.-G. Zhang et al., 2017
SOD1^{G93A}	The microbiota linked to ALS and the vagal nerve link to the disease	Surgical stimulation of the vagal nerve was carried out and the bacteroidales group was seen to be the most numerous microbiome present, apart from Lachnospiraceae and Rikenellaceae	Accessing the vagal nerve through ventral neck surgery and manual stimulation. RNA sequencing, and extraction of DNA	Haney et al., 2018
SOD1^{G93A}	Changes in gut microbiota, epigenome, and disease progression of ALS in the superoxide dismutase mutated mice	High activation of immune response in the mice were observed specially in central nervous system. The animals showed altered gut microbiome in their early life. This study also established the interrelationship between the commensal microbes and the immune system. Gradual reduction of Firmicutes were observed during the lifespan of the mice from 37 days to 60 days and ultimately to 90 days in the mice	Assessment of motor functions using rotarod model. Fecal matter collection and isolation of bacteria for DNA sequencing. Collection of immune cells from blood, spleen, and spinal cord	Figueroa-Romero et al., 2019

TABLE 11.5
Clinical Trial Data on Probiotics for ALS and HD

Study	Sample Size	Method	Key Findings	References
Diversity of microbiological environment in ALS	11 patients, 6 ALS, and 5 healthy volunteers	High-throughput sequencing following genome extraction	Higher number colonies of bacteroidales, dorea, bacteroidia, and Bacteroidetes, while reduction of colonies of Lachnospiraceae, *Anaerostipes*, and *Oscillibacter* in significance was noticed in ALS patients	Fang et al., 2016
Contribution of gut microbiota in the pathogenesis of ALS and their new treatment strategies	100 volunteers, 50 ALS, and 50 healthy	Genomic DNA extraction and, gel electrophoresis, RT-qPCR, and double blinded treatment regimen for the patients	DNA concentration was found to be low in ALS. There was no observed link between yeast colonies and ALS. ALS patients had reduced colonies yeast, *Clostridium* strains, higher colonies of Enterobacteriaceae, and *E. coli* were observed	Martin et al., 2022

11.7 CONCLUSION

In conclusion, the gut–brain axis has a significant role in neurodegenerative disorders. It can modulate the pathogenesis of disease progression. Implementing various probiotic strains and synbiotics can be a promising therapeutic strategy for curing the disease, even though existing conventional therapies are not up to the mark. Recent evidence suggests the importance of probiotic therapeutics and their future perspectives by crafting them with innovative techniques such as nanotechnology, nanomedicine, multi-omics approach, and gene-editing technologies. Thus, merging these might ameliorate the current gaps in microbiota-gut–brain therapy and opportunities that develop precision medicine goals in various NDDs like ALS and HD.

REFERENCES

Al-Fakhrany, O. M., & Elekhnawy, E. (2024). Next-generation probiotics: The upcoming biotherapeutics. *Molecular Biology Reports*, *51*(1), 505. https://doi.org/10.1007/s11033-024-09398-5

Anderson, G. (2023). Amyotrophic lateral sclerosis pathoetiology and pathophysiology: Roles of astrocytes, gut microbiome, and muscle interactions via the mitochondrial melatonergic pathway, with disruption by glyphosate-based herbicides. *International Journal of Molecular Sciences*, *24*(1), Article 1. https://doi.org/10.3390/ijms24010587

Baral, K. C., Bajracharya, R., Lee, S. H., & Han, H.-K. (2021). Advancements in the pharmaceutical applications of probiotics: Dosage forms and formulation technology. *International Journal of Nanomedicine*, *16*, 7535–7556. https://doi.org/10.2147/IJN.S337427

Bates, G. P., Dorsey, R., Gusella, J. F., Hayden, M. R., Kay, C., Leavitt, B. R., Nance, M., Ross, C. A., Scahill, R. I., Wetzel, R., Wild, E. J., & Tabrizi, S. J. (2015). Huntington disease. *Nature Reviews Disease Primers*, *1*(1), 1–21. https://doi.org/10.1038/nrdp.2015.5

Blacher, E., Bashiardes, S., Shapiro, H., Rothschild, D., Mor, U., Dori-Bachash, M., Kleimeyer, C., Moresi, C., Harnik, Y., Zur, M., Zabari, M., Brik, R. B.-Z., Kviatcovsky, D., Zmora, N., Cohen, Y., Bar, N., Levi, I., Amar, N., Mehlman, T., … Elinav, E. (2019). Potential roles of gut microbiome and metabolites in modulating ALS in mice. *Nature*, *572*(7770), 474–480. https://doi.org/10.1038/s41586-019-1443-5

Castelli, V., d'Angelo, M., Quintiliani, M., Benedetti, E., Cifone, M. G., & Cimini, A. (2021). The emerging role of probiotics in neurodegenerative diseases: New hope for Parkinson's disease? *Neural Regeneration Research*, *16*(4), 628. https://doi.org/10.4103/1673-5374.295270

Chakraborty, P., Gamage, H. K. A. H., & Laird, A. S. (2024). Butyrate as a potential therapeutic agent for neurodegenerative disorders. *Neurochemistry International*, *176*, 105745. https://doi.org/10.1016/j.neuint.2024.105745

Chen, D., Yang, X., Yang, J., Lai, G., Yong, T., Tang, X., Shuai, O., Zhou, G., Xie, Y., & Wu, Q. (2017). Prebiotic effect of fructooligosaccharides from Morinda officinalis on Alzheimer's disease in rodent models by targeting the microbiota-gut-brain axis. *Frontiers in Aging Neuroscience*, *9*. https://doi.org/10.3389/fnagi.2017.00403

de Ceglia, R., Ledonne, A., Litvin, D. G., Lind, B. L., Carriero, G., Latagliata, E. C., Bindocci, E., Di Castro, M. A., Savtchouk, I., Vitali, I., Ranjak, A., Congiu, M., Canonica, T., Wisden, W., Harris, K., Mameli, M., Mercuri, N., Telley, L., & Volterra, A. (2023). Specialized astrocytes mediate glutamatergic gliotransmission in the CNS. *Nature*, *622*(7981), 120–129. https://doi.org/10.1038/s41586-023-06502-w

Dhyani, P., Goyal, C., Dhull, S. B., Chauhan, A. K., Singh Saharan, B., Harshita, Duhan, J. S., & Goksen, G. (2024). Psychobiotics for mitigation of neuro-degenerative diseases: Recent advancements. *Molecular Nutrition & Food Research*, *68*(13), 2300461. https://doi.org/10.1002/mnfr.202300461

Duranti, S., Ruiz, L., Lugli, G. A., Tames, H., Milani, C., Mancabelli, L., Mancino, W., Longhi, G., Carnevali, L., Sgoifo, A., Margolles, A., Ventura, M., Ruas-Madiedo, P., & Turroni, F. (2020). Bifidobacterium adolescentis as a key member of the human gut microbiota in the production of GABA. *Scientific Reports*, *10*(1), 14112. https://doi.org/10.1038/s41598-020-70986-z

Esmaeili, M. A., Panahi, M., Yadav, S., Hennings, L., & Kiaei, M. (2013). Premature death of TDP-43 (A315T) transgenic mice due to gastrointestinal complications prior to development of full neurological symptoms of amyotrophic lateral sclerosis. *International Journal of Experimental Pathology*, *94*(1), 56–64. https://doi.org/10.1111/iep.12006

Fang, X., Wang, X., Yang, S., Meng, F., Wang, X., Wei, H., & Chen, T. (2016). Evaluation of the microbial diversity in amyotrophic lateral sclerosis using high-throughput sequencing. *Frontiers in Microbiology*, *7*, 1479. https://doi.org/10.3389/fmicb.2016.01479

Figueroa-Romero, C., Guo, K., Murdock, B. J., Paez-Colasante, X., Bassis, C. M., Mikhail, K. A., Pawlowski, K. D., Evans, M. C., Taubman, G. F., McDermott, A. J., O'Brien, P. D., Savelieff, M. G., Hur, J., & Feldman, E. L. (2019). Temporal evolution of the microbiome, immune system and epigenome with disease progression in ALS mice. *Disease Models & Mechanisms*, *13*(2), dmm041947. https://doi.org/10.1242/dmm.041947

Finkelstein, A., Kunis, G., Seksenyan, A., Ronen, A., Berkutzki, T., Azoulay, D., Koronyo-Hamaoui, M., & Schwartz, M. (2011). Abnormal changes in NKT cells, the IGF-1 axis, and liver pathology in an animal model of ALS. *PloS One*, *6*(8), e22374. https://doi.org/10.1371/journal.pone.0022374

Goutman, S. A., Guo, K., Savelieff, M. G., Patterson, A., Sakowski, S. A., Habra, H., Karnovsky, A., Hur, J., & Feldman, E. L. (2022). Metabolomics identifies shared lipid pathways in independent amyotrophic lateral sclerosis cohorts. *Brain*, *145*(12), 4425–4439. https://doi.org/10.1093/brain/awac025

Haney, M. M., Ericsson, A. C., & Lever, T. E. (2018). Effects of intraoperative vagal nerve stimulation on the gastrointestinal microbiome in a mouse model of amyotrophic lateral sclerosis. *Comparative Medicine*, *68*(6), 452–460. https://doi.org/10.30802/AALAS-CM-18-000039

Hardiman, O., Al-Chalabi, A., Chio, A., Corr, E. M., Logroscino, G., Robberecht, W., Shaw, P. J., Simmons, Z., & van den Berg, L. H. (2017). Amyotrophic lateral sclerosis. *Nature Reviews Disease Primers*, *3*(1), 1–19. https://doi.org/10.1038/nrdp.2017.71

Herdewyn, S., Cirillo, C., Van Den Bosch, L., Robberecht, W., Vanden Berghe, P., & Van Damme, P. (2014). Prevention of intestinal obstruction reveals progressive neurodegeneration in mutant TDP-43 (A315T) mice. *Molecular Neurodegeneration*, *9*, 24. https://doi.org/10.1186/1750-1326-9-24

Kent, S. A., & Miron, V. E. (2024). Microglia regulation of central nervous system myelin health and regeneration. *Nature Reviews Immunology*, *24*(1), 49–63. https://doi.org/10.1038/s41577-023-00907-4

Krause, A., Anderson, D. G., Ferreira-Correia, A., Dawson, J., Baine-Savanhu, F., Li, P. P., & Margolis, R. L. (2024). Huntington disease-like 2: Insight into neurodegeneration from an African disease. *Nature Reviews Neurology*, *20*(1), 36–49. https://doi.org/10.1038/s41582-023-00906-y

Liu, Z., Cheng, X., Zhong, S., Zhang, X., Liu, C., Liu, F., & Zhao, C. (2020). Peripheral and central nervous system immune response crosstalk in amyotrophic lateral sclerosis. *Frontiers in Neuroscience*, *14*. https://doi.org/10.3389/fnins.2020.00575

Loh, J. S., Mak, W. Q., Tan, L. K. S., Ng, C. X., Chan, H. H., Yeow, S. H., Foo, J. B., Ong, Y. S., How, C. W., & Khaw, K. Y. (2024). Microbiota–gut–brain axis and its therapeutic applications in neurodegenerative diseases. *Signal Transduction and Targeted Therapy*, *9*(1), 1–53. https://doi.org/10.1038/s41392-024-01743-1

Ma, Y.-Y., Li, X., Yu, J.-T., & Wang, Y.-J. (2024). Therapeutics for neurodegenerative diseases by targeting the gut microbiome: From bench to bedside. *Translational Neurodegeneration*, *13*(1), 12. https://doi.org/10.1186/s40035-024-00404-1

Martin, S., Battistini, C., & Sun, J. (2022). A gut feeling in amyotrophic lateral sclerosis: Microbiome of mice and men. *Frontiers in Cellular and Infection Microbiology*, *12*. https://doi.org/10.3389/fcimb.2022.839526

Noor Eddin, A., Alfuwais, M., Noor Eddin, R., Alkattan, K., & Yaqinuddin, A. (2024). Gut-modulating agents and amyotrophic lateral sclerosis: Current evidence and future perspectives. *Nutrients*, *16*(5), Article 5. https://doi.org/10.3390/nu16050590

Ogbu, D., Zhang, Y., Claud, K., Xia, Y., & Sun, J. (2022). *Dietary butyrate treatment enhances healthy metabolites by longitudinal untargeted metabolomic analysis in amyotrophic lateral sclerosis mice* (p. 2022.01.15.476456). bioRxiv. https://doi.org/10.1101/2022.01.15.476456

Pandey, R. P., Gunjan, Himanshu, Mukherjee, R., & Chang, C.-M. (2024). Nanocarrier-mediated probiotic delivery: A systematic meta-analysis assessing the biological effects. *Scientific Reports*, *14*(1), Article 1. https://doi.org/10.1038/s41598-023-50972-x

Rooks, M. G., & Garrett, W. S. (2016). Gut microbiota, metabolites and host immunity. *Nature Reviews Immunology*, *16*(6), 341–352. https://doi.org/10.1038/nri.2016.42

Rothstein, S. M., Sen, S., & Mansell, T. J. (2020). Towards high-throughput genome engineering in lactic acid bacteria. *Current Opinion in Biotechnology*, *61*, 181–188. https://doi.org/10.1016/j.copbio.2019.12.015

Sanmarco, L. M., Wheeler, M. A., Gutiérrez-Vázquez, C., Polonio, C. M., Linnerbauer, M., Pinho-Ribeiro, F. A., Li, Z., Giovannoni, F., Batterman, K. V., Scalisi, G., Zandee, S. E. J., Heck, E. S., Alsuwailm, M., Rosene, D. L., Becher, B., Chiu, I. M., Prat, A., & Quintana, F. J. (2021). Gut-licensed IFNγ+ NK cells drive LAMP1+TRAIL+ anti-inflammatory astrocytes. *Nature*, *590*(7846), 473–479. https://doi.org/10.1038/s41586-020-03116-4

Tabrizi, S. J., Flower, M. D., Ross, C. A., & Wild, E. J. (2020). Huntington disease: New insights into molecular pathogenesis and therapeutic opportunities. *Nature Reviews Neurology*, *16*(10), 529–546. https://doi.org/10.1038/s41582-020-0389-4

Tang, Y., Yan, Y., Mao, J., Ni, J., & Qing, H. (2023). The hippocampus associated GABAergic neural network impairment in early-stage of Alzheimer's disease. *Ageing Research Reviews*, *86*, 101865. https://doi.org/10.1016/j.arr.2023.101865

Wallen, Z. D., Demirkan, A., Twa, G., Cohen, G., Dean, M. N., Standaert, D. G., Sampson, T. R., & Payami, H. (2022). Metagenomics of Parkinson's disease implicates the gut microbiome in multiple disease mechanisms. *Nature Communications*, *13*(1), 6958. https://doi.org/10.1038/s41467-022-34667-x

Wang, J., Gu, B. J., Masters, C. L., & Wang, Y.-J. (2017). A systemic view of Alzheimer disease—Insights from amyloid-β metabolism beyond the brain. *Nature Reviews Neurology*, *13*(10), 612–623. https://doi.org/10.1038/nrneurol.2017.111

Wang, L., Hu, J., Li, K., Zhao, Y., & Zhu, M. (2024). Advancements in gene editing technologies for probiotic-enabled disease therapy. *iScience*, *27*(9). https://doi.org/10.1016/j.isci.2024.110791

Wang, R., Cao, S., Bashir, M. E. H., Hesser, L. A., Su, Y., Hong, S. M. C., Thompson, A., Culleen, E., Sabados, M., Dylla, N. P., Campbell, E., Bao, R., Nonnecke, E. B., Bevins, C. L., Wilson, D. S., Hubbell, J. A., & Nagler, C. R. (2023). Treatment of peanut allergy and colitis in mice via the intestinal release of butyrate from polymeric micelles. *Nature Biomedical Engineering*, *7*(1), 38–55. https://doi.org/10.1038/s41551-022-00972-5

Wilson, D. M., Cookson, M. R., Bosch, L. V. D., Zetterberg, H., Holtzman, D. M., & Dewachter, I. (2023). Hallmarks of neurodegenerative diseases. *Cell*, *186*(4), 693–714. https://doi.org/10.1016/j.cell.2022.12.032

Wu, S., Yi, J., Zhang, Y.-G., Zhou, J., & Sun, J. (2015). Leaky intestine and impaired microbiome in an amyotrophic lateral sclerosis mouse model. *Physiological Reports*, *3*(4), e12356. https://doi.org/10.14814/phy2.12356

Yano, J. M., Yu, K., Donaldson, G. P., Shastri, G. G., Ann, P., Ma, L., Nagler, C. R., Ismagilov, R. F., Mazmanian, S. K., & Hsiao, E. Y. (2015). Indigenous bacteria from the gut microbiota regulate host serotonin biosynthesis. *Cell*, *161*(2), 264–276. https://doi.org/10.1016/j.cell.2015.02.047

Yin, L., Azi, F., Xia, X., Jin, Y., Lu, X., Cheng, J., Guan, Y., Cheng, J., Lu, G., & Pang, L. (2024). Microbiome-metabolomics-based insight into the protective effects of dietary fiber from sweetpotato residues on the high-fat diet-induced intestinal integrity damage. *International Journal of Biological Macromolecules*, *275*, 133620. https://doi.org/10.1016/j.ijbiomac.2024.133620

Zhang, W., Guo, Y., Cheng, Y., Yao, W., & Qian, H. (2023). Neuroprotective effects of polysaccharide from *Sparassis crispa* on Alzheimer's disease-like mice: Involvement of microbiota-gut-brain axis. *International Journal of Biological Macromolecules*, *225*, 974–986. https://doi.org/10.1016/j.ijbiomac.2022.11.160

Zhang, Y., Ogbu, D., Garrett, S., Xia, Y., & Sun, J. (2021). Aberrant enteric neuromuscular system and dysbiosis in amyotrophic lateral sclerosis. *Gut Microbes*, *13*(1), 1996848. https://doi.org/10.1080/19490976.2021.1996848

Zhang, Y., Xia, Y., & Sun, J. (2024). Probiotics and microbial metabolites maintain barrier and neuromuscular functions and clean protein aggregation to delay disease progression in TDP43 mutation mice. *Gut Microbes*, *16*(1), 2363880. https://doi.org/10.1080/19490976.2024.2363880

Zhang, Y.-G., Wu, S., Yi, J., Xia, Y., Jin, D., Zhou, J., & Sun, J. (2017). Target intestinal microbiota to alleviate disease progression in amyotrophic lateral sclerosis. *Clinical Therapeutics*, *39*(2), 322–336. https://doi.org/10.1016/j.clinthera.2016.12.014

Zhao, Z., Ning, J., Bao, X., Shang, M., Ma, J., Li, G., & Zhang, D. (2021). Fecal microbiota transplantation protects rotenone-induced Parkinson's disease mice via suppressing inflammation mediated by the lipopolysaccharide-TLR4 signaling pathway through the microbiota-gut-brain axis. *Microbiome*, *9*(1), 226. https://doi.org/10.1186/s40168-021-01107-9

Zhong, Z., Liu, G., Tang, Z., Xiang, S., Yang, L., Huang, L., He, Y., Fan, T., Liu, S., Zheng, X., Zhang, T., Qi, Y., Huang, J., & Zhang, Y. (2023). Efficient plant genome engineering using a probiotic sourced CRISPR-Cas9 system. *Nature Communications*, *14*(1), 6102. https://doi.org/10.1038/s41467-023-41802-9

12 Role of Probiotics in MS, Epilepsy, and Seizures

Kajal Gurow[1], *Deepak Chandra Joshi*[2], *Uttam Singh Baghel*[1], *and Lata Bisht*[3]

[1] Gurukul Pharmacy College, Ranpur, Kota, Rajasthan, India
[2] Department of Pharmacy, Faculty of Pharmaceutical Sciences & Nursing, Vivekananda Global University, Jaipur, Rajasthan, India
[3] School of Pharmaceutical Sciences and Technology, Sardar Bhagwan Singh University, Dehradun, Uttarakhand, India

12.1 INTRODUCTION

12.1.1 Overview of Neurological Disorders: Multiple Sclerosis (MS), Epilepsy, and Seizures

Multiple sclerosis (MS) is recognized as a chronic inflammatory disease (CID) and a debilitating condition that impacts the CNS, resulting in nerve demyelination. MS is a condition that affects about 2.8 million individuals worldwide, with a prevalence of 35.9 per 100,000 population. The prevalence of MS has been increasing in every region on an annual basis (Walton et al., 2020). It typically affects young individuals aged between 20 and 40; however, it can manifest at any age. Furthermore, it is frequently reported that in comparison to men, women are found to be more susceptible, with a ratio of 3:1 or even greater. In approximately 85% of relapsing-remitting multiple sclerosis (RRMS) patients, a progressive stage of MS develops after years and is marked by permanent impairment and continuous disease progression without respite (Cree et al., 2021).

Complex pathogenic pathways result in atrophy of brain and spinal cord, leading to irreversible impairment and malfunction of numerous organs (Ruggieri et al., 1234). The precise etiology of MS remains unidentified; nevertheless, controlling inflammatory pathways and immune system function may positively influence disease management and enhancement. MS leads to immune system's destruction of the protective covering of nerves, therefore disrupting the connection between the brain and the systemic tissues. Finally, this may lead to the irreversible degradation of the nerves themselves. Roughly 15% of MS cases follow a primary progressive MS (PPMS) course, whereas 85% adhere to a relapsing-remitting (RRMS). There are many more phenotypes of MS (Vidal-Jordana & Montalban, 2017). Multiple sclerosis symptoms typically manifest in young individuals aged 20–40. Blood–brain barrier disruption, axonal destruction, demyelination, and immunological dysfunction are characteristic hallmarks of multiple sclerosis.

in the spinal cord and the brain, demyelinated plaques can develop in either the gray matter or the white matter. A neuroinflammatory reaction is initiated by them, which results in the loss of myelin in highly specialized cells, such as oligodendrocytes and neurodegeneration (Frohman et al., 2006). Research using a mouse model of experimental autoimmune encephalomyelitis has shown that CD4+ T lymphocytes participate in the development of MS (Fletcher et al., 2010).

A "seizure" is a temporary disruption of neurological function resulting from the intense and coordinated neuronal activity in the brain. Neurological issues may result from diverse factors, including congenital or acquired brain anomalies, damage to the structure or lesions, and other diseases of the brain. The phrase "epileptic seizure" distinguishes seizures caused by abnormal brain activity from non-epileptic manifestations such as psychogenic seizures. This medical disorder is characterized by repeated, unprovoked episodes of seizures. The causes of epilepsy are manifold, and each one reflects an underlying brain dysfunction. A seizure triggered by a reversible insult (such as fever or hypoglycemia) is not considered epilepsy as it is a transient secondary pathology rather than a long-lasting illness.

Epilepsy is a persistent neurological condition impacting around 65 million individuals globally. At present, even on taking anti-epileptic drug therapy, just 70% of people with epilepsy successfully attain full control over their seizures. Consequently, roughly 33% of individuals with epilepsy experience refractory episodes that disrupt their everyday routines (Arulsamy et al., 2020). A correlation has been proven among epilepsy and intestinal bacterial species, suggesting that genetic modification (GM) may be beneficial in the treatment of epilepsy. In healthy gut microbome, microbial populations influence both pro-inflammatory and anti-inflammatory states, suggesting a potential link

DOI: 10.1201/9781003536802-12

between immune system, and a balanced gut microbiota. Scientific proof has demonstrated that persistent inflammation is a cause of development and advancement of epilepsy. Additionally, genetic modification can influence immune and inflammatory reactions (Zhang et al., 2022).

12.1.2 The Gut–Brain Axis: A Bidirectional Communication Pathway

GBA refers to the interaction that transpires between the brain and the gut microbome. It involves a number of physiological processes that are key players in preserving the microbial, CNS, and GI tract's homeostasis. Despite the structural separation of the stomach and the brain, various mechanisms of communication between gut microbome with the CNS have been suggested. These mechanisms involve the production of neuroactive substances, metabolites, and hormones to regulate the immune system, ENS, vagus nerve, and circulatory system (Forsythe et al., 2014; Strandwitz, 2018). Neuromodulation, energy balance, activation, and regulation of brain activity, and emotions are among the many processes regulated by these GBA systems (Dinan & Cryan, 2017a).

The microbiota has attracted significant scientific interest as a connection between cognitive brain centers, mental health, and peripheral gastrointestinal function.

Research indicates that the GBA disruption may be linked to mood abnormalities and dysbiosis of the gastrointestinal microbome (Dinan & Cryan, 2017b; Huang et al., 2019). Additionally, anxiety and depressive behavior are influenced by gut microbiota through the GBA (Naseribafrouei et al., 2014). Specific microbiota remodeling are observed in autistic patients, contingent upon the disease severity (Qiao et al., 2018).

12.1.3 Emerging Role of Probiotics in Neurological Health

The global consumption of probiotics is on the rise due to their health benefits (Table 12.1).

12.2 THE GUT MICROBIOTA AND NEUROLOGICAL DISORDERS

12.2.1 Composition and Function of the Gut Microbiota

The gastrointestinal tract is home to a consortium of wide variety of microorganisms that coexist in a mutualistic association with hosts, known as gut microbiota. Despite the long-standing recognition of the significance of the gut microbiota, it is only in the past few decades that our comprehension of the gut microbiota has begun to expand as a result of advancements in genomics, metabolomics, and other technologies, such as culturomics (Lagier et al., 2018).

12.2.1.1 Composition

The predominant bacilli in the human gut are derived from five distinct phyla: Firmicutes (60–80%), encompassing the classes Negativicutes, Bacilli, and Clostridium. The Bacteroidetes (20–25%) include Verrucomicrobia, Sphingobacteria, Bacteroidia, Flavobacteria, and Cytophagia.

Actinobacteria (3%), and a small amount of Proteobacteria (5–10%) round out the composition. Finally, the Euryarchaeota is the only Archea phyla (Donaldson et al., 2016).

A recent extensive study revealed that the human gut microflora has around 35,000 bacterial species, in contrast to the previously estimated 500–1,000 species (Ramakrishna, 2007). The Human Microbiome Project and MetaHIT investigations show that the human microbiome that may have over 10 million bacterial genes are nonredundant. The MetaHIT Consortium (Arumugam et al., n.d.) advocates for the classification of gut flora according to species composition, which organizes into stable, well-balanced host-microbial symbiotic relationship that vary by food and medication while remaining consistent across geography and gender. We refer to these clusters as enterotypes (Karlsson et al., 2012).

In general, there are three enterotypes (Hollister et al., 2014; Arumugam et al., 2011).

Enterotype 1: Bacteroides dominates, subsequently succeeded by *Parabacteroides*. The two genera exhibit high levels of enzymes such as galactosidase, hexosaminidase, and protease. People with enterotype-1 are used to Western diets because they may get the greatest energy from fermenting carbs and proteins. This enterotype produces high concentration of pantothenic acid, riboflavin, and biotin (B2–B5–B8) than others. This gut microbiota makeup often indicates a general inflammatory state.

Enterotype 2: Prevotella and *Desulfovibrio* are the dominant genera in enterotype 2, which can generate elevated concentrations of folic acid (B9) and thiamine (B1). Consequently, this enterotype is the consequence of a diet that is entirely vegetarian.

Ruminococcus and *Akkermansia* are the dominant genera in enterotype 3. They specialize in the degradation of mucin but prefer simple carbohydrates. The immune system's typical functionality can be disrupted by these microorganisms.

12.2.2 Mechanisms Linking Gut Microbiota to Neurological Disorders

The CNS and ENS establish communication through several chemical signaling channels, such as direct neuronal, immunological, and endocrine relays.

12.2.2.1 Immune System Modulation

The CNS and gut microbe directly influence the immune system. Gut microbiome affects peripheral immune system development and function (Zheng et al., 2020). The

TABLE 12.1
Role of Probiotics in Neurological Conditions via Modulation of the Gut–Brain Axis

Neurological Condition	Probiotic Strain	Proposed Mechanism of Action	Research Findings	References
Multiple sclerosis	*Lactobacillus casei*, *Bifidobacterium*	Modulation of gut–brain axis, immune regulation, reduction in pro-inflammatory cytokines	Reduced progression and severity of experimental autoimmune encephalomyelitis (EAE) in animal models	Constantinescu et al. (2011)
Epilepsy	*Lactobacillus rhamnosus*, *Bifidobacterium*	Modulation of GABAergic neurotransmission, enhancement of inhibitory signals in the CNS	Increased seizure thresholds and decreased seizure frequency in animal models	Kandratavicius et al. (2014)
Seizures	*Bifidobacterium breve*, *Lactobacillus*	Influence on gut microbiota composition, modulation of neuroinflammatory and oxidative pathways	Reduction in seizure intensity and frequency in rodent models	Garbuz et al. (2022)
Overall Gut–Brain Axis	*Lactobacillus plantarum*, *Lactobacillus reuteri*	Regulation of gut microbiota, improved intestinal barrier function, anti-inflammatory effects	Studies show enhanced neuroprotection, reduction in neuroinflammation, and improved neurological outcomes	Adamu et al. (2024)

brain innate immune cells require microbe for their activation and development. Immune reactions and inflammation may cause psychiatric illnesses (Miller and Raison, 2016). The CNS and cytokine interactions alter neurological circuits that govern motivation, behavior, and motor function (Capuron and Miller, 2011). The gut microbiome and brain are also linked through the immune system and circulating cytokines (Hsiao et al., 2013). Immune cells are capable of infiltering CNS through the BBB or create cytokines and chemokines throughout the CNS (Morais et al., 2021). Intestinal cytokines can pass through the bloodstream, affecting hypothalamus and other brain regions (El Aidy et al., 2014). The blood–brain barrier is a physiological boundary that delineates the neurological isolation of the body's microenvironment. The BBB regulates chemical, ion, and cell movement between the brain and the body (Engelhardt and Liebner, 2014). The blood–brain barrier is crucial for protecting the brain from pathogens and immunological reactions that could destroy neurons and synapses, affecting hypothalamus and other regions of brain (El Aidy et al., 2014). SCFAs directly affect immunomodulators and immune cells for the maintenance of homeostasis. Short-chain fatty acids (SCFAs) affect intestinal mucosal immunity, as Corrêa-Oliveira et al. (2016) describe. They regulate cerebral function via the peripheral immune system. Systemic inflammation may be mitigated by thickening the intestinal boundary and limiting transmission of bacteria and bacterial metabolites or by direct interaction of SCFAs and immune cells, reducing brain neuroinflammation (Dalile et al., 2019). SCFAs control T lymphocyte, macrophage, DC, and neutrophil maturation and activation (Corrêa-Oliveira et al., 2016). Neutrophils, the commonest granulocyte, are produced in bone marrow and are vital to the innate immune system. They are initial responders at the site of inflammation and use cytokines to attract macrophages (Rodrigues et al., 2016). SCFAs induce neutrophils to produce pro-inflammatory cytokines like TNF via inhibiting histone deacetylase (HDAC). They also attract neutrophils by controlling the synthesis of CXC motif chemokines such CXCL1 and CXCL8. SCFAs activate neutrophil FFAR2 to influence chemotaxis (Rodrigues et al., 2016). SCFAs impact T cell growth and proliferation, altering adaptive immune responses (Kim et al., 2014).

12.2.2.2 Endocrine Pathway

SCFAs regulate gut hormones to affect gut–brain function. They activate GPCRs in the colon to produce gut hormones and increase enteroendocrine L cell release of PYY and GLP1. Hormones affect mood, memory, and learning. SCFAs directly activate vagal afferents through free fatty acid receptors (FFARs) to transmit signals to the brain (Dalile et al., 2019). GLP1 affects brain functioning via humoral and neuronal channels in the CNS, PNS, intestines, lungs, heart, and pancreas via numerous receptors (Alvarez et al., 2005; Katsurada and Yada, 2016). GLP1 has been linked to improve memory and learning in mice, hippocampal neuroplasticity and neuroprotection, and decrease the activation microglia and β-amyloid in animal models of AD (Isacson et al., 2011; McClean et al., 2011; Porter et al., 2011). PYY, another anorexic neuropeptide, inhibits stomach motility and appetite. The brain's hypothalamus and pituitary gland are the primary site to express PYY, along with the colon and ileum's L cells (Morimoto et al., 2008). Shortened PYY3-36, which binds primarily with the Y2 neuropeptide Y receptor, is the most prevelant form of PYY in circulation (Murphy and Bloom, 2006).

12.2.2.3 Neurotransmitter Production

Interacting chemically with the neurological system, which encompasses both "direct" and "indirect" communication, gut microbes can assist in regulating the framework of controlling physiological function and the modification of behavior in their animal hosts (Morais et al., 2021). Microorganism synthesize a portion of the neuroactive chemicals, by inducing the host to produce more metabolites, and neurotransmitters by modulating gut–brain signaling (Poutahidis et al., 2013). It seems that cues from microbial metabolism influence microglia immune programming, as the restoration of microglial morphology and function in germ-free mice is achieved through the administration of bacterial-derived SCFAs (Erny et al., 2015). The brain is signaled by molecules derived from microorganisms. The intestinal microbiota produces neurotransmitters, including serotonin, norepinephrine, dopamine, glycine, and gamma-aminobutyric acids, each of which has a distinct effect on cerebral γ-aminobutyric acid (GABA). Alzheimer's disease, Parkinson's disease, ASD, anxiety disorders, and depressive disorders may result from neurotransmitter imbalances.

In a study, bifidobacterium infantis elevates the level of tryptophan in blood, thus influencing the transmission of central serotonin. *Lactobacillus* and *Bifidobacterium* are capable of producing GABA, while *Escherichia*, *Saccharomyces*, *and Bacillus* species can produce noradrenaline. *Streptococcus*, *Escherichia coli*, and *Enterococcus* species can produce serotonin, bacteria can produce dopamine, and *Lactobacillus* synthesises acetycholine. The critical role of bacteria in the synthesis of 5-HT has been established in a study performed in animal mouse model (Clarke et al., 2013). Microbiological metabolites including indole, short-chain fatty acids, secondary bile acids, α-tocopherol,tyramine, and p-aminobenzoate influence the release and production of 5-HT by enteroendocrine cells (EECs).

12.3 PROBIOTICS: DEFINITION AND MECHANISMS OF ACTION

12.3.1 Mechanisms by Which Probiotics Influence the Gut–Brain Axis

12.3.1.1 Modulation of Gut Microbiota Composition

The gut microbiome is composed of bacteria, archae, fungi, viruses, and protozoa, which through mutual influence affect the host's physiological state and overall health. Vitamin B production, improved digestion, and the development of angiogenesis and neuronal function are just a few of the many vital human health tasks supported by the intestinal bacteria. Probiotic *Lactobacillus* strains that strengthen the intestinal barrier may have a variety of effects, including preserving immunological tolerance, reducing bacterial translocation across the mucosa of the intestine, and alleviating symptoms of diseases such GI infections, IBS, and inflammatory bowel disease (Lee et al., 2011).

12.3.1.2 Enhancement of Gut Barrier Function

The luminal contents and the dynamic, variable enteric microbiota are in constant contact with the intestinal epithelium. The gut barrier is an important line of protection that is employed to safeguard the organism originating from environment and to preserve epithelium integrity. The food and bacteria antigens become permeable to submucosa and trigger inflammatory reaction that can lead to intestinal disorders, including IBD, once this barrier function is compromised (Sartor, 2006). Probiotic microorganism have been throughly investigated for their role in preserving intestinal barrier, as ingesting non-pathogenic bacteria can contribute to its function. Conversely, the mechanisms by which beneficial bacteria in the gut enhance the function of intestinal barrier remain incompletely comprehended. Research suggests that increasing tight junction signaling gene expression may strengthen intestinal barrier integrity. Using a cell barrier model T84, lactobacilli regulate genes that encode β-catenin and E-cadherin, the two adhesion junction proteins. Intestinal cells incubated with lactobacilli show variations in adherence junction protein phosphorylation and PKC isoform abundance, positively affecting epithelial barrier function (Bermudez-Brito et al., 2012). The most recent data implies that beneficial bacteria can stimulate the restoration of the protection layer function following injury. *E.coli* Nissle 1917 (EcN1917) has dual effects by preventing injury of mucosal barrier caused by enteropathogenic *E. coli* and by restoring in T84 and Caco-2 cells mucosal integrity. The action is facilitated by the increased production and redistribution of proteins localized within tight junction proteins, namely, the zona occludens (ZO-2) and PKC, ultimately resulting in tight junction complex reconstruction (Zyrek et al., 2007; Šťtinová et al., 2013) .

12.3.1.3 Immune and Neuroprotective Effects

Probiotics mainly modulate the immune system by (i) activating macrophages through probiotic signaling, (ii) stimulating neutrophils and IgA-producing cells, (iii) producing peripheral Ig, (iv) prompting the body to produce more mucin, and (v) preventing cytokines release that promote inflammation. This action has been shown by the capacity of probiotic bacteria to interact with immune cells, including monocytes, lymphocytes, macrophages, and dendritic cells, and also intestinal epithelial cells. Probiotics enhance the function of gut immune system by stimulating β cells to produce IgAs. Orally administrating several probiotics, including lactobacillus casei, acidophilus, rhamnosus, plantarum, lactis, and streptococcus thermophilus, has been shown to exhibit a dose-dependent rise in the population of IgA-producing intestinal cells. Probiotics can promote the proliferation of β cells that are activated to secrete IgAs, without affecting the CD4+ T cells (Perdigón et al., 1999). Orally administrating probiotic bacteria like lactobacillus helveticus R389 and lactobacillus casei CRL 431 elevates the level of IL-6 in a TLR2-dependent manner, leading to an increase in intestinal IgA-producing cells without increasing CD4+ T cells. Evidence indicates lactobacilli can induce B cell clonal proliferation via IL-6 production, leading to IgA secretion (Vinderola et al., 2005).

12.4 PROBIOTICS AND MULTIPLE SCLEROSIS

12.4.1 Pathophysiology of MS and the Role of the Gut Microbiota

The pathogenesis of MS involves a combination of genetic predisposition, environmental factors, and an abnormal immune response. Despite this understanding, the exact cause of the disease remains unclear (Chastain & Miller, 2012). Several studies have suggested that oxidative stress plays a significant role in the progression of MS (Schreibelt et al., 2007). The presence of increased reactive oxygen species (ROS) in the cerebrospinal fluid (CSF) of MS patients indicates a common imbalance between oxidants and antioxidants (Acar et al., 2003). Additionally, elevated oxidative stress markers have been detected in the blood of MS patients, including malondialdehyde (MDA) (Ghonimi et al., 2021), superoxide dismutase (SOD) (Obradovic et al., 2021), and glutathione (GSH) (Choi et al., 2011). In MS, immune dysfunction is likely driven by an imbalance between regulatory cells, such as Th17 and pathogenic T helper cells, as well as regulatory B (Breg) and T (Treg) cells. The immunopathology of MS is characterized by an overactive immune response, particularly an excess of pro-inflammatory cells like macrophages, dendritic cells, monocytes (CD14+/CD16+), and B cells, along with a reduced number of regulatory cells, including Tregs (CD4+/CD25+/FoxP3+) and Bregs (CD19+/CD5+/Cd1d+). Evidence suggests that activated Th1 and Th17 cells, which produce TNF-α, GM-CSF, IFN-γ, IL-17, IL-21, and IL-22, are the main contributors to MS pathogenesis (El-Behi et al., 2011; Hirota et al., 2011). Given the significant role of the gut microbiota in the development and maturation of the immune system, it is plausible that it plays a role in the pathogenesis of MS. Recent findings indicate that the gut microbiome has a profound influence on MS and is susceptible to environmental influences. Studies have shown that MS patients have a higher prevalence of the phylum Firmicutes in their gut microbiota, while the phylum Bacteroidetes is less abundant compared to healthy individuals, as demonstrated by 16S rRNA sequencing (Cosorich et al., 2017). Untreated MS patients also exhibit a greater presence of *Euryachaeota* and *Akkermansia* compared to healthy controls (Jangi et al., 2016). Other research has identified a reduced number of *Prevotella* in patients with RRMS, which may contribute to disease activity. Miyake et al. (2015) also reported a decrease in SCFA production in RRMS patients, linked to a reduced abundance of *Clostridium*. Increased levels of Streptococcus, a higher Firmicutes-to-Bacteroidetes ratio, and lower levels of *Prevotella* have been associated with more severe disease activity and a higher number of intestinal Th17 cells (Miyake et al., 2015).

12.5 PROBIOTICS IN EPILEPSY AND SEIZURES

12.5.1 Pathophysiology of Epilepsy and the Role of the Gut Microbiota

Stress induces seizures; the regulating mechanism in the stress response is HPA axis (Cano-López & González-Bono, 2019). In epilepsy patients, the glucocorticoid levels were increased, while deoxycorticosterone is found beneficial. Corticosterone and CRH influence signaling of excitatory neurotransmitter, which can cause seizures (Reddy & Rogawski, 2002; Davletshin et al., 2023). Recent evidences from studies have shown a link between gut microbiome and the HPA axis. Stressful environments can alter gut microbiota structure and function, affecting gene expression in the colonic CRH pathway. However, additional investigation is needed to ascertain the link between gut microbiome, the HPA axis, and epilepsy. Research suggests that neurotransmitter imbalances are linked to epilepsy, with epileptic foci displaying elevated levels of glutamate, dopamine, and norepinephrine levels and lower GABA levels (Singh & Goel, 2016). In epilepsy sufferers, a disruption in the balance in GABA-glutamate in brain tissue has been identified. GABA is a neurotransmitter that is distributed throughout the nervous system and is inhibitory. Reduced GABA levels can result in alterations in the permeability of neuronal cells to synaptic interstitial chloride ions.

This modification leads to higher epilepsy risk and a reduction in seizure thresholds (Barrett et al., 2012a). It has been discovered that a variety of intestinal microbiomes are able to metabolize their own distinct neurotransmitters. For instance, serotonin can be synthesized by *Enterococcus* spp., *Streptococcus* spp., and *E.coli* spp. *Lactobacillus* spp. and *Bifidobacterium* spp. are responsible for GABA synthesis. On the other hand, *Bacillus* spp. and *E. coli* spp. generate dopamine and noradrenaline (Barrett et al., 2012b; Strandwitz, 2018). Through the regulation of brain neurotransmitters, including GABA, glutamate, the gut microbiome, specifically due to presence of beneficial bacteria like A. mucinophilia and Parabacteroides, may offer protective benefits in seizure frequency. GABA levels may be altered by bacterial dysbiosis, resulting in an exacerbation of seizures.

Firmicutes and Bacteroidetes in the gut microbiome can ferment insoluble fiber, producing SCFAs like acetate, propionate, and butyrate (Den Besten et al., 2013). SCFAs promote microglia development and are connected to seizure vulnerability by altering function and BBB permeability (De Caro et al., 2019). Studies show that people with drug-resistant epilepsy have considerably different amounts of Firmicutes and Bacteroidetes phyla compared to those with drug-sensitive epilepsy and healthy controls (Figure 12.1) (Lindefeldt et al., 2019; H. Lee et al., 2021).

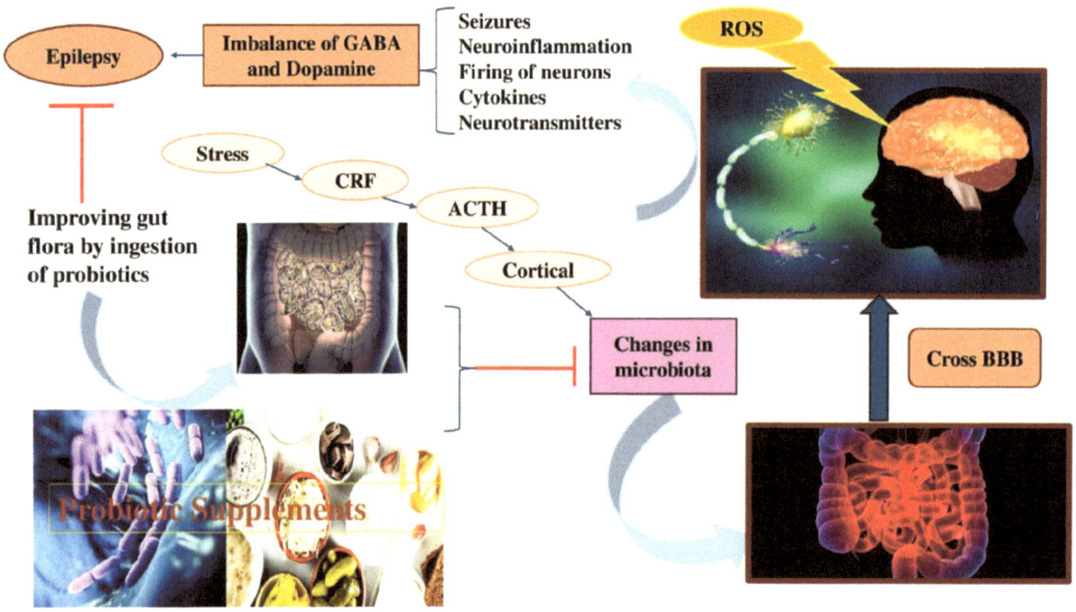

FIGURE 12.1 Mechanism of probiotic action on epilepsy.

12.5.2 Studies on Probiotics in Epilepsy Management (Table 12.2)

TABLE 12.2
Studies on Probiotics Use in Epilepsy

Study	Probiotic Strain	Study Design	Key Findings	Reference
Gomez-Eguilaz et al. (2018)	*Lactobacillus casei Shirota*	Open-label, uncontrolled study	Improvement in seizure frequency and quality of life in drug-resistant epilepsy patients	Gomez-Eguilaz et al., 2018
Skonieczna-Zydecka et al. (2018)	*Lactobacillus rhamnosus GG*	Randomized, double-blind, placebo-controlled trial	Reduced seizure activity in animal models, linked to modulation of the gut–brain axis	Skonieczna-Zydecka et al., 2018
He et al. (2019)	*Bifidobacterium longum, Lactobacillus acidophilus*	Preclinical (rat model)	Significant reduction in seizure susceptibility and inflammatory markers in epileptic rats	He et al., 2017
Varesi et al. (2022)	*Lactobacillus plantarum*	Preclinical (mouse model)	Enhanced neuroprotection, decreased neuroinflammation, and improved seizure thresholds	Varesi et al., 2022
Xie et al. (2021)	*Bifidobacterium breve*	Preclinical (mouse model)	Modulation of gut microbiota, reduction of seizure severity, and improved cognitive function	Xie et al., 2021
Bercik et al. (2019)	*Lactobacillus helveticus*	Randomized, placebo-controlled trial	Reduction in seizure-like activity and improvement in behavioral outcomes in preclinical models	Dalile et al., 2019
Arzimanoglou et al. (2020)	*Lactobacillus rhamnosus*	Case study (drug-resistant epilepsy)	Patient reported improved seizure control and better gastrointestinal health following probiotic use	Arzimanoglou et al., 2020

12.6 FUTURE DIRECTIONS ON USE OF PROBIOTICS IN EPILEPSY AND MULTIPLE SCLEROSIS

Although numerous pathways (inflammation, neuroprotection, and gut–brain axis) have been suggested, additional research should be performed in future to gain a comprehensive understanding of the impact of probiotics on MS and epilepsy. For instance, the refinement of treatment strategies could be facilitated by an understanding of the impact of probiotics on specific neurotransmitter pathways in epilepsy or the modulation of T cell activity in MS.

Identifying biomarkers (e.g., particular cytokines, gut microbial metabolites, or short-chain fatty acids) that suggest probiotic treatment response could assist monitor patient progress and alter treatments.

Diet helps manage epilepsy (ketogenic diet) and MS (anti-inflammatory diet). Dietary treatments and probiotics may work together. Certain diets may boost probiotic efficiency by boosting good bacteria development, while probiotics may boost epilepsy treatment with ketogenic diets. Prebiotics and synbiotics can also be investigated in an integrated method. These may improve gut probiotic stability and colonization.

Genetically modified probiotics may target epilepsy and MS pathways. For instance, strains could increase gut-produced neuroprotective or anti-inflammatory chemicals. Develop broader microbiome-based therapeutics like FMT or modified bacterial consortia to reset the gut–brain axis in epilepsy and MS. Future research could investigate the potential of probiotics to prevent the onset of disease in at-risk populations. For instance, the onset of MS or epilepsy may be delayed or prevented by early probiotic intervention in individuals with a genetic predisposition to these conditions. Additionally, Advanced Probiotic Delivery Systems is a developing field. Novel delivery mechanisms for probiotics, such as encapsulated probiotics that can circumvent stomach acid and more effectively colonize the gut, could be the subject of research. Targeted delivery systems that release probiotics at specific sites within the gastrointestinal tract for maximal efficacy may be developed using nanotechnology.

Personalized, mechanistic, and integrative approaches are the future of probiotics in the treatment of MS and epilepsy. Probiotics may become a potent alternative or adjunct therapy as a result of the progress made in gut microbiome research, precision medicine, and biotechnology. Nevertheless, their effective clinical use will be contingent upon the development of more rigorous clinical trials, a more comprehensive understanding of mechanisms, and optimized delivery methods.

12.7 CONCLUSION

In the past decade, significant theoretical concerns regarding the relationship between the MGB axis and chronic neurological disorders, such as epilepsy and multiple sclerosis, have emerged, highlighting the necessity for innovative research strategies focused on a new class of therapeutic targets, particularly for refractory phenotypes patients. An increasing amount of evidence indicates that gut microbiota is a crucial regulator of immunity, metabolism, and neuroendocrine-signaling pathways. Probiotics may increase GABA, serotonin, and dopamine production, which regulates neuronal excitability and mood. Epilepsy is linked to neurotransmitter imbalances. Certain probiotics can alter inflammatory pathways by lowering pro-inflammatory cytokines (e.g., IL-6, TNF-α) and raising anti-inflammatory cytokines (e.g., IL-10), which are crucial in the etiology of epilepsy. By lowering neuroinflammation, probiotics may reduce seizure risk. Inflammation can compromise the BBB, allowing dangerous compounds to enter and cause seizures. Probiotics may reduce systemic inflammation and protect the blood–brain barrier. The ketogenic diet may reduce epileptic seizures. Probiotics that boost ketogenesis (ketone body formation) may reduce seizures by mimicking the ketogenic diet. Probiotics may regulate tryptophan metabolism, an important amino acid involved in serotonin synthesis. Serotonin regulates mood and neuronal function, and tryptophan metabolism abnormalities can cause epilepsy. Excessive immunological response and chronic inflammation in MS are caused by increased pro-inflammatory cytokines (e.g., IL-17, IL-6, and TNF-α). By suppressing pro-inflammatory cytokines, probiotics can reduce immune-mediated damage in MS.

Maintaining immunological tolerance and avoiding autoimmune assaults requires Tregs. In MS, probiotics may boost Treg function and quantity, suppressing the autoimmune response.

MS imbalances pro-inflammatory Th17 and Th1 cells and anti-inflammatory Treg cells. MS inflammation and demyelination are caused by Th17 and Th1 responses, which probiotics may reduce.

REFERENCES

Acar, G., Idiman, F., Idiman, E., Kirkali, G., Çakmakçi, H., & Özakbaş, S. (2003). Nitric oxide as an activity marker in multiple sclerosis. *Journal of Neurology*, *250*(5), 588–592. https://doi.org/10.1007/S00415-003-1041-0

Adamu, A., Li, S., Gao, F., & Xue, G. (2024). The role of neuroinflammation in neurodegenerative diseases: Current understanding and future therapeutic targets. *Frontiers in Aging Neuroscience, 16*, 1347987. https://doi.org/10.3389/fnagi.2024.1347987

Akkol, S., Doğan, M. C., Esenkar, D., Doğan, H., Karamahmutoğlu, T., & Onat, F. (2017). Effects of probiotic consumption on absence seizures. *Journal of the Turkish Epilepsi Society*, *23*(2), 51. https://doi.org/10.14744/EPILEPSI.2017.59389

Arulsamy, A., Tan, Q. Y., Balasubramaniam, V., O'Brien, T. J., & Shaikh, M. F. (2020). Gut microbiota and epilepsy: A systematic review on their relationship and possible therapeutics. *ACS Chemical Neuroscience*, *11*(21), 3488–3498. https://doi.org/10.1021/ACSCHEMNEURO.0C00431

Arumugam, M., Raes, J., Pelletier, E., Paslier, D. Le, Yamada, T., Mende, D. R., Fernandes, G. R., Tap, J., Bruls, T., Batto, J. M., Bertalan, M., Borruel, N., Casellas, F., Fernandez, L., Gautier, L., Hansen, T., Hattori, M., Hayashi, T., Kleerebezem, M., ... Zeller, G. (2011). Enterotypes of the human gut microbiome. *Nature*, *473*(7346), 174. Retrieved September 6, 2024, from https://www.nature.com/articles/nature09944

Arzimanoglou, A., Brandl, U., Cross, J. H., Gil-Nagel, A., Lagae, L., Landmark, C. J., Specchio, N., Nabbout, R., Thiele, E. A., Gubbay, O., & Panel, C. I. E. (2020). Epilepsy and cannabidiol: a guide to treatment. *PubMed*, *22*(1), 1–14. https://doi.org/10.1684/epd.2020.1141

Barrett, E., Ross, R. P., O'Toole, P. W., Fitzgerald, G. F., & Stanton, C. (2012a). γ-Aminobutyric acid production by culturable bacteria from the human intestine. *Journal of Applied Microbiology*, *113*(2), 411–417. https://doi.org/10.1111/J.1365-2672.2012.05344.X

Bermudez-Brito, M., Plaza-Díaz, J., Muñoz-Quezada, S., Gómez-Llorente, C., & Gil, A. (2012). Probiotic mechanisms of action. *Annals of Nutrition and Metabolism*, *61*(2), 160–174. https://doi.org/10.1159/000342079

Cano-López, I., & González-Bono, E. (2019). Cortisol levels and seizures in adults with epilepsy: A systematic review. *Neuroscience and Biobehavioral Reviews*, *103*, 216–229. https://doi.org/10.1016/J.NEUBIOREV.2019.05.023

Capuron, L., & Miller, A. H. (2011). Immune system to brain signaling: Neuropsychopharmacological implications. *Pharmacology & Therapeutics, 130*(2), 226–238. https://doi.org/10.1016/j.pharmthera.2011.01.014

Chastain, E. M. L., & Miller, S. D. (2012). Molecular mimicry as an inducing trigger for CNS autoimmune demyelinating disease. *Immunological Reviews*, *245*(1), 227–238. https://doi.org/10.1111/J.1600-065X.2011.01076.X

Choi, I. Y., Lee, S. P., Denney, D. R., & Lynch, S. G. (2011). Lower levels of glutathione in the brains of secondary progressive multiple sclerosis patients measured by 1H magnetic resonance chemical shift imaging at 3 T. *Multiple Sclerosis Journal*, *17*(3), 289–296. https://doi.org/10.1177/1352458510384010

Clarke, G., Grenham, S., Scully, P., Fitzgerald, P., Moloney, R. D., Shanahan, F., Dinan, T. G., & Cryan, J. F. (2013). The microbiome–gut–brain axis during early life regulates the hippocampal serotonergic system in a sex-dependent manner. *Molecular Psychiatry, 18*(6), 666–673. https://doi.org/10.1038/mp.2012.

Constantinescu, C. S., Farooqi, N., O'Brien, K., & Gran, B. (2011). Experimental autoimmune encephalomyelitis (EAE) as a model for multiple sclerosis (MS). *British Journal of Pharmacology, 164*(4), 1079–1106. https://doi.org/10.1111/j.1476-5381.2011.01302.x

Corrêa-Oliveira, R., Fachi, J. L., Vieira, A., Sato, F. T., & Vinolo, M. A. (2016). Regulation of immune cell function by short-chain fatty acids. *Clinical & Translational Immunology, 5*(4), e73. https://doi.org/10.1038/cti.2016.1777

Cosorich, I., Dalla-Costa, G., Sorini, C., Ferrarese, R., Messina, M. J., Dolpady, J., Radice, E., Mariani, A., Testoni, P. A., Canducci, F., Comi, G., Martinelli, V., & Falcone, M. (2017). High frequency of intestinal TH17 cells correlates with microbiota alterations and disease activity in multiple sclerosis. *Science Advances*, *3*(7). https://doi.org/10.1126/SCIADV.1700492

Cree, B. A. C., Arnold, D. L., Chataway, J., Chitnis, T., Fox, R. J., Pozo Ramajo, A., Murphy, N., & Lassmann, H. (2021). Secondary progressive multiple sclerosis: New insights. *Neurology*, *97*(8), 378–388. https://doi.org/10.1212/WNL.0000000000012323/ASSET/56DBB7F0-ED66-4C69-A9D7-3E1B8CDB6834/ASSETS/GRAPHIC/14TTU1.JPEG

Dalile, B., Van Oudenhove, L., Vervliet, B., & Verbeke, K. (2019). The role of short-chain fatty acids in microbiota–gut–brain communication. *Nature Reviews Gastroenterology & Hepatology*, *16*(8), 461–478. https://doi.org/10.1038/s41575-019-0157-3

Davletshin, A. I., Matveeva, A. A., Poletaeva, I. I., Evgen'ev, M. B., & Garbuz, D. G. (2023). The role of molecular chaperones in the mechanisms of epileptogenesis. *Cell Stress & Chaperones*, *28*(6), 599. https://doi.org/10.1007/S12192-023-01378-1

De Caro, C., Iannone, L. F., Citraro, R., Striano, P., De Sarro, G., Constanti, A., Cryan, J. F., & Russo, E. (2019). Can we 'seize' the gut microbiota to treat epilepsy? *Neuroscience & Biobehavioral Reviews*, *107*, 750–764. https://doi.org/10.1016/j.neubiorev.2019.10.002

Den Besten, G., Van Eunen, K., Groen, A. K., Venema, K., Reijngoud, D. J., & Bakker, B. M. (2013). The role of short-chain fatty acids in the interplay between diet, gut microbiota, and host energy metabolism. *Journal of Lipid Research*, *54*(9), 2325. https://doi.org/10.1194/JLR.R036012

Dinan, T. G., & Cryan, J. F. (2017a). Gut-brain axis in 2016: Brain-gut-microbiota axis - mood, metabolism and behaviour. *Nature Reviews. Gastroenterology & Hepatology*, *14*(2), 69–70. https://doi.org/10.1038/NRGASTRO.2016.200

Dinan, T. G., & Cryan, J. F. (2017b). Gut-brain axis in 2016: Brain-gut-microbiota axis - mood, metabolism and behaviour. *Nature Reviews. Gastroenterology & Hepatology*, *14*(2), 69–70. https://doi.org/10.1038/NRGASTRO.2016.200

Donaldson, G. P., Lee, S. M., & Mazmanian, S. K. (2016). Gut biogeography of the bacterial microbiota. *Nature Reviews Microbiology*, *14*(1), 20–32. https://doi.org/10.1038/NRMICRO3552

El Aidy, S., Dinan, T. G., & Cryan, J. F. (2015). Gut microbiota: The conductor in the orchestra of immune–neuroendocrine communication. *Clinical Therapeutics, 37*(5), 954–967. https://doi.org/10.1016/j.clinthera.2015.03.002

El-Behi, M., Ciric, B., Dai, H., Yan, Y., Cullimore, M., Safavi, F., Zhang, G. X., Dittel, B. N., & Rostami, A. (2011). The encephalitogenicity of TH 17 cells is dependent on IL-1- and IL-23-induced production of the cytokine GM-CSF. *Nature Immunology*, *12*(6), 568–575. https://doi.org/10.1038/NI.2031

Engelhardt, B., & Liebner, S. (2014). Novel insights into the development and maintenance of the blood–brain barrier. *Cell and Tissue Research, 355*(3), 687–699. https://doi.org/10.1007/s00441-014-1811-2

Erny, D., Hrabě de Angelis, A. L., Jaitin, D., Wieghofer, P., Staszewski, O., David, E., Keren-Shaul, H., Mahlakoiv, T., Jakobshagen, K., Buch, T., Schwierzeck, V., Utermöhlen, O., Chun, E., Garrett, W. S., McCoy, K. D., Diefenbach, A., Staeheli, P., Stecher, B., Amit, I., & Prinz, M. (2015). Host microbiota constantly control maturation and function of microglia in the CNS. *Nature Neuroscience, 18*(7), 965–977. https://doi.org/10.1038/nn.4030

Fletcher, J. M., Lalor, S. J., Sweeney, C. M., Tubridy, N., & Mills, K. H. G. (2010). T cells in multiple sclerosis and experimental autoimmune encephalomyelitis. *Clinical and Experimental Immunology*, *162*(1), 1–11. https://doi.org/10.1111/J.1365-2249.2010.04143.X

Forsythe, P., Bienenstock, J., & Kunze, W. A. (2014). Vagal pathways for microbiome-brain-gut axis communication. *Advances in Experimental Medicine and Biology*, *817*, 115–133. https://doi.org/10.1007/978-1-4939-0897-4_5

Frohman, E. M., Racke, M. K., & Raine, C. S. (2006). Multiple sclerosis--the plaque and its pathogenesis. *The New England Journal of Medicine*, *354*(9), 942–955. https://doi.org/10.1056/NEJMRA052130

Garbuz, M. M., Ovchinnikova, A. A., & Kumeiko, V. V. (2024). Correction: Garbuz et al. Design, optimization and validation of the ARMS PCR protocol for the rapid diagnosis of Wilson's disease using a panel of 14 common mutations for the European population. *Genes, 15*(8), 1070. https://doi.org/10.3390/genes15081070

Ghonimi, N. A. M., Elsharkawi, K. A., Khyal, D. S. M., & Abdelghani, A. A. (2021). Serum malondialdehyde as a lipid peroxidation marker in multiple sclerosis patients and its relation to disease characteristics. *Multiple Sclerosis and Related Disorders, 51.* https://doi.org/10.1016/J.MSARD.2021.102941

Gómez-Eguílaz, M., Ramón-Trapero, J., Pérez-Martínez, L., & Blanco, J. (2018). The beneficial effect of probiotics as a supplementary treatment in drug-resistant epilepsy: a pilot study. *Beneficial Microbes, 9*(6), 875–882. https://doi.org/10.3920/bm2018.0018

He, B., Hoang, T. K., Tian, X., Taylor, C. M., Blanchard, E., Luo, M., Bhattacharjee, M. B., Freeborn, J., Park, S., Couturier, J., Lindsey, J. W., Tran, D. Q., Rhoads, J. M., & Liu, Y. (2019, March). Lactobacillus reuteri reduces the severity of experimental autoimmune encephalomyelitis in mice by modulating gut microbiota. *Frontiers in Immunology, 10.* https://doi.org/10.3389/FIMMU.2019.00385/FULL

He, L., He, T., Farrar, S., Ji, L., Liu, T., & Ma, X. (2017). Antioxidants maintain cellular redox homeostasis by elimination of reactive oxygen species. *Cellular Physiology and Biochemistry, 44*(2), 532–553. https://doi.org/10.1159/000485089

Hirota, K., Duarte, J. H., Veldhoen, M., Hornsby, E., Li, Y., Cua, D. J., Ahlfors, H., Wilhelm, C., Tolaini, M., Menzel, U., Garefalaki, A., Potocnik, A. J., & Stockinger, B. (2011). Fate mapping of IL-17-producing T cells in inflammatory responses. *Nature Immunology, 12*(3), 255–263. https://doi.org/10.1038/NI.1993

Hollister, E., Gao, C., & Gastroenterology, J. V. (2014). Compositional and functional features of the gastrointestinal microbiome and their effects on human health. *Gastroenterology.* Retrieved September 6, 2024, from https://www.sciencedirect.com/science/article/pii/S0016508514001437

Hsiao, E. Y., McBride, S. W., Hsien, S., Sharon, G., Hyde, E. R., McCue, T., Codelli, J. A., Chow, J., Reisman, S. E., Petrosino, J. F., Patterson, P. H., & Mazmanian, S. K. (2013). Microbiota modulate behavioral and physiological abnormalities associated with neurodevelopmental disorders. *Cell, 155*(7), 1451–1463. https://doi.org/10.1016/j.cell.2013.11.024

Huang, T. T., Lai, J. B., Du, Y. L., Xu, Y., Ruan, L. M., & Hu, S. H. (2019, February). Current understanding of gut microbiota in mood disorders: An update of human studies. *Frontiers in Genetics, 10.* https://doi.org/10.3389/FGENE.2019.00098

Jangi, S., Gandhi, R., Cox, L. M., Li, N., Von Glehn, F., Yan, R., Patel, B., Mazzola, M. A., Liu, S., Glanz, B. L., Cook, S., Tankou, S., Stuart, F., Melo, K., Nejad, P., Smith, K., Topçuolu, B. D., Holden, J., Kivisäkk, P., … Weiner, H. L. (2016). Alterations of the human gut microbiome in multiple sclerosis. *Nature Communications, 7*(1), 1–11. https://doi.org/10.1038/ncomms12015

Jiang, C. S., & Gu, C. T. (2022). *Companilactobacillus salsicarnum* Zheng et al. 2020 is a later heterotypic synonym of *Companilactobacillus mishanensis* (Wei and Gu 2019) Zheng et al. 2020. *International Journal of Systematic and Evolutionary Microbiology, 72*(2). https://doi.org/10.1099/ijsem.0.005256

Kandratavicius, L., Balista, P. A., Lopes-Aguiar, C., Ruggiero, R. N., Umeoka, E. H., Garcia-Cairasco, N., Bueno-Junior, L. S., & Leite, J. P. (2014). Animal models of epilepsy: Use and limitations. *Neuropsychiatric Disease and Treatment, 10,* 1693–1705. https://doi.org/10.2147/NDT.S50371

Karlsson, F. H., Fåk, F., Nookaew, I., Tremaroli, V., Fagerberg, B., Petranovic, D., Bäckhed, F., & Nielsen, J. (2012). Symptomatic atherosclerosis is associated with an altered gut metagenome. *Nature Communications.* https://doi.org/10.1038/ncomms2266

Katsurada, K., & Yada, T. (2016). Neural effects of gut- and brain-derived glucagon-like peptide-1 and its receptor agonist. *Journal of Diabetes Investigation, 7*(Suppl 1), 64–69. https://doi.org/10.1111/jdi.12464

Kim, S., Kim, D., Cho, S. W., Kim, J., & Kim, J. S. (2014). Highly efficient RNA-guided genome editing in human cells via delivery of purified Cas9 ribonucleoproteins. *Genome Research, 24*(6), 1012–1019. https://doi.org/10.1101/gr.171322.113

Lagier, J. C., Dubourg, G., Million, M., Cadoret, F., Bilen, M., Fenollar, F., Levasseur, A., Rolain, J. M., Fournier, P. E., & Raoult, D. (2018). Culturing the human microbiota and culturomics. *Nature Reviews Microbiology, 16*(9), 540–550. https://doi.org/10.1038/S41579-018-0041-0

Lee, H., Lee, S., Lee, D. H., & Kim, D. W. (2021). A comparison of the gut microbiota among adult patients with drug-responsive and drug-resistant epilepsy: An exploratory study. *Epilepsy Research, 172.* https://doi.org/10.1016/J.EPLEPSYRES.2021.106601

Lee, J. B., & Bak, Y. T. (2011). Irritable bowel syndrome, gut microbiota and probiotics. *Journal of Neurogastroenterology and Motility, 17*(3). https://doi.org/10.5056/jnm.2011.17.3.252

Lindefeldt, M., Eng, A., Darban, H., Bjerkner, A., Zetterström, C. K., Allander, T., Andersson, B., Borenstein, E., Dahlin, M., & Prast-Nielsen, S. (2019). The ketogenic diet influences taxonomic and functional composition of the gut microbiota in children with severe epilepsy. *NPJ Biofilms and Microbiomes, 5*(1). https://doi.org/10.1038/S41522-018-0073-2

McClean, P. L., Parthsarathy, V., Faivre, E., & Hölscher, C. (2011). The diabetes drug liraglutide prevents degenerative processes in a mouse model of Alzheimer's disease. *The Journal of Neuroscience, 31*(17), 6587–6594. https://doi.org/10.1523/JNEUROS

Miller, A. H., & Raison, C. L. (2016). The role of inflammation in depression: From evolutionary imperative to modern treatment target. *Nature Reviews Immunology, 16*(1), 22–34. https://doi.org/10.1038/nri.2015.5CI.0529-11.2011

Miyake, S., Kim, S., Suda, W., Oshima, K., Nakamura, M., Matsuoka, T., Chihara, N., Tomita, A., Sato, W., Kim, S. W., Morita, H., Hattori, M., & Yamamura, T. (2015). Dysbiosis in the gut microbiota of patients with multiple sclerosis, with a striking depletion of species belonging to clostridia XIVa and IV clusters. *PLoS One, 10*(9), e0137429. https://doi.org/10.1371/JOURNAL.PONE.0137429

Morais, L. H., Schreiber, H. L., 4th, & Mazmanian, S. K. (2021). The gut microbiota-brain axis in behaviour and brain disorders. *Nature Reviews Microbiology, 19*(4), 241–255. https://doi.org/10.1038/s41579-020-00460-0

Morimoto, T., Kadoya, R., Endo, K., Tohata, M., Sawada, K., Liu, S., Ozawa, T., Kodama, T., Kakeshita, H., Kageyama, Y., Manabe, K., Kanaya, S., Ara, K., Ozaki, K., & Ogasawara, N. (2008). Enhanced recombinant protein productivity by genome reduction in *Bacillus subtilis. DNA Research, 15*(2), 73–81. https://doi.org/10.1093/dnares/dsn002

Murphy, K. G., & Bloom, S. R. (2006). Gut hormones and the regulation of energy homeostasis. *Nature, 444*(7121), 854–859.

Naseribafrouei, A., Hestad, K., Avershina, E., Sekelja, M., Linløkken, A., Wilson, R., & Rudi, K. (2014). Correlation between the human fecal microbiota and depression. *Neurogastroenterology and Motility*, *26*(8), 1155–1162. https://doi.org/10.1111/NMO.12378

Obradovic, D., Andjelic, T., Ninkovic, M., Dejanovic, B., & Kotur-Stevuljevic, J. (2021). Superoxide dismutase (SOD), advanced oxidation protein products (AOPP), and disease-modifying treatment are related to better relapse recovery after corticosteroid treatment in multiple sclerosis. *Neurological Sciences*, *42*(8), 3241–3247. https://doi.org/10.1007/S10072-020-04928-Y

Perdigón, G., Vintiñi, E., Alvarez, S., Medina, M., & Medici, M. (1999). Study of the possible mechanisms involved in the mucosal immune system activation by lactic acid bacteria. *Journal of Dairy Science*, *82*(6), 1108–1114. https://doi.org/10.3168/jds.S0022-0302(99)75333-6

Porter, D. L., Levine, B. L., Kalos, M., Bagg, A., & June, C. H. (2011). Chimeric antigen receptor-modified T cells in chronic lymphoid leukemia. *The New England Journal of Medicine, 365*(8), 725–733. https://doi.org/10.1056/NEJMoa1103849

Poutahidis, T., Kleinewietfeld, M., Smillie, C., Levkovich, T., Perrotta, A., Bhela, S., Varian, B. J., Ibrahim, Y. M., Lakritz, J. R., Kearney, S. M., Chatzigiagkos, A., Hafler, D. A., Alm, E. J., & Erdman, S. E. (2013). Microbial reprogramming inhibits Western diet-associated obesity. *PLoS One, 8*(7), e68596. https://doi.org/10.1371/journal.pone.0068596

Qiao, Y., Wu, M., Feng, Y., Zhou, Z., Chen, L., & Chen, F. (2018). Alterations of oral microbiota distinguish children with autism spectrum disorders from healthy controls. *Scientific Reports*, *8*(1), 1–12. https://doi.org/10.1038/s41598-018-19982-y

Ramakrishna, Balakrishnan S. MD, PhD. (2007). The normal bacterial flora of the human intestine and its regulation. *Journal of Clinical Gastroenterology*, *41*(), S2–S6. Retrieved September 6, 2024, from https://journals.lww.com/jcge/fulltext/2007/05001/The_Normal_Bacterial_Flora_of_the_Human_Intestine.2.aspx

Reddy, D. S., & Rogawski, M. A. (2002). Stress-induced deoxycorticosterone-derived neurosteroids modulate GABAA receptor function and seizure susceptibility. *Journal of Neuroscience*, *22*(9), 3795–3805. https://doi.org/10.1523/JNEUROSCI.22-09-03795.2002

Rodrigues, T. M., Castro Caldas, A., & Ferreira, J. J. (2016). Pharmacological interventions for daytime sleepiness and sleep disorders in Parkinson's disease: Systematic review and meta-analysis. *Parkinsonism & Related Disorders, 27*, 25–34. https://doi.org/10.1016/j.parkreldis.2016.03.002

Ruggieri, S., Petracca, M., De Giglio, L., De Luca, F., Giannì · Costanza, Gurreri, F., Petsas, N., Tommasin, S., Pozzilli, C., & Pantano, P. (1234). A matter of atrophy: differential impact of brain and spine damage on disability worsening in multiple sclerosis. *Journal of Neurology*, *268*(12), 4698–4706. https://doi.org/10.1007/s00415-021-10576-9

Sartor, R. B. (2006). Mechanisms of disease: Pathogenesis of Crohn's disease and ulcerative colitis. *Nature Clinical Practice: Gastroenterology & Hepatology*, *3*(7), 390–407. https://doi.org/10.1038/NCPGASTHEP0528

Schreibelt, G., van Horssen, J., van Rossum, S., Dijkstra, C. D., Drukarch, B., & de Vries, H. E. (2007). Therapeutic potential and biological role of endogenous antioxidant enzymes in multiple sclerosis pathology. *Brain Research Reviews*, *56*(2), 322–330. https://doi.org/10.1016/J.BRAINRESREV.2007.07.005

Singh, T., & Goel, R. K. (2016). Evidence in support of using a neurochemistry approach to identify therapy for both epilepsy and associated depression. *Epilepsy and Behavior*, *61*, 248–257. https://doi.org/10.1016/J.YEBEH.2016.05.005/ASSET/171261A6-6110-4EF2-897C-5C896432570C/MAIN.ASSETS/GR5.SML

Skonieczna-Żydecka, K., Marlicz, W., Misera, A., Koulaouzidis, A., & Łoniewski, I. (2018). Microbiome—The missing link in the Gut-Brain axis: Focus on its role in gastrointestinal and mental health. *Journal of Clinical Medicine, 7*(12), 521. https://doi.org/10.3390/jcm7120521

Strandwitz, P. (2018). Neurotransmitter modulation by the gut microbiota. *Brain Research*, *1693*(Pt B), 128. https://doi.org/10.1016/J.BRAINRES.2018.03.015

Šťinová, V., Smetanová, L., Kholová, D., Květina, J., Svoboda, Z., Zídek, Z., & Tlaskalová-Hogenová, H. (2013). The effect of probiotic Escherichia coli strain Nissle 1917 lipopolysaccharide on the 5-aminosalicylic acid transepithelial transport across Caco-2 cell monolayers. *General Physiology and Biophysics*, *32*(3), 371–380. https://doi.org/10.4149/gpb_2013035

Varesi, A., Pierella, E., Romeo, M., Piccini, G. B., Alfano, C., Bjørklund, G., Oppong, A., Ricevuti, G., Esposito, C., Chirumbolo, S., & Pascale, A. (2022). The potential role of gut microbiota in Alzheimer's disease: From diagnosis to treatment. *Nutrients, 14*(3), 668. https://doi.org/10.3390/nu14030668

Vidal-Jordana, A., & Montalban, X. (2017). Multiple sclerosis: Epidemiologic, clinical, and therapeutic aspects. *Neuroimaging Clinics of North America*, *27*(2), 195–204. https://doi.org/10.1016/J.NIC.2016.12.001

Vinderola, G., Matar, C., & Perdigon, G. (2005). Role of intestinal epithelial cells in immune effects mediated by gram-positive probiotic bacteria: Involvement of toll-like receptors. *Clinical and Diagnostic Laboratory Immunology*, *12*(9), 1075–1084. https://doi.org/10.1128/CDLI.12.9.1075-1084.2005

Walton, C., King, R., Rechtman, L., Kaye, W., Leray, E., Marrie, R. A., Robertson, N., La Rocca, N., Uitdehaag, B., van der Mei, I., Wallin, M., Helme, A., Angood Napier, C., Rijke, N., & Baneke, P. (2020). Rising prevalence of multiple sclerosis worldwide: Insights from the Atlas of MS, third edition. *Multiple Sclerosis Journal*, *26*(14), 1816–1821. https://doi.org/10.1177/1352458520970841

Xie, Y., Meijer, A. H., & Schaaf, M. J. M. (2021). Modeling inflammation in zebrafish for the development of anti-inflammatory drugs. *Frontiers in Cell and Developmental Biology, 8*. https://doi.org/10.3389/fcell.2020.620984

Zhang, L., Li, S., Tai, Z., Yu, C., & Xu, Z. (2022). Gut microbes regulate innate immunity and epilepsy. *Frontiers in Neuroscience*, *16*. https://doi.org/10.3389/FNINS.2022.870197

Zyrek, A. A., Cichon, C., Helms, S., Enders, C., Sonnenborn, U., & Schmidt, M. A. (2007). Molecular mechanisms underlying the probiotic effects of Escherichia coli Nissle 1917 involve ZO-2 and PKCzeta redistribution resulting in tight junction and epithelial barrier repair. *Cellular Microbiology*, *9*(3), 804–816. https://doi.org/10.1111/J.1462-5822.2006.00836.X

13 Role of Probiotics in Anxiety and Depression

Mirunalini Gobinath[1], *L Priyanka Dwarampudi*[1], *Shanmugam R*[2], *Parikshit Roychowdhury*[3], *and Subhajit Dutta*[2]

[1] Department of Pharmacognosy, JSS College of Pharmacy, JSS Academy of Higher Education & Research, Ooty, Tamil Nadu, India
[2] TIFAC CORE HD, JSS College of Pharmacy, JSS Academy of Higher Education & Research, Ooty, Tamil Nadu, India
[3] Department of Pharmaceutics, JSS College of Pharmacy, JSS Academy of Higher Education & Research, Ooty, Tamil Nadu, India

13.1 INTRODUCTION

Anxiety represents a devastating neuropsychiatric condition that is characterized by persistent worrying, fearfulness, irritability, panic, or avoidance, which can shift over time. This develops from childhood or early adulthood due to genetic, epigenetic relations, or environmental triggers (Colica et al., 2017). Anxiety disorders have the most impact on the daily social life of the affected individual. Sometimes, anxiety is often comorbid with other psychological conditions such as depression, post-traumatic stress disorders, and other mental disorders. Thus, psychiatric illness has more socioeconomic burden due to complete deviation from normal quality of life (Penninx et al., 2021).

Depression is one of the leading causes of suicide, which is often characterized by persistent sadness with decreased interest or pleasure in daily activities. The individual might lose hope in all things and have a more negative impact on their thoughts and altered emotional status. Subsequently, in both disorders, the prevalence is greater among women than in men. The pathogenesis of the disease is unknown, and it is believed to be due to changes in genetic, environmental, biological, or psychological factors (Marwaha et al., 2023; Verduijn et al., 2015).

Neurotransmitters play a vital role in initiating the pathophysiology of neuropsychiatric illness. These are molecules that help in signaling and the communication of information between each neuron of the brain. Thus, if there's any fault in the transmission of neurotransmitters in a particular part of the brain, it leads to psychiatric disorders, including anxiety, depression, and bipolar disorders. Dysregulation of serotonin, norepinephrine, GABA, and dopamine are the three important neurotransmitters that interplay in the onset of anxiety and depression (Vostanis et al., 2023). These are involved in various cognitive functions such as mood, sleep, appetite, cognition, and attention. All the existing chronic illnesses, including cancer, asthma, lung disease, cardiovascular disorders, obesity, and metabolic syndrome, have the risk of getting anxiety and depression. Current medication therapy like anti-anxiety and antidepressants has drastic side effects, such as sexual dysfunction, dyskinesia, constipation, and vomiting, which might lead to poor social outcomes (Oswald et al., 2019).

Probiotics are derived from living microorganisms that provide several health benefits when consumed in adequate quantity. In ancient civilization, there was a practice of routine consumption of fermented milk or milk products, which is assumed to have positive health results (Hou et al., 2022). Certain groups of bacteria are generally found in the gut microbiome, such as Actinobacteria, Bacteriodetes, Firmicutes, Fusobacteria, Proteobacteria, and Verrucomicrobia. Among these, Bacteroidetes and Firmicutes are predominant in nature. Fungal species also called gut mycobiota are *Candida*, *Cladosporium*, *Malassezia*, and *Saccharomyces*. This composition also depends on the quantity of fibers, proteins, and fats consumed in the diet. These healthy bacteria are also called commensal microbiota. They have vital functions such as nutrient processing, improving brain's anatomy and function, and regulating immune system and host physiology (Valdes et al., 2018).

Usually, after birth, the human gut microbiome develops and assists in the progression of the immune system. The gut microbiota helps in the production and absorption of vital nutrients, including vitamins, minerals, and amino acids, from our diet. Although scientific knowledge on probiotics continues to evolve, dysbiosis of the gut microbiota has been interlinked with various neuropsychiatric disorders (Rogers et al., 2016).

Recent insights in the research of probiotics have garnered interest because gut microbes tend to produce short-chain fatty acids from a source of complex dietary

DOI: 10.1201/9781003536802-13

carbohydrates, which is a process that interplays an essential role in the integration of the gastrointestinal and central nervous systems. This raises a novel treatment strategy called psychobiotics, in which probiotics are used for psychological disorders that have beneficial effects on mental health (Berding & Staudacher, 2022). The mechanism underlying psychobiotics is that they are directly or indirectly linked to neurotransmitters and regulate the production of GABA, serotonin, norepinephrine, etc.

The gut–brain axis allows two directional interactions between autonomic and enteric nervous system, neuroendocrine system, metabolic processes, and immune system pathways to the central nervous system. In physiological conditions, the secondary chemicals like enzymes, proteins, cytokines, or hormones produced by the gut reaches the brain through the bloodstream and passes the blood–brain barrier (Foster & Neufeld, 2013). Additionally, in the same manner, the short-chain fatty acids that are produced by gut microbiota reach the brain and hence they regulate brain-related activities and provide beneficial effects. Thus, the gut influences the CNS through metabolites that are derived from gut microbes, and products, neuroactive compounds, and gut hormones, which pass through the immune system, circulatory system, enteric nervous system, or through the vagus nerve, reach the brain (Liu et al., 2022). For instance, in the case of irritable bowel syndrome, individuals are often affected by mood disorders. This states that the imbalance in the contribution of this gut brain axis in the gut and the nervous system has proved that there are higher chances of comorbidities associated with gastrointestinal and psychiatric disorders (Staudacher et al., 2021).

This chapter deals with comprehensive knowledge regarding how psychobiotics influence mood-related disorders such as anxiety and depression, and their interconnection with gut–brain microbiome axis.

13.2 GUT MICROBIOTA AND THEIR SIGNIFICANT MECHANISMS

The gut microbiome has an intricate influence on several physiological and biological processes, including metabolism, inflammation, immunology, and neurology. Consequently, the gut microbiota has a beneficial interaction with the immune system in modulating the nervous system. The gut microbiota regulates both innate and adaptive types of immunity, where the predominant function involves synthesis of metabolites. The common pathway that involves gut microbiota is through the gut–brain axis. Thus, the correlation between the neuroendocrine system, enteric nervous system, central nervous system, immune system, and gut microbiota are discussed in detail along with their significant mechanisms (Ross, 2023). The important mechanisms that are included in gut–brain axis are given in Figure 13.1.

13.2.1 Mechanism for Neural Pathway

The gastrointestinal system innervates around 300–700 million neurons, in which the enteric nervous system has a strong connection with the central nervous system through the vagus nerve, pelvic nerve, and the sympathetic nervous system. Recent observations in gut microbiota research have stated that psychological conditions like depression, by stimulating the vagus nerve, which is one of the most important cranial nerves, could diminish the consequences associated with depression (Ferrari et al., 2024). In some previous studies, they have encouraged the use of vagus nerve stimulators for the management of treatment-resistant depression. The responses were recorded where it was positive and have decreased the incidence of remission and suicide rates of the disease. Despite being a surgical intervention, it has positive outcomes (Ahmed et al., 2024). Moreover, gut microbiota has proven their mechanism by stimulating vagus nerve and signal transmission from the enteric to central nervous system. This can reduce the surgical intervention of vagus nerve stimulation by incorporating oral intake of psychobiotics. Certain research has proved that the vagus nerve has a crucial role in the interaction between gut bacteria and the brain. They alter the behavioral function of the brain, including anorexia, emotion, hyperalgesia, and lethargy (Mörkl et al., 2023).

In some research, especially on prenatal neurodevelopment, it states that the gut microbiota has a crucial role in regulating brain development and function. They produce various neurotransmitters such as acetylcholine, serotonin, and GABA, which are neuroactive molecules in the brain that involve balancing the communication between the brain and the gut. These neurotransmitters are required to maintain several metabolic functions like mood, sleep, behavior, etc. The gut–brain axis has connection to emotional and cognitive centers of the brain (Vaher et al., 2022). Tryptophan, which is the precursor to serotonin, modulates brain microglial function and the transcriptional factors of astrocytes. The serotonergic system is considered as a major component in regulating mood disorders. They play an essential role in secreting serotonin where dysfunctional serotonergic system might lead to disorders like anxiety and depression. Gut microbiota can also induce the synthesis and production of metabolites and neuroactive molecules, which can alter the functions of the nervous system in various physiological and pathological conditions (Morais et al., 2021). Microbial-derived molecules that alter the host behavior and neuronal function are given in Table 13.1.

13.2.2 Mechanisms of Immune System

Gut microbiota has a significant influence on neurophysiology via direct or indirect pathways. They modulate development of the peripheral immune system and the immune cells of CNS which can be either innate or adaptive type

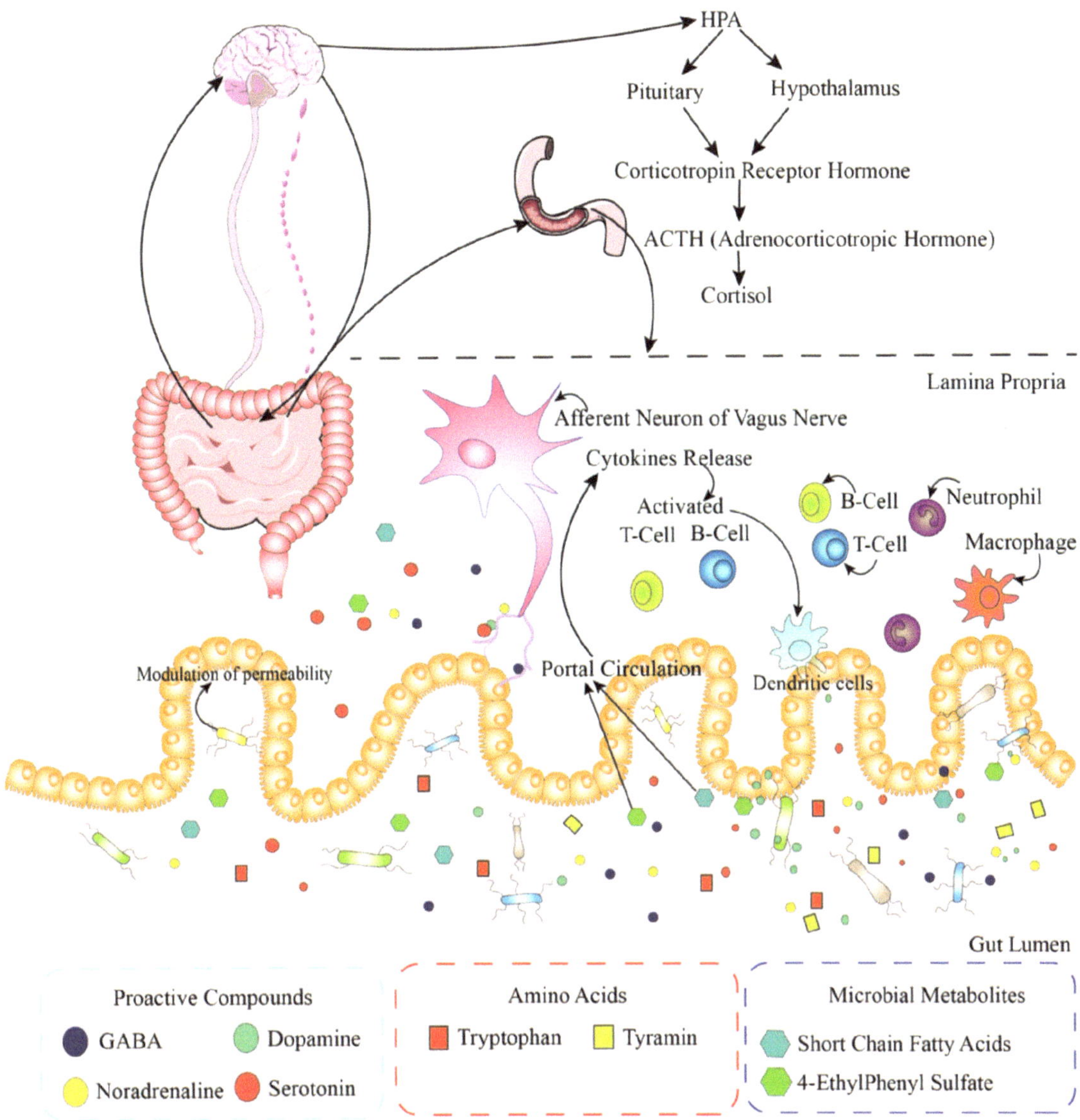

FIGURE 13.1 Important mechanisms between the gut microbiota and the gut–brain axis.

TABLE 13.1
Functions of Gut Microbial Strains toward Neuronal Pathways

S. No.	Microbial Strain	Neuroactive Molecules	Functions
1	*Bifidobacterium longum* NCC3001	Increases brain-derived neurotrophic factors (BDNF) in the vagus nerve	Decreases anxiety behaviors and improves communication between brain and gut
2	*Lactobacillus rhamnosus*	Increased GABA in the vagus nerve and mesolimbic regions of the brain	Diminishes stress response by HPA-axis dysregulation, decreases anxiety and depressive behaviors
3	*Lactobacillus reuteri*	Increases oxytocin in vagus nerve	Helps in regulating neuronal plasticity, increases social behavioral activities and neurons that secrete oxytocin
4	Bacteria that produce short-chain fatty acids	Increased production of short-chain fatty acids (SCFA)	Decreasing response to stress, anxiety, and depressive symptoms. Involves in neuroinflammation process by microglial activation
5	*Bacteriodes fragilis*	Reduces 4-ethylphenylsulfate (4-EPS)	Causes anxiety-like behaviors in mice model

and leads to inflammation that has an unknown etiology till now. Certain inflammatory mediators such as *p*-selectin, which cause monocyte adhesion in cerebral endothelial cells and TNF-α, mainly the TNFR1, have correlated with gut microbiota (Aburto & Cryan, 2024). In germ-free animal models, the gut microorganisms have altered the healthy development of microglia when compared to standard untreated mice. It also seems that they interact with cytokines in the systemic immune system via circulation. Any dysbiosis might increase the risk of peripheral inflammation which is associated with various neuropsychiatric conditions like anxiety, autism spectrum disorder, and depression. In some preclinical reports, the gut microbiome has influenced the permeability of BBB in germ-free mice models. Subsequently, changes in BBB function have a link to the pathogenesis of various neuroinflammatory diseases (Zou et al., 2021). Many cross-sectional studies have reported that the composition of healthy microbes in the gut is different in individuals with neuropsychiatric conditions when compared with healthy individuals (M. Gao et al., 2023). Thus, there are many factors that influence the composition of gut bacteria. These are mentioned in Table 13.2.

13.2.3 Neuroendocrine Mechanism

The gut microbiota has a high influence on neuroendocrine pathways by modulating the hypothalamic–pituitary–adrenal (HPA) axis . Chronic stressors have the tendency to trigger the HPA axis activation and produce the release of glucocorticoids. On the other hand, this can also be induced by gut microbiome due to increased intestinal permeability of the gastrointestinal barrier (Sudo, 2014). Recent studies have shown that peptidoglycans and lipopolysaccharides, which are abundant in most of bacteria, have the potential to stimulate brain activity by activating pattern recognition receptors (PRRs) and toll-like receptors (TLR4), activating innate immune systems by penetrating the intestinal epithelial barrier to modulate HPA axis. These mechanisms, which are also known as stress-induced immune response, occur only when there's an external stimulus such as stress, high-fat diet, and associated psychological symptoms (Makris et al., 2021). Altered microbiota in the gut can lead to increased nucleotide-binding oligomerization domain-like receptors (NOD) and NLRP3 inflammasomes, thereby stimulating innate immunity and production of cytokine. This can disrupt the BBB and intestinal permeability and induce activation of microglia in the CNS. Some instances of beneficial metabolites such as SCFAs can modulate brain function through this pathway. Dysbiosis of these metabolites can lead to clinical depression and anxiety (Clarke, 2020).

13.3 PROBIOTICS AND NEUROTRANSMITTER REGULATION

Probiotics are derived from living microorganisms that have beneficial effects in neurotransmitter regulation. It is demonstrated that gut microorganisms can induce the concentration of neurotransmitters by modulating the synthesis and production. Consequently, altered gut bacteria can substantially increase the risk of anxiety and depression due to the dysregulation of neurotransmitters such as serotonin, norepinephrine, dopamine, GABA, and glutamate (Chen et al., 2021).

TABLE 13.2
Factors that Influence the Composition of Gut Microorganisms

S. No.	Factors	Highlighted Reasons
1	Dietary factors	Changes in intake of dietary fibers and plant-based foods can alter the composition of microbes in the gut. Examples are high intake of sugary beverages and high-fat content diet
2	Genetics	Genetic predisposition of certain individuals has effect on their gut microbial colonization and reduces the composition
3	Environmental factors	In prenatal and postnatal health, studies have reported that if the individual had normal vaginal delivery, they acquire the same bacterial composition from the mother, whereas in case of caesarian section births, there is evidence in altered composition. Occupational health is also correlated with altered gut microbes
4	Stress	Stress has a significant role in dysbiosis of microbiota in the gut due to altered HPA axis and cortisol production, which influences changes in the composition of microbes. In case of depression, decreased serotonin induces hyperactivation of HPA axis due to excess vasopressin and corticotrophin hormones
5	Medications	Use of certain medications like antibiotics, antidepressants, antipsychotics, and other drugs that have activity to suppress or stimulate neurotransmitters have exacerbated the dysbiosis of gut microbiota
6	Inflammation and infections	Infections in the gut and other autoimmune diseases like IBD can lead to dysbiosis of gut microbiota. Patients with chronic inflammation are also prone to risk factors in the gut microbial composition
7	Gut–brain axis communication	Dysfunctional communication between brain and gastrointestinal system can have an impact on the composition of gut microbiota, mainly the SCFAs, which are also modulated by various factors such as immune system and neuroendocrine pathways
8	Circadian rhythm	Unlike humans having biological clock, gut bacteria also possess biological clock that may disrupt the circadian rhythm. Due to its strong relationship with activities including sleep, digestion, and alertness, any changes in these could lead to risk of psychiatric or metabolic diseases

Almost 90% of serotonin (5-hydroxytryptamine) are present in the gastrointestinal system and 10% in the CNS. These are produced by enterochromaffin cells in the gut. Studies have shown that gut microorganisms regulate these cells and maintain the production of serotonin (Oroojzadeh et al., 2022). This can be done by various pathways—by modifying direct serotonin synthesis and production, altering the specialized serotonin transporter (SERT), communication between the immune system and vagus nerve, or via secondary metabolites of gut bacteria such as SCFAs. Dysbiosis can increase bacterial strains, including *Enterococcus* and *Streptococcus*, thus increasing 5-HT production, which affects mood and anxiety (Homan et al., 2015).

GABA (γ-amino butyric acid) is synthesized by glutamate pathway with the help of glutamic acid decarboxylase enzyme (GAD). The gut microbiota has a crucial role in the production of GABA due to which the bacteria contain genes that encode GAD. The reason why gut bacteria produce GABA is to withstand survival in extreme environments in the intestine such as changes in pH of the gut. Certain strains of gut microorganisms that synthesize GABA are *Lactobacillus*, *Bifidobacterium*, and *Bacteroides*. Thus, dysbiosis in these strains have a substantial impact on altered mechanisms of the brain, which can lead to neuropsychiatric conditions like depression and anxiety (Braga et al., 2024).

Catecholamines which are also involved in the pathogenesis of stress-induced psychiatric illness and immune response could be improved by probiotic interventions. Generally, in psychological conditions, especially clinical depression, monoamine deficiency theory has an essential role, which includes the depletion of monoamines such as serotonin, dopamine, and norepinephrine(Cui et al., 2024). These neurotransmitters are involved in behavioral processes like feel-good, attention, pleasure, happiness, motivation, and other cognitive functions. In a crossover design study, they have estimated that both the depletion of serotonergic and the catecholaminergic neuroactive molecules might be the result of depression, mainly the major depressive disorder subtype (Homan et al., 2015).

13.4 PROBIOTICS AND IMMUNE MODULATION AND STRESS RESPONSE

As we already have discussed that the gut microbiota has a substantial influence on immune system and related stress response pathways, administration of probiotics or psychobiotics can have a positive impact in reducing the stress-induced inflammatory or immune responses. Several studies have stated that depression might be induced by hyperactivation of HPA axis and inflammatory responses which are triggered by gut microbiota (Varghese & Brown, 2001). In many preclinical research, they have estimated that the disrupted gut microbiota can lead to altered behaviors, social interactions, and cognitive health in different animal models (Buffington et al., 2016).

When the composition of gut flora is abruptly disturbed, they send signals via the enteric nervous system to the CNS to secrete neuroactive metabolites. This in turn would also increase the fabrication of metabolites, which are 6 carbon atom chain fatty acids like SCFAs, acetate, propionate, and butyrate. These factors modulate T cells and induce immune responses which also dysregulate neurotrophins in hippocampus of the brain (Misra & Mohanty, 2019). These complex mechanisms could initiate chronic stress and stimulate the hyperactivation of HPA axis. This axis has both positive and negative feedback loops in the homeostasis of stress hormones, including cortisol and other glucocorticoids. Finally, the dysfunctional HPA axis pathway can lead to increased cortisol production and increased volume of adrenal cortex, thus precipitating depression and anxiety (Figure 13.2) (Brushett et al., 2023).

13.5 PSYCHOBIOTICS FOR ANXIETY AND DEPRESSION

Psychobiotics is the trending terminology for probiotics which are used in treating various psychiatric illnesses (Figure 13.3). They contain healthy microbes that modulate the gut–brain axis by numerous mechanisms. When consumed, they increase the number of commensal bacteria in the gut, thus promoting their beneficial activity. Psychobiotics exert their activity by neuronal channels that regulate the immune system and the enteric nervous system (Patra, 2016).

Though there is no clear evidence on specific molecular targets of psychobiotics, recent studies have estimated that microbiota-targeted interventions change the internal and external environment of gut microbiota, bringing it to a balanced state, which then stimulates the gut–brain microbial axis to further induce CNS functions and neurotransmitters activity, thus alleviating anxiety and depression (Chao et al., 2020).

Despite being a non-interventional treatment, daily diet routine has tremendous effects over the gut floral profiles. Prebiotics are indigestible fibers that are obtained by nutritious diet, which includes vegetables, fruits, green-leafy vegetables, grains, pulses, and milk. They have high concentrations of oligosaccharides which are fructose and galactose in nature. These provide nutrient media for the gut microorganisms, helping in fulfilling the nutritional values of bacteria (Taylor & Holscher, 2020). For instance, low-fiber diet might reduce the production of butyrate by SCFAs and decrease *Faecalibacterium prausnitzii*, which assists in supporting the immune system and digestive health. In contrast, high-fat diet, especially non-vegetarian diet, which has very less fiber content, might diminish the beneficial bacteria and increase microorganisms that tolerate bile. This is because bile involves in lipid metabolism (Bear et al., 2020). Other approaches like synbiotics, which are the

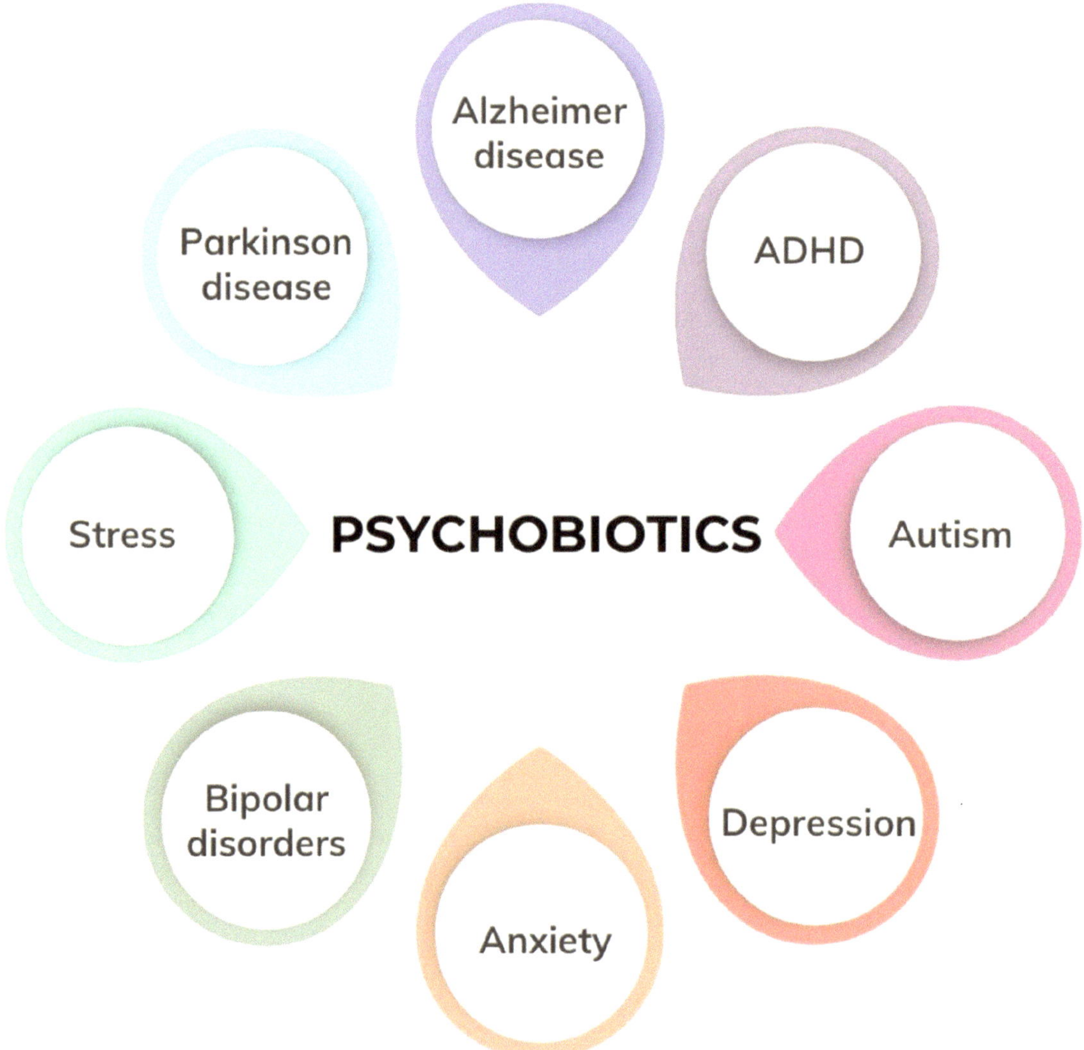

FIGURE 13.2 Significant uses of psychobiotics in various psychiatric diseases.

combination of prebiotics and probiotics, in which prebiotics ameliorates the viability of probiotics. In addition, postbiotics are the secondary metabolites of fermentation of gut bacteria. In pharmaceutical and nutraceutical industry, the utilization of postbiotics has improved due to their longer shelf-life period and safety profiles (Cosme et al., 2022). This is because postbiotics are just the metabolites of live microbes, whereas probiotics contain live microorganisms that might cause harmful effects in immunocompromised individuals, they also have less shelf life due to the presence of viable microbes. Some important examples of postbiotics include short-chain fatty acids, which secrete acetates, propionates, and butyrate. Microbial strains that possess psychobiotic characteristics are *Bifidobacterium*, *Lactobacillus*, and *Lactococcus* (Jarosz et al., 2024). Figure 13.3 shows how psychobiotics modulate several pathways.

13.6 RESULTS AND EVIDENCE FROM PRECLINICAL AND CLINICAL TRIALS AND META-ANALYSIS STUDIES

13.6.1 Efficacy of Probiotics on Anxiety/Depression: A Preclinical Study Analysis

Evaluation of probiotics on anxiety and depression is established by various studies on different animal models; the studies feature organisms such as *Bifidobacterium infantis, Lactobacillus plantarum, Lactobacillus rhamnosus, Lactobacillus paracasei,* and several others (Divyashri & Prapulla, 2022). These are shown to help in stress response by multifaceted pathways such as enhancing serotonin, GABA, and a few essential amino acids such as tryptophan that acts as a precursor for melatonin and serotonin, in turn enhancing general well-being and mental health. In a stress-inducing rat model where the rats are separated

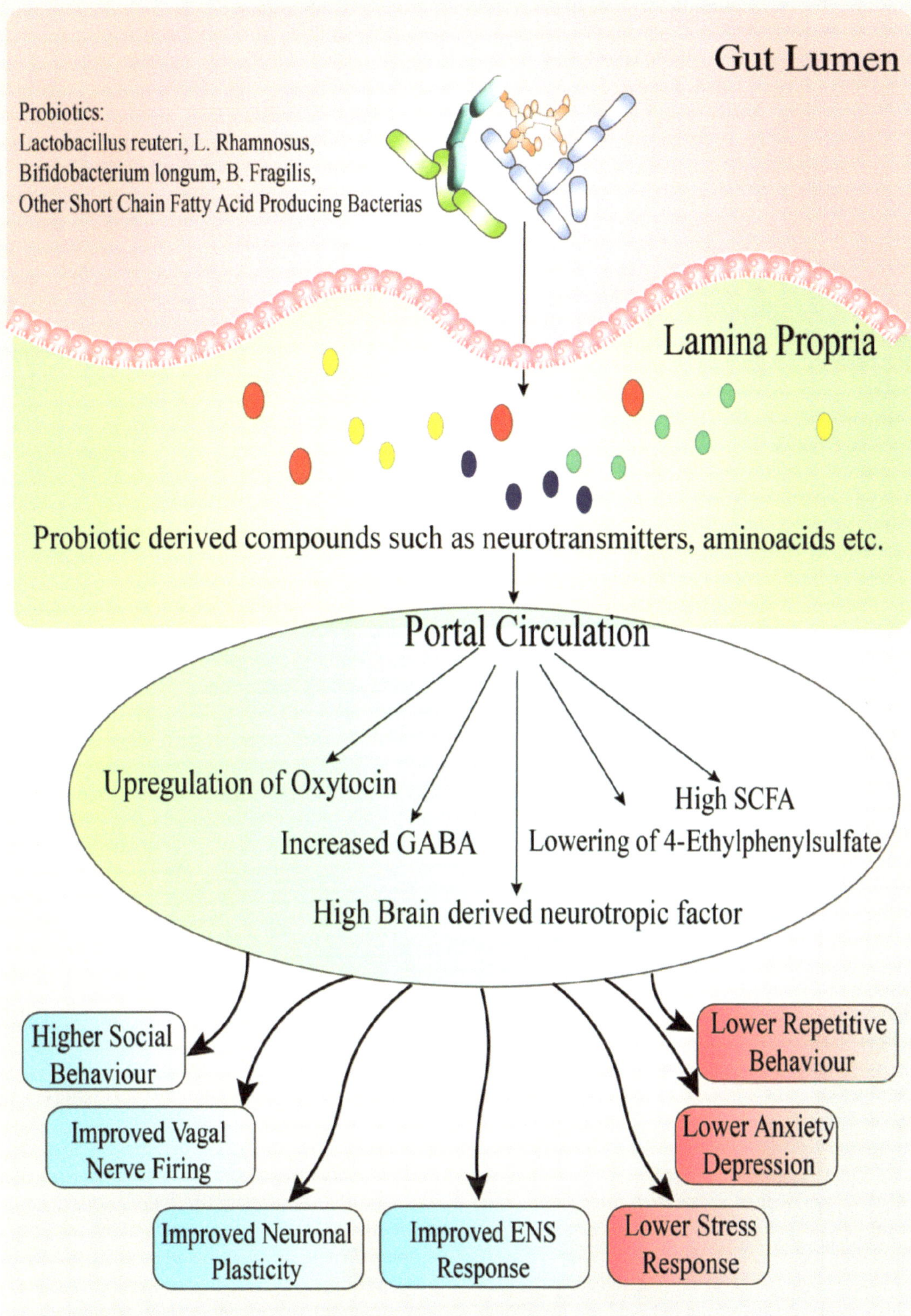

FIGURE 13.3 Mechanisms of psychobiotics in various pathways in psychiatric conditions.

from their mother, *Bifidobacterium infantis* 35624 showed a significant positive intervention. In this early life anxiety and stress that is comparable to citalopram, dosing of 1×10^{10} CFU live bacteria suspended in per milliliter of drinking water showed reduction in IL-6 (previously associated with stress and anxiety) and enhanced the presence of 5-hydroxyindoleactic acid, which is an important precursor for serotonin biosynthesis; the former was seen to be elevated during the maternal separation period (Desbonnet et al., 2010). When the dose of the bacteria is diluted to the same amount of CFUs per 100 ml, it significantly elevates the concentration of tryptophan and reduces the serum levels of tumor necrosis factor-α, interferon-γ, and IL-6, which are major pre-inflammatory cytokines that are also found to be associated with depressive states (Desbonnet et al., 2008). The GABA-A and GABA-B receptor modifications take place and alleviate symptoms of anxious and depressive behaviors with a four-week treatment of *Lactobacillus rhamnosus* in a dose of 1×10^9 CFU in mice. The same regimen also improves the presence of glutamate and *N*-acetyl aspartate (Janik et al., 2016). Both heat-killed and live *L. paracasei* PS23 can improve anxiety and depression states by elevating the levels of dopamine and serotonin in the CNS, especially in the hippocampus; this was evaluated by several studies via the corticosterone-induced anxiety depression rat models, the results were equivalent to that of fluoxetine (Huang et al., 2018). The same strain can also alleviate and leave a promising impact in the anxiety–depression symptoms that are associated with progressive age. Post-administration, the serum levels of brain-derived neurotrophic factor (BDNF), dopamine, and serotonin were seen to be higher, which leads to the delaying of age-related cognitive decline, thus reducing anxiety and depressive situations; the same organism in both live and heat-killed form is also proven effective in the maternal separation early age separation anxiety model (Liao et al., 2019). Another strain of *Lactobacillus*, *Lactobacillus helveticus NS8,* can reduce anxiety, stress, and the inclination of cognitive impairment with progressive age; it is also reported to have reduced inflammation by controlling ILs, TNF, and IFN, apart from controlling the serum levels of corticosterone, adrenocorticotropic hormone, or ACTH. The strain is also reported to have positive impact on the declining concentration of serotonin and norepinephrine in the hippocampal region. A 41-day study on ampicillin-induced anxiety model on rats revealed that the strain *L. fermentum NS9* can alleviate the symptoms and can control the elevated corticosteroid levels post-ampicillin administration. Corticosterone when administered at 40 mg/kg/day can severely deplete the serotonin and dopamine levels, *L. paracasei* PS23 reversed the effect when given in an oral gavage to the treated animals; notably, the anxiolytic effect by restoration of serotonin and dopamine in CNS was higher with the administration of heat-killed bacteria rather than the live ones (Wei et al., 2019). SAMP8 or senescence-accelerated mice show rapid aging after they pass the age of 120 days; these mice when treated with the former organism live, and at a dose of 5×10^9 CFU by oral gavage for 12 weeks showed slower rate of aging and age-related cognitive decline, as well as anxiety and depressive symptoms. Biochemical analysis revealed the improvement of BDNF, serotonin, as well as dopamine and maintenance of inflammatory markers (Huang et al., 2018). *Lactiplantibacillus plantarum* D-9 can yield high amounts of GABA, and when administered to chronic unpredicted mild stress (CUMS) mice , it improved the anxiety symptoms and alleviated depressive behavior. In addition to these, the administration provided modulation and healthy improved colonization of *Ligilactobacillus murinus and Lactobacillus jhonsonii,* with the synergistic effect of lowering the colonies of the *Helicobacter bilis* and *Kineothrix alysoides* in the gut of CUMS mice. These effects in addition to an effective and modulated tryptophan metabolism helped in its antidepressant and anxiolytic action (Jia et al., 2024). *Lactiplantibacillus plantarum* JYLP-326 was evaluated using a C57BL/6 mouse, in association with CUMS, spontaneous behavior that is generally associated with CUMS anxiety is reduced to a significant level, marked improvement in the glial fibrillary acidic protein (GFAP), and Iba1 is also seen, which is directly linked to the neuroprotection activity of the strain (Barrio et al., 2022). The administration increased the metabolism of tryptophan by upregulation of tryptophan hydrolase enzymes, therefore modulating the biosynthesis of serotonin; a significant enhancement was observed in the regulation of serotonin receptor 1 A or 5-HT1AR. In addition, this particular strain showed a reduction in the pro-inflammatory markers, and also improved the gut barrier to provide an all-round depression and anxiety reduction (Zhu et al., 2024). C57BL/6J mice showed improved levels of BDNF, short-chain fatty acids, and serotonin following a five-week oral treatment with 10^9 CFU/ml of *Bifidobacterium breve* CCFM1025. Due to the chronic stress associated with the model hypothalamus, pituitary axis was enhanced, which is observed to be controlled and modulated post treatment; the strain also reduced the inflammatory mediators in the due course of treatment (Tian et al., 2020). A preclinical study containing Sprague Dawley rats and with the intervention of *Faecalibacterium prausnitzii* ATCC 27766 yielded high short-chain fatty acids and IL-10, which lowered IL-6 and corticosterone levels (Hao et al., 2019). Two *Lactococcus* strains—WHH2080 and WHH2078—were evaluated for age-related anxiety and depression in elderly mice. Following a 12-week study on a year-old mice, reversal of low serotonin and 5-HTP, excess of 3-hydroxykynurenine and indolelactic acid in addition to restoration of the gut microbial colonies with Firmicutes, Deferribacterota was observed (K. Gao et al., 2024). A multi-bacterial product, OttaBac containing eight strains of probiotics (lyophilized strains of *Bifidobacterium animalis* BL03, *Lactobacillus acidophilus* BA05, *Lacticaseibacillus paracasei* BP07, *Lactobacillus plantarum, L. helveticus BD08, B. animalis* BI04, *B. breve* BB02, and *Streptococcus thermophilus*) showed a total reversal of postnatal maternal separation anxiety in male mice. Normalcy in behavior was observed along with a pro-neurogenic effect (De Santa et al., 2024). A few notable preclinical experiments are listed out in Table 13.3.

TABLE 13.3
Factors that Influence the Composition of Gut Microorganisms

S. No	Strain of Probiotic	Study Model	Study Duration and Dose	Observations	Reference
1	*Lactiplantibacillus plantarum* PS128	C5BL/6J mice, separation from mother at early age stress	28 days and 10^9 CFU/day	Reduction in depressive and anxious behavior, lowering in anxiety-related hyperactivity and corticosterone levels	Dhyani et al., 2024
2	*L. helveticus NS8*	Sprague Dawley rats subjected to chronic restraint, inducing depression and stress	3 weeks and 10^9 CFU in water	Maintained and enhanced concentrations of serotonin and nor epinephrine in CNS and particularly in hippocampal region along with BDNF levels	Liang et al., 2015
3	*Lacticaseibacillus rhamnosus* NCC4007, in combination with *B. longum* NCC3001	BALB/c male mice infected with whipworm or *Trichuris muris*, causing colonic infection associated with anxiety-like symptoms and BDNF depletion	10 days	Enhanced BDNF mRNA were observed, and the serum levels were improved with normal behavioral traits on treatment with NCC3001, whereas the treatment groups involving NCC4007 did not show any significant improvement of anxiety-like condition or BDNF parameters	Bercik et al., 2010
4	*Bifidobacterium pseudocatenulatum* CECT 7765	Male adult C57BL mice with high fat diet induced neurobehavioral changes	13 weeks with 10^9 CFU	Enhanced weight loss, balanced metabolism, and reduced anxious and compulsive behavioral traits, restoration of the CNS 5HT and dopamine levels were observed. Levels of serum adrenaline was reduced	Agusti et al., 2018
5	*B. longum 1714, and B. breve 1205*	Male adult BALB/c mice with high fat diet induced neurobehavioral changes	2.5 months and 10^9 CFU/ml	1205 triggered weight loss and modulated rate of metabolism, whereas both the strains showed reduction of stress and inhibition of compulsive behavior; however no significant changes on the onset and reversal of depression was observed	Savignac et al., 2014

13.6.2 Effectiveness of Probiotics against Anxiety and Depression in Human Studies

The majorly studied strains as anxiolytics and antidepressants on clinical settings are *Lactobacillus* and *Bifibacterium*, first ever available literature suggests that the success of *Lactobacillus* as an antidepressant dates back to early 1910 (Evrensel et al., 2019). Strains of *B. longum* intervention resulted in the enhancement and maintenance of GABA biosynthesis alleviating symptoms of depression and is also seen to have effects on the myenteric neurons; the strain of *L. helveticus* impacts the CNS and improves the concentration of 5HT, thus reducing the anxious and depressive behavioral traits (Sharma et al., 2021). These two strains are observed to be reducing anger, depression, obsessive and compulsive behavior, etc. when evaluated in a clinical study involving 55 moderately stressed volunteers. The study was validated using tools and techniques like perceived stress scale, Hopkins symptoms checklist-90, etc. (Heidarzadeh-Rad et al., 2020; Messaoudi et al., 2011). In 2019, Kazemi et al. performed a trial comparing *L. Helveticus* and *B. longum* with R0052 and R0175 subspecies, respectively; the trial didn't show any significant difference between the placebo and the treatment group apart from the increase in tryptophan compared to that of the kynurenine. However, when the study was redone with 78 volunteers from the previous study, it was noted that the levels of BDNF were significantly high in the treatment group, which in combination with the former outcome were thought to be the mechanism of action for the antidepressant action of the strains. The responses from the volunteers were validated in the study using Beck's Depression Inventory (BDI) aligned questionnaire (Kazemi et al., 2019). These two strains in combination, in a 9:1 ratio, were seen to improve the symptoms of major depressive disorder. Patients followed by a 28-day treatment. These results were validated using a general anxiety disorder-7-aligned questionnaire. The hypothesized mechanism could be the bacteria's influence on the tryptophan metabolism and serotonin biosynthesis (Wallace & Milev, 2021). Another study explored the combination of those two along with *L. rhamnosus* R0011, polyphenols, ginger root extract, other phytobiotics, fibers, and hydrolyzed guar gum. Followed by a 30-day intervention period, all 32 subjects reported a marked alleviation in depression (55%), fatigue (64%), anger (54%), and tension (45%), the subjects also experienced an improved state of vigor (44%) and by 25% improvement on overall quality of life, mood, and well-being as compared to that of the placebo (Talbott et al., 2019). A significant reduction in depressive behavior and general anxiety was reported from a double-blind-randomized-controlled trial involving 45 volunteers suffering from major depressive disorder. The study reported the improvement in 28 days of intervention with *B. breve*. The causative mechanism was suspected to be enhanced metabolism and production of active metabolites of tryptophan (Dao et al., 2021; Tian et al., 2020).

13.7 CONCLUSIONS AND FUTURE PERSPECTIVES IN RESEARCH

In conclusion, the role of probiotics in anxiety and depression is fascinating. In fact, these probiotics that target gut microbiota are often called psychobiotics. On the other hand, the gut–brain axis plays a significant role in the pathogenesis of mood-related psychiatric conditions, including depression and anxiety. Even though numerous factors contribute to the dysbiosis of the gastrointestinal microbes, these ill-effects can be alleviated through incorporating psychobiotic regimens to ameliorate anxiolytic and antidepressive effects. Future perspectives of administration of probiotics and psychobiotics can be enhanced by modulating the formulation with nanotechnology. These ease the delivery of psychobiotics and improve their bioavailability and other properties. Thus, the importance of psychobiotics and other psychological conditions must be well-studied to such an extent where all the molecular signaling pathways are well-defined, which helps us to develop innovative strategies and to promote the use of probiotics.

REFERENCES

Aburto, M. R., & Cryan, J. F. (2024). Gastrointestinal and brain barriers: Unlocking gates of communication across the microbiota–gut–brain axis. *Nature Reviews Gastroenterology & Hepatology*, *21*(4), 222–247. https://doi.org/10.1038/s41575-023-00890-0

Agusti, A., Moya-Pérez, A., Campillo, I., Montserrat-de la Paz, S., Cerrudo, V., Perez-Villalba, A., & Sanz, Y. (2018). Bifidobacterium pseudocatenulatum CECT 7765 Ameliorates neuroendocrine alterations associated with an exaggerated stress response and anhedonia in obese mice. *Molecular Neurobiology*, *55*(6), 5337–5352. https://doi.org/10.1007/s12035-017-0768-z

Ahmed, G. K., Ramadan, H. K.-A., Elbeh, K., & Haridy, N. A. (2024). Bridging the gap: Associations between gut microbiota and psychiatric disorders. *Middle East Current Psychiatry*, *31*(1), 2. https://doi.org/10.1186/s43045-024-00395-9

Barrio, C., Arias-Sánchez, S., & Martín-Monzón, I. (2022). The gut microbiota-brain axis, psychobiotics and its influence on brain and behaviour: A systematic review. *Psychoneuroendocrinology*, *137*, 105640. https://doi.org/10.1016/j.psyneuen.2021.105640

Bear, T. L. K., Dalziel, J. E., Coad, J., Roy, N. C., Butts, C. A., & Gopal, P. K. (2020). The role of the gut microbiota in dietary interventions for depression and anxiety. *Advances in Nutrition*, *11*(4), 890–907. https://doi.org/10.1093/advances/nmaa016

Bercik, P., Verdu, E. F., Foster, J. A., Macri, J., Potter, M., Huang, X., Malinowski, P., Jackson, W., Blennerhassett, P., Neufeld, K. A., Lu, J., Khan, W. I., Corthesy–Theulaz, I., Cherbut, C., Bergonzelli, G. E., & Collins, S. M. (2010). Chronic gastrointestinal inflammation induces anxiety-like behavior and alters central nervous system biochemistry in mice. *Gastroenterology*, *139*(6), 2102–2112.e1. https://doi.org/10.1053/j.gastro.2010.06.063

Berding, K., & Staudacher, H. M. (2022). Psychobiotics: Feeding the gut to nurture the brain. *The Lancet Gastroenterology & Hepatology*, *7*(5), 393. https://doi.org/10.1016/S2468-1253(22)00097-8

Braga, J. D., Thongngam, M., & Kumrungsee, T. (2024). Gamma-aminobutyric acid as a potential postbiotic mediator in the gut–brain axis. *Npj Science of Food*, *8*(1), 16. https://doi.org/10.1038/s41538-024-00253-2

Brushett, S., Gacesa, R., Vich Vila, A., Brandao Gois, M. F., Andreu-Sánchez, S., Swarte, J. C., Klaassen, M. A. Y., Collij, V., Sinha, T., Bolte, L. A., Wu, J., Swertz, M., de Kroon, M. L. A., Reijneveld, S. A., Wijmenga, C., Weersma, R. K., Fu, J., van Loo, H. M., Kurilshikov, A., & Zhernakova, A. (2023). Gut feelings: The relations between depression, anxiety, psychotropic drugs and the gut microbiome. *Gut Microbes*, *15*(2), 2281360. https://doi.org/10.1080/19490976.2023.2281360

Buffington, S. A., Prisco, G. V. D., Auchtung, T. A., Ajami, N. J., Petrosino, J. F., & Costa-Mattioli, M. (2016). Microbial reconstitution reverses maternal diet-induced social and synaptic deficits in offspring. *Cell*, *165*(7), 1762–1775. https://doi.org/10.1016/j.cell.2016.06.001

Chao, L., Liu, C., Sutthawongwadee, S., Li, Y., Lv, W., Chen, W., Yu, L., Zhou, J., Guo, A., Li, Z., & Guo, S. (2020). Effects of probiotics on depressive or anxiety variables in healthy participants under stress conditions or with a depressive or anxiety diagnosis: A meta-analysis of randomized controlled trials. *Frontiers in Neurology*, *11*. https://doi.org/10.3389/fneur.2020.00421

Chen, Y., Xu, J., & Chen, Y. (2021). Regulation of neurotransmitters by the gut microbiota and effects on cognition in neurological disorders. *Nutrients*, *13*(6), 2099. https://doi.org/10.3390/nu13062099

Clarke, G. (2020). The gut microbiome and depression: Finding a way through troubled waters where the river meets the sea. *Expert Review of Gastroenterology & Hepatology*, *14*(5), 301–304. https://doi.org/10.1080/17474124.2020.1754796

Colica, C., Avolio, E., Bollero, P., Costa de Miranda, R., Ferraro, S., Sinibaldi Salimei, P., De Lorenzo, A., & Di Renzo, L. (2017). Evidences of a new psychobiotic formulation on body composition and anxiety. *Mediators of Inflammation*, *2017*(1), 5650627. https://doi.org/10.1155/2017/5650627

Cosme, F., Inês, A., & Vilela, A. (2022). Consumer's acceptability and health consciousness of probiotic and prebiotic of non-dairy products. *Food Research International*, *151*, 110842. https://doi.org/10.1016/j.foodres.2021.110842

Cui, L., Li, S., Wang, S., Wu, X., Liu, Y., Yu, W., Wang, Y., Tang, Y., Xia, M., & Li, B. (2024). Major depressive disorder: Hypothesis, mechanism, prevention and treatment. *Signal Transduction and Targeted Therapy*, *9*(1), 1–32. https://doi.org/10.1038/s41392-024-01738-y

Dao, V. H., Hoang, L. B., Trinh, T. O., Tran, T. T. T., & Dao, V. L. (2021). Psychobiotics for patients with chronic gastrointestinal disorders having anxiety or depression symptoms. *Journal of Multidisciplinary Healthcare*, *14*, 1395–1402. https://doi.org/10.2147/JMDH.S312316

De Santa, F., Strimpakos, G., Marchetti, N., Gargari, G., Torcinaro, A., Arioli, S., Mora, D., Petrella, C., & Farioli-Vecchioli, S. (2024). Effect of a multi-strain probiotic mixture consumption on anxiety and depression symptoms induced in adult mice by postnatal maternal separation. *Microbiome*, *12*(1), 29. https://doi.org/10.1186/s40168-024-01752-w

Desbonnet, L., Garrett, L., Clarke, G., Bienenstock, J., & Dinan, T. G. (2008). The probiotic *Bifidobacteria infantis*: An assessment of potential antidepressant properties in the rat. *Journal of Psychiatric Research*, *43*(2), 164–174. https://doi.org/10.1016/j.jpsychires.2008.03.009

Desbonnet, L., Garrett, L., Clarke, G., Kiely, B., Cryan, J. F., & Dinan, T. G. (2010). Effects of the probiotic Bifidobacterium infantis in the maternal separation model of depression. *Neuroscience*, *170*(4), 1179–1188. https://doi.org/10.1016/j.neuroscience.2010.08.005

Dhyani, P., Goyal, C., Dhull, S. B., Chauhan, A. K., Singh Saharan, B., Harshita, Duhan, J. S., & Goksen, G. (2024). Psychobiotics for mitigation of neuro-degenerative diseases: Recent advancements. *Molecular Nutrition & Food Research*, *68*(13), 2300461. https://doi.org/10.1002/mnfr.202300461

Divyashri, G., & Prapulla, S. G. (2022). Animal models used for studying the benefits of probiotics in neurodegeneration. In P. K. Deol & S. K. Sandhu (Eds.), *Probiotic research in therapeutics: Volume 4: Probiotics in neurodegenerative disorders* (pp. 237–254). Springer Nature. https://doi.org/10.1007/978-981-16-6760-2_10

Evrensel, A., Önen Ünsalver, B., & Ceylan, M. E. (2019). Therapeutic potential of the microbiome in the treatment of neuropsychiatric disorders. *Medical Sciences*, *7*(2), Article 2. https://doi.org/10.3390/medsci7020021

Ferrari, S., Mulè, S., Parini, F., Galla, R., Ruga, S., Rosso, G., Brovero, A., Molinari, C., & Uberti, F. (2024). The influence of the gut-brain axis on anxiety and depression: A review of the literature on the use of probiotics. *Journal of Traditional and Complementary Medicine*, *14*(3), 237–255. https://doi.org/10.1016/j.jtcme.2024.03.011

Foster, J. A., & Neufeld, K.-A. M. (2013). Gut–brain axis: How the microbiome influences anxiety and depression. *Trends in Neurosciences*, *36*(5), 305–312. https://doi.org/10.1016/j.tins.2013.01.005

Gao, K., Chen, C., Zheng, Z., Fan, Q., Wang, H., Li, Y., & Chen, S. (2024). Lactococcus strains with psychobiotic properties improve cognitive and mood alterations in aged mice. *Frontiers in Nutrition*, *11*. https://doi.org/10.3389/fnut.2024.1439094

Gao, M., Wang, J., Liu, P., Tu, H., Zhang, R., Zhang, Y., Sun, N., & Zhang, K. (2023). Gut microbiota composition in depressive disorder: A systematic review, meta-analysis, and meta-regression. *Translational Psychiatry*, *13*(1), 1–18. https://doi.org/10.1038/s41398-023-02670-5

Hao, Z., Wang, W., Guo, R., & Liu, H. (2019). *Faecalibacterium prausnitzii* (ATCC 27766) has preventive and therapeutic effects on chronic unpredictable mild stress-induced depression-like and anxiety-like behavior in rats. *Psychoneuroendocrinology*, *104*, 132–142. https://doi.org/10.1016/j.psyneuen.2019.02.025

Heidarzadeh-Rad, N., Gökmen-Özel, H., Kazemi, A., Almasi, N., & Djafarian, K. (2020). Effects of a psychobiotic supplement on serum brain-derived neurotrophic factor levels in depressive patients: A *Post Hoc* analysis of a randomized clinical trial. *Journal of Neurogastroenterology and Motility*, *26*(4), 486–495. https://doi.org/10.5056/jnm20079

Homan, P., Neumeister, A., Nugent, A. C., Charney, D. S., Drevets, W. C., & Hasler, G. (2015). Serotonin versus catecholamine deficiency: Behavioral and neural effects of experimental depletion in remitted depression. *Translational Psychiatry*, *5*(3), e532–e532. https://doi.org/10.1038/tp.2015.25

Hou, K., Wu, Z.-X., Chen, X.-Y., Wang, J.-Q., Zhang, D., Xiao, C., Zhu, D., Koya, J. B., Wei, L., Li, J., & Chen, Z.-S. (2022). Microbiota in health and diseases. *Signal Transduction and Targeted Therapy*, *7*(1), 1–28. https://doi.org/10.1038/s41392-022-00974-4

Huang, S.-Y., Chen, L.-H., Wang, M.-F., Hsu, C.-C., Chan, C.-H., Li, J.-X., & Huang, H.-Y. (2018). Lactobacillus paracasei PS23 delays progression of age-related cognitive decline in senescence accelerated mouse prone 8 (SAMP8) mice. *Nutrients*, *10*(7), Article 7. https://doi.org/10.3390/nu10070894

Janik, R., Thomason, L. A. M., Stanisz, A. M., Forsythe, P., Bienenstock, J., & Stanisz, G. J. (2016). Magnetic resonance spectroscopy reveals oral *Lactobacillus* promotion of increases in brain GABA, N-acetyl aspartate and glutamate. *NeuroImage*, *125*, 988–995. https://doi.org/10.1016/j.neuroimage.2015.11.018

Jarosz, Ł. S., Socała, K., Michalak, K., Wiater, A., Ciszewski, A., Majewska, M., Marek, A., Grądzki, Z., & Wlaź, P. (2024). The effect of psychoactive bacteria, Bifidobacterium longum Rosell®-175 and Lactobacillus rhamnosus JB-1, on brain proteome profiles in mice. *Psychopharmacology*, *241*(5), 925–945. https://doi.org/10.1007/s00213-023-06519-z

Jia, L., Xiao, L., Fu, Y., Shao, Z., Jing, Z., Yuan, J., Xie, Y., Guo, J., Wang, Y., & Geng, W. (2024). Neuroprotective effects of probiotics on anxiety- and depression-like disorders in stressed mice by modulating tryptophan metabolism and the gut microbiota. *Food & Function*, *15*(6), 2895–2905. https://doi.org/10.1039/D3FO03897A

Kazemi, A., Noorbala, A. A., Azam, K., Eskandari, M. H., & Djafarian, K. (2019). Effect of probiotic and prebiotic vs placebo on psychological outcomes in patients with major depressive disorder: A randomized clinical trial. *Clinical Nutrition*, *38*(2), 522–528. https://doi.org/10.1016/j.clnu.2018.04.010

Liang, S., Wang, T., Hu, X., Luo, J., Li, W., Wu, X., Duan, Y., & Jin, F. (2015). Administration of *Lactobacillus helveticus* NS8 improves behavioral, cognitive, and biochemical aberrations caused by chronic restraint stress. *Neuroscience*, *310*, 561–577. https://doi.org/10.1016/j.neuroscience.2015.09.033

Liao, J. F., Hsu, C. C., Chou, G. T., Hsu, J. S., Liong, M. T., & Tsai, Y. C. (2019). *Lactobacillus paracasei PS23 reduced early-life stress abnormalities in maternal separation mouse model*. https://doi.org/10.3920/BM2018.0077

Liu, L., Huh, J. R., & Shah, K. (2022). Microbiota and the gut-brain-axis: Implications for new therapeutic design in the CNS. *eBioMedicine*, *77*. https://doi.org/10.1016/j.ebiom.2022.103908

Makris, A. P., Karianaki, M., Tsamis, K. I., & Paschou, S. A. (2021). The role of the gut-brain axis in depression: Endocrine, neural, and immune pathways. *Hormones*, *20*(1), 1–12. https://doi.org/10.1007/s42000-020-00236-4

Marwaha, S., Palmer, E., Suppes, T., Cons, E., Young, A. H., & Upthegrove, R. (2023). Novel and emerging treatments for major depression. *The Lancet*, *401*(10371), 141–153. https://doi.org/10.1016/S0140-6736(22)02080-3

Messaoudi, M., Violle, N., Bisson, J.-F., Desor, D., Javelot, H., & Rougeot, C. (2011). Beneficial psychological effects of a probiotic formulation (Lactobacillus helveticus R0052 and Bifidobacterium longum R0175) in healthy human volunteers. *Gut Microbes*, *2*(4), 256–261. https://doi.org/10.4161/gmic.2.4.16108

Misra, S., & Mohanty, D. (2019). Psychobiotics: A new approach for treating mental illness? *Critical Reviews in Food Science and Nutrition*, *59*(8), 1230–1236. https://doi.org/10.1080/10408398.2017.1399860

Morais, L. H., Schreiber, H. L., & Mazmanian, S. K. (2021). The gut microbiota–brain axis in behaviour and brain disorders. *Nature Reviews Microbiology*, *19*(4), 241–255. https://doi.org/10.1038/s41579-020-00460-0

Mörkl, S., Butler, M. I., & Wagner-Skacel, J. (2023). Gut-brain-crosstalk- the vagus nerve and the microbiota-gut-brain axis in depression. A narrative review. *Journal of Affective Disorders Reports*, *13*, 100607. https://doi.org/10.1016/j.jadr.2023.100607

Oroojzadeh, P., Bostanabad, S. Y., & Lotfi, H. (2022). Psychobiotics: The Influence of Gut Microbiota on the Gut-Brain Axis in Neurological Disorders. *Journal of Molecular Neuroscience*, *72*(9), 1952–1964. https://doi.org/10.1007/s12031-022-02053-3

Oswald, G., Francis, J., & Dhakal, S. (2019). Behavioral and Physiological Interventions for Anxiety and Depression: An Overview of Nontraditional Methods. *Journal of Creativity in Mental Health*, *14*(4), 455–464. https://doi.org/10.1080/15401383.2019.1626314

Patra, S. (2016). Psychobiotics: A paradigm shift in psychopharmacology. *Indian Journal of Pharmacology*, *48*(4), 469. https://doi.org/10.4103/0253-7613.186194

Penninx, B. W., Pine, D. S., Holmes, E. A., & Reif, A. (2021). Anxiety disorders. *The Lancet*, *397*(10277), 914–927. https://doi.org/10.1016/S0140-6736(21)00359-7

Rogers, G. B., Keating, D. J., Young, R. L., Wong, M.-L., Licinio, J., & Wesselingh, S. (2016). From gut dysbiosis to altered brain function and mental illness: Mechanisms and pathways. *Molecular Psychiatry*, *21*(6), 738–748. https://doi.org/10.1038/mp.2016.50

Ross, K. (2023). Psychobiotics: Are they the future intervention for managing depression and anxiety? A literature review. *EXPLORE*, *19*(5), 669–680. https://doi.org/10.1016/j.explore.2023.02.007

Savignac, H. M., Kiely, B., Dinan, T. G., & Cryan, J. F. (2014). Ifidobacteria exert strain-specific effects on stress-related behavior and physiology in BALB/c mice. *Neurogastroenterology & Motility*, *26*(11), 1615–1627. https://doi.org/10.1111/nmo.12427

Sharma, R., Gupta, D., Mehrotra, R., & Mago, P. (2021). Psychobiotics: The Next-Generation Probiotics for the Brain. *Current Microbiology*, *78*(2), 449–463. https://doi.org/10.1007/s00284-020-02289-5

Staudacher, H. M., Mikocka-Walus, A., & Ford, A. C. (2021). Common mental disorders in irritable bowel syndrome: Pathophysiology, management, and considerations for future randomised controlled trials. *The Lancet Gastroenterology & Hepatology*, *6*(5), 401–410. https://doi.org/10.1016/S2468-1253(20)30363-0

Sudo, N. (2014). Microbiome, HPA Axis and Production of Endocrine Hormones in the Gut. In M. Lyte & J. F. Cryan (Eds.), *Microbial Endocrinology: The Microbiota-Gut-Brain Axis in Health and Disease* (pp. 177–194). Springer. https://doi.org/10.1007/978-1-4939-0897-4_8

Talbott, S., Talbott, J., Stephens, B., & Oddou, M. (2019). Effect of Coordinated Probiotic/Prebiotic/Phytobiotic Supplementation on Microbiome Balance and Psychological Mood State in Healthy Stressed Adults. *Functional Foods in Health and Disease*, *9*, 265–275. https://doi.org/10.31989/ffhd.v9i4.599

Taylor, A. M., & Holscher, H. D. (2020). A review of dietary and microbial connections to depression, anxiety, and stress. *Nutritional Neuroscience*, *23*(3), 237–250. https://doi.org/10.1080/1028415X.2018.1493808

Tian, P., O'Riordan, K. J., Lee, Y., Wang, G., Zhao, J., Zhang, H., Cryan, J. F., & Chen, W. (2020). Towards a psychobiotic therapy for depression: *Bifidobacterium breve* CCFM1025 reverses chronic stress-induced depressive symptoms and gut microbial abnormalities in mice. *Neurobiology of Stress*, *12*, 100216. https://doi.org/10.1016/j.ynstr.2020.100216

Vaher, K., Bogaert, D., Richardson, H., & Boardman, J. P. (2022). Microbiome-gut-brain axis in brain development, cognition and behavior during infancy and early childhood. *Developmental Review*, *66*, 101038. https://doi.org/10.1016/j.dr.2022.101038

Valdes, A. M., Walter, J., Segal, E., & Spector, T. D. (2018). Role of the gut microbiota in nutrition and health. *BMJ*, *361*, k2179. https://doi.org/10.1136/bmj.k2179

Varghese, F. P., & Brown, E. S. (2001). The Hypothalamic-Pituitary-Adrenal Axis in Major Depressive Disorder: A Brief Primer for Primary Care Physicians. *Primary Care Companion to the Journal of Clinical Psychiatry*, *3*(4), 151–155. https://doi.org/10.4088/pcc.v03n0401

Verduijn, J., Milaneschi, Y., Schoevers, R. A., van Hemert, A. M., Beekman, A. T. F., & Penninx, B. W. J. H. (2015). Pathophysiology of major depressive disorder: Mechanisms involved in etiology are not associated with clinical progression. *Translational Psychiatry*, *5*(9), e649–e649. https://doi.org/10.1038/tp.2015.137

Vostanis, P., Haffejee, S., Getanda, E., Eruyar, S., Hassan, S., & O'Reilly, M. (2023). Risk factors for mental health and wellness: Children's perspectives from five Majority World Countries. *Health, Risk & Society*, *25*(7–8), 304–323. https://doi.org/10.1080/13698575.2023.2258381

Wallace, C. J. K., & Milev, R. V. (2021). The Efficacy, Safety, and Tolerability of Probiotics on Depression: Clinical Results From an Open-Label Pilot Study. *Frontiers in Psychiatry*, *12*. https://doi.org/10.3389/fpsyt.2021.618279

Wei, C.-L., Wang, S., Yen, J.-T., Cheng, Y.-F., Liao, C.-L., Hsu, C.-C., Wu, C.-C., & Tsai, Y.-C. (2019). Antidepressant-like activities of live and heat-killed *Lactobacillus paracasei* PS23 in chronic corticosterone-treated mice and possible mechanisms. *Brain Research*, *1711*, 202–213. https://doi.org/10.1016/j.brainres.2019.01.025

Zhu, R., Zhao, X., Wu, H., Zeng, X., Wei, J., & Chen, T. (2024). Psychobiotics *Lactiplantibacillus plantarum* JYLP-326: Antidepressant-like effects on CUMS-induced depressed mouse model and alleviation of gut microbiota dysbiosis. *Journal of Affective Disorders*, *354*, 752–764. https://doi.org/10.1016/j.jad.2024.03.136

Zou, R., Tian, P., Xu, M., Zhu, H., Zhao, J., Zhang, H., Chen, W., & Wang, G. (2021). Psychobiotics as a novel strategy for alleviating anxiety and depression. *Journal of Functional Foods*, *86*, 104718. https://doi.org/10.1016/j.jff.2021.104718

14 Role of Probiotics and Prebiotics in Ischemic Stroke in Connection with the Gut–Brain Axis

Aarti Gupta and Alankar Shrivastava

Amity Institute of Pharmacy, Amity University, Raipur, Chhattisgarh, India

14.1 INTRODUCTION

Ischemic stroke (IS), an inadequate supply of blood to a specific area of the cerebral tissue, is reversible loss of tissue function, and eventually there occurs an infarction that results in the damage of neurons and supporting tissue. It leads to loss of electrical function and disturbance of membrane function with calcium input leading to calcium-dependent excitotoxicity, generation of reactive oxygen species, and after the event cell membranes damaged and cell lysis occurred (Paul & Candelario-Jalil, 2021). In 2019, 77.19 million people suffered an ischemic stroke, which resulted in 63.48 million disability-adjusted life years (DALYs) and 3.29 million fatalities. The mortality and DALY rates related with ischemic stroke declined in both genders between 1990 and 2019. International and global predictions suggest that among people aged over 25 years, the lifetime risk of all strokes and ischemic stroke are 24.9% and 18.3%, respectively. In an effort to prevent and treat noncommunicable diseases like stroke, the World Health Organization has suggested a number of primary health care interventions for behavioral and lifestyle risk factors, such as high blood pressure, smoking, and an unhealthy diet (Pu et al., 2023). The enteric and central nervous system are linked by a bidirectional communication network called the gut–brain axis. This network involves anatomical, humoral, endocrine, metabolic, and immunological pathways of communication. The brain can affect intestinal activities such as the activity of functional immune effectors cells, through the autonomic nervous system, the hypothalamic-pituitary-adrenal axis, and nerves within GIT. The gut can also affect mood, cognition, and mental health (Mayer, 2011; Mayer et al., 2014). The gut microbiota plays a vital role in performing specific functions in the host body. It helps in nutrient metabolism, xenobiotics and drug metabolism, maintenance of structural integrity of the gut mucosal barrier, immunomodulation and protection against pathogens, and behaves normally in the intestinal environment (Jandhyala, 2015; Mayer, 2011).

14.2 PATHOPHYSIOLOGY OF ISCHEMIA STROKE VIA THE BRAIN–GUT AXIS

The gut–brain axis consists of the following essential components: the CNS; the autonomic nervous system (sympathetic nerves at T5-L2 level, parasympathetic at S2-S4 level); the enteric nervous system (ENS); neurotransmitter, hormones, and neuropeptides; intestinal microenvironment (the intestinal barrier, gut microbiota, and their metabolic products, enteroendocrine and immune system and); the hypothalamic–pituitary–adrenal (HPA) axis; and the blood–brain barrier (Ferreira et al., 2014). There are three mechanisms by which sensory information is transferred in the gut's information system: primary afferent neurons, immune cells, and enteroendocrine cells (Appleton, 2018; Lake & Heuckeroth, 2013; Mayer, 2011).

Communication involves multiple biological networks, including neural, immune, neuroendocrine, and metabolic pathways. The first is through the vagus nerve, which consists of afferent fiber and efferent fiber. It communicates signals between the intestine and the brain. The afferent role for stimulation by microbial compounds, metabolites, and hormones (for example serotonin, cholecystokinin, glucagon-like peptide, dopamine, serotonin and gamma-aminobutyric acid (GABA) and peptide YY) released by enteroendocrine cells from the epithelial layer of the intestine to initiate signaling from the intestine to the brain. The signals pass to the hypothalamus that regulates the secretion of the pituitary gland (Karemaker, 2017; Petkova, 2018). Second, the immunogenic endotoxin from microbiota, such as lipopolysaccharide, can promote neuronal inflammation either directly or through activation of peripheral immune cells, which can migrate to the brain. Third, the microbiota produces metabolites such as neurotransmitters, short-chain fatty acids, indoles, aryl hydrogen receptors, ligands, and bile acids that enter the circulatory system and travel to brain to the function of neurons, microglia, astrocytes, and the blood–brain barrier (Singh et al., 2016).

Brain physiology and cognitive function can be affected via the changes of microbial action. It goes via direct and

DOI: 10.1201/9781003536802-14

indirect pathways. It includes the enteric nervous system and vagus nerves, the neuroendocrine and hypothalamic–pituitary–adrenal axis, the immune system, and the microbiota-derived neuroactive compounds (Stanley et al., 2018). After IS, these neurotransmitters affect the nervous function, active metabolites such as short-chain fatty acids, and stimulates the innate immunity and development of gut-associated lymphoid tissues and amino acids (tyramine, tryptophan) (Carabotti et al., 2015; X. Zhu et al., 2017). The intestinal epithelium layer secrets hormones via neuroendocrine cells which change the physiological function of the gut. Moreover, it interacts with the host immune cells and metabolism, circulates through blood, and sends direct signals to the brain. Various immune cells and neurons or even blood circulation, which control the signaling molecules, transport from the gut lumen to the lamina propria (Benakis et al., 2020; X. Li et al., 2020). Gut microbiotas stimulate the immune system in which low degree of physiological inflammation helps in defending against pathogens under physiological condition. However, bacterial colony produce nutrition and cytokines for the survival that can inhibit that growth of the microorganism (Rakoff-Nahoum et al., 2004; Sekirov et al., 2010).

14.3 GUT MICROBIOTA COMPOSITION AND METABOLITES IN NORMAL AND ISCHEMIA STROKE CONDITION

Dysbiosis refers to a condition in which changes in composition, distribution, and metabolic activity of intestinal microbiota are based on both qualitative and quantitative terms (Renamastika et al., 2021). The human gut microbiome is highly diverse, encompassing around 552,700 unique genes from 202 complete human gut bacterial genomes which can be analyzed at various taxonomic levels: strain to species, species to genus, and genus. At the level of phylum, the blood microbiota or gut microbiota in normal physiological condition consists of Firmicutes, Bacteriodetes, Proteobacteria, Verrucomicrobia, and Actinobacteria, but Verrucomicrobia, Firmicutes, and Deferribacteres microbes are of higher level than other microbes. In acute IS, Firmicutes, Bacteriodetes, and Verrucomicrobia levels decrease in comparison to normal condition. Actinobacteria, Proteobacteria, OD1, and TM7 levels increase indicating a shift toward a pathological state. At the genus level, *Akkermansia*, *Bacteroides*, *Lactobacillus*, *Ruminococcus* and Oscillospira are present in both conditions but Enterobacteria, *Pseudomonas*, *Flavobacterium*, *Staphylococcus*, *Prevotella*, *Micrococcus*,

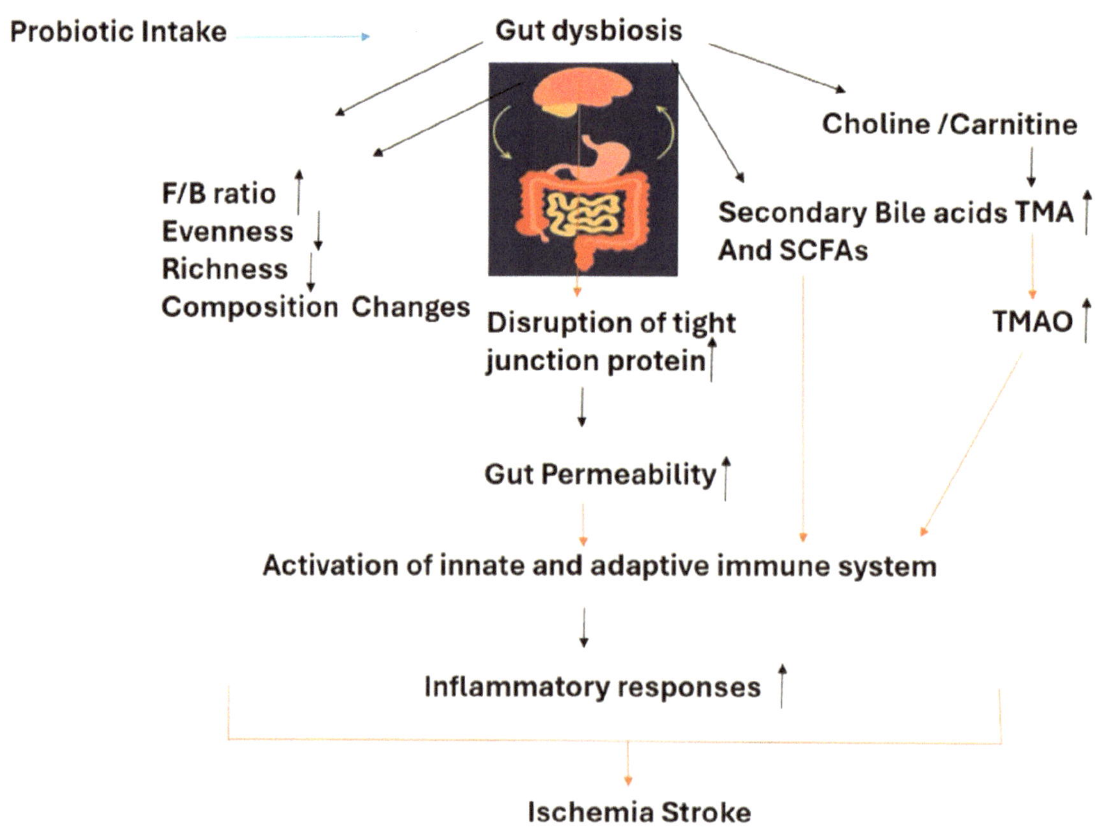

FIGURE 14.1 Mechanisms of the beneficial role of probiotics on IS.

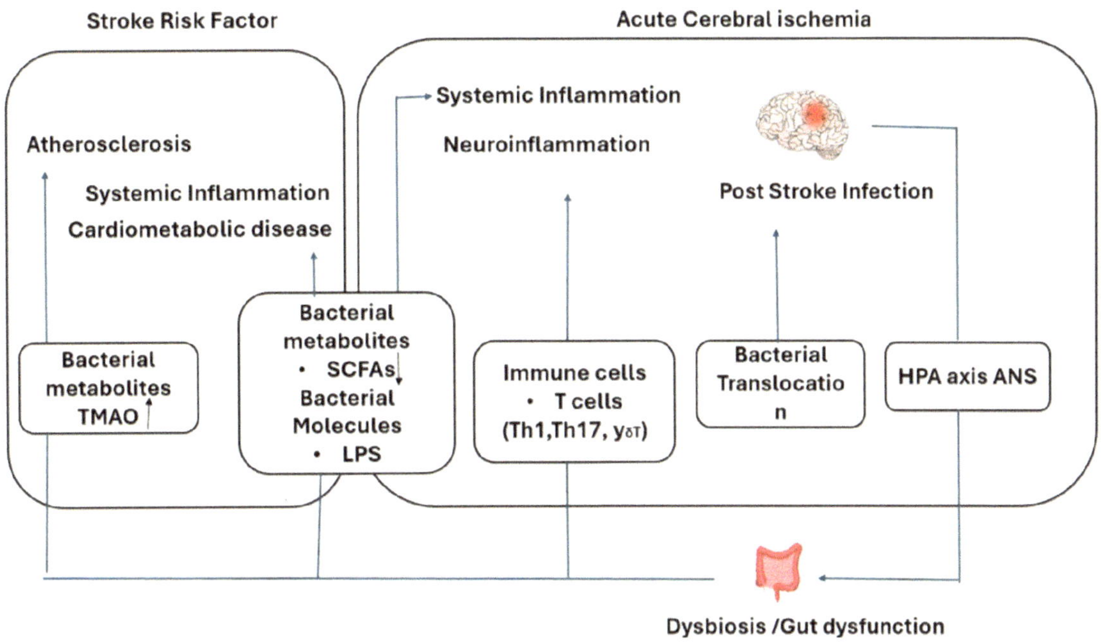

FIGURE 14.2 Role of gut microbiota–mediated pathways in IS.

Corynebacterium, Enhydrobacter, Comamonadaceae, *Collinsella*, *Faecalibacterium*, *Blautia*, *Anaerococcus*, *Finegoldia*, *Dialister*, Neisseriaceae, ZB2, Aerococcaceae, and TM7-1 increase in abundance in stroke. *Bacteroides*, *Akkermansia*, *Clostridiales*, Ruminococcaceae, Lachnospiraceae, *Lactobacillus*, *Ruminococcus*, *Parabacteroides*, *Stenotrophomonas*, and rc4 decrease in IS. Aerococcaceae, ZB2, TM7-1, and Flavobacterium increase drastically, whereas *Mucispirillum*, rc4-4, *Akkermansia*, *Clostridiales*, *Lactobacillus*, and *Stenotrophomonas* decrease immediately (Figures 14.1 and 14.2) (Chang et al., 2021)

After the IS, the intestinal microbiota alpha-diversity and beta-diversity reduce and *Aspergillus*, Enterobacteriaceae, and *Escherichia coli* spp. increase. The change in systemic inflammation, the overgrowth of Enterobacteriaceae, leads to intensify brain infarction and intestinal microbiota. The IS stimulate intestinal ischemia, which produces large number of nitrates in response to free radicals and support nitrate-responsive microbiota of the Enterobacteriaceae, which provoke the inflammatory response and progress of IS (Blasco et al., 2020; Ahnstedt et al., 2020; Xu et al., 2021).

The metabolite elements are trimethylamine oxide (TMAO), SCFAs, and other substances; they affect the level of IS. Choline, carnitine, and phosphatidylcholine produced as trimethylamine (TMA) by the intestinal microbiota get converted into TMAO by hepatics monooxygenase and shows dynamic changes in level during acute phase of IS. TMAO modulating platelets hyperresponsive and thrombosis potential and identify microbial taxa associated with plasma TMAO (Schneider et al., 2020; Craciun & Balskus, 2012; Koeth et al., 2014; W. Zhu et al., 2016; Svingen et al., 2018).

SCFAs play a neuroprotective role, synthesizing the neurotransmitter and modulating the immune system by improving intestinal permeability and the blood–brain barrier(Russo et al., 2018; Brahe et al., 2013). It can be detected as in cerebrospinal fluid. It maintains the integrity of the blood–brain barrier related to the distribution of molecules and nutrition from the circulatory system to the brain. It enters the central nervous system have neuroactive properties that is unknown mechanism. It majorly influences the neurological function and behavior, as well as the improvement in neurological development and neurodegenerative disorder. Neuroinflammation affects the brain function. Sodium butyrate can decrease microglial morphology. It leads to IL-1beta, IL-6, and TNF-alpha, as well as phosphorylation of p38 MAPK, JNF, and NF-κB (Stanisavljević et al., 2019; Patnala et al., 2017).

14.4 PROBIOTICS AND PREBIOTICS USED IN IS

Probiotics must be kept alive in the digestive tract to obtain health benefits. Prebiotics can be employed to keep these bacteria alive. Prebiotics are food components that cannot be digested by the human digestive tract enzymatically. The food is fermented by the microflora present in the large intestine. Prebiotics in the large intestine will support the growth of probiotic bacteria and supress pathogenic bacteria. Some specific mechanisms of intestinal microbiota are (1) production of inhibitory substances, (2) blocking of adhesion sites, (3) competition for nutrients, and (4) degradation of toxin receptor and stimulation of immunity. Restoration of Gut Microbiota, Anti-inflammatory effects, Neuroprotection, Gut Barrier Integrity, Modulation of Neuroinflammation. Prebiotics are nondigestible food

TABLE 14.1
Prebiotics and Role in IS

Prebiotics	Dose	Role in Ischemia Stroke	Reference
Insulin Fructooligosaccharides (FDS) Galactooligosaccharides (GOS)	APOE4 model mouse and old mouse FOS (3g/kg) GOS (4g/kg)	Insulin increases the number of beneficial microbiotas and decreases harmful microbiota. It also increases systematics metabolism as indicated levels of SCFA, tryptophan-derived metabolites, bile acids, glycoltics metabolites, and scylloinositol. Insulin reduces the expression of inflammatory genes in the hippocampus	Hoffman et al., 2019
Chitosan oligosaccharides	COS is given orally in 200, 400, and 800 mg/kg for 14 days	The neuroprotective effect of COS is closely related to its ability to inhibit oxidative stress, which is shown by decreased levels of malondialdehyde, 8-8-hydroxy-2′ deoxyguanosine, and decreased and increased levels of glutathione peroxidase and superoxide dismutase activity. COS suppresses the inflammatory response by decreasing the release of pro-inflammatory cytokines	Jia et al., 2016
Galactooligosaccharides	2% GOS in the amount of 8 ml/kg/day	It slows prolonged disease onset. It increases the concentration of folate, vitamin B12, and lower homocysteine levels. GOS and yogurt prebiotics reduce the decrease in motor neurons, increase atrophy and mitochondrial activity in myocytes. GOS also supresses the activation of astrocytes and microglia and regulates several factors associated with inflammation and apoptosis	Song et al., 2013
Bimunogalaktooligosakarida (B-GOS)	13 g B-GOS powder for 3 weeks	Reduces the pro-inflammatory cytokine IL-1Beta and expression of 5-HT2A cortical receptors	Savignac et al., 2016
Bimunogalaktooligosakarida (B-GOS)	3% B-GOS solution for 3 weeks	It increases plasma acetate levels cortical GluN2B subunits and acetyl Co-A carboxylase mRNA	Gronier et al., 2018
Glucuronomannan oligosaccharides	GMn 10 mg/kg and 20 mg/kg for 7 days	It protects dopaminergic neurons from apoptosis	Y. Liu et al., 2020
Chitosan oligosaccharides	200, 500,1,000 µg/kg for 12 hours	It crosses the blood–brain barrier via GLUT1 transporter. COS reduces Abeta42-mediated cytotoxicity. It reduces amyloid-induced apoptosis, oxidative stress, and the release of inflammatory cytokines	L. Zhu et al., 2020
Fructooligosaccharides from *Morinda officinalis*	50 and 100 mg/kg/day	It improves brain tissues swelling, neuronal apoptosis, and downregulation of expression of the intracellular markers of the disease	D. Chen et al., 2017
Xylooligosaccharides	10% XOS 1 ml/day for 12 weeks	It increases hippocampal plasticity and improved brain mitochondrial dysfunction. Oxidative stress and apoptosis in the hippocampus reduce and microglial activation and improving cognitive function recovery	Chunchai et al., 2018

components that selectively stimulate the growth and activity of beneficial bacteria in the gut. These prebiotics help in the growth of beneficial bacteria such as Bifidobacteria and Lactobacilli, enhancing the growth of probiotics such as insulin, fructooligosaccharides (FOS), and galactooligosaccharides. Production of SCFAs, such as acetate, propionate, and butyrate, improves the metabolic system and gut–brain preamibility (Bugaut, 1987). Propionate and butyrate exert influence on intercellular potassium level involvement of SCFA in the cell signaling system (Tables 14.1 and 14.2) (Tolhurst et al., 2012; Braniste et al., 2014; Duncan et al., 2004).

14.5 CONCLUSION

The importance of MGBA and its mechanism in a person's health status has been highlighted in this chapter. Imbalances in the GI microflora composition should be related to abnormal physiological situation. The biological relationship between the gut microbiota, the CNS, and that immune signaling suggest that systemic microbial signaling or metabolites directly or indirectly impact immunological and neurological activity in IS. For stroke, early entral nutrition combined with probiotics can help in improving the nutritional status, reconstructing the gut microbiota, stabilizing intestinal barrier function, improving immune tolerance, and decreasing the complication of infection and nutritional diarrhoea, making it more effective. The

TABLE 14.2
Some Other Probiotics and Their Roles in IS

Probiotics	Role in Ischemia Stroke	References
Combination of four viable probiotic bacteria strains, namely, *Bifidobacterium breve*, *Lactobacillus casei*, *Lactobacillus bulgaricus* (*Lactobacillus delbrueckii* spp. bulgaricus), and *Lactobacillus acidophilus*	Could not improve neurological function, reduced infarct size 52%, and decreased the malondialdehyde content and TNF alpha level	Akhoundzadeh et al., 2018
Red ginseng and fermented red ginseng	Fermented red ginseng-treated group showed reduction of the infarction area in all region and total infarction volume	Bae et al., 2004
Puerariae Lobatae Radix+ Chuanxiong Rhizoma	Repair neurological impairment, reduce the cerebral infarction, and reverse the dyslipidemia reduce the blood viscosity and thromobotic risk	R. Chen et al., 2019
Panax notoginsenoside extract	Decrease neurological deficit scores, cerebral infract volume, upregulate the expression of GABA receptor in hippocampus, decrease rate of attenuation in BDNF expression	H. Li et al., 2018
Clostridium butyricum	Improve spatial learning ability, ameliorate the morphological changes in the hippocampus, increase in the brain	J. Liu et al., 2015
Shuan tong Ling	Increase neurological scores, reduce infarct volume, and decrease TNF-alpha and IL-1 beta	Mei et al., 2017
Fermented ginseng	Ameliorated loss of hippocampus cornu Ammonis neurons and increase caspase 3/Iba 1	Nagao et al., 2019
Yam gruel	Improve cognitive function, increase relative content of probiotics bacteria and SCFAs in intestinal tract, cerebral cortex reduces oxidative stress and inflammatory response, promoted the expression of neurotransmitter and brain-derived neurotrophic factor	Pang S., 2020
Chungkookjang (fermented soybean)	Prevent symptoms such as drooped eyelid, bristling hair, reduce muscles tone and flexor reflex and abnormal posture, the neuronal cell death, suppress cytokine expression, prevent the impairment of the glucose metabolism	Park et al., 2016
Lactocare capsule, 109 CFU, ZIST, Takhmir (mixture of seven probiotics bacteria strains, including *Lactobacillus casei* ZT-Lca.106, *Lactobacillus acidophilus* ZT-Lac.51, *Lactobacillus rhamnosus* ZT-Lrh.54, *Lactobacillus bulgaricus* ZT-Lbu.90, *Bifidobacterium longum* ZT-La.106, and *Streptococcus thermophilus* ZT-Sth.20)	Reduce spatial memory impairment and neurological dysfunction, reduce neuronal death, malondialdehyde content and brain-derived neurotrophic factor level do not change	Rahmati et al., 2019
Clostridum butyricum	Ameliorate cognitive impairment, histopathologic changes in the hippocampus increase *p*-Atk expression and decrease caspase-3 expression equal inhibit neuronal apoptosis, decrease blood glucose level	Sun et al., 2016
Inactivated *Lactobacillus*	Improve neurobehavioral scores, decrease cerebral infarction volume, neural cells apoptosis, inhibit expression of TLR4, Promote the expression of IkB and A20, reduce oxidate stress	Wanchao et al., 2018
Chamomile	Improve neurological scores, decrease in infarct volume, and increase the activity of HO1 nad Nrf2	Zhang et al., 2019

effects on the SCFA on microglia is still not well defined. In the future, proteomics, metabolomics, and 16S microbiome sequencing should be used for gut microbiome studies and to understand the pathways, microbes, and metabolites involved and to clarify the effects of interaction on IS.

REFERENCES

Ahnstedt, H., Patrizz, A., Chauhan, A., Roy-O'Reilly, M., Furr, J. W., Spychala, M. S., D'Aigle, J., Blixt, F. W., Zhu, L., Bravo Alegria, J., & McCullough, L. D. (2020). Sex differences in T cell immune responses, gut permeability and outcome after ischemic stroke in aged mice. *Brain, Behavior, and Immunity*, *87*, 556–567. https://doi.org/10.1016/j.bbi.2020.02.001

Akhoundzadeh, K., Vakili, A., Shadnoush, M., & Sadeghzadeh, J. (2018). Effects of the oral ingestion of probiotics on brain damage in a transient model of focal cerebral ischemia in mice. *Iranian Journal of Medical Sciences*, *43*(1), 32–40.

Appleton, J. (2018). The gut-brain axis: Influence of microbiota on mood and mental health. *Integrative Medicine (Encinitas, Calif.)*, *17*(4), 28–32.

Bae, E.-A., Hyun, Y.-J., Choo, M.-K., Oh, J. K., Ryu, J. H., & Kim, D.-H. (2004). Protective effect of fermented red ginseng on a transient focal ischemic rats. *Archives of Pharmacal Research*, *27*(11), 1136–1140. https://doi.org/10.1007/BF02975119

Benakis, C., Martin-Gallausiaux, C., Trezzi, J.-P., Melton, P., Liesz, A., & Wilmes, P. (2020). The microbiome-gut-brain axis in acute and chronic brain diseases. *Current Opinion in Neurobiology*, *61*, 1–9. https://doi.org/10.1016/j.conb.2019.11.009

Blasco, M. P., Chauhan, A., Honarpisheh, P., Ahnstedt, H., d'Aigle, J., Ganesan, A., Ayyaswamy, S., Blixt, F., Venable, S., Major, A., Durgan, D., Haag, A., Kofler, J., Bryan, R., McCullough, L. D., & Ganesh, B. P. (2020). Age-dependent involvement of gut mast cells and histamine in post-stroke inflammation. *Journal of Neuroinflammation*, *17*(1), 160. https://doi.org/10.1186/s12974-020-01833-1

Brahe, L. K., Astrup, A., & Larsen, L. H. (2013). Is butyrate the link between diet, intestinal microbiota and obesity-related metabolic diseases? *Obesity Reviews*, *14*(12), 950–959. https://doi.org/10.1111/obr.12068

Braniste, V., Al-Asmakh, M., Kowal, C., Anuar, F., Abbaspour, A., Tóth, M., Korecka, A., Bakocevic, N., Ng, L. G., Kundu, P., Gulyás, B., Halldin, C., Hultenby, K., Nilsson, H., Hebert, H., Volpe, B. T., Diamond, B., & Pettersson, S. (2014). The gut microbiota influences blood-brain barrier permeability in mice. *Science Translational Medicine*, *6*(263). https://doi.org/10.1126/scitranslmed.3009759

Bugaut, M. (1987). Occurrence, absorption and metabolism of short chain fatty acids in the digestive tract of mammals. *Comparative Biochemistry and Physiology Part B: Comparative Biochemistry*, *86*(3), 439–472. https://doi.org/10.1016/0305-0491(87)90433-0

Carabotti, M., Scirocco, A., Maselli, M. A., & Severi, C. (2015). The gut-brain axis: Interactions between enteric microbiota, central and enteric nervous systems. *Annals of Gastroenterology*, *28*(2), 203–209.

Chang, Y., Woo, H. G., Jeong, J. H., Kim, G. H., Park, K. D., & Song, T.-J. (2021). Microbiota dysbiosis and functional outcome in acute ischemic stroke patients. *Scientific Reports*, *11*(1), 10977. https://doi.org/10.1038/s41598-021-90463-5

Chen, D., Yang, X., Yang, J., Lai, G., Yong, T., Tang, X., Shuai, O., Zhou, G., Xie, Y., & Wu, Q. (2017). Prebiotic effect of fructooligosaccharides from Morinda officinalis on Alzheimer's disease in rodent models by targeting the microbiota-gut-brain axis. *Frontiers in Aging Neuroscience*, *9*. https://doi.org/10.3389/fnagi.2017.00403

Chen, R., Wu, P., Cai, Z., Fang, Y., Zhou, H., Lasanajak, Y., Tang, L., Ye, L., Hou, C., & Zhao, J. (2019). Puerariae Lobatae Radix with chuanxiong Rhizoma for treatment of cerebral ischemic stroke by remodeling gut microbiota to regulate the brain–gut barriers. *The Journal of Nutritional Biochemistry*, *65*, 101–114. https://doi.org/10.1016/j.jnutbio.2018.12.004

Chunchai, T., Thunapong, W., Yasom, S., Wanchai, K., Eaimworawuthikul, S., Metzler, G., Lungkaphin, A., Pongchaidecha, A., Sirilun, S., Chaiyasut, C., Pratchayasakul, W., Thiennimitr, P., Chattipakorn, N., & Chattipakorn, S. C. (2018). Decreased microglial activation through gut-brain axis by prebiotics, probiotics, or synbiotics effectively restored cognitive function in obese-insulin resistant rats. *Journal of Neuroinflammation*, *15*(1), 11. https://doi.org/10.1186/s12974-018-1055-2

Craciun, S., & Balskus, E. P. (2012). Microbial conversion of choline to trimethylamine requires a glycyl radical enzyme. *Proceedings of the National Academy of Sciences*, *109*(52), 21307–21312. https://doi.org/10.1073/pnas.1215689109

Duncan, S. H., Holtrop, G., Lobley, G. E., Calder, A. G., Stewart, C. S., & Flint, H. J. (2004). Contribution of acetate to butyrate formation by human faecal bacteria. *British Journal of Nutrition*, *91*(6), 915–923. https://doi.org/10.1079/BJN20041150

Ferreira, C. M., Vieira, A. T., Vinolo, M. A. R., Oliveira, F. A., Curi, R., & Martins, F. dos S. (2014). The central role of the gut microbiota in chronic inflammatory diseases. *Journal of Immunology Research*, *2014*, 1–12. https://doi.org/10.1155/2014/689492

Gronier, B., Savignac, H. M., Di Miceli, M., Idriss, S. M., Tzortzis, G., Anthony, D., & Burnet, P. W. J. (2018). Increased cortical neuronal responses to NMDA and improved attentional set-shifting performance in rats following prebiotic (B-GOS®) ingestion. *European Neuropsychopharmacology*, *28*(1), 211–224. https://doi.org/10.1016/j.euroneuro.2017.11.001

Hoffman, J. D., Yanckello, L. M., Chlipala, G., Hammond, T. C., McCulloch, S. D., Parikh, I., Sun, S., Morganti, J. M., Green, S. J., & Lin, A.-L. (2019). Dietary inulin alters the gut microbiome, enhances systemic metabolism and reduces neuroinflammation in an APOE4 mouse model. *PLoS One*, *14*(8), e0221828. https://doi.org/10.1371/journal.pone.0221828

Jandhyala, S. M. (2015). Role of the normal gut microbiota. *World Journal of Gastroenterology*, *21*(29), 8787. https://doi.org/10.3748/wjg.v21.i29.8787

Jia, S., Lu, Z., Gao, Z., An, J., Wu, X., Li, X., Dai, X., Zheng, Q., & Sun, Y. (2016). Chitosan oligosaccharides alleviate cognitive deficits in an amyloid-β1–42-induced rat model of Alzheimer's disease. *International Journal of Biological Macromolecules*, *83*, 416–425. https://doi.org/10.1016/j.ijbiomac.2015.11.011

Karemaker, J. M. (2017). An introduction into autonomic nervous function. *Physiological Measurement*, *38*(5), R89–R118. https://doi.org/10.1088/1361-6579/aa6782

Koeth, R. A., Levison, B. S., Culley, M. K., Buffa, J. A., Wang, Z., Gregory, J. C., Org, E., Wu, Y., Li, L., Smith, J. D., Tang, W. H. W., DiDonato, J. A., Lusis, A. J., & Hazen, S. L.

(2014). γ-Butyrobetaine Is a proatherogenic intermediate in gut microbial metabolism of L-Carnitine to TMAO. *Cell Metabolism*, *20*(5), 799–812. https://doi.org/10.1016/j.cmet.2014.10.006

Lake, J. I., & Heuckeroth, R. O. (2013). Enteric nervous system development: Migration, differentiation, and disease. *American Journal of Physiology-Gastrointestinal and Liver Physiology*, *305*(1), G1–G24. https://doi.org/10.1152/ajpgi.00452.2012

Li, H., Xiao, J., Li, X., Chen, H., Kang, D., Shao, Y., Shen, B., Zhu, Z., Yin, X., Xie, L., Wang, G., & Liang, Y. (2018). Low cerebral exposure cannot hinder the neuroprotective effects of panax notoginsenosides. *Drug Metabolism and Disposition*, *46*(1), 53–65. https://doi.org/10.1124/dmd.117.078436

Li, X., You, X., Wang, C., Li, X., Sheng, Y., Zhuang, P., & Zhang, Y. (2020). Bidirectional brain-gut-microbiota Axis in increased intestinal permeability induced by central nervous system injury. *CNS Neuroscience & Therapeutics*, *26*(8), 783–790. https://doi.org/10.1111/cns.13401

Liu, J., Sun, J., Wang, F., Yu, X., Ling, Z., Li, H., Zhang, H., Jin, J., Chen, W., Pang, M., Yu, J., He, Y., & Xu, J. (2015). Neuroprotective effects of *clostridium butyricum* against vascular dementia in mice via metabolic butyrate. *BioMed Research International*, *2015*, 1–12. https://doi.org/10.1155/2015/412946

Liu, Y., Jin, W., Deng, Z., Wang, J., & Zhang, Q. (2020). Preparation and neuroprotective activity of glucuronomannan oligosaccharides in an MPTP-induced Parkinson's model. *Marine Drugs*, *18*(9), 438. https://doi.org/10.3390/md18090438

Mayer, E. A. (2011). Gut feelings: The emerging biology of gut–brain communication. *Nature Reviews Neuroscience*, *12*(8), 453–466. https://doi.org/10.1038/nrn3071

Mayer, E. A., Savidge, T., & Shulman, R. J. (2014). Brain–gut microbiome interactions and functional bowel disorders. *Gastroenterology*, *146*(6), 1500–1512. https://doi.org/10.1053/j.gastro.2014.02.037

Mei, Z., Tan, L., Wang, J., Li, X., Huang, W., & Zhou, H. (2017). Fermented Chinese formula Shuan-Tong-Ling attenuates ischemic stroke by inhibiting inflammation and apoptosis. *Neural Regeneration Research*, *12*(3), 425. https://doi.org/10.4103/1673-5374.202946

Nagao, M., Yamano, S., Imagawa, N., Igami, K., Miyazaki, T., Ito, H., Watanabe, T., Kubota, K., Katsurabayashi, S., & Iwasaki, K. (2019). Effect of *Lactobacillus paracasei* A221-fermented ginseng on impaired spatial memory in a rat model with cerebral ischemia and β -amyloid injection. *Traditional & Kampo Medicine*, *6*(2), 96–104. https://doi.org/10.1002/tkm2.1220

Pang, S., & Li, Z. (2020). Effects of Dioscorea polystachya "yam gruel" on the cognitive function of diabetic rats with focal cerebral ischemia-reperfusion injury via the gut-brain axis. *Journal of Integrative Neuroscience*, *19*(2), 273. https://doi.org/10.31083/j.jin.2020.02.69

Park, S., Kim, D. S., Kang, S., & Moon, B. R. (2016). Fermented soybeans, Chungkookjang, prevent hippocampal cell death and β-cell apoptosis by decreasing pro-inflammatory cytokines in gerbils with transient artery occlusion. *Experimental Biology and Medicine*, *241*(3), 296–307. https://doi.org/10.1177/1535370215606811

Patnala, R., Arumugam, T. V, Gupta, N., & Dheen, S. T. (2017). HDAC inhibitor sodium butyrate-mediated epigenetic regulation enhances neuroprotective function of microglia during ischemic stroke. *Molecular Neurobiology*, *54*(8), 6391–6411. https://doi.org/10.1007/s12035-016-0149-z

Paul, S., & Candelario-Jalil, E. (2021). Emerging neuroprotective strategies for the treatment of ischemic stroke: An overview of clinical and preclinical studies. *Experimental Neurology*, *335*, 113518. https://doi.org/10.1016/j.expneurol.2020.113518

Petkova, N. (2018). Characterization inulin from black salsify (Scorzonera Hispanica L.) food and pharmaceutical purposes. *Asian Journal of Pharmaceutical and Clinical Research*, *11*(12), 221. https://doi.org/10.22159/ajpcr.2018.v11i12.28262

Pu, L., Wang, L., Zhang, R., Zhao, T., Jiang, Y., & Han, L. (2023). Projected global trends in ischemic stroke incidence, deaths and disability-adjusted life years from 2020 to 2030. *Stroke*, *54*(5), 1330–1339. https://doi.org/10.1161/STROKEAHA.122.040073

Rahmati, H., Momenabadi, S., Vafaei, A. A., Bandegi, A. R., Mazaheri, Z., & Vakili, A. (2019). Probiotic supplementation attenuates hippocampus injury and spatial learning and memory impairments in a cerebral hypoperfusion mouse model. *Molecular Biology Reports*, *46*(5), 4985–4995. https://doi.org/10.1007/s11033-019-04949-7

Rakoff-Nahoum, S., Paglino, J., Eslami-Varzaneh, F., Edberg, S., & Medzhitov, R. (2004). Recognition of commensal microflora by toll-like receptors is required for intestinal homeostasis. *Cell*, *118*(2), 229–241. https://doi.org/10.1016/j.cell.2004.07.002

Renamastika, S. N., Wahyuningrum, D. R., Putri, V. A., & Aisy, R. (2021). Prebiotics for acute ischemic stroke. *International Journal of Pharmacy and Pharmaceutical Sciences*, 1–10. https://doi.org/10.22159/ijpps.2021v13i5.41110

Russo, R., Cristiano, C., Avagliano, C., De Caro, C., La Rana, G., Raso, G. M., Canani, R. B., Meli, R., & Calignano, A. (2018). Gut-brain Axis: Role of lipids in the regulation of inflammation, pain and CNS diseases. *Current Medicinal Chemistry*, *25*(32), 3930–3952. https://doi.org/10.2174/0929867324666170216113756

Savignac, H. M., Couch, Y., Stratford, M., Bannerman, D. M., Tzortzis, G., Anthony, D. C., & Burnet, P. W. J. (2016). Prebiotic administration normalizes lipopolysaccharide (LPS)-induced anxiety and cortical 5-HT2A receptor and IL1-β levels in male mice. *Brain, Behavior, and Immunity*, *52*, 120–131. https://doi.org/10.1016/j.bbi.2015.10.007

Schneider, C., Okun, J. G., Schwarz, K. V., Hauke, J., Zorn, M., Nürnberg, C., Ungerer, M., Ringleb, P. A., & Mundiyanapurath, S. (2020). Trimethylamine-N-oxide is elevated in the acute phase after ischaemic stroke and decreases within the first days. *European Journal of Neurology*, *27*(8), 1596–1603. https://doi.org/10.1111/ene.14253

Sekirov, I., Russell, S. L., Antunes, L. C. M., & Finlay, B. B. (2010). Gut microbiota in health and disease. *Physiological Reviews*, *90*(3), 859–904. https://doi.org/10.1152/physrev.00045.2009

Singh, V., Roth, S., Llovera, G., Sadler, R., Garzetti, D., Stecher, B., Dichgans, M., & Liesz, A. (2016). Microbiota dysbiosis controls the neuroinflammatory response after stroke. *Journal of Neuroscience*, *36*(28), 7428–7440. https://doi.org/10.1523/JNEUROSCI.1114-16.2016

Song, L., Gao, Y., Zhang, X., & Le, W. (2013). Galactooligosaccharide improves the animal survival and alleviates motor neuron death in SOD1G93A mouse model of amyotrophic lateral sclerosis. *Neuroscience*, *246*, 281–290. https://doi.org/10.1016/j.neuroscience.2013.05.002

Stanisavljević, S., Čepić, A., Bojić, S., Veljović, K., Mihajlović, S., Đedović, N., Jevtić, B., Momčilović, M., Lazarević, M., Mostarica Stojković, M., Miljković, Đ., & Golić, N. (2019). Oral neonatal antibiotic treatment perturbs gut microbiota and aggravates central nervous system autoimmunity in Dark Agouti rats. *Scientific Reports*, *9*(1), 918. https://doi.org/10.1038/s41598-018-37505-7

Stanley, D., Moore, R. J., & Wong, C. H. Y. (2018). An insight into intestinal mucosal microbiota disruption after stroke. *Scientific Reports*, *8*(1), 568. https://doi.org/10.1038/s41598-017-18904-8

Sun, J., Wang, F., Ling, Z., Yu, X., Chen, W., Li, H., Jin, J., Pang, M., Zhang, H., Yu, J., & Liu, J. (2016). Clostridium butyricum attenuates cerebral ischemia/reperfusion injury in diabetic mice via modulation of gut microbiota. *Brain Research*, *1642*, 180–188. https://doi.org/10.1016/j.brainres.2016.03.042

Svingen, G. F. T., Zuo, H., Ueland, P. M., Seifert, R., Løland, K. H., Pedersen, E. R., Schuster, P. M., Karlsson, T., Tell, G. S., Schartum-Hansen, H., Olset, H., Svenningsson, M., Strand, E., Nilsen, D. W., Nordrehaug, J. E., Dhar, I., & Nygård, O. (2018). Increased plasma trimethylamine- N -oxide is associated with incident atrial fibrillation. *International Journal of Cardiology*, *267*, 100–106. https://doi.org/10.1016/j.ijcard.2018.04.128

Tolhurst, G., Heffron, H., Lam, Y. S., Parker, H. E., Habib, A. M., Diakogiannaki, E., Cameron, J., Grosse, J., Reimann, F., & Gribble, F. M. (2012). Short-chain fatty acids stimulate glucagon-like peptide-1 secretion via the G-protein–coupled receptor FFAR2. *Diabetes*, *61*(2), 364–371. https://doi.org/10.2337/db11-1019

Wanchao, S., Chen, M., Zhiguo, S., Futang, X., & Mengmeng, S. (2018). Protective effect and mechanism of Lactobacillus on cerebral ischemia reperfusion injury in rats. *Brazilian Journal of Medical and Biological Research*, *51*(7). https://doi.org/10.1590/1414-431x20187172

Xu, K., Gao, X., Xia, G., Chen, M., Zeng, N., Wang, S., You, C., Tian, X., Di, H., Tang, W., Li, P., Wang, H., Zeng, X., Tan, C., Meng, F., Li, H., He, Y., Zhou, H., & Yin, J. (2021). Rapid gut dysbiosis induced by stroke exacerbates brain infarction in turn. *Gut*, *70*(8), 1486–1494. https://doi.org/10.1136/gutjnl-2020-323263

Zhang, S., Xu, S., Duan, H., Zhu, Z., Yang, Z., Cao, J., Zhao, Y., Huang, Z., Wu, Q., & Duan, J. (2019). A novel, highly-water-soluble apigenin derivative provides neuroprotection following ischemia in male rats by regulating the ERK/Nrf2/HO-1 pathway. *European Journal of Pharmacology*, *855*, 208–215. https://doi.org/10.1016/j.ejphar.2019.03.024

Zhu, L., Li, R., Jiao, S., Wei, J., Yan, Y., Wang, Z. A., Li, J., & Du, Y. (2020). Blood-brain barrier permeable chitosan oligosaccharides interfere with β-Amyloid aggregation and alleviate β-Amyloid protein mediated neurotoxicity and neuroinflammation in a dose- and degree of polymerization-dependent manner. *Marine Drugs*, *18*(10), 488. https://doi.org/10.3390/md18100488

Zhu, W., Gregory, J. C., Org, E., Buffa, J. A., Gupta, N., Wang, Z., Li, L., Fu, X., Wu, Y., Mehrabian, M., Sartor, R. B., McIntyre, T. M., Silverstein, R. L., Tang, W. H. W., DiDonato, J. A., Brown, J. M., Lusis, A. J., & Hazen, S. L. (2016). Gut microbial metabolite TMAO enhances platelet hyperreactivity and thrombosis risk. *Cell*, *165*(1), 111–124. https://doi.org/10.1016/j.cell.2016.02.011

Zhu, X., Han, Y., Du, J., Liu, R., Jin, K., & Yi, W. (2017). Microbiota-gut-brain axis and the central nervous system. *Oncotarget*, *8*(32), 53829–53838. https://doi.org/10.18632/oncotarget.17754

15 Role of Probiotics

Divya Kumari[1], Ayush Madan[2], Priya Chaudhary[3], and Pracheta Janmeda[1]

[1] Department of Bioscience and Biotechnology, Banasthali Vidyapith, Rajasthan, India
[2] Department of Biotechnology, School of Research & Technology, People's University, Madhya Pradesh, India
[3] Centre for Herbal Pharmacology and Environmental Sustainability, Chettinad Hospital and Research Institute, Chettinad Academy of Research and Education, Chennai, Tamil Nadu, India

15.1 INTRODUCTION

Neurodegenerative diseases, as described by Jellinger (2010), involve the progressive decline of neuronal function and structure in the central nervous system (CNS) and the peripheral nervous system (PNS). These disorders, including Alzheimer's disease (AD) and Parkinson's disease (PD), affect various parts of the brain, spinal cord, and nerves, leading to a gradual deterioration of motor, cognitive, and sensory functions (Brown *et al.*, 2005). Recent studies indicate that bacterial probiotics, when consumed in sufficient quantities, provide health benefits and could be beneficial in the prevention and treatment of neurodegenerative diseases (Kechagia *et al.*, 2013). Probiotics are known to enhance gut flora, which can help regulate inflammatory responses and prevent harmful infections from colonizing the body (Valdes *et al.*, 2018).

However, there are several challenges associated with probiotics. Firstly, there is limited evidence for their efficacy in treating specific conditions like dermatitis or irritable bowel syndrome (IBS) (Silva *et al.*, 2020). Secondly, choosing the appropriate dosage and strain can be difficult since different strains have varying effects and the ideal dose depends on the strain (Sun *et al.*, 2021). Thirdly, not all probiotic strains can survive the acidic environment of the stomach, which is essential for colonizing the gut (Hill *et al.*, 2014). Additionally, probiotics are typically used in the short term, and their long-term effects remain unclear (Ojha *et al.*, 2023). Furthermore, while generally safe, probiotics may cause side effects such as gas, bloating, and diarrhea, and in rare cases severe illnesses, especially in immunocompromised individuals (Drago and Vecchi, 2009).

Despite these limitations, probiotics are effective in maintaining a healthy gut microbiome, reducing inflammation, and combating oxidative stress—key factors in neurodegenerative diseases (Cortese *et al.*, 2018). Certain probiotics can also promote the production of brain-derived neurotrophic factor (BDNF), a protein essential for neuronal survival and growth, with deficiencies linked to Alzheimer's and Parkinson's diseases. Further research is needed to determine the optimal strains, doses, and treatment plans for neurodegenerative conditions (Ghassab *et al.*, 2024). This chapter explores the potential of probiotics in neurodegenerative diseases, focusing on their ability to influence inflammation and survival pathways.

15.2 PROBIOTICS: A NEW HOPE FOR GUT-RELATED PROBLEMS

Nutraceutical and therapeutic approaches are essential for improving the microbiome, which often deteriorates with age. Nutraceuticals like fiber-rich foods, indigestible carbohydrates, prebiotics, probiotics, and polyphenols, as well as therapeutic methods such as fecal transplants, help restore the gut microbiome's balance. Studies show that probiotic and prebiotic supplementation enhanced motor function, increased butyrate levels, boosted neurotrophic factors like BDNF, and reduced inflammatory cytokines in the hippocampus of middle-aged male rats (Romo-Araiza *et al.*, 2018). These therapies aim to improve biological outcomes and address gut imbalances in various diseases. Probiotics, biotin, Ginkgo biloba, and curcumin are examples of beneficial nutraceuticals. According to Abd El-Salam and El-Shibiny (2017), nutraceuticals offer medical benefits, including disease prevention. Research indicates that probiotics can improve brain chemistry, reduce symptoms of depression, and enhance neuronal plasticity (Figure 15.1) (Distrutti *et al.*, 2013; Koh *et al.*, 2016).

15.2.1 Role of Probiotics in Parkinson's Disease

Recent research suggests that lifestyle factors, including probiotics, may play a significant role in PD development. Probiotics, as part of nutraceutical interventions, may help address gut dysbiosis, enhance gastrointestinal functions, reduce neuronal inflammation in the enteric nervous system (ENS), and lower gut permeability. While the evidence

DOI: 10.1201/9781003536802-15

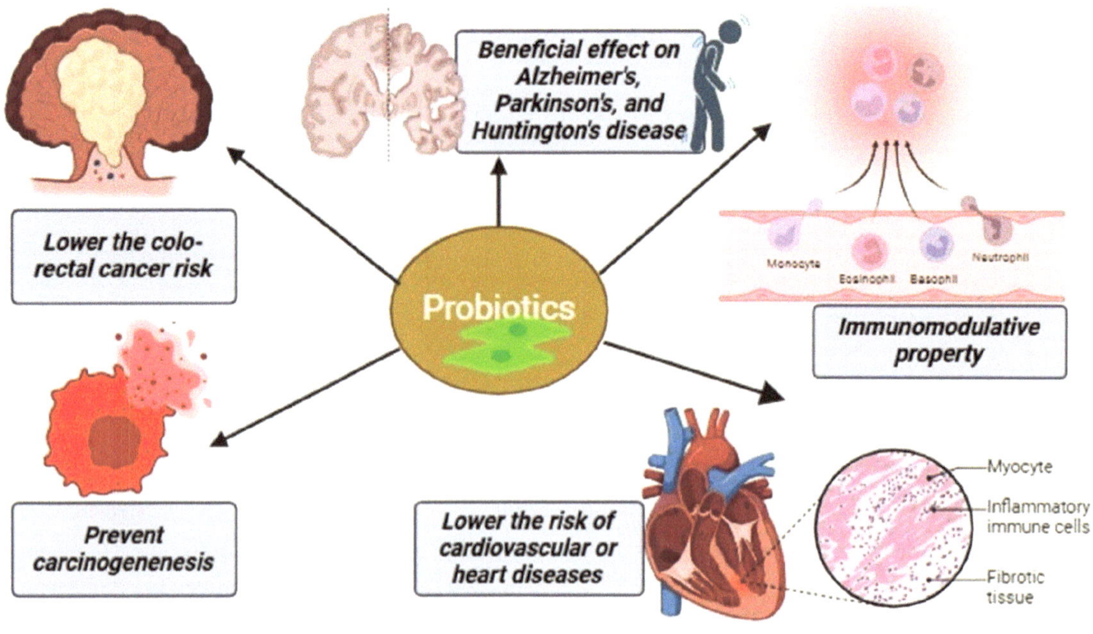

FIGURE 15.1 Role of probiotics in a variety of diseases and disorders.

remains limited, several studies support these benefits. For instance, a 2011 clinical trial showed that *Lactobacillus casei* Shirota-enriched fermented milk alleviated bloating and abdominal pain in PD patients with chronic constipation (Cassani *et al.*, 2011). Additionally, a probiotic tablet improved insulin sensitivity and motor scores in PD individuals (Tamtaji *et al.*, 2019). Another study showed that the liquid probiotic Symprove (K-1803) remained active in the lower gut (Gazerani, 2019). Research also reveals that *Bacillus subtilis*–based probiotics prevent α-synuclein accumulation in a *Caenorhabditis* worm model (Goya *et al.*, 2020). Probiotics have shown potential to reduce pro-inflammatory cytokines and oxidative stress in PD patients. Furthermore, probiotic formulations like SLAB51 and VSL#3 improve neuroprotection, regulate oxidative stress, and enhance cognitive function (Bonfili *et al.*, 2017; Distrutti *et al.*, 2014). SLAB51, in particular, may slow PD progression by restoring gut microbiota balance, indicating probiotics as a promising adjunct therapy for PD (Castelli *et al.*, 2021).

15.2.2 Possible Role of Probiotics in Alzheimer's Disease

Targeting the gut microbiota and its metabolites offers a potential strategy for treating microbiota-related diseases (Ikeda *et al.*, 2022). One approach involves manipulating pathways that drive disease progression, while another option is using probiotics. A meta-analysis by Den *et al.* (2020) suggests that probiotics can improve cognitive function in AD patients by reducing oxidative stress and inflammation. Similarly, Liu *et al.* (2022) reported that probiotics, which balance the gut microbiota, may enhance cognitive performance in AD patients. The gut microbiota produces short-chain fatty acids (SCFAs) like acetic, propionic, and butyric acids through fermentation, which can cross the blood–brain barrier and impact both the gut and brain (Kobayashi *et al.*, 2017). Probiotic formulations containing bifidobacteria and lactic acid bacteria, such as SLAB51, have shown promise in improving gut microbiota composition and boosting SCFA production (Liu *et al.*, 2022).

Recent advances in gene editing technologies like CRISPR-Cas have paved the way for engineered probiotics that could treat inflammatory diseases and deliver therapeutic biomolecules (Ikeda *et al.*, 2022). Additionally, synbiotics and postbiotics may provide therapeutic benefits for AD. Studies also suggest that Helicobacter pylori eradication in AD patients could reduce neurodegeneration (Kobayashi *et al.*, 2017). Moreover, cycloserine showed neuroprotective effects in animal models of AD (Den *et al.*, 2020). Lastly, antibiotics, such as in the ABX cocktail, may reduce Aβ deposition in AD models (Murai and Matsuda, 2023).

15.3 DISTINCT ROLES OF GUT–BRAIN AXIS, GUT MICROBIOTA, AND PROBIOTICS IN NEURODEGENERATION

The gastrointestinal tract (GIT) and brain function are linked by a multitude of interrelated hormonal and metabolic pathways, which underscore the substantial pharmaceutic potential of probiotics in the fight against neurodegeneration. The widespread and powerful anti-inflammatory actions of some bacteria may play an important role in their potential to prevent neurodegeneration. Innate immune cells that present antigens, like dendritic cells and macrophages, are found in the subepithelial lamina propria tissue of the gastrointestinal tract. These immune cells are positioned in close proximity to the gut microbiota, pathogens that invade, and antigens that cross the epithelial barrier to effectively communicate with the body's systemic immune system about external stimuli (Westfall *et al.*, 2017). These immune cells have NOD-like receptors (NLRs) and toll-like receptors (TLRs) that recognize microbe-associated molecular patterns (MAMPs) from bacteria and other microorganisms. These receptors then initiate signaling pathways that cause the release of cytokines that are either pro- or anti-inflammatory. This is critical since neurodegeneration, behavioral problems, and other neurological impairments are closely linked to neuroinflammation (Table 15.1) (Faden and Loane, 2015; Vivekanantham *et al.*, 2015).

15.4 RELATION OF GUT MICROBIOTA WITH NEURODEGENERATIVE DISORDERS

The gut microbiota plays a crucial role in brain health, with its imbalance linked to neurological diseases like AD and PD (Zhang *et al.*, 2022). Communication between the gut and brain occurs through various pathways, including the neuroimmune, neuroendocrine, and autonomic nervous systems (ANS), involving the vagus nerve, metabolites, neurotransmitters, and cell walls (Fedeli *et al.*, 2019). Gut microbiota produces neurotransmitters like tryptophan, GABA, BDNF, and SCFAs, all of which influence brain function and neurodegeneration (Canfora *et al.*, 2015). Neurotrophins, including BDNF, are vital for neuron survival and plasticity, with decreased BDNF levels linked to neurodegenerative diseases (Rieder *et al.*, 2017). SCFAs, produced by gut microbes such as *Eubacterium rectale* and *Faecalibacterium prausnitzii*, reduce inflammation and impact neuronal health by modulating serotonin and neuropeptides (Nagahara and Tuszynski, 2011). However, excessive SCFAs can negatively affect behavior and brain health. Additionally, tryptophan deficiency and GABA imbalances can exacerbate mood disorders and cognitive decline (Bell and Hardingham, 2011; Clarke *et al.*, 2013).

15.4.1 Alzheimer's Disease

Individuals who suffer from gastrointestinal issues are more likely to eventually develop AD. Obesity, diabetes, and inflammatory bowel disease are among the gastrointestinal problems and inflammatory disorders associated with disruptions in complex gut ecosystems. Dysbiosis of the gut microbiota influences metabolism and protein synthesis, which advances disorders such as AD. The microbial balance in the gut changes with age, favoring the proliferation of pro-inflammatory bacteria over anti-inflammatory ones, which impairs GI and blood–brain barrier permeability (Wu and Sun, 2015). Inflammation is associated with pro-inflammatory bacteria, including species like *Pseudomonas aeruginosa*, genera like *Escherichia/Shigella*, and members of the Proteobacteria and Verrucomicrobia phyla. However, species like Ruminococcaceae and *Clostridium* spp. have anti-inflammatory properties. Research indicates that elevated mRNA encoding in *Escherichia/Shigella* causes pro-inflammatory cytokines such NLRP3, CXCL2, and IL6 to be released (Bell and Hardingham, 2011). The presence of Helicobacter pylori in the stomach causes an increase in the serum amyloid β 40/42 ratio because it promotes the release of inflammatory mediators. Other bacteria, such as *Chlamydia pneumoniae* and *Borrelia burgdorferi*, are involved in the hyper phosphorylation of tau, a crucial indicator of AD. On the other hand, the ant-inflammatory taxon *Eubacterium rectale* is linked to the anti-inflammatory cytokine IL-10 (Kesika *et al.*, 2021).

15.4.2 Parkinson's Disease

A multifocal neurological illness, PD is typified by tremors, akinesia, sluggish movement, muscle rigidity, and abnormalities in gait. PD patients frequently have constipation in addition to these symptoms, which are combined with enhanced intestinal permeability and inflammation that directly interact with the small intestine' microbiota (Cattaneo *et al.*, 2017). In PD patients, *Helicobacter pylori* infection and small intestinal bacterial overgrowth are factors that lead to stool-related problems and motor impairment (Roubaud-Baudron *et al.*, 2012). Due to coliform bacteria, many patients also experience increased mucosal permeability and endotoxin exposure (Forsyth *et al.*, 2011). Furthermore, there is a decrease in the good bacteria in feces, including *Blautia faecis*, *Coprococcus eutactus*, *Roseburia intestinalis*, and *Roseburia hominis*, which interfere with the regulation of GABA and lipopolysaccharide production. The brain may be affected by the amyloidogenic proteins released by the Gram-negative bacterium *Escherichia coli* (Ojha *et al.*, 2023).

15.4.3 Huntington's Disease

The overexpression of the huntingtin gene results in the hereditary illness known as Huntington's disease (HD). According to recent study, an imbalance in the gut microbiota might significantly impact gut health by upsetting cytokine levels and causing an excessive hydrogen sulfide production. Bacteroidetes (4%) and Firmicutes (83%), which are more prevalent in HD, are out of balance within the bacterial phylum, leading to both motor impairment and mortality. In addition, abnormal gut microbiota lowers mucosal thickness, interferes with neuropeptide synthesis, and modifies levels of endocrine hormones. *Prevotella scopos* is linked to decreased butyrate generation, as shown with *Blautia producta,* and is associated with ATP levels (Kong *et al.*, 2021; Ojha *et al.*, 2023).

15.4.4 Other Neurological Disorders

Neurological disorders such as Friedreich ataxia, ALS, Lewy body disease, epilepsy, and spinal muscular atrophy involve severe neuronal damage. Epilepsy, for example, is influenced by genetic and environmental factors, with

TABLE 15.1
Effects of Probiotics in Various Diseases

Name of Probiotic	Effects	References
Alzheimer's disease		
Bifidobacterium breve	Inhibited cognitive impairments	Kobayashi *et al.*, 2017
Lactobacillus acidophilus *Lactobacillus fermentum* *Bifidobacterium lactis* *Bifidobacterium longum*	Enhanced memory loss disorder	Athari *et al.*, 2018
Lactobacillus plantarum *Bifidobacteria* and lactic acid bacteria *(SLAB51)*	Reduced amyloid aggregation	Murai and Matsuda, 2023
Lactobacillus casei *Bifidobacterium bifidum* Probiotic cocktail (*Bifidobacterium bifidum, Lactobacillus acidophilus, Lactobacillus casei,* and *Lactobacillus fermentum*)	Enhanced cognitive impairments	Akbari *et al.*, 2016
Microgravity-influenced neurodegenerative diseases		
Lactobacillus helveticus and *Bifdobacterium longum* (R0175)	Hippocampal apoptosis	Sil *et al.*, 2024
Lactobacillus plantarum (DR7) and *Lactobacillus fermentum* (DR9)	Ocular degeneration	
Lactobacillus paracasei PS23 (LPPS23)	Delayed senescence	
Bifdobacterium lactis, Lactobacillus casei, Bifidobacterium bifidum, and *Lactobacillus acidophilus*	Repaired memory loss, synaptic and neuronal injuries	
Neuroprotection		
Clostridium butyricum	Reduced apoptosis, ameliorated neurological impairments, and enhanced antioxidant capacity (superoxide dismutase and malondialdehyde levels decreased)	Sun *et al.*, 2016
Lactobacillus helveticus R0052	Probiotic treatment attenuated the hypothalamic pituitary adrenal (HPA) axis and anterior nervous system (ANS) activities in response to water avoidance stress or WAS	Ait-Belgnaoui *et al.*, 2014
Bacteroides longum R0175	Probiotics increased body weight gain, decreased cytokine expression, and shortened the duration of clinical symptoms	
Multiple sclerosis		
Bacteroides animalis	Probiotics increased body weight gain, decreased cytokine expression, and shortened the duration of clinical symptoms	Ezendam *et al.*, 2008
Bacteroides fragilis PSA	Treatment for prophylaxis postponed the onset of symptoms and lessened their intensity in experimental autoimmune encephalomyelitis (EAE). It also decreased the cytokines-mediated expression such as interleukin and IL-17, interferon gamma (IFN_γ) and retinoic acid-related orphan receptor gamma t ($ROR_\gamma t$), and therapeutic intervention lessened the severity of the disease	Ochoa-Reparaz *et al.*, 2010
Lactobacillus paracasei DSM 13434	Treatment with probiotics decreased neuroinflammation, stimulated T lymphocytes that are regulatory in mesenteric lymph nodes, and increased the expression of TGFβ1	Westfall *et al.*, 2017
Lactobacillus plantarum DSM 15312	Combined inhibitory effect on the disease's course and eliminated its clinical and histological symptoms	
Lactobacillus plantarum DSM 15313 *Lactobacillus casei* *Lactobacillus acidophilus* *Lactobacillus reuteri* *Bacteroides bifidum*	Development and postponed progression, suppressed the polarization of pro-inflammatory T-helper cells (Th1/Th17), and stimulated the production of IL-10 and/or regulatory T cells in forkhead box P3 (Foxp3 (+))	
Anxiety issues and memory deficits		
Lactobacillus helveticus R0052	Reduced inflammatory response and fecal corticosterone in WAS' mice on a western diet, and prevented anxiety-like behavior and memory deficit issues in mice with a pro-inflammatory state and western diet	Ohland *et al.*, 2013

(*Continued*)

TABLE 15.1 (CONTINUED)

Lactobacillus helveticus	Probiotics helped the diabetic animals' poor particular memory	Davari *et al.*, 2013
Lactobacillus rhamnosus	Restored long-term potentiation in the hippocampus and recovered decreased fundamental synaptic transmission	
Lactobacillus rhamnosus R0011	Probiotic medication on a daily basis reduced memory impairment caused by *Clostridium rodentium* infection	Gareau *et al.*, 2011
Lactobacillus helveticus R0052 *Lactobacillus casei* Shirota	Probiotic therapy considerably lessened feelings of anxiety	Rao *et al.*, 2009
Neurodegeneration		
Lactobacillus fermentum NCIMB 5221	Ferulic acid (FA), which possesses strong anti-AD properties, is a chemical that releases strains of *Lactobacillus fermentum*. FA decreased the production of Aβ fibrils, neuroinflammation, and restored the deficits in memory and learning in AD models	Yan *et al.*, 2013
Lactobacillus rhamnosus R0011	Preventive probiotic administration decreased the amygdala and dendritic gyrus's caspase-3 proapoptotic activity and Bax/Bcl-2 ratio	Girard *et al.*, 2009
Lactobacillus helveticus R0052	Similar regions showed an increase in Akt activation	
Lactobacillus helveticus R0052	Probiotic therapy reduced the activation of the ANS and HPA axis in response to WAS	Ait-Belgnaoui *et al.*, 2014
Bacteroides longum R0175	Prevented the reduction in hippocampus neurogenesis brought on by WAS and the alterations in the expression of genes related to synaptic plasticity in the hypothalamus	
Clostridium butyricum	Significantly lessened the histological and cognitive abnormalities, raised Bcl-2 and brain-derived growth factor levels, lowered Bax levels that supported the anti-apoptotic state, activated Akt phosphorylation, and decreased neuronal apoptosis	Liu *et al.*, 2015

TABLE 15.2
Role of Probiotics in Food Products (Sonal *et al.*, 2007; Mishra *et al.*, 2018)

Food Products	Strains	Functions
Milk based	*Bifidobacterium lactis*	Increases the concentration of lactic acid, probiotic numbers, acidification rate, and fermentation time
	Lactobacillus casei	Demonstrates good sensory and nutritional qualities, strengthens the immune system, decreases antibiotic-induced diarrhea, metabolizes fat, and prevents weight gain
Cereal based	Lyophilized probiotic cultures	Uses as an easy-to-use, direct delivery method for probiotics that are dried
	Barley β-glucan	Generates a potent bifidogenic impact
Fruit or vegetable based	*Lactobacillus johnsonii B-2178*	Used to assess fermented cashew and apple juice's probiotic properties
	Lactobacillus rhamnosus GG	Used to create a synbiotic fresh-cut apple wedge by combining prebiotics like inulin and oligofructose with probiotics like *Lactobacillus rhamnosus* GG. Enhances immunity, lessens diarrhea brought on by antibiotics, enhances digestive health, and supports the maintenance of a healthy vaginal flora
	Lactobacillus paraplantarum and one genotype of *Lactobacillus pentosus*	Purple carrots are fermented to make the traditional food "kanji"
Dried probiotics	Spray-dried probiotic bacteria	Utilized in the production of confections, pastries, and baby food
Beverage based		Beverages without alcohol that have been enhanced with vitamins A, C, and E or other useful components. Numerous beverages are available that decrease cholesterol (by combining soy and omega-3), are good for the eyes (including lutein) or the bones (containing inulin and calcium)

research linking it to gut microbiota dysbiosis. This imbalance heightens both pro- and anti-inflammatory responses, contributing to chronic inflammation and disease progression (Mar´c *et al.*, 2022; Gubert *et al.*, 2022). Disruption in lipid and glucose metabolism from certain bacterial overgrowth affects ATP-binding cassette transporters (Cai *et al.*, 2014). ALS, caused by genetic mutations in motor neurons, leads to misfolded proteins and declines in bulbar function, causing complications like weight loss and dysphagia. A leaky gut can trigger pro-inflammatory cytokines, altering the Firmicutes/Bacteroidetes ratio, and contribute to ALS progression (Chio *et al.*, 2009; Ojha *et al.*, 2023). Dysbiosis also plays a role in Lewy body disease progression (Table 15.2) (Rowin *et al.*, 2017).

15.5 RECENT EVIDENCES FOR PROBIOTICS AND INFLAMMATION-MEDIATED NEURODEGENERATIVE DISEASES

Neuroinflammation plays a crucial role in the development of neurodegenerative diseases, accelerating dopaminergic neuron loss. As a result, anti-inflammatory therapies are increasingly important in combating these conditions (Fu *et al.*, 2021). Probiotics like *E. coli*, *Bifidobacterium pseudocatenulatum*, and *Lactiplantibacillus plantarum* show promise in reducing neurodegenerative risks by acting as anti-inflammatory, antioxidant, or anti-pro-inflammatory agents (Eifler *et al.*, 2006). Dopamine and serotonin imbalances are associated with neurological and mental conditions. The VSL#3 probiotic, containing eight bacterial strains, interacts with hematopoietic stem cells to reduce inflammation and neurodegeneration (Pajares *et al.*, 2020). *Bifidobacterium longum* has been shown to improve cognitive function in mice, and a multi-strain probiotic enhanced cognitive benefits in adults (Silva *et al.*, 2020). A recent study involving *Bifidobacterium* strains demonstrated changes in gut microbiota composition, particularly in *Clostridiales* and Prevotellaceae (Kim *et al.*, 2021).

15.6 MECHANISM AND THERAPEUTIC EFFECTS OF PROBIOTICS IN THE PREVENTION OF NEUROLOGICAL DISEASES

Probiotics have garnered significant attention due to their health benefits, particularly in preventing and treating various ailments. While their exact mechanism of action is not fully understood, evidence suggests they impact the gut–brain axis, enhance gut barrier function, regulate immune systems, and prevent neurological issues (Lazarenko *et al.*, 2021). Studies in humans and animals show that probiotics, particularly *Bifidobacterium* and *Lactobacillus*, alleviate Alzheimer's disease (AD) symptoms by improving memory and cognitive dysfunction. In AD, glial cell accumulation and neuroinflammation lead to neuron damage through pro-inflammatory cytokines (Carabotti *et al.*, 2015). Research indicates probiotics can reduce chronic inflammation, abnormal protein buildup, and synaptic dysfunction, preventing neurodegeneration. Clinical trials on PD showed positive results with *Lactobacillus acidophilus*, *Bifidobacterium bifidum*, and others (Bachiller *et al.*, 2018). Furthermore, *Lacticaseibacillus rhamnosus* improved energy balance in rats (Figure 15.2) (Tytgat *et al.*, 2016).

15.7 PHARMACEUTIC ROLE OF PROBIOTICS IN THE GUT–BRAIN AXIS UNDER THE INFLUENCE OF MICROGRAVITY

Recent studies have highlighted the significant impact of probiotics on host immunity. In space, exposure to microgravity leads to changes in the neurological system, including hippocampal apoptosis, altered neuron structure, and cognitive impairments. These effects vary, with some being temporary and others persistent, potentially compromising astronaut health. As space missions grow in frequency and length, strategies are needed to prevent neurological damage. Probiotics have emerged as a potential therapeutic approach for addressing these microgravity-induced neurological disorders. They may help mitigate neurodegenerative conditions by modulating gut microbiota and reducing oxidative stress. The gut–brain axis, which connects the brain and gut, holds promise for treating mental health issues. Microgravity and cosmic radiation affect neuron survival and structure, leading to neuronal damage and cognitive decline. Probiotics, such as *Lactobacillus* strains,

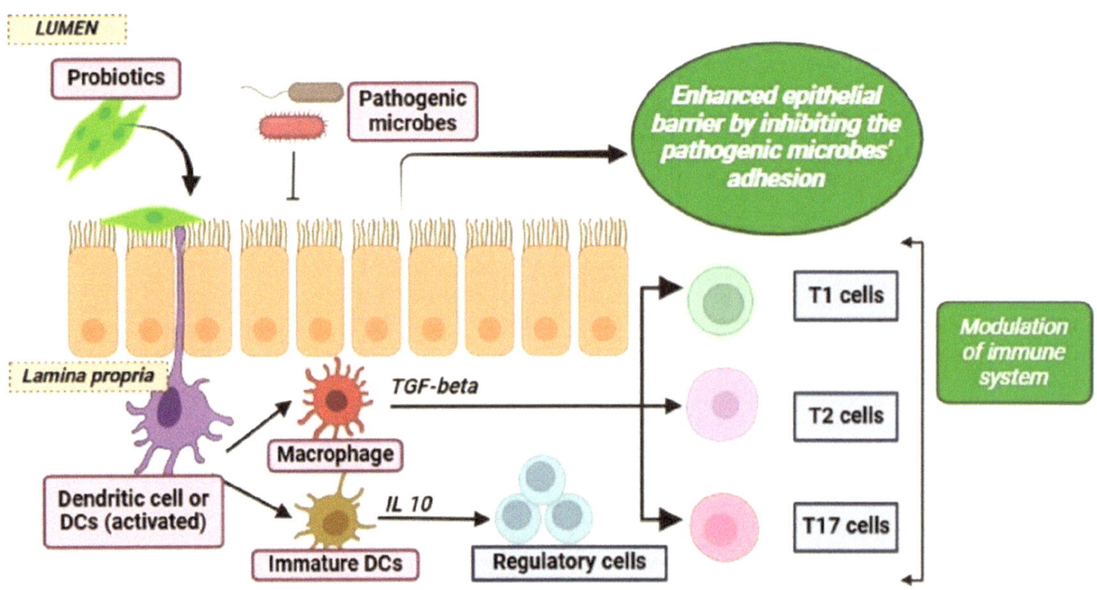

FIGURE 15.2 Mechanistic pathway of probiotics inside the body.

can influence brain gene expression and improve neurological health. The research underscores probiotics' potential in space missions and the need for further studies to aid astronaut recovery (Sil *et al.*, 2024).

15.8 CONCLUSION AND FUTURE ASPECTS

The effectiveness of probiotics in treating neurodegenerative diseases such as PD and AD is still under investigation and cannot be definitively confirmed. Some mentioned studies suggest that probiotics may improve gut microbiota, reduce inflammation, and enhance cognitive function. However, these findings are primarily based on animal models and small human trials, so further research is necessary to validate their safety and effectiveness. Probiotics should not be considered a stand-alone cure for neurodegenerative diseases. Instead, they should complement other treatments, including medications, lifestyle changes, and therapies recommended by health care professionals. Although the potential of probiotics like *Lactobacillus* and *Bifidobacterium* is promising, more extensive randomized controlled trials are needed to fully assess their benefits and limitations in treating neurodegenerative conditions.

REFERENCES

Abd El-Salam, M.H., El-Shibiny, S. (2017). Preparation, properties, and uses of enzymatic milk protein hydrolysates. *Crit Rev Food Sci Nutr.*, 57(6), 1119–1132. https://doi.org/10.1080/10408398.2014.899200

Ait-Belgnaoui, A., Colom, A., Braniste, V., *et al.* (2014). Probiotic gut effect prevents the chronic psychological stress-induced brain activity abnormality in mice. *Neurogastroenterol Motil.*, 26, 510–520. https://doi.org/10.1111/nmo.12295

Akbari, E., Asemi, Z., Daneshvar, K.R., Bahmani, F., Kouchaki, E., Tamtaji, O.R., Hamidi, G.A., Salami, M. (2016). Effect of probiotic supplementation on cognitive function and metabolic status in Alzheimer's disease: A randomized, double-blind and controlled trial. *Front. Aging Neurosci.*, 8, 256.

Athari Nik Azm, S., Djazayeri, A., Safa, M., Azami, K., Ahmadvand, B., Sabbaghziarani, F., Sharifzadeh, M., Vafa, M. (2018). Lactobacilli and bifidobacteria ameliorate memory and learning deficits and oxidative stress in β-amyloid (1–42) injected rats. *Appl. Physiol. Nutr. Metab.*, 43, 718–726.

Bachiller, S., Ferrer, I.J., Paulus, A., Yang, Y., Swanberg, M., Deierborg, T., Boza-Serrano, A. (2018). Microglia in neurological diseases: A road map to brain-disease dependent-inflammatory response. *Front. Cell. Neurosci.*, 12, 488.

Bell, K.F., Hardingham, G.E. (2011). The influence of synaptic activity on neuronal health. *Curr. Opin. Neurobiol.*, 21, 299–305.

Bonfili, L., Cecarini, V., Berardi, S., Scarpona, S., Suchodolski, J.S., Nasuti, C., Fiorini, D., Boarelli, M.C., Rossi, G., Eleuteri, A.M. (2017). Microbiota modulation counteracts Alzheimer's disease progression influencing neuronal proteolysis and gut hormones plasma levels. *Sci. Rep.*, 7, 2426.

Brown, R.C., Lockwood, A.H., Sonawane, B.R. (2005). Neurodegenerative diseases: An overview of environmental risk factors. *Environ. Health Perspect.*, 113, 1250–1256.

Cai, Z., Hussain, M.D., Yan, L.J. (2014). Microglia, neuroinflammation, and beta-amyloid protein in Alzheimer's disease. *Int. J. Neurosci.*, 124, 307–321.

Canfora, E.E., Jocken, J.W., Blaak, E.E. (2015). Short-chain fatty acids in control of body weight and insulin sensitivity. *Nat. Rev. Endocrinol.*, 11, 577–591.

Carabotti, M., Scirocco, A., Maselli, M.A., Severi, C. (2015). The gut-brain axis: Interactions between enteric microbiota, central and enteric nervous systems. *Ann. Gastroenterol.*, 28, 203–209.

Cassani, E., Privitera, G., Pezzoli, G., Pusani, C., Madio, C., Iorio, L., Barichella, M. (2011). Use of probiotics for the treatment of constipation in Parkinson's disease patients. *Minerva Gastroenterol Dietol.*, 57, 117–121.

Castelli, V., d'Angelo, M., Quintiliani, M., Benedetti, E., Cifone, M.G., Cimini, A. (2021). The emerging role of probiotics in neurodegenerative diseases: new hope for Parkinson's disease? *Neural Regen Res.*, 16(4), 628–634.

Cattaneo, A., Cattane, N., Galluzzi, S., Provasi, S., Lopizzo, N., Festari, C., Ferrari, C., Guerra, U.P., Paghera, B., Muscio, C., *et al.* (2017). Association of brain amyloidosis with pro-inflammatory gut bacterial taxa and peripheral inflammation markers in cognitively impaired elderly. *Neurobiol. Aging*, 49, 60–68.

Chio, A., Logroscino, G., Hardiman, O., Swingler, R., Mitchell, D., Beghi, E., Traynor, B. (2009). On behalf of the eurals consortium. Prognostic factors in ALS: A critical review. *ALS*, 10, 310–323.

Clarke, G., Grenham, S., Scully, P., Fitzgerald, P., Moloney, R.D., Shanahan, F., Dinan, T.G., Cryan, J.F. (2013). The microbiome-gut-brain axis during early life regulates the hippocampal serotonergic system in a sex-dependent manner. *Mol. Psychiatry*, 18, 666–673.

Cortese, Y.J., Wagner, V.E., Tierney, M., Devine, D., & Fogarty, A. (2018). Review of catheter-associated urinary tract infections and in vitro urinary tract models. *J. Healthc. Eng.*, 2018, 2986742.

Davari, S., Talaei, S.A., Alaei, H., Salami, M. (2013). Probiotics treatment improves diabetes-induced impairment of synaptic activity and cognitive function: Behavioral and electro-physiological proofs for microbiome–gut–brain axis. *Neuroscience*, 240, 287–296. https://doi.org/10.1016/j.neuroscience.

Den, H., Dong, X., Chen, M., Zou, Z. (2020). Efficacy of probiotics on cognition, and biomarkers of inflammation and oxidative stress in adults with Alzheimer's disease or mild cognitive impairment—A meta-analysis of randomized controlled trials. *Aging*, 12, 4010–4039.

Distrutti, E., Cipriani, S., Mencarelli, A., Renga, B., Fiorucci, S. (2013). Probiotics VSL#3 Protect against development of visceral pain in murine model of irritable bowel syndrome. *PLoS One*, 8, e63893.

Distrutti, E., O'Reilly, J-A., McDonald, C., Cipriani, S., Renga, B., Lynch, M.A., Fiorucci, S. (2014). Modulation of intestinal microbiota by the probiotic VSL#3 resets brain gene expression and ameliorates the age-related deficit in LTP. *PLoS One*, 9, e106503.

Drago, L.O., De Vecchi, E. (2009). Should Lactobacillus sporogenes and Bacillus coagulans have a future? *J. Chemother.*, 21, 371–377.

Eifler, N., Vetsch, M., Gregorini, M., Ringler, P., Chami, M., Philippsen, A., Fritz, A., Müller, S.A., Glockshuber, R., Engel, A., *et al.* (2006). Cytotoxin ClyA from Escherichia coli assembles to a 13-meric pore independent of its redox-state. *EMBO J.*, 25, 2652–2661.

Ezendam, J., de Klerk, A., Gremmer, E.R., van Loveren, H. (2008). Effects of Bifidobacterium animalis administered during lactation on allergic and autoimmune responses in rodents. *Clin Exp Immunol*, 154, 424–431. https://doi.org/10.1111/j.1365-2249.2008.03788.x

Faden, A.I., Loane, D.J. (2015). Chronic neurodegeneration after traumatic brain injury: Alzheimer disease, chronic traumatic encephalopathy, or persistent neuroinflammation? *Neurotherapeutics*, 12, 143–150. https://doi.org/10.1007/s13311-014-0319-5

Fedeli, C., Filadi, R., Rossi, A., Mammucari, C., Pizzo, P. (2019). PSEN2 (presenilin 2) mutants linked to familial Alzheimer disease impair autophagy by altering Ca2+ homeostasis. *Autophagy*, 15, 2044–2062.

Forsyth, C.B., Shannon, K.M., Kordower, J.H., Voigt, R.M., Shaikh, M., Jaglin, J.A., Estes, J.D., Dodiya, H.B., Keshavarzian, A. (2011). Increased intestinal permeability correlates with sigmoid mucosa alpha-synuclein staining and endotoxin exposure markers in early Parkinson's disease. *PLoS One*, 6, e28032.

Fu, J., Wang, T., Xiao, X., Cheng, Y., Wang, F., Jin, M., Wang, Y., Zong, X. (2021). Clostridium butyricum ZJU-F1 benefits the intestinal barrier function and immune response associated with its modulation of gut microbiota in weaned piglets. *Cells*, 10, 527.

Gareau, M.G., Wine, E., Rodrigues, D.M., *et al.* (2011). Bacterial infection causes stress-induced memory dysfunction in mice. *Gut.*, 60, 307–317. https://doi.org/10.1136/gut.2009.202515

Gazerani, P. (2019). Probiotics for Parkinson's disease. *Int. J. Mol. Sci.*, 20, 4121.

Ghassab, F.T., Mahmoudi, F.S., Tinjani, R.T., Meibodi, A.E., Zali, M.R., Yadegar, A. (2024). Probiotics and the microbiota-gut-brain axis in neurodegeneration: Beneficial effects and mechanistic insights. *Life Sci.*, 350, 122748. https://doi.org/10.1016/j.lfs.2024.122748.

Girard, S-A., Bah, T.M., Kaloustian, S., *et al.* (2009). Lactobacillus helveticus and Bifidobacterium longum taken in combination reduce the apoptosis propensity in the limbic system after myocardial infarction in a rat model. *Br. J. Nutr.*, 102, 1420–1425. https://doi.org/10.1017/S0007114509990766.

Goya, M.E., Xue, F., Sampedro-Torres-Quevedo, C., Arnaouteli, S., Riquelme-Dominguez, L., Romanowski, A., Brydon, J., Ball, K.L., Stanley-Wall, N.R., Doitsidou, M. (2020). Probiotic Bacillus subtilis protects against α-synuclein aggregation in C. elegans. *Cell Rep.*, 30, 367–380.

Gubert, C., Choo, J.M., Love, C.J., Kodikara, S., Masson, B.A., Liew, J.J.M., Wang, Y., Kong, G., Narayana, V.K., Renoir, T. (2022). Faecal microbiota transplant ameliorates gut dysbiosis and cognitive deficits in Huntington's disease mice. *Brain Commun.*, 4, fcac205.

Hill, C., Guarner, F., Reid, G., Gibson, G.R., Merenstein, D.J., Pot, B., Morelli, L., Canani, R.B., Flint, H.J., Salminen, S., *et al.* (2014). The international scientific association for probiotics and prebiotics consensus statement on the scope and appropriate use of the term probiotic. *Nat. Rev. Gastroenterol. Hepatol.*, 11, 506–514.

Ikeda, Y., Taniguchi, K., Yoshikawa, S., Sawamura, H., Tsuji, A., Matsuda, S. (2022). A budding concept with certain microbiota, anti-proliferative family proteins, and engram theory for the innovative treatment of colon cancer. *Explor. Med.*, 3, 468–478.

Jellinger, K.A. (2010). Basic mechanisms of neurodegeneration: A critical update. *J. Cell. Mol. Med.*, 14, 457–487.

Kechagia, M., Basoulis, D., Konstantopoulou, S., Dimitriadi, D., Gyftopoulou, K., Skarmoutsou, N., Fakiri, E.M. (2013). Health benefits of probiotics: A review. *ISRN Nutr.*, 2013, 481651.

Kesika, P., Suganthy, N., Sivamaruthi, B.S., Chaiyasut, C. (2021). Role of gut-brain axis, gut microbial composition, and probiotic intervention in Alzheimer's disease. *Life Sci.*, 264, 118627.

Kim, C.-S., Cha, L., Sim, M., Jung, S., Chun, W.Y., Baik, H.W., Shin, D.-M. (2021). Probiotic supplementation improves cognitive function and mood with changes in gut microbiota in community-dwelling older adults: A randomized, double-blind, placebo-controlled, multicenter trial. *J. Gerontol. Ser. A*, 76, 32–40.

Kobayashi, Y., Sugahara, H., Shimada, K., Mitsuyama, E., Kuhara, T., Yasuoka, A., Kondo, T., Abe, K., Xiao, J.Z. (2017). Therapeutic potential of Bifidobacterium breve strain A1 for preventing cognitive impairment in Alzheimer's disease. *Sci. Rep.*, 7, 13510.

Koh, A., De Vadder, F., Kovatcheva-Datchary, P., Bäckhed, F. (2016). From dietary fiber to host physiology: Short-chain fatty acids as key bacterial metabolites. *Cell*, 165, 1332–1345.

Kong, G., Ellul, S., Narayana, V.K., Kanojia, K., Ha, H.T., Li, S., Renoir, T., Le Cao, K.A., Hannan, A.J. (2021). An integrated metagenomics and metabolomics approach implicates the microbiota-gut-brain axis in the pathogenesis of Huntington's disease. *Neurobiol. Dis.*, 148, 105199.

Lazarenko, L.M., Babenko, L.P., Gichka, S.G., Sakhno, L.O., Demchenko, O.M., Bubnov, R.V., Sichel, L.M., Spivak, M.Y. (2021). Assessment of the safety of lactobacillus casei IMV B-7280 probiotic strain on a mouse model. *Probiotics Antimicrob. Proteins*, 13, 1644–1657. 135.

Liu, C., Guo, X., Chang, X. (2022). Intestinal flora balance therapy based on probiotic support improves cognitive function and symptoms in patients with Alzheimer's disease: A systematic review and meta-analysis. *Biomed. Res. Int.*, 2022, 4806163.

Liu, J., Sun, J., Wang, F., *et al.* (2015). Neuroprotective effects of Clostridium butyricum against vascular dementia in mice via metabolic butyrate. *Biomed Res. Int.*, 412946. https://doi.org/10.1155/ 2015/412946

Marć, M.A., Jastrząb, R., Mytych, J. (2022). Does the gut microbial metabolome really matter? The connection between GUT Metabolome and neurological disorders. *Nutrients*, 14, 3967.

Mishra, S., Behera, P., Kar, B., Ray, R. (2018). Advances in probiotics, prebiotics and nutraceuticals. In Probiotic and Prebiotic Foods: Technology, Composition and Health Benefits. Springer. https://doi.org/10.1007/978-3-319-74820-7_7.

Murai, T., Matsuda, S. (2023). Therapeutic implications of probiotics in the gut microbe-modulated neuroinflammation and progression of Alzheimer's disease. *Life*, 13, 1466. https://doi.org/10.3390/ life13071466.

Nagahara, A.H., Tuszynski, M.H. (2011). Potential therapeutic uses of BDNF in neurological and psychiatric disorders. *Nat. Rev. Drug Discov.*, 10, 209–219.

Ochoa-Reparaz, J., Mielcarz, D.W., Wang, Y., *et al.* (2010). A polysaccharide from the human commensal Bacteroides fragilis protects against CNS demyelinating disease. *Mucosal Immunol.*, 3, 487–495. https://doi.org/10.1038/mi.2010.29

Ohland, C.L., Kish, L., Bell, H., *et al.* (2013). Effects of Lactobacillus helveticus on murine behavior are dependent on diet and genotype and correlate with alterations in the gut microbiome. *Psychoneuroendocrinology*, 38, 1738–1747. https://doi.org/10.1016/j. psyneuen.2013.02.008

Ojha, S., Patil, N., Jain, M., Kole, C., Kaushik, P. (2023). Probiotics for neurodegenerative diseases: A systemic review. *Microorganisms*, 11, 1083. https://doi.org/ 10.3390/ microorganisms11041083.

Pajares, M., IRojo, A., Manda, G., Boscá, L., Cuadrado, A. (2020). Inflammation in Parkinson's disease: Mechanisms and therapeutic implications. *Cells*, 9, 1687.

Rao, A.V., Bested, A.C., Beaulne, T.M., *et al.* (2009). A randomized, double-blind, placebo-controlled pilot study of a probiotic in emotional symptoms of chronic fatigue syndrome. *Gut. Pathog.*, 1, 6. https://doi.org/10.1186/1757-4749-1-6

Rieder, R., Wisniewski, P.J., Alderman, B.L., Campbell, S.C. (2017). Microbes and mental health: A review. *Brain Behav. Immun.*, 66, 9–17.

Romo-Araiza, A., Gutiérrez-Salmeán, G., Galván, E.J., Hernández-Frausto, M., Herrera-López, G., Romo-Parra, H., García-Contreras, V., Fernández-Presas, A.M., Jasso-Chávez, R., Borlongan, C.V., Ibarra, A. (2018). Probiotics and prebiotics as a therapeutic strategy to improve memory in a model of middle-aged rats. *Front. Aging. Neurosci.*, 10, 416.

Roubaud-Baudron, C., Krolak-Salmon, P., Quadrio, I., Mégraud, F., Salles, N. (2012). Impact of chronic Helicobacter pylori infection on Alzheimer's disease: Preliminary results. *Neurobiol. Aging*, 33, 1009.e11–1009.e19.

Rowin, J., Xia, Y., Jung, B., Sun, J. (2017). Gut inflammation and dysbiosis in human motor neuron disease. *Physiol. Rep.*, 5, e13443.

Sil, M., Mitra, D., Mukherjee, N., *et al.* (2024). Therapeutic role of probiotics in gut-brain axis under microgravity. *Indian J. Microbiol.* https://doi.org/10.1007/s12088-024-01314-6

Silva, D.R., Sardi, J.D., de Souza Pitangui, N., Roque, S.M., da Silva, A.C., Rosalen, P.L. (2020). Probiotics as an alternative antimicrobial therapy: Current reality and future directions. *J. Funct. Foods*, 73, 104080.

Silva, Y.P., Bernardi, A., Frozza, R.L. (2020). The role of short-chain fatty acids from gut microbiota in gut-brain communication. *Front. Endocrinol.*, 11, 25.

Sonal, S., Suja, A., Lima, T., Aneesh, T. (2007). Probiotics: Friendly microbes for better health. *Internet. J. Nutr. Wellness*, 6(2), 1–6.

Sun, J., Ling, Z., Wang, F., *et al.* (2016). Clostridium butyricum pretreatment attenuates cerebral ischemia/reperfusion injury in mice via anti-oxidation and anti-apoptosis. *Neurosci Lett.*, 613, 30–35. https://doi.org/10.1016/j.neulet.2015.12.047

Sun, P., Su, L., Zhu, H., Li, X., Guo, Y., Du, X., Zhang, L., Qin, C. (2021). Gut microbiota regulation and their implication in the development of neurodegenerative disease. *Microorganisms*, 9, 2281.

Tamtaji, O.R., Taghizadeh, M., Daneshvar Kakhaki, R., Kouchaki, E., Bahmani, F., Borzabadi, S., Oryan, S., Mafi, A., Asemi, Z. (2019). Clinical and metabolic response to probiotic administration in people with Parkinson's disease: A randomized, double-blind, placebo-controlled trial. *Clin. Nutr.*, 38, 1031–1035.

Tytgat, H.L., Douillard, F.P., Reunanen, J., Rasinkangas, P., Hendrickx, A.P., Laine, P.K., Paulin, L., Satokari, R., de Vos, W.M. (2016). Lactobacillus rhamnosus GG outcompetes enterococcus faecium via Mu-cus-Binding Pili: Evidence for a novel and heterospecific probiotic mechanism. *Appl. Environ. Microbiol.*, 82, 5756–5762.

Valdes, A., Walter, J., Segal, E., Spector, T.D. (2018). Role of the gut microbiota in nutrition and health. *BMJ*, 361, k2179.

Vivekanantham, S., Shah, S., Dewji, R., *et al.* (2015). Neuroinflammation in Parkinson's disease: Role in neurodegeneration and tissue repair. *Int. J. Neurosci.*, 125, 717–725. https://doi.org/10.3109/00207454.2014.982795

Westfall, S., Lomis, N., Kahouli, I., Dia, S.Y., Singh, S.P., Prakash, S. (2017). Microbiome, probiotics and neurodegenerative diseases: Deciphering the gut brain axis. *Cell Mol Life Sci.*, *74*(20), 3769–3787. https://doi.org/10.1007/s00018-017-2550-9.

Wu, C., Sun, D. (2015). GABA receptors in brain development, function, and injury. *Metab. Brain Dis.*, 30, 367–379.

Yan, M.H., Wang, X., Zhu, X. (2013). Mitochondrial defects and oxidative stress in Alzheimer disease and Parkinson disease. *Free Radic. Biol. Med.*, 62, 90–101. https://doi.org/10.1016/j.freeradbiomed.2012.11.014

Zhang, H., Chen, Y., Wang, Z., Xie, G., Liu, M., Yuan, B., Chai, H., Wang, W., Cheng, P. (2022). Implications of gut microbiota in neurodegenerative diseases. *Front. Immunol.*, 13, 325.

16 Correlation between Gut Microbiome and Malnutrition and Other Factors in Neurological Disorders

Manikandan Gurusamy

Department of Pharmaceutical Sciences, Tshwane University of Technology, Arcadia Campus, Pretoria, South Africa

16.1 INTRODUCTION

16.1.1 Overview of the Gut Microbiome

Millions of microorganisms representing hundreds of different species of bacteria, fungus, viruses, and protozoa make up the enormous and complex microbial community that inhabits the human gut (Valdes et al., 2018). These microbes are concentrated in the colon, which acts as a hub for them (Valdes et al., 2018). Notably, more than 35,000 different types of bacteria are thought to be present in a healthy gut. Studies have demonstrated a beneficial relationship between gut bacterial diversity and abundance and both general health and illness prevention (Patel, 2021).

It is becoming more well acknowledged that the gut microbiota is an essential organ that creates a sophisticated, reciprocal communication network with other organs. Numerous pathways, including neurological, endocrine, humoral, immunological, and metabolic signals, mediate the complex host–microbe interactions that are facilitated by this gut–brain axis. A vital nexus that is essential to the host's defense against harmful infections is the human gut microbiota (Afzaal et al., 2022).

16.1.2 Significance of the Gut Microbiome in Health

Gut microbiota is considered the most important to maintain our health (Shreiner et al., 2015). Gut bacteria perform many functions such as food fermentation, protection against pathogens, stimulation of the immune response, and production of vitamins (Hillman et al., 2017). Research on the human microbiota has progressed beyond compositional studies and member association explorations due to the microbiota's significant influence on human health, affecting many biological processes and disease development. Notably, the development of advanced high-throughput methodologies, microbiota interaction modeling, and simulation methods has led to a notable change in focus toward clarifying the functional origins of the microbiota (Hou et al., 2022).

Since its discovery, the importance of the microbiota in preserving health and averting disease has been widely documented (Hou et al., 2022). Numerous tasks vital to the host's general health are carried out by the gut microbiota, a varied collection of bacteria (Bander et al., 2020). Numerous microorganisms, such as bacteria, viruses, archaea, and unicellular eukaryotes, live inside the human body. Although microorganisms are found on every surface of the human body, the largest numbers are found in the gut. A complex ecological community known as the gut microbiota is made up of thousands of different microbial species that live in the human gut (Lagier et al., 2016). Interestingly, the human gut microbiome has 150 times as many genes as the human genome. The human body is thought to be home to about 100 trillion microorganisms, which are essential to many biological functions, including both health and illness (Wang et al., 2017). The idea that the gut microbiota has a major impact on immunomodulation, weight control, energy homeostasis, and disorders linked to obesity has been supported by numerous research (Piccioni et al., 2022).

16.1.3 Definition and Scope of Malnutrition

Malnutrition is a condition in which food/energy intake is imbalanced in a person due to insufficient or excessive intake of nutrients. Malnutrition continues to threaten the lives of millions of people worldwide, including children who are affected. Although limited access to food and infectious diseases are the main causes of malnutrition in childhood, the gut microbiota also plays a role (Iddrisu et al., 2021). Malnutrition, also known as undernutrition, is a state of insufficient nutrition that extends from under- to over-nutrition (World Health Organization, 2009; Williams & Suchdev, 2017).

Moreover, half of all deaths in children under 5 are caused by malnutrition, making it the leading cause of

DOI: 10.1201/9781003536802-16

death in this age group. A 2021 report by the World Health Organization (WHO), the World Bank Group, and UNICEF estimates that 149.2 million children under 5 globally are malnourished. Specifically, 45.4 million children are wasted, while 38.9 million are overweight. Africa accounts for 39% of all cases, while Asia accounts for 55%. More than 45% of all deaths in children under 5 are caused by undernutrition. Malnutrition has wide-ranging effects on people and governments around the world, having a major influence on health, cultural, economic, and developmental results (Malnutrition, 2022).

Since malnutrition in children is considered a multifaceted problem, there are many nutritional, infectious, biological, cultural, and psychosocial factors (Seferidi et al., 2022). Each of these malnutrition-related problems must be addressed to reduce overall morbidity and mortality (Chawla et al., 2022).

16.1.4 Neurological Disorders: An Introduction

Over 1 billion people worldwide suffer from neuropsychiatric diseases, which have reached concerning proportions (Chisholm et al., 2016). In particular, the prevalence of anxiety and depression has increased dramatically, affecting roughly 10% of the world's population at this time. Additionally, a startling 30% of the global health burden related to noncommunicable diseases is caused by mental problems (Chisholm et al., 2016). Over the past 15 years, the frequency of autism in children has significantly increased, making it another neurological illness of considerable concern (Ding et al., 2017). Furthermore, neurodegenerative conditions like Alzheimer's disease need to be recognized. Furthermore, it is impossible to ignore eating disorders like anorexia nervosa, which frequently impacts young people (Jayatilleke et al., 2017).

This chapter focused on the gut microbiome: composition and function interplay between gut microbiome and malnutrition, gut microbiome, neurological disorders, nutritional interventions and gut microbiome modulation, other factors influencing the gut microbiome, neurological disorders, and integrative approaches in managing neurological disorders.

16.2 THE GUT MICROBIOME: COMPOSITION AND FUNCTION

The varied community of bacteria that live in the human digestive system is referred to as the gut microbiota. In addition to prokaryotic species, this dynamic ecosystem also includes viruses, fungi, parasites, and archaea. The gut microbiome is the aggregate term for the genetic and functional traits of these microbes (Ursell et al., 2012). Remarkably, people may be roughly categorized into different enterotypes, which are defined by comparable gut microbial assemblages in particular populations (Mobeen et al., 2018). Notably, compared to the 23,000 genes found in the human genome, the 3.3 million genes found in the human gut microbiota are astounding (Qin et al., 2010). The significant influence that the gut microbiota can have on human health is demonstrated by this discrepancy.

Numerous microorganisms, including as bacteria, yeast, and viruses, make up the varied ecosystem known as the gut microbiota. A hierarchical taxonomy comprising phylum, classes, orders, families, genera, and species is used to categorize bacteria in particular. Of the more than 160 species of gut microbiota that have been found so far, only a small number have undergone thorough characterization (Laterza et al., 2016). Firmicutes, Bacteroidetes, Actinobacteria, Proteobacteria, Fusobacteria, and Verrucomicrobia are the six phyla that make up the majority of the gut microbiota. Remarkably, the phyla Firmicutes and Bacteroidetes make up about 90% of the gut microbiota. Firmicutes is further subdivided into more than 200 genera, which together make up 95% of the Firmicutes phylum. These genera include *Lactobacillus, Clostridium, Enterococcus*, and *Ruminicoccus* (Arumugam et al., 2011). Bacteroidetes, on the other hand, are dominated by the genera *Bacteroides* and *Prevotella*. The Actinobacteria phylum, although less abundant, is primarily represented by the *Bifidobacterium* genus (Arumugam et al., 2011).

The breakdown of ingested carbohydrates and host-derived glycans into short-chain fatty acids is one of the gut microbiota's main roles. These fatty acids are essential for the host's energy needs and are also essential for controlling a number of physiological processes. The idea of a "core microbiome," which includes the fundamental roles of a person's microbial population, was developed because of research into the microbiome. The members of the gut microbiota cooperate with the host to create a microbial ecosystem that is both harmonious and productive (Blaut, 2018).

16.2.1 Key Functional Roles of the Gut Microbiome

According to Musso et al. (2010), intestinal bacteria have the enzymatic capacity to break down complex carbohydrates and produce high-energy metabolites like short-chain fatty acids. The stomach mucosa produces the three main short-chain fatty acids: acetate, propionate, and butyrate (Louis et al., 2014). It is noteworthy that the phylum-specific production of these short-chain fatty acids is dominated by Firmicutes, while Bacteroidetes produce the majority of propionate (Louis et al., 2017; Macfarlane & Macfarlane, 2003; Morrison & Preston, 2016). In particular, *Akkermansia muciniphila*, which specializes in mucin breakdown, is a significant propionate generator (Derrien, 2004). Acetate is discharged into peripheral tissues after synthesis, while propionate is mostly absorbed by the liver (Guarner et al., 2003).

Short-chain fatty acids have a major effect on human metabolism, according to recent reviews (Morrison &

Preston, 2016; Lin & Zhang, 2017). Strong anti-inflammatory and anticancer effects have been identified for butyrate in particular (Morrison & Preston, 2016; Lin & Zhang, 2017). Additionally, colonocytes use butyrate as their main energy source (Corrêa-Oliveira et al., 2016). By encouraging colonocyte proliferation at the base of the crypts and causing apoptosis and exfoliation of cells close to the lumen, the gradient of butyrate concentration, which decreases from the lumen to the crypt, has been suggested to control intestinal epithelial homeostasis (Donohoe et al., 2012). Furthermore, by regulating collagen synthesis and mucin production, butyrate has been demonstrated to have positive effects on the intestinal epithelium, including decreased bacterial growth and enhanced barrier function (Morrison & Preston, 2016).

According to Morrison and Preston (2016), short-chain fatty acids are essential for controlling the inflammatory response and immune system. According to Corrêa-Oliveira et al. (2016), they alter the synthesis of cytokines like interleukin-18 (IL-18), which is necessary for preserving and mending epithelial integrity. Notably, butyrate and propionate alter gene expression epigenetically by acting as inhibitors of histone deacetylase (Morrison & Preston, 2016; Lin & Zhang, 2017). Additionally, it has been demonstrated that short-chain fatty acids affect energy consumption and appetite regulation via receptor-mediated pathways (Chambers et al., 2015). Propionate has been shown to suppress reward-based eating behavior through striatal pathways (Byrne et al., 2016) and to have positive effects on β-cell function (Pingitore et al., 2017). Microbial metabolites aside from short-chain fatty acids were pronounced to have an effect on intestinal barrier functions, epithelium proliferation, and the immune gadget (Nagai et al., 2016).

Essential vitamins that the host cannot generate on its own are resynthesized by the gut bacteria (LeBlanc et al., 2013). For example, the synthesis of vitamin B12, which is not generated by plants, animals, or fungi, depends on lactic acid bacteria (LeBlanc et al., 2013; Martens et al., 2002). Contrarily, Bifidobacteria are important providers of folate, a vitamin that is essential for several metabolic functions, including as DNA synthesis and repair (Pompei et al., 2007).

16.2.2 Factors Influencing Gut Microbiome Composition

A vast ecology of microorganisms called the gut microbiota live in the gut. Animal models have been used in research to show that a number of factors can have a major impact on the gut microbiota's makeup:

- Genetics
- Mode of delivery at birth
- Method of infant feeding
- Antibiotics
- Diet

Changes in the makeup of gut microbes can have a significant impact. Notably, gut inflammation and subsequent low-grade inflammation have been connected to a specific gut microbiota signature, which may play a role in the onset of type 2 diabetes (Wen & Duffy, 2017).

16.2.2.1 Genetics

The genetic composition of the host influences the abundance of particular bacteria in the gut microbiota, which in turn impacts metabolism and general health (Goodrich et al., 2014). According to research, family members typically have microbiota communities that are more comparable than those of unrelated people. Additionally, research on twins has shown that monozygotic twins have a more comparable gut microbiome than dizygotic twins (Goodrich et al., 2014). Despite these discoveries, genome-wide research has not yet pinpointed the precise genes and pathways that dictate the gut microbiome's makeup (Blekhman et al., 2015). Nonetheless, a connection between host genetics and microbiota composition has been shown by the identification of certain genes linked to intestinal inflammation (Thompson-Chagoyán et al., 2005; Rehman et al., 2011).

16.2.2.2 Mode of Delivery at Birth

Studies have repeatedly demonstrated that infants born vaginally and those born via cesarean section have different gut bacterial community compositions (Dominguez-Bello et al., 2016; Tamburini et al., 2016). According to Kulas et al. (2013), infants born vaginally are exposed to their mother's germs during birth, which influences their gut microbiota and promotes the growth of their immune system. Research has also looked into how the technique of birth affects children's health over the long run. According to a cross-sectional research of 8,900 preschoolers, children born by caesarean section were 25% more likely to be obese and 35% more likely to be overweight (Rutayisire et al., 2016). Likewise, a smaller cohort research of 672 preschoolers revealed that the prevalence of overweight and obesity among those born via caesarean section was 15.6% and 12.9%, respectively (Portela & TO, 2015). According to the current study, babies born by caesarean section are more likely to acquire diabetes and obesity than babies born vaginally, which is in line with these findings.

16.2.2.3 Method of Infant Feeding

A baby's gut microbiota is greatly influenced by the commensal and maybe probiotic bacteria found in human breast milk (Boix-Amorós et al., 2016). Breast milk offers the growing newborn a varied microbial environment with approximately 700 distinct species of bacteria. Streptococci and Staphylococci are the most often detected bacterial species in breast milk (Fitzstevens et al., 2017). These bacteria are also found in the skin microbiome, suggesting that skin bacteria may be present in breast milk. According to studies, the most common bacteria found in infant colostrum samples are *Weissella*, *Leuconostoc*, *Staphylococcus*, *Streptococcus*, and *Lactococcus* (Boix-Amorós et al., 2016).

Nevertheless, studies have also revealed that the bacterial composition of the milk from obese moms is different, with less variety and more pro-inflammatory qualities (Panagos et al., 2016). Additionally, it has been discovered that the bacteria in the breast milk of moms who had caesarean sections differ from those of mothers who gave birth spontaneously (Cabrera-Rubio et al., 2012). One of the earliest germs to enter an infant's body are the bacteria found in breast milk and the mother's skin. These bacteria may have a major impact on the infant's health outcomes (Cabrera-Rubio et al., 2012). Furthermore, IgA antibodies, which offer defense against a variety of infections, are essential for breast milk (Hanson et al., 1984).

16.2.2.4 Antibiotics

Numerous lives have been saved by antibiotics, which have completely changed the way infections are treated. But the delicate equilibrium of the gut microbiota is also significantly impacted by these life-saving drugs. Antibiotics have quick and perhaps permanent impacts on the gut microbiome. Specifically, broad-spectrum antibiotics change the gut microbiota's makeup by decreasing the number of beneficial bacteria, increasing the number of opportunistic pathogens, and decreasing bacterial diversity (Modi et al., 2014). It has been discovered that the composition of the gut microbiota is permanently altered when broad-spectrum antibiotics, like clindamycin, are administered to newborns and young children (Gough et al., 2014). According to Shaw et al. (2010), early antibiotic exposure in neonates might upset the gut microbiota's equilibrium and cause microbial dysbiosis, which raises the risk of inflammatory bowel disease in later life.

Since female mice tend to gain less weight on a high-fat diet, the majority of mouse studies looking at obesity caused by a high-fat diet have solely utilized male mice. It is important to note, though, that obesity in people does not significantly differ by gender. The relationship between antibiotics and obesity has been studied recently. According to a study by Reijnders et al. (2016), antibiotics change the gut microbiota of the host without influencing the host's general function. Other studies have found that antibiotics can reduce body weight and enhance insulin sensitivity (Membrez et al., 2008). Antibiotics and obesity, however, have a complicated and poorly understood association. The findings of earlier research have been mixed, with some indicating that antibiotics may contribute to obesity, while others have found no discernible impact. To fully understand how antibiotics contribute to obesity, more research is required.

16.2.2.5 Diet

High-calorie diets are clearly linked to the development of obesity and type 2 diabetes, according to extensive research conducted in both humans and animals (Mozaffarian et al., 2011). However, new research indicates that this association is mediated in large part by the gut microbiota (DiBaise et al., 2012). It makes sense to target the gut microbiota for intervention because of its substantial influence on its makeup. Research has repeatedly demonstrated that dietary modifications can cause significant and quick changes in the gut microbiota (David et al., 2014). When compared to a normal diet, studies on mice have shown that a high-fat diet (60% fat) significantly alters the makeup of the gut microbiome (β-diversity) and reduces the richness of bacterial species (α-diversity). Notably, a mouse study revealed that mice with type 2 diabetes and obesity had lower levels of *Akkermansia muciniphila*. However, *A. muciniphila*'s abundance was restored and metabolic parameters improved when it was fed prebiotics. *A. muciniphila* treatment also decreased insulin resistance, inflammation, and fat mass brought on by a high-fat diet. Numerous health benefits of a high-fiber diet have been demonstrated, primarily because of its capacity to alter the gut microbiota (Koh et al., 2016). Overall, the data points to the possibility that a high-fat diet may upset the delicate balance of gut flora, resulting in dysbiosis and raising the risk of illness.

16.3 MALNUTRITION: TYPES, CAUSES, AND EFFECTS

Millions of children worldwide suffer from malnutrition, commonly referred to as undernutrition, a widespread global health concern. Malnutrition is the primary cause of death of children under 5, making for over half of all deaths in this age group, according to the World Health Organization (WHO). The issue is especially severe in nations with poor and moderate incomes. According to a 2021 report by the World Bank Group, the WHO, and the United Nations International Children's Emergency Fund (UNICEF), there were 38.9 million overweight children, 45.4 million wasting children, and 149.2 million stunted children under the age of 5 worldwide. With 39% and 55% of the world's stunting prevalence, respectively, Africa and Asia were shown to be the most impacted continents (Williams & Suchdev, 2017). These figures highlight how urgently comprehensive initiatives are needed to combat malnutrition and guarantee the health and welfare of children everywhere.

16.3.1 Types of Malnutrition

Malnutrition is comprises undernutrition and overnutrition.

16.3.1.1 Undernutrition

It refers to the condition where a person is not getting enough nutrients to meet their physical needs. This can result in stunting, wasting, and underweight. Stunting is a condition characterized by impaired growth and development, particularly in terms of height. Wasting, on the other hand, refers to a rapid and severe weight loss, often associated with acute malnutrition. Underweight refers to a low body weight compared to the individual's age.

16.3.1.2 Overnutrition

Obesity can result from consuming too many nutrients, especially calories. An excessive buildup of body fat is the hallmark of obesity, a chronic illness that has detrimental effects on general health. This illness dramatically raises the risk of getting a number of chronic illnesses, such as diabetes, heart disease, and several types of cancer.

16.3.2 Causes of Malnutrition

UNICEF divides the complicated problem of malnutrition into three primary categories:

- Immediate causes
- Underlying causes
- Basic causes

The direct causes of malnutrition are sickness and food insecurity, which are mostly brought on by inadequate nutrition, which also has an impact on the gut flora. Underlying factors that greatly contribute to children's malnutrition and infections include inadequate maternity and child care, a shortage of food at home, restricted access to healthcare, and poor sanitation. The fundamental causes of malnutrition are more deeply ingrained in a community's or region's political, economic, and cultural structures, which can restrict access to resources, medical care, and wholesome food (Fund UNC, 1991).

16.4 NEUROLOGICAL DISORDERS

According to MoosWalter et al. (2016), noncommunicable diseases have overtaken infectious diseases as the leading cause of morbidity and mortality globally throughout the previous century. More than 1 billion individuals worldwide suffer from neuropsychiatric diseases, which have emerged as a major global health concern (Chisholm et al., 2016). Remarkably, the percentage of people worldwide who suffer from anxiety and depression has risen to about 10%. A significant 30% of the global healthcare burden related to noncommunicable diseases is caused by mental disorders (Chisholm et al., 2016). Alzheimer's disease and autism, whose prevalence has significantly increased in youngsters over the past 15 years (Ding et al., 2017), are other noteworthy neurological illnesses. Furthermore, it is impossible to ignore eating disorders like anorexia nervosa, which primarily afflict young individuals (Jayatilleke et al., 2017).

16.4.1 Common Neurological Disorders Associated with Malnutrition

The immunological, cardiovascular, and neurological systems, all need macronutrients and micronutrients to function at their best. A variety of clinical symptoms may arise from any changes in their levels brought on by illness, malnourishment, or insufficient supplementation. Wernicke's encephalopathy, a brain ailment brought on by a thiamine (vitamin B1) shortage, is a well-known illustration of a nutritional disorder. In certain parts of the brain, thiamine deficiency can result in degenerative lesions and neuronal cell death because it is necessary for a number of neurometabolic processes (Tsalamandris et al., 2023).

Neurological function can be greatly impacted by malnutrition and specific nutritional problems, especially when cerebrovascular diseases like stroke develop (Han et al., 2022; Chen et al., 2022; Liampas et al., 2023). Malnutrition has been linked to an increased risk of stroke; Han et al. (2022) found that patients who experienced an acute ischemic stroke were more likely to suffer from malnutrition than those who did not. Furthermore, Chen et al. (2022) found that in patients who had suffered a major stroke, malnutrition was linked to decreased food intake and weakened immunity. Additionally, changes in brain processes related to metabolism and endocrine function can potentially result in nutritional problems and bad eating patterns (Regensburger et al., 2022). A comprehensive strategy is required to address malnutrition and promote normal neurological performance, as evidenced by the intricate interactions among nutrition, neurological function, and cerebrovascular health.

16.5 INTERPLAY BETWEEN GUT MICROBIOME AND MALNUTRITION

16.5.1 Impact of Malnutrition on Gut Microbiome

Children are disproportionately affected by malnutrition, which is still a major global health concern. According to Iddrisu et al. (2021), children's gut microbiota is extremely vulnerable to dietary impacts, which can affect their nutritional status. Reduced microbial diversity, a decrease in beneficial bacteria like *Faecalibacterium, Bifidobacterium, Clostridium,* and *Ruminococcus*, and an increase in pathogenic microorganisms, especially after 18 months, are characteristics of the immature microbiome linked to severe malnutrition, according to research (Chibuye et al., 2024). Moreover, pathogenic components of the gut microbiota frequently cause diarrhea, a significant cause of malnutrition. In the end, diarrhea perpetuates malnutrition by causing weight loss, decreased energy availability, and malabsorption of vital nutrients. It is crucial for public health to comprehend the causes of malnutrition and create efficient preventative interventions. One promising strategy to reduce malnutrition is to alter the gut microbiota (Pekmez et al., 2019).

16.5.2 Gut Microbiome Dysbiosis and Its Consequences

Dysbiosis, a state of microbial imbalance, is defined by three basic features: reduced microbiota diversity, loss of beneficial microorganisms, and proliferation of dangerous microbes. Notably, changes in macronutrient consumption can cause significant changes in the gut bacterial and fungal flora. The body may experience serious effects from these alterations. For example, eating a lot of simple carbohydrates can cause intestinal inflammation, damage to tissues, and disruption of the intestinal barrier (Vrieze et al., 2012).

Wide-ranging effects of gut microbiota ecosystem disruption can be broadly divided into three categories: disruption of the host metabolic system, immune system imbalance, and gut barrier impairment. Numerous processes, such as direct mucolytic activity, the microbiota's synthesis of acetaldehyde from ethanol sources, and other pathways, might impair the integrity of the gut wall (Llorente & Schnabl, 2015; Png et al., 2010). Microbiota-derived compounds have the ability to influence the host immune system through interactions with the immune system via NOD-like receptor (NLR), toll-like receptor (TLR), and inflammasome signaling. Furthermore, there may be an imbalance between pro-inflammatory and regulatory immune cells. Changes in the composition of bile acids, the synthesis of short-chain fatty acids (SCFAs) from dietary fiber, the conversion of choline to trimethylamine (TMA), and other processes can all have an impact on the human metabolic system, especially glucose and lipid metabolism (Chu et al., 2019).

16.6 GUT MICROBIOME AND NEUROLOGICAL DISORDERS

16.6.1 Gut–Brain Axis: Mechanisms of Interaction

Two main channels enable the intricate communication between the stomach and the brain: blood-borne connections through hormones and other compounds produced by gut bacteria, and physical connections through the autonomic nervous system (Sharon et al., 2016). The "microbiome–gut–brain axis" is the phrase used to describe this complex link (Dalton et al., 2019). The development of the blood–brain barrier, neurogenesis, myelination, and microglia maturation are just a few of the neurological processes that the gut microbiota influences through these interactions. Furthermore, studies have linked the diversity of microbiomes to changes in behavior, mental health, and mood (Sharon et al., 2016; Dalton et al., 2019).

A sophisticated, two-way communication network that unifies the central neurological and gastrointestinal systems is known as the gut–brain axis. This network includes immunological, humoral, metabolic, and endocrine pathways in addition to anatomical ties (Appleton, 2018). The gut microbiota, which generates neurotransmitters like dopamine, serotonin, and γ-aminobutyric acid (GABA), is an essential part of the gut–brain axis. These neurotransmitters have a significant effect on the neurological system, especially during an ischemic stroke (IS), and they control nerve function. Additionally, microbial metabolites such as aryl hydrocarbon receptor (AhR) ligands and SCFAs are produced by the gut microbiota. Following IS, these metabolites experience substantial alterations that result in the maturation of gut-associated lymphoid tissues and the activation of the innate immune system. Furthermore, amino acids like tryptophan and tyramine are produced by the gut microbiota and cause both systemic and local immunological reactions. Additionally, these amino acids are essential for the removal of toxins and medications from the body (Li et al., 2020).

16.6.2 Influence of Gut Microbiome on Brain Function

The neurological system in particular is significantly impacted by the gut microbiota. The construction and organization of the nervous system over time correspond with the early colonization of the microbiome during fetal development. The active regulation of early brain development by the microbiota and its by-products has been brought to light by recent studies. On the other hand, interference with this early developmental stage can have long-term effects, impacting brain function and playing a role in the emergence of a number of neurological and neuropsychiatric disorders (NPD) (Dash et al., 2022).

The complex interaction of several systems, including the immunological, endocrine, and enteric neurological systems, as well as the central, autonomic, and enteric nervous systems (ENSs), is necessary for the microbiota-gut–brain axis to work at its best (Cryan & Dinan, 2012). In particular, the sympathetic and parasympathetic branches of the autonomic nervous system, the ENS, and the central nervous system (CNS) all communicate with the gut microbiota in both directions (Cryan & Dinan, 2012). Efferent and afferent nerves allow for neural communication between the brain and the digestive system, enabling this intricate conversation (Furness, 2012). As a result, the CNS regulates a number of sensory and secretory processes, including gastrointestinal motility (Badawy, 2017).

16.6.3 Gut Microbiome Alterations in Neurological Disorders

Through hormone synthesis and the control of physiological processes, the gut microbiota communicates with different organs and plays a crucial role in the host's physiology and disease. Complex chemical, neurological, immunological, and endocrine processes interact with gastrointestinal signaling pathways in the central nervous system. By generating microbial metabolites and neurotransmitters and promoting interactions with host cells, including intestinal epithelial cells and immune cells, a healthy gut microbiota helps the host. On the other hand, dysbiosis or alterations in the intestinal microbiota can result in a number of gastrointestinal issues, which can then have an impact on the host's body because of the aberrant metabolites produced by the microbes. The pathophysiology of neurological and neurodevelopmental problems can be influenced by gut dysbiosis, which can upset the delicate balance between the gut microbiota and the central nervous system. Numerous clinical studies indicate that gut microorganisms could be a risk factor for the onset of neurological conditions such as multiple sclerosis, Parkinson's disease, Alzheimer's disease, and autism spectrum disorder. Numerous immune-related neurological conditions, such as developmental disorders, neurodegeneration, and emotional dysregulation, have been connected to changes in the gut microbiota and the synthesis of microbial metabolites (Ullah et al., 2023).

16.7 NUTRITIONAL INTERVENTIONS AND GUT MICROBIOME MODULATION

16.7.1 Dietary Interventions to Modulate the Gut Microbiome

The importance of the gut microbiome in human health and disease is widely accepted. Disturbance of the balance between different microbial populations living in the gut is related to the development of many diseases. Modulation of the gut microbiome through dietary intervention is a novel therapeutic and preventive strategy for many conditions. However, individual differences in response to clinical treatments or diets are observed during clinical trials, and recent studies have suggested that subject-specific characteristics of the gut microbiota are the responsibility (Purdel et al., 2023).

Human metabolic function has been demonstrated to be improved by dietary interventions and treatments targeted at resolving microbial abnormalities (Purdel et al., 2023). In contrast to animals fed an isocaloric high-fat diet rich in n-3 PUFA from fish oil, preclinical research has shown that animals fed a diet high in saturated lipids have lower abundances and diversity of beneficial bacteria, including *Lactobacillus*, *Bifidobacterium*, and *Akkermansia muciniphila* (Flach et al., 2018). Increased insulin resistance, inflammation, and metabolic impairment are linked to this decline in good bacteria. These results demonstrate how nutrition and the host microbiota interact intricately, with each factor affecting the other. Diet can change the microbiota, and the different metabolic products that come from the microbiome's conversion of nutrients can change the host's metabolism. Interestingly, the gut microbiota can carry out tasks that the host cannot (Caesar et al., 2015).

The effects of dietary modifications on the intestinal microbiota of people and animals are still not well understood, despite the field's increasing interest. Furthermore, it is uncertain if the translatability of preclinical findings is limited by variations in the makeup and dynamics of the gut microbiota between human and animal models. It is unclear whether the effects of calorie restriction on gut microbiota composition are more noticeable than those of time-restricted fasting, despite the fact that it has been demonstrated that these regimens are effective in lowering body weight and adiposity, as well as the risk of noncommunicable diseases (Fontana & Partridge, 2015). Furthermore, it is unknown how these modifications may affect intestinal mucosa renewal and the composition of the gut microbiota in the long run. Notably, there has been little research done on the intricate link between the immune system and gut microbiota under various fasting circumstances (Mohr et al., 2021).

16.7.2 Impact of Nutritional Rehabilitation on Neurological Outcomes

A stroke is a neurological emergency that necessitates either rapid medical care or surgery. It is brought on by acute brain damage from either blood artery rupture or blockage, which are the two main vascular insults. Stroke is still a major cause of mortality and disability globally, even with major improvements in treatment approaches. During the acute and rehabilitative phases, survivors of their first stroke undergo varying degrees of neurological recovery (Hendricks et al., 2002). However, according to Roberts and Counsell (1998), 34–54% of survivors will suffer from moderate-to-severe disability for the rest of their lives. During the rehabilitation phase, functional outcomes may benefit from a multidisciplinary team approach that includes dietary support (Cifu & Stewart, 1999). Elderly patients, who frequently have concomitant conditions and are at risk of malnutrition, comprise the majority of stroke victims (FOOD Trial Collaboration, 2003). Research indicates that nutritional support is essential for the best possible recovery, even though the exact effect of post-stroke nutritional status on long-term survival is yet unknown. However, the confusion surrounding this issue has been exacerbated by the disparity in nutritional assessment methodologies and the time of post-stroke assessments (Martineau et al., 2005).

For many nutrients, the lowest nutrition group lived shorter than the highest nutrition stroke subgroup. In addition, stroke patients with the highest 25% intake of each nutrient, excluding moisture and total fat, had shorter survival than nonstroke patients with the lowest 25% intake. Eating disorders are very common in stroke patients and are associated with high morbidity. The drastic change in energy needs during illness requires a change in diet to ensure adequate nutrient intake (Lu et al., 2023).

16.8 CONCLUSION

The gut microbiota is essential to preserving human health. Food fermentation, pathogen defence, immune response activation, and vitamin synthesis are just a few of the vital tasks carried out by the diverse variety of gut bacteria. Research on the human microbiota has progressed beyond compositional studies and membership links due to the microbiota's critical importance in human health, including its involvement in many biological processes and the development of disease. Malnutrition continues to be a serious worldwide problem that endangers the lives of millions of people, especially children. Childhood malnutrition is mostly caused by viral illnesses and malnutrition, but the gut microbiota also has a big impact. Through hormone synthesis and the control of physiological processes, the gut microbiota interacts with several organs. Complex chemical, neurological, immunological, and endocrine processes interact with gastrointestinal signaling pathways in the central nervous system. This chapter emphasizes the complex interrelationships between the gut microbiota, central nervous system, and general human health, as well as the crucial role the gut microbiota plays in controlling neurological diseases.

REFERENCES

Afzaal, M., Saeed, F., Shah, Y. A., Hussain, M., Rabail, R., Socol, C. T., ... & Aadil, R. M. (2022). Human gut microbiota in health and disease: Unveiling the relationship. *Frontiers in Microbiology*, *13*, 999001.

Appleton, J. (2018). The gut-brain axis: Influence of microbiota on mood and mental health. *Integrative Medicine: A Clinician's Journal*, *17*(4), 28.

Arumugam, M., Raes, J., Pelletier, E., Le Paslier, D., Yamada, T., Mende, D. R., ... & Bork, P. (2011). Enterotypes of the human gut microbiome. *Nature*, *473*(7346), 174–180.

Badawy, A. A. (2017). Tryptophan availability for kynurenine pathway metabolism across the life span: control mechanisms and focus on aging, exercise, diet and nutritional supplements. *Neuropharmacology*, *112*, 248–263.

Bander, Z., Nitert, M. D., Mousa, A., & Naderpoor, N. (2020). The gut microbiota and inflammation: An overview. *International Journal of Environmental Research and Public Health*, *17*(20), 7618.

Blaut, M. (2018). Composition and function of the gut microbiome. In D. Haller (Ed.), *The Gut Microbiome in Health and Disease* (pp. 5–30), Springer.

Blekhman, R., Goodrich, J. K., Huang, K., Sun, Q., Bukowski, R., et al. (2015). Host genetic variation impacts microbiome composition across human body sites. *Genome Biology*, *16*, 191.

Boix-Amorós, A., Collado, M. C., & Mira, A. (2016). Relationship between milk microbiota, bacterial load, macronutrients, and human cells during lactation. *Frontiers in Microbiology*, *7*, 492.

Byrne, C. S., Chambers, E. S., Alhabeeb, H., Chhina, N., Morrison, D. J., Preston, T., ... & Frost, G. S. (2016). Increased colonic propionate reduces anticipatory reward responses in the human striatum to high-energy foods. *The American Journal of Clinical Nutrition*, *104*(1), 5–14.

Cabrera-Rubio, R., Collado, M. C., Laitinen, K., Salminen, S., Isolauri, E., & Mira, A. (2012). The human milk microbiome changes over lactation and is shaped by maternal weight and mode of delivery. *The American Journal of Clinical Nutrition*, *96*(3), 544–551.

Caesar, R., Tremaroli, V., Kovatcheva-Datchary, P., Cani, P. D., & Bäckhed, F. (2015). Crosstalk between gut microbiota and dietary lipids aggravates WAT inflammation through TLR signaling. *Cell Metabolism*, *22*(4), 658–668.

Chambers, E. S., Morrison, D. J., & Frost, G. (2015). Control of appetite and energy intake by SCFA: What are the potential underlying mechanisms? *Proceedings of the Nutrition Society*, *74*(3), 328–336.

Chawla, M., Gupta, R., & Das, B. (2022). Gut microbiome dysbiosis in malnutrition. *Progress in Molecular Biology and Translational Science*, *192*(1), 205–229.

Chen, Y., Yang, X., Zhu, Y., Zhang, X., Ni, J., & Li, Y. (2022). Malnutrition defined by geriatric nutritional risk index predicts outcomes in severe stroke patients: A propensity score-matched analysis. *Nutrients*, *14*(22), 4786.

Chibuye, M., Mende, D. R., Spijker, R., Simuyandi, M., Luchen, C. C., Bosomprah, S., ... & Harris, V. C. (2024). Systematic review of associations between gut microbiome composition and stunting in under-five children. *npj Biofilms and Microbiomes*, *10*(1), 46.

Chisholm, D., Sweeny, K., Sheehan, P., Rasmussen, B., Smit, F., Cuijpers, P., & Saxena, S. (2016). Scaling-up treatment of depression and anxiety: A global return on investment analysis. *The Lancet Psychiatry*, *3*(5), 415–424.

Chu, H., Duan, Y., Yang, L., & Schnabl, B. (2019). Small metabolites, possible big changes: A microbiota-centered view of non-alcoholic fatty liver disease. *Gut*, *68*(2), 359–370.

Cifu, D. X., & Stewart, D. G. (1999). Factors affecting functional outcome after stroke: A critical review of rehabilitation interventions. *Archives of Physical Medicine and Rehabilitation*, *80*(5), S35–S39.

Corrêa-Oliveira, R., Fachi, J. L., Vieira, A., Sato, F. T., & Vinolo, M. A. R. (2016). Regulation of immune cell function by short-chain fatty acids. *Clinical & Translational Immunology*, *5*(4), e73.

Cryan, J. F., & Dinan, T. G. (2012). Mind-altering microorganisms: The impact of the gut microbiota on brain and behaviour. *Nature Reviews Neuroscience*, *13*(10), 701–712.

Dalton, A., Mermier, C., & Zuhl, M. (2019). Exercise influence on the microbiome–gut–brain axis. *Gut Microbes*, *10*(5), 555–568.

Dash, S., Syed, Y. A., & Khan, M. R. (2022). Understanding the role of the gut microbiome in brain development and its association with neurodevelopmental psychiatric disorders. *Frontiers in Cell and Developmental Biology*, *10*, 880544.

David, L. A., Maurice, C. F., Carmody, R. N., Gootenberg, D. B., Button, J. E., Wolfe, B. E., ... & Turnbaugh, P. J. (2014). Diet rapidly and reproducibly alters the human gut microbiome. *Nature*, *505*(7484), 559–563.

Derrien, M., Vaughan, E. E., Plugge, C. M., & de Vos, W. M. (2004). Akkermansia muciniphila gen. nov., sp. nov., a human intestinal mucin-degrading bacterium. *International Journal of Systematic and Evolutionary Microbiology*, *54*(5), 1469–1476.

DiBaise, J. K., Frank, D. N., & Mathur, R. (2012). Impact of the gut microbiota on the development of obesity: current concepts. *The American Journal of Gastroenterology Supplements*, *1*(1), 22.

Ding, H. T., Taur, Y., & Walkup, J. T. (2017). Gut microbiota and autism: Key concepts and findings. *Journal of Autism and Developmental Disorders*, *47*, 480–489. https://doi.org/10.1007/s10803-016-2960-9

Dominguez-Bello, M. G., De Jesus-Laboy, K. M., Shen, N., Cox, L. M., Amir, A., Gonzalez, A., ... & Clemente, J. C. (2016). Partial restoration of the microbiota of cesarean-born infants via vaginal microbial transfer. *Nature Medicine*, *22*(3), 250–253.

Donohoe, D. R., Collins, L. B., Wali, A., Bigler, R., Sun, W., & Bultman, S. J. (2012). The Warburg effect dictates the mechanism of butyrate-mediated histone acetylation and cell proliferation. *Molecular Cell*, *48*(4), 612–626.

Fitzstevens, J. L., Smith, K. C., Hagadorn, J. I., Caimano, M. J., Matson, A. P., & Brownell, E. A. (2017). Systematic review of the human milk microbiota. *Nutrition in Clinical Practice*, *32*(3), 354–364.

Flach, J., Van Der Waal, M. B., Kardinaal, A. F. M., Schloesser, J., Ruijschop, R. M. A. J., & Claassen, E. (2018). Probiotic research priorities for the healthy adult population: A review on the health benefits of Lactobacillus rhamnosus GG and Bifidobacterium animalis subspecies lactis BB–12. *Cogent Food & Agriculture*, *4*(1), 1452839.

Fontana, L., & Partridge, L. (2015). Promoting health and longevity through diet: from model organisms to humans. *Cell*, *161*(1), 106–118.

FOOD Trial Collaboration. (2003). Poor nutritional status on admission predicts poor outcomes after stroke: Observational data from the FOOD trial. *Stroke*, *34*(6), 1450–1456.

Fund UNC. (1991). United Nations Children's fund. Strategy for improved nutrition of children and women in developing countries. *Indian Journal of Pediatrics*, *58*(1), 13–24. https://doi.org/10.1007/bf02810402

Furness, J. B. (2012). The enteric nervous system and neurogastroenterology. *Nature Reviews Gastroenterology & Hepatology*, *9*(5), 286–294.

Goodrich, J. K., Waters, J. L., Poole, A. C., Sutter, J. L., Koren, O., Blekhman, R., ... & Ley, R. E. (2014). Human genetics shape the gut microbiome. *Cell*, *159*(4), 789–799.

Gough, E. K., Moodie, E. E., Prendergast, A. J., Johnson, S. M., Humphrey, J. H., Stoltzfus, R. J., ... & Manges, A. R. (2014). The impact of antibiotics on growth in children in low and middle income countries: Systematic review and meta-analysis of randomised controlled trials. *BMJ*, *348*, g2267.

Guarner, F., & Malagelada, J. R. (2003). Gut flora in health and disease. *The Lancet*, *361*(9356), 512–519.

Han, X., Cai, J., Li, Y., Rong, X., Li, Y., He, L., ... & Tang, Y. (2022). Baseline objective malnutritional indices as immune-nutritional predictors of long-term recurrence in patients with acute ischemic stroke. *Nutrients*, *14*(7), 1337.

Hanson, L. A., Ahlstedt, S., Andersson, B., Cruz, J. R., Dahlgren, U., Fällström, S. P., ... & Wettergren, B. (1984). The immune response of the mammary gland and its significance for the neonate. *Annals of Allergy*, *53*(6 Pt. 2), 576–582.

Hendricks, H. T., Van Limbeek, J., Geurts, A. C., & Zwarts, M. J. (2002). Motor recovery after stroke: A systematic review of the literature. *Archives of Physical Medicine and Rehabilitation*, *83*(11), 1629–1637.

Hillman, E. T., Lu, H., Yao, T., & Nakatsu, C. H. (2017). Microbial ecology along the gastrointestinal tract. *Microbes and Environments*, *32*(4), 300–313.

Hou, K., Wu, Z. X., Chen, X. Y., Wang, J. Q., Zhang, D., Xiao, C., ... & Chen, Z. S. (2022). Microbiota in health and diseases. *Signal Transduction and Targeted Therapy*, *7*(1), 1–28.

Iddrisu, I., Monteagudo-Mera, A., Poveda, C., Pyle, S., Shahzad, M., Andrews, S., & Walton, G. E. (2021). Malnutrition and gut microbiota in children. *Nutrients*, *13*(8), 2727.

Jayatilleke, N., Hayes, R. D., Dutta, R., Shetty, H., Hotopf, M., Chang, C. K., & Stewart, R. (2017). Contributions of specific causes of death to lost life expectancy in severe mental illness. *European Psychiatry*, *43*, 109–115. https://doi.org/10.1016/j.eurpsy.2017.02.487

Koh, A., De Vadder, F., Kovatcheva-Datchary, P., & Bäckhed, F. (2016). From dietary fiber to host physiology: Short-chain fatty acids as key bacterial metabolites. *Cell*, *165*(6), 1332–1345.

Kulas, T., Bursac, D., Zegarac, Z., Planinic-Rados, G., & Hrgovic, Z. (2013). New views on cesarean section, its possible complications and long-term consequences for children's health. *Medical Archives*, *67*(6), 460.

Lagier, J. C., Khelaifia, S., Alou, M. T., Ndongo, S., Dione, N., Hugon, P., et al. (2016). Culture of previously uncultured members of the human gut microbiota by culturomics. *Nature Microbiology*, *1*, 16203. https://doi.org/10.1038/nmicrobiol.2016.203

Laterza, L., Rizzatti, G., Gaetani, E., Chiusolo, P., & Gasbarrini, A. (2016). The gut microbiota and immune system relationship in human graft-versus-host disease. *Mediterranean Journal of Hematology and Infectious Diseases*, *8*(1), e2016025

LeBlanc, J. G., Milani, C., De Giori, G. S., Sesma, F., Van Sinderen, D., & Ventura, M. (2013). Bacteria as vitamin suppliers to their host: A gut microbiota perspective. *Current Opinion in Biotechnology*, *24*(2), 160–168.

Li, X. J., You, X. Y., Wang, C. Y., Li, X. L., Sheng, Y. Y., Zhuang, P. W., & Zhang, Y. J. (2020). Bidirectional brain-gut-microbiota axis in increased intestinal permeability induced by central nervous system injury. *CNS Neuroscience & Therapeutics*, *26*(8), 783–790.

Liampas, A., Zis, P., Hadjigeorgiou, G., & Vavougios, G. D. (2023). Selenium, stroke, and infection: A threefold relationship; where do we stand and where do we go? *Nutrients*, *15*(6), 1405.

Lin, L., & Zhang, J. (2017). Role of intestinal microbiota and metabolites on gut homeostasis and human diseases. *BMC Immunology*, *18*, 1–25.

Llorente, C., & Schnabl, B. (2015). The gut microbiota and liver disease. *Cellular and Molecular Gastroenterology and Hepatology*, *1*(3), 275–284.

Louis, P., & Flint, H. J. (2017). Formation of propionate and butyrate by the human colonic microbiota. *Environmental Microbiology*, *19*(1), 29–41.

Louis, P., Hold, G. L., & Flint, H. J. (2014). The gut microbiota, bacterial metabolites and colorectal cancer. *Nature Reviews Microbiology*, *12*(10), 661–672.

Lu, H. Y., Ho, U. C., & Kuo, L. T. (2023). Impact of nutritional status on outcomes of stroke survivors: A post hoc analysis of the NHANES. *Nutrients*, *15*(2), 294.

Macfarlane, S., & Macfarlane, G. T. (2003). Regulation of short-chain fatty acid production. *Proceedings of the Nutrition Society*, *62*(1), 67–72.

Malnutrition, W. H. O. (2022). Who/int/news-room/fact-sheets/detail/malnutrtiion. Accessed March 3, 2021.

Martens, J. H., Barg, H., Warren, M. A., & Jahn, D. (2002). Microbial production of vitamin B 12. *Applied Microbiology and Biotechnology*, *58*, 275–285.

Martineau, J., Bauer, J. D., Isenring, E., & Cohen, S. (2005). Malnutrition determined by the patient-generated subjective global assessment is associated with poor outcomes in acute stroke patients. *Clinical Nutrition*, *24*(6), 1073–1077.

Membrez, M., Blancher, F., Jaquet, M., Bibiloni, R., Cani, P. D., Burcelin, R. G., ... & Chou, C. J. (2008). Gut microbiota modulation with norfloxacin and ampicillin enhances glucose tolerance in mice. *The FASEB Journal*, *22*(7), 2416–2426.

Mobeen, F., Sharma, V., & Tulika, P. (2018). Enterotype variations of the healthy human gut microbiome in different geographical regions. *Bioinformation*, *14*(9), 560.

Modi, S. R., Collins, J. J., & Relman, D. A. (2014). Antibiotics and the gut microbiota. *The Journal of Clinical Investigation*, *124*(10), 4212–4218.

Mohr, A. E., Gumpricht, E., Sears, D. D., & Sweazea, K. L. (2021). Recent advances and health implications of dietary fasting regimens on the gut microbiome. *American Journal of Physiology-Gastrointestinal and Liver Physiology*, *320*(5), G847–G863.

MoosWalter, H., FallerDouglas, V., HarppDavid, N., & PowersWhitney, R. (2016). Microbiota and neurological disorders: A gut feeling. *BioResearch Open Access*, *5*(1), 137–145.

Morrison, D. J., & Preston, T. (2016). Formation of short chain fatty acids by the gut microbiota and their impact on human metabolism. *Gut Microbes*, *7*(3), 189–200.

Mozaffarian, D., Hao, T., Rimm, E. B., Willett, W. C., & Hu, F. B. (2011). Changes in diet and lifestyle and long-term weight gain in women and men. *New England Journal of Medicine*, *364*(25), 2392–2404.

Musso, G., Gambino, R., & Cassader, M. (2010). Obesity, diabetes, and gut microbiota: The hygiene hypothesis expanded? *Diabetes Care*, *33*(10), 2277–2284.

Nagai, M., Obata, Y., Takahashi, D., & Hase, K. (2016). Fine-tuning of the mucosal barrier and metabolic systems using the diet-microbial metabolite axis. *International Immunopharmacology*, *37*, 79–86.

Panagos, P. G., Vishwanathan, R., Penfield-Cyr, A., Matthan, N. R., Shivappa, N., Wirth, M. D., ... & Sen, S. (2016). Breastmilk from obese mothers has pro-inflammatory properties and decreased neuroprotective factors. *Journal of Perinatology*, *36*(4), 284–290.

Patel, V. (2021). The gut microbiome. In E. Short (Ed.), *A Prescription for Healthy Living* (pp. 165–175). Academic Press.

Pekmez, C. T., Dragsted, L. O., & Brahe, L. K. (2019). Gut microbiota alterations and dietary modulation in childhood malnutrition – The role of short chain fatty acids. *Clinical Nutrition*, *38*(2), 615–630.

Piccioni, A., Cicchinelli, S., Valletta, F., De Luca, G., Longhitano, Y., Candelli, M., et al. (2022). Gut microbiota and autoimmune diseases: A charming real world together with probiotics. *Current Medicinal Chemistry, 29*, 3147–3159. https://doi.org/10.2174/ 0929867328666210922161913

Pingitore, A., Chambers, E. S., Hill, T., Maldonado, I. R., Liu, B., Bewick, G., ... & Persaud, S. J. (2017). The diet-derived short chain fatty acid propionate improves beta-cell function in humans and stimulates insulin secretion from human islets in vitro. *Diabetes, Obesity and Metabolism*, *19*(2), 257–265.

Png, C. W., Lindén, S. K., Gilshenan, K. S., Zoetendal, E. G., McSweeney, C. S., Sly, L. I., ... & Florin, T. H. (2010). Mucolytic bacteria with increased prevalence in IBD mucosa augmentin vitroutilization of mucin by other bacteria. *Official Journal of the American College of Gastroenterology| ACG*, *105*(11), 2420–2428.

Pompei, A., Cordisco, L., Amaretti, A., Zanoni, S., Matteuzzi, D., & Rossi, M. (2007). Folate production by bifidobacteria as a potential probiotic property. *Applied and Environmental Microbiology*, *73*(1), 179–185.

Portela, D. S., Vieira T. O., Matos, S. M., de Oliveira, N. F., & Vieira, G. O. (2015). Maternal obesity, environmental factors, cesarean delivery and breastfeeding as determinants of overweight and obesity in children: Results from a cohort. *BMC Pregnancy Childbirth*, *15*(1), 94.

Purdel, C., Margină, D., Adam-Dima, I., & Ungurianu, A. (2023). The beneficial effects of dietary interventions on gut microbiota—An up-to-date critical review and future perspectives. *Nutrients*, *15*(23), 5005.

Qin, J., Li, R., Raes, J., Arumugam, M., Burgdorf, K. S., Manichanh, C., ... & Wang, J. (2010). A human gut microbial gene catalogue established by metagenomic sequencing. *Nature*, *464*(7285), 59–65.

Regensburger, M., Krumm, L., Schmidt, M. A., Schmid, A., Spatz, I. T., Marterstock, D. C., ... & Winkler, J. (2022). Neurometabolic dysfunction in SPG11 hereditary spastic paraplegia. *Nutrients*, *14*(22), 4803.

Rehman, A., Sina, C., Gavrilova, O., Häsler, R., Ott, S., Baines, J. F., ... & Rosenstiel, P. (2011). Nod2 is essential for temporal development of intestinal microbial communities. *Gut*, *60*(10), 1354–1362.

Reijnders, D., Goossens, G. H., Hermes, G. D., Neis, E. P., van der Beek, C. M., Most, J., ... & Blaak, E. E. (2016). Effects of gut microbiota manipulation by antibiotics on host metabolism in obese humans: A randomized double-blind placebo-controlled trial. *Cell Metabolism*, *24*(1), 63–74.

Roberts, L., & Counsell, C. (1998). Assessment of clinical outcomes in acute stroke trials. *Stroke*, *29*(5), 986–991.

Rutayisire, E., Wu, X., Huang, K., Tao, S., Chen, Y., & Tao, F. (2016). Cesarean section may increase the risk of both overweight and obesity in preschool children. *BMC Pregnancy and Childbirth*, *16*, 1–8.

Seferidi, P., Hone, T., Duran, A. C., Bernabe-Ortiz, A., & Millett, C. (2022). Global inequalities in the double burden of malnutrition and associations with globalisation: A multilevel analysis of Demographic and Health Surveys from 55 low-income and middle-income countries, 1992–2018. *The Lancet Global Health*, *10*(4), e482–e490.

Sharon, G., Sampson, T. R., Geschwind, D. H., & Mazmanian, S. K. (2016). The central nervous system and the gut microbiome. *Cell*, *167*(4), 915–932.

Shaw, S. Y., Blanchard, J. F., & Bernstein, C. N. (2010). Association between the use of antibiotics in the first year of life and pediatric inflammatory bowel disease. *Official Journal of the American College of Gastroenterology| ACG*, *105*(12), 2687–2692.

Shreincr, A. B., Kao, J. Y., & Young, V. B. (2015). The gut microbiome in health and in disease. *Current Opinion in Gastroenterology*, *31*(1), 69–75.

Tamburini, S., Shen, N., Wu, H. C., & Clemente, J. C. (2016). The microbiome in early life: implications for health outcomes. *Nature Medicine*, *22*(7), 713–722.

Thompson-Chagoyán, O. C., Maldonado, J., & Gil, A. (2005). Aetiology of inflammatory bowel disease (IBD): Role of intestinal microbiota and gut-associated lymphoid tissue immune response. *Clinical Nutrition*, *24*(3), 339–352.

Tsalamandris, G., Hadjivassiliou, M., & Zis, P. (2023). The role of nutrition in neurological disorders. *Nutrients*, *15*(22), 4713.

Ullah, H., Arbab, S., Tian, Y., Liu, C. Q., Chen, Y., Qijie, L., ... & Li, K. (2023). The gut microbiota–brain axis in neurological disorder. *Frontiers in Neuroscience*, *17*, 1225875.

Ursell, L. K., Metcalf, J. L., Parfrey, L. W., & Knight, R. (2012). Defining the human microbiome. *Nutrition Reviews*, *70*(suppl_1), S38–S44.

Valdes, A. M., Walter, J., Segal, E., & Spector, T. D. (2018). Role of the gut microbiota in nutrition and health. *BMJ, 361*, k2179. https://doi.org/10.1136/bmj.k2179

Vrieze, A., Van Nood, E., Holleman, F., Salojärvi, J., Kootte, R. S., Bartelsman, J. F., ... & Nieuwdorp, M. (2012). Transfer of intestinal microbiota from lean donors increases insulin sensitivity in individuals with metabolic syndrome. *Gastroenterology*, *143*(4), 913–916.

Wang, B., Yao, M., Lv, L., Ling, Z., & Li, L. (2017). The human microbiota in health and disease. *Engineering, 3*, 71–82. https://doi.org/10.1016/J.ENG.2017.01.008

Wen, L., & Duffy, A. (2017). Factors influencing the gut microbiota, inflammation, and type 2 diabetes. *The Journal of Nutrition*, *147*(7), 1468S–1475S.

Williams, A. M., & Suchdev, P. S. (2017). Assessing and improving childhood nutrition and growth globally. *Pediatric Clinics*, *64*(4), 755–768.

World Health Organization. (2009). WHO child growth standards and the identification of severe acute malnutrition in infants and children: Joint statement by the World Health Organization and the United Nations Children's Fund.

World Health Organization. ISBN: 9789241598163. https://www.who.int/publications/i/item/9789241598163

17 Tools and Bioinformatic Approaches to Study Gut Microbiota and Neurodegeneration

Rideb Chakraborty [a], *Naureen Afrose*[a], *Pratibha Bhowmick*[a], and *Mithun Bhowmick*[a]

[a] Bengal College of Pharmaceutical Sciences and Research, Durgapur, West Bengal, India

17.1 INTRODUCTION

The human body is comprised of various organs and, of course, the gut and the brain are partners that top the list. Gut to brain communication is information flow between the two organs and includes the nervous, the endocrinal, and the immunologic systems. This complex has a role to do with several functions of the body, including the digestive mechanism, the breakdown of nutrients, and even emotions and thoughts, among others. Biomedical science has discovered that the relation between the digestive system and the brain is quite complex and this has led many researches to focus on in this field. The main element of this axis is the gut microbiota, which concerns the inhabitants that live in the abdomen and intestines. This microbial is necessary for the human body and plays role in digestion and absorption, immunology, and ward off many pathogens. Recent studies have pointed out that this microbial plays a role in various aspects of human life, such as neurological [1–5].

17.2 TOOLS FOR GUT MICROBIOTA ANALYSIS

17.2.1 16S rRNA Sequencing

Specific to the molecular biology approach that is used in the identification and phenotypic grouping of bacterial organisms by targeting their genetic markers is the technique known as 16S rRNA sequencing [6, 7]. It is specifically made for identifying a particular gene known as the 16S rRNA gene that is unique to all the varying kinds of bacteria. There are some defined regions that are very much conserved and there are also some regions that are mutable within this gene. It is present in more or less all the bacterial and archeal species. Some of these segments integrate the universal primers in the amplification of the gene, while in others you have the so-called hypervariable regions that would help in the identification of the microorganisms involved. It has taken the role of configuring and distributing the composition of gut microbiota as an important element of analysis [8]. This technique is based on identification of the conserved gene in most of the bacterial organisms known as 16S rRNA gene. It means that specificity of different bacterial species in the given sample is defined by an amount of the gene amplification and sequencing whether the given particular method of amplification is used. The approaches of next-generation sequencing for the identification of intestinal microbiota include the 16S rRNA sequencing [9], as described here and in the subsequent Figure 17.1.

1. *Sample Collection:* Regarding stool samples in relation to the gut microbioma, there are differences in relations to techniques.
2. *RNA Extraction:* The sample mentioned above is subjected to isolation of RNA from the microbe population [10].
3. *PCR Amplification:* The 16S rRNA gene is then precipitated and then amplified with specific primers which will match the conserved region of the gene 16S rRNA.
4. *Sequencing:* The obtained amplified DNA is then once again submitted for sequencing using next-generation sequencing technologies [11].
5. *Data Analysis:* The subsequent sequences are again measured with the primary overall objective of establishing and/or identifying the proportions of bacterial taxa [].

Table 17.1 briefly describes the advantages and disadvantages of 16S rRNA sequencing to investigate the gut microbiome and neurodegenerative diseases.

DOI: 10.1201/9781003536802-17

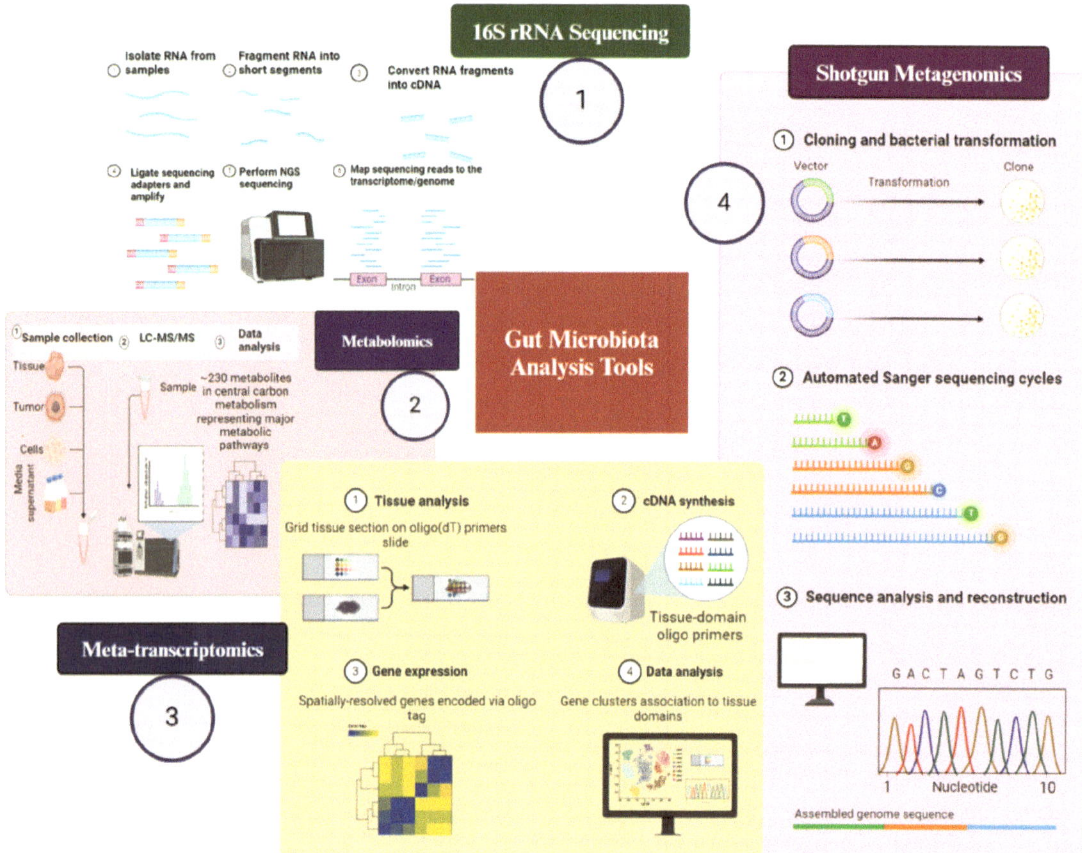

FIGURE 17.1 Tools for gut microbiota analysis.

17.2.2 Shotgun Metagenomics

Shotgun metagenomics is an effective method of analyze shooting the entire DNA content of a microbial community with isolating the organisms in its turn [16, 17]. It offers the analysts a view of the microbial composition and putative capabilities within a sample. Shotgun metagenomics can give better view of the gut microbiota compared to 16S rRNA sequencing due to the approach that it uses [18]. As it identifies and quantifies all the genetic material in a sample, it yields a view of the microbial system encompassing of density of different species as well as their metabolic possibilities.

Shotgun metagenomics is applied to the below detailed procedures in order to accomplish the shoot of describing the microbial community of the gut [19]. Figure 17.1 highlights and gives a more expanded understanding of the procedure by which shotgun metagenomics helps in deciphering the gut microbiome.

1. *Sample Collection and Preparation:* Fecal samples are the type of sample material that is most commonly used to determine the composition of gut microbiota. Other kinds of samples which could be useful may include mucosal biopsies. It is written at the bottom of the tube samples and these samples are preserved at −80°C. An appropriate method for isolating microbial DNA is applied in this step when conducting the experiment.
2. *Library Preparation:* The genomic DNA gets further fragmented or else diluted to even smaller units suitable for sequencing. Following amplification, the DNA fragments are attached to adapters and barcodes to get it ready for sequencing. Some of the process are end repair, A-tailing, adapter ligation, and PCR amplification, among others.
3. *Sequencing:* The prepared library is then subjected to high-throughput-sequencing platforms, including the Illumina, Pacific Bioscience's PacBio, or Oxford Nanopore's Nanopore [18]. These channels create tens or hundreds of millions or even billions of short DNA-like sequences.
4. *Data Preprocessing:* Sequencing generates direct nucleotide sequence as a result, which is further preprocessed to remove the low quality reads; residual adaptor sequences, and DNA of host origin by trimming.
5. *Sequence Assembly:* Short reads are then grouped from longer, nonoverlapping segments known as Contigs, and then forms a new microbial genome, which is known as De Novo Assembly. If they are a reference genome in place, these reads are then

TABLE 17.1
Strengths and limitations of 16S rRNA Sequencing

Strengths	Scenario	Limitations	Scenario
High- throughput	Useful in testing of several samples at once which is very important in population-related large studies [14].	**Limited taxonomic resolution**	However, the identification of Microbial species has some disadvantages such as; 16S rRNA has low power of resolution ; power of resolution reduces the opportunities to distinguish between two relatively close microbial species; this means that there may be no differences in microbial community [14, 15].
Cost-effective	In general, 16S rRNA sequencing is cheaper than the other methods for metagenomic analysis , although, here also the price may vary depending on many factors.	**Bias**	As to the choice of the primer the latter influences the result to the extent that concentrations of individual microorganisms of the microbial population shall not be discernible.
Sensitive	One of the benefits of this test is the feasibility of identification of a very low quantity of microorganisms that will help to calculate the density of the population in the stools.	**No functional information**	Consists about the features of the type of microorganisms in the sample but never mentioned that microorganisms can perform the given procedures.
Established methodology	This question actually does not include any specific instruction; however, the approaches to the analysis of the data appear to be clear as bioinformatics has been stated to have found prosperous territories.	**Inability to capture complex interactions**	For this reason it might be impossible to notice the existence of all those effects which definite groups of the microbiota can influence the state of the host.
Identification of microbial biomarkers	Stated further that its capability was in the identification of certain microbial signatures associated with the neurodegenerative diseases for confirmation and prognosis of the diseases [12]		

aligned to the reference genome to learn about the variants and gene that is present.

6. *Taxonomic Profiling:* Sequencing runs are aligned to reference databases (e.g., of NCBI, EMBL, and SILVA) and the one's abundance is determined, for example, with qPCR [22].
7. *Functional Profiling:* These might have been suggested from the contigs or anyhow postulated from the read data in which case they are known as the open reading frames or ORFs. Subsequently, problems of the predicted gene, including intron–exon division and function, can be stated in detail through for a number of databases such as in KEGG, COG, and Pfam. Table 17.2 presents a tabular form of disadvantage and advantage of 16S rRNA sequencing.

17.2.3 Metabolomics

Metabolomics is the scientific study of the entire complement of metabolites (endogenous and exogenous bioactive) within a biological system [23]. It gives a picture of the activity of the metabolic processes taking place at a certain time and can be used for assessing the condition of a system at that time. The study of the fecal metabolome is helpful to understand the metabolic activity of microbes, the interaction between the host and microbes, and changes of the microbiota with health status [24, 25]. Gut microbiota metabolomics, as distinguished from the general topic of metabolomics, is a subfield and contains the following steps (Figure 17.1).

1. *Sample Collection and Preparation:* Fecal sample is taken to be cultured on the culture media of the gut microbes. Other samples that be can be used include, for example, mucosal biopsies, stored at temperature below –80°C to prevent degradation followed by processing by homogenization, extraction, and filtration to get the metabolites.
2. *Metabolite Extraction:* Methanol, acetonitrile, water, and other solvents are used and procedures include approaches like liquid–liquid extraction, solid-phase extraction, or protein precipitation.
3. *Metabolite Analysis:* With regard to the assays employed to obtain the data, some of them includes nuclear magnetic resonance (NMR) spectrometer,

TABLE 17.2
Strengths and Limitations of Shotgun Metagenomics

Strengths		Limitations	
Comprehensive analysis	Provides a holistic view of acterial, archaeal, viral surrogates, fungi, and microeukaryote's tRNA for the improvement	Data complexity	Gives huge amounts of info and it has been understood that thousands of pieces of info can be processed only with the help of the up-to-date programs
Functional insights	Uses the functional potential of the microbiome designed with the oriented genes of certain procedures	High cost	Gauging can be time-consuming and this means it can be expensive, so is data analysis
Novel microorganism discovery	Permits the identification of the other microorganisms which were not taken into consideration earlier [22]	Assembly challenges	The process of rarefying the short reads to entire genomes can, therefore, be quite a tedious affair in microbial conservation analysis
Quantitative analysis	Accordingly, it can be stated that it is can be clearly seen that one can possibly obtain a rather approximate idea of the contents of a different type of microorganism	Host DNA contamination	In microbial sequence analysis, one can contaminate host DNA interference before sequencing, thus the DNA has to be extracted
Strain-level resolution	Originally designed to describe how one could possibly conduct a comparison of distensions of the same species	Limited temporal resolution	Provides only a "snapshot" of the microbiome hence it cannot show the conditions of the microbiome after a week or a month

mass spectrometer, gas chromatography–mass spectrometry (GC-MS), and liquid chromatography–mass spectrometry (LC-MS) [26].

4. *Data Preprocessing and Analysis:* Data is aligned to remove baseline shifts, normalize and clean data with the removal of any noise. Bands/peaks that correspond to metabolites are detected and their intensity is measured. Thus, the identified metabolites are aligned against the metabolic pathways in order to deliberate potential biological indications. Metabolomics data can also be compared/merged with other –omics data such as, but not limited to, genomics data, transcriptomics data, etc.
5. *Interpretation and Biological Validation:* Metabolites are spotted on the basis of reference databases or by comparison with authentic standards. The variations in the metabolite levels are analyzed with reference to existing knowledge about the metabolic sequence and biological events. Data obtained can be further substantiated by targeted analysis such as metabolomics or other experiments.

There are several specific uses of metabolomics in contextualizing the gut microbiota research. An outline of metabolomics and its potential as well as the niches in exploring gut microbiota and neurodegeneration are depicted in Table 17.3.

17.2.4 Meta-Transcriptomics

Meta-transcriptomics is a very effective method of investigation that offers data on the functional activity of the microbial communities based on the information about RNA transcripts contained in the sample [30, 31]. This approach can be complementary to metagenomics that define possible functions of a community based on the genes' sequences, comparing it to the identification of actual functions in a given condition. Meta-transcriptomics is helpful for looking at the functional aspect of the gut microbiome [32]. Meta-transcriptomics consists of some steps elaborated below, to decipher the functional state of gut microbiome as shown in Figure 17.1.

1. *Sample Collection and Preservation:* These can be utilized here also; fecal samples or mucosal biopsies, whichever be the case. RNA is highly unstable, hence must be immobilized immediately on the matrix. There are some methods available like RNAprotect Bacteria Reagent, and the reagents and solutions should be stored at –80°C.
2. *RNA Extraction:* The stock is then washed and RNA extraction in high quality obtained. RNA extraction kits are used in the process. Some of the RNA may be from the host microbe associations, hence its removal may be necessary when dealing with such samples.
3. *cDNA Library Preparation:* This is done with the help of either the RNA integrity number (RIN) or

TABLE 17.3
Strengths and Limitations of Metabolomics

	Strengths		Limitations
Functional readout	In particular, metabolomics represents the functional analysis of microbial activity that is expressed in terms of its metabolic final products in the gut microbiota [26]	Complexity	The metabolite concentrations are rather labile and can be altered by several factors depending on diet among others
Phenotypic relevance	Metabolites are normally tightly associated with the phenotypes, therefore metabolomics is highly useful for studying the disease processes as well as biomarkers determination [27, 28]	Analytical challenges	Metabolomics is inclined toward analysis and requires specific skills and expertise quite often
Complementary to genomics	Metabolomics can perform the functions of expressing the behavior of microbial genes and their interactions with other microorganisms and the surrounding environment	Data interpretation	Difficulties can be observed when the analysis of metabolomics data is conducted due to the fact that numerous features might be correlated with each other and might resemble each other [29]
Noninvasive	The use of metabolomics analysis involves the use of fecal samples and therefore the process is quite invasive	Limited taxonomic resolution	The latter procedure demonstrated herein-known provides no information immediately pointing at specified microorganisms responsible for formation of metabolic products, while metabolomics provides such information

the Bioanalyzer profiles that have been acquired. RNA is reversed transcribed to get cDNA: the latter being a stable molecule for this preceding sequencing.

4. *Sequencing:* Afterward, the libraries need to be prepared for sequencing and adding adapters and barcodes to the cDNA libraries. NGS is then done to get the sequence data using one of the commonly used platforms: the Illumina, the PacBio, or the Nanopore sequencer [33–35].
5. *Data Analysis:* Sequenced raw reads are further cleaned and any of them which are of low quality or contain adapters are excluded. Sequences are sorted based to specific reference genome or assembled into contigs for the samples which have no reference genome available. The gene expression levels are in turn assessed using the relevant software. Comparison methods are used to search for such genes within which the average intensity in groups is increased or, on the contrary, reduced. The descriptions for genes are added into the annotations by adopting the use of databases such as KEGG COG or Pfam [36–38]. The possibility and the limitation associated with the metatranscriptomics approach concerning gut microbiota and neurodegeneration are presented in Table 17.4.

17.3 BIOINFORMATIC APPROACHES FOR DATA ANALYSIS

17.3.1 Data Preprocessing

High-throughput sequencing technologies have resulted in a shift in the field of microbiome analysis because of the incorporation of massive high-dimensional data [40]. Processing of data is therefore very important so as to get considerable insight. Discussed sequencing reads are transformed with the help of several bioinformatic tools in order to be read easily by a biologist. The types of data processing include quality control, clustering, normalization, and data imputation, among others. Concerning metagenomics data, it may be predicted for gene content and the metabolic pathways utilizing tools, including KEGG and COG [41, 42].

17.3.2 Functional Profiling

The functional profiling focuses on the opportunities of the gut microbiome that functional genomics provides, which can be described as giving a profound view on neurodegenerative diseases [43]. Unlike physiology-based systematics, this method reveals metabolic pathways and uses various genes and proteins present in the microbial genome. PICRUSt and HUMAnN that use 16S rRNA data for comprehensive analysis give a primary functional annotation [44–46]. However, by using reference databases, their

TABLE 17.4
Strengths and Limitations of Meta-Transcriptomics

	Strengths		Limitations
Functional insights	Gives a quantitative measure of gene expression differential to the reference, gives a functional end point of the microbiome	RNA degradation	RNA is less stable than DNA and thus RNA samples should be obtained and processed in a proper manner
Dynamic analysis	Is a good example of the microbial activity which is useful for assessing the changes that take place in a certain period of time	Low RNA abundance	It is also generally appreciated that some transcripts might be present at very low levels, and thus difficult to detect
Complementary to metagenomics	Conveys the functional component to the metagenomic data or information	Data analysis complexity	In other words, meta transcriptomic data analysis and the interpretation of the results involve intricate computation; therefore, they should be performed using an appropriate bioinformatics approach [39]
Potential for new discoveries	It enables the identification of many unknown gene and other factors that are helpful for the assessment of the intestinal health and disease condition	Host RNA contamination	The efficient removal of host RNA is very pertinent for right identification of microbes' existence

efficiency is somewhat compromised. Shotgun metagenomics, however, targets the microbial DNA and hence provides functional profile that is more extensive than the culture-based approach. Sophisticated approaches like the metagenomic binning and the assembly constitute nearly complete genomes adding to the functional genomics.

17.3.3 Microbe-Gut–Brain Axis Analysis

Consequently, bioinformatics can be regarded as the foundation for understanding the intricate interaction between the gut microbiota and the brain [47–52]. This strategy as depicted in Figure 17.2 has given new details on the complex microbial societies along with their role in functionality in neurodegenerative disorders. With vast sizes of the data produced in microbiome investigations, machine learning has proved crucial in the analysis of such data.

17.4 APPLICATIONS IN NEURODEGENERATION RESEARCH

17.4.1 Identifying Microbiome Signatures in Neurodegenerative Diseases

Recent research about the gut microbiome is related to neurological diseases and new developments have emerged to describe microbial features linked to neurodegenerative diseases. In this case, modern science employs methods such as 16S rRNA sequencing, shotgun metagenomics to uncover one of a kind microbial biomarkers associated with neurological disorders like Alzheimer's, Parkinson's, and ALS. The present findings reveal the shifts in the relative abundance of particular bacterial taxa associated with the diseases state and encompass particularly the changes in the relative numbers of several bacterial taxa [53–55]. For instance, it has been reported that the density of healthy bacteria discovered as *Bifidobacterium* was low in patients diagnosed with Alzheimer's, while the probability of having inflammatory bacteria of *Escherichia coli* was high. In the similar manner, Guess et al. have found that Parkinson's disease patients had reduced *Prevotella* count; increased Enterobacteriaceae count, and suggested that dysbiosis may be involved in the development of the disease.

17.4.2 Exploring Functional Links between Gut Microbiota and Neurodegeneration

In the latest research works related to neurodegenerative diseases, the link between the lack of favorable bacteria in the intestinal niche and the emergence of a disease was revealed. Researchers have decided to intervene with specific microbial taxa associated with specific ailments, including Alzheimer's disease, Parkinson's disease, and ALS. It is suggested that a decrease in the quantity of Bacteroides and Firmicutes and an increase in the level of Proteobacteria are linked with Alzheimer's illness; moreover, dysbiosis is related to amyloid plaque formation and neuroinflammation [56, 57]. Here they have been also succeeded in determining that other microbial metabolites such as SCFAs along with tryptophan metabolites are effective in neurodegeneration. Thus, butyrate on example affects microglial activity, and contributes to the blood–brain barrier; nonetheless, alterations in SCFAs carry effects on neuroinflammation and synaptic plasticity. Meta-transcriptomics gives data on the functional gene expression regulation status of gut microbiota that may impact neurodegeneration [58–61].

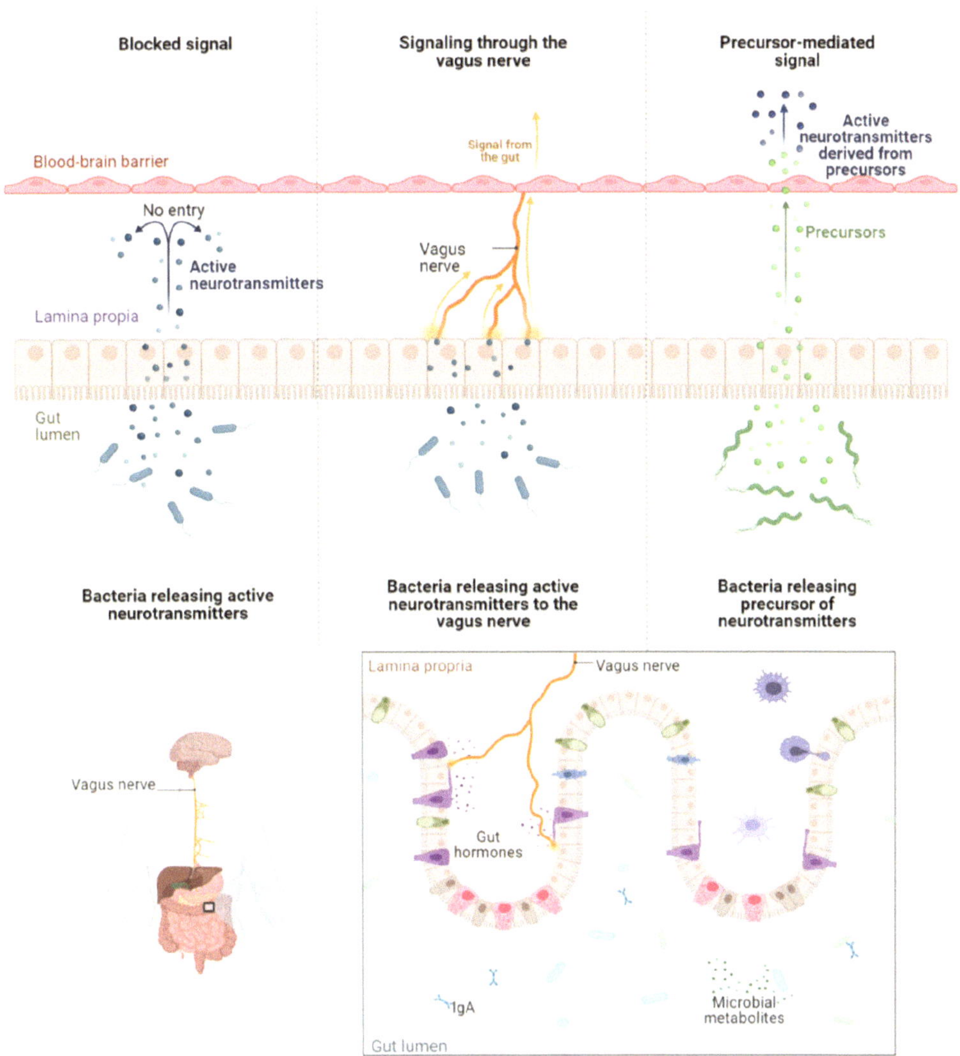

FIGURE 17.2 Microbe-gut–brain axis analysis.

17.4.3 Potential Therapeutic Strategies Targeting the Gut Microbiome

New discoveries in the area of the gut–brain axis have provided new concepts involving the use of the gut microbiota in prevention or alleviation of neurodegenerative diseases. *Lactobacillus* and *Bifidobacterium* strains and fructooligosaccharide have demonstrated possible benefits in the improvement of the gut barrier integrity and the production of anti-inflammatory SCFAs [62, 63]. Fresh fecal microbiota administered in animal models for Parkinson's disease have led to functional recovery, highlighting the fact that microbiota transplant from healthy donors has therapeutic potential in PD. Although ketogeni and Mediterranean diets are successful in decreasing gut microbiota, the anti-inflammatory action of both diets has been observed for the improvement of gut health [64].

17.5 CONCLUSION

The multiple links between the gut microbiome and neurodegenerative diseases constitute a rather complicated yet promising field of research. This chapter has explored the methodological repertoire that is crucial in explaining this dynamic relationship. In the course of the reviews focused on the advantages and challenges of 16S rRNA sequencing, shotgun metagenomics, metabolomics, and meta-transcriptomics, the priority was given to the highest level of evidence that could be obtained only through the synoptic approach to the identification and quantification of gut microbiota.

REFERENCES

1. Sorboni, S.G., Moghaddam, H.S., Jafarzadeh-Esfehani, R. and Soleimanpour, S., 2022. A comprehensive review on the role of the gut microbiome in human neurological disorders. *Clinical Microbiology Reviews*, *35*(1), pp.e00338–20.
2. Needham, B.D., Kaddurah-Daouk, R. and Mazmanian, S.K., 2020. Gut microbial molecules in behavioural and neurodegenerative conditions. *Nature Reviews Neuroscience*, *21*(12), pp.717–731.
3. Chen, M., Ruan, G., Chen, L., Ying, S., Li, G., Xu, F., Xiao, Z., Tian, Y., Lv, L., Ping, Y. and Cheng, Y., 2022. Neurotransmitter and intestinal interactions: Focus on the microbiota-gut-brain axis in irritable bowel syndrome. *Frontiers in Endocrinology*, *13*, p.817100.
4. Guo, Y., Zhu, X., Zeng, M., Qi, L., Tang, X., Wang, D., Zhang, M., Xie, Y., Li, H., Yang, X. and Chen, D., 2021. A diet high in sugar and fat influences neurotransmitter metabolism and then affects brain function by altering the gut microbiota. *Translational Psychiatry*, *11*(1), p.328.
5. Chakraborty, R., Afrose, N., Bhowmick, P. and Bhowmick, M., 2025. Biomedical sensors in wearable health technologies. In S. Mahajan and A.K., Pandit (Eds.), *Innovations in Biomedical Engineering* (pp. 159–184). Academic Press.
6. Yadav, S., Dwivedi, A., Tripathi, A. and Tripathi, A.K., 2022. Therapeutic potential of short-chain fatty acid production by gut microbiota in neurodegenerative disorders. *Nutrition Research*, *106*, pp.72–84.
7. Weinroth, M.D., Belk, A.D., Dean, C., Noyes, N., Dittoe, D.K., Rothrock Jr, M.J., Ricke, S.C., Myer, P.R., Henniger, M.T., Ramírez, G.A. and Oakley, B.B., 2022. Considerations and best practices in animal science 16S ribosomal RNA gene sequencing microbiome studies. *Journal of Animal Science*, *100*(2), p.skab346.
8. Church, D.L., Cerutti, L., Gürtler, A., Griener, T., Zelazny, A. and Emler, S., 2020. Performance and application of 16S rRNA gene cycle sequencing for routine identification of bacteria in the clinical microbiology laboratory. *Clinical Microbiology Reviews*, *33*(4), pp.10–1128.
9. Gao, B., Chi, L., Zhu, Y., Shi, X., Tu, P., Li, B., Yin, J., Gao, N., Shen, W. and Schnabl, B., 2021. An introduction to next generation sequencing bioinformatic analysis in gut microbiome studies. *Biomolecules*, *11*(4), p.530.
10. Gallardo-Becerra, L., Cornejo-Granados, F., García-López, R., Valdez-Lara, A., Bikel, S., Canizales-Quinteros, S., López-Contreras, B.E., Mendoza-Vargas, A., Nielsen, H. and Ochoa-Leyva, A., 2020. Metatranscriptomic analysis to define the Secrebiome, and 16S rRNA profiling of the gut microbiome in obesity and metabolic syndrome of Mexican children. *Microbial Cell Factories*, *19*, pp.1–18.
11. Średnicka, P., Roszko, M.Ł., Popowski, D., Kowalczyk, M., Wójcicki, M., Emanowicz, P., Szczepańska, M., Kotyrba, D. and Juszczuk-Kubiak, E., 2023. Effect of in vitro cultivation on human gut microbiota composition using 16S rDNA amplicon sequencing and metabolomics approach. *Scientific Reports*, *13*(1), p.3026.
12. Olovo, C.V., Huang, X., Zheng, X. and Xu, M., 2021. Faecal microbial biomarkers in early diagnosis of colorectal cancer. *Journal of Cellular and Molecular Medicine*, *25*(23), pp.10783–10797.
13. Liu, Y., Li, W., Yang, H., Zhang, X., Wang, W., Jia, S., Xiang, B., Wang, Y., Miao, L., Zhang, H. and Wang, L., 2021. Leveraging 16S rRNA microbiome sequencing data to identify bacterial signatures for irritable bowel syndrome. *Frontiers in Cellular and Infection Microbiology*, *11*, p.645951.
14. Rizal, N.S.M., Neoh, H.M., Ramli, R., Hanafiah, A., Samat, M.N.A., Tan, T.L., Wong, K.K., Nathan, S., Chieng, S., Saw, S.H. and Khor, B.Y., 2020. Advantages and limitations of 16S rRNA next-generation sequencing for pathogen identification in the diagnostic microbiology laboratory: Perspectives from a middle-income country. *Diagnostics*, *10*(10), pp.816–834.
15. Weinroth, M.D., Belk, A.D., Dean, C., Noyes, N., Dittoe, D.K., Rothrock Jr, M.J., Ricke, S.C., Myer, P.R., Henniger, M.T., Ramírez, G.A. and Oakley, B.B., 2022. Considerations and best practices in animal science 16S ribosomal RNA gene sequencing microbiome studies. *Journal of Animal Science*, *100*(2), p.skab346.
16. Sunny, J.S. and Saleena, L.M., 2020. Metagenomics: A vital source of information for modeling interaction networks in bacterial communities. In S. de Mandal and P. Bhatt (Eds.), *Recent Advancements in Microbial Diversity* (pp. 507–535). Academic Press.
17. Ramadurai, S., Moorthy, A. and Balasundaram, U., 2021. Metagenomic approach in relation to plant–microbe and microbe–microbe interactions. *Omics Science for Rhizosphere Biology*, pp.21–40.
18. Bars-Cortina, D., Ramon, E., Rius-Sansalvador, B., Guinó, E., Garcia-Serrano, A., Mach, N., Khannous-Lleiffe, O., Saus, E., Gabaldón, T., Ibáñez-Sanz, G. and Rodríguez-Alonso, L., 2024. Comparison between 16S rRNA and shotgun sequencing in colorectal cancer, advanced colorectal lesions, and healthy human gut microbiota. *BMC Genomics*, *25*(1), p.730.
19. Pérez-Cobas, A.E., Gomez-Valero, L. and Buchrieser, C., 2020. Metagenomic approaches in microbial ecology: An update on whole-genome and marker gene sequencing analyses. *Microbial Genomics*, *6*(8), p.e000409.
20. Wright, I.A., Delaney, K.E., Katusiime, M.G.K., Botha, J.C., Engelbrecht, S., Kearney, M.F. and van Zyl, G.U., 2021. NanoHIV: A bioinformatics pipeline for producing accurate, near full-length HIV proviral genomes sequenced using the oxford nanopore technology. *Cells*, *10*(10), p.2577.
21. Furlani, B., Kouter, K., Rozman, D. and Paska, A.V., 2021. Sequencing of nucleic acids: From the first human genome to next generation sequencing in COVID-19 pandemic. *Acta Chimica Slovenica*, *68*(2), pp.268–278.
22. Tully, B.J., Finkel, S.E. and Corzett, C.H., 2023. Benchmarking a novel quantitative PCR-based microbiome profiling platform against sequencing-based methods. *bioRxiv*, pp.2023–12.
23. Liu, R., Bao, Z.X., Zhao, P.J. and Li, G.H., 2021. Advances in the study of metabolomics and metabolites in some species interactions. *Molecules*, *26*(11), p.3311.
24. Wu, R., Xiong, R., Li, Y., Chen, J. and Yan, R., 2023. Gut microbiome, metabolome, host immunity associated with inflammatory bowel disease and intervention of fecal microbiota transplantation. *Journal of Autoimmunity*, *141*, p.103062.
25. Misheva, M., Ilott, N.E. and McCullagh, J.S., 2021. Recent advances and future directions in microbiome metabolomics. *Current Opinion in Endocrine and Metabolic Research*, *20*, p.100283.
26. Zhou, L., Yu, D., Zheng, S., Ouyang, R., Wang, Y. and Xu, G., 2021. Gut microbiota-related metabolome analysis based on chromatography-mass spectrometry. *TrAC Trends in Analytical Chemistry*, *143*, p.116375.

27. Li, X., Fan, X., Yang, H. and Liu, Y., 2022. Review of metabolomics-based biomarker research for Parkinson's disease. *Molecular Neurobiology*, pp.1–17.
28. Ristori, M.V., Mortera, S.L., Marzano, V., Guerrera, S., Vernocchi, P., Ianiro, G., Gardini, S., Torre, G., Valeri, G., Vicari, S. and Gasbarrini, A., 2020. Proteomics and metabolomics approaches towards a functional insight onto AUTISM spectrum disorders: Phenotype stratification and biomarker discovery. *International Journal of Molecular Sciences*, *21*(17), p.6274.
29. Perez De Souza, L., Alseekh, S., Brotman, Y. and Fernie, A.R., 2020. Network-based strategies in metabolomics data analysis and interpretation: From molecular networking to biological interpretation. *Expert Review of Proteomics*, *17*(4), pp.243–255.
30. Kumar, A. and Yadav, A., 2024. Next generation sequencing in metagenomics and metatranscriptomics. In I. Mani and V. Singh (Eds.), *Multi-Omics Analysis of the Human Microbiome: From Technology to Clinical Applications* (pp. 49–75). Singapore: Springer Nature.
31. Chakraborty, R., Jana, B., Koner, S., Maiti, S.K., Afrose, N., Bhowmick, P. and Bhowmick, M., 2024. Exploring the transcriptomic landscape of rheumatoid arthritis using next-generation RNA sequencing. *INNOSC Theranostics and Pharmacological Sciences*, p.4449.
32. Afrose, N., Chakraborty, R., Bhowmick, P. and Bhowmick, M., 2025. Advancing drug delivery systems through biomedical engineering: Innovations and future directions. In S. Mahajan and A.K. Pandit (Eds.), *Innovations in Biomedical Engineering* (pp. 1–32). Academic Press.
33. Cuber, P., Chooneea, D., Geeves, C., Salatino, S., Creedy, T.J., Griffin, C., Sivess, L., Barnes, I., Price, B. and Misra, R., 2023. Comparing the accuracy and efficiency of third generation sequencing technologies, Oxford Nanopore Technologies, and Pacific Biosciences, for DNA barcode sequencing applications. *Ecological Genetics and Genomics*, *28*, p.100181.
34. Segerman, B., 2020. The most frequently used sequencing technologies and assembly methods in different time segments of the bacterial surveillance and RefSeq genome databases. *Frontiers in Cellular and Infection Microbiology*, *10*, p.527102.
35. Hu, T., Chitnis, N., Monos, D. and Dinh, A., 2021. Next-generation sequencing technologies: An overview. *Human Immunology*, *82*(11), pp.801–811.
36. Galperin, M.Y., Wolf, Y.I., Makarova, K.S., Vera Alvarez, R., Landsman, D. and Koonin, E.V., 2021. COG database update: Focus on microbial diversity, model organisms, and widespread pathogens. *Nucleic Acids Research*, *49*(D1), pp.D274–D281.
37. Djemiel, C., Maron, P.A., Terrat, S., Dequiedt, S., Cottin, A. and Ranjard, L., 2022. Inferring microbiota functions from taxonomic genes: A review. *GigaScience*, *11*, p.giab090.
38. Ferreira, C.R., de Lima Gomes, P.C.F., Robison, K.M., Cooper, B.R. and Shannahan, J.H., 2024. Implementation of multiomic mass spectrometry approaches for the evaluation of human health following environmental exposure. *Molecular Omics*, pp.296–321.
39. Costa-Silva, J., Domingues, D.S., Menotti, D., Hungria, M. and Lopes, F.M., 2023. Temporal progress of gene expression analysis with RNA-Seq data: A review on the relationship between computational methods. *Computational and Structural Biotechnology Journal*, *21*, pp.86–98.
40. Regueira-Iglesias, A., Balsa-Castro, C., Blanco-Pintos, T. and Tomás, I., 2023. Critical review of 16S rRNA gene sequencing workflow in microbiome studies: From primer selection to advanced data analysis. *Molecular Oral Microbiology*, *38*(5), pp.347–399.
41. Gwak, H.J., Lee, S.J. and Rho, M., 2021. Application of computational approaches to analyze metagenomic data. *Journal of Microbiology*, *59*, pp.233–241.
42. Frioux, C., Singh, D., Korcsmaros, T. and Hildebrand, F., 2020. From bag-of-genes to bag-of-genomes: Metabolic modelling of communities in the era of metagenome-assembled genomes. *Computational and Structural Biotechnology Journal*, *18*, pp.1722–1734.
43. Chen, H., Li, L., Yu, F. and Ma, Z., 2024. A comprehensive diversity analysis on the gut microbiomes of ASD patients: From alpha, beta to gamma diversities. *FEMS Microbiology Letters*, *371*, p.fnae014.
44. Sulit, A.K., Kolisnik, T., Frizelle, F.A., Purcell, R. and Schmeier, S., 2023. MetaFunc: taxonomic and functional analyses of high throughput sequencing for microbiomes. *Gut Microbiome*, *4*, p.e4.
45. Wemheuer, F., Taylor, J.A., Daniel, R., Johnston, E., Meinicke, P., Thomas, T. and Wemheuer, B., 2020. Tax4Fun2: prediction of habitat-specific functional profiles and functional redundancy based on 16S rRNA gene sequences. *Environmental Microbiome*, *15*, pp.1–12.
46. Ismaeil, M., Alsharif, S.M., Saeed, A.M. and El-Sayed, W.S., 2024. Metagenomic 16S rRNA analysis and predictive functional profiling revealed intrinsic organohalides respiration and bioremediation potential in mangrove sediment. *BMC Microbiology*, *24*(1), p.176.
47. Chakrabarti, A., Geurts, L., Hoyles, L., Iozzo, P., Kraneveld, A.D., La Fata, G., Miani, M., Patterson, E., Pot, B., Shortt, C. and Vauzour, D., 2022. The microbiota–gut–brain axis: Pathways to better brain health. Perspectives on what we know, what we need to investigate and how to put knowledge into practice. *Cellular and Molecular Life Sciences*, *79*(2), p.80.
48. Guo, F., Cai, D., Li, Y., Gu, H., Qu, H., Zong, Q., Bao, W., Chen, A. and Liu, H.Y., 2021. How early-life gut microbiota alteration sets trajectories for health and inflammatory bowel disease? *Frontiers in Nutrition*, *8*, p.690073.
49. Qian, X.H., Song, X.X., Liu, X.L. and Tang, H.D., 2021. Inflammatory pathways in Alzheimer's disease mediated by gut microbiota. *Ageing Research Reviews*, *68*, p.101317.
50. Nagu, P., Parashar, A., Behl, T. and Mehta, V., 2021. Gut microbiota composition and epigenetic molecular changes connected to the pathogenesis of Alzheimer's disease. *Journal of Molecular Neuroscience*, *71*(7), pp.1436–1455.
51. Wang, Y., 2023. The role of the gut microbiota and microbial metabolites in the pathogenesis of Alzheimer's disease. *CNS & Neurological Disorders-Drug Targets (Formerly Current Drug Targets-CNS & Neurological Disorders)*, *22*(4), pp.577–598.
52. Liu, R.T., Rowan-Nash, A.D., Sheehan, A.E., Walsh, R.F., Sanzari, C.M., Korry, B.J. and Belenky, P., 2020. Reductions in anti-inflammatory gut bacteria are associated with depression in a sample of young adults. *Brain, Behavior, and Immunity*, *88*, pp.308–324.
53. Neu, A.T., Allen, E.E. and Roy, K., 2021. Defining and quantifying the core microbiome: challenges and prospects. *Proceedings of the National Academy of Sciences*, *118*(51), p.e2104429118.

54. Xing, M., Gao, H., Yao, L., Wang, L., Zhang, C., Zhu, L. and Cui, D., 2023. Profiles and diagnostic value of intestinal microbiota in schizophrenia patients with metabolic syndrome. *Frontiers in Endocrinology*, *14*, p.1190954.
55. Liu, X., Zhao, D., Wu, X., Zhao, S., Li, Z. and Qin, X., 2022. Comprehensive 16S rRNA sequencing based microbiomes and 1H NMR based metabolomics reveal the relationships of aging and constipation. *Experimental Gerontology*, *166*, p.111882.
56. Liu, S., Gao, J., Zhu, M., Liu, K. and Zhang, H.L., 2020. Gut microbiota and dysbiosis in Alzheimer's disease: Implications for pathogenesis and treatment. *Molecular Neurobiology*, *57*, pp.5026–5043.
57. Shabbir, U., Arshad, M.S., Sameen, A. and Oh, D.H., 2021. Crosstalk between gut and brain in Alzheimer's disease: The role of gut microbiota modulation strategies. *Nutrients*, *13*(2), p.690.
58. Delanote, J., Correa Rojo, A., Wells, P.M., Steves, C.J. and Ertaylan, G., 2024. Systematic identification of the role of gut microbiota in mental disorders: A TwinsUK cohort study. *Scientific Reports*, *14*(1), p.3626.
59. Barchi, A., Massimino, L., Mandarino, F.V., Vespa, E., Sinagra, E., Almolla, O., Passaretti, S., Fasulo, E., Parigi, T.L., Cagliani, S. and Spanò, S., 2023. Microbiota profiling in esophageal diseases: Novel insights into molecular staining and clinical outcomes. *Computational and Structural Biotechnology Journal*, pp.626–637.
60. Turco, L., Opallo, N., Buommino, E., De Caro, C., Pirozzi, C., Mattace Raso, G., Lembo, F. and Coretti, L., 2023. Zooming into gut dysbiosis in Parkinson's disease: New insights from functional mapping. *International Journal of Molecular Sciences*, *24*(11), p.9777.
61. Rana, S., Singh, P., Bhardwaj, T. and Somvanshi, P., 2024. A comprehensive metagenome study identifies distinct biological pathways in asthma patients: An in-silico approach. *Biochemical Genetics*, pp.1–16.
62. Roy, S. and Dhaneshwar, S., 2023. Role of prebiotics, probiotics, and synbiotics in management of inflammatory bowel disease: Current perspectives. *World Journal of Gastroenterology*, *29*(14), p.2078.
63. Sharma, S., Bhatia, R., Devi, K., Rawat, A., Singh, S., Bhadada, S.K., Bishnoi, M., Sharma, S.S. and Kondepudi, K.K., 2023. A synbiotic combination of Bifidobacterium longum Bif10 and Bifidobacterium breve Bif11, isomaltooligosaccharides and finger millet arabinoxylan prevents dextran sodium sulphate induced ulcerative colitis in mice. *International Journal of Biological Macromolecules*, *231*, p.123326.
64. Ross, F.C., Patangia, D., Grimaud, G., Lavelle, A., Dempsey, E.M., Ross, R.P. and Stanton, C., 2024. The interplay between diet and the gut microbiome: implications for health and disease. *Nature Reviews Microbiology*, pp.1–16.

18 Brain–Gut Axis Focusing on High-Throughput Technologies Like Gene Sequencing, Brain Imaging, and Metabolomics

*Vaibhav Dagaji Aher[1], Abhishek Sharma[2], Simanchal Panda[3], Krishna Chandra Panda[4], and *Pankaj Sharma[5]*

[1] Department of Pharmaceutical Medicine, Maharashtra University of Health Sciences, Nashik, Maharashtra, India
[2] Department of Pharmacology, Amity Institute of Pharmacy, Department of Pharmacology, Amity University, Greater Noida Campus, Greater Noida, Uttar Pradesh, India
[3] Department of Pharmacy, School of Pharmaceutical Sciences, Jaipur National University, Jaipur, India
[4] Department of Pharmacy, Roland Institute of Pharmaceutical Sciences, Khodasingi, Berhampur, India
[5] Department of Pharmaceutics, ShriRam College of Pharmacy, Morena, Madhya Pradesh, India

18.1 INTRODUCTION

The brain–gut axis (BGA) involves the central nervous system (CNS) as well as the enteric nervous system (ENS) that is located in the gastrointestinal (GI) tract wall. The schematic diagram depicts how the brain and the gut communicate and elaborate their reciprocal relation where the brain can make change in motility secretion and immune function in the intestine and similarly the intestine can influence the brain, behavior, and function. The ENS, also known as the "second brain," is a greatly developed nerve tissue consisting of neurons, neurotransmitters, and glial cells with its own functional independence, yet it is in close relation with the CNS via the vagus nerve, spinal cord, and hormones and cytokines [1].

The most significant part of the body and a bundle of nerves in this email exchange system is the gut microbiota, which is a collection of microorganisms present in the gastrointestinal tract. It communicates with the ENS and the CNS by short-chain fatty acids (SCFAs) and tryptophan metabolites, neuroactive components, and microbes. These interactions can act on all the brain functions such as mood, cognition, and even pains [2]. This bidirectional cross talk between the gut and the brain is critical for sustaining a healthy body and normal physiological functions, and it is understood that alteration of this axis results in functional gastrointestinal disorders, psychological disorders, neurodegenerative diseases, and others.

There are references to the relation between the brain and the gut dating back thousands of years. Other primitive forms of practice, like the ones documented in ancient Greece and China, stressed the impact of emotions and mental conditions on the digestive system. However, the scientific research of this relation became established in the 19th century and the beginning of 20th century by recognizing the autonomic part of the nervous system and its function of overseeing the internal functions of the human body, such as digestion [3].

During the middle of the 20th century, neurogastroenterology as a discipline that deals with the ENS and the interaction between ENS and CNS was developed. Scholars started investigating the interaction between stress and psychological factors with GI functions, thereby having realization about the existence of the complex communication system between the brain, hormones, and immune system present in the gut. This bidirectional communication system is known as the "brain–gut axis" [4].

Modern technological developments have greatly enhanced knowledge on the brain–gut link. Advanced technical procedures, like the next-generation sequencing (NGS), has made studies for the gut microbiome much more elaborate for the complexity and interactivity of the microbes with the host [5]. This has resulted in the recognition of certain microbial signatures correlated with IBS, IBD, alongside mental disorders such as depression and anxiety.

DOI: 10.1201/9781003536802-18

Magnetic resonance imaging (MRI) and positron emission tomography (PET) studies on humans have helped researchers to map the brain's response to gut stimuli, and also to discover which parts of the brain are involved in the control of gut functions. The combination of using these imaging studies has helped to reveal how the brain and the gut interact, as well as showing the impact that the CNS has on the function of the gut [6]. Omics, which investigates the detailed information about the metabolites existing in the samples of biological origin, have given new insights into the metabolic crosstalk between the gut and the brain. This approach has revealed relevant metabolic pathways and metabolites in the brain–gut cross talk, for instance, SCFAs synthesized by our gut bacteria affecting the brain and behavior [7].

These technologies have helped to increase knowledge on the brain–gut axis and its interaction in relation to health and disease. These studies have described the relationships between the gut microbiotic and the enteric nervous system and the central nervous system both on configuration and functions, and they have highlighted a key role for this gut microbe, ENS/CNS balance in physiological homeostasis, and disease initiation. This is gradually building up knowledge that is defining new diagnostics and therapies of the brain–gut axis to address several diseases.

18.2 THE BRAIN–GUT AXIS: MECHANISMS OF COMMUNICATION

These connective tracks are central to establishing CNS–ENS interaction as a part of the nervous system interface (Figure 18.1). These pathways include sensory and motor fibers that are mainly of the vagus nerve, spinal afferents, and the autonomic nervous system (ANS).

18.2.1 Neural Pathways

The neural pathways are a part of the brain-gut axis that provides the essential link between the CNS and the ENS, thereby promoting two-way traffic. It is understood that about 8% of the fibers of the vagal nerve are afferent, while the remaining 2% are efferent in nature. It passes on information received from the gut to the brain and instructions from the brain to the gut, thus being critical in regulating the contraction of the gut's muscles, secretion, and inflammation [8]. Spinal afferents likewise play a role in that it conveys nociceptive and non-nociceptive information from the gut to the brain, which is essential in visceral pain [9]. The ANS, which has branches, namely, sympathetic and parasymphathetic divisions, continues to regulate this axis using tonic inhibitory and excitatory influences on the gastrointestinal activities [10]. Innovations in neuroimaging involving functional MRI (fMRI) and PET have finally offered better ideas about these pathways; these are the areas of the brain which are turned on by gut stimuli and the neural networks associating the gut and the brain [11]. Knowledge of these neural pathways informs therapeutics focused on manipulating the brain–gut axis for IBS, and/or depression [12].

18.2.2 Hormonal Signals

Obviously, hormonal signaling with such agents as ghrelin, leptin cholecystokinin (CCK), and glucagon-like peptide-1 (GLP-1) occupy a very significant role in the brain–gut connection. Ghrelin is known as Hunger hormone. It is produced in the stomach. It increases appetite by affecting the hypothalamus before meals [13]. It is secreted by adipocytes and has an action on the brain to decrease the food's intake, aiming at controlling energy balance; its dysfunction is associated with obesity [14]. CCK, which is secreted in the small intestine in response to fat and proteins, inhibits further food intake and also stimulates the pancreatic enzyme secretion by stimulating the vagal afferent [15]. The incretin hormone GLP-1 secreted from the intestines increases the secretion of insulin, suppresses glucagon release, and promotes satiety, which has attracted attention as a diabetes and obesity drug candidate [16]. Through various feedback signals, these hormones become a component of the brain–gut axis, thus, exemplifying their function of regulating homeostasis and providing information on possible therapy for the metabolic and gastrointestinal diseases [17].

18.2.3 Immune System Involvement

There is a growing focus on the regulation of the brain–gut relationship by the immune system; immune cells and cytokines being influential in this regulation. The M cell system of the Peyer's patches and the lamina propria lymphocytes can inhibit the invasions of the pathogens [18]. Cytokines, which are released in inflammation such as TNF-α, IL-1ē, and IL-6, can penetrate through the blood–brain barrier (BBB) or stimulate through the vagal afferents and affect the brain, possibly causing neuroinflammation and disorders like depression and anxiety [19]. Derangement of the gut microbiota unleashes immune responses; for instance, LPS from pathogenic bacteria stimulate inflammances, inflammation, and breakdown of the BBB [20]. Gut symbionts release SCFAs such as butyrate, which has an anti-inflammatory effect and preserves the BBB [21]. The immune cells also communicate with the ENS, which is important in controlling the movement and secretion in the gut [22]. Stress becomes chronic through time and it impairs the immune system, making the permeability of gut higher and thus increases general inflammation and affects the brain and behavior. Familiarity with these interactions has therapeutic implications, including anti-cytokine therapies of the neuropsychiatric disorders and probiotic intervention to regulate the gut microbiotic aliments and consequent neuroinflammation [23].

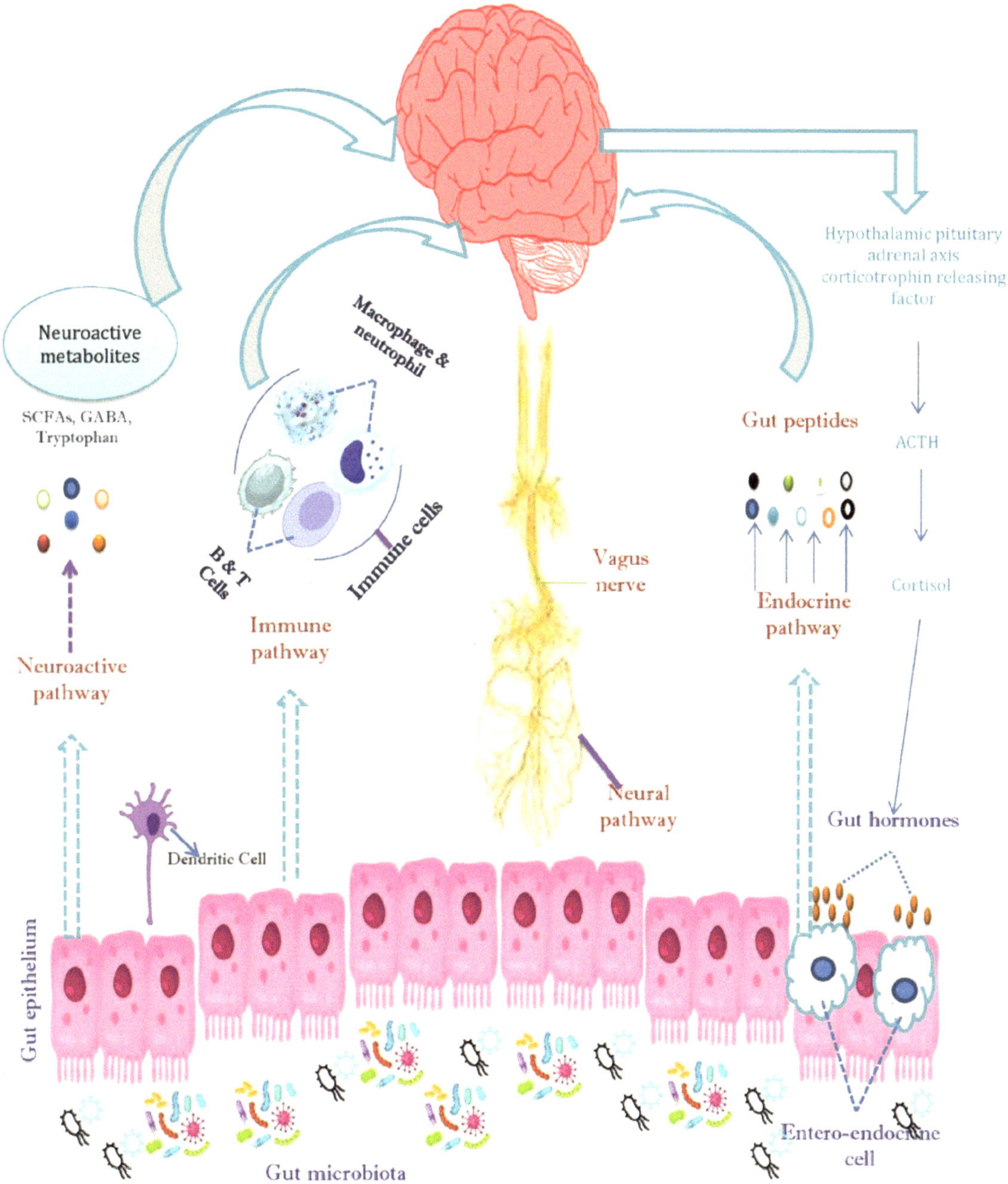

FIGURE 18.1 The brain–gut axis: mechanisms of communication.

18.2.4 Microbial Metabolites and Their Impact

These products also greatly affect the communication between the brain and the gut, therefore balancing the gut and the brain. They are the short-chain fatty acids, neurotransmitters, as well as the bile acids. The SCFAs such as butyrate, which are synthesized from the fermentation of the dietary fibers, exhibit anti-inflammatory properties and help in maintaining the gut barrier function, thus decreasing the systemic inflammation and protecting the BBB [24]. Specific neurotransmitters like serotonin and dopamine synthesized by the bacteria in the gut are influenced directly with emotion, thinking, and activity. For instance, 90% of the body's serotonin is created in the gut, which determines the movement of food through the digestive tract, as well as effect the brain and it's processes [25]. Gut bacteria further change bile acids, which affect metabolism and immune system; hence they have an indirect impact on the brain [26].

Common neurological disorders are related to changes in microbial metabolites. Beutler et al.'s documents show that diarrhea-predominant IBS patients have reduced butyrate-producing bacteria, which is associated with autism spectrum disorder (ASD) and gut–brain connections in neurodevelopment [27]. Likewise, alterations of gut microbiota and SCFAs have been linked with Parkinson's disease, which also points toward the context of neurodegeneration [28]. Another application of microbiotics includes the use of probiotics and prebiotics to strengthen the wholesome gut bacteria and their by-products which have been viewed to support barrier functionality of the gut wall and reduce inflammation presumably favorable for mental health [29].

18.3 GENE SEQUENCING IN BRAIN–GUT AXIS RESEARCH

18.3.1 Introduction to Gene Sequencing Technologies

Modern approaches in gene sequencing have greatly advanced the scientific knowledge about the brain–gut axis because it is now possible to have the genomics analysis of the genes and the microbes that compose this intricate signaling system. Various high-throughput sequencing techniques like the NGS are now unavoidable tools in the study of the BGA. These technologies enable the identification of the specific facets of the users' gut microbiota, host genetic differences, and gene activity, which offer understanding of how genetics and microbiome impact the brain and the gut [30].

There are several platforms of next-generation sequencing, some of which include Illumina, Ion Torrent, and PacBio, each with their own merits of aspect such as read length, accuracy, and throughput. For example, Illumina sequencing is noted for its high precision and productivity and it is recommendable for research projects that involve a big number of samples of microbial population and host genes [31]. When determining a part of the microbial community within the gut by the sequencing of 16S ribosomal RNA (rRNA) genes, such important components of the gut microbiota, including the degree of its diversification and organization that might be reflected in the interaction with brain functions and behavior, can be revealed [32].

Whole genome sequencing (WGS) and whole exome sequencing (WES) are also critical in studies targeting the brain–gut axis as well since the latter is a complex connectivity between the gastrointestinal and central nervous systems. The integration of WGS results in the study of the whole genome gives better chances to consider the genetic variation applied to SNPs or other structural variations that might be associated with the regulation of the BGA [33]. WES, however, mostly targets the protein-coding regions of the genome, which can be cheaper and yield relevant information on genetic factors that modulate the brain–gut connection [34].

RNA sequencing (RNA-seq) not only extends our knowledge by indicating differential gene expression in the brain and gut tissues. RNA-seq enables one to study the changes in gene expression due to differences in microbial conveyor belts, diet, or disease, and thus the intricate process of molecular interactions between the gut and the brain [35]. For instance, it has been demonstrated that the regulation of genes concerned with neurotransmitter biosynthesis and immune response to different stimuli is related to modifications in the microbiota of the gut and has been implicated in neurological and psychological disorders [36].

The combination of these gene sequencing technologies has boosted the research in brain–gut axis research immensely. For example, assessing the microbial composition in the gut, NGS application in discovering biomarkers or signatures linked to IBS or ASD for the diagnosis and therapeutics [37]. Moreover, genetic research has established microbial genotypes that are associated with host variations of gut microbiota, pointing toward the relevance of applying personal genomic techniques for treating the physiological schemes of the brain–gut axis illnesses [38].

18.3.2 Characterizing the Gut Microbiome

There are several sophisticated tools in sequencing that help in describing the array of microorganisms and functions that occur in the human GI tracts. Another molecular technique that is widely used is 16S ribosomal RNA (rRNA) gene sequencing that targets the gene of this molecule that exists in all bacteria. This method makes it possible to determine the species composition of bacteria and their numbers by comparing the variable parts of the gene, which vary across species [39]. The sequences that are acquired from the samples taken from the gut can then be compared with a reference database and this then creates a representation of the bacterial richness and density which aids in the determination of the changes in the bacterial composition of the community when health conditions are under consideration. For instance, changes in Firmicutes and Bacteroidetes phyla are known to be related to obesity and metabolic diseases [40].

Another term which often resonates closely with metagenomics is genomics; however, in metagenomics, all the genetic materials in a sample are sequenced, thus offering a broad perspective of species other than bacteria, such as viruses, fungi, etc. This approach also enables the functional analysis [41] because it highlights genes linked with the metabolic processes and microbial activities. For instance, the metagenomics approach has offered details on how alterations in the composition of gut microbes impact the fermentation of fiber and production of SCFAs associated with IBD and other related ailments [42].

In methodological comparison, compared to 16S rRNA sequencing, shotgun metagenomics gives more comprehensive data as the metagenomic data yields the complete DNA sequences of microorganisms in the sample, thus enabling identification of low abundance species and new genes in the sample [43]. This method has greatly contributed in identifying the microbial richness and capacity that is useful in establishing how the composition of gut microbial consortia influence host health and sickness. For instance, shotgun metagenomics has described certain bacterial species and features of their functional roles that are dysregulated in diseases such as Crohn's disease and ulcerative colitis [44].

Complementarily to genomic analysis, the functional metagenomics, which involves cloning and expressing the environmental DNA in the host systems for the testing of functional property, gives more information on the functional capacity of the gut microbiota. This technique has helped in identifying genes that are involved in the breakdown of the consumed nutrients and the formation of new metabolites that the host microbiota has an effect on host physiology and disease [45].

The integration of these sequencing technologies has therefore enhanced knowledge on the gut microbiome in health and disease. For instance, dysbiosis, which means the disruption of gut bacteria, has been scientifically associated with metabolic syndrome, Autism spectrum disorders, and neurological disorders, among others [46, 47]. Moreover, by describing the gut microbiome in greater detail, the researchers can establish biomarkers of a particular disease, design-specific treatments, and other forms of medicine based on an individual's genetic makeup.

18.3.3 Genomic Insights into Microbiota Composition and Function

Among these methods, the metagenomic and the shotgun metagenomic have greatly enhanced the knowledge regarding the composition and function of gut microbiotas. The microbiome composition and genomics give information about microbial species and gene repertoires, including SCFAs' synthesis from GFs, essential for gut homeostasis and the anti-inflammatory effect [48, 49]. Shotgun metagenomics (SMPG) lets one study the microbial community and genes, respectively, which helps in identifying the correlation between the microbial gene and diseases such as colorectal cancer and obesity [50, 51]. These genomic instruments assist in outlining how such microbiota affect host metabolism and immune systems; these can be targeted for treatments.

18.3.4 Functional Implications of Microbiome Diversity on Brain Health

Microbiome complexity in the gut is considered to have direct impacts on the brain and its functions. It was also observed that the diverse gut microbiota has a well-moderated production of metabolite and neurotransmitter involved in the functioning of the brain (Table 18.1) [52]. Gut bacterial metabolites, the SCFAs like butyrate, acetate, and propionate resulting from the breakdown of the dietary fibers also primarily help in BBB maintenance and other inflammatory processes. This has been explained by the fact that a healthy microbiome produces these SCFAs that are known to have neuroprotective properties together with mood and cognitive functions [53].

On the other hand, it was established that lower levels of microbiota diversity negatively affect the brain. The research has shown that lowered microbiota diverseness increases the level of circulating inflammatory markers and changes the production of SCFAs, which can compromise the structure and function of the BBB and underlying neurodegenerative pathologies, including Alzheimer's and Parkinson's diseases [54, 55]. For example, dysbiosis, an imbalance in the composition of the gut microbes, was found to be connected to the signs of anxiety and depression; therefore, the microbiome influences mental health [56].

Furthermore, the complexity of gut microbiota is in charge of the production of neurotransmitters such as serotonin and dopamine; it is in charge of moods and emotions [57]. A diverse microbial community promotes the better and healthier production of these neurotransmitters and enhances the psychological health of a person, thus reducing the possibilities of mood disorders.

Thus, the general health and diversification of the microbiome is crucial to brain health as it aids in the formation of metabolic products, the preservation of the BBB, and the increased production of neurotransmitters; this shows the significance of the microbiome to neuropsychological well-being.

18.4 BRAIN IMAGING TECHNIQUES

18.4.1 Overview of Brain Imaging Technologies

18.4.1.1 Magnetic Resonance Imaging

MRI entails a method of imaging the brain without causing harm and entails the use of magnetic waves to give details of the inside structures of the brain. The MRI is known to work through the application of strong magnetic fields as well as radiofrequency waves to provide clear images of the organ of interest [58]. It offers very good contrast between different soft tissues and thus it is very useful in demonstrating structural, for example, tumors, lesions, and other illnesses. MRI is used systematically in clinics to diagnose and manage the neurological disorders such as multiple sclerosis and epilepsy [59]. On the same note, newer MRI examined shapes and sizes of the brain, tissue quality and changes in it, and presence of various types of lesions in depth.

18.4.1.2 Functional MRI

Functional MRI is an enhancement of the MRI as it is a method that assesses the changes in the blood flow that indicates the amount of activity in the brain. fMRI measures changes in blood oxygenation levels, detected as the blood oxygen level-dependent (BOLD) signal [60]. This technique enables the researcher to map the functions of the brain and get to know which part of the brain is involved in a certain function, for example, memory, feeling, and sensation. fMRI is helpful in the analysis of brain connections and in the research of the neural substrate of mental disorders like major depressive disorder and schizophrenia [61]. It is also very useful in planning for surgery by pointing out areas of the brain that should not be operated on.

TABLE 18.1
Functional Implications of Microbiome Diversity on Brain Health

Aspect	Implications	Example	Mechanism
Neurotransmitter production	The state of gut microbiota affects synthesis of hormones such as serotonin and dopamine that control one's temperance	Increased serotonin production linked to specific gut bacteria	Bacterial metabolites act on enterochromaffin cells in the gut to stimulate serotonin synthesis [56, 57]
Immune system modulation	Intestinal bacteria influence the immune response neuroinflammation and the development of neurological diseases	Less inflammation in the brain of germ-free mice, which were re-colonized with particular gut bacteria	Microbial metabolites as reported in the previous section exert effects on the systemic immune system and decrease neuroinflammation [53, 55]
Metabolite production	Metabolites generated from the microbial metabolism include the short-chain fatty acids and other metabolites that affect the brain	Higher circulating levels of SCFAs also had benefits on neuronal function on mice	SCFAs are able to penetrate the blood–brain barrier and affect the brain activity and its development [49, 52]
Gut–brain communication	The communication is bidirectional and is carried through the vagus nerve and other neural pathways and affects mood and cognitive functions	Vagotomy minimizes the effects elicited by the changes in the microbiota on the behavior of animal models	It means that the signals which are sent through the vagal nerve impact the brain [48, 56]
Microbiome and cancer	There are findings which pointed to dysbiosis as a trigger of colorectal cancer and which may furthermore impact on the brain	Colorectal cancer patients' gut dysbiosis related to systemic inflammation	This immunomodulatory process leads to the mentioned microbial dysbiosis, and as a result inflammation negatively impacts the brain [50, 51]
Developmental influences	Research has revealed that early-life microbiota composition influences the brain's functions and has potential long-lasting consequences on mental health and cognitive function	Disruptions to the bacterial populations in the gut as a young mouse are associated with behavioral differences in adulthood	Prod plaint BC influences the brain development, and the effects are mediated by immune and metabolic processes [48, 52]
Mood and behavior	Imbalances in the composition of the bacterial flora affect mood, anxiety, and depression	In clinical trials, probiotics' treatment has been seen to enhance the moods and decrease anxiety	These gut microbiota communicate with the brain via modulating the neurotransmitters and the immune system [56, 57]
Neurodegenerative diseases	Reduced bacterial diversity has already been associated with neurodegenerative disorders such as Parkinson's and Alzheimer's	New research details particular microbial associations with Parkinson's disease	The disturbed gut–brain axis is involved in the process of neurodegenerative diseases' progression [53, 55]
Serotonin biosynthesis	Some bacteria in the gut hold the ability of controlling the synthesis of serotonin by the host body, which in aspect results in mood and gastrointestinal regulation	Higher amounts of serotonin found with certain bacteria in the gut	Some bacterial strains are known to raise the levels of serotonin found in the gut. Largely, bacterial metabolites help increase serotonin in the gut, which affects the brain [57]
Cognitive functions	The present study showed that the shift in gut microbiota leads to cognitive dysfunction and impaired learning and memory function	There are additional improvements in learning and memory when using animal models undergoing probiotic treatment	For this reason, microbial metabolites and modulation of the immune system affect cognitive functions [48, 56]

18.4.1.3 Positron Emission Tomography

Resonance magnetic imaging is a technique that produces detailed images of soft tissue and organs of the body, while PET is an imaging technique that measures the metabolic and biochemical activity in the body or the brain especially. They include PET scans, which entail use of a radiotracer, a radioactive compound that attaches itself to, and certain molecules administered into the bloodstream. Positron-emitting radiotracers are used; where the tracer decays, it releases positrons that are picked up by PET scanner in constructing images of the brain activity [62]. PET is useful in determining metabolism of the brain, neurotransmitter systems, and receptor binding. It is usually applied in studying the pathogenesis of neurodegenerative diseases, including Alzheimer's disease, based on the alteration of glucose utilization and amyloid deposition [63]. PET abilities help stimulate an understanding of molecular processes that cause the disease, and contributes to the development of new drugs [64].

18.4.1.4 Diffusion Tensor Imaging (DTI)

Diffusion tensor imaging (DTI) is an MRI-based method applied for the purpose of tracing the diffusion of water in brain tissues and obtaining data on the direction of fibers in the white matter [65]. DTI quantifies the extent to which water diffusion occurs in a given direction and is influenced by the orientation of neurfibers; hence, it facilitates visualization of the white matter paths and connections [65]. This technique is especially applied to investigate brain connections, evaluate the effects of brain damage, and research developmental disorders. For example, DTI has applied in researching modifications in white matter linking traumatic brain injury and developmental disorders such as the ASD [66].

18.4.2 Structural and Functional Connectivity between the Gut and the Brain

The bidirectional communication between the gut, including structural and functional neural connections, is through the neural pathways, hormones, and immune system. New developments in neuroscience, particularly in neuroimaging, have helped to clarify how these connections affect the brain, and the gut.

18.4.2.1 Structural Connectivity

Structural connectivity relates to the direct linkages of the brain areas with input of gut sensations. Diffusion tensor imaging gives information about the white matter tractography though which brain regions are interconnected. Studies conducted with the help of DTI have indicated that changes in the composition of the gut microbiota can affect the white matter tracts that are associated with cognitive and emotive functions [67]. For example, it has been demonstrated that shifts in microbial profile affect the wiring of two vital brain regions for stress and emotion regulation, namely, the prefrontal cortex and the amygdala [68].

18.4.2.2 Functional Connectivity

Functional connectivity is a study on how networks of the brain are functionally connected to other networks in regard to gut-derived signals. These interactions can be investigated using fMRI by determining the level of correspondences in the activity of the different areas. In the fMRI analysis, it has been shown that changes in the composition of the gut microbiota can affect the activity of the brain regions associated with the regulation of mood, pain sensations, and cognitive processes [69]. For instance, published studies carried out on the effects of probiotics exhibited the ability of gut microbiota affecting the functioning of the brain regions with a role in regulating emotions which include the insula and the anterior cingulate cortex [70].

18.4.2.3 Gut–Brain Axis Mechanisms

The healthy communication between the gut and the brain is made possible by these factors. Our nervous system also plays a significant part of relaying gut signals to the brain through what they called the vagus nerve. This two-way process enables the brain to send signals to the gut and at the same time receive signals from it. These interactions are also carried by hormonal signals like gut-derived peptides including ghrelin and leptin that influence the parts of the brain responsible for hunger and satiety signals [71]. Also, this connectivity is maintained by immune cytokines and other signaling molecules that modulate the inflammation of the brain and other neurological processes [72].

18.4.2.4 Clinical Implications

Knowledge of the relationship between the structural and functional connectivity between the gastrointestinal tract and CNS is therefore quite important. Abnormality in this connectivity is associated with several neurological and psychological diseases such as depression, anxiety, and neurodegenerative diseases [70]. For instance, imbalances in the gut bacteria composition were found to be linked with changes in the neural network that is implicated in mood disorders and dementia [73]. Exploring these links could result in identifying less conventional treatments, for example, using certain bacterial strains or adjusting person's diet to alter the interaction between the gut and the brain and thus promote better mental health.

18.4.3 Brain Imaging Biomarkers of Gut Microbiome Influence

Imaging derived biomarkers offers important information on how the gut microbiome modulates the brain, and its structure. These biomarkers are useful in determining the alterations in brain function, networks, and structure that are related to the changes in the gut bacteria. The body–mind dualism is evidently a complex one; the following imaging biomarkers are relevant to this relationship.

18.4.3.1 Metabolic Activity

Functional imaging that includes PET scans looks at the metabolic and biochemical differences between the normal and diseased brain. Molecular effects of PET biomarkers such as the alteration in glucose metabolism or number of neurotransmitter receptor binding can show how the gut microbiota alters brain biochemistry. For example, PET examination revealed that the changes in gut microbiota have significant effects on the brain glucose metabolism and neurotransmitter systems that are instrumental in the examination of the effect on mental health and cognition [74]. Changes in PET biomarkers may help reveal how products of the gut microbiota affect the brain and its functions and pathologies.

18.4.3.2 White Matter Integrity

From the respective citations it can be understood that DTI is used to identify white matter tracts fiber integrity in the brain. DTI biomarkers are fractional anisotropy and mean diffusivity which can represent the alterations of white matter tracts due to the shifts of gut microbiota. Existing research employing DTI has shown that the changes of gut microbiota may affect the white matter tract that come with the key roles in regulating emotions and cognition [75, 76]. These observations raise the understanding of the microbiota impact on brain networks and its consequences for the psychological state.

18.4.3.3 Neuroinflammation

Neuroinflammation, which is a typical side product of the dysbiosis, can also be diagnosed using MRI as well as PET. Molecular imaging biomarkers of inflammation as measured by radiotracer uptake in PET or signal changes in the MRI can show how the microbiota act on the brain's inflammation. Since increased neuroinflammation is associated with many neurological and psychiatric disorders, and targeting gut health has a link to brain inflammation, imaging biomarkers are useful in studying this relationship [77].

18.5 METABOLOMICS IN THE STUDY OF THE BRAIN–GUT AXIS

Thus, metabolomics play a significant role in the context of understanding the impact of the products formed by the gut microbiota on the brain and other organs. Here is a breakdown of its uses.

18.5.1 Analytical Techniques in Metabolomics

Several analytical approaches exist in metabolomics, which strive for the identification of metabolites derived from biological samples.

18.5.1.1 Mass Spectrometry (MS)

This technique involves determination of the mass to charge ratio of ions for the purpose of identifying and even quantifying metabolites. MS is very selective and is capable of identifying numerous metabolites, thus ideal for metabolic profiling [78].

18.5.1.2 Nuclear Magnetic Resonance (NMR) Spectroscopy

NMR spectroscopy gives specific data of structure and form of the metabolites present in the sample. It is especially valuable for characterizing structure and concentration of metabolites in the complex matrix like blood plasma or feces [79].

18.5.1.3 Gas Chromatography-Mass Spectrometry (GC-MS)

GC-MS is based on separation of volatile metabolites on a column before an analysis by mass spectrometer. It works for small and rather fluctuant metabolites and gives high-resolution data [80].

18.5.1.4 Liquid Chromatography-Mass Spectrometry (LC-MS)

LC-MS links liquid chromatography and mass spectrometry to detect a large variety of metabolites, ranging from relatively nonpolar to polar and having large molecular masses [81].

18.5.2 Metabolic Profiles of Gut Microbiota

The biomethod used in metabolic profiling of gut microbiota entails assessing the metabolites that are generated by gut bacteria.

18.5.2.1 Short-Chain Fatty Acids

Butyrate, acetate, and propionate are some of the metabolic generated from the breakdown of dietary fibers by gut microbiome. SCFAs affect the function of the brain through the control of the concentrations of neurotransmitters and the integrity of the blood–brain barrier [82].

18.5.2.2 Tryptophan Metabolism

Bacteria control the ability to metabolite tryptophan and therefore serotonin, which is significantly involved in moods. An association between psychiatric diseases and changes in tryptophan metabolism have been established [83].

18.5.2.3 Phenolic Compounds

Polyphenols are eaten from food and these are fermented by the gut microbiota and produce bioactive products that influence the brain. It is well known that these metabolites were antioxidant and anti-inflammatory, which could affect neuroinflammation and cognition [84].

18.5.3 Microbial Metabolites and Effects on Brain Function

18.5.3.1 Neurotransmitter Modulation

There seems to be interaction between the metabolites SCFAs, and tryptophan-derived metabolites affect synthesis and release of neurotransmitters, serotonin, and dopamine, which are a vital component of mood, memory, and behavior [85].

18.5.3.2 Blood–Brain Barrier Integrity

The butyrate among the SCFAs has been shown to play a role in the modulation of blood–brain barrier and afford the brain the much-needed protection from inflammatory and neurotoxic substances [86].

18.5.3.3 Neuroinflammation

Bacteria within the gut can release products that influence systemic inflammation, which impacts the brains' inflammation and function. The dysbiosis and the changes in the metabolites were found to be linked to higher neuroinflammation and neurological disorders [87].

18.5.4 Combining Metabolomics Data with Genomic and Imaging Data

Combinations of metabolomics with genomic and imaging data offer a robust insight into the brain–gut physiology.

18.5.4.1 Genomic Data

Combining metabolomics with genomic information allows one to discover relations between specific genetic differences and metabolic activity and its impact on the brain's function. For example, genetic differences can be revealed regarding the metabolism of the gut microbiota, which in turn affects the transformation of metabolites that are crucial for the overall brain function [88].

18.5.4.2 Imaging Data

Metabolomics integrated with brain imaging data – structural and functional MRI data, fMRI – enables researchers to link changes in metabolism with structural and functional brain changes. This integration assists in establishing the biomarkers of the disease and the effect of gut-derived metabolites on the brain activity and network [89].

18.5.4.3 Systems Biology Approaches

Systems biology combines various datasets to depict the interconnectivity between the gut microorganisms, metabolites, and neuronal activities. This framework enables the determination of main links and possible therapeutic interventions for neurodegenerative and psychiatric disorders [90].

18.6 INTEGRATIVE MULTI-OMICS APPROACHES

The concept of multi-omics involves the integration of several "omics platforms," including genomics, transcriptomics, proteomics, metabolomics, imaging, etc. This integrative strategy improves the chances of understanding the relationships between genes, metabolites, and the brain better.

18.6.1 Definition of Multi-Omics Integration

Systems biology is defined as the study of interactions within cells, organisms, and populations through a combination of data from several omics layers. When genomics, transcriptomics, proteomics, and metabolomics data are combined, all the layers of data which are independent of each other will help to identify new interactions and regulatory mechanisms, which cannot be observed in isolation. The aim is to predict the current and potential stages of health across the biological hierarchy and disease process [91].

It has been possible through integrative approaches that give the different aspects of genetics variations, gene/mRNA expression changes, proteins, and changes in metabolites that are entailing in biological functions [92].

18.6.2 Using Genome Sequencing, Magnetic Resonance Imaging, and Metabolomics

Combining gene sequencing, brain imaging, and metabolomics provides a powerful framework for understanding the brain–gut axis and other complex biological systems. Combining gene sequencing, brain imaging, and metabolomics provides a powerful framework for understanding the brain–gut axis and other complex biological systems.

18.6.2.1 Gene Sequencing

It gives complete information of related genetic variations, gene profiles, and possible mutations. It also assists in determining the patients' gene features that make them susceptible to certain illnesses and genetic modifications that may influence metabolite formation and mind activity [93].

18.6.2.1 Brain Imaging

MRI fMRI and PET scans, for instance, show structural and functional changes in the body's most important organ. Combining brain image information with additional types of information, such as genomic and/or metabolomic data, assists in being able to associate genetic and metabolic alterations with brain and connectivity data [94].

18.6.2.2 Metabolomics

Review and analyze the metabolic end products of the gut microbiota and evaluate their physiological and pathological impacts. When metabolomic data is linked with gene and brain imaging data, new insights can be gained from metabolites produced by the gut bacteria and their effects on the brain and neurological disorders [95].

18.6.3 Case Studies and Examples of Multi-Omics Research

18.6.3.1 Alzheimer's Disease

Scientific investigation of the genomic and metabolomic profile along with neuroimaging has helped in understanding the pathophysiology of Alzheimer's disease. For instance, there are publications indicating specific genes that link to changes in metabolite, and neuroimaging biomarkers that are related to progression of the disease (Table 18.2) [96].

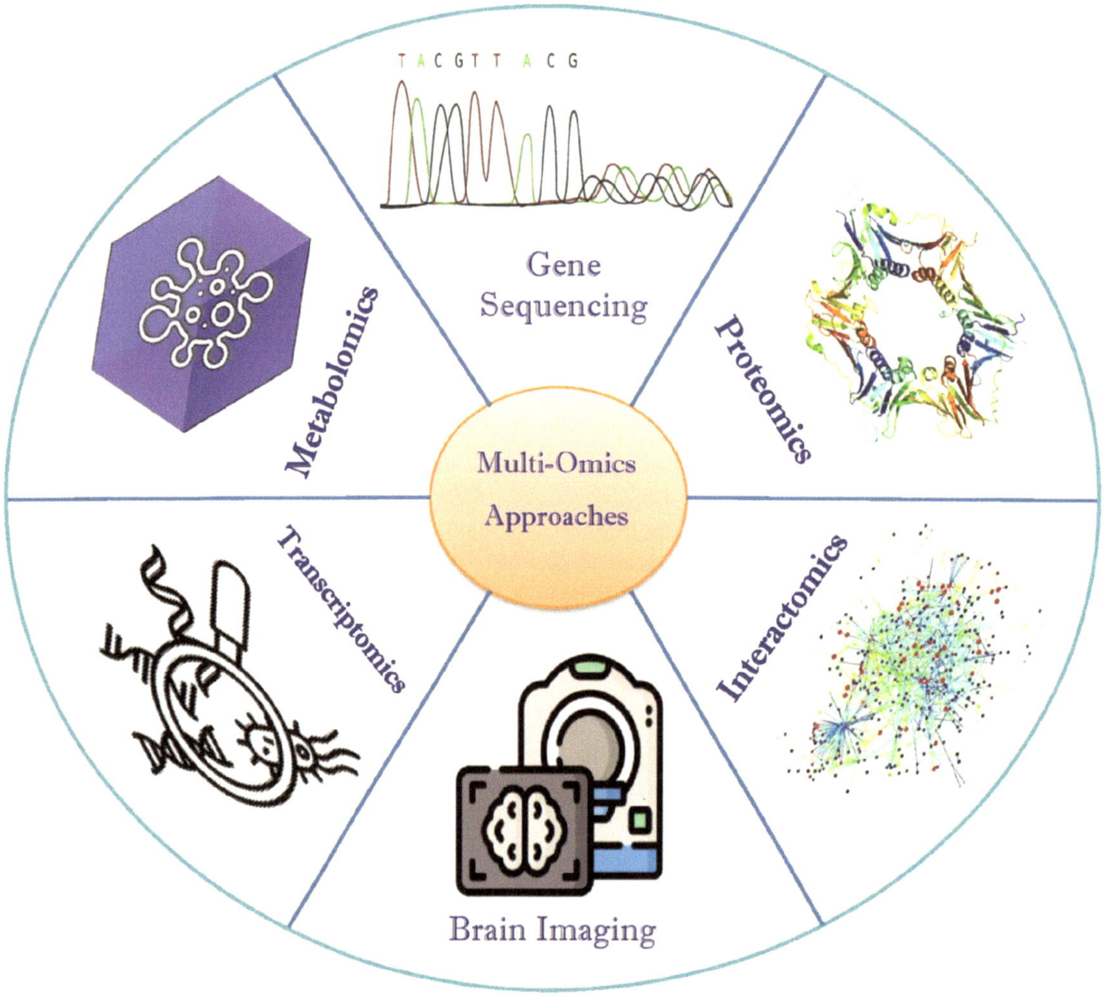

FIGURE 18.2 Concept of multi-omics integration.

TABLE 18.2
Case Studies and Examples of Multi-Omics Research

Case Study	Research Focus	Omics Integrated	Key Findings
Alzheimer's disease neuroimaging initiative (ADNI)	Neuroimaging in Alzheimer's disease	Genomics, proteomics, metabolomic	Found several biomarkers for screening and evaluation of the intensity of Alzheimer's disease [96]
Gut microbiome and brain function	Brain–gut axis	Genomics, metabolomics	Explained how gut microbiota affects the brain and mental health [97]
Cancer research	Multi-omics integration in cancer	Genomics, transcriptomics, proteomics	Provided new information about the pathogenesis of cancer and possible goals of treatment [98]
Resting-state fMRI functional connectivity	Brain network analysis	Transcriptomics, neuroimaging	Discussed connectivity in the brain and the optics in neuropsychiatric disorders [94]
Gut microbiome and metabolomics	Health and disease	Metabolomics, microbiomics	Emphasized the role of gut microbiota on metabolic health and disease status [95]

18.6.3.2 Gut–Brain Axis

Whole-genome shotgun sequencing of the gut microbiome and integrative multimodal neuroimaging and metabolomics have illustrated the communication between the metabolites in the gut and brain functions and mental health. These researches have shown that there are definite sets of metabolic processes that are related to mood and cognition [97].

18.6.3.3 Cancer Research

In diagnostics and prognosis of cancer, a range of multi-omics approaches have been looked at to determine biomarkers. The inclusion of genomic, proteomic, and metabolomic databases has provided new cancer biomarkers and therapies [98].

18.6.4 Integration of Multi-Omics Data and Its Pros and Cons

18.6.4.1 Challenges

Data Complexity: Interactomics is the study of the relationship between different omes and thus entails working with massive but complex datasets that must be handled and analyzed using powerful computational tools [99].

Data Standardization: This is because different omics technologies produce information in numerous styles and with coefficients that are easy to compare and combine [100].

Interpretation: Interactions between layers and effects of changes at one layer on other layers and the overall biological system may not always be easily comprehensible and thus often needs the use of complex modeling and interpretation [101].

18.6.4.2 Opportunities

Holistic Understanding: This multi-omics integration is more comprehensive than the single-omics and leads to the discovery of novel information that is not obtainable from one dimension alone [102].

Personalized Medicine: Thus, the integration of different types of data could allow for the introduction of individualized treatment working with an individual's genetic, metabolic, and brain information [103].

Novel Discoveries: Therefore, integration of different omics data supports research into novel biomarkers and targets in a disease and helps to enhance knowledge of multifactorial diseases [104].

18.7 IMPLICATIONS FOR NEUROPSYCHIATRIC AND GASTROINTESTINAL DISORDERS

To sum up, gut microbiota and its interaction with diverse illnesses opens up important research directions for neuropsychiatric and gastrointestinal diseases. The subsequent sections describe the effects of microbes' presence and the resulting metabolomic shifts and opportunities for the development of precision medicine solutions.

18.7.1 GI Microbiome and Mental Well-Being

Some key interactions between gut microbes and mental health are as follows: The synthesis of neurotransmitters, which are used in the management of brain disorders, influence the immune system, and straight interaction with the brain through the vagus nerve [105]. Prescribing a psychological disorder, there are changes in the balance of the bacterial population in the gut, known as dysbiosis, that have been associated with depression, anxiety, and bipolar disorder. For instance, one piece of research has captured the attention of scientists where it was highlighted that people suffering from depression have different compositions of the gut biota than the normal population, hence a possibility of microbial therapy to treat depression [106, 107]. Furthermore, the application of probiotics and prebiotics as treatment interventions for altering the microbiota with the intention of enhancing mental health has also been looked at [108].

18.7.2 Microbial Mechanisms and Neurological/Psychiatric Diseases

In this section, the given paper also reveals how microbial factors are relevant to different neurological and psychiatric conditions.

18.7.2.1 Autism Spectrum Disorders

The findings made in the studies reveal that children and adults diagnosed with ASD are characterized by changes in gut microbiota composition. A change in the composition of microbiota may impact on behavioral and cognitive manifestations in regard to ASD [109]. Former research shows that microbiota in the gut may partially influence the symptoms and/or cognitive performance of the ASD patients [110].

18.7.2.2 Parkinson's Disease

Given data lead to the conclusion about the possible role of the changes in the composition of the gut microbiota in the development or progression of Parkinson's disease. Alteration in the composition of gut microbiome can alter the synthesis of neurotoxic metabolites, which has effects on neuronal function, development of neuronal diseases [111]. It has also become more common to explore that probiotic treatments might be able to influence the further course of the disease or reduce the symptoms [112].

18.7.2.3 Schizophrenia

The role of gut microbiota in schizophrenia has commenced to attract the attention of researchers. Studies show that microbiota imbalance can affect the development and manifestation of psychotic features using immune regulation and neurotransmitter synthesis [113, 114].

18.7.3 Metabolomic Alterations in Gastrointestinal Afflictions

Thus, metabolomics provides information about the related biochemical changes occurring in GI disorders.

18.7.3.1 Crohn's Disease

In IBD like Crohn's disease and ulcerative colitis, alterations in the levels of SCFAs, bile acids, and other metabolites have been demonstrated with the help of metabolomic profiling. Such changes are associated with alterations in gut microbial metabolism and can be used for providing biomarkers of the disease's activity and progression [115, 116].

18.7.3.2 IBS Patients

Based on metabolomic analysis, there are variations in the levels of the different metabolites between patients with IBS and the control subjects. These include shift in the reversible amino acids, SCFAs, and other metabolates where changes are seen to have symptoms such as abdominal pain and bloating [117].

18.7.3.3 Colorectal Cancer

In the field of biomarkers, metabolomic profiling in CRC has been conducted for the identification of biomolecules, which are either upregulated or downregulated in CRC patient and involved in cellular metabolism and inflammation pathway [118]. For this reason, these biomarkers also have the potential of having a role in screening and assessing the languish of ailment.

18.7.4 Opportunities for the Use of Individualized Therapies

The joining of the two platforms of microbiome and metabolomic data is expected to improve on the idea of personalized medicine.

18.7.4.1 Tailored Therapies

Practically, when the specialist investigates the contents of the human gut and the metabolite profiles, more particular targeted treatment plans can be designed. This includes the use of probiotic or prebiotic therapy that involves the maintenance of this balance through the prescription of a unique gear to suit the patient's conditions [119].

18.7.4.2 Precision Diagnostics

Arguably, metabolomic and microbiome data can improve the diagnostic accuracy due to the unique biomarkers in the study of diseases. This enables more early and precise monitor of the treatment responses [120].

18.7.4.3 Personalised Nutritional and Lifestyle Modifications

Biomedical recommendations of microbiotics and metabolic profiles in patients offer an efficient way of preventing diseases. For instance, regarding those constructed to enhance the positive microbial species or decrease negative by-products, specific diets beneficial for people can be developed [121, 122].

18.8 FUTURE DIRECTIONS AND EMERGING TECHNOLOGIES

Current research involving the brain–gut axis is a comparatively growing area of focus with new technologies such as sequencing, imaging, metabolomics, as well as the artificial intelligence and machine learning. These developments offer hope for enhanced understanding of the brain and gut relationship and for better practice. Other current developments in the sequencing technologies include the high-throughput and low-cost technologies like the single-cell RNA-seq and long-read sequencing. Analyzing the gene expressed by single cells enables making conclusions about cellular variations and the certain types of cells engaged in the brain–gut connection with the use of single-cell RNA sequencing [123]. From the result, long-read sequencing enables the improvement of the genome assembly and the identification of structural variation in aiding the genetic variations that affect the gut–brain interactions [124].

Diagnostic imaging has also undergone improvement by way of technology. Better pictures of the brain in terms of structure and function have been made possible by better techniques such as high-resolution MRI and fMRI. Some of them include the ultrahigh-field MRI, also known as 7T, and more, which give a very clear insight on how connectivity of the brain is affected and changes resulting from the alteration of gut microbiome [125]. Positron emission tomography scan with state-of-art radiotracers allows the visualization of the neuroreceptors and neurotransmitter systems that demonstrate the functional effects of microbial metabolites influencing the brain [126]. This is because metabolomics is growing with the progress in more powerful and through analytical methods. Spatial metabolomics thus enables organization of form metabolite reference within desired tissues which show specific metabolic changes within ended tiny places that may be related to brain function [127]. Moreover, developments in Multiple Metabolite Profiling, which combines information from more than one kind of mass spectrometry, and NMR help improve a larger comprehensiveness of the metabolites besides the interactions [128]. These techniques help in understanding the role of the metabolites originated in the gut in relation to the brain and diseases affecting it.

AI and ML are often used in contemporary studies involving the analysis of large volumes of data in relation to the brain–gut axis. Several features based on genomics, transcriptomics, metabolomics, and imaging can be inputted into AI algorithms and the ensuing patterns and correlations are almost impossible to spot when viewed individually [129]. For instance, it is possible to predict the development or otherwise of certain diseases or the effectiveness of certain treatments through the analysis of data from diverse information sources through machine learning models [130]. Machine learning is also considered to assist the real-time analysis of imaging data of the brain to enhance diagnostic and therapeutic methods [131].

Potential of these emerging technologies for clinical applications is witnessed in the integration of these technologies.

Personalized Medicine: New technologies in sequencing, imaging, and metabolomics allow for the prescription of gene, and metabolic and brain-specific targeted plans of treatment. For example, like in the case of probiotics or prebiotics, specific treatment or nutritional approaches based on the person's microbial profiles may be prescribed to enhance mood or treat gastrointestinal diseases [132].

Early Diagnosis and Monitoring: Imaging and metabolomic studies have been used to improve the diagnostic potential and intensify the surveillance of the diseases associated with the brain–gut axis. The biomarkers discovered through such technologies may result in early diagnosis, therefore improving methods of managing diseases like depression, IBS, and neurodegenerative diseases [133].

Targeted Therapies: AI and machine learning models can help in identification of new targets for the development of drugs by leveraging multi-omics data to understand the molecular pathways that are involved in the disease process. From this, one could potentially arrive at the dawn of customized medications that focus on the root causes of conditions concerning the brain–gut axis [134].

18.9 CONCLUSION

Exploring the concept of brain–gut axis along with the help of sequencing, imaging techniques, metabolomics, and AI provides deep indications concerning the linkages between gut microbiome and brain. New approaches are changing the way researchers study gut-mentioned messages with regard to the brain and the diseases of the body and mind. New and improved genetic, structural, and functional techniques to sequence and image the brains and gastrointestinal tracts offer progressively finer perspectives of the connection between the two. The advanced methods of the high-resolution sequencing techniques and new generation scanners help the researchers better understand how the gut and the brain interact at the molecular and physiological levels. They help in the discovery and understanding of biomarkers associated with several diseases and conditions. Originally defined as the analysis of small molecules, metabolomics is now revolutionizing views on the metabolic alterations that occur in conjunction with gut microbiota and their effects on the brain. Technological advances in spatial and multi-omics metabolomics help investigate how the microbial metabolites regulate the system and/subsequent regional brain activity. This progress increases the capacity to discover new biomarkers and/or targets of treatment. This is sometimes worsened by the fact that a single patient can have multiple omics data, which presents a complex challenge to deciphering; this is where the incorporation of artificial intelligence and machine learning comes into play. Smartness of AI methods improve our skill to forecast outcomes and identify solutions to medical issues on the basis of large amount of information accumulating over the time making treatment plans more sensitive to individual's organism. This integration is believed to quicken the pace of developing personalised treatment plans that meet each patient's trend analysis depending on the health profile materials collected. Thus, looking to the future, the potential of clinical usage is rather large. A new perspective in neuropsychiatric and gastrointestinal diseases' treatment is aimed at the precision approach approximated through the integrated diagnostics of multiple aspects of a patient's state with the help of multi-omics data. Accurate and timely methods of diagnosing, monitoring, and specific therapeutic approaches may help to change the fate of patients and expand the knowledge about the brain and gut relationship.

FUNDING

There was no funding for this project from any source.

ACKNOWLEDGMENTS

The authors acknowledge in particular ShriRam College of Pharmacy, Morena, Madhya Pradesh, India, for encouraging and providing necessary facilities.

REFERENCES

1. Furness, J. B. (2012). The enteric nervous system and neurogastroenterology. *Nature Reviews Gastroenterology & Hepatology, 9*(5), 286–294.
2. Cryan, J. F., & Dinan, T. G. (2012). Mind-altering microorganisms: the impact of the gut microbiota on brain and behaviour. *Nature Reviews Neuroscience, 13*(10), 701–712.
3. Mayer, E. A. (2011). Gut feelings: the emerging biology of gut–brain communication. *Nature Reviews Neuroscience, 12*(8), 453–466.
4. Collins, S. M., Surette, M., & Bercik, P. (2012). The interplay between the intestinal microbiota and the brain. *Nature Reviews Microbiology, 10*(11), 735–742.
5. Qin, J., Li, R., Raes, J., Arumugam, M., Burgdorf, K. S., Manichanh, C., ... & Wang, J. (2010). A human gut microbial gene catalogue established by metagenomic sequencing. *Nature, 464*(7285), 59–65.

6. Tillisch, K., Labus, J., Kilpatrick, L., Jiang, Z., Stains, J., Ebrat, B., ... & Mayer, E. A. (2013). Consumption of fermented milk product with probiotic modulates brain activity. *Gastroenterology, 144*(7), 1394–1401.
7. Nicholson, J. K., Holmes, E., Kinross, J. M., Burcelin, R., Gibson, G., Jia, W., & Pettersson, S. (2012). Host-gut microbiota metabolic interactions. *Science, 336*(6086), 1262–1267.
8. Bonaz, B., Bazin, T., & Pellissier, S. (2018). The vagus nerve at the interface of the microbiota-gut-brain axis. *Frontiers in Neuroscience, 12*, 49.
9. Vermeulen, W., De Man, J. G., & Pelckmans, P. A. (2014). The complex interplay between inflammation, pain and the vagal nerve: a role for calcitonin-gene related peptide (CGRP). *Neuropeptides, 48*(1), 67–80.
10. Goyal, R. K., & Hirano, I. (1996). The enteric nervous system. *New England Journal of Medicine, 334*(17), 1106–1115.
11. Mayer, E. A., Tillisch, K., & Bradesi, S. (2006). Review article: modulation of the brain–gut axis as a therapeutic approach in gastrointestinal disease. *Alimentary Pharmacology & Therapeutics, 24*(6), 919–933.
12. Berthoud, H. R., & Neuhuber, W. L. (2000). Functional and chemical anatomy of the afferent vagal system. *Autonomic Neuroscience, 85*(1–3), 1–17.
13. Cummings, D. E., & Overduin, J. (2007). Gastrointestinal regulation of food intake. *Journal of Clinical Investigation, 117*(1), 13–23.
14. Dockray, G. J. (2014). Cholecystokinin. *Current Opinion in Endocrinology, Diabetes, and Obesity, 21*(1), 46–51.
15. Holst, J. J., & Deacon, C. F. (2005). Glucagon-like peptide-1 mediates the therapeutic actions of DPP-IV inhibitors. *Diabetologia, 48*(4), 612–615.
16. Morton, G. J., Meek, T. H., & Schwartz, M. W. (2014). Neurobiology of food intake in health and disease. *Nature Reviews Neuroscience, 15*(6), 367–378.
17. Mowat, A. M. (2003). Anatomical basis of tolerance and immunity to intestinal antigens. *Nature Reviews Immunology, 3*(4), 331–341.
18. Dantzer, R., O'Connor, J. C., Freund, G. G., Johnson, R. W., & Kelley, K. W. (2008). From inflammation to sickness and depression: when the immune system subjugates the brain. *Nature Reviews Neuroscience, 9*(1), 46–56.
19. Banks, W. A. (2005). Blood-brain barrier transport of cytokines: A mechanism for neuropathology. *Current Pharmaceutical Design, 11*(8), 973–984.
20. Macia, L., Tan, J., Vieira, A. T., Leach, K., Stanley, D., Luong, S., ... & Mackay, C. R. (2015). Metabolite-sensing receptors GPR43 and GPR109A facilitate communication between the gut microbiota and inflammatory cells to control intestinal inflammation and extra-gastrointestinal inflammation. *Immunity, 42*(5), 989–1001.
22. Gabanyi, I., Muller, P. A., Feighery, L., Oliveira, T. Y., Costa-Pinto, F. A., & Mucida, D. (2016). Neuro-immune interactions drive tissue programming in intestinal macrophages. *Cell, 164*(3), 378–391.
22. Gareau, M. G., Silva, M. A., & Perdue, M. H. (2008). Pathophysiological mechanisms of stress-induced intestinal damage. *Current Molecular Medicine, 8*(4), 274–281.
23. Slyepchenko, A., Carvalho, A. F., Cha, D. S., Kasper, S., & McIntyre, R. S. (2014). Gut emotions—Mechanisms of action of probiotics as novel therapeutic targets for depression and anxiety disorders. *CNS & Neurological Disorders-Drug Targets, 13*(10), 1770–1786.
24. Koh, A., De Vadder, F., Kovatcheva-Datchary, P., & Bäckhed, F. (2016). From dietary fiber to host physiology: Short-chain fatty acids as key bacterial metabolites. *Cell, 165*(6), 1332–1345.
25. Sharma, P. (2021). Modification of human behavior due to coronavirus outbreak: A brief study on current scenario. *Asian Journal of Pharmaceutics (AJP), 15*(3), 392–398.
26. Wahlström, A., Sayin, S. I., Marschall, H. U., & Bäckhed, F. (2016). Intestinal crosstalk between bile acids and microbiota and its impact on host metabolism. *Cell Metabolism, 24*(1), 41–50.
27. Kang, D. W., Park, J. G., Ilhan, Z. E., Wallstrom, G., LaBaer, J., Adams, J. B., & Krajmalnik-Brown, R. (2013). Reduced incidence of Prevotella and other fermenters in intestinal microflora of autistic children. *PLOS ONE, 8*(7), e68322.
28. Sampson, T. R., & Mazmanian, S. K. (2015). Control of brain development, function, and behavior by the microbiome. *Cell Host & Microbe, 17*(5), 565–576.
29. Dinan, T. G., & Cryan, J. F. (2017). Gut-brain axis in 2016: Brain-gut-microbiota axis—Mood, metabolism and behaviour. *Nature Reviews Gastroenterology & Hepatology, 14*(2), 69–70.
30. Wang, Z., & Gerstein, M. (2008). Next-generation sequencing technology and its applications. *Nature Reviews Genetics, 10*(1), 57–63.
31. Metzker, M. L. (2010). Sequencing technologies—The next generation. *Nature Reviews Genetics, 11*(1), 31–46.
32. Jovel, J., Patterson, J., Wang, W., Hotte, N., O'Keefe, S., & Madsen, K. L. (2016). Characterization of the gut microbiome using 16S or shotgun metagenomics. *Frontiers in Microbiology, 7*, 459.
33. Meyerson, M., Gabriel, S., & Getz, G. (2010). Advances in understanding cancer genomes through second-generation sequencing. *Nature Reviews Genetics, 11*(10), 685–696.
34. Sharma, P., Jain, V., & Tailang, M. (2022). Selection and role of polymers for designing of a drug carrier. In *Drug Carriers* (pp. 1–23). London: IntechOpen.
35. Wang, Z., Gerstein, M., & Snyder, M. (2009). RNA-Seq: a revolutionary tool for transcriptomics. *Nature Reviews Genetics, 10*(1), 57–63.
36. Hsiao, E. Y., McBride, S. W., Hsien, S., Sharon, G., Hyde, E. R., McCue, T., ... & Mazmanian, S. K. (2013). Microbiota modulate behavioral and physiological abnormalities associated with neurodevelopmental disorders. *Cell, 155*(7), 1451–1463.
37. Jeffery, I. B., O'Toole, P. W., Ohman, L., Claesson, M. J., Deane, J., Quigley, E. M., & Simrén, M. (2012). An irritable bowel syndrome subtype defined by species-specific alterations in faecal microbiota. *Gut, 61*(7), 997–1006.
38. Sharma, P., & Tailang, M. (2018). In-vivo study of orodispersible tablet of primaquine. *International Journal of Pharmaceutical Sciences and Research, 9*(8), 3506–3510.
39. Jovel, J., Patterson, J., Wang, W., Hotte, N., O'Keefe, S., & Madsen, K. L. (2016). Characterization of the gut microbiome using 16S or shotgun metagenomics. *Frontiers in Microbiology, 7*, 459.
40. Turnbaugh, P. J., Ley, R. E., Mahowald, M. A., Magrini, V., Mardis, E. R., & Gordon, J. I. (2006). An obesity-associated gut microbiome with increased capacity for energy harvest. *Nature, 444*(7122), 1027–1031.
41. Quince, C., Walker, A. W., Simpson, J. T., Loman, N. J., & Segata, N. (2017). Shotgun metagenomics, from sampling to analysis. *Nature Biotechnology, 35*(9), 833–844.

42. Morgan, X. C., Tickle, T. L., Sokol, H., Gevers, D., de Vries, M., & Ward, D. V. (2012). Dysfunction of the intestinal microbiome in inflammatory bowel disease and treatment. *Genome Biology, 13*(9), R79.
43. Sharma, P., & Tailang, M. (2022). Primaquine-loaded transdermal patch for treating malaria: design, development, and characterization. *Future Journal of Pharmaceutical Sciences, 8*(1), 43.
44. Schirmer, M., Garber, J., & Vlamakis, H. (2018). Microbial functional capacity and gut microbiota composition are linked in inflammatory bowel disease. *Journal of Crohn's and Colitis, 12*(3), 257–267.
45. Handelsman, J., Rondon, M. R., Brady, S. F., Clardy, J., & Goodman, R. M. (1998). Molecular biological access to the chemistry of unknown soil microbes: A new frontier. *Chemical Reviews, 105*(2), 586–597.
46. Kang, D. W., Park, J. G., Ilhan, Z. E., Wallstrom, G., LaBaer, J., Adams, J. B., & Krajmalnik-Brown, R. (2013). Reduced incidence of Prevotella and other fermenters in intestinal microflora of autistic children. *PLOS ONE, 8*(7), e68322.
47. Zhang, X., Zhang, D., Wang, H., Wang, S., & Zhao, L. (2015). Human gut microbiota changes reveal the progression of colorectal cancer. *Frontiers in Microbiology, 6*, 478.
48. Qin, J., Li, Y., Cai, Z., Li, S., Zhu, J., & Zhang, F. (2012). A human gut microbial gene catalogue established by metagenomic sequencing. *Nature, 490*(7418), 56–65.
49. Sharma, P., Jain, V., & Jain, S. (2021). Fabrication, optimization and evaluation of chronotropic drug delivery system of captopril. *International Journal of Pharmaceutical Sciences and Research, 12*, 2203–2210.
50. Montalban-Arques, A., & Llopis, M. (2016). Microbiota and its role in cancer. *Current Opinion in Oncology, 28*(3), 256–263.
51. Sharma, P., & Tailang, M. (2020). Design, optimization and evaluation of buccal drug delivery system of propranolol for hypertension treatment. *International Journal of Pharmaceutical Sciences and Research, 11*(1), 301–311.
52. Louis, P., & Flint, H. J. (2017). Development of the gut microbiota in infants and children. *Food & Function, 8*(8), 2281–2292.
53. Mörkl, S., Salvatore, S., & Kaser, A. (2021). The role of the gut microbiome in the development of neurological disorders. *Frontiers in Neurology, 12*, 752897.
54. Zhang, X., Zhang, D., Wang, H., Wang, S., & Zhao, L. (2015). Human gut microbiota changes reveal the progression of colorectal cancer. *Frontiers in Microbiology, 6*, 478.
55. Parker, A., & O'Malley, D. (2022). The role of the microbiome in neurodegenerative diseases. *Brain Research, 1781*, 147765.
56. Dinan, T. G., & Cryan, J. F. (2017). Gut-brain axis in 2016: Brain-gut-microbiota axis—Mood, metabolism and behaviour. *Nature Reviews Gastroenterology & Hepatology, 14*(2), 69–70.
57. Yano, J. M., Yu, K., Donaldson, G. P., Shastri, G. G., Ann, P., Ma, L., ... & Hsiao, E. Y. (2015). Indigenous bacteria from the gut microbiota regulate host serotonin biosynthesis. *Cell, 161*(2), 264–276.
58. Sharma, P., & Jain, V. (2023). An overview of current advances and pharmaceutical uses of 3D and 4D printing. *Exploration of Medicine, 4*, 560–75.
59. Woolrich, M. W., Behrens, T. E., & Smith, S. M. (2004). Constrained linear basis sets for HRF modelling in EEG and fMRI. *NeuroImage, 21*(1), 174–184.
60. Ogawa, S., Lee, T. M., Kay, A. R., & Tank, D. W. (1990). Brain magnetic resonance imaging with contrast dependent on blood oxygenation. *Nature, 348*(6298), 599–604.
61. Logan, J., & Fowler, J. S. (1993). Positron emission tomography: Theory and applications. *Science, 259*(5100), 1426–1433.
62. Clark, J. C., & Smith, E. F. (2011). Advances in imaging technology: PET and SPECT. *Clinical Imaging, 35*(4), 328–334.
63. Sharma, P., Jain, V., & Tailang, M. (2023). Advancement of nanocarrier-based engineering for specific drug delivery for cancer therapy. In *Targeted Cancer Therapy in Biomedical Engineering* (pp. 465–486). Singapore: Springer Nature.
64. Basser, P. J., & Pierpaoli, C. (1996). Microstructural and physiological features of tissues elucidated by quantitative-diffusion-tensor MRI. *Journal of Magnetic Resonance, 111*(3), 209–219.
65. Sharma, P. (2022). Applications of statistical tools for optimization and development of smart drug delivery system. In *Smart Drug Delivery* (p. 183). London: Intech Open.
66. Schmithorst, V. J., & Wilke, M. (2008). The relationship between white matter and cognitive ability in children. *Developmental Cognitive Neuroscience, 2*(4), 413–426.
67. Matsuoka, K., & Kato, K. (2018). Brain-gut microbiota interactions: Implications for neurodegenerative diseases. *Neuropsychobiology, 77*(2), 87–97.
68. Fan, L., Li, X., & Wang, H. (2017). Microbiota-gut-brain axis and its role in neuropsychiatric disorders. *Brain Research Bulletin, 129*, 1–11.
69. Tillisch, K., Labus, J. S., & Kilpatrick, L. A. (2013). Consumption of fermented milk product with probiotic modulates brain activity. *Gastroenterology, 144*(7), 1394–1401.
70. Desbonnet, L., Clarke, G., & Traplin, A. (2015). Gut microbiota depletion alters the function of the blood-brain barrier. *Frontiers in Neuroscience, 9*, 17.
71. Cryan, J. F., & Dinan, T. G. (2012). Mind-altering microorganisms: The impact of the gut microbiota on brain and behavior. *Nature Reviews Neuroscience, 13*(10), 701–712.
72. Zhang, X., Zhang, D., & Wang, H. (2015). Gut microbiota and brain function: insights from functional imaging studies. *Current Opinion in Neurology, 28*(6), 707–714.
73. Kharbouch, A., Jha, S., & Kumar, S. (2020). PET imaging and its role in understanding brain-gut microbiota interactions. *Journal of Nuclear Medicine, 61*(9), 1346–1354.
74. Mörkl, S., Salvatore, S., & Kaser, A. (2021). The role of the gut microbiome in the development of neurological disorders. *Frontiers in Neurology, 12*, 752897.
75. Zhang, X., Zhang, D., & Wang, H. (2015). Gut microbiota and brain function: insights from functional imaging studies. *Current Opinion in Neurology, 28*(6), 707–714.
76. Sharma, P., Bhargava, S., Parashar, D., & Mangal, A. (2017). Formulation and evaluation of nanoparticles containing cyclophosphamide. *WJPR, 7*(1), 785–798.
77. Nakatani, K., & Kinoshita, H. (2019). Neuroimaging biomarkers of neuroinflammation: A review. *Frontiers in Neuroscience, 13*, 711.
78. Wishart, D. S., Feunang, Y. D., & Marcu, A. (2018). HMDB 4.0: The human metabolome database for 2018. *Nucleic Acids Research, 46*(D1), D608–D617.
79. Markley, J. L., Bruschweiler–Li, L., & Edison, A. S. (2017). The future of NMR-based metabolomics. *Current Opinion in Biotechnology, 43*, 34–40.

80. Pankaj, S., Ashish, D., Dheerendra, R., Yogendra, R., Ramakant, J., & Kuldeep, S. (2015). To formulate and evaluate orodispersible tablets of primaquine. *Indo American Journal of Pharmaceutical Research, 5*(5), 1625–32.
81. Wang, M., Carr, P. W., & Rejtar, T. (2010). Liquid chromatography-mass spectrometry for the analysis of complex biological samples. *Journal of Chromatography A, 1217*(25), 3827–3839.
82. Louis, P., & Flint, H. J. (2017). Development of the gut microbiota in infants and children. *Food & Function, 8*(8), 2281–2292.
83. Mayer, E. A., & Gutner, J. (2021). The role of the gut microbiome in psychiatric and neurodegenerative disorders. *Nature Reviews Neuroscience, 22*(10), 570–581.
84. Roda, A., & Sartor, R. B. (2021). The role of microbial metabolites in neurodegenerative disease. *Trends in Neurosciences, 44*(7), 580–591.
85. Ridaura, V. K., & Ley, R. E. (2013). Gut microbiota from twins discordant for obesity modulate metabolism in mice. *Science, 341*(6150), 1241214.
86. Zheng, P., & Zhang, J. (2016). Gut microbiome and its role in neurodevelopment and neurodegeneration. *Frontiers in Neuroscience, 10*, 468.
87. Duca, F. A., & Zhang, L. (2018). The role of the gut microbiome in metabolic disease. *Current Diabetes Reports, 18*(7), 36.
88. Gupta, A., & Singh, P. (2022). Integrating genomics and metabolomics for personalized medicine. *Journal of Personalized Medicine, 12*(4), 613.
89. Sharma, P., Jain, V., & Tailang, M. (2022). Swarm intelligence and machine learning algorithms for cancer diagnosis. In *Swarm Intelligence and Machine Learning* (pp. 34–50). CRC Press.
90. Choi, M., & Edden, R. A. (2017). Metabolomics and systems biology approaches to understand brain function. *Trends in Neurosciences, 40*(6), 447–461.
91. Cordero, F., & Fernandez, M. (2020). Multi-omics approaches to understanding complex biological systems. *Nature Reviews Molecular Cell Biology, 21*(12), 759–774.
92. Patti, G. J., Yanes, O., & Siuzdak, G. (2012). Metabolomics: The apogee of the omics trilogy. *Nature Reviews Molecular Cell Biology, 13*(4), 263–269.
93. Adams, R. D., & Montgomery, J. (2018). Integrative genomics for understanding complex diseases. *Journal of Human Genetics, 63*(7), 751–759.
94. van den Heuvel, M. P., & Hulshoff Pol, H. E. (2010). Exploring the brain network: Areview on resting-state fMRI functional connectivity. *European Neuropsychopharmacology, 20*(8), 519–534.
95. Mardinoglu, A., & Boren, J. (2018). The gut microbiome and metabolomics in health and disease. *Nature Reviews Gastroenterology & Hepatology, 15*(6), 345–357.
96. Kantarci, K., & Weiner, M. W. (2020). Alzheimer's disease neuroimaging initiative (ADNI): A review of neuroimaging studies in Alzheimer's disease. *Journal of Alzheimer's Disease, 74*(3), 737–758.
97. Mörkl, S., & Kaser, A. (2021). Gut microbiome and brain function: A review of the literature. *Frontiers in Neurology, 12*, 752897.
98. Yao, J., & Zhang, X. (2020). Multi–omics integration in cancer research. *Frontiers in Oncology, 10*, 538.
99. van den Berg, R. A., & Hoefsloot, H. C. J. (2018). The challenges of integrating multi-omics data. *Trends in Biotechnology, 36*(3), 202–210.
100. Misra, B. B., & Langefeld, C. D. (2018). Data integration and multi-omics approaches in personalized medicine. *Bioinformatics, 34*(24), 4106–4117.
101. Batra, R., Jain, V., & Sharma, P. (2021). Adenosine: A partially discovered medicinal agent. *Future Journal of Pharmaceutical Sciences, 7*, 1–10.
102. Mo, Y., & Chen, J. (2018). Multi-omics approaches for discovering biomarkers in complex diseases. *Bioinformatics, 34*(15), 2517–2523.
103. Hood, L., & Eberwine, J. (2018). Systems biology and personalized medicine. *Science Translational Medicine, 10*(423), eaao3798.
104. Pankaj, S., & Mukul, T. (2019). PH dependent release potential of natural polymers in sustained release of ornidazole from colon targeted delivery system. *Magnesium, 10*(10), 10.
105. Cryan, J. F., & Dinan, T. G. (2012). Mind-altering microorganisms: The gut microbiota and mental health. *Nature Reviews Microbiology, 10*(9), 678–688.
106. Sharma, P., & Tailang, M. (2017). Opportunities and challenges to develop transdermal patch with nanocarrier. *WJPR, 6*(15), 392–400.
107. Wallace, C. J. K., & Milev, R. (2017). The effects of probiotics on depressive symptoms in humans: A systematic review. *Annals of General Psychiatry, 16*, 14.
108. Mörkl, S., & Kaser, A. (2021). Probiotics and prebiotics in the management of mental health conditions: A review of current evidence. *Journal of Psychiatric Research, 139*, 255–267.
109. Sharma, P. (2023). Design, optimization, and characterization of herbal gel for facial application. *Asian Journal of Pharmaceutics (AJP), 17*(3), 486–494.
110. Finegold, S. M., & Molitoris, D. R. (2008). Gastrointestinal microbiota studies in autism. *Clinical Infectious Diseases, 45*(6), 727–733.
111. Unger, M. M., & Spiegel, J. (2016). Alteration of the gut microbiome in Parkinson's disease. *Movement Disorders, 31*(5), 756–759.
112. van den Bogaard, E., & Paulissen, S. M. J. (2017). Probiotic supplementation in Parkinson's disease: A review. *Movement Disorders Clinical Practice, 4*(5), 532–542.
113. Sharma, P., Jain, V., & Tailang, M. (2023). How artificial intelligence is transforming medicine: The future of pharmaceutical research. In *Industry 4.0 and Healthcare: Impact of Artificial Intelligence* (pp. 133–150). Singapore: Springer Nature.
114. Müller, M., & Funder, J. W. (2020). Gut microbiota and schizophrenia: A review of the literature. *Journal of Neuropsychiatry and Clinical Neurosciences, 32*(2), 82–89.
115. Sharma, P., Jain, V., Jain, A., & Tailang, M. (2022). Role of swarm intelligence for health monitoring and related actions. In *Internet of Things and Fog Computing-Enabled Solutions for Real-Life Challenges* (pp. 95–118). CRC Press.
116. Sokol, H., & Seksik, P. (2020). Metabolomics in inflammatory bowel disease. *Journal of Crohn's and Colitis, 14*(3), 345–357.
117. Parkes, G. C., & Pritchard, S. A. (2012). Metabolomics of irritable bowel syndrome: A review. *World Journal of Gastroenterology, 18*(27), 3687–3696.
118. Kuehne, A., & Hsu, C. (2019). Metabolomic profiling of colorectal cancer. *Cancer Metabolism, 7*, 7.
119. Kuehne, A., & Martinez, M. (2020). Personalized microbiome-based approaches in the treatment of gastrointestinal diseases. *Nature Reviews Gastroenterology & Hepatology, 17*(12), 723–735.

120. Sharma, P. (2023). Nanostructured biomaterials in drug delivery: Current trends and upcoming possibilities. In *Engineered biomaterials: synthesis and applications* (pp. 261–280). Singapore: Springer Nature.
121. David, L. A., & Maurice, C. F. (2014). Diet rapidly and reproducibly alters the gut microbiome of mice. *Nature, 505*(7484), 559–563.
122. Rooks, M. G., & Garrett, W. S. (2016). From the gut microbiota to the gut-brain axis: Dietary modulation of the microbiome. *Journal of Neuroimmune Pharmacology, 11*(2), 261–273.
123. Dixit, A., Jain, A. K., Sharma, N., Saluja, G. S., Acharya, M., Rathore, D., ... & Singh, Y. (2015). Hepatoprotective activity of wrightia tinctoria leaves against d-galactosamine induced hepatotoxicity in rats. *Indo American Journal of Pharmaceutical Research, 5*(1), 422–426.
124. Oxford Nanopore Technologies. (2020). Advances in long-read sequencing technologies. *Nature Reviews Genetics, 21*(8), 467–480.
125. Marques, J. P., & Turner, R. (2019). High-resolution brain imaging: Advances and applications. *NeuroImage, 195*, 55–70.
126. Ikoma, Y., & Toyama, H. (2020). Advances in PET imaging technologies and radiotracers. *Frontiers in Nuclear Medicine, 10*, 274.
127. Gibbons, M. A., & Aguiar, M. (2020). Spatial metabolomics for brain research: New horizons and challenges. *Metabolomics, 16*(7), 67.
128. Kuehnbaum, N. L., & Britz-McKibbin, P. (2013). Advanced mass spectrometry techniques in metabolomics. *Annual Review of Analytical Chemistry, 6*, 215–234.
129. Shen, X., & Xu, X. (2021). Integrating multi-omics data with machine learning for personalized medicine. *Bioinformatics, 37*(12), 1761–1770.
130. Liu, Y., & Wang, Y. (2021). Machine learning approaches in analyzing multi-omics data for disease prediction. *Journal of Biomedical Informatics, 116*, 103727.
131. Lee, H., & Lee, S. (2020). Real-time analysis of brain imaging data using AI: Current trends and future directions. *Neuroinformatics, 18*(1), 19–36.
132. Goyal, M. S., & McCarthy, C. (2022). Personalized probiotics for mental health: A future perspective. *Journal of Clinical Gastroenterology, 56*(5), 410–422.
133. Fiehn, O. (2021). Metabolomics for early detection and monitoring of disease. *Clinical Chemistry, 67*(3), 348–359.
134. Schadt, E. E., & Zhang, B. (2021). Targeted therapies through multi-omics data integration. *Nature Reviews Drug Discovery, 20*(1), 11–29.

19 Signaling Pathways and Gut–Brain Axis *Relationship in Neurological Disorders*

Papia Acharjee[1], *Vishal Vikram Singh*[1], *Shambhu Kumar Prasad*[1], *Priyanka Thakur*[1], *and Arup Acharjee*[2*]

[1] Biochemistry and Molecular Biology Unit, Department of Zoology, Institute of Science, Banaras Hindu University, Varanasi, India
[2] Molecular Omics Laboratory, Department of Zoology, University of Allahabad, Prayagraj, India

19.1 INTRODUCTION

19.1.1 Overview of Gut–Brain Axis

The gastrointestinal tract (GIT) and the central nervous system (CNS) are connected by the GBA, a bidirectional reciprocal communication network that facilitates information exchange and preserves physiological homeostasis, especially in neurological disorders. This network is made up of neural, immune, endocrine, and microbial pathways. The neurological components include the enteric nervous system (ENS), vagus nerve, and autonomic nervous system (ANS), with the ENS managing gastrointestinal functions and the vagus nerve mediates gut brain communication associated with digestion, mood, and inflammation (1). The immune pathway involves cytokines and immune signaling molecules from the gut, responsible for chronic gut inflammation led blood–brain barrier (BBB) disruption and neurodegeneration (2). Endocrine pathways involve hormones like cortisol and ghrelin that regulate stress responses, appetite, and immune functions. The hypothalamic–pituitary–adrenal (HPA) axis serves as the central stress response system (3). Metabolites like short-chain fatty acids (SCFAs) and tryptophan derivatives released by gut microbiota play a crucial role in gut–brain signaling, via microbial pathway and in neurological functions.

19.1.2 Importance of Signaling Pathways in Gut–Brain Communication

The GBA signaling pathways, involving neural, immune, endocrine, and microbial mechanisms, are vital for gut–brain communication. The ENS, vagus nerve, and autonomic nervous system regulate gastrointestinal motility, mood, and behavior (1). Gut-derived immune signals affect brain function, and chronic gut inflammation is associated with multiple sclerosis (MS) and autism (2). Gut-produced hormones like cortisol and ghrelin cross the BBB and modulate neural circuits and neurotransmitter systems essential for stress responses and emotional regulation (3). The gut microbiota-produced metabolites (SCFAs and tryptophan derivatives) influence physiological functioning of brain and behavior. Microbial imbalance or dysbiosis causes neuroinflammation, a primary pathological cause in several neurological disorders (4). Understanding these pathways is imperative to comprehend the role of GBA disruption in neurological disorders and the development of targeted therapies.

19.1.3 Relevance to Neurological Disorders

In recent times, impact of GBA in brain disorders has received significant attention. Various neurological conditions, like Alzheimer's disease (AD), multiple sclerosis (MS), Parkinson's disease (PD), depression, ASD, epilepsy, and Huntington's disease (HD) show distinct interactions with GBA. In AD, gut dysbiosis is linked to amyloid plaque formation and neuroinflammation, suggesting that gut microbiota may influence its pathogenesis (5). PD is characterized by alpha-synuclein aggregation, modulated by gut microbiota, with the vagus nerve serving as a crucial transmission route for these signals (4). In MS, immune-mediated pathways affected by gut microbiota play a significant role, with peripheral inflammation affecting brain health (2). Besides, gut microbiota modulates the HPA stress center and inflammatory pathways, playing a definitive role in depression and anxiety (6). In ASD, the gut microbiota influences neurodevelopment, and dysbiosis is associated with behavioral symptoms (7). Epilepsy is caused by microbial dysbiosis, which can influence seizure susceptibility and severity, implicating the pathophysiology (4). Lastly, in HD, microbial metabolites affect neuroinflammation and neurodegeneration, highlighting the influence of (8). The complex relation of the gut and the brain underscores the importance of GBA in understanding and treating neurological disorders.

DOI: 10.1201/9781003536802-19

19.2 COMPONENTS OF THE GUT–BRAIN AXIS

19.2.1 Neural Pathways

19.2.1.1 Enteric Nervous System

The ENS (also known as second brain) comprises of an immense network of neurons lining the GIT and can function independently and with the CNS. It regulates gastrointestinal activities such as motility, secretion, and absorption using neurotransmitters such as serotonin and acetylcholine to execute complex reflexes autonomously and connect the CNS through the vagus nerve (1).

19.2.1.2 Vagus Nerve

The vagus nerve is a primary pathway connecting the gut with the brain, extending from the brainstem to the thoracic and abdominal organs, such as the heart, lungs, and digestive tract. It primarily transmits afferent signals from the gut to the brain, conveying information on nutrient levels and inflammation and affecting brain areas related to mood, stress, and satiety. Efferent signals from the brain to the gut through the vagus nerve regulate gastrointestinal activities such as peristalsis and secretion (9).

19.2.1.3 Autonomic Nervous System

The cornerstone of the GBA is the ANS, which is comprised of the parasympathetic and sympathetic nervous systems. It controls involuntary processes, including respiration, heartbeat, bowel movements, and pupillary reflex. The SNS, linked with the "fight or flight" response, can slow gastrointestinal motility and reduce digestive secretions during acute stress, which is potentially harmful if chronic (4). The PNS, including the vagus nerve, supports "rest and digest" activities, promotes gastrointestinal motility, enhances digestive secretions, and aids nutrient absorption.

19.2.2 Immune Signaling Pathways

The immune pathways of GBA involve the gastrointestinal immune and nervous systems, collaborating to detect and respond to the changing intestinal environment. Maintaining immune homeostasis and preventing disease development is dependent on a variety of both adaptive and innate immune cell types, which include macrophages, dendritic cells, T cells, microglia, and astrocytes, as well as signaling molecules like cytokines, gut-derived nutrients, and gut microbiota. These cells also affect gut and brain function. Neuroinflammation, ischemic brain damage, neurogenesis, and plasticity are all associated with immune system activation in conjunction with GBA. Gut immune cells generate cytokines, which are tiny proteins that control and regulate immunity, inflammation, and hematopoiesis. Pro- and anti-inflammatory cytokines include interleukins (ILs), tumor necrosis factor-alpha (TNF-α), and interferons (IFNs). These cytokines affect the vagus nerve, transmit signals from the gut to the brain, regulating mood and behavior. When gut inflammation occurs, they can pass across the BBB, which lets immune cells and signaling chemicals into the CNS and induce neuroinflammation (10) (Figure 19.1).

19.2.3 Endocrine Signaling Pathways

Endocrine pathways in GBA involve hormone release and regulation, linking the GIT, CNS, and the HPA axis. Serotonin (5-HT), a gut-produced hormone, regulates intestinal movements, mood, behavior, and brain function via

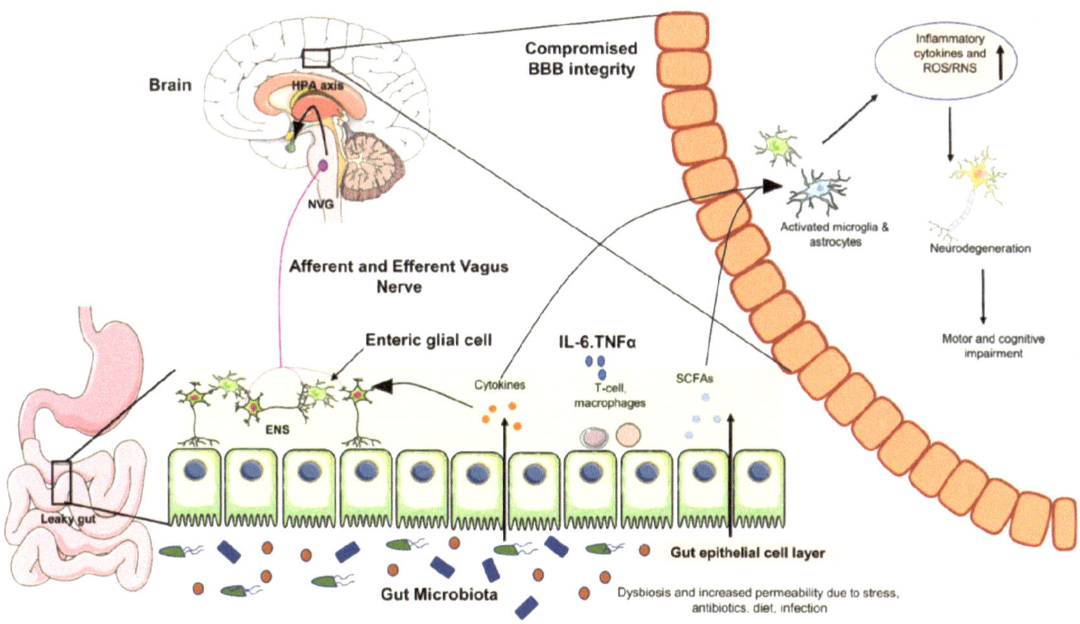

FIGURE 19.1 Pathways of gut–brain communication and the impact of dysbiosis on neural and immune signaling.

vagus nerve signaling and immune system influences (2). Cortisol, a stress hormone from the adrenal glands, controls metabolism, immune responses, and stress (1). Ghrelin, produced in the stomach, stimulates appetite (11), while Peptide YY (PYY), secreted post-food intake, decreases appetite, regulates gut motility and nutrient absorption, and affects brain function via vagus nerve signaling (12). Glucagon-like peptide-1 (GLP-1), released during food intake, enhances insulin secretion, offers neuroprotection, promotes satiety, and reduces food intake (13).

19.2.3.1 Hypothalamic–Pituitary–Adrenal Axis

The hypothalamus, pituitary, and adrenal glands are linked with GBA by the HPA axis, a major stress response mechanism. When under stress, the hypothalamus produces corticotropin-releasing hormone (CRH), which leads to secretion of adrenocorticotropic hormone (ACTH) by the pituitary gland. In order to help manage stress, ACTH stimulates the adrenal glands to create cortisol, which increases energy availability and suppresses non-essential processes, including immunological reactions and digestion (14). Chronic HPA axis activation can cause "leaky gut," alter gut microbiota, affect brain function, and contribute to depression and anxiety (1) (Figure 19.2).

19.2.4 Microbial Signaling Pathways

The microbial pathways in the GBA include gut microbiota and their metabolites, which significantly impact gut brain functions. The gut microbiota consisting of trillions of microorganisms in the GIT generates various metabolites that affect GBA. Gut bacteria degrade dietary fiber to create SCFAs (butyrate, propionate, and acetate). SCFAs are crucial for maintaining gut barrier integrity, modulating immune responses, and influencing brain function via their anti-inflammatory and neuroactive properties (15). Tryptophan, an essential aromatic amino acid, is metabolized into bioactive substances, including kynurenine and serotonin by gut microorganisms. These molecules affect behavior and brain function via influencing immunological responses and neurotransmitter systems (16). Gut bacteria also modify bile acids, which then serve as signaling molecules that affect metabolism and brain function. Altered bile acid profiles are associated with various neurological disorders (17).

19.3 NEURAL SIGNALING PATHWAYS

19.3.1 Enteric Nervous System and Brain Communication

The ENS consists of a complex network of intrinsic primary afferent neurons (IPANs), interneurons, and motor neurons embedded within the walls of GIT. The whole gastrointestinal system is home to enteric nerve fibers, arranged in form of ganglionated networks that encircle the intestinal tube. These neurons are physically classified into two layers: the submucosal plexus in the submucosa and the myenteric plexus, located between the circular and longitudinal muscle layers. Enteric glial cells, motor neurons, and interneurons are tightly linked to these two plexi. Reflex circuits that regulates peristalsis and secretory functions are also formed by them. Furthermore, they combine inputs from sympathetic, parasympathetic, and extrinsic sensory neurons to govern the physiology of the GIT (18).

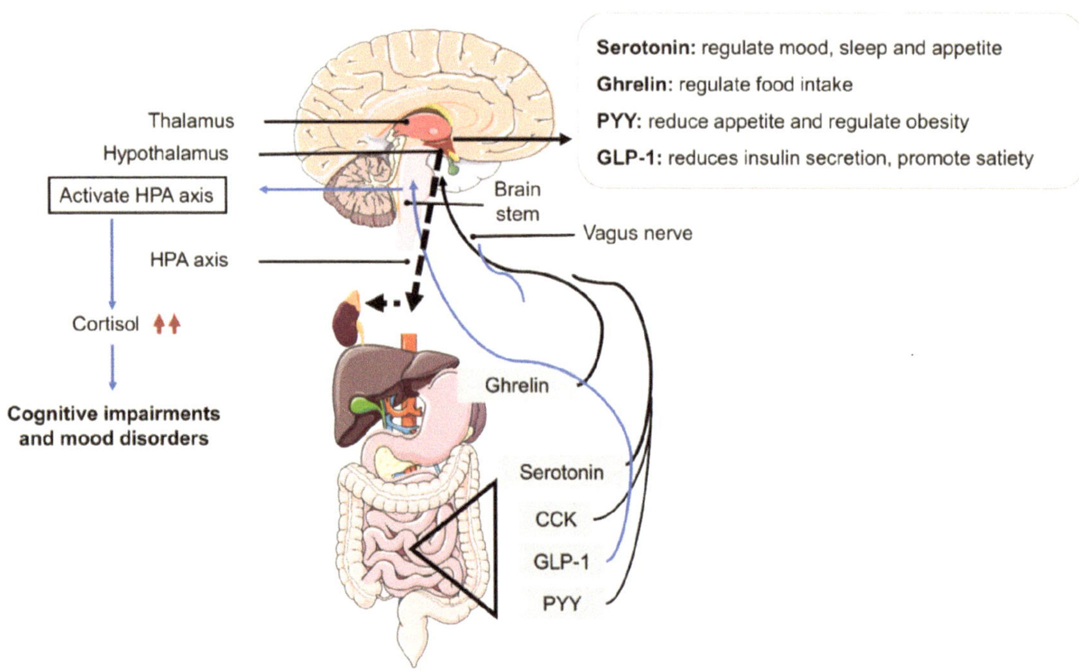

FIGURE 19.2 Bidirectional communication between the gut and the brain via the vagus nerve and HPA axis.

It operates independent of the CNS, but can also interact with the brain via various neural pathways, including the vagus nerve and spinal afferents (1). The ENS regulates numerous gut functions, including motility, secretion, blood flow, and absorption. It uses neurotransmitters such as serotonin, acetylcholine, and nitric oxide to coordinate these activities. For example, serotonin plays a vital role in gut motility and mood regulation, linking gut function to psychological states (6). Altered status of the ENS has been linked to conditions such as irritable bowel syndrome (IBS) and other functional gastrointestinal disorders, often correlated with anxiety and depression comorbidities (6).

19.3.2 Vagus Nerve: The Major Gut–Brain Signaling Pathway

The vagus nerve, with 80% afferent and 20% efferent fibers, is the primary conduit for gut–brain communication, extending from the brainstem to various organs (9). It transmits sensory information about the gut's nutrient content and inflammatory status to the brain and mediates anti-inflammatory reflexes, production of cytokines, and protection against systemic inflammation in conditions such as sepsis and IBD (19). Vagal afferent signals influence brain regions regulating mood, stress, and appetite, whereas efferent signals modulate gastrointestinal functions such as peristalsis and secretion, which are essential for digestion and gut–brain communication, particularly in response to stress and dietary changes (6, 20).

19.3.3 Role of the Sympathetic and Parasympathetic Nervous System

The SNS (part of the ANS), is linked to the "fight or flight" response and significantly influences gut function, particularly under stress. SNS activation can decelerate gastrointestinal motility and reduce digestive secretions, which are beneficial during acute stress by reallocating energy to critical functions. However, chronic SNS activation can cause gastrointestinal disorders, such as functional dyspepsia and IBS, and contribute to systemic inflammation and metabolic disturbances (21). The SNS also interacts with the HPA axis to mediate stress responses. Elevated levels of cortisol (stress hormones) affect permeability of gut and microbiota composition, influencing gut–brain communication and potentially worsening conditions during depression and anxiety (14). Dysregulation of the PNS is linked to gastrointestinal and psychological disorders, highlighting the need for balanced autonomic function (5). Balancing sympathetic and parasympathetic activity is crucial for gastrointestinal homeostasis and overall health. Imbalances can result in gastrointestinal disorders with psychological symptoms, illustrating the interconnectedness (22).

19.3.4 Neural Responses to Gastrointestinal Stimuli

Neural responses to gastrointestinal stimuli involve complex signaling mechanisms that enable the brain to interpret and react to various gut signals, including mechanical signals detected by stretch receptors in the gut wall (23), chemical signals such as gut hormones triggered by nutrients (11), and microbial metabolites produced by the gut microbiota that influence neural signaling (15). The interplay between these neural responses and gut stimuli is crucial for GBA function and disruptions can lead to gastrointestinal and neurological disorders (5) (for more detail, see Table 19.1)

19.4 IMMUNE SIGNALING PATHWAYS

In order for the gut and brain GBA to communicate, immune signaling pathways are essential. This interaction involves various immune cells and signaling molecules that help maintain homeostasis and influence both gastrointestinal and neurological health.

19.4.1 Immune Cell Communication in the Gut and Brain

Immune cells, such as macrophages, dendritic cells, T cells, microglia, and astrocytes, monitor the gut and brain environments and relay information that affects physiological states. In gut-associated lymphoid tissue (GALT), dendritic cells and macrophages detect pathogens, maintain immune homeostasis, and regulate T cells (24). During infections, they produce cytokines that influence brain function via the BBB or vagus nerve signaling (10). Regulatory T cells (Tregs) formed from naïve T cells in the gut maintain immune tolerance, reduce inflammation, and decrease BBB permeability by secreting IL-10 and TGF-β (25, 26).

19.4.2 Impact of Inflammation on Neurological Health

Inflammation in the gut severely impacts neurological health via GBA. Chronic inflammation (hallmark of most neurological disorders) can be exacerbated by factors related to the gut. BBB disruption by inflammatory cytokines from the gut allows entry of detrimental elements in the brain and causes neuroinflammation, possibly contributing to conditions such as MS and AD (27). Additionally, systemic inflammation caused by gut-derived inflammation can affect the brain and lead to psychiatric comorbidities such as depression and anxiety in conditions like IBD (14). Moreover, gut inflammation activates microglia and astrocytes (CNS immune cells) contributing to neurodegenerative diseases (28).

TABLE 19.1
Gut Microbiota and Brain Communication via Neural Pathways

Gut Microbiota	Key Findings	Conclusion	References
Campylobacter jejuni	Increased c-Fos expression in vagal neurons without elevated pro-inflammatory cytokines	Sensory neurons signal early to the brain about pathogens	(81)
Lactobacillus rhamnosus JB-1	Increased firing rate of vagal afferents and response to gut distension	Probiotics affect gut–brain communication	(82)
Lactobacillus reuteri	Enhanced social behavior via synaptic plasticity in ASD models; dependent on oxytocin receptors	Influences social behavior through vagal pathways	(83)
Gut flora	Microbiome status affects gut sensory neuron function and calbindin expression	Intact microbiome is crucial for gut sensory function	(84)
Microbial community (ENS)	ENS dysfunction leads to inflammation due to dysbiosis in zebrafish	ENS modulates gut microbiota, crucial for intestinal health	(85)
Intestinal microbiota	Dysbiosis causes ENS structural defects and altered neurotransmission	Microbiota is integral to ENS development	(86, 87)
Gut microbiota (BDNF/TrkB)	Dysbiosis alters BDNF/TrkB expression, affecting IBS and psychiatric conditions	Dysbiosis impacts BDNF-TrkB in ENS and CNS differentially	(88)
Gut microbiota (TLR4/MyD88)	TLR4/MyD88 deficiency delays gut motility and reduces nitrergic neurons	TLR4/MyD88 signaling is crucial for neuron survival and gut function	(89)
Lactobacillus johnsonii La1 (LJLa1)	LJLa1 lowers blood pressure via central histaminergic pathways	Modulates autonomic neurotransmission	(90)
Gut microbiome	IBD fecal supernatants increase vagal neuron excitability via cysteine proteases	Dysbiosis during IBD leads to vagal neuron excitability	(91)
Gut microbiota (Indole/GLP-1)	Indole stimulates GLP-1 secretion, increasing vagal nerve activity	L-cells transmit microbial signals to the nervous system	(92)
Lactobacillus rhamnosus (JB-1), *Bacteroides fragilis*	JB-1 activates sensory neurons rapidly, similar to *Bacteroides fragilis* polysaccharide A	Bacterial carbohydrates signal to sensory neurons	(93)
Gut microbiota (fatty acids)	Long-chain and SCFAs activate vagal afferents via different mechanisms	Fatty acids influence gut function and satiety	(94)
Gut microbiota (sympathetic neurons)	Microbiota depletion increases cFos in gut-extrinsic sympathetic neurons; bacterial colonization suppresses it	Microbiota modulates sympathetic activity	(95)

19.4.3 Cross Talk between Gut Immune Cells and CNS

The gut immune system and CNS interact through complex signaling mechanisms. Cytokine signaling is pivotal in this communication because gut-produced cytokines easily cross the BBB as well as act on the vagus nerve. Pro-inflammatory cytokines (TNF-α, IL-1β, and IL-6) affect mood and cognitive functions (9). Microbial metabolites, such as SCFAs and tryptophan derivatives, also modulate immune responses and brain function, thereby influencing both gut and brain health (15). The vagus nerve plays an important role in neuroimmune interactions, and its activation releases acetylcholine, which binds to immune cell receptors and reduces pro-inflammatory cytokine production (29) (Figure 19.1).

19.4.4 Role of Microglia and Astrocytes in Brain Inflammation

Microglia and astrocytes (immune cells of the brain) are vital for neuronal health and response to injuries or infections. Activated microglia, the CNS resident macrophages, initiate immune responses to brain injuries or infections. It leads to release of pro-inflammatory cytokines and reactive oxygen species (ROS) that can cause neuroinflammation and neuronal damage (1). Astrocytes support neurons by preserving their extracellular environment, offering metabolic support, and regulating synaptic activity. Inflammation can make astrocytes reactive, alter their morphology and function, and contribute to neuroinflammation via cytokine release (30). Gut-derived signals such as microbial metabolites and cytokines can influence microglia and astrocytes. SCFAs from gut bacteria can modulate the inflammatory response of these cells, thereby affecting brain health (31). Use of probiotics, prebiotics, and dietary changes can modulate gut microbiota and reduce inflammation, thereby benefiting brain health (23). (For more details, see Table 19.1.)

19.5 ENDOCRINE SIGNALING PATHWAYS

19.5.1 Hormonal Regulation of Gut–Brain Axis

Serotonin, ghrelin, glucagon-like peptide-1, and peptide YY influence GBA, affecting brain function and behavior. Serotonin, mainly gut-produced, affects motility, nausea, vomiting, secretion, peristalsis, metabolism, mood, sleep, and appetite (6). Children with autism exhibit increased level of plasma serotonin, which is inversely related to reduced serotonergic signaling (32). Ghrelin, produced by gastric X/A-like cells and influenced by the gut microbiome and SCFAs, stimulates appetite and interacts with brain areas associated with reward and motivation (11, 33). Peptide YY from digestive endocrine cells inhibits gastrointestinal motility and secretion, reduces appetite, regulates obesity (12), and affects brain function via the vagus nerve. Glucagon-like peptide-1, released after food intake, enhances insulin secretion, offers neuroprotection, and promotes satiety (13). Cortisol, part of the HPA axis, manages stress; however, chronic elevation can lead to cognitive impairments and mood disorders, necessitating balanced hormone levels for optimal mental health (2) (Figure 19.2) (Table 19.2).

19.5.2 Stress Responses and HPA Axis Activation

The hypothalamus, pituitary, and adrenal glands are connected by the HPA axis, the key biological mechanism for stress response. By controlling stress reactions and preserving homeostasis, the HPA axis is essential to GBA.

Activation and Regulation: ACTH is secreted by the pituitary gland in response to the production of CRH by hypothalamus under stress. The adrenal glands then create cortisol in response to ACTH, which helps regulate stress by enhancing energy availability and inhibiting non-essential processes, including immunological responses and digestion. (14).

Chronic Stress and HPA Dysregulation: Persistent HPA axis activation can cause dysregulation, resulting in elevated cortisol levels and chronic inflammation. This affects gut permeability and microbiota composition, influences gut–brain communication, and potentially worsens conditions of depression and anxiety (1).

TABLE 19.2
Endocrine and Microbial Signaling Pathways

Topic	Key Findings	Involved Elements	References
Hormonal regulation of gut–brain axis	EECs in the gut play a crucial role in gut–brain communication, affecting conditions like IBS and obesity. Dysbiosis contributes to hormonal dysfunction in depression	EECs, VIP, GLP-1, cortisol, serotonin, and ghrelin	(96–98)
Stress responses and HPA axis	Dysregulation of the HPA axis due to stress leads to metabolic, cardiovascular, and psychiatric disorders. Gut–brain signaling impacts energy and glucose homeostasis	HPA axis, CRF, glucocorticoid receptors, and gut peptides	(99–103)
Gut hormones and brain function	Early-life stress impacts GI physiology, with sex differences in response. Brain regions respond differently to gut hormone levels after macronutrient consumption	β-estradiol, GPER receptor, and BOLD signal	(104, 105)
Impact of gut-derived hormones	Gut bacteria modulate neurotransmitter production, influencing digestion, immunity, and gut permeability. Hormones like GLP-1, ghrelin, and PYY regulate metabolism	Serotonin, GABA, GLP-1, ghrelin, and PYY	(106–108))
Role of gut microbiota in neurological health	The gut microbiome influences brain health through the GBA, affecting neurodevelopmental, psychiatric, and neurodegenerative disorders	Gut microbiota, vagal nerve, neurotransmitters, and immune system	(109–111)
Metabolites produced by gut microbes	SCFAs, tryptophan metabolites, and polyamines from gut microbes regulate immune responses, stress, and neurotransmitter synthesis. Dysbiosis impacts neurological health	SCFAs, polyamines, tryptophan, TLR2, and B6 vitamin	(112–122)
Tryptophan metabolites and neurotransmitter synthesis	Tryptophan is a precursor for serotonin and kynurenine, influencing neuroendocrine functions. Vitamin B6 is crucial for tryptophan metabolism	Tryptophan, serotonin, kynurenine, and Vitamin B6	(123–127)
Impact of dysbiosis on neurological disorders	Dysbiosis reduces SCFA levels, contributing to neurological disorders like AD, PD, and MS	SCFAs, gut microbiota, immune system, and CNS	(128, 129)

19.5.3 Impact of Gut-Derived Hormones on Neurotransmission

Gut-derived hormones considerably affect neurotransmission and brain functions. Serotonin, which is crucial for mood, cognition, and gastrointestinal function, can affect brain serotonin levels and influence mood and behavior. Dysregulated serotonin pathways are implicated in psychiatric disorders (6). GABA, an inhibitory neurotransmitter, regulates neuronal excitability. It is produced by the gut microbiota and influences brain function by modulating stress and anxiety responses, affecting neural circuits involved in mood and behavior (21). The gut microbiota also regulates production and signaling pathways of Dopamine, required for reward, motivation, and motor control. Dysregulation of these pathways is linked to conditions such as PD and addiction (5). Understanding endocrine signaling pathways within the GBA is crucial for developing therapies for gastrointestinal and neurological disorders, as targeting these pathways can modulate gut–brain communication and improve health outcomes. (For more detail see Table 19.2)

19.6 Exploring Microbial Signaling Pathways

The gut microbiota (a community of bacteria, fungi, viruses, and archaea) significantly influences neurological and gastrointestinal health through intricate interactions with the host. Gut microbiota helps maintain balance between pro-inflammatory and anti-inflammatory responses, which is crucial for preventing neuroinflammation (34). Changes in the intestinal microbiota composition is linked to behavioral and cognitive alterations, including anxiety, depression, autism spectrum disorders, and other neuropsychiatric conditions (4).

19.6.1 Metabolites Produced by Gut Microbes

Gut microorganisms create metabolites such as SCFAs (through the fermentation of dietary fibers), which have anti-inflammatory qualities and affect brain function via modulating the BBB and neurotransmitter systems (15). It also produces tryptophan metabolites (serotonin and kynurenine) that are vital for neurotransmitter synthesis and immune regulation (16). Bile acids are modified by gut bacteria and act as signaling molecules that affect metabolism and brain function, with altered profiles associated with neurological disorders (17). SCFAs enhance gut barrier integrity, prevent leaky gut syndrome, and reduce systemic inflammation (15). Moreover, SCFAs stimulate the development of Tregs and suppress production of pro-inflammatory cytokine, both of which are critical for immunological tolerance and the avoidance of neuroinflammation (35). Moreover, SCFAs cross the BBB, modulate microglial activity, and promote neurotrophic factors such as BDNF (36) .

19.6.2 Tryptophan Metabolites and Neurotransmitter Synthesis

Tryptophan metabolism by gut microbes produces several essential metabolites that are vital for neurotransmitter synthesis and brain function. The following are some of the significant roles tryptophan metabolism plays in brain function:

Serotonin Production: Serotonin is a key neurotransmitter that regulates mood, cognition, and gastrointestinal function. Serotonin produced in the gut can influence brain serotonin levels, affecting mood and behavior (6).

Kynurenine Pathway: Tryptophan can be metabolized to kynurenine, which has neuroactive properties. A balance between kynurenine and its neurotoxic derivatives is critical for maintaining brain health. Dysregulation of this pathway has been linked to neurodegenerative diseases (2).

Indole Derivatives: The indoles generated by tryptophan metabolism by gut bacteria alter immunological responses and functioning of the gut barrier, influencing behavior and brain function (37) (for detail see Table 19.2).

19.7 Specific Neurological Disorders and Gut–Brain Axis

19.7.1 Alzheimer's Disease

The link between AD and gut dysbiosis has garnered significant attention owing to amyloid-beta pathology. The gut microbiota can influence amyloid-beta aggregation via systemic inflammation and metabolic pathways. Increased intestinal permeability caused by dysbiosis makes it possible for bacterial endotoxins to enter the circulation, escalating neuroinflammation and causing accumulation of amyloid-beta in the brain (38, 39). Research indicates that gut microbiota composition significantly differs in AD patients compared to healthy individuals, with an increase in prevalence of pro-inflammatory bacteria like *Escherichia/Shigella* over beneficial bacteria such as *Eubacterium* and *Bacteroides* (39). CCAAT/enhancer-binding protein beta (C/EBPβ) and asparagine endopeptidase (AEP) signaling pathways are pivotal in neuroinflammation and neurodegeneration in AD. The activation of this pathway by gut-derived inflammatory mediators can worsen amyloid-beta pathology. Inhibition of the C/EBPβ/AEP pathway has shown promise in reducing neuroinflammation and amyloid-beta deposition, highlighting potential therapeutic targets for AD (40) (Figure 19.3).

19.7.2 Parkinson's Disease

PD is characterized by alpha-synuclein aggregates, which is possibly influenced by the gut microbiota. Dysbiosis in PD patients, correlates with increased intestinal permeability and systemic inflammation, promoting alpha-synuclein aggregation in the gut and its transmission to the brain through the vagus nerve (41). Research suggests that the

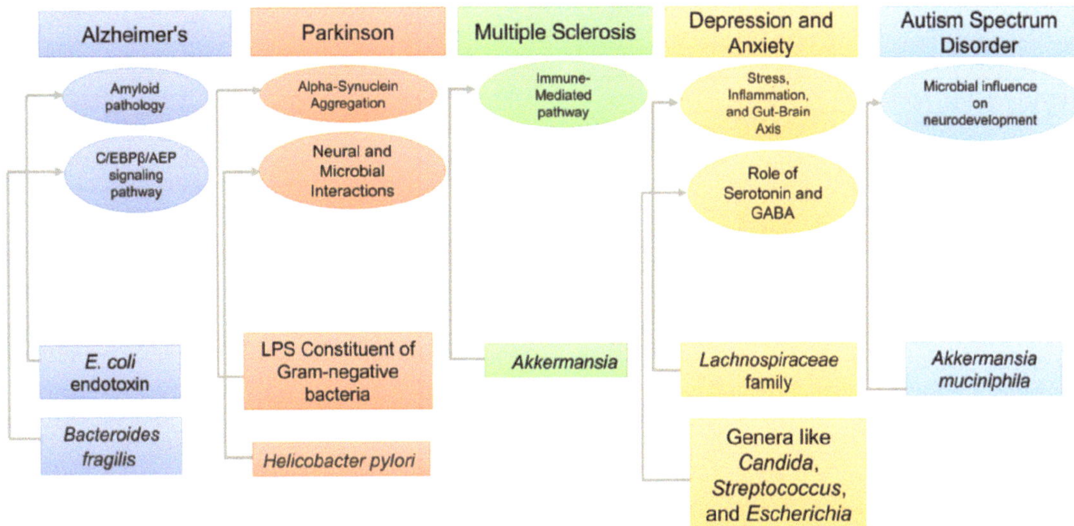

FIGURE 19.3 Microbes and their metabolites in the gut that leads to progression of the certain neurological disorder.

gut microbiome of PD patients is altered, with elevated Enterobacteriaceae and reduced Prevotellaceae levels, linked to motor symptoms and gastrointestinal dysfunction (42). The bidirectional gut–brain communication in PD involves complex neural–gut microbiota interactions. The vagus nerve is crucial, with vagotomy potentially lowering the PD risk (43). SCFAs (gut metabolite) can affect brain function and neuroinflammation, further connecting gut health to PD pathogenesis (41) (Figure 19.3).

19.7.3 Multiple Sclerosis

MS, an autoimmune disease, is affected by the gut microbiota through immune pathways. Dysbiosis in MS patients results in reduced diversity and altered gut microbiota composition, impacting immune regulation and promoting inflammation (44). Specific bacterial taxa, such as *Akkermansia* and *Methanobrevibacter*, are associated with MS and linked to the illness's course and intensity. These bacteria modulate immune responses, influencing the balance between pro-inflammatory and regulatory T cells, thus affecting disease outcomes (45) (Figure 19.3).

19.7.4 Depression and Anxiety

Depression and anxiety have been reported to be associated with GBA dysfunction and are impacted by stress and inflammation. Chronic stress led gut microbiota alterations, increased intestinal permeability and systemic inflammation, and caused changes in brain function and mood (46). Changes in the gut microbiota promotes increase in pro-inflammatory bacteria declining availability of beneficial bacteria like *Bifidobacterium* and *Lactobacillus* (47) and is reported to cause depression and anxiety. These gut microbiota significantly affects neurotransmitter production of serotonin and GABA, involved in mood regulation. Their reduction in the gut can decrease the levels of serotonin and GABA in the brain, resulting in depression and anxiety (48) (Figure 19.3).

19.7.5 Autism Spectrum Disorders (ASD)

Dysregulation of gut microbiota is implicated in ASD, with children often exhibiting altered composition, decreased diversity, and an imbalance between useful and harmful bacteria (49). The development and function of the brain are impacted by microbial metabolites, and is associated with behavioral and cognitive impairments in ASD patients (50). The GBA affects ASD behavior through various pathways, including immune modulation and neurotransmitter production. Intestinal permeability and systemic inflammation can be exacerbated by gut dysbiosis, which can exacerbate behavioral symptoms associated with ASD (51) (Figure 19.3).

19.7.6 Epilepsy

Epilepsy has been linked to gut microbiota dysbiosis, which influences seizures via the inflammatory and metabolic pathways. Specific bacterial taxa are altered in patients with epilepsy, affecting the GBA and seizure susceptibility (52–54). Interventions such as ketogenic diets and probiotics have shown potential in reducing seizure frequency and severity, underscoring the therapeutic value of modulating the GBA in epilepsy (55).

19.7.7 Huntington's Disease

HD is (neurodegenerative disorder) influenced by the gut microbiota through microbial metabolites. Dysbiosis in HD patients can lead to altered levels of SCFAs, which can affect neuroinflammation and neuronal health (56).

According to research, probiotics and dietary modifications of the gut microbiota may have therapeutic advantages for HD, perhaps delaying the course of the illness and enhancing quality of life (57). For more details on impact of microbiota on HD, see Table 19.3.

19.8 THERAPEUTIC IMPLICATIONS

GBA represents a complex communication between GIT and CNS, which is essential for health and disease. This understanding has catalyzed therapeutic interventions to modulate the gut microbiota for better neurological and psychiatric outcomes, including probiotics, prebiotics, dietary changes, pharmacological treatments, and fecal microbiota transplantation (FMT).

19.8.1 Probiotics and Prebiotics

Probiotics and prebiotics create bioactive molecules like SCFAs, impacts make up of gut microbiota, thereby modulating immune system and improve the function of the gut barrier. These actions influence brain function and behavior via the GBA. Clinical evidence supports probiotics usage in therapeutic management of psychiatric disorders, like depression and anxiety. For example, a systematic review found that probiotics are beneficial in major depressive disorder and Alzheimer's disease, reducing symptoms and potentially delaying disease progression (58,59). Prebiotics also show promise in modulating the GBA. A review suggested that prebiotics might mitigate depression symptoms by influencing the gut microbiota and neuroinflammatory pathways (60).

19.8.2 Dietary Interventions

Dietary interventions are important in modulating the gut microbiota and the GBA. Specific dietary forms and nutrients influence microbial composition, gut health, and brain function.

19.8.2.1 Impact of Diet on Gut Microbiota and Brain Function

Fiber-rich, fermented, and omega-3 fatty acid diets enhance gut health. Fiber is a prebiotic, fostering beneficial bacteria that produces SCFAs, which is anti-inflammatory and has neuroprotective effects (61). Omega-3s in fish and flaxseeds also have anti-inflammatory benefits and promote brain health. The Mediterranean diet consisting of lots of fruits, vegetables, whole grains, and healthy fats enhances the makeup of the gut microbiota and reduces the risk of neurological illnesses (7).

19.8.3 Pharmacological Interventions

Pharmacological interventions targeting the GBA aim to modify the gut microbiota or influence signaling pathways between the gut and brain.

19.8.3.1 Drugs Targeting Gut–Brain Signaling Pathways

Several drugs have been investigated for their potential to modulate the GBA. Antibiotics, primarily for infections, can alter gut microbiota composition and affect neuroinflammatory pathways. However, their use is limited by antibiotic resistance and adverse effects on gut health (62).

TABLE 19.3
Neurological Disorders and Gut–Brain Axis

Disorder	Key Findings	Microbiota	References
Alzheimer's disease	Gut bacteria release amyloids and LPS, impacting signaling pathways and inflammation. Dysbiosis precedes amyloid plaque development in AD models	*Escherichia coli*, *Salmonella enterica*, *Akkermansia*, *Blautia*	130–132
Parkinson's disease	LPS from gut bacteria influences α-synuclein aggregation, a key event in PD. Curli-producing bacteria enhance αSyn disease in mice	*Escherichia coli*, *Helicobacter pylori*	133–135
Multiple sclerosis	Gut microbiota alteration reduces severity of EAE, an MS model. Variations in microbiome influence immune signaling pathways in MS	*Methanobrevibacter*, *Akkermansia*	136, 137
Depression and anxiety	Stress-induced GBA disruption leads to inflammation and mood impairments. Gut microbiota modulates serotonin and GABAergic systems, affecting mood	*Candida*, *Streptococcus*, *Lactobacillus rhamnosus*	138–140
Autism spectrum disorder	Imbalance in gut flora linked to gastrointestinal issues in ASD. Reduced beneficial bacteria in ASD patients, with potential behavioral implications	*Clostridium*, *Bifidobacteria*, *Akkermansia muciniphila*	141–144
Epilepsy	Dysbiosis linked to seizure activity. Probiotic therapy shows potential in reducing seizure severity	*Bacteroides fragilis*, *Lactobacillus rhamnosus*	145–147
Huntington's disease	Microbial metabolites influence the severity of HD symptoms. Dysbiosis is associated with altered plasma metabolites affecting the GBA	*Bacteroides*, *Christensenella*	148, 149

A family of probiotics called psychobiotics is being investigated for the treatment of psychiatric diseases. These include *Lactobacillus* and *Bifidobacterium* strains producing GABA and serotonin, which influence mood and anxiety (63).

19.8.4 Fecal Microbiota Transplantation

To rebuild a healthy gut microbiome, FMT entails transferring feces from a healthy donor to the recipient's digestive system. It is being investigated for several ailments, including those connected to the GBA. Studies have shown promising results in treating *Clostridium difficile* infection. Clinical trials have examined FMT's efficacy in the treatment of psychiatric and neurological disorders. For instance, decline in cognitive function of hepatic encephalopathy patients improved by FMT and is under study for ASD (64). Research shows that FMT can modify gut microbiota and reduce neuroinflammation, leading to better neurological outcomes (65). However, more rigorous trials are required to fully understand FMT's therapeutic potential and safety concerning the GBA.

19.9 CASE STUDIES AND EXPERIMENTAL MODELS

Understanding GBA and its implications for neurological disorders has led to several clinical and experimental studies exploring potential treatments. This section highlights the key findings of human case studies and animal models.

19.9.1 Human Case Studies

Human case studies furnish valuable insights into the potency and safety of therapeutic interventions targeting the GBA. Two notable interventions are FMT for PD, and probiotic interventions for AD.

19.9.1.1 FMT in Parkinson's Disease

FMT has shown promise in alleviating gastrointestinal and motor dysfunction in PD patients. With an increased plethora of beneficial bacteria and improvements in the Hoehn-Yahr grade, Unified Parkinson's Disease Rating Scale (UPDRS), and Non-Motion Symptom Questionnaire (NMSS) scores, a study involving 11 PD patients (with constipation) showed significant improvements in gastrointestinal and motor symptoms post-FMT (66). A review summarized both clinical and basic studies, indicating that FMT could effectively reconstruct the gut microbiota and alleviate PD symptoms, although more extensive clinical trials are required (67). FMT improved autonomic symptoms as compared to the placebo group in a randomized, placebo-controlled trial involving 56 patients with PD. This suggests that FMT may be useful in treating PD by augmenting the complexity of the microecological system(68).

Thus, studies suggest that FMT is a possible viable supportive treatment for PD (addressing both motor and cognitive symptoms) by modulating gut microbiota.

19.9.1.2 Probiotic Interventions in Alzheimer's Disease

Probiotics have shown promise as a therapeutic option for AD by modulating the GBA (69). After 12 weeks of supplementation, a randomized, double-blind, placebo-controlled study using strains of *Lactobacillus rhamnosus* and *Bifidobacterium longum* showed significant improvements in cognitive function, improved Mini-Mental State Examination scores, and decline in anxiety level and physical activity in patients with AD (70). According to a meta-analysis report, probiotics considerably enhance cognitive function by lowering oxidative stress and inflammatory biomarkers in people with AD and moderate cognitive impairment (71). These studies showcase the usage of probiotics as a non-pharmacological intervention to improve GBA health and slow cognitive impairment in AD.

19.9.2 Animal Models

Testing possible therapies and understanding the mechanics behind the GBA dysfunction in neurological disorders have benefited from animal models. Germ-free and antibiotic-treated models have revealed the role of the gut microbiome on neurological health. Studies with germ-free mice revealed that the absence of gut microbiota worsens PD symptoms, whereas antibiotic-treated mice exhibit altered motor functions and increased inflammation, suggesting a protective role of the microbiota against PD (41). In AD models, antibiotics changed the gut microbiota composition and worsened cognitive deficits, indicating that restoring the gut microbiota could alleviate AD symptoms, highlighting its therapeutic potential (38). These models allow the controlled study of gut–brain interactions and microbiota-targeted therapies. Zebrafish models, with their transparent anatomy and well-documented genetics, benefit GBA research. Modifying the gut microbiota in zebrafish affects PD-like symptoms, making it valuable for testing therapies (72). In Alzheimer's models, probiotic-treated mice showed improved cognitive function and reduced amyloid-beta accumulation, supporting their use in GBA studies (73). Animal models offer a cost-effective platform for exploring gut–brain interactions and new therapeutic strategies. FMT and probiotics show promise for treating PD and AD, with germ-free mouse and zebrafish models providing key insights into the mechanisms of these treatments.

19.10 FUTURE DIRECTIONS AND RESEARCH GAPS

The study of GBA has grown rapidly in recent years, revealing complex interactions between the gastrointestinal tract and CNS. Despite significant advancements, many areas still require further exploration. This section outlines the emerging areas of research, challenges, limitations, and potential for new therapeutic strategies.

19.10.1 Emerging Areas of Research

Emerging research areas are enhancing our understanding of GBA and exploring new mechanisms, potential biomarkers, and innovative therapies.

1. *Metabolomics and GBA:* The metabolic pathways impacted by gut bacteria can be understood by metabolomics. Studying metabolites' effects on neurological health and illness, SCFAs, tryptophan derivatives, and bile acids are critical for GBA (4).
2. *Immune System Interactions:* The immune system significantly mediates gut–brain interactions. Research has investigated how immune cells, cytokines, and inflammatory pathways contribute to neurodegenerative and psychiatric disorders, potentially leading to new treatments for conditions, such as AD and MS (23).
3. *Microbiota and Neurodevelopment:* Study reports suggests that gut microbiota influences brain and behavior development in the early years of life. Studies have concentrated on the effects of food, early childhood antibiotic exposure, and maternal microbiome on neurodevelopment and mental health (74).
4. *Psychobiotics and Mental Health:* Psychobiotics, live bacteria that benefit patients with psychiatric illnesses, is a promising field. Research has explored specific probiotic strains that can treat conditions such as depression, anxiety, and schizophrenia by modulating gut microbiota (75).
5. *Gut–Liver–Brain Axis:* This emerging field studies the interactive relationship between gut microbiota, liver function, and brain health, particularly in metabolic diseases, such as NAFLD and hepatic encephalopathy (76).

19.10.2 Challenges and Limitations

Though progress has been made, a number of obstacles and constraints still stand in the way of GBA's impact on research on neurological disorders. These include the complexity and interindividual variation of the gut microbiota, which are subject to various influences and make standardizing research findings and treatment approaches more difficult (77). Although correlations between the brain function and gut microbiota is established, the precise mechanisms of interaction remain unclear and require further research (2). Translating findings from animal models to humans is challenging because of species-specific differences in the microbiota composition and physiology (78). GBA research requires an interdisciplinary approach, posing significant technical and methodological integration challenges (79). Standardized methodologies in GBA research are required to reduce variability and enhance reproducibility (80).

Although many challenges and limitations remain, the potential for novel therapeutic strategies is immense. Emerging areas of research, such as metabolomics, immune interactions, and personalized medicine, hold promise in advancing our understanding and treatment of neurological and psychiatric disorders. Continued interdisciplinary efforts and methodological advancements are essential to unlock the full potential of GBA research and translate the findings into effective therapies.

REFERENCES

1. Post Z, Manfready RA, Keshavarzian A. Overview of the gut–brain axis: from gut to brain and back again. *Semin Neurol.* 2023 Aug;43(4):506–17. Available from: http://www.thieme-connect.de/DOI/DOI?10.1055/s-0043-1771464.
2. Gheorghe CE, Martin JA, Manriquez FV, Dinan TG, Cryan JF, Clarke G. Focus on the essentials: tryptophan metabolism and the microbiome-gut-brain axis. *Curr Opin Pharmacol.* 2019 Oct;48:137–45. Available from: https://linkinghub.elsevier.com/retrieve/pii/S1471489218301279.
3. Malagelada JR. The brain-gut team. *Dig Dis.* 2020;38(4):293–8. Available from: https://karger.com/DDI/article/doi/10.1159/000505810.
4. Cryan JF, O'Riordan KJ, Cowan CSM, Sandhu KV, Bastiaanssen TFS, Boehme M, et al. The microbiota-gut-brain axis. *Physiol Rev.* 2019 Oct 1;99(4):1877–2013. Available from: https://www.physiology.org/doi/10.1152/physrev.00018.2018.
5. Zheng Y, Bonfili L, Wei T, Eleuteri AM. Understanding the gut–brain axis and its therapeutic implications for neurodegenerative disorders. *Nutrients.* 2023 Oct 31;15(21):4631. Available from: https://www.mdpi.com/2072-6643/15/21/4631.
6. Margolis KG, Cryan JF, Mayer EA. The microbiota-gut-brain axis: from motility to mood. *Gastroenterology.* 2021 Apr;160(5):1486–501. Available from: https://linkinghub.elsevier.com/retrieve/pii/S0016508521002687.
7. Suganya K, Koo BS. Gut–brain axis: role of gut microbiota on neurological disorders and how probiotics/prebiotics beneficially modulate microbial and immune pathways to improve brain functions. *IJMS.* 2020 Oct 13;21(20):7551. Available from: https://www.mdpi.com/1422-0067/21/20/7551.
8. Ekwudo MN, Gubert C, Hannan AJ. The microbiota–gut–brain axis in Huntington's disease: pathogenic mechanisms and therapeutic targets. *FEBS J*;n/a(n/a). Available from: https://onlinelibrary.wiley.com/doi/abs/10.1111/febs.17102.
9. Romijn JA, Corssmit EP, Havekes LM, Pijl H. Gut???brain axis. *Curr Opin Clin Nutr Metab Care.* 2008 Jul;11(4):518–21. Available from: http://journals.lww.com/00075197-200807000-00026.
10. Agirman G, Yu KB, Hsiao EY. Signaling inflammation across the gut-brain axis. *Science.* 2021 Nov 26;374(6571):1087–92. Available from: https://www.science.org/doi/10.1126/science.abi6087.
11. Bistoletti M, Bosi A, Banfi D, Giaroni C, Baj A. The microbiota-gut-brain axis: Focus on the fundamental communication pathways. In: *Progress in Molecular Biology*

and Translational Science. Elsevier; 2020. p. 43–110. Available from: https://linkinghub.elsevier.com/retrieve/pii/S1877117320301381.
12. Wu Y, He H, Cheng Z, Bai Y, Ma X. The role of Neuropeptide Y and Peptide YY in the development of obesity via gut-brain axis. *CPPS*. 2019 Jun 27;20(7):750–8. Available from: http://www.eurekaselect.com/169346/article.
13. Sun EWL, Martin AM, Young RL, Keating DJ. The regulation of peripheral metabolism by gut-derived hormones. *Front Endocrinol*. 2019 Jan 4;9:754. Available from: https://www.frontiersin.org/article/10.3389/fendo.2018.00754/full.
14. Makris AP, Karianaki M, Tsamis KI, Paschou SA. The role of the gut-brain axis in depression: Endocrine, neural, and immune pathways. *Hormones*. 2021 Mar;20(1):1–12. Available from: https://link.springer.com/10.1007/s42000-020-00236-4.
15. Silva YP, Bernardi A, Frozza RL. The role of short-chain fatty acids from gut microbiota in gut-brain communication. *Front Endocrinol*. 2020 Jan 31;11:25. Available from: https://www.frontiersin.org/article/10.3389/fendo.2020.00025/full.
16. Gao K, Mu C long, Farzi A, Zhu W yun. Tryptophan metabolism: A link between the gut microbiota and brain. *Adv Nutr*. 2020 May;11(3):709–23. Available from: https://linkinghub.elsevier.com/retrieve/pii/S2161831322002952.
17. Deng W, Yi P, Xiong Y, Ying J, Lin Y, Dong Y, et al. Gut metabolites acting on the gut-brain axis: Regulating the functional state of microglia. *Aging Dis*. 2024;15(2):480. Available from: https://www.aginganddisease.org/EN/10.14336/AD.2023.0727.
18. Jacobson A, Yang D, Vella M, Chiu IM. The intestinal neuro-immune axis: crosstalk between neurons, immune cells, and microbes. *Mucosal Immunol*. 2021 May;14(3):555–65. Available from: https://www.nature.com/articles/s41385-020-00368-1.
19. Teratani T, Mikami Y, Nakamoto N, Suzuki T, Harada Y, Okabayashi K, et al. The liver–brain–gut neural arc maintains the Treg cell niche in the gut. *Nature*. 2020 Sep 24;585(7826):591–6. Available from: https://www.nature.com/articles/s41586-020-2425-3.
20. Yu CD, Xu QJ, Chang RB. Vagal sensory neurons and gut-brain signaling. *Curr Opin Neurobiol*. 2020 Jun;62:133–40. Available from: https://linkinghub.elsevier.com/retrieve/pii/S0959438820300635.
21. Pusceddu MM, Barboza M, Schneider M, Stokes P, Sladek JA, Torres-Fuentes C, et al. Nod-like receptors are critical for gut-brain axis signaling. 2019. Available from: http://biorxiv.org/lookup/doi/10.1101/647032.
22. Abdullah N, Defaye M, Altier C. Neural control of gut homeostasis. *Am J Physiol Gastrointest Liver Physiol*. 2020 Dec 1;319(6):G718–32. Available from: https://journals.physiology.org/doi/10.1152/ajpgi.00293.2020.
23. Günther C, Rothhammer V, Karow M, Neurath M, Winner B. The gut-brain axis in inflammatory bowel disease—Current and future perspectives. *IJMS*. 2021 Aug 18;22(16):8870. Available from: https://www.mdpi.com/1422-0067/22/16/8870.
24. Filardy AA, Ferreira JRM, Rezende RM, Kelsall BL, Oliveira RP. The intestinal microenvironment shapes macrophage and dendritic cell identity and function. *Immunol Lett*. 2023 Jan 1;253:41–53. Available from: https://www.sciencedirect.com/science/article/pii/S0165247823000032.
25. Izcue A, Coombes J, Powrie F. Regulatory T cells suppress systemic and mucosal immune activation to control intestinal inflammation. *Immunol Rev*. 2006;212. Available from: https://consensus.app/papers/regulatory-cells-mucosal-activation-control-izcue/f44d7eebb1555cc0a46f0d57871309e4/.
26. Omenetti S, Pizarro T. The Treg/Th17 axis: A dynamic balance regulated by the gut microbiome. *Front Immunol*. 2015;6. Available from: https://consensus.app/papers/tregth17-axis-dynamic-balance-regulated-microbiome-omenetti/55972d15653f59cba159173445dc441d/.
27. Battaglini D, Pimentel-Coelho PM, Robba C, Dos Santos CC, Cruz FF, Pelosi P, et al. Gut microbiota in acute ischemic stroke: From pathophysiology to therapeutic implications. *Front Neurol*. 2020 Jun 25;11:598. Available from: https://www.frontiersin.org/article/10.3389/fneur.2020.00598/full.
28. Tyler Patterson T, Grandhi R. Gut microbiota and neurologic diseases and injuries. In: Chen P, editor. *Gut Microbiota and Pathogenesis of Organ Injury*. Singapore: Springer; 2020. p. 73–91. (Advances in Experimental Medicine and Biology; vol. 1238). Available from: http://link.springer.com/10.1007/978-981-15-2385-4_6.
29. Long-Smith C, O'Riordan KJ, Clarke G, Stanton C, Dinan TG, Cryan JF. Microbiota-gut-brain axis: new therapeutic opportunities. *Annu Rev Pharmacol Toxicol*. 2020 Jan 6;60(1):477–502. Available from: https://www.annualreviews.org/doi/10.1146/annurev-pharmtox-010919-023628.
30. Dehhaghi M, Kazemi Shariat Panahi H, Heng B, Guillemin GJ. The gut microbiota, kynurenine pathway, and immune system interaction in the development of brain cancer. *Front Cell Dev Biol*. 2020 Nov 19;8:562812. Available from: https://www.frontiersin.org/articles/10.3389/fcell.2020.562812/full.
31. Sultan S, Yousuf B, Yeo J, Ahmed T, Bouhlel NE, Hassan H, et al. Comprehensive multi-omics characterization of gut microbiome extracellular vesicles reveals a connection to gut-brain axis signaling. 2022. Available from: http://biorxiv.org/lookup/doi/10.1101/2022.10.28.514259.
32. Appleton J. The gut-brain axis: Influence of microbiota on mood and mental health. *Integr Med (Encinitas)*. 2018 Aug;17(4):28–32. Available from: https://www.ncbi.nlm.nih.gov/pmc/articles/PMC6469458/
33. Schalla MA, Stengel A. Effects of microbiome changes on endocrine ghrelin signaling – A systematic review. *Peptides*. 2020 Nov;133:170388.
34. Mitrea L, Nemeş SA, Szabo K, Teleky BE, Vodnar DC. Guts imbalance imbalances the brain: a review of gut microbiota association with neurological and psychiatric disorders. *Front Med*. 2022 Mar 31;9:813204. Available from: https://www.frontiersin.org/articles/10.3389/fmed.2022.813204/full.
35. De Caro C, Iannone LF, Citraro R, Striano P, De Sarro G, Constanti A, et al. Can we 'seize' the gut microbiota to treat epilepsy? *Neurosci Biobehav Rev*. 2019 Dec;107:750–64. Available from: https://linkinghub.elsevier.com/retrieve/pii/S0149763419306517.
36. Cerdó T, Diéguez E, Campoy C. Impact of gut microbiota on neurogenesis and neurological diseases during infancy. *Curr Opin Pharmacol*. 2020 Feb;50:33–7. Available from: https://linkinghub.elsevier.com/retrieve/pii/S147148921930116X.

37. Banerjee A, Pradhan LK, Sahoo PK, Jena KK, Chauhan NR, Chauhan S, et al. Unravelling the potential of gut microbiota in sustaining brain health and their current prospective towards development of neurotherapeutics. *Arch Microbiol.* 2021 Aug;203(6):2895–910. Available from: https://link.springer.com/10.1007/s00203-021-02276-9.
38. Minter MR, Zhang C, Leone V, Ringus DL, Zhang X, Oyler-Castrillo P, et al. Antibiotic-induced perturbations in gut microbial diversity influences neuro-inflammation and amyloidosis in a murine model of Alzheimer's disease. *Sci Rep.* 2016 Jul 21;6(1):30028. Available from: https://www.nature.com/articles/srep30028.
39. Vogt NM, Kerby RL, Dill-McFarland KA, Harding SJ, Merluzzi AP, Johnson SC, et al. Gut microbiome alterations in Alzheimer's disease. *Sci Rep.* 2017 Oct 19;7(1):13537. Available from: https://www.nature.com/articles/s41598-017-13601-y.
40. Xiong J, Zhang Z, Ye K. C/EBPβ/AEP signaling drives Alzheimer's disease pathogenesis. *Neurosci Bull.* 2023 Jul 1;39(7):1173–85. Available from: https://doi.org/10.1007/s12264-023-01025-w.
41. Sampson TR, Debelius JW, Thron T, Janssen S, Shastri GG, Ilhan ZE, et al. Gut microbiota regulate motor deficits and neuroinflammation in a model of Parkinson's disease. *Cell.* 2016 Dec 1;167(6):1469–80.e12. Available from: https://www.cell.com/cell/abstract/S0092-8674(16)31590-2.
42. Scheperjans F, Aho V, Pereira PAB, Koskinen K, Paulin L, Pekkonen E, et al. Gut microbiota are related to Parkinson's disease and clinical phenotype. *Movement Disorders.* 2015 Mar;30(3):350–8. Available from: https://movementdisorders.onlinelibrary.wiley.com/doi/10.1002/mds.26069.
43. Liu B, Fang F, Pedersen NL, Tillander A, Ludvigsson JF, Ekbom A, et al. Vagotomy and Parkinson disease: A Swedish register–based matched-cohort study. *Neurology.* 2017 May 23;88(21):1996–2002. Available from: https://www.neurology.org/doi/10.1212/WNL.0000000000003961.
44. Berer K, Gerdes LA, Cekanaviciute E, Jia X, Xiao L, Xia Z, et al. Gut microbiota from multiple sclerosis patients enables spontaneous autoimmune encephalomyelitis in mice. *Proc Natl Acad Sci U S A.* 2017 Oct 3;114(40):10719–24.
45. Jangi S, Gandhi R, Cox LM, Li N, Von Glehn F, Yan R, et al. Alterations of the human gut microbiome in multiple sclerosis. *Nat Commun.* 2016 Jun 28;7(1):12015. Available from: https://www.nature.com/articles/ncomms12015.
46. Foster JA, Rinaman L, Cryan JF. Stress & the gut-brain axis: Regulation by the microbiome. *Neurobiol Stress.* 2017 Dec;7:124–36. Available from: https://linkinghub.elsevier.com/retrieve/pii/S2352289516300509.
47. Kelly JR, Borre Y, O' Brien C, Patterson E, El Aidy S, Deane J, et al. Transferring the blues: Depression-associated gut microbiota induces neurobehavioural changes in the rat. *J Psych Res.* 2016 Nov;82:109–18. Available from: https://linkinghub.elsevier.com/retrieve/pii/S0022395616301571.
48. Dinan TG, Cryan JF. Brain–gut–microbiota axis — Mood, metabolism and behaviour. *Nat Rev Gastroenterol Hepatol.* 2017 Feb;14(2):69–70. Available from: https://www.nature.com/articles/nrgastro.2016.200.
49. Strati F, Cavalieri D, Albanese D, De Felice C, Donati C, Hayek J, et al. New evidences on the altered gut microbiota in autism spectrum disorders. *Microbiome.* 2017 Dec;5(1):24. Available from: http://microbiomejournal.biomedcentral.com/articles/10.1186/s40168-017-0242-1.
50. Kang DW, Ilhan ZE, Isern NG, Hoyt DW, Howsmon DP, Shaffer M, et al. Differences in fecal microbial metabolites and microbiota of children with autism spectrum disorders. *Anaerobe.* 2018 Feb;49:121–31.
51. Hsiao EY, McBride SW, Hsien S, Sharon G, Hyde ER, McCue T, et al. Microbiota modulate behavioral and physiological abnormalities associated with neurodevelopmental disorders. *Cell.* 2013 Dec;155(7):1451–63. Available from: https://linkinghub.elsevier.com/retrieve/pii/S0092867413014736.
52. Gómez-Eguílaz M, Ramón-Trapero JL, Pérez-Martínez L, Blanco JR. The beneficial effect of probiotics as a supplementary treatment in drug-resistant epilepsy: a pilot study. *Benef Microbes.* 2018 Dec 7;9(6):875–81.
53. Iannone LF, Gómez-Eguílaz M, Citaro R, Russo E. The potential role of interventions impacting on gut-microbiota in epilepsy. *Expert Rev Clin Pharmacol.* 2020 Apr;13(4):423–35.
54. Iannone LF, Gómez-Eguílaz M, De Caro C. Gut microbiota manipulation as an epilepsy treatment. *Neurobiol Dis.* 2022 Nov;174:105897.
55. Lindefeldt M, Eng A, Darban H, Bjerkner A, Zetterström CK, Allander T, et al. The ketogenic diet influences taxonomic and functional composition of the gut microbiota in children with severe epilepsy. *NPJ Biofilms Microbiomes.* 2019;5(1):5.
56. Wasser CI, Mercieca EC, Kong G, Hannan AJ, McKeown SJ, Glikmann-Johnston Y, et al. Gut dysbiosis in Huntington's disease: associations among gut microbiota, cognitive performance and clinical outcomes. *Brain Commun.* 2020 Jul 1;2(2):fcaa110. Available from: https://doi.org/10.1093/braincomms/fcaa110.
57. Wasser CI, Mercieca EC, Kong G, Hannan AJ, Allford B, McKeown SJ, et al. A randomized controlled trial of probiotics targeting gut dysbiosis in Huntington's disease. *J Huntingtons Dis.* 2023;12(1):43–55.
58. Barbosa RSD, Vieira-Coelho MA. Probiotics and prebiotics: focus on psychiatric disorders – A systematic review. *Nutri Rev.* 2020 Jun 1;78(6):437–50. Available from: https://academic.oup.com/nutritionreviews/article/78/6/437/5643897
59. Zagórska A, Marcinkowska M, Jamrozik M, Wiśniowska B, Paśko P. From probiotics to psychobiotics – The gut-brain axis in psychiatric disorders. *BM.* 2020 Dec 2;11(8):717–32. Available from: https://brill.com/view/journals/bm/11/8/article-p717_1.xml.
60. Radford-Smith DE, Anthony DC. Prebiotic and probiotic modulation of the microbiota–gut–brain axis in depression. *Nutrients.* 2023 Apr 13;15(8):1880. Available from: https://www.mdpi.com/2072-6643/15/8/1880.
61. Dalile B, Verbeke K, Van Oudenhove L, Vervliet B. Nourishing the gut microbiota: the potential of prebiotics in microbiota-gut-brain axis research. *Behav Brain Sci.* 2019;42:e69. Available from: https://www.cambridge.org/core/product/identifier/S0140525X18002856/type/journal_article.
62. Jacobs JP, Mayer EA. Psychobiotics: shaping the mind with gut bacteria. *Am J Gastroenterol.* 2019 Jul;114(7):1034–5. Available from: https://journals.lww.com/00000434-201907000-00013.

63. Forssten SD, Ouwehand AC, Griffin SM, Patterson E. One giant leap from mouse to man: the microbiota–gut–brain axis in mood disorders and translational challenges moving towards human clinical trials. *Nutrients*. 2022 Jan 27;14(3):568. Available from: https://www.mdpi.com/2072-6643/14/3/568.
64. Baldi S, Mundula T, Nannini G, Amedei A. Microbiota shaping — the effects of probiotics, prebiotics, and fecal microbiota transplant on cognitive functions: A systematic review. *WJG*. 2021 Oct 21;27(39):6715–32. Available from: https://www.wjgnet.com/1007-9327/full/v27/i39/6715.htm.
65. Tan Q, Orsso CE, Deehan EC, Kung JY, Tun HM, Wine E, et al. Probiotics, prebiotics, synbiotics, and fecal microbiota transplantation in the treatment of behavioral symptoms of autism spectrum disorder: A systematic review. *Autism Res*. 2021 Sep;14(9):1820–36. Available from: https://onlinelibrary.wiley.com/doi/10.1002/aur.2560.
66. Kuai X yi, Yao X han, Xu L juan, Zhou Y qing, Zhang L ping, Liu Y, et al. Evaluation of fecal microbiota transplantation in Parkinson's disease patients with constipation. *Microb Cell Fact*. 2021 Dec;20(1):98. Available from: https://microbialcellfactories.biomedcentral.com/articles/10.1186/s12934-021-01589-0.
67. Kang Y, Kang X, Zhang H, Liu Q, Yang H, Fan W. Gut microbiota and Parkinson's disease: Implications for faecal microbiota transplantation therapy. *ASN Neuro*. 2021 Jan 1;13. Available from: https://doi.org/10.1177/17590914211016217.
68. Cheng Y, Tan G, Zhu Q, Wang C, Ruan G, Ying S, et al. Efficacy of fecal microbiota transplantation in patients with Parkinson's disease: Clinical trial results from a randomized, placebo-controlled design. *Gut Microbes*. 2023 Dec 18;15(2):2284247. Available from: https://www.tandfonline.com/doi/full/10.1080/19490976.2023.2284247.
69. Naomi R, Embong H, Othman F, Ghazi HF, Maruthey N, Bahari H. Probiotics for Alzheimer's disease: A systematic review. *Nutrients*. 2021 Dec 22;14(1):20. Available from: https://www.mdpi.com/2072-6643/14/1/20.
70. Akhgarjand C, Vahabi Z, Shab-Bidar S, Etesam F, Djafarian K. Effects of probiotic supplements on cognition, anxiety, and physical activity in subjects with mild and moderate Alzheimer's disease: A randomized, double-blind, and placebo-controlled study. *Front Aging Neurosci*. 2022 Oct 31;14:1032494. Available from: https://www.frontiersin.org/articles/10.3389/fnagi.2022.1032494/full.
71. Zhu X, Hu J, Deng S, Tan Y, Qiu C, Zhang M, et al. Bibliometric and visual analysis of research on the links between the gut microbiota and depression from 1999 to 2019. *Front Psychiatry*. 2021 Jan 8;11:587670. Available from: https://www.frontiersin.org/articles/10.3389/fpsyt.2020.587670/full.
72. Manfready RA, Goetz CG, Keshavarzian A. Chapter Four - Intestinal microbiota and neuroinflammation in Parkinson's disease: At the helm of the gut-brain axis. In: Sampson TR, editor. *International Review of Neurobiology*. Academic Press; 2022 p. 81–99. (Microbiome in Neurological Disease; vol. 167). Available from: https://www.sciencedirect.com/science/article/pii/S0074774222000794.
73. Cao J, Amakye WK, Qi C, Liu X, Ma J, Ren J. Bifidobacterium Lactis Probio-M8 regulates gut microbiota to alleviate Alzheimer's disease in the APP/PS1 mouse model. *Eur J Nutr*. 2021 Oct;60(7):3757–69. Available from: https://link.springer.com/10.1007/s00394-021-02543-x.
74. Cowan CSM, Dinan TG, Cryan JF. Annual research review: critical windows – The microbiota-gut-brain axis in neurocognitive development. *J Child Psychol Psychiatry*. 2020 Mar;61(3):353–71.
75. Taylor VH. The microbiome and mental health: hope or hype? *jpn*. 2019 Jul 1;44(4):219–22. Available from: http://www.jpn.ca/lookup/doi/10.1503/jpn.190110.
76. Ding JH, Jin Z, Yang XX, Lou J, Shan WX, Hu YX, et al. Role of gut microbiota via the gut-liver-brain axis in digestive diseases. *WJG*. 2020 Oct 28;26(40):6141–62. Available from: https://www.wjgnet.com/1007-9327/full/v26/i40/6141.htm.
77. Zyoud SH, Smale S, Waring WS, Sweileh WM, Al-Jabi SW. Global research trends in microbiome-gut-brain axis during 2009–2018: A bibliometric and visualized study. *BMC Gastroenterol*. 2019 Dec;19(1):158. Available from: https://bmcgastroenterol.biomedcentral.com/articles/10.1186/s12876-019-1076-z.
78. Moysidou CM, Owens RM. Advances in modelling the human microbiome–gut–brain axis *in vitro*. *Biochem Soc Trans*. 2021 Feb 26;49(1):187–201. Available from: https://portlandpress.com/biochemsoctrans/article/49/1/187/227811/Advances-in-modelling-the-human-microbiome-gut.
79. Bajinka O, Simbilyabo L, Tan Y, Jabang J, Saleem SA. Lung-brain axis. *Crit Rev Microbiol*. 2022 May 4;48(3):257–69. Available from: https://www.tandfonline.com/doi/full/10.1080/1040841X.2021.1960483.
80. Ou Y, Belzer C, Smidt H, De Weerth C. Methodological recommendations for human microbiota-gut-brain axis research. *Microbiome Res Rep*. 2024 Jan 1;3(1). Available from: https://www.oaepublish.com/articles/mrr.2023.33.
81. Goehler LE, Gaykema RPA, Opitz N, Reddaway R, Badr N, Lyte M. Activation in vagal afferents and central autonomic pathways: Early responses to intestinal infection with *Campylobacter jejuni*. *Brain Behav Immunity*. 2005 Jul 1;19(4):334–44.
82. Perez-Burgos A, Wang B, Mao YK, Mistry B, Neufeld KAM, Bienenstock J, et al. Psychoactive bacteria Lactobacillus rhamnosus (JB-1) elicits rapid frequency facilitation in vagal afferents. *Am J Physiol Gastrointest Liver Physiol*. 2013 Jan 15;304(2):G211–20.
83. Sgritta M, Dooling SW, Buffington SA, Momin EN, Francis MB, Britton RA, et al. Mechanisms underlying microbial-mediated changes in social behavior in mouse models of autism spectrum disorder. *Neuron*. 2019 Jan 16;101(2):246–59.e6.
84. McVey Neufeld KA, Perez-Burgos A, Mao YK, Bienenstock J, Kunze WA. The gut microbiome restores intrinsic and extrinsic nerve function in germ-free mice accompanied by changes in calbindin. *Neurogastroenterol Motil*. 2015 May;27(5):627–36.
85. Rolig AS, Mittge EK, Ganz J, Troll JV, Melancon E, Wiles TJ, et al. The enteric nervous system promotes intestinal health by constraining microbiota composition. *PLoS Biol*. 2017 Feb 16;15(2):e2000689.
86. Caputi V, Marsilio I, Filpa V, Cerantola S, Orso G, Bistoletti M, et al. Antibiotic-induced dysbiosis of the microbiota impairs gut neuromuscular function in juvenile mice. *Br J Pharmacol*. 2017;174(20):3623–39.
87. Collins J, Borojevic R, Verdu EF, Huizinga JD, Ratcliffe EM. Intestinal microbiota influence the early postnatal development of the enteric nervous system. *Neurogastroenterol Motility*. 2014;26(1):98–107.

88. Bistoletti M, Caputi V, Baranzini N, Marchesi N, Filpa V, Marsilio I, et al. Antibiotic treatment-induced dysbiosis differently affects BDNF and TrkB expression in the brain and in the gut of juvenile mice. *PLoS One.* 2019 Feb 22;14(2):e0212856.
89. Anitha M, Vijay–Kumar M, Sitaraman SV, Gewirtz AT, Srinivasan S. Gut microbial products regulate murine gastrointestinal motility via toll-like receptor 4 signaling. *Gastroenterology.* 2012 Oct 1;143(4):1006–16.e4.
90. Tanida M, Yamano T, Maeda K, Okumura N, Fukushima Y, Nagai K. Effects of intraduodenal injection of *Lactobacillus johnsonii* La1 on renal sympathetic nerve activity and blood pressure in urethane-anesthetized rats. *Neurosci Lett.* 2005 Dec 2;389(2):109–14.
91. Tashtush AA, Reed D, Lomax AE. Excitation of vagal afferent neurons by fecal supernatant from inflammatory bowel disease patients. Available from: https://faseb.onlinelibrary.wiley.com/doi/10.1096/fasebj.2022.36.S1.R2894.
92. Buckley MM, O'Brien R, Brosnan E, Ross RP, Stanton C, Buckley JM, et al. Glucagon-like peptide-1 secreting L-cells coupled to sensory nerves translate microbial signals to the host rat nervous system. *Front Cell Neurosci.* 2020 Apr 30;14. Available from: https://www.frontiersin.org/journals/cellular-neuroscience/articles/10.3389/fncel.2020.00095/full.
93. Mao YK, Kasper DL, Wang B, Forsythe P, Bienenstock J, Kunze WA. Bacteroides fragilis polysaccharide A is necessary and sufficient for acute activation of intestinal sensory neurons. *Nat Commun.* 2013 Feb 12;4(1):1465.
94. Lal S, Kirkup AJ, Brunsden AM, Thompson DG, Grundy D. Vagal afferent responses to fatty acids of different chain length in the rat. *Am J Physiol Gastrointest Liver Physiol.* 2001 Oct;281(4):G907–15.
95. Muller PA, Schneeberger M, Matheis F, Wang P, Kerner Z, Ilanges A, et al. Microbiota modulate sympathetic neurons via a gut–brain circuit. *Nature.* 2020 Jul;583(7816):441–6.
96. Barton JR, Londregan AK, Alexander TD, Entezari AA, Covarrubias M, Waldman SA. Enteroendocrine cell regulation of the gut-brain axis. *Front Neurosci.* 2023 Nov 7;17. Available from: https://www.frontiersin.org/journals/neuroscience/articles/10.3389/fnins.2023.1272955/full.
97. Siba IP, Martynhak BJ, Pereira M. When gut hormones influence brain function in depression. *Appl Biosci.* 2023 Mar;2(1):31–51.
98. Wachsmuth HR, Weninger SN, Duca FA. Role of the gut–brain axis in energy and glucose metabolism. *Exp Mol Med.* 2022 Apr;54(4):377–92.
99. Hinds JA, Sanchez ER. The role of the Hypothalamus–Pituitary–Adrenal (HPA) axis in test-induced anxiety: assessments, physiological responses, and molecular details. *Stresses.* 2022 Mar;2(1):146–55.
100. Karin O, Raz M, Tendler A, Bar A, Kohanim YK, Milo T, et al. A new model for the HPA axis explains dysregulation of stress hormones on the timescale of weeks. *Mol Syst Biol.* 2020 Jul 16. Available from: https://www.embopress.org/doi/10.15252/msb.20209510.
101. Reul JMHM Holsboer F. On the role of corticotropin-releasing hormone receptors in anxiety and depression. *Dialogues Clin Neurosci.* 2002 Mar;4(1):31–46.
102. Risbrough VB, Stein MB. Role of corticotropin releasing factor in anxiety disorders: A translational research perspective. *Horm Behav.* 2006 Nov;50(4):550–61.
103. Tsigos C, Chrousos GP. Hypothalamic–pituitary–adrenal axis, neuroendocrine factors and stress. *J Psychosom Res.* 2002 Oct 1;53(4):865–71.
104. Accarie A, Toth J, Wauters L, Farré R, Tack J, Vanuytsel T. Estrogens play a critical role in stress-related gastrointestinal dysfunction in a spontaneous model of disorders of gut–brain interaction. *Cells.* 2022 Jan;11(7):1214.
105. Li J, An R, Zhang Y, Li X, Wang S. Correlations of macronutrient-induced functional magnetic resonance imaging signal changes in human brain and gut hormone responses123. *Am J Clin Nutr.* 2012 Aug 1;96(2):275–82.
106. Dicks LMT. Gut bacteria and neurotransmitters. *Microorganisms.* 2022 Sep;10(9):1838.
107. Sun EWL, Martin AM, Young RL, Keating DJ. The regulation of peripheral metabolism by gut-derived hormones. *Front Endocrinol.* 2019 Jan 4;9. Available from: https://www.frontiersin.org/journals/endocrinology/articles/10.3389/fendo.2018.00754/full.
108. Zanchi D, Depoorter A, Egloff L, Haller S, Mählmann L, Lang UE, et al. The impact of gut hormones on the neural circuit of appetite and satiety: A systematic review. *Neurosci Biobehav Rev.* 2017 Sep 1;80:457–75.
109. Liu L, Wang H, Chen X, Xie P. Gut microbiota: A new insight into neurological diseases. *Chin Med J.* 2023 Jun 5;136(11):1261.
110. Tiwari P, Dwivedi R, Bansal M, Tripathi M, Dada R. Role of gut microbiota in neurological disorders and its therapeutic significance. *J Clin Med.* 2023 Feb 19;12(4):1650.
111. Ullah H, Arbab S, Tian Y, Liu C qing, Chen Y, Qijie L, et al. The gut microbiota–brain axis in neurological disorder. *Front Neurosci.* 2023 Aug 4;17. Available from: https://www.frontiersin.org/journals/neuroscience/articles/10.3389/fnins.2023.1225875/full.
112. Chen J, Rao JN, Zou T, Liu L, Marasa BS, Xiao L, et al. Polyamines are required for expression of Toll-like receptor 2 modulating intestinal epithelial barrier integrity. *Am J Physiol-Gastroint Liver Physiol.* 2007 Sep;293(3):G568–76.
113. Feng C, Jin C, Liu K, Yang Z. Microbiota-derived short chain fatty acids: Their role and mechanisms in viral infections. *Biomed Pharmacother.* 2023 Apr 1;160:114414.
114. He J, Zhang P, Shen L, Niu L, Tan Y, Chen L, et al. Short-chain fatty acids and their association with signalling pathways in inflammation, glucose and lipid metabolism. *Int J Mol Sci.* 2020 Sep 2;21(17):6356.
115. Kim CH. Complex regulatory effects of gut microbial short-chain fatty acids on immune tolerance and autoimmunity. *Cell Mol Immunol.* 2023 Apr;20(4):341–50.
116. Lee JH, Lee J. Indole as an intercellular signal in microbial communities. *FEMS Microbiol Rev.* 2010 Jul 1;34(4):426–44.
117. Miller TL, Wolin MJ. Pathways of acetate, propionate, and butyrate formation by the human fecal microbial flora. *Appl Environ Microbiol.* 1996 May;62(5):1589–92.
118. Oldendorf W. Carrier-mediated blood-brain barrier transport of short-chain monocarboxylic organic acids. *Am J Physiol-Legacy Content.* 1973 Jun;224(6):1450–3.
119. Ríos-Covián D, Ruas-Madiedo P, Margolles A, Gueimonde M, de los Reyes-Gavilán CG, Salazar N. Intestinal short chain fatty acids and their link with diet and human health. *Front Microbiol.* 2016 Feb 17;7. Available from: https://www.frontiersin.org/journals/microbiology/articles/10.3389/fmicb.2016.00185/full.
120. Ufnal M, Zadlo A, Ostaszewski R. TMAO: A small molecule of great expectations. *Nutrition.* 2015 Nov 1;31(11):1317–23.
121. Vijay N, Morris ME. Role of monocarboxylate transporters in drug delivery to the brain. *Curr Pharmaceut Design.* 2014 Mar 1;20(10):1487–98.

122. Zhang D, Jian YP, Zhang YN, Li Y, Gu LT, Sun HH, et al. Short-chain fatty acids in diseases. *Cell Commun Signal.* 2023 Aug 18;21(1):212.
123. Agus A, Planchais J, Sokol H. Gut microbiota regulation of tryptophan metabolism in health and disease. *Cell Host Microbe.* 2018 Jun 13;23(6):716–24.
124. Ciorba MA. Kynurenine pathway metabolites: relevant to vitamin B-6 deficiency and beyond123. *Am J Clin Nutr.* 2013 Oct 1;98(4):863–4.
125. Flux MC, Lowry CA. Finding intestinal fortitude: Integrating the microbiome into a holistic view of depression mechanisms, treatment, and resilience. *Neurobiol Dis.* 2020 Feb 1;135:104578.
126. Mesripour A, Hajhashemi V, Kuchak A. Effect of concomitant administration of three different antidepressants with vitamin B6 on depression and obsessive compulsive disorder in mice models. *Res Pharmaceut Sci.* 2017 Feb;12(1):46.
127. O'Mahony SM, Clarke G, Borre YE, Dinan TG, Cryan JF. Serotonin, tryptophan metabolism and the brain-gut-microbiome axis. *Behav Brain Res.* 2015 Jan 15;277:32–48.
128. Hazan SC, Daniels J. S159 The role of dysbiosis in neurological disorders. *Am J Gastroenterology.* 2021 Oct;116:S70.
129. Wu W, Kong Q, Tian P, Zhai Q, Wang G, Liu X, et al. Targeting gut microbiota dysbiosis: Potential intervention strategies for neurological disorders. *Engineering.* 2020 Apr 1;6(4):415–23.
130. Pistollato F, Sumalla Cano S, Elio I, Masias Vergara M, Giampieri F, Battino M. Role of gut microbiota and nutrients in amyloid formation and pathogenesis of Alzheimer disease. *Nutr Rev.* 2016 Oct 1;74(10):624–34.
131. Chen Y, Fang L, Chen S, Zhou H, Fan Y, Lin L, et al. Gut microbiome alterations precede cerebral amyloidosis and microglial pathology in a mouse model of Alzheimer's disease. *BioMed Res Int.* 2020;2020(1):8456596.
132. Marizzoni M, Mirabelli P, Mombelli E, Coppola L, Festari C, Lopizzo N, et al. A peripheral signature of Alzheimer's disease featuring microbiota-gut-brain axis markers. *Alz Res Therapy.* 2023 May 31;15(1):101.
133. Holmqvist S, Chutna O, Bousset L, Aldrin-Kirk P, Li W, Björklund T, et al. Direct evidence of Parkinson pathology spread from the gastrointestinal tract to the brain in rats. *Acta Neuropathol.* 2014 Dec 1;128(6):805–20.
134. Bhattacharyya D, Mohite GM, Krishnamoorthy J, Gayen N, Mehra S, Navalkar A, et al. Lipopolysaccharide from gut microbiota modulates α-synuclein aggregation and alters its biological function. *ACS Chem Neurosci.* 2019 May 15;10(5):2229–36.
135. Sampson TR, Challis C, Jain N, Moiseyenko A, Ladinsky MS, Shastri GG, et al. A gut bacterial amyloid promotes α-synuclein aggregation and motor impairment in mice. Chiu IM, Garrett WS, Desjardins M, editors. *eLife.* 2020 Feb 11;9:e53111.
136. Ochoa-Repáraz J, Mielcarz DW, Ditrio LE, Burroughs AR, Foureau DM, Haque-Begum S, et al. Role of gut commensal microflora in the development of experimental autoimmune encephalomyelitis1. *J Immunol.* 2009 Nov 15;183(10):6041–50.
137. Jangi S, Gandhi R, Cox LM, Li N, von Glehn F, Yan R, et al. Alterations of the human gut microbiome in multiple sclerosis. *Nat Commun.* 2016 Jun 28;7(1):12015.
138. Bravo JA, Forsythe P, Chew MV, Escaravage E, Savignac HM, Dinan TG, et al. Ingestion of *Lactobacillus* strain regulates emotional behavior and central GABA receptor expression in a mouse via the vagus nerve. *Proc Natl Acad Sci USA.* 2011 Sep 20;108(38):16050–5.
139. Makris AP, Karianaki M, Tsamis KI, Paschou SA. The role of the gut-brain axis in depression: endocrine, neural, and immune pathways. *Hormones.* 2021 Mar 1;20(1):1–12.
140. Varanoske AN, McClung HL, Sepowitz JJ, Halagarda CJ, Farina EK, Berryman CE, et al. Stress and the gut-brain axis: Cognitive performance, mood state, and biomarkers of blood-brain barrier and intestinal permeability following severe physical and psychological stress. *Brain Behav Immunity.* 2022 Mar 1;101:383–93.
141. Parracho HM, Bingham MO, Gibson GR, McCartney AL. Differences between the gut microflora of children with autistic spectrum disorders and that of healthy children. *J Med Microbiol.* 2005;54(10):987–91.
142. Heijtz RD, Wang S, Anuar F, Qian Y, Björkholm B, Samuelsson A, et al. Normal gut microbiota modulates brain development and behavior. *PNAS.* 2011 Feb 15;108(7):3047–52.
143. Wang L, Christophersen CT, Sorich MJ, Gerber JP, Angley MT, Conlon MA. Low relative abundances of the mucolytic bacterium akkermansia muciniphila and bifidobacterium spp. in feces of children with autism. *Appl Environ Microbiol.* 2011 Sep 15;77(18):6718–21.
144. Damiani F, Cornuti S, Tognini P. The gut-brain connection: Exploring the influence of the gut microbiota on neuroplasticity and neurodevelopmental disorders. *Neuropharmacology.* 2023 Jun 15;231:109491.
145. Tang AT, Choi JP, Kotzin JJ, Yang Y, Hong CC, Hobson N, et al. Endothelial TLR4 and the microbiome drive cerebral cavernous malformations. *Nature.* 2017 May;545(7654):305–10.
146. Zhang Y, Zhou S, Zhou Y, Yu L, Zhang L, Wang Y. Altered gut microbiome composition in children with refractory epilepsy after ketogenic diet. *Epilepsy Res.* 2018 Sep;145:163–8.
147. Bagheri S, Heydari A, Alinaghipour A, Salami M. Effect of probiotic supplementation on seizure activity and cognitive performance in PTZ-induced chemical kindling. *Epilepsy Behav.* 2019 Jun 1;95:43–50.
148. Kong G, Ellul S, Narayana VK, Kanojia K, Ha HTT, Li S, et al. An integrated metagenomics and metabolomics approach implicates the microbiota-gut-brain axis in the pathogenesis of Huntington's disease. *Neurobiol Dis.* 2021 Jan 1;148:105199.
149. Qian SX, Bao YF, Li XY, Dong Y, Zhang XL, Wu ZY. Multi-omics analysis reveals key gut microbiota and metabolites closely associated with huntington's disease. *Mol Neurobiol.* 2024 Jun 8. Available from: https://doi.org/10.1007/s12035-024-04271-9

20 Prebiotic as Functional Food in Neurological Disorder

Pankaj Sharma[1], Sarvesh Bhargava[1], Vinay Jain[2], Deepesh Parashar[3], and Wasim Akram[4]*

[1] Department of Pharmaceutics, ShriRam College of Pharmacy, Morena, Madhya Pradesh, India
[2] Department of Pharmacognosy, ShriRam College of Pharmacy, Morena, Madhya Pradesh, India
[3] Department of Medicinal Chemistry, Amity Institute of Pharmacy, Amity University, Gwalior, Madhya Pradesh, India
[4] Department of Pharmaceutics, Amity Institute of Pharmacy, Amity University, Gwalior, Madhya Pradesh, India

20.1 INTRODUCTION

Prebiotics can be understood as food components that act on the host's benefits by selectively encouraging the growth of desirable intestinal bacteria, mainly bifidobacteria and lactobacilli. They usually are oligosaccharides, for example, inulin, fructooligosaccharides (FOS), and galactooligosaccharides (GOS). Prebiotics can be obtained from foods like chicory root, garlic, onions, leeks, asparagus, and bananas, among others. These compounds remain insoluble in the upper gut and reach the colon, where their fermentation through the action of gut microbes produces SCFAs, mainly butyrate, propionate, and acetate; these have various health effects on the body [1, 2].

Consumers' understanding of functional foods is much more extensive and incorporates a far wider range of products, prebiotics being a subgroup of functional foods. They are mostly fortified with micronutrients, including vitamins, essential minerals, live active bacterial cultures, and organic compounds referred to as prebiotics. Within the context of functional foods, products can be categorically described as the real or whole foods or as the foods which have been fortified, enriched, or enhanced in some manner to help minimize the risk of disease and boost health status. Being a rich source of phytochemicals they are useful in the control and prevention of chronic diseases like cardiovascular diseases, diabetes, obesity, and even neurological disorders due to their bioactive compounds that show multifaceted physiological effects [3, 4].

In the mid-1990s, the term prebiotics was defined by Gibson and Roberfroid, which can be considered as a landmark in human nutritional science. Drawing from this observation, they described prebiotics as "non-digestible food ingredients that selectively stimulate the growth and or activity of one or limited number of bacteria in the colon to directly and/or indirectly enhance the host's health" [5]. The first investigations were aimed at the impact on the gut; it was acknowledged that prebiotics could stimulate the proliferation of favorable bacteria inside the gut, as well as promote regular bowel movements and strengthen the barriers against infections.

There has been growth in the knowledge about prebiotics since they were discovered with subsequent research extending to the immune response and metabolic functions and mental health. It has also been established that prebiotics can affect immunological reactions and decrease inflammation, and also affect lipid balance. Due to the growing body of evidence on the impact of prebiotics on health, global research and the development of such products have expanded significantly [6, 7].

However, studies on the neurological effect of prebiotics have gained more attention in the recent past owing to development in the knowledge of the gut–brain axis. Such cross talk can be constituted by neural, endocrine, immune signaling between the gut and the brain, or through the signaling by gut microbiota. Studies also revealed that the gut microbiota can modulate the central nervous system (CNS) and behavior through several different pathways; for example, SCFAs, neurotransmitters, cytokines. etc. [8].

In this trial, on account of prebiotics, the averagely favorable gut microbiota community has helped to boost the beneficial nutrients that yield these neuroactive ingredients. Research has also shown that prebiotic consumption has an effect of elevating mood, depression reduction, better cognition, and signs of neurodegenerative diseases such as Alzheimer's and Parkinson's. For example, inulin and GOS raise the concentration of *Bifidobacterium breve* and *Limosilactobacillus reuteri* in the gut, which is connected with better mental health. These findings about prebiotics suggest that they have the ability to affect the

 DOI: 10.1201/9781003536802-20

neurological health, thus expanding the possibility of the prevention and treatment of neurological diseases through diet [9, 10].

20.2 UNDERSTANDING PREBIOTICS

20.2.1 Definition and Classification of Prebiotics

Prebiotics are determined as selectively stimulated substances that alter in a proven beneficial manner the composition and/or activity of the GI microbiota. It came to existence from the works of Gibson and Roberfroid in the year 1995 and the idea has since expanded to other compounds. Prebiotics are usually oligosaccharides and carbohydrates that are not digested and are different from probiotics which are live and ajax microorganisms that confer health benefits when consumed in specific amounts [11, 12].

The classification of prebiotics is mainly in terms of chemical structure and the ability to turn them into usable substrates by the gut microbiota. The prebiotics chemical characteristics that have been paid the most attention in research are inulin-type fructans which comprise inulin and FOS, and GOS. Inulin and FOS are present in foods such as chicory root, raw garlic cloves, onions, as well as leeks, while GOS is mainly produced from lactose and widely incorporated in baby foods. Other prebiotics include oligosaccharides like xylooligosaccharides (XOS), soy oligosaccharides, and isomaltooligosaccharides (IMOS) [13, 14].

Inulin-type fructans are synthesized by accumulation of fructose with the terminal glucose moiety, and their DP ranges from 2 to 60. This variation affects their fermentability as well as their functionality in the gastrointestinal tract. FOS, with generally a DP of 2–10, have a tendency to be more readily fermented by the beneficial gut bacteria such as *Bifidobacterium* and *Lactobacillus*, improving their growth and activity. On the other hand, inulin, having a higher DP, is fermented at a slower rate, thus has a longer prebiotic impact in the body [15].

GOSs, which are made up of galactose units with a terminal glucose, are also stimulating the growth of the beneficial bacteria. It is specifically efficient in increasing the populations of *Bifidobacterium* that is beneficial in regulating gut health. GOS are added to the formulas to copy the impact of human milk oligosaccharides, which helps to form and maintain a proper and healthy bacterial milieu in the gut of an infant [16].

Evidently, the utilitarian values of prebiotics transcend the control of gut health. Prebiotics affect the gut microbiota and, consequently, can alter systemic immune response, decrease inflammation, enhance mineral absorption, and some outcomes of the gut–brain axis. These characteristics of prebiotics that make it possible to stimulate the growth and reproduction and suppress the pathogenic bacteria are some of the reasons why it is believed that prebiotics could be used in the management of various health challenges [17, 18].

20.2.2 Common Sources of Prebiotics

20.2.2.1 Inulin

Thus, inulin is a fructosan natural polysaccharide, generally with a terminal fructose unit at the nonreducing end. It is present in many plant products, especially in the root of the chicory, Jerusalem artichoke, garlic, onion, and leeks. Inulin is also resistant to the human enzymes in the small intestine and gets to the large intestine where it is consumed by the friendly bacteria, including *Bifidobacterium* and *Lactobacillus*. This process leads to the generation of the short-chain fatty acids (SCFAs), which have positive health impacts such as a healthy gut, boosted immunity, and a lower probability of contracting colon cancer. Because inulin enhances the prolific of favorable bacteria over pathogenic ones, it is a prominent dietary prebiotic [19, 20].

20.2.2.2 Fructooligosaccharides

Fructooligosaccharides are oligosaccharides that are made mostly of fructose molecules linked the way in which fructose molecules are linked together, FOSs are short chains and occur naturally in various plants such as bananas, onions, chicory root, garlic, asparagus, and wheat. FOS are similar to inulin; however, they are shorter in-chain molecules manufactured from 2 to 10 fructose units. They are not broken down and absorbed by enzymes that humans use, and are instead received, metabolized, and absorbed in the colon by bacteria. Several research have found that FOS stimulate the growth of healthy bacteria, especially *Bifidobacteria*, and enhance bowel movement through the production of larger amount of stool and increased frequency. Furthermore, it has been suggested that FOS intake is beneficial for enhancing the absorption of minerals, improving the immune system and lowering the level of pathogenic bacteria [21, 22].

20.2.2.3 Galactooligosaccharides

Galactooligosaccharides are produced from galactose connected to each other, where one of the ends connects to glucose, and are usually synthesized from lactose. It exists in human breast milk and can be incorporated into infant formulas as one of the marketing strategies to mimic HMOs. GOS selectively stimulate specific advantageous resident gut bacteria such as *Bifidobacterium* that is beneficial in the creation of a healthy ENT in infants. In the context to the human subjects, GOS have proven to have positive impacts on the gastrointestinal health, immune system as well as cognition and mental health. It is found in foods such as milk, yoghurt, soybean products, and beans, peas, and some tubers [23, 24].

20.2.2.4 Resistant Starch

Resistant starch is, basically, a starch that is not digested in the small intestine but is instead analyzed in the large intestine with most of the characteristics of dietary fiber. It is classified into four types based on its source and structure: RS1 like physically inaccessible starch, RS2 or resistant granules, RS3 or retrograded starch, and RS4 which is chemically modified starch. RS is present in foods that include green plantains, cooked and cold potatoes, native beans, and whole grain products among others. Professionally, it is fermented by gut microbiota, where it forms SCFAs such as butyrate that have multiple knock-on health benefits, including enhancing gut congestion, insulin sensitivity, and decrease in the risk of colorectal cancer. Consuming resistant starch has also been associated with increase in the levels of insatiety and a decrease in appetite, thus considering for obesity control [25, 26].

20.2.3 Mechanisms of Action

20.2.3.1 Modulation of Gut Microbiota

Prebiotics' ability to positively affect the body has been attributed to the ability to alter the population of active microorganisms within the gut. Prebiotics are fermentable by the beneficial gut bacteria such as bifidobacteria and lactobacilli enhancing their numbers and activity in the gut (Figure 20.1). This selective fermentation results in increase in the number of these advocate bacteria hence competing with pathogenic bacteria for space and resources and also through the production of substances that inhibit the growth of pathogenic bacteria [27, 28]. For instance, earlier research carried out indicates that inulin and FOS enhance the population of bifidobacteria in the gut and at the same time reduce the susceptibility to various kinds of gut infections [29, 30].

20.2.3.2 Production of Short-Chain Fatty Acids

Another critical mechanism is the production of short-SCFAs during the fermentation of prebiotics by gut microbiota (Figure 20.2). SCFAs, such as acetate, propionate, and butyrate, have various beneficial effects on host health. Butyrate, in particular, is the primary energy source for colonocytes and plays a vital role in maintaining colonic health. It has anti-inflammatory properties, promotes the integrity of the gut barrier, and has been shown to induce apoptosis in colon cancer cells, thereby reducing the risk of colorectal cancer [31, 32]. SCFAs also influence systemic metabolic processes; for example, propionate has been shown to regulate gluconeogenesis in the liver and reduce lipogenesis, contributing to improved metabolic health [33].

20.2.3.3 Immune Modulation

Prebiotics also play a significant role in modulating the immune system (Figure 20.1). The interaction between prebiotics, gut microbiota, and the host immune system can enhance the body's defense mechanisms. Prebiotics have been shown to increase the production of anti-inflammatory cytokines and decrease the production of pro-inflammatory cytokines, thereby helping to regulate immune responses and reduce chronic inflammation [34, 35]. Additionally, SCFAs produced from prebiotic fermentation can influence the activity of immune cells, such as macrophages and regulatory T cells, promoting a balanced immune response and enhancing the body's ability to fight infections [36].

20.2.3.4 Impact on the Gut–Brain Axis

The gut–brain axis is a complex communication network that links the gut and the brain (Figure 20.1), involving neural, hormonal, and immune pathways. Prebiotics can impact brain function and behavior through their effects on the gut microbiota and the production of neuroactive compounds. SCFAs, for instance, can cross the blood–brain barrier and influence brain function by modulating neurotransmitter synthesis and signaling pathways. Additionally, the modulation of gut microbiota by prebiotics can influence the production of metabolites such as tryptophan, which is a precursor to serotonin, a key neurotransmitter involved in mood regulation [37, 38]. Studies have demonstrated that prebiotics like GOS and FOS can reduce anxiety and improve cognitive function, highlighting their potential in managing stress-related and neurodegenerative disorders [39, 40].

20.3 THE GUT–BRAIN AXIS: A PATHWAY FOR NEUROLOGICAL HEALTH

The CNS–GI axis is the two-way communication system that links the CNS with the gastrointestinal system. It also encompasses neural hormonal and immune communication systems that make the gut and the brain have direct talks and affect each other. Novel data have revealed importance of this axis for the neurological homeostasis, on mood, cognitive, and even neurodegenerative diseases [41, 42].

There are several neural connections linking up the gut and the brain. Most Internal communication to and from the brain is accomplished through the vagus nerve, which serves as a communication link between the gut and the brain. The immune system also possesses immune messengers called the cytokines and other immune mediators that may affect the brain and behavior. The endocrine thereby involves releasing hormones like cortisol and serotonin, which influence both the gut and the mind. All together these pathways are vital for maintaining gut–brain communication, which is always dynamic and functional for handling stress and achieving homeostasis [43, 44].

The studies on the gut microbiota show that it has a direct impact on the functioning of the brain and behavior through the GBA. For instance, SCFAs, which are by-products of gut bacteria metabolism; tryptophan, a precursor of neurotransmitter serotonin; and neurotransmitters, which directly influence brain chemistry and neuronal directness. Research has also revealed that changes in the composition

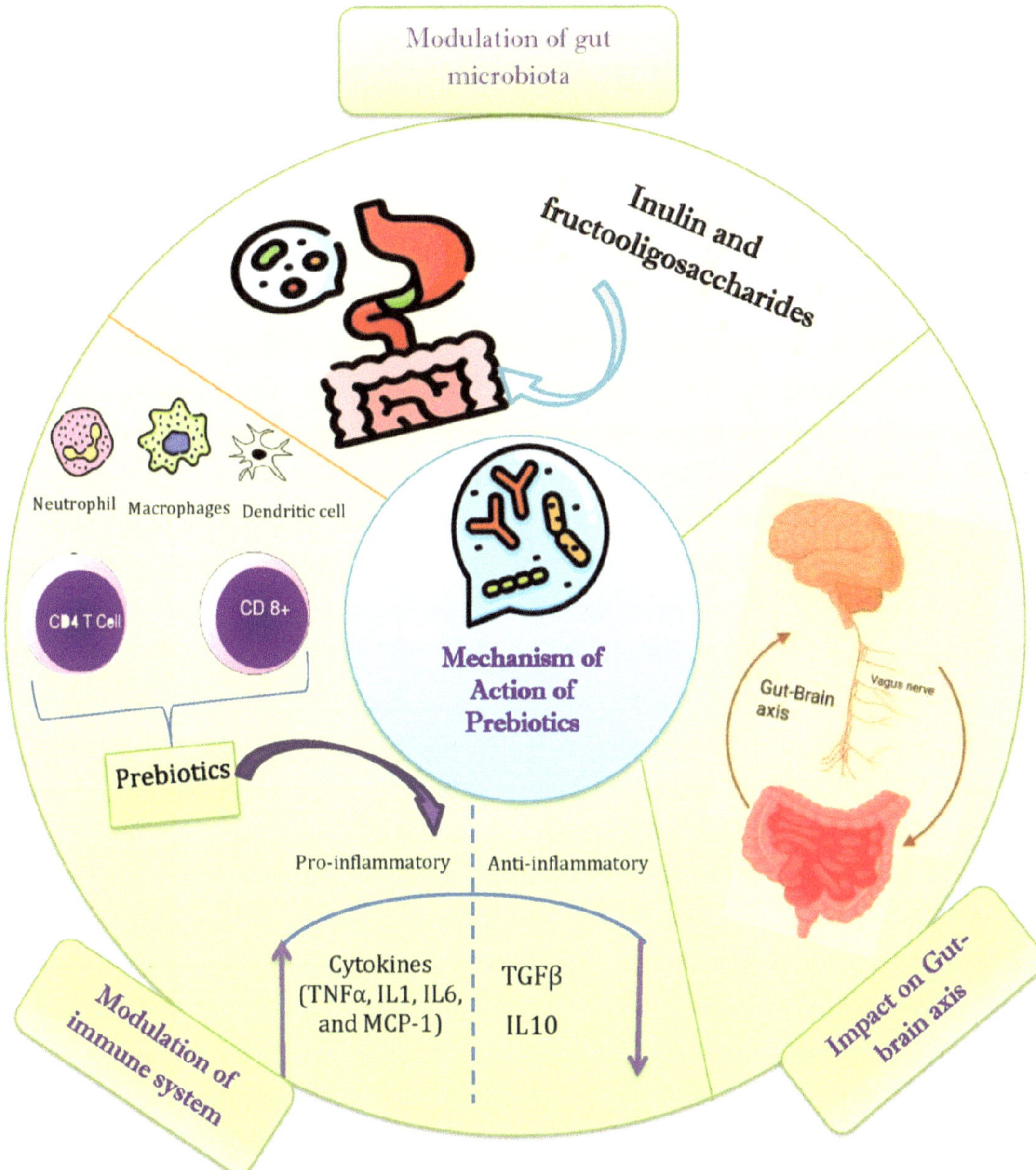

FIGURE 20.1 Mechanisms of actions of prebiotics.

of the people's gut microbes are linked to anxiety, depression, and autism spectrum disorders. Due to the effects of prebiotics on the gut microbiome, these compounds can significantly contribute to improved brain function and mental status, constituting a fundamental component of neurological disorder treatments [45, 46].

20.4 IMPACT OF PREBIOTICS ON GUT MICROBIOTA

20.4.1 Modulation of Microbiota Composition

Prebiotics only favor the growth and activity of a good beneficial bacteria which in turn will create a healthy population. People consume them and reach various gut bacteria, which leads to change of microbiota. For instance, prebiotic substances like inulin and FOS can help in escalating the population of useful bacteria and, at the same time, decreasing the range of pathogenic bacteria that are detrimental to the organism's gut health and function [47, 48].

20.4.2 Enhancement of Beneficial Microbes (e.g., Bifidobacteria, Lactobacilli)

Specific prebiotics like GOS and FOS especially have a positive effect on the beneficial microbes that are Bifidobacteria and Lactobacilli. These bacteria are good for the human body in many ways; they help in digestion, boost the immune system, and help to prevent chances of one getting infected with intestinal diseases. Research has shown

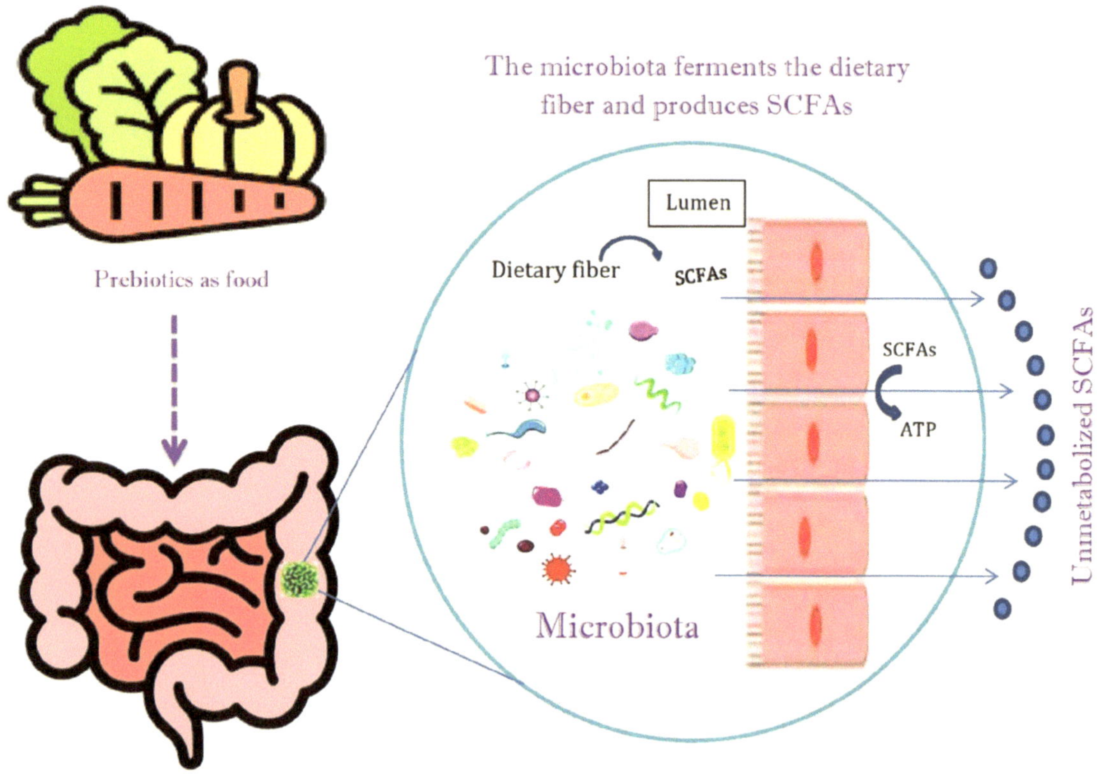

FIGURE 20.2 The microbiota ferments the dietary fiber and produces SCFAs.

that the ingestion of prebiotics leads to a marked enhancement of the level of the above beneficial bacteria in the gut, hence health [49, 50].

20.4.3 Production of Short-Chain Fatty Acids and Their Effects

The breakdown of prebiotics by the gut bacteria yields SCFAs, among them being acetate, propionate, and butyrate. The SCFAs are very essential to the health of the gut because they act as an energy source to the colonocytes, increase the mucus layer, and have anti-inflammatory properties. Concisely, butyrate, which is a SCFA, protects against CRC and enhances gut barrier function. Furthermore, SCFAs can affect systemic health via changing the immune system activity and metabolism [51, 52].

20.5 Prebiotics and Neuroinflammation

20.5.1 Understanding Neuroinflammation

Neuroinflammation is a type of inflammation affecting the site in the body commonly known as the CNS, which comprises the brain and spinal cord. This response is generally achieved through the involvement of microglia and astrocytes that are the immune cells of the central nervous system. Thus, acute neuroinflammation is neuroprotective; however, chronic neuroinflammation is linked with neurodegenerative diseases, including the Alzheimer's disease, Parkinson's disease, and multiple sclerosis (Figure 20.3). Until now, it has been identified that prolonged inflammation in CNS can result in neuronal damage, altered synapse, and neural network, and cognitive impairment [53, 54].

20.5.2 Mechanisms by Which Prebiotics Modulate Inflammation

Prebiotics affect neuroinflammation through a direct action as well as the effects produced in the gut microbiota and SCFAs. Foods rich in prebiotics such as inulin as well as galactooligosaccharides enhance the growth of the friendly bacteria in the gut that in turn produce SCFAs, including butyrate. Butyrate has anti-inflammatory effect and can suppress the synthesis of cytokines that are pro-inflammatory and that can adversely affect the CNS by triggering inflammation throughout the whole body. Similarly, prebiotics are also known to improve the intestinal barrier integrity that in turn reduces the movement of molecules such as lipopolysaccharides (LPS) that instigate inflammation in the brain [55, 56].

20.5.3 Preclinical and Clinical Evidence

Experimental work links indicate positive impacts of prebiotics on neuroinflammation based on laboratory tests. For example (Table 20.1), in vitamin- and minerals-enriched

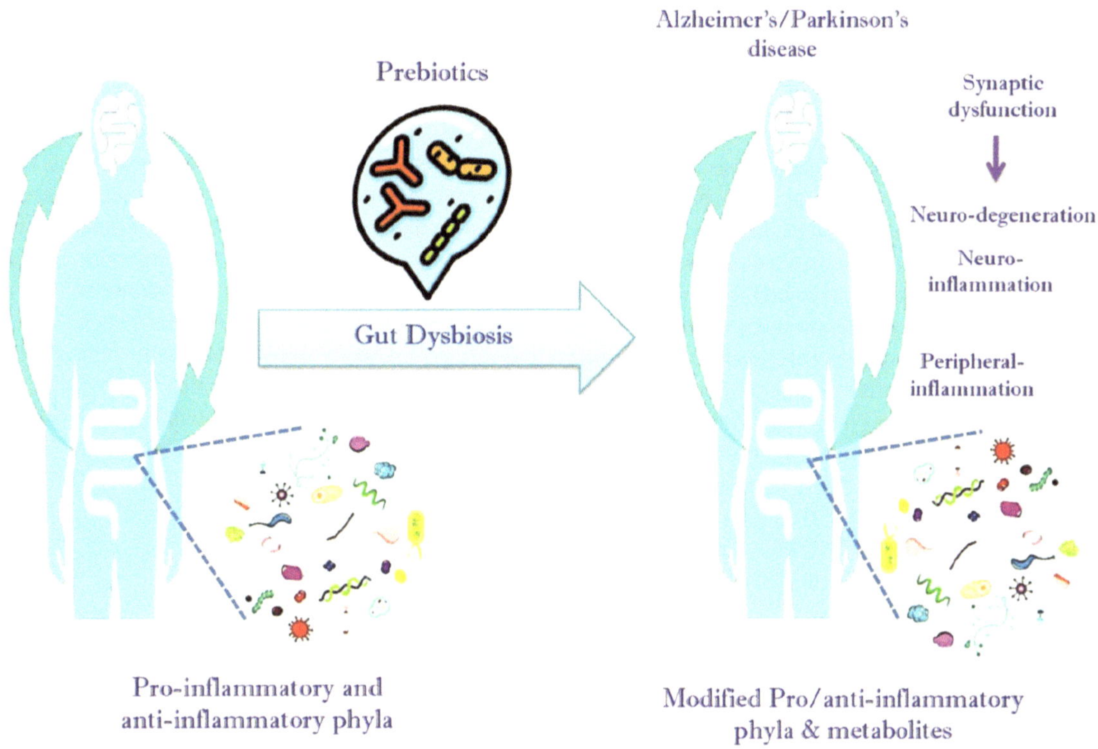

FIGURE 20.3 Gut dysbiosis and neuroinflammation.

TABLE 20.1
Preclinical and Clinical Evidences for Prebiotic Uses

Study Type	Prebiotics Used	Model/System	Key Findings
Preclinical	Inulin, GOS	Animal models (e.g., mice)	Decrease in neuroinflammation, memory enhancement, decreased level of pro-inflammatory cytokines, increased level of anti-inflammatory cytokines and neurotrophic factors [57, 58]
Clinical	GOS	Healthy volunteers	Waking salivary cortisol levels, enhanced affect regulation, signifying depressive status, and suggestion of lower inflammation [46]
Preclinical	Various prebiotics	Animal models (e.g., rats)	Reversal of the diabetes-influenced negative impact on synaptic activity to restore its normal level, cognition enhancement [57]
Clinical	Prebiotic supplement	Healthy volunteers	Regulation of the immune response, decrease in the level of inflammatory indices are identified [60]
Preclinical	Oligosaccharides	Mouse models	Enhanced anxiety-like symptoms decreased the extent of neuroinflammation in the hippocampus, Baumgardner's microbiota alterations [58]

animal diet that contain nondigestible compounds, including inulin and GOS, it was observed that neuroinflammation is decreased and cognition is enhanced in models of Alzheimer's disease and other related diseases. These studies suggest that prebiotics may lower the CON of the pro-inflammatory cytokines in the brain and elevate the CON of anti-inflammatory cytokines and neurotrophic factors [57, 58]. Yet, much clinician knowledge is still borrowed as some human trials indicated that prebiotics are capable of influencing the immune system and decrease inflammation. For instance, an empirical research among a group of healthy subjects found that prebiotic intake prevails to decrease waking cortisol response that is a sign of less systemic inflammation and enhance the process of emotional regulation implying possible positive effects for health of mind [59, 60].

20.6 PREBIOTICS AND NEUROTRANSMITTER PRODUCTION

20.6.1 Influence on Serotonin, GABA, and Other Neurotransmitters

Prebiotics also has a major role in the synthesis and management of neurotransmitters that is vital in the brain and our mental health. Serotonin, which is associated with mood, is part of the gut microbiota. It is also well established that 90% of the body's serotonin is produced in the gut, while prebiotics can also increase the production of serotonin through the stimulation of bacteria that produce tryptophan [61]. Also, it was revealed that prebiotics could raise the levels of gamma-aminobutyric acid (GABA), which is an inhibitory neurotransmitter that helps to regulate neuronal hyperexcitability and is involved in stress and anxiety reactions [62]. The fermentation of the prebiotics by the gut microbiota results in formation of SCFAs and it is known to impact the CNS and neurotransmitter biosynthesis [63, 64].

20.6.2 Impact on Mood and Cognitive Functions

Prebiotics, thus, affect mood and cognitive capabilities because they are involved in the regulation of neurotransmitter synthesis. Higher levels of serotonin and GABA lead to a better mood, diminished anxiety, and a more efficient way of handling stress. FOS and GOS, types of prebiotics, have been shown to positively alter mood and sleep by stimulating the secretion of these neurotransmitters. For instance, prebiotics can help in the reduction of the signs of depression and anxiety through regulation of neurotransmitters in the brain, which are typically imbalanced in individuals experiencing such ailments [65, 66]. Also, the favorable changes in the gut microbiota associated with prebiotic intake are emphasized as the effects that can influence the brain functions such as memory, learning, and word recalled [67].

20.6.3 Evidence from Animal and Human Studies

Animal studies also back the effectiveness of prebiotics in neurotransmitter synthesis while human studies also back the prebiotics as important in mental health. In traditional animal models, the addition of prebiotics has been shown to enhance the brain neurotransmitter such as serotonin and GABA, thereby reducing the anxiety level and enhancing the cognitive function [68, 69]. According to the clinical studies in humans, it is evident that prebiotic intake has a positive impact on mood and cognitive functions. For instance, one study identifying healthy volunteers reported that they witnessed a decreased cortisol level, which refers to stress with the supplementation of the prebiotic [70, 71]. The conclusion of the study underlines the prospect of prebiotics in the treatment of mental diseases and cognitive disorders.

20.7 PREBIOTICS IN NEURODEGENERATIVE DISEASES

20.7.1 Alzheimer's Disease

Alzheimer's disease (AD) or senile dementia is a disease that is chronic in nature and affects the neurons in the brain causing deterioration of mental ability, short-term memory loss, and changes in behavior. Some of the current research has proposed that there is an alterations in the gut microbiota that have an impact in AD development by enhancing the neuroinflammation and amyloid-beta plaques deposition [72, 73]. Due to ability to alter the gut microbiota, prebiotics can be used for the treatment of AD. For example, use of FOS and inulin as prebiotics have proven to increase the friendly gut flora, thus decreasing neuroinflammation and improving the cognitive ability of rat models of AD [74, 75]. Thus, the positive role of GBA in the attenuation of the development of AD can be explained by the modulation of the gut–brain axis.

20.7.2 Parkinson's Disease

Another neurodegenerative illness is the Parkinson's disease (PD), which mainly concerns motor function since dopaminergic neurons in substantia nigra degrade. Interest in the relationship between gut microbiota and PD has risen recently, where the disruption of gut microbiota is thought to worsen neuroinflammation and the progression of the disease [76, 77]. Research has demonstrated that prebiotics can help improve PD's symptoms through aiding in modulation of the balance of gut flora, thus sharply decreasing inflammation. In regard to animals with PD, earlier experiments revealed that GOS can positively affect motor coordination and has neuroprotective effects, decreasing the expression of pro-inflammatory cytokines; all these are due to augmentation of the presence of Bifidobacteria and Lactobacilli [78, 79].

20.7.3 Mechanistic Insights and Experimental Evidence

The mechanism through which the useful impact of prebiotic substances in the treatment of NDD has been determined is that prebiotics encourage the growth of helpful gut bacteria that in return synthesize by-products such as SCFAs. SCFAs, especially butyrate, have anti-inflammatory effects, while some can pass through the blood–brain barrier to exhibit neuroprotection [80, 81]. In addition, prebiotics help in maintaining gut health not only by improving the barrier function which has effects on the system inflammation that can reach the CNS. This is backed by experimental data from preclinical and clinical investigations of the role of antioxidants. For instance, a study performed on a mouse model of AD showed that prebiotic supplement intervention reduced amyloid-beta plaques and enhanced memory competence [82]. Similar to the mentioned effect of prebiotic supplementation, clinical trials in patients with neurodegenerative diseases indicated that prebiotics can affect the gut microbiota and have an anti-inflammatory effect on the human brain [83].

20.8 PREBIOTICS IN MENTAL HEALTH DISORDERS

20.8.1 Depression and Anxiety

Depression and anxiety are common mental disorders that have a huge effect on the patient's well-being. Increasing data from animal and human studies shared that the GMI has a direct role in creating these disorders through the GB axis. Some of these SCFAs such as prebiotics when able to influence the composition of gut microbiota has proved to help manage symptoms of depression and anxiety. For instance, FOS and GOS have been suggested to cause the proliferation of Bifidobacteria and Lactobacilli that generate SCFAs; these are neurotransmitters that can impact the brain and mood [84, 85]. Data on the effects of prebiotic agents have also demonstrated that supplementation leads to a decrease in cortisol levels, which is an indicator of stress, and an enhancement of the sufferers' emotional processing in depression and anxiety cases [86, 87].

20.8.2 Autism Spectrum Disorders

ASD is neurodevelopmental disorder that mainly affects social interaction and communication and is accompanied by repetitive behaviors. Gut microbiota/microbiome dysbiosis has been associated with the ASD and therefore uses microbiota manipulation as intervention for the condition could be a viable option. The literature review located illustrates that prebiotics have been explored for their capability in aiding gastrointestinal dysfunction and behavioral abnormalities associated with ASD [88]. For example, experimental animal research proves that GOS, a type of prebiotics, can alter the actinomyces community, decrease the permeability of the gut, and enhance social behaviors. There are not many clinical studies present, but the available [89] studies point toward a possibility of potentially using prebiotics to treat this disorder; prebiotic supplementation was shown to enhance GI symptoms and the social interactions of children with ASD [27, 90, 91].

20.8.3 Clinical Trials and Outcomes

Some recent clinical studies regarding the impact of prebiotics on specific mental health disorders are presented in Table 20.2. Another study conducted on healthy adults in a double-blind, cross-over, placebo-controlled manner that used GOS prebiotic indicated that, after a day's supplement consumption, there was lowering of waking cortisol levels and perceived anxiety, indicating improvement in the emotional well-being capacity of the supplements [92]. Yet another study revealed that prebiotic use in people with depression caused serious changes in mood as well as cognitive performance, which definitively points that prebiotics can be used as an additional method of the treatment of depressive disorders [93]. Clinical studies involving kids with ASID also mentioned that the use of prebiotics helped in the diversification of the gut microbiota, decrease GI symptoms, as well as the positive behavioral change pointing to the likelihood of its positive effects [94, 95]. Thus, these concerns evidence the uses of prebiotics for mental states and neurodevelopmental disorders: a justified need for further investigations and cures.

20.9 SYNERGISTIC EFFECTS: PREBIOTICS, PROBIOTICS, AND SYNBIOTICS

Synbiotics is therefore the integration of prebiotics and probiotics with the intention of, at the same time, promoting intestinal and brain health. While prebiotics are nondigestible food products that encourage the cultures of advantageous microorganisms in the gastrointestinal tract for growth and activeness, probiotics are useful live microorganisms given in sufficient doses in order to improve the host's health. Synbiotics combine elements of both since they create a favorable environment for members of the probiotic bacteria, thus enhancing the impact of the favorable changes in the composition of the gut microbiota and improved health. Specific actions of synbiotics affecting the gut and the brain are enhanced gut barrier function, immuno-modulations, synthesis of neurotransmitters, and SCFAs with neuroprotective and anti-inflammatory attributes. For example, a study done to show the effects of synbiotic supplementation in the management of IBS evidenced that the interventions led to a significant lesser mean score of gastrointestinal symptoms, and a decrease in the levels of anxiety [96]. Another example is the application of synbiotics in children with ASD; when given both prebiotics and probiotics, there was positive alteration in the child subjects' behaviors, as well as improvement in their gut microbiota [97]. In light of these case studies, it is evident that synbiotics have the capacity of being effective in offering a comprehensive remedy for the issues related with the gut–brain axis and further investigation into the application of synbiotics in therapy is advised.

20.10 DIETARY STRATEGIES AND PRACTICAL APPLICATIONS

20.10.1 Incorporating Prebiotics into the Diet

To include prebiotics in the diet, one is supposed to take foods containing these natural fibers in their diet plans. Some of the foods containing prebiotics include chicory root, garlic, onions, leeks, asparagus, bananas, and whole grain products like barley and oat. These foods rich in inulin, FOS, and other unabsorbed fibers selectively encouraging the bifidogenic and lactogenic properties in the gut helpful for improvement and prolonged beneficial fermentation by the

TABLE 20.2
Clinical Studies on Prebiotics for Treatment of Mental Health Disorders

Study Type	Participants/Population	Prebiotics Used	Key Findings
Randomized controlled trial	Healthy volunteers	GOS	Decreased waking cortisol, enhanced mood, therefore provided evidence that practice may cause a decrease in anxiety and stress [92]
Randomized controlled trial	Individuals with depression	Multispecies probiotics containing prebiotics	Larger magnitudes of improvement both for mood and emanating cognitive ratings suggesting that the compound may be useful in further augmenting depressive disorders [93]
Clinical trial	Children with autism spectrum disorders (ASD)	Prebiotic supplement	Enhanced gut microbiota diversity, reduced gastrointestinal symptoms, improved behavioral outcomes [94]
Pilot study	Children with ASD	Probiotics with prebiotics	Literature reported that enhanced social skills, decrease in gastrointestinal complaint, and modulation of gut microbe populations [95]
Review of multiple clinical trials	Children with ASD	Prebiotics and probiotics	Altering the symbiotic microbiota, decrease in ASD symptoms, enhancing the gastrointestinal function and behavior [91]

favorable set comprising of Bifidobacteria and Lactobacilli [98]. For example, chicory root contains inulin, which is a prebiotic fiber that has been shown to increase the beneficial microbes in the digestive tract [99]. Likewise, bananas and onions contain FOS and assist in the enhancement of the digestion of foods by beneficial bacteria [100]. Thus, their inclusion in daily meals can substantially increase the consumption of prebiotics and therefore have a positive impact on digestion and well-being.

20.10.2 Food Sources and Supplementation

Of course, the best way of enhancing prebiotic intake is by consuming prebiotic-rich foods. All the same, supplements are a practical option for people who may not have adequate prebiotic from their foods. Prebiotics come in forms of functional foods, vitamins, minerals, capsules, enzymes, and even beverages such as prebiotic fortified yogurt, smoothies, cereals, and baked products. Some of the usual prebiotics you can consume are GOS, FOS, and resistant starch. Research has also established that prebiotics such as GOS could be useful in adjusting the gut microbiota's population and composition, boosting immunity, and promoting metabolic syndromes [101]. For instance, a study on elder individuals concluded that supplementation with GOS enhanced the level of the beneficial *Bifidobacteria* spp. without any negative effect on its participants' physiologic parameters, indicating gut health [102]. This shows that there is possibility of the prebiotic supplements to act as source of prebiotics besides the diet, so that a good and convenience dose of the prebiotics could be obtained.

20.10.3 Guidelines for Effective Use

To add prebiotics to one's diet, there are certain principles which if followed would help in the absorption of prebiotics without compromising on the side effects involved. It is advised to begin with a small amount of prebiotics and increase the daily dosage gradually to give the digestive system a chance to adapt because prebiotics as well as the bacterial mass in the gut are likely to grow; this can cause bloating and gas [103]. The amount of prebiotics that should be consumed also differs, though it has been recommended that 5–20 g of prebiotic should be taken daily based on the ability of the individual and his/her objectives [104]. In this case, it is similarly noted that the synergistic action of various prebiotics may also increase the efficiency of utilization, and if different raw materials are used in the diet, this promotes various forms of beneficial bacteria in the intestine [105]. Also, the combination of pre- and probiotics known as synbiotics enhances gut health more than when using one of them. For example, incorporating prebiotic fibers to the probiotics containing foods like the yoghurt increases the growth and efficiency of the probiotic bacteria, thus enhancing the general health of the gut [106]. The following practical tips can also be used over a period of time to assist anyone in their quest to incorporate prebiotics into their meals.

20.11 CHALLENGES AND FUTURE DIRECTIONS

Several challenges and limitations can be drawn from current research on prebiotics and their effects in an individual's general and neurological well-being. A few limitations, which are specific to this study, include the following: An important factor, which often presents variability to the intervention's outcomes, is the response of the subjects to prebiotic supplements, and this is often occasioned by factors such as genetic makeup, current composition of the gut microbiota, and the comprehensive dietary pattern of the subjects [107]. Also, most of the current experiments employ animals as subjects, although valuable as they are, they do not always respond to stimuli in the same manner as people do [108]. As high-throughput DNA sequencing and metabolomics enter the scene for a detailed description of GMS and their relationships with prebiotics, the facet of functional improvement of the diets gets illuminated [109]. The uses of information technology in such personalized nutrition and microbiome-based therapeutic applications may similarly be a good predictor for the use of individualized prebiotic administration to improve on the effectiveness and reduce on the side effects [110]. Further studies should focus on large-scale and long-term human clinical trials in order to determine the correct recommended dosages and to define the list of specific prebiotics that have the best impact on the different health-related indicators. However, basic science research involving disciplines such as microbiology, neurology, and nutrition together with the integration of the gut–brain axis as well as prebiotic outcome on neurological receptors are still complex and require more research, hence extensive studies are required [111]. Mitigating these issues employing the participatory and creative research strategies will be vital in realizing the optimized pharmacological potentials of prebiotics.

20.12 CONCLUSION

Thus, the emerging science of prebiotics and their impact in the neurological area is a prospective direction that may improve human psychological and cognitive condition using food strategies. Since prebiotics help to increase the population of beneficial bacteria in the gastrointestinal tract and modulate the composition of gut microbiota, which is considered to be an essential component of the mentioned gut–brain axis, prebiotics are highly important. It is through this connection that it is possible to originate the ability of prebiotics to offer CNS command, mood, and conduct, as well as offering medication-free approaches to the actualization and deterrence of neuropsychiatric and mental health conditions. The latest scientific studies look into the correlation between prebiotics and neuroinflammation, neurotransmitter synthesis, and brain health generally. Experimental data have demonstrated the immunomodulatory effects of prebiotics and the lesser level of neuroinflammation that can be beneficial in such diseases as Alzheimer's and Parkinson's. Also, other clinical studies have revealed that prebiotics can result in mood enhancement, reduction in anxiety and depression, and better cognition, particularly for individuals with mental illness like depression and autism spectrum disorders. However, the field is not without its difficulties, such as fluctuations in the manner in which individuals react to treatment, the requirement for better human clinical trials, and not forgetting the immense challenges of converting data from animal studies to human usage. There is much potential in new technologies and biochemical individuality of nutrition to utilize them and eliminate the above-mentioned drawbacks, which means the future of prebiotics is going to be individualized at last. In general, the inclusion of prebiotics as part of dietary approaches present a rather logical, noninvasive, and potentially effective concept to enhance neurological conditions. As more knowledge accumulated, prebiotics might be recognized as an important part of the nutritional guidelines for treating the mental and cognitive disorders to ensure the proper integration of nutrition and neuroscience in the advancement of medicine.

FUNDING

There was no funding for this project from any source.

ACKNOWLEDGMENTS

The authors acknowledge in particular ShriRam College of Pharmacy, Banmore, Morena, Madhya Pradesh, India, for encouraging us and providing necessary facilities.

REFERENCES

1. Gibson, G. R., & Roberfroid, M. B. (1995). Dietary modulation of the human colonic microbiota: Introducing the concept of prebiotics. *The Journal of Nutrition, 125*(6), 1401–1412.
2. Slavin, J. (2013). Fiber and prebiotics: Mechanisms and health benefits. *Nutrients, 5*(4), 1417–1435.
3. Martirosyan, D. M., & Singh, J. (2015). A new definition of functional food by FFC: What makes a new definition unique? *Functional Foods in Health and Disease, 5*(6), 209–223.
4. Roberfroid, M. B. (2000). Prebiotics and probiotics: Are they functional foods? *The American Journal of Clinical Nutrition, 71*(6), 1682S–1687S.
5. Gibson, G. R., & Roberfroid, M. B. (1995). Dietary modulation of the human colonic microbiota: Introducing the concept of prebiotics. *The Journal of Nutrition, 125*(6), 1401–1412.
6. Roberfroid, M. B. (2007). Prebiotics: The concept revisited. *The Journal of Nutrition, 137*(3), 830S–837S.
7. Slavin, J. L. (2013). Fiber and prebiotics: Mechanisms and health benefits. *Nutrients, 5*(4), 1417–1435.
8. Cryan, J. F., & Dinan, T. G. (2012). Mind-altering microorganisms: The impact of the gut microbiota on brain and behaviour. *Nature Reviews Neuroscience, 13*(10), 701–712.
9. Schmidt, K., Cowen, P. J., Harmer, C. J., Tzortzis, G., & Errington, S. (2015). Prebiotic intake reduces the waking cortisol response and alters emotional bias in healthy volunteers. *Psychopharmacology, 232*(10), 1793–1801.
10. Misiak, B., Łoniewski, I., Marlicz, W., Frydecka, D., Szulc, A., & Rymaszewska, J. (2020). The HPA axis dysregulation in severe mental illness: Can we shift the blame to gut microbiota? *Progress in Neuro-Psychopharmacology and Biological Psychiatry, 102*, 109951.

11. Gibson, G. R., & Roberfroid, M. B. (1995). Dietary modulation of the human colonic microbiota: Introducing the concept of prebiotics. *The Journal of Nutrition, 125*(6), 1401–1412.
12. Roberfroid, M. (2007). Prebiotics: The concept revisited. *The Journal of Nutrition, 137*(3), 830S–837S.
13. Slavin, J. (2013). Fiber and prebiotics: Mechanisms and health benefits. *Nutrients, 5*(4), 1417–1435.
14. Patel, S., & Goyal, A. (2012). The current trends and future perspectives of prebiotics research: A review. *3 Biotech, 2*(2), 115–125.
15. Kolida, S., Meyer, D., & Gibson, G. R. (2007). A double-blind placebo-controlled study to establish the bifidogenic dose of inulin in healthy humans. *European Journal of Clinical Nutrition, 61*(10), 1189–1195.
16. Davis, L. M. G., Martínez, I., Walter, J., Goin, C., & Hutkins, R. (2010). Barcoded pyrosequencing reveals that consumption of galactooligosaccharides results in a highly specific Bifidogenic response in humans. *PLoS One, 6*(9), e25200.
17. Roberfroid, M. B., Gibson, G. R., Hoyles, L., McCartney, A. L., Rastall, R. A., Rowland, I. R., ... & Cummings, J. H. (2010). Prebiotic effects: Metabolic and health benefits. *British Journal of Nutrition, 104*(S2), S1–S63.
18. Sanz, Y., & De Palma, G. (2009). Gut microbiota and probiotics in modulation of epigenetic mechanisms involved in the immune function. *Journal of Clinical Gastroenterology, 43*(S1), S49–S54.
19. Kolida, S., Meyer, D., & Gibson, G. R. (2007). A double-blind placebo-controlled study to establish the bifidogenic dose of inulin in healthy humans. *European Journal of Clinical Nutrition, 61*(10), 1189–1195.
20. Slavin, J. L. (2013). Fiber and prebiotics: Mechanisms and health benefits. *Nutrients, 5*(4), 1417–1435.
21. Hidaka, H., Eida, T., Takizawa, T., Tokunaga, T., & Tashiro, Y. (1986). Effects of fructooligosaccharides on intestinal flora and human health. *Bifidobacteria and Microflora, 5*(1), 37–50.
22. Roberfroid, M. B. (2005). Introducing inulin-type fructans. *British Journal of Nutrition, 93*(S1), S13–S25.
23. Davis, L. M. G., Martínez, I., Walter, J., Goin, C., & Hutkins, R. (2010). Barcoded pyrosequencing reveals that consumption of galactooligosaccharides results in a highly specific Bifidogenic response in humans. *PLoS One, 6*(9), e25200.
24. Van Loo, J., Cummings, J., Delzenne, N., Englyst, H., Franck, A., Hopkins, M., ... & Roberfroid, M. (1999). Functional food properties of non-digestible oligosaccharides: A consensus report from the ENDO project (DGXII AIRII-CT94-1095). *British Journal of Nutrition, 81*(2), 121–132.
25. Nugent, A. P. (2005). Health properties of resistant starch. *Nutrition Bulletin, 30*(1), 27–54.
26. Sharma, P., & Tailang, M. (2022). Primaquine-loaded transdermal patch for treating malaria: Design, development, and characterization. *Future Journal of Pharmaceutical Sciences, 8*(1), 43.
27. Gibson, G. R., & Roberfroid, M. B. (1995). Dietary modulation of the human colonic microbiota: Introducing the concept of prebiotics. *The Journal of Nutrition, 125*(6), 1401–1412.
28. Slavin, J. (2013). Fiber and prebiotics: Mechanisms and health benefits. *Nutrients, 5*(4), 1417–1435.
29. Kolida, S., Meyer, D., & Gibson, G. R. (2007). A double-blind placebo-controlled study to establish the bifidogenic dose of inulin in healthy humans. *European Journal of Clinical Nutrition, 61*(10), 1189–1195.
30. Hidaka, H., Eida, T., Takizawa, T., Tokunaga, T., & Tashiro, Y. (1986). Effects of fructooligosaccharides on intestinal flora and human health. *Bifidobacteria and Microflora, 5*(1), 37–50.
31. Hamer, H. M., Jonkers, D., Venema, K., Vanhoutvin, S., Troost, F. J., & Brummer, R. J. (2008). Review article: The role of butyrate on colonic function. *Alimentary Pharmacology & Therapeutics, 27*(2), 104–119.
32. Pryde, S. E., Duncan, S. H., Hold, G. L., Stewart, C. S., & Flint, H. J. (2002). The microbiology of butyrate formation in the human colon. *FEMS Microbiology Letters, 217*(2), 133–139.
33. Wong, J. M. W., de Souza, R., Kendall, C. W. C., Emam, A., & Jenkins, D. J. A. (2006). Colonic health: Fermentation and short-chain fatty acids. *Journal of Clinical Gastroenterology, 40*(3), 235–243.
34. Lindsay, J. O., Whelan, K., Stagg, A. J., Gobin, P., Al-Hassi, H. O., Rayment, N., ... & Forbes, A. (2006). Clinical, microbiological, and immunological effects of fructo-oligosaccharide in patients with Crohn's disease. *Gut, 55*(3), 348–355.
35. Lomax, A. R., & Calder, P. C. (2009). Prebiotics, immune function, infection and inflammation: A review of the evidence. *British Journal of Nutrition, 101*(5), 633–658.
36. Sharma, P., & Tailang, M. (2020). Design, optimization and evaluation of buccal drug delivery system of propranolol for hypertension treatment. *International Journal of Pharmaceutical Sciences and Research, 11*(1), 301–311.
37. Cryan, J. F., & Dinan, T. G. (2012). Mind-altering microorganisms: The impact of the gut microbiota on brain and behaviour. *Nature Reviews Neuroscience, 13*(10), 701–712.
38. Misiak, B., Łoniewski, I., Marlicz, W., Frydecka, D., Szulc, A., & Rymaszewska, J. (2020). The HPA axis dysregulation in severe mental illness: Can we shift the blame to gut microbiota? *Progress in Neuro-Psychopharmacology and Biological Psychiatry, 102*, 109951.
39. Schmidt, K., Cowen, P. J., Harmer, C. J., Tzortzis, G., & Errington, S. (2015). Prebiotic intake reduces the waking cortisol response and alters emotional bias in healthy volunteers. *Psychopharmacology, 232*(10), 1793–1801.
40. Smith, A. P. (2014). The concept of well-being: Relevance to nutrition research. *British Journal of Nutrition, 111*(S1), S1–S5.
41. Carabotti, M., Scirocco, A., Maselli, M. A., & Severi, C. (2015). The gut-brain axis: Interactions between enteric microbiota, central and enteric nervous systems. *Annals of Gastroenterology, 28*(2), 203–209.
42. Cryan, J. F., & Dinan, T. G. (2012). Mind-altering microorganisms: The impact of the gut microbiota on brain and behaviour. *Nature Reviews Neuroscience, 13*(10), 701–712.
43. Mayer, E. A. (2011). Gut feelings: The emerging biology of gut-brain communication. *Nature Reviews Neuroscience, 12*(8), 453–466.
44. Rhee, S. H., Pothoulakis, C., & Mayer, E. A. (2009). Principles and clinical implications of the brain-gut-enteric microbiota axis. *Nature Reviews Gastroenterology & Hepatology, 6*(5), 306–314.
45. Foster, J. A., & McVey Neufeld, K.-A. (2013). Gut-brain axis: How the microbiome influences anxiety and depression. *Trends in Neurosciences, 36*(5), 305–312.

46. Sharma, P., & Tailang, M. (2018). In-vivo study of orodispersible tablet of primaquine. *International Journal of Pharmaceutical Sciences and Research, 9*(8), 3506–3510.
47. Roberfroid, M. B. (2007). Prebiotics: The concept revisited. *The Journal of Nutrition, 137*(3), 830S–837S.
48. Slavin, J. L. (2013). Fiber and prebiotics: Mechanisms and health benefits. *Nutrients, 5*(4), 1417–1435.
49. Davis, L. M. G., Martínez, I., Walter, J., Goin, C., & Hutkins, R. (2010). Barcoded pyrosequencing reveals that consumption of galactooligosaccharides results in a highly specific Bifidogenic response in humans. *PLoS One, 6*(9), e25200.
50. Gibson, G. R., & Roberfroid, M. B. (1995). Dietary modulation of the human colonic microbiota: Introducing the concept of prebiotics. *The Journal of Nutrition, 125*(6), 1401–1412.
51. Hamer, H. M., Jonkers, D., Venema, K., Vanhoutvin, S., Troost, F. J., & Brummer, R. J. (2008). Review article: The role of butyrate on colonic function. *Alimentary Pharmacology & Therapeutics, 27*(2), 104–119.
52. Sharma, P. (2021). Modification of human behavior due to coronavirus outbreak: A brief study on current scenario. *Asian Journal of Pharmaceutics (AJP), 15*(3), 392–398.
53. Perry, V. H., Nicoll, J. A. R., & Holmes, C. (2010). Microglia in neurodegenerative disease. *Nature Reviews Neurology, 6*(4), 193–201.
54. Heneka, M. T., Golenbock, D. T., & Latz, E. (2015). Innate immunity in Alzheimer's disease. *Nature Immunology, 16*(3), 229–236.
55. Patterson, E., Cryan, J. F., Fitzgerald, G. F., Ross, R. P., & Dinan, T. G. (2014). Gut microbiota, the pharmabiotics they produce and host health. *Proceedings of the Nutrition Society, 73*(4), 477–489.
56. Dalile, B., Van Oudenhove, L., Vervliet, B., & Verbeke, K. (2019). The role of short-chain fatty acids in microbiota–gut–brain communication. *Nature Reviews Gastroenterology & Hepatology, 16*(8), 461–478.
57. Davari, S., Talaei, S. A., Alaei, H., & Salami, M. (2013). Probiotics treatment improves diabetes-induced impairment of synaptic activity and cognitive function: Behavioral and electrophysiological proofs for microbiome-gut-brain axis. *Neuroscience, 240*, 287–296.
58. Jang, H. M., Lee, K. E., Kim, D. H., & Shim, J. J. (2018). Chronic ingestion of oligosaccharides improves anxiety-like behavior, hippocampal neuroinflammation, and gut dysbiosis in mice. *International Journal of Molecular Sciences, 19*(10), 3441.
59. Schmidt, K., Cowen, P. J., Harmer, C. J., Tzortzis, G., & Errington, S. (2015). Prebiotic intake reduces the waking cortisol response and alters emotional bias in healthy volunteers. *Psychopharmacology, 232*(10), 1793–1801.
60. Besedovsky, L., Lange, T., & Haack, M. (2019). The sleep-immune crosstalk in health and disease. *Physiological Reviews, 99*(3), 1325–1380.
61. Yano, J. M., Yu, K., Donaldson, G. P., Shastri, G. G., Ann, P., Ma, L., ... & Hsiao, E. Y. (2015). Indigenous bacteria from the gut microbiota regulate host serotonin biosynthesis. *Cell, 161*(2), 264–276.
62. Clarke, G., Stilling, R. M., Kennedy, P. J., Stanton, C., Cryan, J. F., & Dinan, T. G. (2014). Minireview: Gut microbiota: The neglected endocrine organ. *Molecular Endocrinology, 28*(8), 1221–1238.
63. Sharma, P., Jain, V., & Jain, S. (2021). Fabrication, optimization and evaluation of chronotropic drug delivery system of captopril. *International Journal of Pharmaceutical Sciences and Research, 12*, 2203–2210.
64. Dalile, B., Van Oudenhove, L., Vervliet, B., & Verbeke, K. (2019). The role of short-chain fatty acids in microbiota–gut–brain communication. *Nature Reviews Gastroenterology & Hepatology, 16*(8), 461–478.
65. Schmidt, K., Cowen, P. J., Harmer, C. J., Tzortzis, G., & Errington, S. (2015). Prebiotic intake reduces the waking cortisol response and alters emotional bias in healthy volunteers. *Psychopharmacology, 232*(10), 1793–1801.
66. Sharma, P., & Jain, V. (2023). An overview of current advances and pharmaceutical uses of 3D and 4D printing. *Exploratory Medicine, 4*, 560–575.
67. Savignac, H. M., Corona, G., Mills, H., Chen, L., Spencer, J. P. E., Tzortzis, G., ... & Burnet, P. W. J. (2013). Prebiotic feeding elevates central brain derived neurotrophic factor, N-methyl-d-aspartate receptor subunits and D-serine. *Neurochemistry International, 63*(8), 756–764.
68. Burokas, A., Arboleya, S., Moloney, R. D., Peterson, V. L., Murphy, K., Clarke, G., ... & Cryan, J. F. (2017). Targeting the microbiota-gut-brain axis: Prebiotics have anxiolytic and antidepressant-like effects and alter gene expression in the hippocampus. *Psychopharmacology, 234*(3), 603–619.
69. Messaoudi, M., Lalonde, R., Violle, N., Javelot, H., Desor, D., Nejdi, A., ... & Cazaubiel, M. (2011). Assessment of psychotropic-like properties of a probiotic formulation (Lactobacillus helveticus R0052 and Bifidobacterium longum R0175) in rats and human subjects. *British Journal of Nutrition, 105*(5), 755–764.
70. Vulevic, J., Juric, A., Tzortzis, G., & Gibson, G. R. (2013). A mixture of trans-galactooligosaccharides reduces markers of metabolic syndrome and modulates the fecal microbiota and immune function of overweight adults. *The Journal of Nutrition, 143*(3), 324–331.
71. Steenbergen, L., Sellaro, R., van Hemert, S., Bosch, J. A., & Colzato, L. S. (2015). A randomized controlled trial to test the effect of multispecies probiotics on cognitive reactivity to sad mood. *Brain, Behavior, and Immunity, 48*, 258–264.
72. Harach, T., Marungruang, N., Duthilleul, N., Cheatham, V., Mc Coy, K. D., Frisoni, G., ... & Bolmont, T. (2017). Reduction of Abeta amyloid pathology in APPPS1 transgenic mice in the absence of gut microbiota. *Scientific Reports, 7*(1), 41802.
73. Zhuang, Z. Q., Shen, L. L., Li, W. W., Fu, X., Zeng, F., Gui, L., ... & Zhu, C. (2018). Gut microbiota is altered in patients with Alzheimer's disease. *Journal of Alzheimer's Disease, 63*(4), 1337–1346.
74. Burokas, A., Moloney, R. D., Dinan, T. G., & Cryan, J. F. (2015). Microbiota regulation of the Mammalian gut-brain axis. *Advances in Applied Microbiology, 91*, 1–62.
75. Kobayashi, Y., Kuhara, T., Oki, M., & Xiao, J. Z. (2019). Effects of Bifidobacterium breve A1 on the cognitive function of older adults with mild cognitive impairment: A randomized, double-blind, placebo-controlled trial. *Bioscience of Microbiota, Food and Health, 38*(1), 1–10.
76. Unger, M. M., Spiegel, J., Dillmann, K. U., Grundmann, D., Philippeit, H., Bürmann, J., ... & Schäfer, K. H. (2016). Short chain fatty acids and gut microbiota differ between patients with Parkinson's disease and age-matched controls. *Parkinsonism & Related Disorders, 32*, 66–72.

77. Hill-Burns, E. M., Debelius, J. W., Morton, J. T., Wissemann, W. T., Lewis, M. R., Wallen, Z. D., ... & Knight, R. (2017). Parkinson's disease and Parkinson's disease medications have distinct signatures of the gut microbiome. *Movement Disorders, 32*(5), 739–749.
78. Sharma, P., Jain, V., & Tailang, M. (2023). Advancement of nanocarrier-based engineering for specific drug delivery for cancer therapy. In R. Malviya & S. Sundram (Eds.), *Targeted Cancer Therapy in Biomedical Engineering* (pp. 465–486). Springer.
79. Sun, M. F., Shen, Y. Q., Zhu, Y. L., Xu, P. W., Shen, Y. Q., Yang, Q., ... & Fang, Y. Q. (2018). The probiotics Lactobacillus and Bifidobacterium attenuate the body weight loss, loss of motor functions, and dopaminergic neuron injury in a Parkinson's disease mouse model. *Food & Function, 9*(9), 4312–4321.
80. Stilling, R. M., van de Wouw, M., Clarke, G., Stanton, C., Dinan, T. G., & Cryan, J. F. (2016). The neuropharmacology of butyrate: The bread and butter of the microbiota-gut-brain axis? *Neurochemistry International, 99*, 110–132.
81. Silva, Y. P., Bernardi, A., & Frozza, R. L. (2020). The role of short-chain fatty acids from gut microbiota in gut-brain communication. *Frontiers in Endocrinology, 11*, 25.
82. Nimgampalle, M., & Kuna, Y. (2017). Anti-Alzheimer properties of probiotic, Lactobacillus plantarum MTCC 1325 in Alzheimer's disease induced albino rats. *Journal of Clinical and Diagnostic Research, 11*(8), KC01–KC05.
83. Sochocka, M., Donskow-Łysoniewska, K., Sahajpal, N. S., & Woźniak, M. (2019). The gut microbiome alterations and inflammation-driven pathogenesis of Alzheimer's disease—A critical review. *Molecular Neurobiology, 56*(3), 1841–1851.
84. Schmidt, K., Cowen, P. J., Harmer, C. J., Tzortzis, G., & Errington, S. (2015). Prebiotic intake reduces the waking cortisol response and alters emotional bias in healthy volunteers. *Psychopharmacology, 232*(10), 1793–1801.
85. Collins, S. M., Surette, M., & Bercik, P. (2012). The interplay between the intestinal microbiota and the brain. *Nature Reviews Microbiology, 10*(11), 735–742.
86. Cryan, J. F., & Dinan, T. G. (2012). Mind-altering microorganisms: The impact of the gut microbiota on brain and behaviour. *Nature Reviews Neuroscience, 13*(10), 701–712.
87. Desbonnet, L., Garrett, L., Clarke, G., Bienenstock, J., & Dinan, T. G. (2008). The probiotic Bifidobacteria infantis: An assessment of potential antidepressant properties in the rat. *Journal of Psychiatric Research, 43*(2), 164–174.
88. Sharma, P., Bhargava, S., Parashar, D., & Mangal, A. (2017). Formulation and evaluation of nanoparticles containing cyclophosphamide. *World Journal of Pharmaceutical Research, 7*(1), 785–798.
89. Sharma, P. (2023). Nanostructured biomaterials in drug delivery: Current trends and upcoming possibilities. In R. Malviya & S. Sundram (Eds.), *Engineered Biomaterials: Synthesis and Applications* (pp. 261–280). Springer.
90. Parracho, H. M., Bingham, M. O., Gibson, G. R., & McCartney, A. L. (2005). Differences between the gut microflora of children with autistic spectrum disorders and that of healthy children. *Journal of Medical Microbiology, 54*(10), 987–991.
91. Grimaldi, R., Camilleri, G., & D'Inca, R. (2020). Prebiotics and probiotics in the management of children with autism and neurodevelopmental disorders: A review. *Nutrients, 12*(8), 2335.
92. Schmidt, K., & Cowen, P. J. (2015). Prebiotics and emotional bias in healthy volunteers. *Psychopharmacology, 232*(10), 1793–1801.
93. Steenbergen, L., Sellaro, R., van Hemert, S., Bosch, J. A., & Colzato, L. S. (2015). A randomized controlled trial to test the effect of multispecies probiotics on cognitive reactivity to sad mood. *Brain, Behavior, and Immunity, 48*, 258–264.
94. Grimaldi, R., & Costabile, A. (2018). A prebiotic intervention study in children with autism spectrum disorders (ASDs). *Microbiome, 6*(1), 133.
95. Sanctuary, M. R., Kain, J. N., Chen, S. Y., Kalanetra, K., Lemay, D. G., Rose, D. R., & Mills, D. A. (2019). Pilot study of probiotics for dietary mitigation of autism-related social behavior in the BTBR T+ Itpr3tf/J mouse model. *PLOS ONE, 14*(12), e0226745.
96. Krumbeck, J. A., Maldonado-Gomez, M. X., Ramer-Tait, A. E., & Hutkins, R. W. (2016). Prebiotics and synbiotics: Dietary strategies for improving gut health. *Current Opinion in Gastroenterology, 32*(2), 110–119.
97. Grimaldi, R., Gibson, G. R., Vulevic, J., Giallourou, N., Castro-Mejía, J. L., Hansen, L. H., ... & Costabile, A. (2018). A prebiotic intervention study in children with autism spectrum disorders (ASDs). *Microbiome, 6*(1), 133.
98. Slavin, J. (2013). Fiber and prebiotics: Mechanisms and health benefits. *Nutrients, 5*(4), 1417–1435.
99. Roberfroid, M. B. (2005). Introducing inulin-type fructans. *British Journal of Nutrition, 93*(S1), S13–S25.
100. Gibson, G. R., & Roberfroid, M. B. (1995). Dietary modulation of the human colonic microbiota: Introducing the concept of prebiotics. *The Journal of Nutrition, 125*(6), 1401–1412.
101. Gibson, G. R., Hutkins, R., Sanders, M. E., Prescott, S. L., Reimer, R. A., Salminen, S. J., & Reid, G. (2017). Expert consensus document: The international scientific association for probiotics and prebiotics (ISAPP) consensus statement on the definition and scope of prebiotics. *Nature Reviews Gastroenterology & Hepatology, 14*(8), 491–502.
102. Vulevic, J., Juric, A., Walton, G. E., Claus, S. P., Tzortzis, G., Toward, R. E., & Gibson, G. R. (2015). Influence of galactooligosaccharides on gut microbiota, immune parameters and metabonomics in elderly persons. *British Journal of Nutrition, 114*(4), 586–595.
103. Slavin, J. (2013). Fiber and prebiotics: Mechanisms and health benefits. *Nutrients, 5*(4), 1417–1435.
104. Sharma, P., Dasputre, N., & Jain, V. (2023). Nanomedicine: Concepts, characteristics, and regulatory challenges. In Pankaj Sharma, Wasim Akram, Vinay Jain, & Mukul Tailang (Eds.), *Smart Nanocarrier for Effective Drug Delivery* (pp. 1–16). CRC Press.
105. Roberfroid, M. B. (2007). Prebiotics: The concept revisited. *The Journal of Nutrition, 137*(3), 830S–837S.
106. Swanson, K. S., Gibson, G. R., Hutkins, R., Reimer, R. A., Reid, G., Verbeke, K., & Sanders, M. E. (2020). The International Scientific Association for Probiotics and Prebiotics (ISAPP) consensus statement on the definition and scope of synbiotics. *Nature Reviews Gastroenterology & Hepatology, 17*(11), 687–701.

107. Hill, C., Guarner, F., Reid, G., Gibson, G. R., Merenstein, D. J., Pot, B., ... & Sanders, M. E. (2014). Expert consensus document: The International Scientific Association for Probiotics and Prebiotics consensus statement on the scope and appropriate use of the term probiotic. *Nature Reviews Gastroenterology & Hepatology, 11*(8), 506–514.
108. Dalile, B., Van Oudenhove, L., Vervliet, B., & Verbeke, K. (2019). The role of short-chain fatty acids in microbiota–gut–brain communication. *Nature Reviews Gastroenterology & Hepatology, 16*(8), 461–478.
109. Patterson, E., Ryan, P. M., Cryan, J. F., Dinan, T. G., Ross, R. P., & Fitzgerald, G. F. (2016). Gut microbiota, the pharmabiotics they produce and host health. *Proceedings of the Nutrition Society, 75*(2), 207–215.
110. Smolinska, A., Tedjo, D. I., Blanchet, L., Swann, J. R., & Zhernakova, A. (2018). Nontargeted metabolomics in human disease diagnosis: A review of validation strategies. *Journal of Proteome Research, 17*(12), 4423–4441.
111. Cryan, J. F., & Dinan, T. G. (2012). Mind-altering microorganisms: the impact of the gut microbiota on brain and behaviour. *Nature Reviews Neuroscience, 13*(10), 701–712.

21 Postbiotics and Psychobiotics
Therapeutic Interventions in Neurodegenerative Diseases

Anas Islam[1], Badruddeen[1], Mohammad Irfan Khan[1], Mohsin Vahid Khan[2], and Yusuf Asad[3]

[1] Faculty of Pharmacy, Integral University, Lucknow, India
[2] Department of Biosciences, Integral University, Lucknow, India
[3] Department of Biomedical Engineering, Rowan University, New Jersey, United States

21.1 INTRODUCTION

21.1.1 Overview of Neurodegenerative Diseases

Neurodegenerative diseases encompass a range of conditions marked by the gradual deterioration and reduction of neurons. Neurons are the fundamental building blocks of the nervous system and the neural network that makes up the brain. These disorders frequently have a long duration and cause serious limitations in cognitive, motor, and behavioral abilities, ultimately leading to substantial illness and death. Alzheimer's disease (AD), Parkinson's disease (PD), Huntington's disease (HD), and amyotrophic lateral sclerosis (ALS) are some of the most common neurodegenerative illnesses. Although they have different clinical characteristics, they all have similar underlying pathogenic processes (Islam et al., 2023).

Alzheimer's disease is the predominant etiology of dementia responsible for around 60–80% of all cases globally. The key indicators of this illness are the buildup of amyloid-beta plaque and the development of neurofibrillary tangles in the brain caused by hyperphosphorylated tau protein. These deteriorating characteristics result in the demise of neurons, namely, in areas such as the cerebral cortex and the hippocampus, which have a significant impact on memory and cognitive function (Scheltens et al., 2021). The clinical manifestation comprises a gradual decline in memory, difficulty in speaking, confusion, and alterations in behavior and personality. While the precise origin of AD remains uncertain, it is widely accepted that the development of the illness is influenced by a variety of variables, including genetic, environmental, and lifestyle-related factors (Masters et al., 2015).

Parkinson's disease is a prevalent neurological disorder that mostly impacts motor control and affects many individuals. The degeneration of dopaminergic neurons in the substantia nigra, a brain area crucial for movement control, is the underlying cause of the condition. One of the defining characteristics of PD is the lack of dopamine, which is an essential neurotransmitter for control of motor function. Tremors, bradykinesia (slow movement), stiffness, and postural instability are some of the different symptoms that can be described as a consequence of this insufficiency (Poewe et al., 2017). People who have Parkinson's disease generally exhibit symptoms that are not related to movement, such as cognitive deterioration, emotional issues, and autonomic dysfunction. These symptoms often have a substantial and harmful impact on their quality of life. While most occurrences of PD occur randomly, there are some genetic anomalies that have been associated with family variants of the ailment. These abnormalities include those found in the LRRK2 and SNCA genes (Nalls et al., 2014).

Huntington's disease is an inherited neurological disorder caused by an elevated number of CAG repeats in the HTT gene. The development of a huntingtin protein that is mutated is brought about by this expansion. The aggregation of this protein within neurons, namely, in the striatum and cortex, leads to a decline in their functioning and ultimately causes cell death. From a clinical perspective, HD is distinguished by a combination of three main symptoms: motor abnormalities, cognitive impairment, and mental problems. These symptoms encompass involuntary, dance-like movements known as chorea, a loss in cognitive function, and mood disorders such as despair and irritability (Walker, 2007). Huntington's disease is distinctive among neurodegenerative disorders due to its autosomal dominant inheritance pattern. This means that individuals who possess a single copy of the mutated gene would inevitably acquire the condition, usually after middle age (Bates et al., 2015).

The degenerative neurological disease known as ALS mostly affects motor neurons in the brain and spinal cord. It's also frequently referred to as Lou Gehrig's sickness.

DOI: 10.1201/9781003536802-21

Neuronal degeneration results in muscular weakening, atrophy, and ultimately, paralysis. Individuals diagnosed with ALS commonly have a swift deterioration in their physical abilities, and the majority perish from the illness within a span of three to five years as a result of respiratory insufficiency (Brown and Al-Chalabi, 2017). Although most instances of ALS occur randomly, over 10% are hereditary, including mutations in genes such as TARDBP, SOD1, and C9orf72 (Chia et al., 2018). Despite extensive research, the exact mechanisms underlying neuronal death in ALS remain incompletely understood, though abnormal protein aggregation, excitotoxicity, and neuroinflammation are thought to play central roles (Hardiman et al., 2017).

The underlying cause of neurodegenerative disorders is complex and diverse, encompassing a blend of hereditary susceptibility, environmental influences, and age-related elements. Neurodegenerative diseases frequently display an abnormal accumulation of proteins that are distorted in some way. These proteins include amyloid-beta in AD, alpha-synuclein in PD, mutant huntingtin in HD, and TDP-43 in ALS (Goedert et al., 2017). These proteins have the ability to create harmful clumps that interfere with the normal functioning of cells, resulting in impaired brain cell activity and eventual death. Moreover, it is well known that the development of these diseases is much influenced by neuroinflammation, mitochondrial malfunction, oxidative stress, and poor autophagy (Heneka et al., 2015; Wyss-Coray and Rogers, 2012). Understanding the complex interplay between these processes is essential for developing effective therapeutic interventions.

21.1.2 The Gut–Brain Axis: A Bidirectional Communication System

The gut–brain axis (GBA), a massive two-way communication network, connects the gastrointestinal tract to the central nervous system (CNS). Among the interconnected systems that come under this axis are the immunological system, the endocrine system, the vagus nerve of the autonomic nervous system, and the enteric neural system. Importantly, the gut microbiome is part of this axis (Carabotti et al., 2015). The GBA is becoming more well acknowledged as a crucial element in maintaining homeostasis and exerting an impact on both gut and brain health. Malfunctions in this axis have been linked to several neurological and mental problems, including as neurodegenerative diseases (Cryan and Dinan, 2012).

Human health greatly depends on the gut microbiota, which is mostly located in the large intestine and is made up of many species including bacteria, viruses, fungus, and archaea. These microbes have a role in immune system control, digestion, metabolism, and the manufacture of neurotransmitters and neuroactive substances (Dinan and Cryan, 2017). The types and abundance of bacteria in the gastrointestinal tract can affect the development and operation of the central nervous system, which in turn affects emotions, cognition, and behavior. The gut's microbiota composition and function can be influenced by the central nervous system (CNS) through neuronal, endocrine, and immunological pathways (Foster et al., 2017).

The vagus nerve serves as an essential information channel between the brainstem and the stomach by directly connecting the two organ systems. The vagus nerve transports neural impulses from the gastrointestinal tract to the brain, where they have an impact on behavior and cognitive processes. For example, studies have shown that vagus nerve stimulation may impact mood and cognitive abilities while reducing neuroinflammation (Bonaz et al., 2018). Because of this, researchers have been looking at the application of vagus nerve stimulation as a therapeutic strategy for neurological and mental disorders like Alzheimer's disease, depression, and epilepsy (Johnson and Wilson, 2018).

Additionally, the immune system participates in the GBA in a significant way. The gut-associated lymphoid tissue (GALT) reflects the tight link between the gut bacteria and the central nervous system (CNS). This tissue is responsible for housing a considerable proportion of the immune cells that are found in the body. Inflammation that is broad can be caused by a disruption of immune responses in the gastrointestinal tract. This inflammation is recognized as a contributing component to the onset of neurodegenerative diseases (Harach et al., 2017). Gut inflammation can initiate the production of pro-inflammatory cytokines and immunological mediators, which have the potential to penetrate the blood–brain barrier and cause neuroinflammation and neuronal harm (Dantzer et al., 2008).

Furthermore, it is worth noting that the GBA is influenced by the endocrine system, namely the hypothalamic-pituitary-adrenal (HPA) axis, which plays a crucial role in regulating the body's stress response. In addition to affecting gut permeability, microbiota composition, and immunological function, chronic stress can disrupt the gut–brain axis (GBA), which can result in an increased susceptibility to neurodegenerative illnesses or conditions (O'Mahony et al., 2015). Hormones and neurotransmitters essential for brain function and mood control, including gamma-aminobutyric acid (GABA), serotonin, and dopamine, can also be produced and modulated by the gut bacteria (Strandwitz, 2018). Therefore, with important ramifications for mental and neurological health, the GBA acts as a crucial mediator of the interactions between the stomach and the brain.

21.1.3 Introduction to Postbiotics and Psychobiotics

Bioactive substances known as postbiotics are created in the gut during fermentation by probiotics. Among these substances are enzymes, peptides, polysaccharides, organic acids, cell surface proteins, and short-chain fatty acids (SCFAs) (Aguilar-Toalá et al., 2018). Postbiotics are nonviable and do not depend on the life of the bacteria that created them to function, in contrast to probiotics, which are living microorganisms. Because postbiotics do not involve the danger of infection or translocation that comes with living bacteria, they may be safer and more stable than probiotics (Tsilingiri and Rescigno, 2013).

Understanding the diverse roles of prebiotics, probiotics, and postbiotics is crucial for comprehending their impact on health. Prebiotics are indigestible dietary components, such specific types of fibers, that stimulate the proliferation and function of advantageous intestinal bacteria. Probiotics are living microorganisms that, when given in sufficient quantities, exert positive effects on the health of the host organism. On the other hand, postbiotics refer to the substances produced as a result of probiotic activity. These substances can have many positive impacts on the host without the need for live bacteria (Gibson et al., 2017).

Studies have shown that postbiotics possess various biological properties that can potentially impact the well-being of the brain and gastrointestinal system. For instance, scientific evidence has demonstrated that short-chain fatty acids (SCFAs) possess anti-inflammatory, immunomodulatory, and neuroprotective properties. Short-chain fatty acids (SCFAs) are generated through the process of food fiber digestion by gut bacteria (Stilling et al., 2014). Numerous mechanisms, including modifying gene expression, turning on G-protein-coupled receptors, and preventing histone deacetylases, are affected in order to produce these effects. These pathways have a role in controlling inflammation and neuroplasticity (Dalile et al., 2019).

Psychobiotics are a specific type of probiotic that specifically targets the connection between the stomach and the brain. They have the ability to enhance mental well-being by influencing the microorganisms in the stomach. The term "psychobiotic" was initially used by Dinan et al. (2013) to refer to probiotics that, when consumed in sufficient quantities, can provide a therapeutic advantage for individuals with mental disorders. Psychobiotics can impact the brain through several methods, such as the synthesis of neurotransmitters, the control of inflammation, and the regulation of the HPA axis (Dinan et al., 2013).

Although the study of psychobiotics is still in its early stages, a growing amount of data suggests that they may be useful in treating mental health issues. Various strains of *Lactobacillus* and *Bifidobacterium* have demonstrated potential in alleviating symptoms of anxiety and depression, as observed in both animal models and human clinical studies (Bravo et al., 2011; Dhyani et al., 2024). These benefits are believed to be mediated via immune system and gut microbiome regulation, as well as by the synthesis of neurotransmitters including GABA and serotonin (Wallace and Milev, 2017).

Psychobiotics may have therapeutic promise in neurodegenerative illnesses, in addition to their potential significance in mental health. Psychobiotics have the potential to safeguard against neurodegenerative illnesses like Alzheimer's and Parkinson's by altering the gut microbiota and decreasing neuroinflammation (Bonnechère et al., 2022). Moreover, the ability of psychobiotics to enhance neuroplasticity and cognitive function suggests that they could be used to support cognitive health in aging populations (Grochowska et al., 2020).

21.2 MECHANISMS OF ACTION: HOW POSTBIOTICS AND PSYCHOBIOTICS AFFECT THE BRAIN

For a detailed visual representation of how postbiotics and psychobiotics influence brain function via the gut–brain axis, refer to Figure 21.1, which illustrates the key pathways including immune system modulation and vagus nerve stimulation, pivotal in understanding the therapeutic implications in neurodegenerative diseases.

21.2.1 Microbiota-Derived Metabolites and Neurotransmitter Modulation

Metabolites called short-chain fatty acids (SCFAs), generated by the microbiota, play a crucial role in facilitating interaction among the microbes in the gut and the brain. Acetate, propionate, and butyrate are specific instances of short-chain fatty acids (SCFAs) that are generated as a result of the breakdown of dietary fibers by gut bacteria. It has been demonstrated that these metabolites may pass across the blood–brain barrier and have a major impact on behavior and brain function. Butyrate in particular has attracted attention because to its neuroprotective properties. It functions as a histone deacetylase inhibitor, affecting the expression of genes related to neuronal survival, inflammation, and neuroplasticity (Stilling et al., 2014).

The modulation of neurotransmitters by SCFAs is another critical mechanism through which these metabolites influence brain function. For example, by making tryptophan, the precursor of serotonin, more readily available in the brain, SCFAs can improve the synthesis of serotonin, a neurotransmitter that is essential for mood control (Strandwitz, 2018). Serotonin, predominantly produced in the gut, is crucial for maintaining mood balance, and alterations in its levels are associated with mood disorders such as anxiety and depression (Yano et al., 2015). Furthermore, research has demonstrated that short-chain fatty acids have an impact on the production of gamma-aminobutyric acid, the main inhibitory neurotransmitter in the central nervous system. GABA's role in reducing neuronal excitability is essential for preventing the onset of anxiety and other neuropsychiatric conditions. SCFAs accomplish this via modulating the expression of enzymes responsible for GABA production in the gastrointestinal tract, hence affecting the amounts of GABA in the brain (Cryan and Dinan, 2012).

In addition to short-chain fatty acids, other metabolites produced by the microbiota, such as bile acids and tryptophan metabolites, also have important effects on regulating brain function. Bile acids, historically acknowledged for their function in lipid metabolism, are presently acknowledged for their capacity to function as signaling molecules

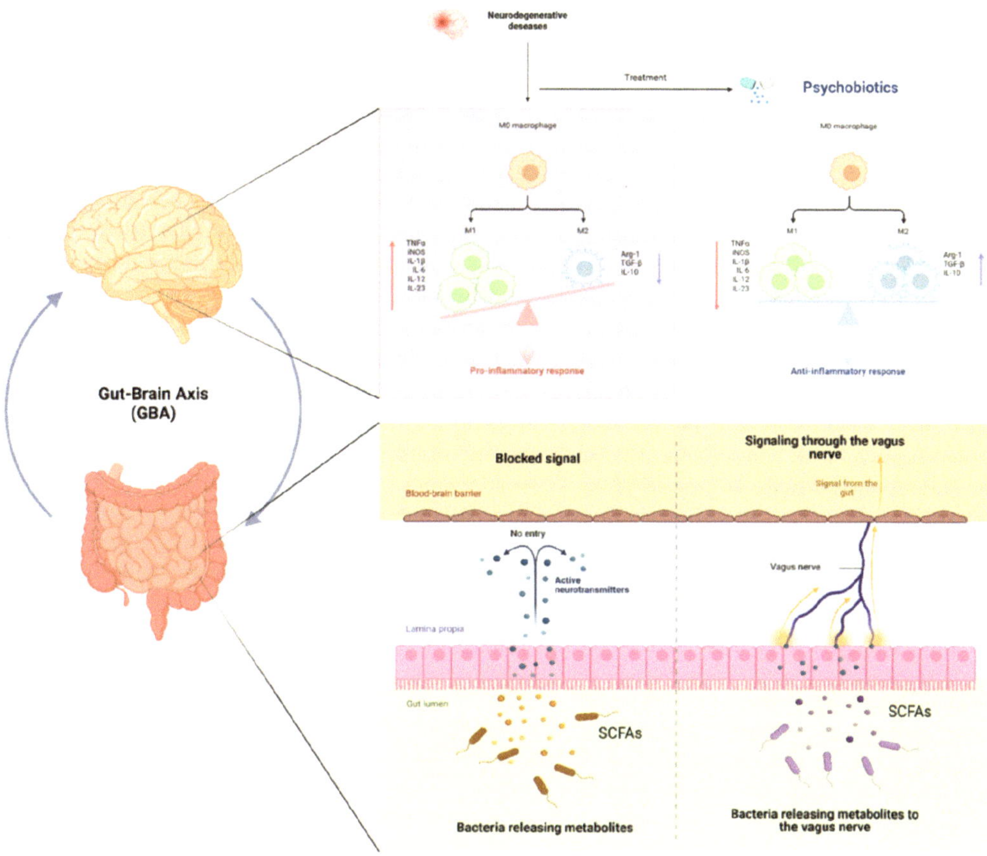

FIGURE 21.1 A comprehensive illustration of the mechanisms by which postbiotics and psychobiotics impact brain function through the gut–brain axis, particularly in the context of neurodegenerative diseases. Central to the diagram is the depiction of the gut–brain axis, which highlights the bidirectional communication between the gastrointestinal system and the brain. Detailed panels within the image focus on two primary mechanisms: immune system modulation and vagus nerve stimulation. In the immune system modulation panels, the transition from a pro-inflammatory state to an anti-inflammatory state under the influence of psychobiotics is shown. This anti-inflammatory state is depicted by the conversion of M0 macrophages to M2 macrophages that secrete anti-inflammatory cytokines, demonstrating the therapeutic potential of psychobiotics in reducing neuroinflammation. The vagus nerve stimulation segment is split into two subpanels. The first shows a scenario where the blood–brain barrier blocks neurotransmitter signals, and the second illustrates active signaling through the vagus nerve. This pathway is activated by metabolites from gut bacteria that influence brain functions related to mood and cognition via signals transmitted through the vagus nerve.

in the gut–brain axis. These acids have the capacity to control the activity of various brain receptors, which in turn influences behavior and the brain's capacity for adaptation and change (neuroplasticity) (Kiriyama and Nochi, 2019). Tryptophan metabolites, such as indoles and kynurenine, are substances produced by microorganisms that may get through the protective barrier around the brain and impact the systems responsible for transmitting signals between nerve cells, including those related to serotonin and dopamine. As a result, they can influence mood and cognitive abilities (Agus et al., 2018).

21.2.2 Immune System Modulation

The immune system has a significant impact on the gut–brain axis, especially in relation to neuroinflammation, which is a prevalent characteristic of several neurodegenerative disorders. The gut microbiota and its metabolites have the ability to affect the immune system in ways that either exacerbate or alleviate neuroinflammation. Dysbiosis, or an unbalanced population of gut microbes, has been associated with increased intestinal permeability, or "leaky gut." This condition makes it possible for microbial compounds to enter the bloodstream, including lipopolysaccharides (LPS). This causes the immune cells called microglia, which are found in the brain, to become activated and causes extensive inflammation throughout the body (Fung et al., 2017).

When microglia are activated, they have the ability to generate cytokines that promote inflammation. These cytokines include tumor necrosis factor-alpha (TNF-α), interleukin-6 (IL-6) and interleukin-1β (IL-1β). These cytokines may intensify neuroinflammation and accelerate the development of neurodegenerative illnesses, such as Alzheimer's and Parkinson's diseases (Heneka et al., 2015). Psychobiotics can restore balance and decrease intestinal permeability by modifying the gut microbiota. This lessens neuroinflammation and keeps the immune system from being too engaged. Certain strains of *Lactobacillus* and *Bifidobacterium* have been demonstrated to regulate immune responses and strengthen the intestinal barrier. As a result, there are less pro-inflammatory cytokines in the blood and microglial cells are activated in a more balanced manner (Matenchuk et al., 2020).

By boosting the development of regulatory T cells (Tregs), which are essential for maintaining immunological tolerance and lowering excessive inflammation, postbiotics can also have a direct impact on the immune system. Butyrate, specifically, has been demonstrated to improve the generation and performance of Tregs, thereby exerting an anti-inflammatory impact that can safeguard against neuroinflammation and its linked neurodegenerative processes (Furusawa et al., 2013).

21.2.3 Vagus Nerve Stimulation

The vagus nerve, which is an essential part of the parasympathetic nervous system, is the primary nerve that connects the brain to the stomach. This neuron controls a wide range of physiological processes, such as immunological response, digestion, and heart rate. It has a significant role in mood regulation and cognitive function as well. The impact of signals originating from the stomach on the brain via the vagus nerve is a subject of extensive investigation, especially regarding its significance for mental health and neurodegenerative disorders (Bonaz et al., 2018).

Through complex interactions between the brain, enteric nervous system, and gut bacteria, the vagus nerve enables these effects. Short-chain fatty acids (SCFAs) and other signaling molecules are examples of microbial metabolites that can activate the stomach's vagal afferent neurons. The brainstem receives messages from these neurons and uses them to influence the amygdala and hippocampus, two regions of the brain involved in mood regulation (Breit et al., 2018).

Vagus nerve stimulation (VNS) has been investigated as a potential treatment option for a wide range of neurological and psychiatric conditions, such as epilepsy, Alzheimer's disease, and depression, among others. It is thought that VNS reduces neuroinflammation and affects the way brain regions linked to mood and cognition operate (Manta et al., 2009). The anti-inflammatory properties of VNS are especially significant when it comes to neurodegenerative illnesses since persistent inflammation plays a significant role in the damage that neurons endure. Research indicates that inhibiting the synthesis of pro-inflammatory cytokines using vagus nerve stimulation might reduce neuroinflammation and perhaps halt the course of neurodegenerative disorders (Pavlov and Tracey, 2012).

The ramifications of gut–brain communication mediated by the vagus nerve go beyond the treatment of neurodegenerative illnesses. This route also has a substantial impact on mood management and cognitive performance. For instance, research has suggested that the gut microbiota's impact on the vagus nerve is connected to anxiety and depression. Studies have demonstrated that severing the vagus nerve, a procedure known as vagotomy, can eliminate the calming and mood-lifting effects of some psychobiotics (Bravo et al., 2011). These results suggest that the transmission of the gut microbiota's effects to the brain involves the vagus nerve. Consequently, focusing on this route might be a potentially effective approach for treating mood disorders. Table 21.1 provides a concise overview of the primary mechanisms through which postbiotics and psychobiotics influence brain function, including the modulation of neurotransmitters, reduction of neuroinflammation, and the role of the vagus nerve in gut–brain communication.

TABLE 21.1
Mechanisms of Action: How Postbiotics and Psychobiotics Affect the Brain

Mechanism	Description	Key Compounds/Examples	References
Microbiota-derived metabolites and neurotransmitter modulation	*SCFAs and brain function:* SCFAs (e.g., butyrate) cross the blood–brain barrier, influencing neuroplasticity and reducing neuroinflammation	SCFAs (butyrate, acetate)	Stilling et al., 2014
Psychobiotics modulate neurotransmitter levels, such as serotonin and GABA, impacting mood and cognition	*Lactobacillus*, *Bifidobacterium*	Strandwitz, 2018; Bravo et al., 2011	
Immune system modulation	*Neuroinflammation reduction:* postbiotics reduce neuroinflammation by modulating microglial activation and cytokine production	SCFAs, exopolysaccharides	Heneka et al., 2014; Kim et al., 2017
Vagus nerve stimulation	*Gut–brain communication:* Signals from gut-derived metabolites activate the vagus nerve, influencing mood and cognitive functions	SCFAs, neuroactive compounds from gut bacteria	Breit et al., 2018; Bonaz et al., 2018

21.3 POSTBIOTICS IN NEURODEGENERATIVE DISEASE MANAGEMENT

Postbiotics, which refer to the bioactive compounds produced by probiotics during fermentation processes, have shown promise in influencing various physiological processes, including those related to neurodegeneration. These compounds, which include microbial metabolites, enzymes, cell wall fragments, and other bioactive molecules, offer unique advantages over live probiotics, such as enhanced safety, stability, and targeted effects (Aguilar-Toalá et al., 2018).

Postbiotics refer to the non-living substances produced by probiotic microbes, which provide health advantages to the host. Unlike probiotics, which require live bacteria to exert their effects, postbiotics can provide similar benefits without the risks associated with introducing live bacteria into the host, such as the potential for infection or translocation across the gut barrier (Tsilingiri and Rescigno, 2013). The sources of postbiotics are diverse, including microbial fermentation of dietary fibers, production of bioactive compounds during the metabolism of prebiotics, and extraction from probiotic cultures. All of the postbiotics—short-chain fatty acids (SCFAs), polysaccharides, lipoteichoic acids, exopolysaccharides, enzymes, and cell wall fragments—have important functions in preserving gut health and adjusting immune responses (Aguilar-Toalá et al., 2018).

In animal models, postbiotics have been shown to influence several key markers of neurodegeneration, including the reduction of amyloid-beta plaques, attenuation of neuroinflammation, and improvement in cognitive functions. Studies have shown, for example, that a common form of postbiotic, SCFAs, may pass the blood–brain barrier and modulate neuroinflammation pathways to lower oxidative stress and increase neurogenesis, hence exerting neuroprotective benefits (Stilling et al., 2014). Butyrate is one particular SCFA that has been demonstrated in Alzheimer's disease models to decrease amyloid-beta deposition and enhance cognitive function, indicating its potential as a therapeutic agent to slow the course of neurodegenerative illnesses (Frost et al., 2014).

Cellular studies provide further mechanistic insights into how postbiotics exert their neuroprotective effects. In vitro experiments have demonstrated that postbiotics can modulate the expression of genes involved in inflammation, oxidative stress, and apoptosis, which are critical pathways implicated in neurodegeneration. For example, butyrate has been shown to inhibit HDACs, leading to increased acetylation of histones, which in turn modulates the expression of neuroprotective genes (Soliman and Rosenberger, 2011). Additionally, it has been found that a particular type of postbiotic called exopolysaccharides increases the production of cytokines that are anti-inflammatory and decreases the expression of mediators that encourage inflammation. This offers more proof of their capacity to lower neuroinflammation and increase neuronal survival (Chen et al., 2019).

The variation in postbiotic preparation composition and dose, which might produce contradictory results, is one of the difficulties in doing clinical research. Additionally, the complexity of the human microbiome and its interactions with postbiotics poses a significant challenge in understanding how these compounds may exert different effects across individuals. The heterogeneity of neurodegenerative diseases, each with its unique pathophysiology, further complicates the identification of specific postbiotics that could be universally effective (Ouwehand et al., 2002).

Moreover, the majority of clinical trials on postbiotics have focused on general markers of gut health and immune modulation, with limited direct assessment of their impact on neurodegenerative disease outcomes. This emphasizes the necessity for more focused research that explicitly examines the impact of postbiotics on cognitive performance, neuroinflammation, and other pertinent outcomes in individuals suffering from neurodegenerative disorders (Scott et al., 2015).

21.4 PSYCHOBIOTICS: POTENTIAL THERAPEUTIC AGENTS FOR NEURODEGENERATIVE DISEASES

The concept of psychobiotics has emerged as an innovative approach in the field of neuropsychiatry, with significant implications for the control of neurodegenerative diseases. Psychobiotics are a special kind of probiotics that help with mental health by controlling the flow of information from the stomach to the brain. The bidirectional communication mechanism between the gut microbiota and the brain is crucial for preserving neuronal function and resistance against stress and neurodegenerative disorders. Psychobiotics are being studied as potential therapeutic agents for neurodegenerative diseases because of their ability to affect important systems like neuroinflammation, neuroplasticity, and stress resilience, all of which are important in the progression of neurodegenerative diseases (Dhyani et al., 2024).

Psychobiotics are live microorganisms that alter the gut–brain axis to enhance the host's mental health when taken in adequate amounts (Dinan and Cryan, 2013). Numerous species of *Lactobacillus* and *Bifidobacterium* are among the most often researched psychobiotic strains; these bacteria have shown great promise in modifying behavior, lowering anxiety, and enhancing cognitive abilities. For instance, research has demonstrated that *Lactobacillus rhamnosus* affects the expression of GABA receptors in the central nervous system, which play a vital role in the regulation of mood and anxiety (Bravo et al., 2011). Similarly, *Bifidobacterium* longum has been associated with reduced anxiety-like behavior and improved cognitive function in animal models, further supporting its potential as a psychobiotic (Savignac et al., 2015).

Psychobiotics affect the brain through several pathways that impact mental health and neurodegenerative processes, so their mechanisms are varied and complicated. One of the main processes is the improvement of stress resistance, which is very essential in reducing the consequences of persistent stress—a recognized risk factor for neurodegenerative illnesses. The dysregulation of the hypothalamic–pituitary–adrenal (HPA) axis linked to chronic stress is responsible for increased cortisol and other stress chemicals that could impair brain function (O'Mahony et al., 2015). Psychobiotics have been demonstrated to alter the HPA axis, therefore lowering the physiological stress response and fostering resilience. In both animal models and human research, for example, supplements including *Lactobacillus helveticus* and *Bifidobacterium longum* have been observed to lower cortisol levels and enhance stress-related behaviors (Messaoudi et al., 2011). This modulation of the stress response may help protect against the cognitive decline and neurodegeneration associated with chronic stress.

Apart from their function in stress resilience, psychobiotics also affect neuroinflammation and neuroplasticity, two important mechanisms engaged in the course of neurodegenerative disorders. Common in disorders including Alzheimer's, Parkinson's, and multiple sclerosis, neuroinflammation is defined by microglia activating and pro-inflammatory cytokines releasing action. Chronic neuroinflammation contributes to neuronal damage, synaptic dysfunction, and ultimately, the degeneration of brain tissue (Heneka et al., 2015). Psychobiotics have the ability to regulate neuroinflammatory pathways by affecting the gut microbiota, which then impacts both the immune responses in the body and the immunological responses in the central nervous system. For instance, studies have demonstrated that *Lactobacillus reuteri* can decrease the synthesis of pro-inflammatory cytokines like TNF-α and IL-6, therefore mitigating neuroinflammation in animal models (Zolfaghari et al., 2021).

Moreover, psychobiotics have the potential to augment neuroplasticity, which refers to the brain's capacity to restructure itself through the creation of novel connections among neurons. Learning, memory formation, and the recovery from neuronal injury are all heavily reliant on neuroplasticity. One hallmark of neurodegenerative diseases is the decline in neuroplasticity. Psychobiotics influence the communication pathway between the stomach and the brain, which subsequently impacts the synthesis of brain-derived neurotrophic factor (BDNF), a crucial chemical involved in the brain's ability to change and adapt (Bercik et al., 2011).

Animal models in preclinical investigations have shown significant data supporting the cognitive and behavioral advantages of psychobiotics, so providing a clearer understanding of the processes by which they work. Psychobiotics have demonstrated efficacy in mouse models by enhancing cognitive function, reducing anxiety-related behaviors, and improving social relationships. For instance, a research conducted on mice with Alzheimer's disease showed that adding *Lactobacillus plantarum* to their diet greatly enhanced their memory and decreased the build-up of amyloid-beta in the brain, which is a characteristic feature of Alzheimer's disease (Song et al., 2022). Further investigation found that in an animal model of Parkinson's disease, the administration of *Bifidobacterium breve* resulted in enhanced cognitive performance and decreased neuroinflammation. This highlights how psychobiotics may slow the progression of neurodegenerative diseases (Kobayashi et al., 2017).

Further investigating the function of psychobiotics in the gut–brain axis and offering understanding of how these microbes affect brain activity, mechanistic investigations have One important mechanism is the direct influence of gut bacteria' synthesis of neuroactive chemicals and neurotransmitters on brain activity. For example, some strains of *Lactobacillus* and *Bifidobacterium* are known to generate GABA, serotonin, and other neurotransmitters that are very vital for controlling mood, cognition, and behavior (Barrett et al., 2012).

Furthermore, crucial is the regulation of the makeup of the gut flora by psychobiotics, which might affect the blood–brain barrier and gut barrier permeability. Gut permeability that is elevated is associated with dysbiosis, which refers to an instability in the gut flora. This allows harmful substances such as lipopolysaccharides (LPS) to enter the bloodstream and trigger systemic inflammation, hence exacerbating neuroinflammation and neurodegeneration (Fung et al., 2017). Psychobiotics have the ability to reestablish a balanced and healthy community of microorganisms in the gut, which in turn reduces the permeability of the gut and prevents the movement of chemicals that cause inflammation. Consequently, this process aids in safeguarding the brain against harm caused by inflammation and promotes cognitive well-being.

Preclinical research has prompted the start of many clinical trials to assess the efficiency and safety of psychobiotics in human subjects, specifically in relation to neurodegenerative illnesses. Although most clinical research on psychobiotics has mostly examined their impact on mood disorders including depression and anxiety, recent studies are now investigating their potential in neurodegenerative illnesses. A study was conducted to examine the impact of a probiotic mixture including several species on the cognitive abilities of older individuals with mild cognitive impairment. The findings demonstrated that individuals who got the probiotic had notable enhancements in their memory and executive function, as compared to the placebo group (Akbari et al., 2016). This suggests that psychobiotics may hold promise as a therapeutic intervention for early-stage neurodegenerative diseases.

Despite these promising findings, there are challenges and limitations in the clinical research on psychobiotics. One major challenge is the heterogeneity of neurodegenerative diseases, which can make it difficult to identify specific psychobiotic strains that are effective across different conditions. Additionally, individual differences in gut microbiota composition and the gut–brain axis may influence the efficacy of psychobiotics, necessitating personalized approaches to treatment (Quigley, 2017). Furthermore, while psychobiotics are generally considered safe, potential side effects such as gastrointestinal discomfort or interactions with other medications must be carefully monitored, particularly in elderly or immunocompromised individuals.

21.5 INTEGRATIVE APPROACHES: COMBINING POSTBIOTICS AND PSYCHOBIOTICS WITH TRADITIONAL THERAPIES

The management of neurodegenerative diseases is complex and multifaceted, often requiring a combination of pharmacotherapy, dietary modifications, lifestyle interventions, and emerging therapeutic approaches such as the use of postbiotics and psychobiotics. An increasing number of researchers are interested in investigating how these innovative medicines might be combined with established treatments to improve clinical results as our knowledge of the gut–brain axis and its impact on neurological health grows. The potential for synergy between postbiotics, psychobiotics, and conventional neurodegenerative disease therapies holds promise for developing more effective, personalized, and holistic treatment strategies.

Combining postbiotics and psychobiotics with current pharmacotherapies for neurodegenerative illnesses is one of the most promising research directions. The main goals of conventional pharmaceutical therapies for illnesses including ALS, Parkinson's disease, and Alzheimer's disease are to control symptoms and halt the course of the disease. These treatments often include cholinesterase inhibitors, dopamine agonists, and other drugs aimed at modulating neurotransmitter systems or protecting neurons from further damage (Cummings et al., 2019). However, these therapies are often limited by their side effects, lack of efficacy in certain patient populations, and the inability to halt or reverse disease progression.

The integration of postbiotics and psychobiotics into these treatment regimens may offer a novel approach to overcoming some of these limitations. For instance, research has shown that postbiotics, such as short-chain fatty acids (SCFAs), have neuroprotective and anti-inflammatory qualities that may augment the effects of neuroprotective drugs. By influencing the generation of anti-inflammatory cytokines and controlling the activity of microglia, the immune cells found in the brain, short-chain fatty acids (SCFAs) can reduce neuroinflammation (Stilling et al., 2016). This anti-inflammatory effect could enhance the efficacy of drugs designed to protect neurons from inflammatory damage, potentially improving outcomes for patients with neurodegenerative diseases.

Psychobiotics, on the other hand, have been shown to influence neurotransmitter systems that are often targeted by traditional pharmacotherapies. For example, certain strains of *Lactobacillus* and *Bifidobacterium* can modulate the levels of serotonin, dopamine, and GABA in the brain, which are critical neurotransmitters involved in mood regulation, cognition, and motor control (Barrett et al., 2012). By influencing these neurotransmitter systems, psychobiotics may enhance the effectiveness of drugs like selective serotonin reuptake inhibitors (SSRIs) or dopamine agonists, potentially reducing the required dosage and minimizing side effects. Additionally, the anxiolytic and antidepressant effects of psychobiotics could complement pharmacological treatments for mood disorders that often accompany neurodegenerative diseases, such as depression and anxiety (Dinan and Cryan, 2013).

The mechanisms of synergy between postbiotics, psychobiotics, and traditional therapies are complex and multifaceted. The increase of gut–brain axis communication is a crucial mechanism that is involved in both the development of neurodegenerative illnesses and the therapeutic benefits of these therapies. Postbiotics and psychobiotics have the ability to modulate the gut microbiota and affect the creation of neuroactive chemicals. This, in turn, can improve the brain's response to medication. For example, by improving gut barrier function and reducing systemic inflammation, these compounds may enhance the bioavailability and efficacy of neuroprotective drugs, potentially leading to better clinical outcomes (Fung et al., 2017).

Another mechanism of synergy is the potential for postbiotics and psychobiotics to mitigate some of the side effects associated with traditional pharmacotherapies. Many drugs used to treat neurodegenerative diseases can disrupt the gut microbiota, leading to gastrointestinal side effects and further complicating disease management (Koppel et al., 2017). By restoring a healthy balance of gut bacteria and supporting gut barrier integrity, postbiotics and psychobiotics may help prevent or alleviate these side effects, improving patient adherence to treatment and overall quality of life.

A holistic strategy to addressing neurodegenerative disorders must include dietary and lifestyle changes, since they significantly increase the efficacy of postbiotics and psychobiotics. The efficacy of therapies that focus on the gut–brain axis can be affected by certain dietary patterns, as the gut microbiota is highly responsive to dietary variables. Ingesting a diet rich in fiber and fermented foods can stimulate the proliferation of advantageous gut microbiota and the synthesis of short-chain fatty acids (SCFAs), which are crucial for gastrointestinal and cognitive processes (Morrison and Preston, 2016). Conversely, eating a diet high in processed foods, sugars, and bad fats can upset the delicate balance of gut microbes and exacerbate inflammation, which can reduce the benefits of postbiotic and psychobiotic therapies.

Prebiotics, which are specific dietary components, might augment the efficacy of psychobiotics by supplying the necessary substances for the proliferation and functioning of advantageous bacteria. Prebiotics are dietary fibers that the body does not digest, such as inulin, fructooligosaccharides, and galactooligosaccharides. They have the unique capacity to promote the growth of advantageous bacteria in the gut, which raises postbiotic synthesis and eventually enhances the general wellbeing of the gut microbiota (Slavin, 2013). By including prebiotics in the diet, patients may experience greater benefits from psychobiotic supplementation, including improved mood, cognitive function, and neuroprotection.

Lifestyle factors, such as exercise and sleep, also interact with gut–brain axis therapies and can significantly influence their effectiveness. Scientific evidence has demonstrated that engaging in regular physical exercise has positive effects on the gut microbiota, enhances the synthesis of neurotrophic factors such as BDNF, and decreases neuroinflammation. These effects are particularly advantageous in the setting of neurodegenerative illnesses (Gubert et al., 2020). Furthermore, sufficient sleep is crucial for preserving intestinal health and facilitating the brain's restorative mechanisms. Disturbed sleep patterns can cause changes in the gut microbiota, heightened inflammation, and worsened symptoms of neurodegenerative disorders (Wang et al., 2021). Therefore, incorporating regular exercise and promoting good sleep hygiene are important strategies for enhancing the therapeutic effects of postbiotics and psychobiotics.

The use of postbiotics and psychobiotics into personalized medicine strategies is becoming more acknowledged as crucial for enhancing the treatment of neurodegenerative disorders. The gut microbiota composition exhibits significant inter-individual variation, which is modulated by variables including genetics, dietary patterns, environmental conditions, and health status (Rothschild et al., 2018). As a result, the effectiveness of postbiotic and psychobiotic interventions can vary depending on the individual's unique microbiome profile.

Tailoring postbiotic and psychobiotic interventions to the individual's microbiome could enhance their efficacy and minimize the risk of adverse effects. Advances in microbiome sequencing and analysis technologies have made it possible to characterize an individual's gut microbiota with a high degree of precision, allowing for the identification of specific microbial imbalances or deficiencies that may be addressed through targeted interventions (Knight et al., 2017). For example, an individual with low levels of butyrate-producing bacteria may benefit from postbiotic supplementation with butyrate or prebiotics that promote the growth of these bacteria, leading to improved gut–brain axis function and better neurological outcomes.

Future directions in personalized gut–brain axis therapies may also involve the development of novel psychobiotic strains or postbiotic compounds specifically designed to target the unique needs of patients with neurodegenerative diseases. New therapeutic targets will probably be found as studies into the complex connections between the gut microbiota and the brain progress, which will hopefully result in the creation of more individualized and efficient treatment plans (Mimee et al., 2016).

21.6 FUTURE DIRECTIONS AND RESEARCH GAPS

The study of postbiotics and psychobiotics is progressing quickly, but there are various obstacles that need to be overcome in order to properly harness their therapeutic capabilities. A significant obstacle that researchers face is the methodological constraints found in existing studies. Many studies rely on small sample sizes, short durations, and varied protocols, leading to inconsistencies in findings (Ouwehand et al., 2002). Furthermore, the variability in human microbiome responses presents a significant obstacle. Differences in diet, genetics, environment, and existing health conditions can lead to varied responses to postbiotic and psychobiotic interventions, making it difficult to generalize findings across populations (Rothschild et al., 2018).

Emerging trends and technologies offer promising avenues to overcome these challenges. Recent progress in microbiome sequencing and analysis is allowing for more accurate descriptions of individual microbiomes, which can result in personalized therapies that are customized to meet the particular requirements of patients (Knight et al., 2017). In addition, there is a possibility for the creation of novel psychobiotic strains and postbiotic drugs that specifically target different elements of the gut–brain axis. This might possibly result in more efficient therapies for neurodegenerative illnesses (Mimee et al., 2016).

Further investigation should prioritize the implementation of longterm studies to evaluate the enduring impacts of postbiotics and psychobiotics on neurodegenerative disorders. These studies are essential for understanding the sustained impacts of these interventions and for identifying any long-term risks or benefits (Binda et al., 2024). Moreover, exploring novel therapeutic targets within the gut–brain axis could lead to breakthroughs in treatment strategies. Identifying new pathways and mechanisms by which gut microbiota influence brain function could pave the way for innovative therapies that offer better outcomes for patients with neurodegenerative diseases (Cryan et al., 2020).

21.7 CONCLUSION

In conclusion, the integration of postbiotics and psychobiotics into the management of neurodegenerative diseases represents a promising frontier in therapeutic strategies. These treatments have possible advantages via changing the gut–brain axis: lower neuroinflammation, increase neuroplasticity, and strengthen stress resilience. The synergistic effects when combined with traditional pharmacotherapies, alongside the influence of diet and lifestyle factors, underscore the importance of a holistic approach. Personalized medicine, guided by advances in microbiome analysis, holds the key to optimizing these treatments. In order to fully understand the promise and constraints of these novel medicines, further study is necessary, especially long-term and focused investigations. The future of neurodegenerative disease management may well lie in the precise and integrative application of gut–brain axis therapies.

REFERENCES

Aguilar-Toalá, J. E., Garcia-Varela, R., Garcia, H. S., Mata-Haro, V., González-Córdova, A. F., Vallejo-Cordoba, B., & Hernández-Mendoza, A. (2018). Postbiotics: An evolving term within the functional foods field. *Trends in Food Science & Technology*, 75, 105–114.

Agus, A., Planchais, J., & Sokol, H. (2018). Gut microbiota regulation of tryptophan metabolism in health and disease. *Cell Host & Microbe*, 23(6), 716–724.

Akbari, E., Asemi, Z., Daneshvar Kakhaki, R., Bahmani, F., Kouchaki, E., Tamtaji, O. R., ... & Salami, M. (2016). Effect of probiotic supplementation on cognitive function and metabolic status in Alzheimer's disease: A randomized, double-blind and controlled trial. *Frontiers in Aging Neuroscience*, 8, 256.

Barrett, E., Ross, R. P., O'Toole, P. W., Fitzgerald, G. F., & Stanton, C. (2012). γ -Aminobutyric acid production by culturable bacteria from the human intestine. *Journal of Applied Microbiology*, 113(2), 411–417.

Bates, G. P., Dorsey, R., Gusella, J. F., Hayden, M. R., Kay, C., Leavitt, B. R., ... & Tabrizi, S. J. (2015). Huntington disease. *Nature Reviews Disease Primers*, 1(1), 1–21.

Bercik, P., Park, A. J., Sinclair, D., Khoshdel, A., Lu, J., Huang, X., ... & Verdu, E. F. (2011). The anxiolytic effect of Bifidobacterium longum NCC3001 involves vagal pathways for gut–brain communication. *Neurogastroenterology & Motility*, 23(12), 1132–1139.

Binda, S., Tremblay, A., Iqbal, U. H., Kassem, O., Le Barz, M., Thomas, V., ... & Parker, J. A. (2024). Psychobiotics and the microbiota–gut–brain axis: Where do we go from here? *Microorganisms*, 12(4), 634.

Bonaz, B., Bazin, T., & Pellissier, S. (2018). The vagus nerve at the interface of the microbiota-gut-brain axis. *Frontiers in Neuroscience*, 12, 336468.

Bonnechère, B., Amin, N., & van Duijn, C. (2022). The role of gut microbiota in neuropsychiatric diseases–creation of an atlas-based on quantified evidence. *Frontiers in Cellular and Infection Microbiology*, 12, 831666.

Bravo, J. A., Forsythe, P., Chew, M. V., Escaravage, E., Savignac, H. M., Dinan, T. G., ... & Cryan, J. F. (2011). Ingestion of Lactobacillus strain regulates emotional behavior and central GABA receptor expression in a mouse via the vagus nerve. *Proceedings of the National Academy of Sciences*, 108(38), 16050–16055.

Breit, S., Kupferberg, A., Rogler, G., & Hasler, G. (2018). Vagus nerve as modulator of the brain–gut axis in psychiatric and inflammatory disorders. *Frontiers in Psychiatry*, 9, 44.

Brown, R. H., & Al-Chalabi, A. (2017). Amyotrophic lateral sclerosis. *New England Journal of Medicine*, 377(2), 162–172.

Carabotti, M., Scirocco, A., Maselli, M. A., & Severi, C. (2015). The gut-brain axis: interactions between enteric microbiota, central and enteric nervous systems. *Annals of Gastroenterology*, 28(2), 203.

Chen, Y. H., Tsai, W. H., Wu, H. Y., Chen, C. Y., Yeh, W. L., Chen, Y. H., ... & Lai, C. H. (2019). Probiotic Lactobacillus spp. act against Helicobacter pylori-induced inflammation. *Journal of Clinical Medicine*, 8(1), 90.

Chia, R., Chiò, A., & Traynor, B. J. (2018). Novel genes associated with amyotrophic lateral sclerosis: Diagnostic and clinical implications. *The Lancet Neurology*, 17(1), 94–102.

Cryan, J. F., & Dinan, T. G. (2012). Mind-altering microorganisms: The impact of the gut microbiota on brain and behaviour. *Nature Reviews Neuroscience*, 13(10), 701–712.

Cryan, J. F., O'Riordan, K. J., Sandhu, K., Peterson, V., & Dinan, T. G. (2020). The gut microbiome in neurological disorders. *The Lancet Neurology*, 19(2), 179–194.

Cummings, J. L., Tong, G., & Ballard, C. (2019). Treatment combinations for Alzheimer's disease: Current and future pharmacotherapy options. *Journal of Alzheimer's Disease*, 67(3), 779–794.

Dalile, B., Van Oudenhove, L., Vervliet, B., & Verbeke, K. (2019). The role of short-chain fatty acids in microbiota–gut–brain communication. *Nature Reviews Gastroenterology & Hepatology*, 16(8), 461–478.

Dantzer, R., O'connor, J. C., Freund, G. G., Johnson, R. W., & Kelley, K. W. (2008). From inflammation to sickness and depression: When the immune system subjugates the brain. *Nature Reviews Neuroscience*, 9(1), 46–56.

Dhyani, P., Goyal, C., Dhull, S. B., Chauhan, A. K., Singh Saharan, B., Harshita, ... & Goksen, G. (2024). Psychobiotics for mitigation of neuro-degenerative diseases: Recent advancements. *Molecular Nutrition & Food Research*, 68(13), 2300461.

Dinan, T. G., & Cryan, J. F. (2013). Melancholic microbes: A link between gut microbiota and depression?. *Neurogastroenterology & Motility*, 25(9), 713–719.

Dinan, T. G., & Cryan, J. F. (2017). The microbiome-gut-brain axis in health and disease. *Gastroenterology Clinics*, 46(1), 77–89.

Dinan, T. G., Stanton, C., & Cryan, J. F. (2013). Psychobiotics: A novel class of psychotropic. *Biological Psychiatry*, 74(10), 720–726.

Foster, J. A., Rinaman, L., & Cryan, J. F. (2017). Stress & the gut-brain axis: Regulation by the microbiome. *Neurobiology of Stress*, 7, 124–136.

Frost, G., Sleeth, M. L., Sahuri-Arisoylu, M., Lizarbe, B., Cerdan, S., Brody, L., ... & Bell, J. D. (2014). The short-chain fatty acid acetate reduces appetite via a central homeostatic mechanism. *Nature Communications*, 5(1), 3611.

Fung, T. C., Olson, C. A., & Hsiao, E. Y. (2017). Interactions between the microbiota, immune and nervous systems in health and disease. *Nature Neuroscience*, 20(2), 145–155.

Furusawa, Y., Obata, Y., Fukuda, S., Endo, T. A., Nakato, G., Takahashi, D., ... & Ohno, H. (2013). Commensal microbe-derived butyrate induces the differentiation of colonic regulatory T cells. *Nature*, 504(7480), 446–450.

Gibson, G. R., Hutkins, R., Sanders, M. E., Prescott, S. L., Reimer, R. A., Salminen, S. J., ... & Reid, G. (2017). Expert consensus document: The International Scientific Association for Probiotics and Prebiotics (ISAPP) consensus statement on the definition and scope of prebiotics. *Nature Reviews Gastroenterology & Hepatology*, 14(8), 491–502.

Goedert, M., Eisenberg, D. S., & Crowther, R. A. (2017). Propagation of tau aggregates and neurodegeneration. *Annual Review of Neuroscience*, 40(1), 189–210.

Grochowska, M., Wojnar, M., & Radkowski, M. (2018). The gut microbiota in neuropsychiatric disorders. *Acta Neurobiologiae Experimentalis*, 78(2), 69–81.

Gubert, C., Kong, G., Renoir, T., & Hannan, A. J. (2020). Exercise, diet and stress as modulators of gut microbiota: Implications for neurodegenerative diseases. *Neurobiology of Disease*, 134, 104621.

Harach, T., Marungruang, N., Duthilleul, N., Cheatham, V., Mc Coy, K. D., Frisoni, G., ... & Bolmont, T. (2017). Reduction of Abeta amyloid pathology in APPPS1 transgenic mice in the absence of gut microbiota. *Scientific Reports*, 7(1), 41802.

Hardiman, O., Al-Chalabi, A., Chio, A., Corr, E. M., Logroscino, G., Robberecht, W., Shaw, P. J., Simmons, Z., & Van Den Berg, L. H. (2017). Amyotrophic lateral sclerosis. *Nature Reviews Disease Primers*, 3(1), 1–9.

Heneka, M. T., Golenbock, D. T., & Latz, E. (2015). Innate immunity in Alzheimer's disease. *Nature Immunology*, 16(3), 229–236.

Islam, A., Mishra, A., Ahsan, R., & Fareha, S. (2023). Phytopharmaceuticals and herbal approaches to target neurodegenerative disorders. *Drug Research*, 73(7), 388–407.

Johnson, R. L., & Wilson, C. G. (2018). A review of vagus nerve stimulation as a therapeutic intervention. *Journal of Inflammation Research*, 11, 203–213.

Kiriyama, Y., & Nochi, H. (2019). The biosynthesis, signaling, and neurological functions of bile acids. *Biomolecules*, 9(6), 232.

Knight, R., Callewaert, C., Marotz, C., Hyde, E. R., Debelius, J. W., McDonald, D., & Sogin, M. L. (2017). The microbiome and human biology. *Annual Review of Genomics and Human Genetics*, 18(1), 65–86.

Kobayashi, Y., Sugahara, H., Shimada, K., Mitsuyama, E., Kuhara, T., Yasuoka, A., ... & Xiao, J. Z. (2017). Therapeutic potential of Bifidobacterium breve strain A1 for preventing cognitive impairment in Alzheimer's disease. *Scientific Reports*, 7(1), 13510.

Koppel, N., Maini Rekdal, V., & Balskus, E. P. (2017). Chemical transformation of xenobiotics by the human gut microbiota. *Science*, 356(6344), eaag2770.

Manta, S., Dong, J., Debonnel, G., & Blier, P. (2009). Optimization of vagus nerve stimulation parameters using the firing activity of serotonin neurons in the rat dorsal raphe. *European Neuropsychopharmacology*, 19(4), 250–255.

Masters, C. L., Bateman, R., Blennow, K., Rowe, C. C., Sperling, R. A., & Cummings, J. L. (2015). Alzheimer's disease. *Nature Reviews Disease Primers*, 1(1), 1–18.

Matenchuk, B. A., Mandhane, P. J., & Kozyrskyj, A. L. (2020). Sleep, circadian rhythm, and gut microbiota. *Sleep Medicine Reviews*, 53, 101340.

Messaoudi, M., Lalonde, R., Violle, N., Javelot, H., Desor, D., Nejdi, A., ... & Cazaubiel, J. M. (2011). Assessment of psychotropic-like properties of a probiotic formulation (Lactobacillus helveticus R0052 and Bifidobacterium longum R0175) in rats and human subjects. *British Journal of Nutrition*, 105(5), 755–764.

Mimee, M., Citorik, R. J., & Lu, T. K. (2016). Microbiome therapeutics—Advances and challenges. *Advanced Drug Delivery Reviews*, 105, 44–54.

Morrison, D. J., & Preston, T. (2016). Formation of short chain fatty acids by the gut microbiota and their impact on human metabolism. *Gut Microbes*, 7(3), 189–200.

Nalls, M. A., Pankratz, N., Lill, C. M., Do, C. B., Hernandez, D. G., Saad, M., ... & Singleton, A. B. (2014). Large-scale meta-analysis of genome-wide association data identifies six new risk loci for Parkinson's disease. *Nature Genetics*, 46(9), 989–993.

O'Mahony, S. M., Clarke, G., Borre, Y. E., Dinan, T. G., & Cryan, J. F. (2015). Serotonin, tryptophan metabolism and the brain-gut-microbiome axis. *Behavioural Brain Research*, 277, 32–48.

Ouwehand, A. C., Salminen, S., & Isolauri, E. (2002). Probiotics: An overview of beneficial effects. In R. J. Siezen, J. Kok, T. Abee, & G. Schasfsma (Eds.), *Lactic Acid Bacteria: Genetics, Metabolism and Applications: Proceedings of the Seventh Symposium on Lactic Acid Bacteria: Genetics, Metabolism and Applications, 1–5 September 2002, Egmond aan Zee, the Netherlands* (pp. 279–289). Springer.

Pavlov, V. A., & Tracey, K. J. (2012). The vagus nerve and the inflammatory reflex—Linking immunity and metabolism. *Nature Reviews Endocrinology*, 8(12), 743–754.

Poewe, W., Seppi, K., Tanner, C. M., Halliday, G. M., Brundin, P., Volkmann, J., ... & Lang, A. E. (2017). Parkinson disease. *Nature Reviews Disease Primers*, 3(1), 1–21.

Quigley, E. M. (2017). Microbiota-brain-gut axis and neurodegenerative diseases. *Current Neurology and Neuroscience Reports*, 17, 1–9.

Rothschild, D., Weissbrod, O., Barkan, E., Kurilshikov, A., Korem, T., Zeevi, D., ... & Segal, E. (2018). Environment dominates over host genetics in shaping human gut microbiota. *Nature*, 555(7695), 210–215.

Savignac, H. M., Tramullas, M., Kiely, B., Dinan, T. G., & Cryan, J. F. (2015). Bifidobacteria modulate cognitive processes in an anxious mouse strain. *Behavioural Brain Research*, 287, 59–72.

Scheltens, P., De Strooper, B., Kivipelto, M., Holstege, H., Chételat, G., Teunissen, C. E., ... & van der Flier, W. M. (2021). Alzheimer's disease. *The Lancet*, 397(10284), 1577–1590.

Scott, K. P., Jean-Michel, A., Midtvedt, T., & van Hemert, S. (2015). Manipulating the gut microbiota to maintain health and treat disease. *Microbial Ecology in Health and Disease*, 26(1), 25877.

Slavin, J. (2013). Fiber and prebiotics: Mechanisms and health benefits. *Nutrients*, 5(4), 1417–1435.

Soliman, M. L., Puig, K. L., Combs, C. K., & Rosenberger, T. A. (2012). Acetate reduces microglia inflammatory signaling in vitro. *Journal of Neurochemistry*, 123(4), 555–567.

Soliman, M. L., & Rosenberger, T. A. (2011). Acetate supplementation increases brain histone acetylation and inhibits histone deacetylase activity and expression. *Molecular and Cellular Biochemistry*, 352(1), 173–180.

Song, X., Zhao, Z., Zhao, Y., Wang, Z., Wang, C., Yang, G., & Li, S. (2022). Lactobacillus plantarum DP189 prevents cognitive dysfunction in D-galactose/AlCl3 induced mouse model of Alzheimer's disease via modulating gut microbiota and PI3K/Akt/GSK-3β signaling pathway. *Nutritional Neuroscience*, 25(12), 2588–2600.

Stilling, R. M., Dinan, T. G., & Cryan, J. F. (2014). Microbial genes, brain & behaviour–epigenetic regulation of the gut–brain axis. *Genes, Brain and Behavior*, 13(1), 69–86.

Strandwitz, P. (2018). Neurotransmitter modulation by the gut microbiota. *Brain Research*, 1693, 128–133.

Tsilingiri, K., & Rescigno, M. (2013). Postbiotics: What else? *Beneficial Microbes*, 4(1), 101–107.

Walker, F. O. (2007). Huntington's disease. *The Lancet*, 369(9557), 218–228.

Wallace, C. J., & Milev, R. (2017). The effects of probiotics on depressive symptoms in humans: A systematic review. *Annals of General Psychiatry*, 16, 1–10.

Wang, Z., Chen, W. H., Li, S. X., He, Z. M., Zhu, W. L., Ji, Y. B., ... & Han, Y. (2021). Gut microbiota modulates the inflammatory response and cognitive impairment induced by sleep deprivation. *Molecular Psychiatry*, 26(11), 6277–6292.

Wyss-Coray, T., & Rogers, J. (2012). Inflammation in Alzheimer disease—A brief review of the basic science and clinical literature. *Cold Spring Harbor Perspectives in Medicine*, 2(1), a006346.

Yano, J. M., Yu, K., Donaldson, G. P., Shastri, G. G., Ann, P., Ma, L., ... & Hsiao, E. Y. (2015). Indigenous bacteria from the gut microbiota regulate host serotonin biosynthesis. *Cell*, 161(2), 264–276.

Zolfaghari, S. I., Khorasgani, M. R., & Noorbakhshnia, M. (2021). The effects of lactobacilli (L. rhamnosus, L. reuteri, L. Plantarum) on LPS-induced memory impairment and changes in CaMKII-α and TNF-α genes expression in the hippocampus of rat. *Physiology & Behavior*, 229, 113224.

22 Gut Microbiota-Derived Bacterial Metabolites

Role in Neurological Disorders

Karan Goel[1], Isha Chawla[2], Garima[1], and Meenakshi Dhanawat[3]

[1] Chitkara College of Pharmacy, Chitkara University, Rajpura, Punjab, India
[2] M.M College of Pharmacy, Maharishi Markandeshwar (Deemed to be University), Ambala, Haryana, India
[3] Amity Institute of Pharmacy, Amity University Haryana, Amity Education Valley, Gurugram, Haryana, India

22.1 INTRODUCTION TO GUT MICROBIOTA AND BACTERIAL METABOLITES

The microbiota of the human digestive system is composed of trillions of microorganisms that are necessary for numerous metabolic functions related to health and disease prevention. It influences various biological processes such as food absorption and digestion, metabolism control mechanism, immune function, and mental health. It turns out that gut bacteria produce a class of substances called bacterial metabolites, which are instrumental in complex feedback loops between the gut and other sites throughout the body, particularly with respect to those systems connected with central nervous system.

The gut microbiota comprises many different types of bacteria, most of which are members of the phyla Firmicutes, Bacteroidetes, Actinobacteria, and Proteobacteria (Chandel et al., 2024; Chen et al., 2024). Diet, age, inheritance, and various environmental exposures are some of the factors affecting this microbial diversity. Dysbiosis can be linked with many diseases, like inflammatory bowel disease (IBD), obesity, diabetes, and some neurological disorders. Still, homeostasis is maintained when the gut microbiota has a balanced composition. In order to determine the involvement of gut microbiota regarding health as well as disease, it is essential to know its composition and functionality. The association between the gut and the brain shows two-way communication highway to which we send input from stomach bacteria back into our brains. It has recently become a key area of study. This is a discussion on the metabolites: SCFAs, bile acids, and neurotransmitter precursors released through gut bacteria (Ahmadi et al., 2024, Khaledi et al., 2024). These effectors could participate in neuroinflammation and the appearance of neurological diseases like Parkinson´s models, amyotrophic lateral sclerosis (ALS), Alzheimer´s, and autism spectrum models. Not least, the burgeoning evidence connecting gut microbiota to brain disease highlights an urgent need for detailed investigations into how bacterial metabolites access, influence, and regulate central nervous system function. Bacteria produce numerous bioactive compounds that range from bile acids, SCFAs, along with indole derivatives to even some neurotransmitter-associated metabolites. These are metabolites created from the fermentation of food fibres, amino acids, and other microbial phenomena.

Metabolites have consequences on host physiology due to their functionalities and properties. Some metabolites, such as cut-up and trimethylamine-N-oxide (TMAO), have been connected to neurological disorders. In contrast, others, consisting of butyrate and an SCFA, have anti-inflammatory and neuroprotective properties. That being the case, these opening sections have a vast potential to provide readers with an informative snapshot of what is known regarding gut microbiota and their bacterial-derived metabolites, both health and disease. In the upcoming sections, we will also delve deeper into gut microbiota and learn about some of the bacterial metabolites existing in it as well as their functions in physiological processes. We will also explore how these metabolites could impact our brain health. Our research in these fields aims to further our understanding of gut–brain axis along with potential remedial applications for neurodegenerative disorders and neurodevelopment (Zhou et al., 2023, 2024).

22.2 NEUROLOGICAL DISORDERS: AN OVERVIEW

Various illnesses involving poor functioning of the central and peripheral nervous systems represent the category of neurological illnesses. They lead to the disruption of controls such as motor, sensory, and cognitive ones. These

DOI: 10.1201/9781003536802-22

diseases tend to possess intricate and diverse pathologies that some sources can trigger; that is, environmental factors, infections, and certain lifestyle choices can contribute to the development of these disorders. Due to the low quality of life, the high levels of comorbidities, and irreversibility, neurological illnesses are considered to have profound health implications. The essential role of the peripheral as well as central nervous systems, which the peripheral is a part of, the brain and the spinal cord, is in communication between different parts of the body and controls various physiological processes.

The symptoms due to the dysfunction of these systems vary from slight cognitive impairments to mild movement problems and can go up to severe neurodegenerative diseases and developmental neuropsychiatric disorders (Ashique et al., 2024; Li et al., 2023). Neurological diseases such as multiple targeted approaches or requirements that must be met for both scans and experiments (Figure we feel it is paramount to strike a balance between acquiring the highest quality data) are among Alzheimer's, Parkinsonism, stroke, epilepsy, and MS, other than autism spectrum disorder. Emerging evidence from neuroscience fields, particularly studies on gut microbiota and its metabolites, implicates systemic factors—beyond the well-established role of brain-specific abnormalities—in neurological disorders. According to the hypothesized gut–brain axis, modifications to the structure and activity of the gut microbiota might play a role in the development of neurological conditions. This axis highlights the bidirectional contact among the gut and the brain. To devise effective therapies for neurological disorders, it is essential to comprehend their fundamental aetiology (Y. Wang et al., 2023; Yang et al., 2023; Zhang et al., 2023).

22.3 THE GUT MICROBIOTA AND GUT–BRAIN AXIS

22.3.1 Composition and Function

The gut microbiota refers to a varied collection of bacteria inside the gastrointestinal (GI) tract. The microorganisms include bacteria, viruses, fungi, and archaea. The gut microbiota provides a vital element of human well-being. An individual's body composition may vary significantly according to several factors, such as diet, genetics, age, environment, and drug use. The most abundant species of bacteria are Firmicutes, Bacteroidetes, Actinobacteria, and Proteobacteria. A diverse microbiome is often associated with enhanced health outcomes (Dhanawat et al., 2020). These microorganisms play an essential part in several important metabolic activities, including breaking down complex carbohydrates and generating SCFAs such as acetate, butyrate, and propionate (Figure 22.1). These SCFAs act as a source of energy for colonocytes and possess anti-inflammatory properties that impact the whole body. Additionally, some gut-dwelling bacteria provides a vital tactic for synthesising essential vitamins and amino acids, including vitamins K and B, influencing the host's nutritional status (Ashique et al., 2024; Dai et al., 2024; Y. Kang et al., 2022).

22.3.2 Interactions with the Host

The gut microbiota is also important for immune system development as well as its function by influencing systemic and local immunological responses. It controls mucus production and antimicrobial peptide levels, which in turn helps keep the intestinal barrier intact and prevents toxins and infections from entering the circulation. The communicaton among the microbiota, the nervous system, and the mind go beyond only the production of neurotransmitters and the regulation of neuronal pathways to impact emotions, thoughts, and actions (Y. Kang et al., 2022; Yang et al., 2023).

22.3.3 Communication Pathways between Gut and Brain

The link between the gastrointestinal system and the brain is facilitated by many interconnected routes, such as neurological, endocrine, immunological, and metabolic pathways. Furthermore, the vagus nerve, a vital component of the neural pathway, is responsible for relaying signals from the stomach to the brain. These signals provide information on the movement of the digestive system, feeling of fullness, and presence of inflammation. The ENS is the network of neurons located inside the gut, which can communicate with our CNS. It is even called the "second brain". This network can affect digestion and communication. Gut hormones Ghrelin, Leptin, and Peptide YY are released from the endocrine system. These hormones influence hunger, metabolism, and response to stress within the body. The hypothalamic-pituitary-adrenal (HPA) axis possesses a role in stabilizing cortisol levels, which is mainly responsible for stress management and modulates how we respond to it as well; however, there is also a robust interface between the microbiome and this system. The HPA axis has a role in stress reduction, and the microbiome affects both cortisol levels and how we respond to stress (Kim et al., 2024; X. Wang et al., 2024).

The gut microbiota functions to control the cytokine release in immune pathways. Such signalling molecules may either pass blood–brain barrier or send their messages through a different route, like via one of the primary nerves in the body, such as the vagus nerve, to indicate some influence on brain function. The microbiota interacts with the gut-associated lymphoid tissue (GALT), leading to immune-mediated effects systemically, some of which may reach the brain. Furthermore, the metabolites generated by the microbiota—such as bile acids and tryptophan derivatives or SCFAs—prove essential for several physiological mechanisms. These metabolites can affect the behaviour and functions of the brain. SCFAs might influence the production of neurotrophic factors and neurotransmitters,

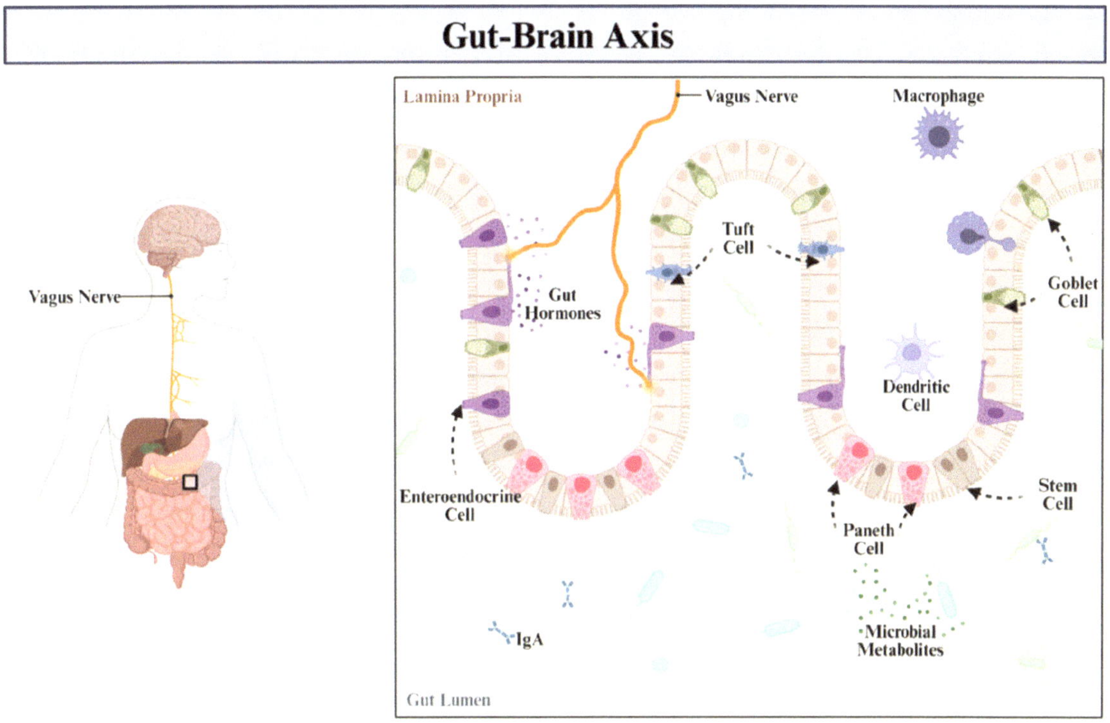

FIGURE 22.1 Gut–brain axis.

demonstrating an intricate bond between gut health as well as psychological well-being. To understand the effect of gut–brain axis on physical and mental well-being and create targeted treatment approaches, it is crucial to have a complete comprehension of intricate connections and communication routes engaged (Y. Lin et al., 2024; X. Wang et al., 2024; Ye et al., 2024).

22.4 BACTERIAL METABOLITES AND NEUROLOGICAL HEALTH

Digesting dietary components and engaging in diverse metabolic functions result in the synthesis of biologically active compounds, such as bacterial metabolites, by gut microbiota. These metabolites could influence neuronal health in the host through multiple pathways. The binding of a range of gut-produced molecules, neurotransmitters, bile acids, tryptophan metabolites, lipopolysaccharides (LPS), and SCFAs have been shown to adjust the crosstalk between the brain–gut axis (Lu et al., 2024) (Table 22.1).

22.5 NEUROTRANSMITTER PRODUCTION BY GUT BACTERIA

Neurotransmitters are chemical messengers that play essential functions in communication within the neurological system. Gut bacteria may create and influence numerous neurotransmitters. Neurotransmitters produced in the stomach may affect cognitive processes, emotional states, and behavioural patterns via the gut–brain axis, which arbitrates interactions among the gut microbiota along with CNS (Zheng et al., 2024). Important neurotransmitters produced by gut bacteria and the effects they have are as follows (Table 22.2):

22.6 BACTERIAL METABOLITES AND BLOOD–BRAIN BARRIER (BBB) INTEGRITY

Several different types of microorganisms living in the gut can produce neurotransmitters. One-way they may impact brain function is by producing these messengers. While most neurotransmitters like GABA, serotonin, and dopamine typically cannot cross the BBB directly, there are pathways for them to reach the brain. For instance, neurotransmitters generated through gut microbes affecting the brain by the vagus nerve as well as its sensory neurons. It is also possible that active neurotransmitters are made from precursor molecules that can cross the blood–brain barrier. Take tryptophan, a precursor to serotonin, for example; gut microorganisms could influence the presence and processing of tryptophan in the body. The levels of tryptophan in the bloodstream have been linked to levels in the brain, potentially impacting how serotonin signals inside nervous system (Fornelos et al., 2020; Gonzalez-Santana & Diaz Heijtz, 2020; Zheng et al., 2024).

The BBB plays a role in allowing essential nutrients to enter the brain while blocking substances from getting through. It's essential to maintain a balance to support brain function. When the BBB is disrupted, it can lead to issues like inflammation and degenerative disorders. The gut microbiota produces metabolites that can affect the BBB integrity through mechanisms. Bile acids, LPS, and SCFAs

TABLE 22.1
List of Some Bacterial Metabolites and Their Neurological Effects

Metabolite	Sources	Neurological Effects	Mechanisms	Reference
Short-chain fatty acids (SCFAs)	Obtained by dietary fibers fermentation through gut bacteria (e.g., Bacteroides, *Faecalibacterium, and Roseburia*)	• Regulate neurotransmitter production (e.g., serotonin) • Anti-inflammatory effects • Maintain blood–brain barrier integrity	• Modulate gene expression through epigenetic mechanisms (e.g., HDAC inhibition) • Influence production of neurotrophic factors (e.g., BDNF)	Huang et al., 2024
Neurotransmitters	Produced by gut bacteria (e.g., *Lactobacillus and Bifidobacterium*)	• Modulate mood, behavior, and cognition	• Direct production of neurotransmitters (e.g., GABA, serotonin, dopamine) • Indirect modulation through precursor availability	Lu et al., 2024; Qu et al., 2024
Bile acids	Generated through liver and modulated by gut bacteria	• Influence brain function and behavior • Modulate immune responses	• Pass blood–brain barrier and undergoes interaction with receptors (e.g., FXR, TGR5) • Modulate gut inflammation and systemic immune responses	Ferrell & Chiang, 2021
Tryptophan metabolites	Generated through gut bacteria by the metabolism of dietary tryptophan	• Affects mood and cognition • Modulate immune and inflammatory responses	• Metabolism into kynurenine and serotonin pathways • Production of indole derivatives, which can influence immune function and gut barrier integrity	Labus et al., 2016
Lipopolysaccharides (LPS)	Gram-negative bacteria's outer membrane components	• Pro-inflammatory effects • Activation of systemic and neuroinflammatory responses	• Signaling pathways associated with Toll-like receptor 4 (TLR4) activation • Induction of pro-inflammatory cytokines and chemokines	Babaei et al., 2024

are examples of metabolites that significantly impact the BBB integrity (Gonzalez-Santana & Diaz Heijtz, 2020; Johnson & Foster, 2018) (Figure 22.2).

22.6.1 Mechanisms of Blood–Brain Barrier Regulation

The BBB is a gate that regulates substance entry from your blood to your brain. The CNS is responsible for its basic functions and protecting the brain against poisonous substances. It takes many different approaches to keep the BBB intact. Endothelial cells form a barrier by interacting with proteins, preventing substances from passing easily (Occluding claudins (claudin 5) Y. Lin et al., 2024; JAMs Ye et al., 2024). These proteins are required to form the BBB and its structural strength and permeability. Although endothelial cells provide this function to maintain a BBB, they can take care of the production of junction proteins and respond to inflammation-associated signals. In addition, they are involved in helping nutrients and waste materials to cross the blood–brain barrier. These interacting cells, pericytes as well as astrocytic extensions surround blood capillaries, forming the BBB. The BBB function is modulated by pericytes (in green) and supported by astrocytes help. This maintains and repairs the BBB by maintaining junction protein production, aiding in endothelial cell growth. Cytokines are inflammation mediators that can disrupt the permeability of the BBB. Metabolites and

TABLE 22.2
Effects of Neurotransmitters Produced by Gut Bacteria

Neurotransmitter	Producing Bacteria	Functions and Effects	Mechanisms	Reference
Serotonin (5-HT)	*Escherichia*, *Enterococcus*, *Streptococcus*, *Lactobacillus*, and *Bifidobacterium*	• Mood regulation • Gut motility • Appetite control • Sleep cycles	• Gut bacteria can produce serotonin or modulate its synthesis by influencing the availability of its precursor, tryptophan	Sancho-Alonso et al., 2024
Gamma-aminobutyric acid (GABA)	*Lactobacillus* and *Bifidobacterium*	• Inhibition of neural activity • Anxiety and stress regulation	• Direct synthesis of GABA by gut bacteria • Modulation of GABA receptor expression	Devi et al., 2023
Dopamine	*Bacillus* and *Serratia*	• Reward and pleasure pathways • Motor control • Regulation of mood	• Synthesis and release of dopamine by gut bacteria • Influence on dopamine levels and receptor activity	Gutman & Litvak, 2024
Norepinephrine	*Escherichia*, *Bacillus*, and *Saccharomyces*	• Fight-or-flight response • Regulation of attention and arousal	• Production of norepinephrine by gut bacteria • Modulation of sympathetic nervous system activity	Tian et al., 2024
Acetylcholine	*Lactobacillus* and *Bacillus*	• Muscle activation • Memory and learning • Regulation of autonomic functions	• Synthesis of acetylcholine by gut bacteria • Modulation of cholinergic signaling	Ndjim et al., 2024
Histamine	*Lactobacillus* and *Enterobacter*	• Immune response • Regulation of sleep-wake cycles • Gastric acid secretion	• Production of histamine by gut bacteria • Modulation of histaminergic signaling and immune response	W. Lin et al., 2023
N-Butyryltyramine	*Bacteroides* and *Prevotella*	• Modulation of mood • Potential impact on depression and anxiety	• Metabolite of dopamine and involved in regulating mood and neuroinflammation	Fornelos et al., 2020
Phenylethylamine (PEA)	*Bacteroides* and *Clostridium*	• Mood enhancement • Cognitive function and alertness	• Acts as a neuromodulator that can influence mood and cognitive functions • Modulates the release of other neurotransmitters	Zhai et al., 2023
Tryptamine	*Lactobacillus* and *Bifidobacterium*	• Regulation of mood • Involvement in the sleep–wake cycle	• A precursor to serotonin; can influence mood and behavior through the modulation of serotonergic pathways	Zhai et al., 2023
5-Hydroxytryptophan (5-HTP)	*Bifidobacterium* and *Lactobacillus*	• Precursor to serotonin • Mood regulation and anxiety reduction	• Directly influences serotonin levels by serving as a precursor	Hu et al., 2023

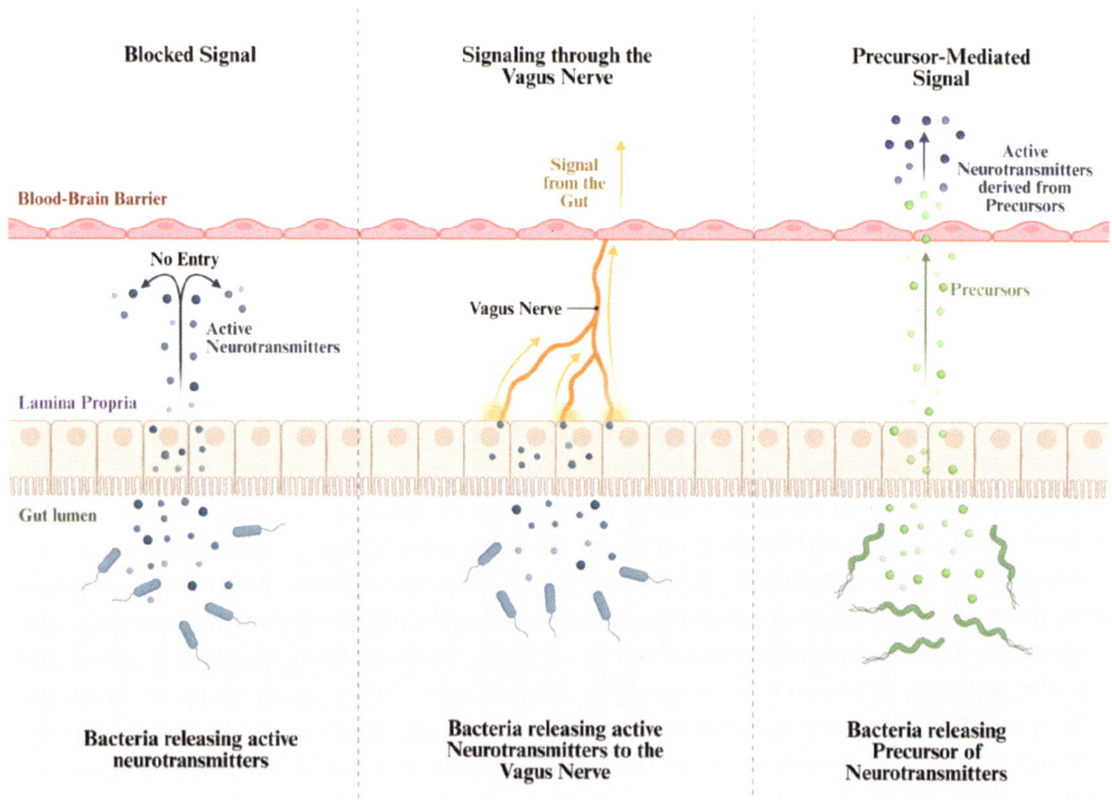

FIGURE 22.2 Signaling between the blood–brain barrier (BBB) and bacterial metabolites via gut–brain neurotransmitters.

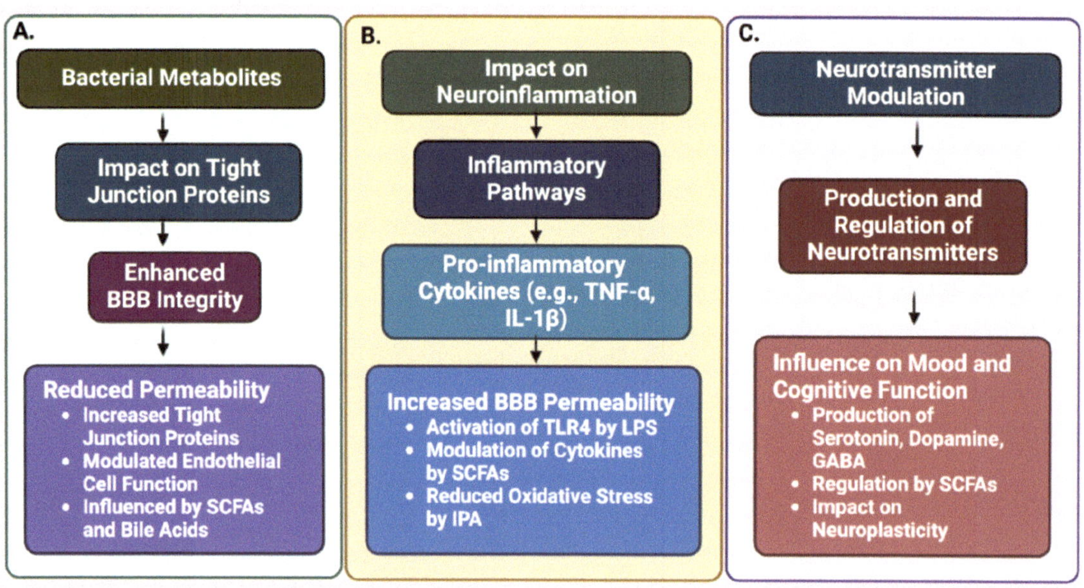

FIGURE 22.3 Modulation of BBB regulation.

anti-inflammatory cytokines can maintain or restore the integrity of BBB (Fornelos et al., 2020; Ye et al., 2024). On the other hand, pro-inflammatory cytokines like TNF-α and IL-1β may weaken tight junctions of the BBB and increase its permeability (Figure 22.3). Intestinal Bacterial metabolites generated by the gut microbiota might potentially have positive and negative influences on the integrity of the BBB. Bile acids and SCFAs are metabolites that can potentially affect the tight junction protein expression, the endothelial cells function, and the body's inflammatory responses (Ferrell & Chiang, 2021; Huang et al., 2024; Lu et al., 2024).

22.6.2 Neuroinflammation

The brain is subjected to inflammation by infections, injuries, or metabolic disorders affecting the nervous system. The impact of this dietetic-induced modulates the accompanying brain inflammation, which can be in part related to bacterial released byproducts. Endothelial cells could be activated by certain substances, such as LPS, and generate molecules like TNF-α and IL-1β. These molecules can produce brain inflammation, increase BBB thickness, and muddle junctions (Fornelos et al., 2020; Ye et al., 2024). Certain substances produced by bacteria, like SCFAs, have properties that can help reduce inflammation and influence the production of cytokines. Propionate and butyrate, for example, may play a role in lowering neuroinflammation by decreasing inflammatory cytokines and increasing anti-inflammatory ones. This process can contribute to the protection of the BBB integrity. SCFAs can activate the system's microglia and immune cells while influencing their functions. Regulating microglial activation and reducing responses to SCFAs may help safeguard the BBB. A specific bacterial metabolite called Indole 3 acid (IPA) has been found to decrease stress in the brain. These metabolites can potentially lessen neuroinflammation and preserve BBB integrity by mitigating harm (Kim et al., 2024; X. Wang et al., 2024).

22.6.3 Neurotransmitter Modulation

Bacteria by-products impact both the BBB. Brain well-being affects BBB function and neurotransmitter production and control (Qu et al., 2024). Our gut bacteria can produce neurotransmitters like serotonin, dopamine, and GABA. Some bacteria, like Lactobacillus and Bifidobacterium, also generates GABA, a neurotransmitter that is an inhibitor—inhibitors act the opposite of exciters—and could influence BBB permeability. The metabolites produced by these bacteria influence the abundance and activity of neurotransmitters. SCFAs, like butyrate, might influence how neurotransmitters are made and released by affecting the activity of enzymes involved in their production. Additionally, bacterial metabolites could impact how neurotransmitter receptors are expressed and function in our bodies (Ashique et al., 2024; Huang et al., 2024; Y. Wang et al., 2023). SCFAs could influence emotions. We were thinking about the ability to manage the activation of GABA and serotonin receptors. Consequently, this could indirectly affect the strength of the BBB. Products such as butyrate can potentially regulate neurotransmitter networks, thus impacting brain flexibility and cognitive performance. This impact on neuroplasticity can potentially disrupt both the brain's functioning and the BBB's integrity (Qu et al., 2024; Tian et al., 2024).

22.7 ROLE OF BACTERIAL METABOLITES IN NEURODEGENERATIVE DISEASES

The gut microbiota synthesizes compounds that could influence many neurological and neurodegenerative diseases. Several processes, including the gut–brain axis, neuroinflammation, metabolic alterations, and neurotransmitter modulation, influence these connections. We examine the potential consequences of specific bacterial metabolites obtained by gut microbiota for the progression of Alzheimer's disease, Parkinson's disease (Dhanawat et al., 2024), and autism spectrum disorders (Ashique et al., 2024; Dai et al., 2024; Y. Wang et al., 2023) (Table 22.3).

22.8 RESEARCH METHODS AND TECHNIQUES

Numerous research projects have delved into the relationship between the gut microbiota along with central nervous system within the fields of the gut–brain axis and bacterial metabolites. Essential research methods involve tools such as microbiome and metabolomics analysis, often utilising animal models. These methodologies can help us grasp how bacterial metabolites influence brain health and unravel the mechanisms underlying the gut–brain connection (Das et al., 2024; Verma et al., 2022).

22.8.1 Animal Models in Gut–Brain Research

Studying how the gut microbiota interacts with the system (CNS) involves delving into physiology, genetics, and biochemistry. This exploration often relies on animal models, which play a role in research that investigates the connection between the stomach and the brain (Moossavi et al., 2022). Here are some common examples of animal models used in this field (Table 22.4).

22.8.2 Metabolomics and Microbiome Analysis

Cutting-edge techniques, like metabolomics and studying the microbiome, could offer clues about the digestive system's bacteria and compounds and their functions (Table 22.5). These tools help us delve into how the gut interacts with the brain by collecting information about its chemical and microbial makeup (Labus et al., 2016; Zhai et al., 2023).

22.9 THERAPEUTIC POTENTIAL OF BACTERIAL METABOLITES

Studying the gut–brain axis along with bacterial metabolites role has revealed new therapy methods for maintaining healthy neurological systems. Neurological diseases could be treated using techniques that regulate the microbiota, including dietary changes, the administration of probiotics

TABLE 22.3
Role of Various Bacterial Metabolites on Neurodegenerative Conditions, Including their Potential Mechanisms of Action, Effects on Brain Health, and Implications for Therapeutic Strategies

Disorder	Bacterial Metabolites	Source	Potential Impact	Mechanisms	Reference
Alzheimer's disease	SCFAs	Fermentation of dietary fibers by gut bacteria	• Neuroprotective effects • Modulation of neuroinflammation and amyloid-beta pathology	• Butyrate acts as an HDAC inhibitor • Regulates gene expression related to inflammation	Huang et al., 2024
	LPS	The outer membrane of Gram-negative bacteria	• Promotes neuroinflammation • Exacerbates amyloid-beta pathology	• Activation of TLR4 and microglia • Induction of pro-inflammatory cytokines	Y. Lin et al., 2024
	Trimethylamine-*N*-oxide (TMAO)	Metabolized from dietary choline by gut bacteria	• Associated with amyloid-beta aggregation • Increases oxidative stress and inflammation	• Modulates cholesterol and lipid metabolism • Promotes plaque formation	Zhou et al., 2024
	Bile acids	Liver-derived, modified by gut bacteria	• Alters neuronal signaling • Influences cognitive function and neuroinflammation	• Interaction with FXR and TGR5 receptors • Influence on lipid metabolism and inflammation	Ferrell & Chiang, 2021
Parkinson's disease	SCFAs	Produced by gut bacteria from dietary fibers	• Modulates neuroinflammation • Influences motor and non-motor symptoms	• Butyrate's neuroprotective properties • Altered SCFA levels affecting gut–brain axis	Sancho-Alonso et al., 2024
	LPS	The outer membrane of Gram-negative bacteria	• Induces neuroinflammation • Contributes to degeneration of dopaminergic neurons	• Activation of microglia and astrocytes • Increases intestinal permeability (leaky gut)	Huang et al., 2024; Sancho-Alonso et al., 2024
	Neuroactive compounds (e.g., dopamine)	Produced or metabolized by gut bacteria	• Affects motor and non-motor symptoms • Modulates neurotransmitter systems	• Production of dopamine by gut bacteria • Influence on synaptic function and neurotransmitter regulation	Gutman & Litvak, 2024
Autism spectrum disorders	SCFAs	Produced by gut bacteria from dietary fibers	• Potentially contributes to ASD-like behaviors • Modulates neuroinflammation and neurotransmission	• Elevated propionate levels affect behavior • Influence on neurotransmitter systems	Dai et al., 2024
	Amines (e.g., histamine)	Produced by gut bacteria from amino acids	• Influences ASD symptoms • Affects neurotransmission and immune responses	• Modulation of neurotransmitter receptors	Yang et al., 2023
	Indole derivatives	Metabolized from tryptophan by gut bacteria	• Affects neurotransmitter systems • Potentially contributes to neurodevelopmental features of ASD	• Influence on gut–brain axis • odulation of serotonin and other neurotransmitter systems	Zhou et al., 2023
	Gut microbiota dysbiosis	Inequality in gut microbial composition	• Alters metabolite production • Contributes to ASD symptoms	• Imbalance in beneficial vs. harmful bacteria • Impact on CNS function through gut–brain axis	Dai et al., 2024; Yang et al., 2023

TABLE 22.4
Effects of Animal Models in Gut–Brain Research

Category	Method/ Technique	Details	Applications	Reference
Germ-free (GF) animals	Animals raised in sterile conditions without microbiota	• GF animals lack any microbiota, allowing the study of host functions without microbial influence • Used to introduce specific bacteria to observe their effects on the host	• Investigate how the gut microbiome impacts immunological responses, cognitive function, and development • Investigate how colonization with specific microbes affects behavior and physiology	Moossavi et al., 2022; Verma et al., 2022
Gnotobiotic animals	Selectively colonized with known bacterial strains	• Gnotobiotic animals are germ-free animals colonized with specific, known bacteria or microbial communities • Allows for controlled studies of individual bacterial species or consortia	• Understand the impact of specific bacteria on host physiology and behavior • Explore microbial interactions and their role in gut–brain communication	Wilson et al., 2023
Conventional animals with altered microbiota	Animals with microbiota altered by antibiotics, diet, or environment	• Conventional animals are subjected to changes in their microbiota through various means, such as antibiotic treatment, dietary changes, or environmental factors • Used to model dysbiosis and its effects	• Study the importance of microbiota changes (dysbiosis) over gut–brain axis and related diseases. • Investigate microbiota that alterations affect neurological and immune responses	X. Kang et al., 2023
Transgenic and knockout models	Genetically modified animals (e.g., transgenic or knockout mice)	• Animals genetically engineered to overexpress, underexpress, or lack specific genes, allow the study of gene function in microbiota-host interactions	• Investigate the role of specific genes in regulating gut–brain interactions • Explore genetic factors that influence microbial composition and host immune and neurological functions	W. Liu et al., 2018
Behavioral assessments	Cognitive, emotional, and social behavior tests	• Animals are subjected to a variety of behavioral tests, such as maze navigation, social interaction tasks, and anxiety or depression assays • These assessments help link microbiota or metabolite changes to specific behavioral outcomes	• Correlate changes in gut microbiota or bacterial metabolites with alterations in behavior • Study microbiota influences cognitive and emotional functions relevant to neurological as well as psychiatric disorders	Gonzalez-Santana & Diaz Heijtz, 2020; Johnson & Foster, 2018

and prebiotics, and other methods. These strategies aim to enhance health by decreasing disease-related mechanisms, boosting the production of beneficial metabolites, and altering the gut microbiota composition and functionality (Acharya et al., 2024; X. Lin et al., 2024; Papa et al., 2023).

22.9.1 Probiotics and Prebiotics in Neurological Disorders

22.9.1.1 Probiotics

Probiotics are live microorganisms that improve a host's health when given in adequate amounts. Probiotics may generate several impacts on neurological disorders. Probiotics may modify the makeup and gut microbiota operation, potentially leading to a favorable change in plenty of helpful bacteria and a decrease in harmful ones. Modifying these settings may reduce the intestinal barrier translocation and subsequent bloodstream entry of inflammatory substances, such as LPS (Y. Lin et al., 2024; X. Liu et al., 2024). Some probiotic strains can create neuroactive and neurotransmitter substances, like serotonin, GABA, and short-chain fatty acids (SCFAs) (Huang et al., 2024; X. Liu et al., 2024; Papa et al., 2023). These substances may alter how the brain works and how people act. Probiotics reduce systemic and neuroinflammation via their immunomodulatory and anti-inflammatory impact on the immune response. Increasing

TABLE 22.5
Effects of Metabolomics and Microbiome in the Gut–Brain Axis

	Method/Technique	Details	Applications	Reference
Metabolomics	Comprehensive analysis of metabolites in biological samples (e.g., blood, feces, urine)	• Metabolomics involves the extensive investigation of tiny molecules (metabolites) identified in tissues, cells, biofluids, and living things • Utilizes technologies like nuclear mMagnetic resonance (NMR) and mass spectrometry (MS) for identifying and quantifying metabolites	• Identify specific bacterial metabolites involved in gut–brain communication • Study metabolic pathways affected by microbiota and correlate them with disease states or behavioral phenotypes • Discover potential biomarkers for neurological diseases	Zhai et al., 2023
Microbiome analysis	Examination of gut microbial community composition and function	• Microbiome analysis explores the diversity, abundance, and function of microorganisms in the gut • Techniques include 16S rRNA gene sequencing for identifying bacterial species, whole-genome shotgun sequencing for functional potential, and metatranscriptomics for active microbial gene expression	• Determine the structural arrangement and functional capabilities of gut microbiota • Study microbial communities participation in health and disease, including their role in neurological disorders • Identify microbial genes involved in metabolite production affecting the CNS	Y. Kang et al., 2022
Integration of metabolomics and microbiome data	Combining metabolomic and microbiome datasets for a comprehensive analysis	• Integrating metabolomics with microbiome data allows for an enlarged holistic comprehension of gut–brain axis • Helps link specific microbes to the production of metabolites that influence brain function	• Provide a detailed picture of how gut microbiota and their metabolites correlate with host biology • Identify novel therapeutic targets and biomarkers by linking microbial functions with metabolic outcomes in neurological diseases	Chen et al., 2024; Li et al., 2023

anti-inflammatory cytokine production and decreasing pro-inflammatory cytokine production are potential advantages in conditions such as Alzheimer's (Adıgüzel et al., 2022; Ciernikova et al., 2023; X. Liu et al., 2024).

22.9.1.2 Prebiotics

Prebiotics are indigestible dietary components that encourage the development and activities of beneficial bacteria inside the digestive tract. They may also affect brain health in other ways, like boosting the generation of good metabolites. Intestinal bacteria decompose prebiotics, which triggers fermentation and the production of SCFAs like acetate, butyrate, and propionate. Numerous studies shows that SCFAs have anti-inflammatory, neuroprotective, and immunomodulatory properties. Enhanced gut barrier activity increases the mucin production and tight junction proteins, prebiotics may fortify the gut wall, reducing the permeability of the gut lining and preventing potentially harmful substances from entering the bloodstream. By altering the composition of gut flora and metabolites, prebiotics may influence behavioral and cognitive processes. The release of more of the feel-good neurotransmitters serotonin and similar ones is one way they do this (Adıgüzel et al., 2022; Ciernikova et al., 2023; Huang et al., 2024; X. Liu et al., 2024).

22.9.2 Harnessing the Power of Probiotics and Prebiotics

The optimal strains of probiotic bacteria and prebiotic compounds for brain function are now the subject of intense investigation. *Lactobacillus* and *Bifidobacterium* are two examples of good bacteria that have shown promise in reducing depressive and anxious symptoms. When prebiotics (synbiotics), medicines, or dietary modifications are added to the mix, the health benefits of probiotics may be amplified. Antibiotics may provide an optimal environment for probiotic bacteria, allowing them to survive and colonize more easily. Intestinal microbial compositions vary greatly across individuals. By tailoring prebiotic and probiotic treatments to each patient's unique microbiota, for example, it may be possible to improve treatment outcomes.

Although there is hope for the effective advantages of probiotics and prebiotics for treating neurological diseases, sufficient clinical trials are mandatory to confirm their safety as well as efficacy. Dosages, strains, and treatment durations must be standardized for reliable results (Adıgüzel et al., 2022; Ciernikova et al., 2023; X. Liu et al., 2024; Papa et al., 2023).

22.9.3 Dietary Interventions

The bacteria that live in one's stomach and the substances they create are greatly affected by the food that a person eats. Nutritional therapies have the potential to enhance neurological function and regulate the gut–brain axis. Those who adhere to the Mediterranean diet which is rich in plant-based foods, healthy fats, almonds, and olive oil are at a reduced risk of developing neurodegenerative diseases. The high polyphenol and fiber content of the Mediterranean diet promotes healthy gut flora and the generation of short-chain fatty acids. The ketogenic diet is characterized by a high-fat, low-carbohydrate diet that has two consequences: it alters the gut flora's makeup and increases the synthesis of ketone bodies. The latter have the ability to pass blood–brain barrier and provide neurons with an alternate source of energy. Potentially useful as a treatment for epilepsy and other neurological disorders. Diets high in fiber promote the growth of bacteria that create SCFAs, leading to an increase in the synthesis of butyrate and other SCFAs. The anti-inflammatory and neuroprotective characteristics of these metabolites may be useful in the treatment of diseases like MS and Parkinson's. Foods that undergo Fermented foods including yogurt, kefir, kimchi, and sauerkraut include good bacteria. If you suffer from mood swings or brain fog, consuming fermented foods regularly may help restore harmony to your gut bacteria. Some people may find relief from gastrointestinal issues or cognitive dysfunction by adhering to a gluten-free or casein-free diet (Adıgüzel et al., 2022; Ciernikova et al., 2023; Dai et al., 2024; X. Liu et al., 2024).

22.10 CHALLENGES AND FUTURE DIRECTIONS

Numerous potential and obstacles exist in the area of gut–brain axis research along with the impact of bacterial metabolites on neurological health. Although there has been a lot of headway in figuring out how gut microbes relate to neurological disorders, a full mechanistic understanding is still out of reach. It is difficult to determine whether changes in gut microbiota cause or result in neurological illnesses, and our understanding of the specific mechanisms by which bacterial metabolites impact brain function is limited. The intricacy of this topic is further compounded by the fact that each person's gut microbiota is unique in terms of richness and functioning, shaped by their unique combination of genetic, environmental, and lifestyle variables. To determine causation and provide efficient treatment strategies, further clinical trials, mechanistic research, and longitudinal studies are urgently required (Dave et al., 2012; Lu et al., 2024; Verma et al., 2022).

New technological resources provide encouraging means of closing these disparities. Latter generation and third-generation sequencing are examples of high-throughput technologies that allow for the thorough characterization of genetic potential and microbiota composition (Zhou et al., 2023). The metabolite profiles linked to different microbiota may be better understood with the use of metabolomics tools like nuclear magnetic resonance spectroscopy and mass spectrometry. The development of predictive models, pattern recognition in massive amounts of multi-omics data, and other related tasks requires cutting-edge bioinformatics and machine learning. Recent advances in in vitro and in vivo models have greatly boosted our capacity to research disease causes and mimic human physiology. These models include organ-on-a-chip systems and complex animal models (Adıgüzel et al., 2022; Jia et al., 2021).

Regulatory and ethical concerns are also present in the sector. Probiotics, prebiotics, and fecal microbiota transplantation are all treatments based on the microbiota; hence, it is crucial to establish clear standards for their safety and effectiveness. As part of this process, quality control procedures, formulations, and doses will be standardized. Concerns about off-target consequences, informed permission, and the use of microbiota originating from humans are of the utmost importance from an ethical standpoint. Finding a happy medium between innovation and access requires thoughtful deliberation on intellectual property concerns, such as the patenting of living creatures or particular bacterial strains (Ashique et al., 2024; Y. Wang et al., 2023; Zhang et al., 2023).

Embracing a multidisciplinary approach that includes microbiology, neuroscience, medicine, and ethics is vital as research advances. The gut–brain axis may be better understood and new biomarkers and treatment targets can be found via the integration of multi-omics methodologies. These approaches combine genomics, transcriptomics, proteomics, and metabolomics. More effective therapies may be possible with personalized medicine techniques that take into account each person's genetic makeup and microbiome. Targeted therapies for neurological illnesses may be possible via the exploration of innovative microbiota manipulation methods, such as engineered probiotics, and phage therapy. Overcoming these obstacles will require gut–brain researchers to work together, think creatively, and conduct their work ethically if they want to find novel ways to cure and prevent neurological illnesses (Dave et al., 2012; Xie et al., 2024).

22.11 CONCLUSION

Neurological illnesses are caused by the importance of bacterial metabolites on brain function. Although our comprehension of the gut–brain axis has expanded, the precise mechanisms via which it operates remain unidentified (Qu et al., 2024). Conduct longitudinal research to investigate the effective roles of alterations in gut microbiota for the development or manifestation of neurological diseases. The intricate and reciprocal nature of the gut–brain axis poses difficulties in demonstrating causation (Li et al., 2023). Examine the gut microbiota and bacterial metabolites using customized methods as a possible treatment for neurological disorders. The microbiome of a person is influenced by genetics, upbringing, and everyday behaviors. Bacterial metabolites have an impact on the brain via the use of in vitro and in vivo models that accurately mimic human physiology (Xie et al., 2024). Advancements in technology have facilitated the accessibility of new research equipment. Ensuring the safe and ethical use of microbiome-based medications, such as probiotics, prebiotics, and fecal transplants, requires the establishment of clear regulatory and ethical frameworks. By using several multi-omics techniques such as transcriptomics, metabolomics, genomics, and proteomics, we may enhance our comprehension of the gut–brain axis and uncover new biomarkers and therapeutic targets (Zhai et al., 2023; Zhou et al., 2024). Explore novel approaches to modulating the microbiome, such as phage treatment and modified probiotics, as possible interventions for neurological disorders. Examine the impact of bacterial metabolites like bile acids, tryptophan derivatives, and SCFAs on neurotransmitter systems and brain function. The ketogenic, fermented, and Mediterranean diets are dietary interventions that impact gut microbiota, and the production of bacterial metabolites, and have potential advantages for neurological disorders (Acharya et al., 2024; Dai et al., 2024; Huang et al., 2024).

REFERENCES

Acharya, A., Shetty, S. S., & Kumari N, S. (2024). Role of gut microbiota derived short chain fatty acid metabolites in modulating female reproductive health. *Human Nutrition & Metabolism*, *36*, 200256. https://doi.org/10.1016/j.hnm.2024.200256

Adıgüzel, E., Çiçek, B., Ünal, G., Aydın, M. F., & Barlak-Keti, D. (2022). Probiotics and prebiotics alleviate behavioral deficits, inflammatory response, and gut dysbiosis in prenatal VPA-induced rodent model of autism. *Physiology & Behavior*, *256*, 113961. https://doi.org/10.1016/j.physbeh.2022.113961

Ahmadi, S., Taghizadieh, M., Mehdizadehfar, E., Hasani, A., Khalili Fard, J., Feizi, H., Hamishehkar, H., Ansarin, M., Yekani, M., & Memar, M. Y. (2024). Gut microbiota in neurological diseases: Melatonin plays an important regulatory role. *Biomedicine & Pharmacotherapy*, *174*, 116487. https://doi.org/10.1016/j.biopha.2024.116487

Ashique, S., Mohanto, S., Ahmed, M. G., Mishra, N., Garg, A., Chellappan, D. K., Omara, T., Iqbal, S., & Kahwa, I. (2024). Gut-brain axis: A cutting-edge approach to target neurological disorders and potential synbiotic application. *Heliyon*, *10*(13), e34092. https://doi.org/10.1016/j.heliyon.2024.e34092

Babaei, F., Navidi-Moghaddam, A., Naderi, A., Ghafghazi, S., Mirzababaei, M., Dargahi, L., Mohammadi, G., & Nassiri-Asl, M. (2024). The preventive effects of Saccharomyces boulardii against oxidative stress induced by lipopolysaccharide in rat brain. *Heliyon*, *10*(9), e30426. https://doi.org/10.1016/j.heliyon.2024.e30426

Chandel, P., Thapa, K., Kanojia, N., Rani, L., Singh, T. G., & Rohilla, P. (2024). Exploring therapeutic potential of phytoconstituents as a gut microbiota modulator in the management of neurological and psychological disorders. *Neuroscience*, *551*, 69–78. https://doi.org/10.1016/j.neuroscience.2024.05.002

Chen, C.-Y., Wang, Y.-F., Lei, L., & Zhang, Y. (2024). Impacts of microbiota and its metabolites through gut-brain axis on pathophysiology of major depressive disorder. *Life Sciences*, *351*, 122815. https://doi.org/10.1016/j.lfs.2024.122815

Ciernikova, S., Sevcikova, A., Drgona, L., & Mego, M. (2023). Modulating the gut microbiota by probiotics, prebiotics, postbiotics, and fecal microbiota transplantation: An emerging trend in cancer patient care. *Biochimica et Biophysica Acta (BBA) – Reviews on Cancer*, *1878*(6), 188990. https://doi.org/10.1016/j.bbcan.2023.188990

Dai, H., Jiang, Y., Liu, S., Li, D., & Zhang, X. (2024). Dietary flavonoids modulate the gut microbiota: A new perspective on improving autism spectrum disorder through the gut-brain axis. *Food Research International*, *186*, 114404. https://doi.org/10.1016/j.foodres.2024.114404

Dhanawat, M., Mehta, DK., Gupta, S., & Das, R. (2020). Understanding the pathogenesis involved in Parkinson's disease and potential therapeutic treatment strategies. *Central Nervous System Agents in Medicinal Chemistry*, *20*, 88–102. https://doi.org/10.2174/1871524920666200705222842

Dhanawat, M., Malik, G., Wilson, K., Gupta, S., Gupta, N., & Sardana, S. (2024). The gut microbiota-brain axis: A new Frontier in Alzheimer's disease pathology. *CNS & Neurological Disorders – Drug Targets*. https://doi.org/10.2174/0118715273302508240613114103

Das, M., Thajuddin, N., Patra, S., & Pundir, M. (2024). Mind-body techniques on stress-induced gut microbiota dysbiosis in Asthmatics: A narrative review. *Brain Behavior and Immunity Integrative*, *5*, 100040. https://doi.org/10.1016/j.bbii.2023.100040

Dave, M., Higgins, P. D., Middha, S., & Rioux, K. P. (2012). The human gut microbiome: current knowledge, challenges, and future directions. *Translational Research*, *160*(4), 246–257. https://doi.org/10.1016/j.trsl.2012.05.003

Devi, P. B., Rajapuram, D. R., Jayamanohar, J., Verma, M., Kavitake, D., Meenachi Avany, B. A., Rani, P. U., Ravi, R., & Shetty, P. H. (2023). Gamma-aminobutyric acid (GABA) production by potential probiotic strains of indigenous fermented foods origin and RSM based production optimization. *LWT*, *176*, 114511. https://doi.org/10.1016/j.lwt.2023.114511

Ferrell, J. M., & Chiang, J. Y. L. (2021). Bile acid receptors and signaling crosstalk in the liver, gut and brain. *Liver Research*, *5*(3), 105–118. https://doi.org/10.1016/j.livres.2021.07.002

Fornelos, N., Franzosa, E. A., Bishai, J., Annand, J. W., Oka, A., Lloyd-Price, J., Arthur, T. D., Garner, A., Avila-Pacheco, J., Haiser, H. J., Tolonen, A. C., Porter, J. A., Clish, C. B., Sartor, R. B., Huttenhower, C., Vlamakis, H., & Xavier, R. J. (2020). Growth effects of N-acylethanolamines on gut bacteria reflect altered bacterial abundances in inflammatory bowel disease. *Nature Microbiology*, *5*(3), 486–497. https://doi.org/10.1038/s41564-019-0655-7

Gonzalez-Santana, A., & Diaz Heijtz, R. (2020). Bacterial Peptidoglycans from Microbiota in Neurodevelopment and Behavior. *Trends in Molecular Medicine*, *26*(8), 729–743. https://doi.org/10.1016/j.molmed.2020.05.003

Gutman, E., & Litvak, Y. (2024). Gut dopamine kick: How gut microbes turn on host receptors to fight pathogens. *Cell Host & Microbe*, *32*(5), 623–624. https://doi.org/10.1016/j.chom.2024.04.011

Hu, G., Zhu, Y., Ding, S., & Zheng, L. (2023). Role of gut microbiota in the 5-hydroxytryptamine signal transduction mechanism. *Metabolism and Translational Medicine*, *1*. https://doi.org/10.54844/mtm.2023.0344

Huang, Y., Wang, Y. F., Miao, J., Zheng, R. F., & Li, J. Y. (2024). Short-chain fatty acids: Important components of the gut-brain axis against AD. *Biomedicine & Pharmacotherapy*, *175*, 116601. https://doi.org/10.1016/j.biopha.2024.116601

Jia, P.-P., Junaid, M., Wen, P.-P., Yang, Y.-F., Li, W.-G., Yang, X.-G., & Pei, D.-S. (2021). Role of germ-free animal models in understanding interactions of gut microbiota to host and environmental health: A special reference to zebrafish. *Environmental Pollution*, *279*, 116925. https://doi.org/10.1016/j.envpol.2021.116925

Johnson, K. V.-A., & Foster, K. R. (2018). Why does the microbiome affect behaviour? *Nature Reviews Microbiology*, *16*(10), 647–655. https://doi.org/10.1038/s41579-018-0014-3

Kang, X., Ng, S.-K., Liu, C., Lin, Y., Zhou, Y., Kwong, T. N. Y., Ni, Y., Lam, T. Y. T., Wu, W. K. K., Wei, H., Sung, J. J. Y., Yu, J., & Wong, S. H. (2023). Altered gut microbiota of obesity subjects promotes colorectal carcinogenesis in mice. *EBioMedicine*, *93*, 104670. https://doi.org/10.1016/j.ebiom.2023.104670

Kang, Y., Li, L., Kang, X., Zhao, Y., & Cai, Y. (2022). Gut microbiota and metabolites in myasthenia gravis: Early diagnostic biomarkers and therapeutic strategies. *Clinical Immunology*, *245*, 109173. https://doi.org/10.1016/j.clim.2022.109173

Khaledi, M., Sameni, F., Gholipour, A., Shahrjerdi, S., Golmohammadi, R., Gouvarchin Ghaleh, H. E., Poureslamfar, B., Hemmati, J., Mobarezpour, N., Milasi, Y. E., Rad, F., Mehboodi, M., & Owlia, P. (2024). Potential role of gut microbiota in major depressive disorder: A review. *Heliyon*, *10*(12), e33157. https://doi.org/10.1016/j.heliyon.2024.e33157

Kim, Y., Lim, J., & Oh, J. (2024). Taming neuroinflammation in Alzheimer's disease: The protective role of phytochemicals through the gut–brain axis. *Biomedicine & Pharmacotherapy*, *178*, 117277. https://doi.org/10.1016/j.biopha.2024.117277

Labus, J. S., Tillisch, K., Oezguen, N., Savidge, T., Versalovic, J., & Mayer, E. A. (2016). Su1573 gut-derived metabolites involved in tryptophan metabolism associated with brain morphometry. *Gastroenterology*, *150*(4), S530. https://doi.org/10.1016/S0016-5085(16)31826-1

Li, L., Yang, J., Liu, T., & Shi, Y. (2023). Role of the gut-microbiota-metabolite-brain axis in the pathogenesis of preterm brain injury. *Biomedicine & Pharmacotherapy*, *165*, 115243. https://doi.org/10.1016/j.biopha.2023.115243

Lin, W., Li, K., Qin, Y., Han, X., Chen, X., & Ren, Y. (2023). Flunitrazepam induces neurotoxicity in zebrafish through microbiota-gut-brain axis. *Science of The Total Environment*, *901*, 165974. https://doi.org/10.1016/j.scitotenv.2023.165974

Lin, X., He, K., Gu, Z., & Zhao, X. (2024). Emerging chemophysiological diversity of gut microbiota metabolites. *Trends in Pharmacological Sciences*. https://doi.org/10.1016/j.tips.2024.07.006

Lin, Y., Xie, Z., Li, Z., Yuan, C., Zhang, C., Li, Y., Xie, K., & Wang, K. (2024). The microbiota-gut-brain axis: A crucial immunomodulatory pathway for Bifidobacterium animalis subsp. lactis' resilience against LPS treatment in neonatal rats. *International Journal of Biological Macromolecules*, *266*, 131255. https://doi.org/10.1016/j.ijbiomac.2024.131255

Liu, W., Pan, H., Wang, Q., & Zhao, Z. (2018). The application of transgenic and gene knockout mice in the study of gastric precancerous lesions. *Pathology – Research and Practice*, *214*(12), 1929–1939. https://doi.org/10.1016/j.prp.2018.10.022

Liu, X., Su, S., Yao, J., Zhang, X., Wu, Z., Jia, L., Liu, L., Hou, R., Farag, M. A., & Liu, L. (2024). Research advance about plant polysaccharide prebiotics, benefit for probiotics on gut homeostasis modulation. *Food Bioscience*, *59*, 103831. https://doi.org/10.1016/j.fbio.2024.103831

Lu, S., Zhao, Q., Guan, Y., Sun, Z., Li, W., Guo, S., & Zhang, A. (2024). The communication mechanism of the gut-brain axis and its effect on central nervous system diseases: A systematic review. *Biomedicine & Pharmacotherapy*, *178*, 117207. https://doi.org/10.1016/j.biopha.2024.117207

Moossavi, S., Arrieta, M.-C., Sanati-Nezhad, A., & Bishehsari, F. (2022). Gut-on-chip for ecological and causal human gut microbiome research. *Trends in Microbiology*, *30*(8), 710–721. https://doi.org/10.1016/j.tim.2022.01.014

Ndjim, M., Gasmi, I., Herbert, F., Joséphine, C., Bas, J., Lamrani, A., Coutry, N., Henry, S., Zimmermann, V. S., Dardalhon, V., Campillo Poveda, M., Turtoi, E., Thirard, S., Forichon, L., Giordano, A., Ciancia, C., Homayed, Z., Pannequin, J., Britton, C., … Jay, P. (2024). Tuft cell acetylcholine is released into the gut lumen to promote anti-helminth immunity. *Immunity*, *57*(6), 1260–1273.e7. https://doi.org/10.1016/j.immuni.2024.04.018

Papa, A., Santini, P., De Lucia, S. S., Maresca, R., Porfidia, A., Pignatelli, P., Gasbarrini, A., Violi, F., & Pola, R. (2023). Gut dysbiosis-related thrombosis in inflammatory bowel disease: Potential disease mechanisms and emerging therapeutic strategies. *Thrombosis Research*, *232*, 77–88. https://doi.org/10.1016/j.thromres.2023.11.005

Qu, S., Yu, Z., Zhou, Y., Wang, S., Jia, M., Chen, T., & Zhang, X. (2024). Gut microbiota modulates neurotransmitter and gut-brain signaling. *Microbiological Research*, *287*, 127858. https://doi.org/10.1016/j.micres.2024.127858

Sancho-Alonso, M., Sarriés-Serrano, U., Miquel-Rio, L., Yanes Castilla, C., Paz, V., Meana, J. J., Perello, M., & Bortolozzi, A. (2024). New insights into the effects of serotonin on Parkinson's disease and depression through its role in the gastrointestinal tract. *Spanish Journal of Psychiatry and Mental Health*. https://doi.org/10.1016/j.sjpmh.2024.07.002

Tian, D., Zhang, W., Lu, L., Yu, Y., Yu, Y., Zhang, X., Li, W., Shi, W., & Liu, G. (2024). Enrofloxacin exposure undermines gut health and disrupts neurotransmitters along the microbiota-gut-brain axis in zebrafish. *Chemosphere*, *356*, 141971. https://doi.org/10.1016/j.chemosphere.2024.141971

Verma, A., Sharma, A., Kumar, M., Bansal, S., & Shrivastava, R. (2022). Techniques and challenges in studies related with human gut microbiome. In *Human-Gut Microbiome* (pp. 37–57). Elsevier. https://doi.org/10.1016/B978-0-323-91313-3.00006-4

Wang, X., Wen, X., Yuan, S., & Zhang, J. (2024). Gut-brain axis in the pathogenesis of sepsis-associated encephalopathy. *Neurobiology of Disease*, *195*, 106499. https://doi.org/10.1016/j.nbd.2024.106499

Wang, Y., Du, W., Hu, X., Yu, X., Guo, C., Jin, X., & Wang, W. (2023). Targeting the blood–brain barrier to delay aging-accompanied neurological diseases by modulating gut microbiota, circadian rhythms, and their interplays. *Acta Pharmaceutica Sinica B*, *13*(12), 4667–4687. https://doi.org/10.1016/j.apsb.2023.08.009

Wilson, N. G., Hernandez-Leyva, A., Rosen, A. L., Jaeger, N., McDonough, R. T., Santiago-Borges, J., Lint, M. A., Rosen, T. R., Tomera, C. P., Bacharier, L. B., Swamidass, S. J., & Kau, A. L. (2023). The gut microbiota of people with asthma influences lung inflammation in gnotobiotic mice. *IScience*, *26*(2), 105991. https://doi.org/10.1016/j.isci.2023.105991

Xie, W., Sharma, A., Kaushik, H., Sharma, L., Nistha, Anwer, M. K., Sachdeva, M., Elossaily, G. M., Zhang, Y., Pillappan, R., Kaur, M., Behl, T., Shen, B., & Singla, R. K. (2024). Shaping the future of gastrointestinal cancers through metabolic interactions with host gut microbiota. *Heliyon*, *10*(15), e35336. https://doi.org/10.1016/j.heliyon.2024.e35336

Yang, Y., Zhou, S., Xing, Y., Yang, G., & You, M. (2023). Impact of pesticides exposure during neurodevelopmental period on autism spectrum disorders – A focus on gut microbiota. *Ecotoxicology and Environmental Safety*, *260*, 115079. https://doi.org/10.1016/j.ecoenv.2023.115079

Ye, X., Zhang, M., Zhang, N., Wei, H., & Wang, B. (2024). Gut-brain axis interacts with immunomodulation in inflammatory bowel disease. *Biochemical Pharmacology*, *219*, 115949. https://doi.org/10.1016/j.bcp.2023.115949

Zhai, L., Xiao, H., Lin, C., Wong, H. L. X., Lam, Y. Y., Gong, M., Wu, G., Ning, Z., Huang, C., Zhang, Y., Yang, C., Luo, J., Zhang, L., Zhao, L., Zhang, C., Lau, J. Y.-N., Lu, A., Lau, L.-T., Jia, W., … Bian, Z.-X. (2023). Gut microbiota-derived tryptamine and phenethylamine impair insulin sensitivity in metabolic syndrome and irritable bowel syndrome. *Nature Communications*, *14*(1), 4986. https://doi.org/10.1038/s41467-023-40552-y

Zhang, S., Jin, M., Ren, J., Sun, X., Zhang, Z., Luo, Y., & Sun, X. (2023). New insight into gut microbiota and their metabolites in ischemic stroke: A promising therapeutic target. *Biomedicine & Pharmacotherapy*, *162*, 114559. https://doi.org/10.1016/j.biopha.2023.114559

Zheng, B., Li, M., Zhang, T., Li, B., Li, Q., Saiding, Q., Chen, W., Guo, M., Koo, S., Ji, X., & Tao, W. (2024). Functional modification of gut bacteria for disease diagnosis and treatment. *Med*, *5*(8), 863–885. https://doi.org/10.1016/j.medj.2024.06.003

Zhou, Y., Chen, Y., He, H., Peng, M., Zeng, M., & Sun, H. (2023). The role of the indoles in microbiota-gut-brain axis and potential therapeutic targets: A focus on human neurological and neuropsychiatric diseases. *Neuropharmacology*, *239*, 109690. https://doi.org/10.1016/j.neuropharm.2023.109690

Zhou, Y., Zhang, Y., Jin, S., Lv, J., Li, M., & Feng, N. (2024). The gut microbiota derived metabolite trimethylamine N-oxide: Its important role in cancer and other diseases. *Biomedicine & Pharmacotherapy*, *177*, 117031. https://doi.org/10.1016/j.biopha.2024.117031

23 Therapeutic Applications of Prebiotics, Probiotics, and Postbiotics

Surendra Agrawal, Pranita Jirvankar, Sunita Vaidya, Nikhita Chambhare, and Pravina Gurjar

Datta Meghe College of Pharmacy, Datta Meghe Institute of Higher Education & Research, Wardha, Maharashtra, India

23.1 INTRODUCTION

The prebiotics concept is not new in the pharmaceutical domain. It was introduced long ago in 1995 by Glenn Gibson and M. Roberfroid. A prebiotic is defined as "a non-digestible food ingredient that selectively stimulates the growth and/or activity of one or a limited number of bacteria in the colon, thereby improving host health and having a beneficial effect on the host" (Zawistowska-Rojek & Tyski, 2022).

This definition remained generally unchanged for over 15 years. Prebiotics mainly include short- and long-chain β-fructans (inulin as well as fructooligosaccharides (FOS), galactooligosaccharides (GOS), and lactulose. The International Scientific Association of Probiotics and Prebiotics (ISAPP) defined "dietary prebiotics" as "a selectively fermented ingredient that results in specific changes in the composition and/or activity of the gastrointestinal microbiota, thus conferring benefit(s) upon host health" at its sixth meeting in 2008 (Davani-Davari et al., 2019). The prebiotics, probiotics, and postbiotics are compared in Table 23.1.

23.1.1 Prebiotics

Probiotics are supplements or food containing live microorganisms that modify the host's microbiota. Prebiotics are not easily digested like food but benefits the host by selectively increasing bacteria in the colon. These chemicals serve as food for probiotics, promoting their development and activity. Dietary fibers such as GOS, FOS, and inulin are popular prebiotics. Prebiotics may be found in many foods, including whole grains, onions, and garlic. Their primary goal is to enhance gut flora, which promotes nutrient absorption, digestion, and overall gastrointestinal health (Okburan & Kızıler, 2023).

23.1.2 Probiotics

The appropriate administration of probiotics, which are live microbes, gives health advantages to the host. These beneficial bacteria or yeasts occur naturally in yoghurt, sauerkraut, kimchi, and kefir. It can also be found as dietary supplements in other fermented foods. Probiotics serve to balance the intestinal microbiota, limit the growth of harmful microorganisms, boost the immune system, and promote gut health. *Lactobacillus*, *Bifidobacterium*, and *Saccharomyces boulardii* are among the most common probiotic strains. Probiotics can help cure a variety of gastrointestinal problems, boost immunological function, and even improve mental health via the gut–brain axis (Rad et al., 2020; Taghizadeh Ghassab et al., 2024).

23.1.3 Postbiotics

Postbiotics are bacterial products metabolic byproducts that are non-viable and obtained from probiotic microorganisms. Derived from fermentation, postbiotics are secreted by live bacteria or released during bacterial lysis. These exogenous compounds contain microbial cells, cellular constituents, and metabolites (e.g., short-chain fatty acids, peptides, and polysaccharides), as well as cell wall fragments. Postbiotics are not live microorganisms, which offer probiotics some advantages in terms of stability and safety. Postbiotics can have diverse positive effects on human health, such as the ability to enhance immune response capacity and promote decreased inflammation in addition to maintaining gut barrier integrity. They are found abundantly in fermented food and can be used as dietary supplements or pharmacologic agents to prevent or manage a wide range of health conditions (Odriozola et al., 2024; Zolkiewicz et al., 2020).

DOI: 10.1201/9781003536802-23

TABLE 23.1
Comparison of Prebiotics, Probiotics, and Postbiotics

Aspects	Prebiotics	Probiotics	Postbiotics	Ref
Definition	Dietary components that selectively stimulate the growth of beneficial gut microorganisms	Live microorganisms that are administered for improving gut microflora	Non-viable bacterial products or metabolic byproducts produced by probiotics during fermentation	Davani-Davari et al., 2019; Zolkiewicz et al., 2020
Nature and composition	Nondigestible carbohydrates and fibers	Live beneficial microorganisms	Nonliving bioactive compounds produced during fermentation	Odriozola et al., 2024
Mechanism of action	Serve as food for probiotics, enhancing growth	Colonize intestinal tract, modulate immune system, inhibit harmful bacteria	Exerts health benefits through bioactive compounds affecting gut microbiota, immune responses, and cellular functions	Asif et al., 2023
Sources	Fruits, vegetables, whole grains (e.g., garlic, onions, bananas, oats)	Fermented foods (e.g., yogurt, kefir, sauerkraut, kimchi), dietary supplements	Fermented foods can be isolated and purified	Arora et al., 2024
Health benefits	Enhance gut health, boost immune function, improve digestion, and increase mineral absorption	Maintain gut microbiota balance, prevent/treat GI disorders, enhance immune responses, and improve mental health	Provide anti-inflammatory effects, support gut barrier function, modulate immune system, and promote overall health	Zolkiewicz et al., 2020
Stability	Not affected by temperature or acid	Sensitive to temperature and stomach acid	Stable	Federici et al., 2020
Safety	Overconsumption can lead to gastrointestinal discomfort	It can cause infections in immunocompromised individuals	Safe, however, need more studies for administration of large amounts	Canton et al., 2024
Examples	Inulin, galactooligosaccharides (GOS), fructooligosaccharides (FOS), foods like garlic, asparagus, onions, and whole grains	*Lactobacillus*, *Bifidobacterium*, *Saccharomyces*, yogurt, kefir, fermented foods, certain supplements	Propionate, butyrate, and acetate that are short-chain fatty acids (SCFAs) as well as peptides, and polysaccharides	Abd El-Ghany, 2020
Key functions	Increase growth of beneficial bacteria, improve gut health, enhance mineral absorption, and reduce risk of colon cancer	Restore gut microbiome balance, reduce inflammation, boost immune system, and improve mental health	Enhance intestinal barrier function, reduce inflammation, promote mucosal healing, and affect immune system	Chowdhury et al., 2024

23.2 MECHANISM OF ACTION

23.2.1 Prebiotics

Prebiotics have diverse actions mainly mediated by changes in microbial flora of the gut, modulating the action of probiotics and postbiotics, and thereby altering immune responses. These mechanisms are mainly responsible for their therapeutic outcomes in systemic and gastrointestinal disorders, as shown in Figure 23.1.

23.2.2 Probiotics

Probiotics when administered in adequate amounts, confer enrichment in the healthy microflora. Their mechanisms of action are shown in Figure 23.2.

23.2.3 Postbiotics

Postbiotics are bioactive compounds produced by probiotics during fermentation. Their mechanisms of action are shown in Figure 23.3.

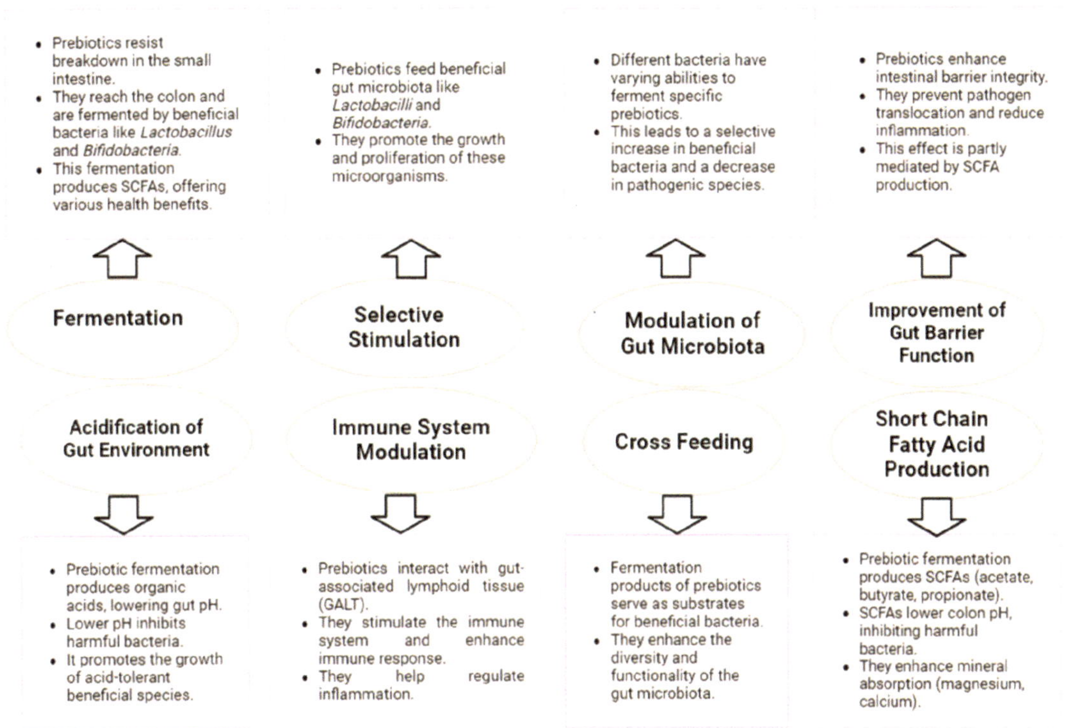

FIGURE 23.1 Mechanism of action of prebiotics.

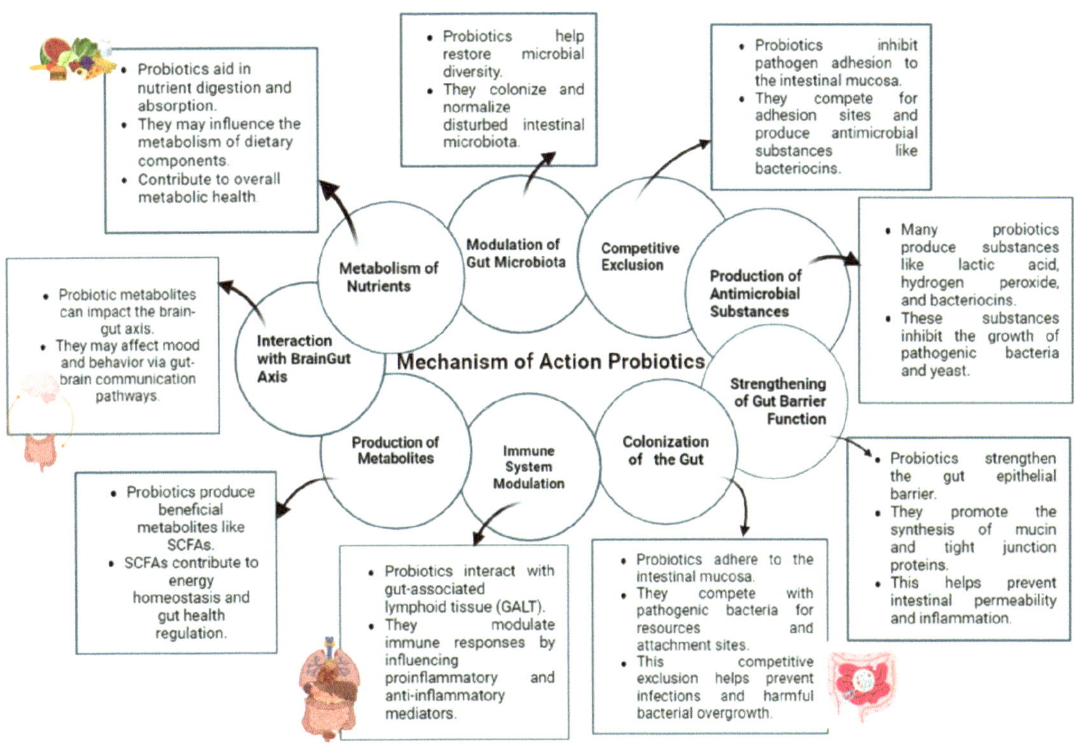

FIGURE 23.2 Mechanism of action of probiotics.

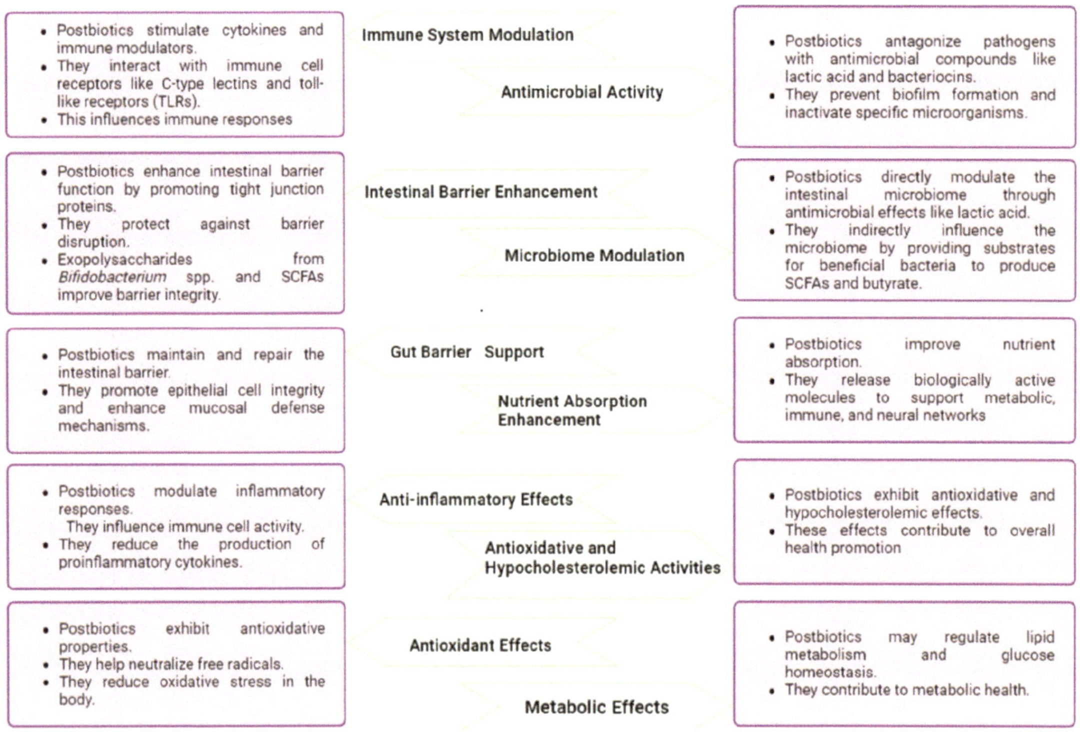

FIGURE 23.3 Mechanism of action of postbiotics.

23.3 SYNERGISTIC EFFECTS OF PREBIOTICS, PROBIOTICS, AND POSTBIOTICS

A balanced microbiome is responsible for overall health, which can be built through the use of prebiotics, probiotics, and postbiotics. These agents help improve the body's physiological processes. The combination of all three is preferable because each of them improves the function of the other. The role of prebiotics provides important substrates for probiotics, stimulating their growth and activity in the gut. In addition, the prebiotics are metabolized to release some byproducts after fermentation in the presence of probiotics, for example, *Lactobacillus* and *Bifidobacterium*. The main metabolites include SCFAs and lactate. These metabolites nourish the gut, improve intestinal integrity, and regulate immune responses. SCFAs and other bioactive compounds inhibit inflammation by giving a protective effect. Studies have demonstrated the synergistic effects of this multifaceted approach in the treatment of inflammatory bowel disease (IBD) and metabolic disorders.

Probiotics produce postbiotics that further support gut health, resulting in a more effective modulation of gut microbiota and improved gastrointestinal function. Postbiotics are stable as compared to live microorganisms, which can be challenging to maintain in some formulations offering sustained therapeutic effects. Prebiotics promote beneficial bacteria that produce postbiotics, enhancing mucosal immunity. The synergistic effects of prebiotics, probiotics, and postbiotics can influence various aspects of health, such as intestinal homeostasis, immune system regulation, and metabolic processes, thus enabling a holistic approach to therapeutic interventions (Chowdhury et al., 2024; Sharma et al., 2023; Wei et al., 2022).

23.4 HEALTH BENEFITS OF PREBIOTICS, PROBIOTICS, AND POSTBIOTICS

23.4.1 Gut Health

Maintaining intestinal health and a balanced and functional gastrointestinal system relies heavily on prebiotics, probiotics, and postbiotics. Digestion can be optimized by promoting the growth of beneficial microorganisms. Symptoms of constipation and diarrhea are alleviated and regular bowel movements are promoted through improved nutrient absorption and increased digestive efficiency. Prebiotics help symbiotic intestinal microbes proliferate and maintain the microbiota by acting as a substrate for these microbes. Probiotics deliver colony-forming units to the intestine, which directly contributes to maintaining homeostasis. Probiotics have been studied as potential treatment options for gastrointestinal disorders that lead to a variety of intestinal symptoms, including stomach ulcers, IBD, and diarrhea in children. To treat these diseases, their ability to control inflammatory processes and restore microbial balance is crucial (Cuevas-Sierra et al., 2019).

23.4.2 Metabolic Health

The body's ability to regulate lipid levels and energy balance is an inherent part of metabolic health. The body weight and fat composition of the body are altered by modulating appetite-stimulating hormones and improving fat metabolism. The energy expenditure is grossly affected and subsequently the fat storage. Probiotics and prebiotics lower LDL cholesterol and triglyceride levels. Postbiotics also contribute to cardiovascular health and lipid metabolism. They regulate blood sugar levels by improving insulin sensitivity. The person developing metabolic disorders and having diabetes is beneficiary of prebiotics, probiotics, and postbiotics (Park et al., 2022).

23.4.3 Immune Boost

The complex immune system regulates the physiological processes and the diseases caused by infections. The gut-associated lymphoid tissue (GALT) plays a crucial role and is regulated by prebiotics and postbiotics. The mucosal surface of the intestinal mucosa is protected mainly by the development of T cells and macrophages and increased production of immunoglobulin A (IgA). This is particularly beneficial in inflammatory diseases. They reduce the inflammatory cytokines while increasing the anti-inflammatory cytokines, thereby alleviating chronic inflammatory conditions and autoimmune diseases. Probiotics produce antimicrobial substances (Odriozola et al., 2024; Zawistowska-Rojek & Tyski, 2022).

23.4.4 Skin Health

Postbiotics and probiotics promote healthy skin appearance as they possess anti-inflammatory properties. It reduces skin inflammation and irritation and is thereby useful in treating acne, eczema, and psoriasis. They synthesize skin-protective molecules and modulate the systemic inflammatory processes, nourish the skin by increasing moisture content and providing elasticity (Callewaert et al., 2021).

23.4.5 Other Benefits

Probiotics and prebiotics can alleviate mood and symptoms of depression and anxiety. This is possible through their interaction with the gut microbiota and thereby alteration in the production of neurotransmitters. They are helpful in gingivitis and periodontitis due to their protective action against harmful bacteria in the oral cavity. Certain probiotics and postbiotics have the ability to modulate immune responses and reduce the severity of allergic reactions and asthma symptoms. By promoting a balanced microbiota, improving metabolic functions and optimizing immune responses, prebiotics, probiotics, and postbiotics contribute to overall health and well-being (Wuethrich et al., 2021).

23.5 THERAPEUTIC APPLICATIONS OF PREBIOTICS, PROBIOTICS, AND POSTBIOTICS

Prebiotics, probiotics, and postbiotics are used to treat numerous diseases and disorders, as shown in Figure 23.4. They are useful in treating chronic inflammatory diseases such as rheumatoid arthritis and IBD. They inhibit the production of inflammatory cytokines. Production of protective metabolites such as SCFAs reduces the risk of cancer. Reduction in blood pressure has been observed due to reduction in systemic inflammation. Reduction in lipid levels is associated with usefulness in cardiovascular health and prevention of cardiovascular diseases. The significant contribution in irritable bowel syndrome (IBS) and IBD is noted in restoring balanced gut microbiota. The

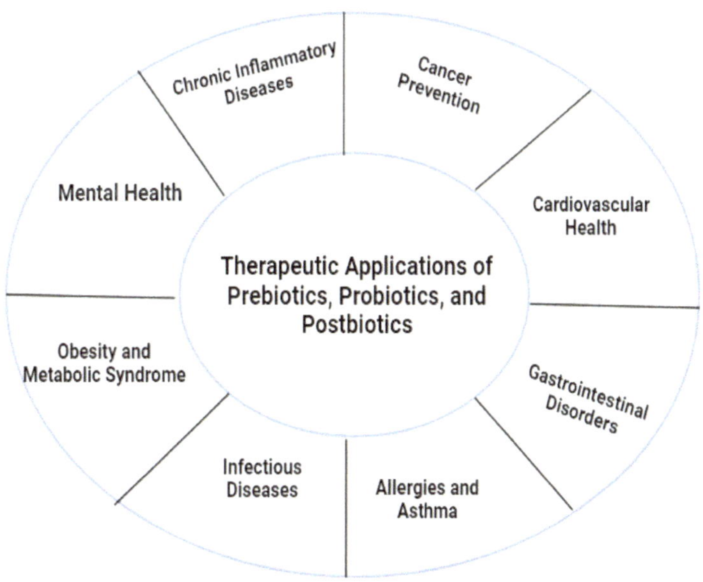

FIGURE 23.4 Therapeutic applications of prebiotics, probiotics, and postbiotics.

TABLE 23.2
Safety and Appropriate Dosage of Prebiotics, Probiotics, and Postbiotics

Types	Safety/Dosage	Description	Reference
Prebiotics	Safety	Generally considered safe, prebiotics can cause gastrointestinal discomfort, such as bloating and gas, if consumed in excessive amounts	Marteau & Seksik, 2004
	Dosage	The recommended dosage varies depending on the type of prebiotic. For example, inulin is typically consumed at doses 5–20 grams per day	
Probiotics	Safety	Most probiotics are safe for the general population. However, immunocompromised individuals and those with severe illnesses should consult healthcare professionals before use	Marteau & Seksik, 2004
	Dosage	Probiotic dosage is measured in colony-forming units (CFUs). Effective doses usually range from 1 billion to 10 billion CFUs per day, depending on the strain and condition being treated	
Postbiotics	Safety	Postbiotics are generally safe and well-tolerated as they do not contain live microorganisms, reducing the risk of infections	Cuevas-González et al., 2020
	Dosage	There is no standardized dosage for postbiotics yet, as research is still emerging. However, doses similar to those of probiotics are often used in studies	

anti-inflammatory markers such as cytokines help in treating allergies and asthma. The protection and restoration of the mucosal lining is associated with the production of antimicrobial peptides that inhibit the growth of pathogenic microorganisms. Probiotics and prebiotics can help regulate weight by reducing hunger. They also increase insulin sensitivity and burn excess lipid by promoting gluconeogenesis. The gut–brain axis helps modulate mood and cognitive functions by altering neurotransmitter levels. This leads to a reduction in symptoms of depression and anxiety (Cang et al., 2022; Gao et al., 2021; Lee et al., 2024).

23.6 EFFICACY AND SAFETY OF PREBIOTICS, PROBIOTICS, AND POSTBIOTICS

The multiple clinical trials are being conducted to standardize the efficacy and safety of prebiotics, probiotics, and postbiotics. Although these are safe but studies are still ongoing and reported for the dosage for achieving better therapeutic outcomes, as listed in Table 23.2.

23.7 CLINICAL EVIDENCE, RESEARCH, AND OVERVIEW OF CLINICAL TRIALS

Clinical trials investigating prebiotics, probiotics, and postbiotics, as listed in Table 23.3, typically involve controlled experiments on human subjects to assess the effects of these compounds on specific outcomes.

23.8 SYNBIOTICS

Synbiotics are a combination of prebiotics and probiotics designed to synergistically enhance each other's effects. Synbiotics support the growth and activity of beneficial gut bacteria. The use of rationale synbiotics (deciding combinations) is rapidly evolving as there is availability of a large number of prebiotics and probiotics. The mode of delivery to the patient for better efficacy is a major challenge that can be resolved with innovations in encapsulation.

23.9 CONCLUSION AND FUTURE DIRECTIONS

The prebiotics, probiotics, and postbiotics are emerging in the prophylaxis and treatment of a number of disorders and diseases, respectively. The advancements in their rational combination with delivery modes can enhance the stability, viability, and bioavailability. The use of nanotechnology can be an alternative and ray of hope in providing better therapeutic outcomes. The tailormade interventions with the use of these agents can assist in relieving multiple disorders that are not treated with medicines. The new strains and associated metabolites could be better therapeutic interventions and need more research.

TABLE 23.3
Clinical Evidence for Prebiotics, Probiotics, and Postbiotics as Microbial Therapeutics

			Prebiotics		
Study Design	**Condition**	**Population**	**Prebiotic**	**Results**	**Reference**
Clinical trial	Asthma	Children with asthma (4–7 years old)	*Lactobacillus rhamnosus* GG (LGG)	Reduced concentration of exhaled nitric oxide; no symptom reduction observed with other strains like *Lactobacillus reuteri*	Wu et al., 2019
	Eczema, atopic dermatitis	Pregnant women and their children (up to age 2)	*Bifidobacterium breve, Lactobacillus rhamnosus, Propionibacterium freudenreichii*	Reduction in the chances of atopic dermatitis in children up to age 2	Sivamaruthi et al., 2018
	Food allergy (cow's milk)	Confirmed cow's milk allergy (220 children)	*Lactobacillus rhamnosus* GG with highly hydrolyzed casein	Accelerated adaption to tolerance toward cow's milk protein; increased threshold for allergies	Canani et al., 2017
	Alzheimer's disease (AD)	Alzheimer's disease (adults).	*Lactobacillus casei, Lactobacillus acidophilus, Lactobacillus fermentum, Bifidobacterium bifidum*	Improved cognitive functions after three months of supplementation; mixed results depending on disease stage	Mancuso & Santangelo, 2018
	Parkinson's disease (PD)	Parkinson's patients	*Lactobacillus reuteri, Lactobacillus acidophilus, Bifidobacterium bifidum, Lactobacillus fermentum*	Improved movement, increased glutathione concentration, reduced CRP, and improved insulin sensitivity after 12 weeks	Amani & Mehrabani, 2023
	Peanut allergy	Children	*Lactobacillus rhamnosus* CGMCC1.3724 with peanut oral immunotherapy (PPOIT)	89.7% desensitization in PPOIT group versus 7.1% in placebo group; 67% could safely eat peanuts after four years	Reinwald et al., 2022
	Multiple sclerosis (MS)	Patients with MS	Various strains of *Bifidobacterium* and *Lactobacillus*	Reduced dysbiosis, induced anti-inflammatory response, and reduced pro-inflammatory immune pathway gene expression	Altieri et al., 2023
Meta-analysis	Atopic dermatitis	Prenatal and postnatal women	Prebiotics (*Lactobacillus rhamnosus* GG)	Effective in preventing, but not treating, atopic dermatitis	Voigt & Lele, 2022
	Hypercholesterolemia	High, borderline, and normal cholesterol levels (1971 patients)	*Lactobacillus acidophilus, Lactobacillus plantarum, Lactococcus lactis*	Significantly reduced total serum cholesterol and LDL fraction	Abdi et al., 2022
Double-masked, placebo-controlled trial	Common acute pediatric illnesses	140 healthy infants and toddlers (4–24 months)	Oligofructose (0.55 g per 15 g of dry cereal)	Similar cereal consumption, normal growth, well-tolerated. Lower emesis, regurgitation, and bowel discomfort. Reduced fever during diarrhea episodes	Nakamura et al., 2006
	Response to measles vaccination	Healthy infants (50) (79 months)	Oligofructose and inulin (1 g per 25 g dry cereal)	Significant increase in anti-measles IgG levels post-vaccination. No significant difference in growth or anti-measles IgM levels	Saavedra & Tschernia, 2002
Double-masked trial	Acute gastroenteritis	Healthy children (58)	*Lactobacillus rhamnosus* and oligofructose	Significantly shortened duration of diarrhea in the supplemented group	Saavedra & Tschernia, 2002
Multicenter, double-masked controlled trial	Mild to moderate malnutrition	children (1–6 years old) (626)	*Bifidobacterium infantis, Lactobacillus acidophilus*, and oligofructose.	Significantly fewer sick days, particularly in children aged 3–5 years. Overall decrease in sick days as nutritional status improved	Heuven et al., 2021

(Continued)

TABLE 23.3 (CONTINUED)

Probiotic

Study Design	Condition	Population	Probiotic	Results	
Randomized controlled trial (RCT)	Atopic dermatitis	Pregnant women, infants (prenatal and postnatal)	*Lactobacillus paracasei, Lactobacillus rhamnosus, Lactobacillus reuteri, Bifidobacterium, Lactobacillus acidophilus, Propionibacterium*	Reduced incidence of dermatitis; effectiveness varied by strain; *Lactobacillus rhamnosus* and *Lactobacillus paracasei* were effective	Xie et al., 2023
Meta-analysis	Pediatric acute infectious diarrhea	Children	*Saccharomyces boulardii, Lactobacillus rhamnosus* GG (LGG)	Shortened diarrhea duration by one day; most effective at high doses (≥10^{10} CFU/day); strain-dependent effectiveness	Huang et al., 2021
Systematic review	Antibiotic-associated diarrhea (AAD)	Children, younger adults (18–64 years)	*Saccharomyces boulardii, Lactobacillus rhamnosus* GG (LGG)	Reduced AAD risk; more effective when started within two days of antibiotic use; not effective in adults ≥65 years	Szajewska & Kołodziej, 2015
Systematic review with meta-analysis	IBD	Patients suffering from Crohn's disease and ulcerative colitis.	*Saccharomyces boulardii, Escherichia coli* Nissle 1917, LGG, *Lactobacillus johnsonii*, VSL#3	Modest benefits in ulcerative colitis when combined with conventional therapy; not effective for Crohn's disease	Chen et al., 2021
Systematic review	IBS	Adults with IBS	*B. breve, B. longum, S., salivarius, L. acidophilus, L. plantarum, L. casei*	Modest symptom reduction; effectiveness varied by strain and symptom; multi-strain products more effective	Jonkers et al., 2012
Randomized controlled trials (RCTs)	Hypercholesterolemia	Adults with elevated cholesterol levels	*Lactobacillus acidophilus, Bifidobacterium lactis, Lactobacillus plantarum, Enterococcus faecium*	Significantly reduced LDL and total cholesterol; greater benefits with multi-strain products, longer duration, and higher baseline cholesterol	Mo et al., 2019
RCTs	Obesity	Overweight/obese adults	*Lactobacillus rhamnosus, Lactobacillus gasseri*	*Lactobacillus gasseri* reduced visceral fat, BMI, waist/hip circumference; *Lactobacillus rhamnosus* reduced body weight; mixed results overall	Barengolts, 2016
RCT	Acute infectious diarrhea	Children	Probiotic strains like *S. boulardii, Lactobacillus* GG	Reduced infection duration and intensity; accepted therapy in Europe for children	Canani et al., 2007
Systematic review	Antibiotic-associated diarrhea	Various populations	*Lactobacillus* GG; *Saccharomyces boulardii*	Probiotics may reduce AAD risk; mixed results on efficacy for *Clostridium difficile* infection	Blaabjerg et al., 2017
Meta-analysis	Antibiotic-associated diarrhea	Children and elderly	*Lactobacillus casei, Lactobacillus* GG, DN-114 001	Probiotics may reduce risk of AAD; inconsistent findings for CDAD prevention	McFarland & Goh, 2013
RCT	Rotavirus diarrhea	Children	Probiotic strains *L. rhamnosus* GG, *L. casei Shirota, Lactobacillus reuteri*	Reduced duration of rotavirus diarrhea by approximately one day	Guandalini & Sansotta, 2017
	Travellers' diarrhea	Travelers	*Lactobacillus* GG, galactooligosaccharide, prebiotics like fructo-oligosaccharide	Mixed results; some studies show prevention benefits, others do not	McFarland & Goh, 2019
	Allergic diseases	Preterm infants	Neutral and acid oligosaccharides	No difference in incidence of allergic diseases compared to placebo	Westerbeek et al., 2008
	Allergic diseases	Infants at risk for allergy	Short-chain GOS and long-chain FOS	Reduced cumulative incidence of atopic dermatitis	Rao et al., 2009
	Respiratory tract infections	Healthy infants	Prebiotic formula with fructo oligosaccharide and galactooligosaccharide	Fewer episodes of infections; no impact on severity or duration of symptomatic rhinovirus infections	Rashidi et al., 2021
	Respiratory tract infections	Preterm infants	Probiotic *L. rhamnosus* GG, prebiotics galacto- and polydextrose	Lower incidence of respiratory infections; no effect on severity or duration	Rashidi et al., 2021

(Continued)

TABLE 23.3 (CONTINUED)

	Cardiovascular disease (CVD)	Smokers	Probiotic *Lactobacillus plantarum* 299v	Reduced systolic blood pressure, leptin, and fibrinogen levels	Sun & Buys, 2015
	Hypercholesterolemia	Hypercholesterolemic individuals	Yogurt with *Lactobacillus acidophilus* and *B. lactis*	Significant reduction in blood cholesterol rates	Chaiyasut et al., 2019
	CVD	Normocholesterolemic men	*B. animalis* Bb-12, *Lactobacillus acidophilus, S. thermophilus*	Reduced LDL-cholesterol levels and improved LDL ratio	da Silva Pontes et al., 2021
	Hypercholesterolemia	Obese and hypercholesterolemic subjects	Inulin (7g/day)	Increase in good cholesterol and reduction in bad lipids	Talukdar et al., 2022
	Hypercholesterolemia	Mildly hypercholesterolemic individuals	Short-chain FOS (10.6 g/day)	No significant effects on lipid metabolism	Liu et al., 2017
Double-blind RCT	Cardiovascular disease (CVD)	Overweight and obese patients	Probiotic cultures CAUSIDO® (*Enterococcus faecium, S. thermophilus*)	Reduced LDL-cholesterol levels; increased fibrinogen rates (within normal range)	Mohammadi et al., 2021
Randomized, double-blind, placebo-controlled	Calcium absorption and bone mineralization	Calcium absorption and bone mineralization	40 g inulin daily for 26 days	Calcium absorption	Franck, 2006
	Helicobacter pylori infection	Patients with *H. pylori*	*L. johnsonii* La1	Reduced gastritis intensity and pathogen density; effects persisted after treatment	Szajewska et al., 2009
	Helicobacter pylori infection	Patients with *H. pylori*	*L. reuteri* ATCC 55730	Significant reduction in gastrointestinal symptoms during *H. pylori* eradication therapy	Francavilla et al., 2014
	Calcium absorption	Menopausal women	10 g FOS daily for five weeks	Consumption of sc-FOS for 36 days stimulated magnesium absorption in girls but did not affect calcium absorption	van den Heuvel et al., 2009
	Allergic diseases	113 preterm infants	Neutral and acid oligosaccharides		Westerbeek et al., 2010
	Osteoporosis	Menopausal women	10 g of FOS daily for five weeks	No significant improvement in Ca absorption compared to placebo group	Geraghty, 2006
	Urogenital infections	Non-pregnant women	*L. rhamnosus* and *B. longum* (oral), *L. rhamnosus, L. delbrueckii, L. acidophilus, S. thermophilus* (intravaginal)	No effectiveness in preventing infections compared to control	Ballini et al., 2019
	Periodontal disease	Volunteers with gingivitis	*L. reuteri* strains (LR-1 or LR-2) in chewing gum	Reduced dental plaque index scores, colonization observed	
	Periodontal disease	Patients with chronic periodontitis	*L. brevis* CD2	Improved plaque and gingival index scores, reduced bleeding	Lee et al., 2015
	Ulcerative colitis (UC)	Patients with acute UC	VSL#3® (*L. casei, L. plantarum, L. delbrueckii* ssp. *bulgaricus, L. acidophilus, B. longum, B. breve, B. infantis, S. thermophilus*)	Improved intestinal dendritic cell function, increased regulatory cytokines, reduced pro-inflammatory cytokines, and TLR expression	Panetta et al., 2020
Randomized, placebo-controlled	Halitosis	Patients with malodorous breath	*S. salivarius* K12	Reduced levels of volatile sulfur compounds and bacterial populations responsible for halitosis	He et al., 2020
	Halitosis	Patients with malodorous breath	*L. reuteri* DSM 17938 and *L. reuteri* PTA 5289 (chewing gum)	No change in volatile sulfur compounds, but improved organoleptic scores	Musić, 2021
Meta-analysis of eight clinical trials	Crohn's disease (CD)	Patients with CD	Probiotics (general)	Probiotics were not effective in maintaining remission or preventing recurrence of CD	Jia et al., 2018

(Continued)

TABLE 23.3 (CONTINUED)

Meta-analysis of 23 RCTs	Ulcerative colitis (UC)	Total of 1,763 subjects	VSL#3®	Most effective for UC management; less effective for CD	Mardini & Grigorian, 2014
Review of clinical trials	Irritable bowel syndrome (IBS)	Various studies	*Lactobacillus* strains (general)	Probiotics showed mixed results; some improvement in symptoms was observed	Dale et al., 2019
Multicentre, randomized study	IBS	Patients with IBS	Flortec (prebiotics + *Lactobacillus paracasei* B21060)	Reduced bowel movements, abdominal pain, and IBS scores in IBS-predominant diarrhea	Andriulli et al., 2008
Various trials (23 trials with 24 unique effects)	Depression	Participants in multiple studies	Lactobacilli, *Bifidobacterium longum, Bifidobacterium, Bacillus coagulans*	Lower depression scores with probiotics compared to placebo. Significant effect in trials lasting >1 month ($d = -0.28$), but not ≤1 month ($d = -0.10$). Larger effect in clinical/medical samples ($d = -0.45$) compared to community samples ($d = -0.09$). No publication bias detected	Romijn et al., 2017
Various trials (22 trials with 23 unique effects)	Anxiety	Participants in multiple studies	Lactobacilli, *Bifidobacterium, Bacillus coagulans*	Lower anxiety scores with probiotics compared to placebo. No significant heterogeneity or publication bias. Overall effect size ($d = -0.11$) remained stable with slight asymmetry	Liu et al., 2019
Postbiotics					
Randomized, double-blind, placebo-controlled trial	IBS	Patients with IBS (443)	*Bifidobacterium bifidum* MIMBb75	Compared to placebo: risk ratio 1.7, 95% CI 1.3–2.4; $p = 0 \cdot 0007$) *B. bifidum* reduced IBS and its symptoms	Andresen et al., 2020
	Seasonal rhinitis due to allergy	Subjects (adults and children) (90)	Heat-killed *Lactobacillus paracasei* LP33	Following a 30-day treatment period, the participants who received either live or heat-killed LP33 showed a significant enhancement in their quality of life. This enhancement was observed in relation to the frequency and intensity of discomfort, as compared to the group that received a placebo. The effectiveness of HK-LP33 was similar to that of the live form	Costa et al., 2014
	Upper respiratory tract infections (URTI)	Healthy subjects with high psychological stress levels (78)	Heat-killed *Lactobacillus plantarum* L-137	The HK L-137 group had a significantly decreased incidence of URTI compared to the control group ($p = 0.011$). An interesting correlation has been shown between the length of time that HK L-137 is taken and the occurrence, length, intensity, and duration of URTI. The HK L-137 group exhibited a significantly higher proliferation of peripheral blood mononuclear cells triggered by Concanavalin A compared to the control group ($p = 0.048$)	Hirose et al., 2013

(Continued)

TABLE 23.3 (CONTINUED)

	Influence on levels of salivary secretory IgA	Elderly participants (80)	Heat-killed *Lactobacillus pentosus* b240	By week 4, salivary secretory immunoglobulin A (SIgA) levels had increased by 20% and remained steady until week 12. The B240 group had a considerably faster growth rate of SIgA in comparison to the placebo group	Shimizu et al., 2014
	Obesity-related disorders	Overweight/obese adults with insulin resistance (32)	Pasteurized *Akkermansia muciniphila*	In the group that received pasteurized *A. muciniphila*, there was a significant increase in insulin sensitivity by 28.62%. Additionally, insulinemia decreased by 34.08%, plasma total cholesterol decreased by 8.68%, and body weight decreased by 2.27 kg. These findings indicate positive changes in various health markers. The hip circumference decreased by 2.63 cm and the fat mass reduced by 1.37 kg compared to the initial measurements	Payahoo et al., 2019
	Viral respiratory tract infections	Children aged 3–6 years (172)	Heat-killed *Pediococcus acidilactici* K15	When compared to the placebo group, the K15 group demonstrated a substantial maintenance of higher salivary IgA levels. K15 revealed that youngsters who ate little to no fermented food had a significantly shorter fever	(Hishiki et al., 2020)
Multicenter, randomized, double-blind, placebo-controlled trial	Prevention of diarrhea	Children aged 12–48 months (126)	Cow's milk fermented with heat-treated *Lactobacillus paracasei*	Children receiving *L. paracasei* had a significantly lower proportion of acute gastroenteritis episodes (p <0.05) compared to the placebo group and were 22% less likely to experience diarrhea (p <0.01)	(Kumar et al., 2023)
	Common infectious diseases	Children aged 12–48 months (126).	Cow's milk fermented with *Lactobacillus paracasei* CBA L74.	In the L74 group, a significant number of children encountered common infectious diseases, whereas in the placebo group, the occurrence of such diseases was higher. There was a significant decrease in the occurrence of URTI among children in the L74 group compared to the placebo group (51% vs. 74%, $p < 0.05$)	Corsello et al., 2017
Preclinical, animal model	Microbiota composition and function	Animal model	Heat-inactivated *Lacticaseibacillus paracasei* N1115	Partially restored gut microbiota disrupted by antibiotic exposure	Vinderola et al., 2022
	Feed efficiency, cholesterol, and creatinine levels	Yellow-feathered broilers	Heat-inactivated *Bacillus subtilis* (strain not declared) and *Lactobacillus acidophilus* BFI	Enhanced feed efficiency, decreased plasma cholesterol and creatinine, and changes in cecal microbiota	Vinderola et al., 2022

(Continued)

TABLE 23.3 (CONTINUED)

RCT	Stress-associated symptoms, gut symptoms	Healthy young adults and IBS patients	Heat-inactivated *Lactobacillus gasseri* CP2305	Reduced anxiety and sleep disturbances; attenuated stress-induced changes in microbiota	Nishida et al., 2017; Vinderola et al., 2022
	Respiratory tract infections	Healthy children aged 3–6 years	Heat-killed *Pediococcus acidilactici* K15	Not effective for preventing respiratory tract infections	Hishiki et al., 2020; Vinderola et al., 2022
	Rhinovirus load	Healthy adults (aged 18–65 years)	Juice with live or heat-inactivated *L. rhamnosus* GG	No effect on rhinovirus load	Vinderola et al., 2022; Wegh et al., 2019
	Influenza incidence and symptoms	Healthy adults aged 20–59 years	Heat-killed L*evilactobacillus brevis* KB290 with β-carotene	Reduced influenza incidence in subjects aged <40 years	Satomi et al., 2021; Vinderola et al., 2022
	Innate and acquired immunity	Healthy adults	Heat-killed *L. plantarum* L-137	Enhanced innate immunity and Th1-related immune functions	(Nishihira et al., 2016; Vinderola et al., 2022)

BIBLIOGRAPHY

Abd El-Ghany, W. A. (2020). Paraprobiotics and postbiotics: Contemporary and promising natural antibiotics alternatives and their applications in the poultry field. *Open Veterinary Journal*, *10*(3), 323–330. https://doi.org/10.4314/ovj.v10i3.11

Abdi, M., Ghaleh, H. E. G., & Ranjbar, R. (2022). Lactobacilli and Bifidobacterium as anti-atherosclerotic agents. *Iranian Journal of Basic Medical Sciences*, *25*(8), 934. https://doi.org/10.22038/IJBMS.2022.63860.14073

Altieri, C., Speranza, B., Corbo, M. R., Sinigaglia, M., & Bevilacqua, A. (2023). Gut-microbiota, and multiple sclerosis: background, evidence, and perspectives. *Nutrients*, *15*(4), 942. https://doi.org/10.3390/nu15040942

Amani, R., & Mehrabani, S. (2023). Prebiotics and probiotics and Parkinson's disease. In *Treatments, Nutraceuticals, Supplements, and Herbal Medicine in Neurological Disorders* (pp. 641–673). Elsevier. https://doi.org/10.1016/B978-0-323-90052-2.00034-2

Andresen, V., Gschossmann, J., & Layer, P. (2020). Heat-inactivated Bifidobacterium bifidum MIMBb75 (SYN-HI-001) in the treatment of irritable bowel syndrome: A multicentre, randomised, double-blind, placebo-controlled clinical trial. *The Lancet Gastroenterology & Hepatology*, *5*(7), 658–666. https://doi.org/10.1016/S2468-1253(20)30056-X

Andriulli, A., Neri, M., Loguercio, C., Terreni, N., Merla, A., Cardarella, M. P., Federico, A., Chilovi, F., Milandri, G. L., & De Bona, M. (2008). Clinical trial on the efficacy of a new symbiotic formulation, Flortec, in patients with irritable bowel syndrome: A multicenter, randomized study. *Journal of Clinical Gastroenterology*, *42*, S218–S223. https://doi.org/10.1097/MCG.0b013e31817fadd6

Arora, R., Kaur, R., Babbar, R., Dhingra, S., Dhingra, A. K., & Grewal, A. S. (2024). Evolving advances in the cosmetic use of probiotics and postbiotics: Health, regulatory and marketing aspects. *Current Pharmaceutical Biotechnology*, *25*(11), 1349–1361. https://doi.org/10.2174/1389201024666230703115732

Asif, A., Afzaal, M., Shahid, H., Saeed, F., Ahmed, A., Shah, Y. A., Ejaz, A., Ghani, S., Ateeq, H., & Khan, M. R. (2023). Probing the functional and therapeutic properties of postbiotics in relation to their industrial application. *Food Science & Nutrition*, *11*(8), 4472–4484. https://doi.org/10.1002/fsn3.3465

Ballini, A., Gnoni, A., De Vito, D., Dipalma, G., Cantore, S., Gargiulo Isacco, C., Saini, R., Santacroce, L., Topi, S., & Scarano, A. (2019). Effect of probiotics on the occurrence of nutrition absorption capacities in healthy children: A randomized double-blinded placebo-controlled pilot study. *European Review for Medical and Pharmacological Sciences*, *23*(19), 8645–8657. https://dx.doi.org/10.26355/eurrev_201910_19182

Barengolts, E. (2016). Gut microbiota, prebiotics, probiotics, and synbiotics in management of obesity and prediabetes: Review of randomized controlled trials. *Endocrine Practice*, *22*(10), 1224–1234. https://doi.org/10.4158/EP151157.RA

Blaabjerg, S., Artzi, D. M., & Aabenhus, R. (2017). Probiotics for the prevention of antibiotic-associated diarrhea in outpatients—A systematic review and meta-analysis. *Antibiotics*, *6*(4), 21. https://doi.org/10.3390/antibiotics6040021

Callewaert, C., Knodlseder, N., Karoglan, A., Guell, M., & Paetzold, B. (2021). Skin microbiome transplantation and manipulation: Current state of the art. *Computational and Structural Biotechnology Journal*, *19*, 624–631. https://doi.org/10.1016/j.csbj.2021.01.001

Canani, R. B., Cirillo, P., Terrin, G., Cesarano, L., Spagnuolo, M. I., De Vincenzo, A., Albano, F., Passariello, A., De Marco, G., & Manguso, F. (2007). Probiotics for treatment of acute diarrhoea in children: Randomised clinical trial of five different preparations. *BMJ*, *335*(7615), 340. https://doi.org/10.1136/bmj.39272.581736.55

Canani, R. B., Di Costanzo, M., Bedogni, G., Amoroso, A., Cosenza, L., Di Scala, C., Granata, V., & Nocerino, R. (2017). Extensively hydrolyzed casein formula containing Lactobacillus rhamnosus GG reduces the occurrence of other allergic manifestations in children with cow's milk allergy: 3-year randomized controlled trial. *Journal of Allergy and Clinical Immunology*, *139*(6), 1906–1913.e1904. https://doi.org/10.1016/j.jaci.2016.10.050

Cang, W., Wu, J., Ding, R., Wang, W., Li, N., Shi, H., Shi, L., Lee, Y., & Wu, R. (2022). Potential of probiotics as an adjunct for patients with major depressive disorder. *Molecular Nutrition & Food Research*, *66*(10), e2101057. https://doi.org/10.1002/mnfr.202101057

Canton, R., De Lucas Ramos, P., Garcia-Botella, A., Garcia-Lledo, A., Hernandez-Sampelayo, T., Gomez-Pavon, J., Gonzalez Del Castillo, J., Martin-Delgado, M. C., Martin Sanchez, F. J., Martinez-Selles, M., Molero Garcia, J. M., Moreno Guillen, S., Rodriguez-Artalejo, F. J., Reigadas, E., Del Campo, R., Serrano, S., Ruiz-Galiana, J., & Bouza, E. (2024). Human intestinal microbiome: Role in health and disease. *Revista Española de Quimioterapia*. https://doi.org/10.37201/req/056.2024

Chaiyasut, C., Sivamaruthi, B. S., Sirilun, S., Tirawat, Y., & Sittiprapaporn, P. (2019). A preliminary study on effect of lactobacillus paracasei on high-density and low-density lipoproteins level in hypercholesterolemic patients. In *2019 16th International Conference on Electrical Engineering/Electronics, Computer, Telecommunications and Information Technology (ECTI-CON)*. https://doi.org/10.1109/ECTI-CON47248.2019.8955200

Chen, M., Feng, Y., & Liu, W. (2021). Efficacy and safety of probiotics in the induction and maintenance of inflammatory bowel disease remission: A systematic review and meta-analysis. *Annals of Palliative Medicine*, *10*(11), 118211829. https://doi.org/10.21037/apm-21-2996

Chowdhury, M., Raj Chaudhary, N., Kaur, P., Goyal, A., & Sahu, S. K. (2024). Different strategies targeting gut microbiota for the management of several disorders: A sustainable approach. *Infectious Disorders - Drug Targets*, *24*(5), e160124225675. https://doi.org/10.2174/0118715265267536231121095634

Corsello, G., Carta, M., Marinello, R., Picca, M., De Marco, G., Micillo, M., Ferrara, D., Vigneri, P., Cecere, G., & Ferri, P. (2017). Preventive effect of cow's milk fermented with Lactobacillus paracasei CBA L74 on common infectious diseases in children: A multicenter randomized controlled trial. *Nutrients*, *9*(7), 669. https://doi.org/10.3390/nu9070669

Costa, D., Marteau, P., Amouyal, M., Poulsen, L. K., Hamelmann, E., Cazaubiel, M., Housez, B., Leuillet, S., Stavnsbjerg, M., & Molimard, P. (2014). Efficacy and safety of the probiotic Lactobacillus paracasei LP-33 in allergic rhinitis: A double-blind, randomized, placebo-controlled trial (GA2LEN Study). *European Journal of Clinical Nutrition*, *68*(5), 602–607. https://doi.org/10.1038/ejcn.2014.13

Cuevas-González, P., Liceaga, A., & Aguilar-Toalá, J. (2020). Postbiotics and paraprobiotics: From concepts to applications. *Food Research International*, *136*, 109502. https://doi.org/10.1016/j.foodres.2020.109502

Cuevas-Sierra, A., Ramos-Lopez, O., Riezu-Boj, J. I., Milagro, F. I., & Martinez, J. A. (2019). Diet, gut microbiota, and obesity: Links with host genetics and epigenetics and potential applications. *Advances in Nutrition*, *10*(suppl_1), S17–S30. https://doi.org/10.1093/advances/nmy078

da Silva Pontes, K. S., Guedes, M. R., da Cunha, M. R., de Souza Mattos, S., Silva, M. I. B., Neves, M. F., Marques, B. C. A. A., & Klein, M. R. S. T. (2021). Effects of probiotics on body adiposity and cardiovascular risk markers in individuals with overweight and obesity: A systematic review and meta-analysis of randomized controlled trials. *Clinical Nutrition*, *40*(8), 4915–4931. https://doi.org/10.1016/j.clnu.2021.06.023

Dale, H. F., Rasmussen, S. H., Asiller, Ö. Ö., & Lied, G. A. (2019). Probiotics in irritable bowel syndrome: An up-to-date systematic review. *Nutrients*, *11*(9), 2048. https://doi.org/10.3390/nu11092048

Davani-Davari, D., Negahdaripour, M., Karimzadeh, I., Seifan, M., Mohkam, M., Masoumi, S. J., Berenjian, A., & Ghasemi, Y. (2019). Prebiotics: Definition, types, sources, mechanisms, and clinical applications. *Foods*, *8*(3), 92. https://doi.org/10.3390/foods8030092

Federici, S., Suez, J., & Elinav, E. (2020). Our microbiome: On the challenges, promises, and hype. *Results and Problems in Cell Differentiation*, *69*, 539–557. https://doi.org/10.1007/978-3-030-51849-3_20

Francavilla, R., Polimeno, L., Demichina, A., Maurogiovanni, G., Principi, B., Scaccianoce, G., Ierardi, E., Russo, F., Riezzo, G., & Di Leo, A. (2014). Lactobacillus reuteri strain combination in Helicobacter pylori infection: A randomized, double-blind, placebo-controlled study. *Journal of Clinical Gastroenterology*, *48*(5), 407–413. https://doi.org/10.1097/MCG.0000000000000007

Franck, A. (2006). Oligofructose-enriched inulin stimulates calcium absorption and bone mineralisation. *Nutrition Bulletin*, *31*(4), 341–345. https://doi.org/10.1111/j.1467-3010.2006.00584.x

Gao, Z. Y., Cui, Z., Yan, Y. Q., Ning, L. J., Wang, Z. H., & Hong, J. (2021). Microbe-based management for colorectal cancer. *Chinese Medical Journal*, *134*(24), 2922–2930. https://doi.org/10.1097/CM9.0000000000001887

Geraghty, M. E. (2006). *Physiological and molecular indicators of change in the intestinal microflora of postmenopausal women consuming soy and fructooligosaccharides (FOS)*. The Ohio State University.

Guandalini, S., & Sansotta, N. (2017). Probiotics use in infectious disease (respiratory, diarrhea, and antibiotic-associated diarrhea). In *The Microbiota in Gastrointestinal Pathophysiology* (pp. 299–313). Elsevier. https://doi.org/10.1016/B978-0-12-804024-9.00033-1

He, L., Yang, H., Chen, Z., & Ouyang, X. (2020). The effect of Streptococcus salivarius K12 on halitosis: A double-blind, randomized, placebo-controlled trial. *Probiotics and Antimicrobial Proteins*, *12*, 1321–1329. DOI10.1007/s12602-020-09646-7

Heuven, L. A., Pyle, S., Greyling, A., Melse-Boonstra, A., & Eilander, A. (2021). Gut microbiota–targeted nutritional interventions improving child growth in low-and middle-income countries: A systematic review. *Current Developments in Nutrition*, *5*(11), nzab124. https://doi.org/10.1093/cdn/nzab124

Hirose, Y., Yamamoto, Y., Yoshikai, Y., & Murosaki, S. (2013). Oral intake of heat-killed Lactobacillus plantarum L-137 decreases the incidence of upper respiratory tract infection in healthy subjects with high levels of psychological stress. *Journal of Nutritional Science*, *2*, e39. doi:10.1017/jns.2013.35

Hishiki, H., Kawashima, T., Tsuji, N. M., Ikari, N., Takemura, R., Kido, H., & Shimojo, N. (2020). A double-blind, randomized, placebo-controlled trial of heat-killed Pediococcus acidilactici K15 for prevention of respiratory tract infections among preschool children. *Nutrients*, *12*(7), 1989. https://doi.org/10.3390/nu12071989

Huang, R., Xing, H.-Y., Liu, H.-J., Chen, Z.-F., & Tang, B.-B. (2021). Efficacy of probiotics in the treatment of acute diarrhea in children: A systematic review and meta-analysis of clinical trials. *Translational Pediatrics*, *10*(12), 3248. https://doi.org/10.21037/tp-21-511

Jia, K., Tong, X., Wang, R., & Song, X. (2018). The clinical effects of probiotics for inflammatory bowel disease: A meta-analysis. *Medicine*, *97*(51), e13792. https://doi.org/10.1097/MD.0000000000013792

Jonkers, D., Penders, J., Masclee, A., & Pierik, M. (2012). Probiotics in the management of inflammatory bowel disease: A systematic review of intervention studies in adult patients. *Drugs*, *72*, 803–823. https://doi.org/10.2165/11632710-000000000-00000

Kumar, H., Schütz, F., Bhardwaj, K., Sharma, R., Nepovimova, E., Dhanjal, D. S., Verma, R., Kumar, D., Kuča, K., & Cruz-Martins, N. (2023). Recent advances in the concept of paraprobiotics: Nutraceutical/functional properties for promoting children health. *Critical Reviews in Food Science and Nutrition*, *63*(19), 3943–3958. https://doi.org/10.1080/10408398.2021.1996327

Lee, J. K., Kim, S. J., Ko, S. H., Ouwehand, A. C., & Ma, D.-S. (2015). Modulation of the host response by probiotic L actobacillus brevis CD 2 in experimental gingivitis. *Oral Diseases*, *21*(6), 705–712. https://doi.org/10.1111/odi.12332

Lee, S. H., Lee, J. H., & Lee, S. W. (2024). Application of microbiome-based therapies in chronic respiratory diseases. *Journal of Microbiology*, *62*(3), 201–216. https://doi.org/10.1007/s12275-024-00124-1

Liu, F., Prabhakar, M., Ju, J., Long, H., & Zhou, H. (2017). Effect of inulin-type fructans on blood lipid profile and glucose level: A systematic review and meta-analysis of randomized controlled trials. *European Journal of Clinical Nutrition*, *71*(1), 9–20. https://doi.org/10.1038/ejcn.2016.156

Liu, R. T., Walsh, R. F., & Sheehan, A. E. (2019). Prebiotics and probiotics for depression and anxiety: A systematic review and meta-analysis of controlled clinical trials. *Neuroscience & Biobehavioral Reviews*, *102*, 13–23. https://doi.org/10.1016/j.neubiorev.2019.03.023

Mancuso, C., & Santangelo, R. (2018). Alzheimer's disease and gut microbiota modifications: The long way between preclinical studies and clinical evidence. *Pharmacological Research*, *129*, 329–336. https://doi.org/10.1016/j.phrs.2017.12.009

Mardini, H. E., & Grigorian, A. Y. (2014). Probiotic mix VSL# 3 is effective adjunctive therapy for mild to moderately active ulcerative colitis: A meta-analysis. *Inflammatory Bowel Diseases*, *20*(9), 1562–1567. https://doi.org/10.1097/MIB.0000000000000084

Marteau, P., & Seksik, P. (2004). Tolerance of probiotics and prebiotics. *Journal of Clinical Gastroenterology*, *38*, S67–S69. https://doi.org/10.1097/01.mcg.0000128929.37156.a7

McFarland, L. V., & Goh, S. (2013). Preventing pediatric antibiotic-associated diarrhea and Clostridium difficile infections with probiotics: A meta-analysis. *World Journal of Meta-Analysis*, *1*(3), 102–120. https://doi.org/10.13105/wjma.v1.i3.102

McFarland, L. V., & Goh, S. (2019). Are probiotics and prebiotics effective in the prevention of travellers' diarrhea: A systematic review and meta-analysis. *Travel Medicine and Infectious Disease*, *27*, 11–19. https://doi.org/10.1016/j.tmaid.2018.09.007

Mo, R., Zhang, X., & Yang, Y. (2019). Effect of probiotics on lipid profiles in hypercholesterolaemic adults: A meta-analysis of randomized controlled trials. *Medicina Clínica (English Edition)*, *152*(12), 473–481. https://doi.org/10.1016/j.medcle.2018.09.013

Mohammadi, H., Ghavami, A., Faghihimani, Z., Sharifi, S., Nattagh-Eshtivani, E., Ziaei, R., & Miraghajani, M. (2021). Effects of probiotics fermented milk products on obesity measure among adults: A systematic review and meta-analysis of clinical trials. *Journal of Functional Foods*, *82*, 104494. https://doi.org/10.1016/j.jff.2021.104494

Musić, L. (2021). *Effect of probiotic Lactobacillus reuteri as an adjunct to nonsurgical treatment on halitosis of patients with periodontitis: A randomised placebo-controlled trial* University of Zagreb. School of Dental Medicine. Department of Periodontology. https://urn.nsk.hr/urn:nbn:hr:127:817040

Nakamura, S., Sarker, S. A., Wahed, M. A., Wagatsuma, Y., Oku, T., & Moji, K. (2006). Prebiotic effect of daily fructooligosaccharide intake on weight gain and reduction of acute diarrhea among children in a Bangladesh urban slum: A randomized double-masked placebo-controlled study. *Tropical Medicine and Health*, *34*(3), 125–131. https://doi.org/10.2149/tmh.34.125

Nishida, K., Sawada, D., Kawai, T., Kuwano, Y., Fujiwara, S., & Rokutan, K. (2017). Para-psychobiotic Lactobacillus gasseri CP2305 ameliorates stress-related symptoms and sleep quality. *Journal of Applied Microbiology*, *123*(6), 1561–1570. https://doi.org/10.1111/jam.13594

Nishihira, J., Moriya, T., Sakai, F., Kabuki, T., Kawasaki, Y., & Nishimura, M. (2016). Lactobacillus gasseri SBT2055 stimulates immunoglobulin production and innate immunity after influenza vaccination in healthy adult volunteers: A randomized, double-blind, placebo-controlled, parallel-group study. *Functional Foods in Health and Disease*, *6*(9), 544–568. https://doi.org/10.31989/ffhd.v6i9.284

Odriozola, A., Gonzalez, A., Odriozola, I., Alvarez-Herms, J., & Corbi, F. (2024). Microbiome-based precision nutrition: Prebiotics, probiotics and postbiotics. *Advanced Genetics*, *111*, 237–310. https://doi.org/10.1016/bs.adgen.2024.04.001

Okburan, G., & Kızıler, S. (2023). Human milk oligosaccharides as prebiotics. *Pediatrics & Neonatology*, *64*(3), 231–238. https://doi.org/10.1128/9781555815462.ch10

Panetta, V., Bacchieri, A., Papetti, S., De Stefani, E., & Navarra, P. (2020). The safety profile of probiotic VSL# 3 (R). A meta-analysis of safety data from double-blind, randomized, placebo-controlled clinical trials. *European Review for Medical and Pharmacological Sciences*, *24*, 963–973. https://dx.doi.org/10.26355/eurrev_202001_20082

Park, M., Joung, M., Park, J. H., Ha, S. K., & Park, H. Y. (2022). Role of postbiotics in diet-induced metabolic disorders. *Nutrients*, *14*(18). https://doi.org/10.3390/nu14183701

Payahoo, L., Khajebishak, Y., Alivand, M. R., Soleimanzade, H., Alipour, S., Barzegari, A., & Ostadrahimi, A. (2019). Investigation the effect of oleoylethanolamide supplementation on the abundance of Akkermansia muciniphila bacterium and the dietary intakes in people with obesity: A randomized clinical trial. *Appetite*, *141*, 104301. https://doi.org/10.1016/j.appet.2019.05.032

Rad, A. H., Abbasi, A., Kafil, H. S., & Ganbarov, K. (2020). Potential pharmaceutical and food applications of postbiotics: A review. *Current Pharmaceutical Biotechnology*, *21*(15), 1576–1587. https://doi.org/10.2174/1389201021666200516154833

Rao, S., Srinivasjois, R., & Patole, S. (2009). Prebiotic supplementation in full-term neonates: A systematic review of randomized controlled trials. *Archives of Pediatrics & Adolescent Medicine*, *163*(8), 755–764. https://doi.org/10.1001/archpediatrics.2009.94

Rashidi, K., Darand, M., Garousi, N., Dehghani, A., & Alizadeh, S. (2021). Effect of infant formula supplemented with prebiotics and probiotics on incidence of respiratory tract infections: A systematic review and meta-analysis of randomized clinical trials. *Complementary Therapies in Medicine*, *63*, 102795. https://doi.org/10.1016/j.ctim.2021.102795

Reinwald, S., Rolland, J. M., O'Hehir, R. E., & Van Zelm, M. C. (2022). Peanut oral immunotherapy: Current trends in clinical trials. *Immunotherapy Advances*, *2*(1), ltac004. https://doi.org/10.1093/immadv/ltac004

Romijn, A. R., Rucklidge, J. J., Kuijer, R. G., & Frampton, C. (2017). A double-blind, randomized, placebo-controlled trial of Lactobacillus helveticus and Bifidobacterium longum for the symptoms of depression. *Australian & New Zealand Journal of Psychiatry*, *51*(8), 810–821. https://doi.org/10.1177/0004867416686669

Saavedra, J., & Tschernia, A. (2002). Human studies with probiotics and prebiotics: Clinical implications. *British Journal of Nutrition*, *87*(S2), S241–S246. https://doi.org/10.1079/BJN/2002543

Satomi, S., Waki, N., Arakawa, C., Fujisawa, K., Suzuki, S., & Suganuma, H. (2021). Effects of heat-killed Levilactobacillus brevis KB290 in combination with β-carotene on influenza virus infection in healthy adults: A randomized controlled trial. *Nutrients*, *13*(9), 3039. https://doi.org/10.3390/nu13093039

Sharma, N., Kang, D. K., Paik, H. D., & Park, Y. S. (2023). Beyond probiotics: A narrative review on an era of revolution. *Food Science and Biotechnology*, *32*(4), 413–421. https://doi.org/10.1007/s10068-022-01212-x

Shimizu, K., Sato, H., Suga, Y., Yamahira, S., Toba, M., Hamuro, K., Kakumoto, K., Kohda, N., Akama, T., & Kono, I. (2014). The effects of Lactobacillus pentosus strain b240 and appropriate physical training on salivary secretory IgA levels in elderly adults with low physical fitness: A randomized, double-blind, placebo-controlled trial. *Journal of Clinical Biochemistry and Nutrition*, *54*(1), 61–66. https://doi.org/10.3164/jcbn.13-62

Sivamaruthi, B. S., Kesika, P., & Chaiyasut, C. (2018). Probiotic based therapy for atopic dermatitis: Outcomes of clinical studies. *Asian Pacific Journal of Tropical Bomedicine*, *8*(6), 328–332. https://doi.org/10.4103/2221-1691.235328

Sun, J., & Buys, N. (2015). Effects of probiotics consumption on lowering lipids and CVD risk factors: A systematic review and meta-analysis of randomized controlled trials. *Annals of Medicine*, *47*(6), 430–440. https://doi.org/10.3109/07853890.2015.1071872

Szajewska, H., Albrecht, P., & Topczewska-Cabanek, A. (2009). Randomized, double-blind, placebo-controlled trial: Effect of lactobacillus GG supplementation on Helicobacter pylori eradication rates and side effects during treatment in children. *Journal of Pediatric Gastroenterology and Nutrition*, *48*(4), 431–436.

Szajewska, H., & Kołodziej, M. (2015). Systematic review with meta-analysis: Lactobacillus rhamnosus GG in the prevention of antibiotic-associated diarrhoea in children and adults. *Alimentary Pharmacology & Therapeutics*, *42*(10), 1149–1157. https://doi.org/10.1111/apt.13404

Taghizadeh Ghassab, F., Shamlou Mahmoudi, F., Taheri Tinjani, R., Emami Meibodi, A., Zali, M. R., & Yadegar, A. (2024). Probiotics and the microbiota-gut-brain axis in neurodegeneration: Beneficial effects and mechanistic insights. *Life Sciences*, *350*, 122748. https://doi.org/10.1016/j.lfs.2024.122748

Talukdar, J. R., Cooper, M. A., Lyutvyn, L., Zeraatkar, D., Ali, R., Bierbrier, R., Janes, S., Ha, V., Darling, P. B., Sievenpiper, J. L., Jenkins, D. J. A., Banfield, L., Mbuagbaw, L., & de Souza, R. J. (2022). Effects of inulin-type fructans supplementation on cardiovascular disease risk factors: A protocol for a systematic review and meta-analysis of randomised controlled trials. *BMJ Open*, *12*(7), e058875. https://doi.org/10.1136/bmjopen-2021-058875

van den Heuvel, E. G., Muijs, T., Brouns, F., & Hendriks, H. F. (2009). Short-chain fructo-oligosaccharides improve magnesium absorption in adolescent girls with a low calcium intake. *Nutrition research*, *29*(4), 229–237. https://doi.org/10.1016/j.nutres.2009.03.005

Vinderola, G., Sanders, M. E., Salminen, S., & Szajewska, H. (2022). Postbiotics: The concept and their use in healthy populations. *Frontiers in Nutrition*, *9*, 1002213. https://doi.org/10.3389/fnut.2022.1002213

Voigt, J., & Lele, M. (2022). Lactobacillus rhamnosus used in the perinatal period for the prevention of atopic dermatitis in infants: A systematic review and meta-analysis of randomized trials. *American Journal of Clinical Dermatology*, *23*(6), 801–811. https://doi.org/10.1007/s40257-022-00723-x

Wegh, C. A., Geerlings, S. Y., Knol, J., Roeselers, G., & Belzer, C. (2019). Postbiotics and their potential applications in early life nutrition and beyond. *International Journal of Molecular Sciences*, *20*(19), 4673. https://doi.org/10.3390/ijms20194673

Wei, L., Duan, L., Zhao, H., & Tai, J. (2022). Role of a mixed probiotic product, VSL#3, in the prevention and treatment of colorectal cancer. *European Journal of Pharmacology*, *930*, 175152. https://doi.org/10.1016/j.ejphar.2022.175152

Westerbeek, E. A., van den Berg, J. P., Lafeber, H. N., Fetter, W. P., Boehm, G., Twisk, J. W., & van Elburg, R. M. (2010). Neutral and acidic oligosaccharides in preterm infants: A randomized, double-blind, placebo-controlled trial. *The American Journal of Clinical Nutrition*, *91*(3), 679–686. https://doi.org/10.3945/ajcn.2009.28625

Westerbeek, E. A., van Elburg, R. M., Berg, A. v. d., Berg, J. v. d., Twisk, J. W., Fetter, W. P., & Lafeber, H. N. (2008). Design of a randomised controlled trial on immune effects of acidic and neutral oligosaccharides in the nutrition of preterm infants: Carrot study. *BMC pediatrics*, *8*, 1–6. https://doi.org/10.1186/1471-2431-8-46

Wu, C.-T., Lin, F.-H., Lee, Y.-T., Ku, M.-S., & Lue, K.-H. (2019). Effect of Lactobacillus rhamnosus GG immunopathologic changes in chronic mouse asthma model. *Journal of Microbiology, Immunology and Infection*, *52*(6), 911–919. https://doi.org/10.1016/j.jmii.2019.03.002

Wuethrich, I., B, W. P., Khodamoradi, Y., & Vehreschild, M. (2021). The role of the human gut microbiota in colonization and infection with multidrug-resistant bacteria. *Gut Microbes*, *13*(1), 1–13. https://doi.org/10.1080/19490976.2021.1911279

Xie, A., Chen, A., Chen, Y., Luo, Z., Jiang, S., Chen, D., & Yu, R. (2023). Lactobacillus for the treatment and prevention of atopic dermatitis: Clinical and experimental evidence. *Frontiers in Cellular and Infection Microbiology*, *13*, 1137275. https://doi.org/10.3389/fcimb.2023.1137275

Zawistowska-Rojek, A., & Tyski, S. (2022). How to improve health with biological agents-narrative review. *Nutrients*, *14*(9). https://doi.org/10.3390/nu14091700

Zolkiewicz, J., Marzec, A., Ruszczynski, M., & Feleszko, W. (2020). Postbiotics – A step beyond pre- and probiotics. *Nutrients*, *12*(8). https://doi.org/10.3390/nu12082189

24 Nanocarriers Encapsulated Polyphenols, Functional Foods to Modulate Gut Microbiota

In Neurological Disorders

Ameya Sharma[1], Malkiet Kaur[2], Divya Dheer[1], Aditya Gautam[1], and Vivek Puri[1]

[1] Chitkara University School of Pharmacy, Chitkara University, Himachal Pradesh, India
[2] Maharishi Markandeshwar College of Pharmacy, Maharishi Markandeshwar University, 2 Ambala, Haryana, India

24.1 INTRODUCTION

The gut microbiota's role in human health and disease has recently attracted a lot of interest from researchers and healthcare professionals. The gut microbiota is a complex network of microbes that exists within the intestines and is vital for digestive health, immunity, and homeostasis (Williams et al., 2023) (Bao et al., 2024). Multiple sclerosis, autism spectrum disorders, Alzheimer's disease, Parkinson's disease, and dysbiosis, an imbalance in the composition of the gut microbiota, all have been associated with neurological disorders. A family of bioactive substances called polyphenols is present in plant-based diets and has grown in popularity due to its possible health benefits, which include neuroprotective, anti-inflammatory, and antioxidant properties (Tao et al., 2024) (Riaz et al., 2023). Nevertheless, polyphenols frequently have poor therapeutic benefits due to their low bioavailability and stability. Using nanocarriers like micelles, liposomes, or nanoparticles could be an excellent strategy to transport polyphenols to specific tissues, increase their bioavailability, and protect them from the gastrointestinal tract's harsh conditions. New functional foods including polyphenols in nanocarriers offer a promising strategy to alter gut microbiota and slow the development of neurological diseases (Jaberi et al., 2024). Innovative formulations that regulate the gut microbiota could possess positive impacts on brain function and cognitive health by targeting the gut–brain axis, a bidirectional communication mechanism between the gut and the brain. Dietary polyphenol metabolism is significantly impacted by the gut flora, which in turn affects the bioavailability and bioactivity of these compounds (Cocean & Vodnar, 2024) (Fekete et al., 2024) (Singh et al., 2024) (El-Saadony et al., 2024). Scientists have found a way to make polyphenols more stable and absorbable in the stomach by using nanocarriers. This allows for more precise delivery of the compounds to the colon, which is the repository of most gut flora. These nanocarriers have the ability to regulate the release of polyphenols once they reach the colon. This allows them to interact with certain gut microorganisms and alter the composition of the gut microbiota (Kassem et al., 2024) (Dasriya et al., 2024).

Several studies have shown that functional foods containing nanocarrier-encapsulated polyphenols have the ability to alter the gut flora and enhance neurological outcomes. For instance, in neuroinflammation animal models, resveratrol-loaded nanoparticles increased the proliferation of good gut bacteria like *Lactobacillus* and *Bifidobacterium* and decreased the prevalence of bad bacteria. In a similar vein, preclinical models of Alzheimer's disease have shown that curcumin-loaded liposomes can modulate the composition of the gut microbiota to reduce neuroinflammation and improve cognitive function (Li et al., 2023) (Chu et al., 2024). An encouraging approach to use these bioactive substances for the treatment of neurological diseases is the development of polyphenol formulations in functional foods based on nanocarriers. These novel formulations provide an unprecedented opportunity to influence brain health and function by modulating gut microbial communities by addressing the gut microbiota (Rosales & Fabi, 2023) (Sasso et al., 2023). To further understand how nanocarrier-encapsulated polyphenols impact the gut–brain axis and apply this knowledge into practice to treat and prevent neurological diseases, a little more research is required. With an emphasis on polyphenols as functional nutrients, this chapter explores the complex relationship between neurological disorders and the gut flora. One potential approach to improving the distribution of polyphenols with the goal of modulating the gut flora is the use of nanocarriers. This system works by releasing compounds into the colon at certain

DOI: 10.1201/9781003536802-24

times in order to affect neuroinflammatory pathways and brain function through interactions with specific strains of gut bacteria. The stability and biocompatibility of nanocarriers are important concerns to address. Looking ahead, it's important to understand how this works and then use this knowledge to treat neurological disorders. One promising strategy is to use functional foods that contain nanocarrier-encapsulated polyphenols to improve brain health by acting on the axis between the brain and the gut.

24.2 ROLE OF GUT MICROBIOTA IN NEUROLOGICAL HEALTH

Recent advancements emphasize polyphenol-encapsulated nanocarriers as a promising mechanism for the targeted delivery of bioactive compounds, especially in the field of neurological health through gut–brain axis modulation (Grabska-Kobyłecka et al., 2023) (Ozkan et al., 2024). Despite being well-known for their anti-inflammatory and antioxidant qualities, polyphenols have challenges with stability and bioavailability. Encapsulation in nanocarriers, including micelles, nanoparticles, and liposomes, improves their absorption and prevents them from gastrointestinal distribution (Barchitta et al., 2024) (Y. Li et al., 2023). These nanocarriers may reach the intestines, affect gut microbiota composition, and alter brain function (Jalil et al., 2023). Preclinical investigations indicate their ability to enhance cognitive function, reduce neuroinflammation, and prevent neurodegenerative disorders (Grabska-Kobyłecka et al., 2023) (Gandhi et al., 2023) (Garbati et al., 2024). Neurological disorders such as ischemic stroke and perioperative neurocognitive dysfunction (PND) are greatly influenced by the gut flora. According to Que et al. (2024), research indicates that modulating microbiota could potentially be used as a therapeutic approach (Que et al., 2024). As an example, Acorus tatarinowii (AT) oils were studied in an ischemic stroke model by Huang et al. (2024). The use of AT oils decreased levels of pro-inflammatory cytokines (TNF-α, IL-17, IL-6, IFN-γ), protected neurons against apoptosis, and decreased cerebral infarction. Furthermore, they reduced detrimental species (*Verrucomicrobia*, *Akkermansia*) and increased good bacteria (*Prevotella copri*, Tenericutes) to restore the chemical composition of the gut microbiota. Results from a study by Huang and team in 2024 revealed that AT oils, when introduced through fecal microbiota transplantation, decreased microglial activation without impacting gut permeability or blood–brain barrier integrity (Y. Huang et al., 2024). Similarly, Gastrodin (GAS) reduced depressive symptoms in a CUMS model and changed the composition of the gut microbiota, according to Zhao et al. (2024). LDA showed that there was a rise in *Bacteroides* in the CUMS-GAS group (Zhao et al., 2024). After treating GV20, LI4, and PC6 acupoints with electroacupuncture, Tang and team in 2024 showed enhanced memory in a study on postpartum depression. Hippocampal inflammation (NSE, S-100β, IL-6) was reduced and gut microbiota restoration was validated by 16S rRNA sequencing and LC-MS analysis. Transplantation of fecal microbiota has been reported to alter gut microbiota and decrease neuroinflammation (Tang et al., 2024).

24.3 DYSBIOSIS IN NEUROLOGICAL DISORDERS

Neurological disorders associated with dysbiosis have potential treatments in the form of nanocarrier-encapsulated polyphenols. Despite their neuroprotective, anti-inflammatory, and antioxidant properties, polyphenols found in plant-based foods are not very bioavailable. Liposomes and nanoparticles are nanocarriers that promote the stability, protection, and delivery of polyphenols to the intestines and brain (Sangnim et al., 2023). The gut microbiota can be regulated by nanocarriers to restore microbial balance, which helps relieve neuroinflammation, oxidative stress, and neurodegeneration associated with diseases such as Alzheimer's and Parkinson's (Puri, Kanojia, et al., 2022) (Puri, Nagpal, et al., 2022). Neurodegenerative disorders associated with dysbiosis are an area of ongoing research. Nanocarrier design, safety, and therapeutic application can all be improved with more studies. A study conducted by Venkidesh et al. (2023) revealed that rats were subjected to a single 6 Gy pelvic irradiation, which disrupted the brain–gut axis and caused neuroinflammation, decreased locomotion, intestinal damage, neuronal death, and dysbiosis (Venkidesh et al., 2023). Similarly, Xu et al. (2024) showed anxiety-like symptoms in mice following hormone-triggered pseudopregnancy and ovarian steroid withdrawal, showing a connection between gut microbiota imbalance and postpartum depression (PPD). The gut–brain axis relationship was further supported when fecal microbiota transplantation in antibiotic-treated and PPD animals showed comparable behavioral and hippocampal impairments (Xu et al., 2024).

24.4 NANOCARRIERS FOR POLYPHENOL DELIVERY

Polyphenols exhibit antioxidant properties, showing their potential in health improvement and disease prevention. Along with their beneficial properties, they also exhibit low solubility, poor permeability, and instability (Bayraktar et al., 2017). The nanoformulation of polyphenols is one potential alternative to this challenge and can enhance bioavailability, extend circulation, improve localization, boost efficacy, and lowering multiple drug resistance (Tabrez et al., 2020). Since nanocarriers are minute in size, they can easily pass through tiny capillaries and cells, maximizing polyphenol accumulation and allowing for prolonged release. The enhanced biological properties, including toxicities of nanoformulations (nanoemulsion, nanoencapsulation, polymeric nanoparticles, nanoliposomes, solid lipid nanoparticles) are due to their physicochemical characteristics, which include shape, size, and surface characteristics (Nedovic et al., 2011). Polyphenols can be effectively given using nanoformulation to treat neurological, cardiovascular, and cancerous conditions as they can be better encapsulated, protected, and delivered using bio-based nanocarriers, increasing their oral bioavailability. These formulations are widely used in health and nutrition goods such as nutraceuticals (Z. Zhang et al., 2022).

24.4.1 Types of Nanocarriers for Polyphenol Encapsulation

Enhancing the distribution of bioactive compounds like vitamins, minerals, fatty acids, lycopene, and antioxidants can be accomplished with the use of nanocarrier systems. Using encapsulation, one or more substances are immobilized in a matrix or wall. This matrix may take the shape of a homogeneous or heterogeneous solid or liquid phase (Acosta, 2009). The process of encapsulation and its advantages are shown in Figure 24.1. The exterior of the capsule is referred to as the shell, wall, encapsulate, or carrier material, while the component is immobilized within the matrix, which is typically referred to as the core or active part of the material. Depending on the increased surface area per unit volume, the nutraceutical delivery qualities, solubility, and bioavailability can be enhanced by decreasing the particle size (Bayraktar et al., 2017).

24.5 NANOSPHERES

Using solid polymer matrices, nanospheres are able to transport hydrophobic drugs on their exterior and hydrophilic ones on the interior. They can be altered with a variety of biocompatible and biodegradable polymers, such as chitosan, PEG, and PLGA. The three main methods for drug incorporation are encapsulation, dissolution, and adsorption (H. Wang et al., 2020). The release of phenolic compounds was investigated by Mohamed et al. (2024) using microbeads manufactured from alginate and acetone extracted from peanut meal. Highlighting the benefits of peanut byproducts, the mixture exhibited neuroprotective potential against aluminum-induced Alzheimer's (Mohamed et al., 2024).

Nanocapsules: Nanocapsules (NCs) have a similar composition but with a core-shell construction in which the medicine is encased in a polymer membrane. Nanocapsules can carry active ingredients on their surfaces or in their layers (Qi et al., 2023). *Hibiscus sabdariffa* L. nanocapsules were prepared using carboxymethyl chitosan (CMC), chitosan hydrochloride (CHC), and β-lactoglobulin (β-Lg). Peonidin-3-glucoside and petunidin-3-glucoside most likely interact with and bind to the outer β-Lg layer of the nanocapsules, enhancing their release during in vitro digestion, according to simulated digestion experiments and molecular

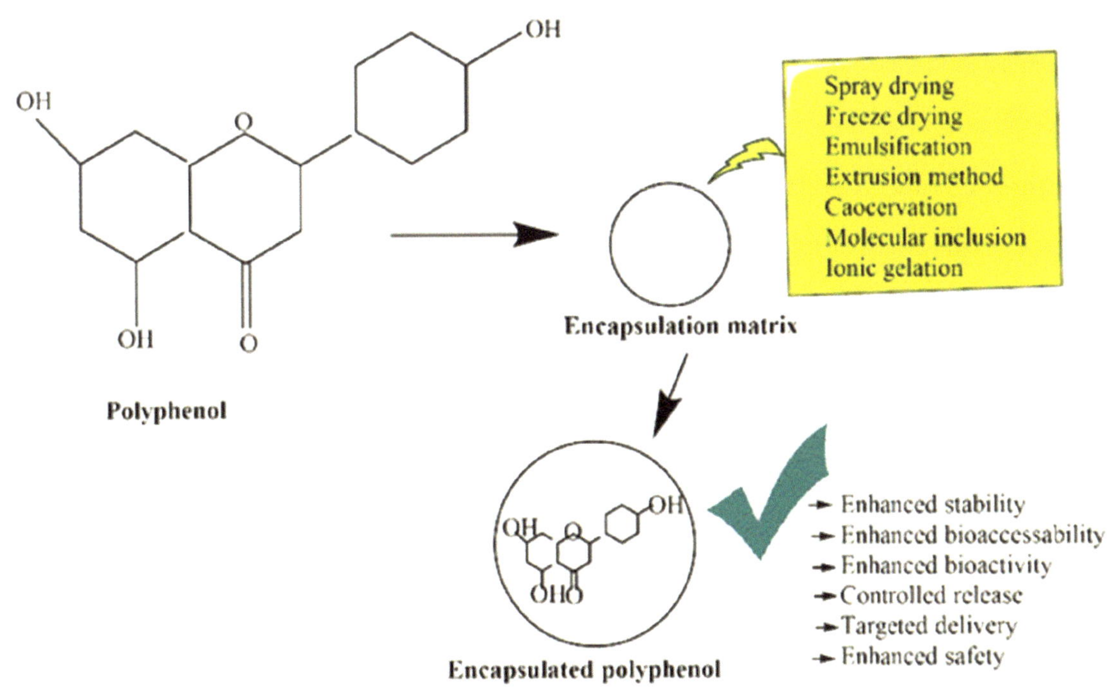

FIGURE 24.1 Process of polyphenol encapsulation.

docking. This study shows that roselle anthocyanin's bioaccessibility is greatly increased when they are encapsulated in CMC, CHC, and β-Lg nanocapsules (Song et al., 2024).

Cyclodextrins: Bacteria break down cellulose to produce cyclodextrins (CDs). They are hydrophilic from the outside and lipophilic in the core, making them cyclic oligosaccharides. These substances are not soluble in water. Effective nanodrug delivery systems are produced using water-soluble cyclodextrin derivatives, such as hydroxypropyl CD and CD, randomly methylated β-cyclodextrin, and sulfobutylether β-cyclodextrin sodium salt. Through hydrogen bonding, hydrophobic bonding, or van der Waals interactions, CDs combine molecules to form compounds. Low temperature and an opposing charge promote this complexation (Lombardo et al., 2019).

Micelles: Amphiphilic chemicals are used in controlled temperatures and concentrations to form colloidal dispersions called micelles (MCs), which have a size range of 5–100 nm. As a result, they restore hydrogen bonds and remove hydrophobic particles, lowering free energy. Solubilization of drugs is made possible by their hydrophobic core, which also ensures stability and biocompatibility. Amphiphilic agents like pluronic, poly(ethylene glycol), poly(D,L-lactide-*co*-glycolide), and polycaprolactone (PCL) are used in microcapsule development (Fan et al., 2021). MCs enhance drug solubility, prolong circulation, improve targeting, and reduce toxicity. Polymeric micelles were developed to improve the intestinal permeability of an *Olea europaea* L. leaf extract rich in polyphenols (49% w/w, 41% oleuropein). Their stability, release profile, encapsulation efficiency, and physicochemical properties were analyzed. Caco-2 and PAMPA assays confirmed enhanced polyphenol penetration (Bergonzi et al., 2023).

Solid Lipid nanoparticles: Aqueous surfactants and lipids with a high melting point form solid lipid nanoparticles (SLNs), which are used as drug carriers. High biocompatibility, stability, and bioavailability are characteristics of SLNs, which are composed of core lipids like fatty acids and stabilizers like phospholipids and sterols (Kumar et al., 2022; Fan et al., 2021). Curcumin was used to create solid lipid nanoparticles (SLNs) by emulsifying sodium caseinate–lactose (NaCas-Lac) Maillard conjugate with equal masses of propylene glycol monopalmitate and glyceryl monostearate as the lipid matrix. Additionally, SLNs significantly increased curcumin's storage stability and antioxidant activity when it was encapsulated. To distribute lipophilic chemicals in functional meals and beverages, the current SLNs may find use (Madureira et al., 2016).

Liposomes: Drug delivery systems called liposomes (LS) develop when lipid powder is hydrated in an aqueous medium. Aqueous units are surrounded by one or more lamellae (phospholipidic bilayer membranes) in liposomes, which range in size from 30 nm to several micrometers. Depending on the temperature, dispersion circumstances, and shape of the molecules, polar lipids self-assemble into colloidal particles (Negahdari et al., 2021). Researchers used enteropathogenic *Escherichia coli* to challenge mice with nanoliposome-encapsulated phenolic-rich fraction (PRF) from *Alcea rosea* leaves as a dietary phytobiotic. The nanoliposome-encapsulated PRF is a promising phytobiotic that may be particularly effective against *E. coli* infection in mice because of its increased intestinal solubility, bioavailability, and absorption of phenolic compounds encapsulated in the nanoliposome carrier (Hassirian et al., 2022).

Nanoemulsions: Nanoemulsions that have droplet sizes of 20–200 nm are heterogeneous systems that are thermodynamically unstable. They consist of dispersions of nanoscale droplets produced by the rupture produced by shear. The resulting emulsion is known as an oil-in-water (O/W) nanoemulsion when the dispersal phase is oil and the continuous phase is water. However, the resulting emulsion is known as a water-in-oil (W/O) nanoemulsion when the water phase is disseminated in the oil phase (Torchilin, 2006).

24.6 MECHANISMS OF ACTION IN MODULATING GUT MICROBIOTA

24.6.1 Interactions between Polyphenols and Gut Microbiota

Polyphenols regulate gut microbiota, restore microbial balance, and strengthen intestinal barrier integrity. Their antibacterial properties disrupt bacterial metabolism, biofilms, and nucleic acid production. They are thought to offer protection against several diseases, including cancer, stroke and myocardial infarction, cardiovascular diseases, and certain immunological and neurological disorders (Silva & Pogačnik, 2020). The intestinal microbiota can produce a variety of metabolites, which can subsequently impact central nervous system function (Erny et al., 2021). Polyphenols impacts on the gut microbiota usually arise from their combined antibacterial and growth-stimulating (perhaps prebiotic) properties (Makarewicz et al., 2021). Bacteria having important biological roles, like probiotic species from the *Lactobacilli* and *Bifidobacteria* genera, are stimulated by polyphenol-rich extracts. This stimulation results from either the inhibition of potentially dangerous bacteria, the re-establishment of the mucosal pro- and anti-inflammatory balance, the direct use of polyphenols by gut bacteria, or a shift in the microbial ecological niches caused by polyphenols (Rodríguez-Daza et al., 2021). Proteins and polyphenols can interact to limit bacterial nucleic acid production, change the fluidity and function of cell membranes, alter the integrity and synthesis of cell walls, impact cell metabolism, and stop the formation of biofilms (Firrman et al., 2016). The influence of polyphenols on bacterial growth and metabolism depends on the polyphenol structure, the dosage assayed and the microorganism strain. For instance, Gram-negative bacteria are more resistant to polyphenols than Gram-positive bacteria, possibly due to the differences found in their wall composition. Polyphenols, such as catechins, act on different bacterial species (*E. coli, Bordetella bronchiseptica, Serratia marcescens, Klebsiella pneumoniae, Salmonella choleraesuis, Pseudomonas aeruginosa, Staphylococcus aureus,* and *Bacillus subtilis*) by generating hydrogen peroxide and by altering the permeability of the microbial membrane (S. Huang et al., 2020).

24.6.2 Effects on Gut Microbiota Composition

Approximately 10^{14} microbial cells inhabit the human gastrointestinal tract, with about a third composed of Bacteroides. Other prominent genera include *Bifidobacterium, Eubacterium, Clostridium, Prevotella, Peptococcus, Lactobacillus*, and others. For instance, an increase in *Clostridium difficile* in the gut can lead to infectious diarrhea (Ma & Chen, 2020). A small percentage of dietary polyphenols (5–10% of the total intake, primarily those with monomeric and dimeric structures) can be directly absorbed in the small intestine, typically following deconjugation reactions like deglycosylation (Manach et al., 2005).

After passing through the small intestine, polyphenols are converted into soluble metabolites for excretion and circulation through Phase I and Phase II biotransformations in hepatocytes and enterocytes, respectively. Polyphenols, which make up 90–95% of total intake, undergo structural and functional modifications in the large intestine as a result of their breakdown by colonic bacteria (Cardona et al., 2013).

24.7 THERAPEUTIC POTENTIAL IN NEUROLOGICAL DISORDERS

The gut microbiota is an important aspect of human health, and changes in gut microbiota can affect different biological parameters of the human body (Figure 24.2). Henceforth, microbiota regulation is being explored for its biological potential in neurological disorders employing nanomaterials (W. Li et al., 2024).

24.7.1 Neuroprotective Effects of Polyphenols

In the last few years, natural polyphenolic agents have shown tremendous potential for medicinal drugs especially in Parkinson's disease (PD), Alzheimer's disease (AD), and mental disorders such as depression, anxiety, and stress (Grabska-Kobyłecka et al., 2023). These agents have observed neuroprotective action by altering pathophysiological responses, such as oxidative stress as well as neuroinflammation. These two mechanistic parameters are crucial in the management of PD as their mutation may intensifiy ROS generation and initiate neurodegradation (Gahtani et al., 2024). Polyphenols have exhibited prominent neuroprotective action owing to their anti-inflammatory and antioxidant effects including flavonoids and non-flavonoids, such as resveratrol, curcumin, coumarin, and phenolic acids (Ashwin et al., 2021). Interestingly, polyphenols are present in fruits, vegetables, cereals, and beverages as tea and are often recommended as a diet plan for patients with depression. Polyphenols restrict the pro-inflammatory cytokines mechanism, transcription factors, as well as protein complexes responsible for neuroinflammatory responses, henceforth reducing behavioral symptoms occurring during depression. Alteration in the microbiome has been explored in depressed patients consuming flavonoid-rich orange juice and resulted in better health (Winiarska-Mieczan et al., 2023). In fact, dietary polyphenols are present in the form of glycosylation or esterification in plant foods or are combined with polymers, and later undergo Phase I and Phase II metabolic reactions with the help of intestinal enzymes. These polyphenols ameliorate intestinal microbiota disorders by altering the structure as well as role of intestinal microbiota, activating good bacteria to synthesize short-chain fatty acids, increasing the production of hormones and neurotransmitters, thus playing a pivotal role in the prevention and treatment of brain-related diseases involving microbe–gut–brain axis (Riegelman et al., 2024).

24.7.2 Clinical Applications in Neurological Disorders

In an interesting preclinical study, he and coworkers designed an orally administered polyphenol-based nanocarrier for targeting gut microbiota along with its interactions with the brain in IBD (inflammatory bowel diseases). The delivery system comprised of a chitosan and tannic acid-based multilayer preventing antioxidation stress conditions as well as from prebiotic environment, thus influencing the varied gut microbiota (He et al., 2023). Another study reported a polyphenolic-based biomimetic nanocarrier alleviating cognitive dysfunctions by modulating the gut–brain axis (Gong et al., 2023). Nanomaterials have proven to be an excellent carrier for transporting certain microbiota-modulating agents into the intestine for efficient therapeutic response. The polyphenolic substances can modulate the gut microbiota, and some of them may change its structure and functions promoting the growth of useful bacteria and decreasing harmful bacteria (W. Li et al., 2024). Well-planned clinical trials are imperative to obtain the efficiency as well as safety of polyphenol-based

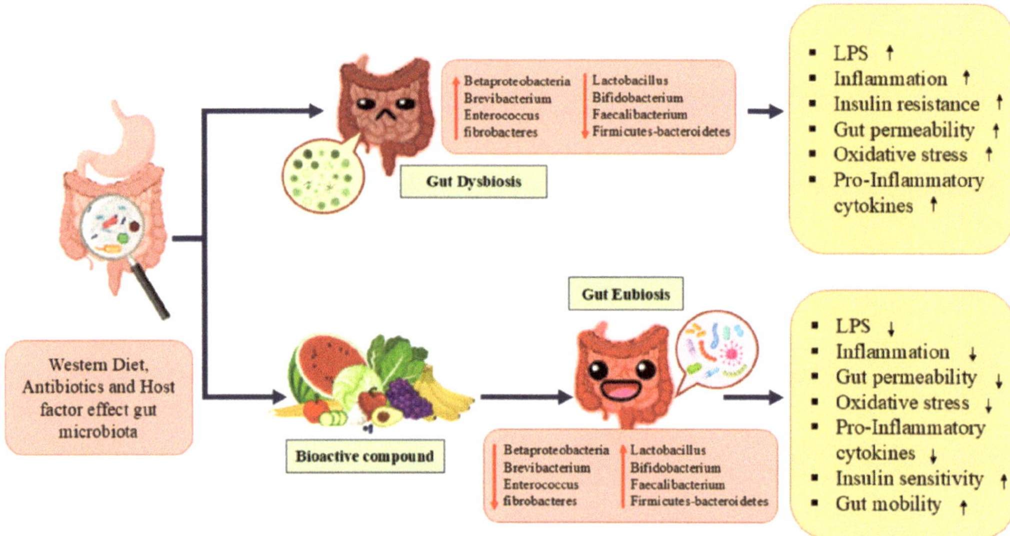

FIGURE 24.2 Factors affecting gut microbiota.

interventions in the management of neurological diseases (Gahtani et al., 2024). Oliveira et al. observed the response of Brazilian depressed women consuming lesser polyphenols than healthy women, indicating the significance of food ingredients in the management of depressive disorders (Mrozek et al., 2023). A double-blind controlled randomized human trial of 60 individuals exhibited that a 42-day rich-antioxidant diet plan (blueberries) diminished symptoms of depression (Velichkov et al., 2024). In AD, PD or epilepsy-based animal assays, polyphenols have shown to alter the composition of the intestinal microbiota and its metabolic constituents thus reducing the progression of these diseases through the microbiota–gut–brain (MGB) axis (Y. Zhang et al., 2022) (H. Zhu et al., 2024). Different polyphenolic flavonoids and non-flavonoids, such as resveratrol, curcumin, coumarin, and other dietary compounds are being explored in clinical trials for their neuroprotective therapeutic efficiency (Condezo-Hoyos et al., 2021). Interestingly, various combination therapies are being explored incorporating polyphenolic components with chemotherapy in preclinical as well as clinical treatment (Jakobušić Brala et al., 2023) and can be explored in neuronal diseases.

24.8 CHALLENGES AND OPPORTUNITIES FOR CLINICAL TRANSLATION

In a few other studies, polyphenolic supplements were found to be effective in augmenting the patient's health in depression, but phenolic compounds do not show the same action. Henceforth, the type of polyphenolic agent has to be carefully selected for the patient. It has also been proven that the Western eating styles may magnify the risk and complexity of depression in young people owing to a lesser plant-based diet and an increased fat and processed food diet. The therapeutic potential of nanocarriers encapsulated with polyphenols for regulating gut microbiota in neurological conditions must be realized; however, there are numerous challenges that must be addressed on the path to clinical translation (Vivarelli et al., 2022). The effectiveness of polyphenols in modifying gut microbiota can be limited by degradation during storage and digestion, making it extremely challenging to ensure their bioavailability and stability within nanocarriers. Furthermore, it is complex to determine the appropriate administration methods and dosage regimens for these new treatments, taking into account patient variability, treatment duration, as well as potential interactions with drugs. Extensive preclinical and clinical investigations are required to evaluate the long-term consequences and reduce negative responses associated with nanocarriers and polyphenols due to safety issues, such as toxicity risks (Sangnim et al., 2024). Another challenge is the regulatory environment, which requires substantial evidence of safety, effectiveness, and quality for clinical applications in order to get nanotechnology-based therapies approved. In addition, patients may not be able to afford these advanced therapeutic drugs due to the high costs of research, development, and production. Sustainable funding systems and cost-effective manufacturing procedures are needed to address this issue (Dri et al., 2023). Advancements in the use of nanocarrier-encapsulated polyphenols for neurological disorders are possible, regardless of limitations. It is possible to improve results and decrease side effects by customizing therapies according to gut microbiota. Clinical trials can proceed more quickly when regulators, researchers, and clinicians work together; awareness improves persistence, and monitoring ensures safety. These developments hold great potential for the treatment of neurological diseases on a worldwide scale (Qi et al., 2023).

24.9 FUTURE PERSPECTIVES

There is tremendous potential for developments in personalized medicine, nanotechnology, clinical translation, regulatory frameworks, and multi-omics integration in the area of using polyphenol-encapsulated nanocarriers to modify gut microbiota in neurological diseases (M. Li et al., 2024). Nanocarrier design advancements will allow for the development of next-generation delivery systems with improved properties for targeted delivery and modulation of gut microbiota, while precision medicine approaches tailored to individual variations in gut microbiota composition and response to polyphenol therapy can optimize treatment outcomes (Wellens et al., 2023). To better understand treatment mechanisms, it is necessary to integrate multi-omics techniques. This will allow us to get full insights into the intricate relationships between gut microbiota, polyphenols, and neurological health. The safety and effectiveness of nanocarrier-encapsulated polyphenols in neurological illnesses must be confirmed through large-scale research before they can be employed in clinical practice (Baysoy et al., 2023) (A. Li et al., 2023). The development, approval, and commercialization of these innovative therapies will be heavily influenced by regulatory considerations. In order to improve patient care and neurological health outcomes, researchers, industry partners, and regulatory bodies must work together to ensure safe and effective translation (Borton et al., 2020).

24.10 CONCLUSION

A potential approach to the treatment of neurological disorders could be the use of functional foods that contain nanocarriers encapsulated with polyphenols that regulate the gut microbiota. Dysbiosis should be considered as a possible therapeutic target due to the complex relationship between gut microbiota and neurological health. The bioactive characteristics and wide range of health advantages associated with polyphenols make them a promising natural approach to controlling the overall composition of gut microbes and, consequently, limiting the development of neurological diseases. Nanocarriers increase the therapeutic potential of polyphenols by making them more stable, increasing their bioavailability, and allowing for targeted release in the GI tract. Encapsulating polyphenols and promoting their interaction with gut microbiota can be achieved by the use of several nanocarriers, including solid lipid nanoparticles, polymeric nanoparticles, and liposomes. The importance of a well-balanced gut microbial environment to brain function has been illuminated by the molecular insights into the interactions between polyphenols and gut bacteria. Polyphenols may offer neuroprotective benefits and alleviate neurological symptoms by altering the composition and function of the gut flora. In the future, researchers should try to figure out exactly how polyphenols affect cerebral function and gut flora. In order to evaluate the therapeutic efficacy of nanocarrier-encapsulated polyphenols in neurological illnesses and to guide their application in clinical practice, additional clinical trials need to be conducted. Nanotechnology, polyphenol research, and gut microbiota modification all are areas that show great potential for enhancing brain health and well-being and for developing neurology as a whole.

REFERENCES

Acosta, E. (2009). Bioavailability of nanoparticles in nutrient and nutraceutical delivery. *Current Opinion in Colloid & Interface Science*, *14*(1), 3–15. https://doi.org/10.1016/J.COCIS.2008.01.002

Ashwin, K., Pattanaik, A. K., & Howarth, G. S. (2021). Polyphenolic bioactives as an emerging group of nutraceuticals for promotion of gut health: A review. *Food Bioscience*, *44*, 101376. https://doi.org/10.1016/J.FBIO.2021.101376

Bao, H., Wang, Y., Xiong, H., Xia, Y., Cui, Z., & Liu, L. (2024). Mechanism of iron ion homeostasis in intestinal immunity and gut microbiota remodeling. *International Journal of Molecular Sciences*, *25*(2), 727. https://doi.org/10.3390/IJMS25020727

Barchitta, M., Van Zonneveld, S. M., Van Den Oever, E. J., Haarman, B. C. M., Grandjean, E. L., Nuninga, J. O., Van De Rest, O., & Sommer, I. E. C. (2024). An anti-inflammatory diet and its potential benefit for individuals with mental disorders and neurodegenerative diseases—A narrative review. *Nutrients*, *16*(16), 2646. https://doi.org/10.3390/NU16162646

Bayraktar, O., Erdogan, I., Köse, M. D., & Kalmaz, G. (2017). Nanocarriers for plant-derived natural compounds. In *Nanostructures for Antimicrobial Therapy: Nanostructures in Therapeutic Medicine Series* (pp. 395–412). https://doi.org/10.1016/B978-0-323-46152-8.00017-2

Baysoy, A., Bai, Z., Satija, R., & Fan, R. (2023). The technological landscape and applications of single-cell multi-omics. *Nature Reviews Molecular Cell Biology*, *24*(10), 695–713. https://doi.org/10.1038/s41580-023-00615-w

Bergonzi, M. C., De Stefani, C., Vasarri, M., Ivanova Stojcheva, E., Ramos-Pineda, A. M., Baldi, F., Bilia, A. R., & Degl'Innocenti, D. (2023). Encapsulation of olive leaf polyphenol-rich extract in polymeric micelles to improve its intestinal permeability. *Nanomaterials*, *13*(24), 3147. https://doi.org/10.3390/NANO13243147/S1

Borton, D. A., Dawes, H. E., Worrell, G. A., Starr, P. A., & Denison, T. J. (2020). Developing collaborative platforms to advance neurotechnology and its translation. *Neuron*, *108*(2), 286–301. https://doi.org/10.1016/j.neuron.2020.10.001

Cardona, F., Andrés-Lacueva, C., Tulipani, S., Tinahones, F. J., & Queipo-Ortuño, M. I. (2013). Benefits of polyphenols on gut microbiota and implications in human health. *The Journal of Nutritional Biochemistry*, *24*(8), 1415–1422. https://doi.org/10.1016/J.JNUTBIO.2013.05.001

Chu, J., Zhang, W., Liu, Y., Gong, B., Ji, W., Yin, T., Gao, C., Liangwen, D., Hao, M., Chen, C., Zhuang, J., Gao, J., & Yin, Y. (2024). Biomaterials-based anti-inflammatory treatment strategies for Alzheimer's disease. *Neural Regeneration Research*, *19*(1), 100–115. https://doi.org/10.4103/1673-5374.374137

Cocean, A. M., & Vodnar, D. C. (2024). Exploring the gut-brain Axis: Potential therapeutic impact of Psychobiotics on mental health. *Progress in Neuro-Psychopharmacology and Biological Psychiatry*, *134*, 111073. https://doi.org/10.1016/J.PNPBP.2024.111073

Condezo-Hoyos, L., Gazi, C., & Pérez-Jiménez, J. (2021). Design of polyphenol-rich diets in clinical trials: A systematic review. *Food Research International*, *149*, 110655. https://doi.org/10.1016/J.FOODRES.2021.110655

Dasriya, V. L., Samtiya, M., Ranveer, S., Dhillon, H. S., Devi, N., Sharma, V., Nikam, P., Puniya, M., Chaudhary, P., Chaudhary, V., Behare, P. V., Dhewa, T., Vemuri, R., Raposo, A., Puniya, D. V., Khedkar, G. D., Vishweswaraiah, R. H., Vij, S., Alarifi, S. N., … Puniya, A. K. (2024). Modulation of gut-microbiota through probiotics and dietary interventions to improve host health. *Journal of the Science of Food and Agriculture*, *104*(11), 6359–6375. https://doi.org/10.1002/JSFA.13370

Dri, D. A., Rinaldi, F., Carafa, M., & Marianecci, C. (2023). Nanomedicines and nanocarriers in clinical trials: Surfing through regulatory requirements and physico-chemical critical quality attributes. *Drug Delivery and Translational Research*, *13*(3), 757–769. https://doi.org/10.1007/S13346-022-01262-Y/TABLES/3

El-Saadony, M. T., Yang, T., Saad, A. M., Alkafaas, S. S., Elkafas, S. S., Eldeeb, G. S., Mohammed, D. M., Salem, H. M., Korma, S. A., Loutfy, S. A., Alshahran, M. Y., Ahmed, A. E., Mosa, W. F. A., Abd El-Mageed, T. A., Ahmed, A. F., Fahmy, M. A., El-Tarabily, M. K., Mahmoud, R. M., AbuQamar, S. F., … Lorenzo, J. M. (2024). Polyphenols: Chemistry, bioavailability, bioactivity, nutritional aspects and human health benefits: A review. *International Journal of Biological Macromolecules*, *277*, 134223. https://doi.org/10.1016/J.IJBIOMAC.2024.134223

Erny, D., Dokalis, N., Mezö, C., Castoldi, A., Mossad, O., Staszewski, O., Frosch, M., Villa, M., Fuchs, V., Mayer, A., Neuber, J., Sosat, J., Tholen, S., Schilling, O., Vlachos, A., Blank, T., Gomez de Agüero, M., Macpherson, A. J., Pearce, E. J., & Prinz, M. (2021). Microbiota-derived acetate enables the metabolic fitness of the brain innate immune system during health and disease. *Cell Metabolism*, *33*(11), 2260–2276.e7. https://doi.org/10.1016/J.CMET.2021.10.010/ATTACHMENT/B7D2F808-4760-43F6-9D9B-663C6A3E5A29/MMC20.XLSX

Fan, Y., Marioli, M., & Zhang, K. (2021). Analytical characterization of liposomes and other lipid nanoparticles for drug delivery. *Journal of Pharmaceutical and Biomedical Analysis*, *192*, 113642. https://doi.org/10.1016/J.JPBA.2020.113642

Fekete, M., Lehoczki, A., Major, D., Fazekas-Pongor, V., Csípő, T., Tarantini, S., Csizmadia, Z., & Varga, J. T. (2024). Exploring the influence of gut–brain axis modulation on cognitive health: A comprehensive review of prebiotics, probiotics, and symbiotics. *Nutrients*, *16*(6), 789. https://doi.org/10.3390/NU16060789

Firrman, J., Liu, L. S., Zhang, L., Arango Argoty, G., Wang, M., Tomasula, P., Kobori, M., Pontious, S., & Xiao, W. (2016). The effect of quercetin on genetic expression of the commensal gut microbes Bifidobacterium catenulatum, Enterococcus caccae and Ruminococcus gauvreauii. *Anaerobe*, *42*, 130–141. https://doi.org/10.1016/J.ANAEROBE.2016.10.004

Gahtani, R. M., Shoaib, S., Hani, U., Jayachithra, R., Alomary, M. N., Chauhan, W., Jahan, R., Tufail, S., & Ansari, M. A. (2024). Combating Parkinson's disease with plant-derived polyphenols: Targeting oxidative stress and neuroinflammation. *Neurochemistry International*, *178*, 105798. https://doi.org/10.1016/J.NEUINT.2024.105798

Gandhi, G. R., Hillary, V. E., Athesh, K., da Cruz Ramos, M. L. C., de Oliveira Krauss, G. P., Jothi, G., Sridharan, G., Sivasubramanian, R., Hariharan, G., Vasconcelos, A. B. S., Montalvão, M. M., Ceasar, S. A., da Silva Calisto, V. K., & Gurgel, R. Q. (2023). The use of nanocarriers to enhance the anti-neuroinflammatory potential of dietary flavonoids in animal models of neurodegenerative diseases: A systematic review. *Mini-Reviews in Medicinal Chemistry*, *24*(13), 1293–1305. https://doi.org/10.2174/1389557523666230907093441

Garbati, P., Picco, C., Magrassi, R., Signorello, P., Cacopardo, L., Dalla Serra, M., Faticato, M. G., De Luca, M., Balestra, F., Scavo, M. P., & Viti, F. (2024). Targeting the gut: A systematic review of specific drug nanocarriers. *Pharmaceutics*, *16*(3), 431. https://doi.org/10.3390/PHARMACEUTICS16030431

Gong, Y., Liu, Z., Zhou, P., Li, J., & Miao, Y.-B. (2023). Biomimetic nanocarriers harnessing microbial metabolites usher the path for brain disease therapy. *Nano TransMed*, *2*(4), 100020. https://doi.org/10.1016/J.NTM.2023.100020

Grabska-Kobyłecka, I., Szpakowski, P., Król, A., Książek-Winiarek, D., Kobyłecki, A., Głąbiński, A., & Nowak, D. (2023). Polyphenols and their impact on the prevention of neurodegenerative diseases and development. *Nutrients*, *15*(15), 3454. https://doi.org/10.3390/NU15153454

Hassirian, N., Karimi, E., & Oskoueian, E. (2022). Nanoliposome-encapsulated phenolic-rich fraction from Alcea rosea as a dietary phytobiotic in mice challenged by Escherichia coli. *Annals of Microbiology*, *72*(1), 1–11. https://doi.org/10.1186/S13213-022-01665-9/TABLES/8

He, H., Qin, Q., Xu, F., Chen, Y., Rao, S., Wang, C., Jiang, X., Lu, X., & Xie, C. (2023). Oral polyphenol-armored nanomedicine for targeted modulation of gut microbiota-brain interactions in colitis. *Science Advances*, *9*(21). https://doi.org/10.1126/SCIADV.ADF3887/SUPPL_FILE/SCIADV.ADF3887_SM.PDF

Huang, S., He, J., Cao, L., Lin, H., Zhang, W., & Zhong, Q. (2020). Improved physicochemical properties of curcumin-loaded solid lipid nanoparticles stabilized by sodium caseinate-lactose maillard conjugate. *Journal of Agricultural and Food Chemistry*, *68*(26), 7072–7081. https://doi.org/10.1021/ACS.JAFC.0C01171/ASSET/IMAGES/MEDIUM/JF0C01171_0009.GIF

Huang, Y., Li, Y., Guan, D., Pan, Y., Yang, C., Liu, H., Chen, C., Chen, W., Liu, J., Wan, T., Zhuang, L., Wang, Q., & Zhang, Y. (2024). Acorus tatarinowii oils exert protective effects on microglia-mediated inflammatory injury via restoring gut microbiota composition in experimental stroke rats. *Brain Research Bulletin*, *213*, 110990. https://doi.org/10.1016/J.BRAINRESBULL.2024.110990

Jaberi, K. R., Alamdari-palangi, V., Savardashtaki, A., Vatankhah, P., Jamialahmadi, T., Tajbakhsh, A., & Sahebkar, A. (2024). Modulatory effects of phytochemicals on gut–brain axis: Therapeutic implication. *Current Developments in Nutrition*, *8*(6), 103785. https://doi.org/10.1016/J.CDNUT.2024.103785

Jakobušić Brala, C., Karković Marković, A., Kugić, A., Torić, J., & Barbarić, M. (2023). Combination chemotherapy with selected polyphenols in preclinical and clinical studies—An update overview. *Molecules*, *28*(9), 3746. https://doi.org/10.3390/MOLECULES28093746

Jalil, A. T., Thabit, S. N., Hanan, Z. K., Alasheqi, M. Q., Al-Azzawi, A. K. J., Zabibah, R. S., & Fadhil, A. A. (2023). Modulating gut microbiota using nanotechnology to increase anticancer efficacy of the treatments. *Macromolecular Research*, *31*(8), 739–752. https://doi.org/10.1007/S13233-023-00168-Z/METRICS

Kassem, A. M., Almukainzi, M., Faris, T. M., Ibrahim, A. H., Anwar, W., Elbahwy, I. A., El-Gamal, F. R., Zidan, M. F., Akl, M. A., Abd-ElGawad, A. M., Elshamy, A. I., & Elmowafy, M. (2024). A pH-sensitive silica nanoparticles for colon-specific delivery and controlled release of catechin: Optimization of loading efficiency and in vitro release kinetics. *European Journal of Pharmaceutical Sciences*, *192*, 106652. https://doi.org/10.1016/J.EJPS.2023.106652

Li, A., Kou, R., Liu, H., Chen, M., Wang, J., Liu, Q., Xing, X., Zhang, B., Dong, L., & Wang, S. (2023). Multi-omics analyses reveal relationships among polyphenol-rich oolong tea consumption, gut microbiota, and metabolic profile: A pilot study. *Food Chemistry*, *426*, 136653. https://doi.org/10.1016/J.FOODCHEM.2023.136653

Li, M., Wang, L. L., Lin, D., Liu, Z., Wang, H., Yang, Y., Sun, C., Ye, J., & Liu, Y. (2024). Advanced bioinspired multifunctional platforms focusing on gut microbiota regulation. *ACS Nano*. https://doi.org/10.1021/ACSNANO.4C05013/ASSET/IMAGES/MEDIUM/NN4C05013_0023.GIF

Li, W., Zhan, M., Wen, Y., Chen, Y., Zhang, Z., Wang, S., Tian, D., & Tian, S. (2024). Recent progress of oral functional nanomaterials for intestinal microbiota regulation. *Pharmaceutics*, *16*(7), 921. https://doi.org/10.3390/PHARMACEUTICS16070921

Li, Y., Peng, Y., Shen, Y., Zhang, Y., Liu, L., & Yang, X. (2023). Dietary polyphenols: Regulate the advanced glycation end products-RAGE axis and the microbiota-gut-brain axis to prevent neurodegenerative diseases. *Critical Reviews in Food Science and Nutrition*, *63*(29), 9816–9842. https://doi.org/10.1080/10408398.2022.2076064

Lombardo, D., Kiselev, M. A., & Caccamo, M. T. (2019). Smart nanoparticles for drug delivery application: Development of versatile nanocarrier platforms in biotechnology and nanomedicine. *Journal of Nanomaterials*, *2019*(1), 3702518. https://doi.org/10.1155/2019/3702518

Lv, H., Wang, Y., Yang, X., Ling, G., & Zhang, P. (2023). Application of curcumin nanoformulations in Alzheimer's disease: Prevention, diagnosis and treatment. *Nutritional Neuroscience*, *26*(8), 727–742. https://doi.org/10.1080/1028415X.2022.2084550

Ma, G., & Chen, Y. (2020). Polyphenol supplementation benefits human health via gut microbiota: A systematic review via meta-analysis. *Journal of Functional Foods*, *66*, 103829. https://doi.org/10.1016/J.JFF.2020.103829

Madureira, A. R., Nunes, S., Campos, D. A., Fernandes, J. C., Marques, C., Zuzarte, M., Gullón, B., Rodríguez-Alcalá, L. M., Calhau, C., Sarmento, B., Gomes, A. M., Pintado, M. M., & Reis, F. (2016). Safety profile of solid lipid nanoparticles loaded with rosmarinic acid for oral use: In vitro and animal approaches. *International Journal of Nanomedicine*, *11*, 3621–3640. https://doi.org/10.2147/IJN.S104623

Makarewicz, M., Drożdż, I., Tarko, T., & Duda-Chodak, A. (2021). The interactions between polyphenols and microorganisms, especially gut microbiota. *Antioxidants*, *10*(2), 188. https://doi.org/10.3390/ANTIOX10020188

Manach, C., Williamson, G., Morand, C., Scalbert, A., & Rémésy, C. (2005). Bioavailability and bioefficacy of polyphenols in humans. I. Review of 97 bioavailability studies. *The American Journal of Clinical Nutrition*, *81*(1), 230S-242S. https://doi.org/10.1093/AJCN/81.1.230S

Mohamed, R. S., Fouda, K., Salama, A., & Akl, E. M. (2024). Peanut meal-derived bioactive compounds: Extraction, co-extrusion encapsulation and neuroprotection against aluminum-induced Alzheimer's disease via in silico and in vivo studies. *Phytomedicine Plus*, *4*(3), 100588. https://doi.org/10.1016/J.PHYPLU.2024.100588

Mrozek, W., Socha, J., Sidorowicz, K., Skrok, A., Syrytczyk, A., Piątkowska-Chmiel, I., & Herbet, M. (2023). Pathogenesis and treatment of depression; role of diet in prevention and therapy. Review. *Nutrition*, *115*. https://doi.org/10.1016/J.NUT.2023.112143

Nedovic, V., Kalusevic, A., Manojlovic, V., Levic, S., & Bugarski, B. (2011). An overview of encapsulation technologies for food applications. *Procedia Food Science*, *1*, 1806–1815. https://doi.org/10.1016/J.PROFOO.2011.09.265

Negahdari, R., Bohlouli, S., Sharifi, S., Maleki Dizaj, S., Rahbar Saadat, Y., Khezri, K., Jafari, S., Ahmadian, E., Gorbani Jahandizi, N., & Raeesi, S. (2021). Therapeutic benefits of rutin and its nanoformulations. *Phytotherapy Research*, *35*(4), 1719–1738. https://doi.org/10.1002/PTR.6904

Ozkan, G., Ceyhan, T., Çatalkaya, G., Rajan, L., Ullah, H., Daglia, M., & Capanoglu, E. (2024). Encapsulated phenolic compounds: Clinical efficacy of a novel delivery method. *Phytochemistry Reviews*, 1–39. https://doi.org/10.1007/S11101-023-09909-5

Puri, V., Kanojia, N., Sharma, A., Huanbutta, K., Dheer, D., & Sangnim, T. (2022). Natural product-based pharmacological studies for neurological disorders. *Frontiers in Pharmacology*, *13*, 1011740. https://doi.org/10.3389/FPHAR.2022.1011740/BIBTEX

Puri, V., Nagpal, M., Singh, I., Singh, M., Dhingra, G. A., Huanbutta, K., Dheer, D., Sharma, A., & Sangnim, T. (2022). A comprehensive review on nutraceuticals: Therapy support and formulation challenges. *Nutrients*, *14*(21), 4637. https://doi.org/10.3390/NU14214637

Qi, G., Shi, G., Wang, S., Hu, H., Zhang, Z., Yin, Q., Li, Z., & Hao, L. (2023). A novel pH-responsive iron oxide core-shell magnetic mesoporous silica nanoparticle (M-MSN) system encapsulating doxorubicin (DOX) and glucose oxidase (Gox) for pancreatic cancer treatment. *International Journal of Nanomedicine*, *18*, 7133–7147.

Que, M., Li, S., Xia, Q., Li, X., Luo, X., Zhan, G., & Luo, A. (2024). Microbiota-gut-brain axis in perioperative neurocognitive and depressive disorders: Pathogenesis to treatment. *Neurobiology of Disease*, *200*, 106627. https://doi.org/10.1016/J.NBD.2024.106627

Riaz, M., Khalid, R., Afzal, M., Anjum, F., Fatima, H., Zia, S., Rasool, G., Egbuna, C., Mtewa, A. G., Uche, C. Z., & Aslam, M. A. (2023). Phytobioactive compounds as therapeutic agents for human diseases: A review. *Food Science & Nutrition*, *11*(6), 2500–2529. https://doi.org/10.1002/FSN3.3308

Riegelman, E., Xue, K. S., Wang, J. S., & Tang, L. (2024). Gut–brain axis in focus: Polyphenols, microbiota, and their influence on α-synuclein in Parkinson's disease. *Nutrients*, *16*(13), 2041. https://doi.org/10.3390/NU16132041

Rodríguez-Daza, M. C., Pulido-Mateos, E. C., Lupien-Meilleur, J., Guyonnet, D., Desjardins, Y., & Roy, D. (2021). Polyphenol-mediated gut microbiota modulation: Toward prebiotics and further. *Frontiers in Nutrition*, *8*, 689456. https://doi.org/10.3389/FNUT.2021.689456/BIBTEX

Rosales, T. K. O., & Fabi, J. P. (2023). Valorization of polyphenolic compounds from food industry by-products for application in polysaccharide-based nanoparticles. *Frontiers in Nutrition*, *10*, 1144677. https://doi.org/10.3389/FNUT.2023.1144677/BIBTEX

Sangnim, T., Dheer, D., Jangra, N., Huanbutta, K., Puri, V., & Sharma, A. (2023). Chitosan in oral drug delivery formulations: A review. *Pharmaceutics*, *15*(9), 2361. https://doi.org/10.3390/PHARMACEUTICS15092361

Sangnim, T., Puri, V., Dheer, D., Venkatesh, D. N., Huanbutta, K., & Sharma, A. (2024). Nanomaterials in the wound healing process: New insights and advancements. *Pharmaceutics*, *16*(3), 300. https://doi.org/10.3390/PHARMACEUTICS16030300

Sasso, J. M., Ammar, R. M., Tenchov, R., Lemmel, S., Kelber, O., Grieswelle, M., & Zhou, Q. A. (2023). Gut microbiome-brain alliance: A landscape view into mental and gastrointestinal health and disorders. *ACS Chemical Neuroscience*, *14*(10), 1717–1763. https://doi.org/10.1021/ACSCHEMNEURO.3C00127/ASSET/IMAGES/LARGE/CN3C00127_0017.JPEG

Silva, R. F. M., & Pogačnik, L. (2020). Polyphenols from food and natural products: Neuroprotection and safety. *Antioxidants*, *9*(1), 61. https://doi.org/10.3390/ANTIOX9010061

Singh, H., Chopra, C., Singh, H., Malgotra, V., Khurshid Wani, A., Singh Dhanjal, D., Sharma, I., Nepovimova, E., Alomar, S., Singh, R., Sharma, V., & Kuca, K. (2024). Gut-brain axis and Alzheimer's disease: Therapeutic interventions and strategies. *Journal of Functional Foods*, *112*, 105915. https://doi.org/10.1016/J.JFF.2023.105915

Song, W., Yuan, Q., Xie, Y., Wang, Y., Deng, D., & Guo, H. (2024). Formulation and characterization of nanocapsules loaded with roselle anthocyanins extract and enhancement of anthocyanins bioaccessibility. *Food Chemistry*, *459*, 140446. https://doi.org/10.1016/J.FOODCHEM.2024.140446

Tabrez, S., Jabir, N. R., Adhami, V. M., Khan, M. I., Moulay, M., Kamal, M. A., & Mukhtar, H. (2020). Nanoencapsulated dietary polyphenols for cancer prevention and treatment: Successes and challenges. *Nanomedicine*, *15*(11), 1147–1162. https://doi.org/10.2217/NNM-2019-0398

Tang, L., Zhang, X., Zhang, B., Chen, T., Du, Z., Song, W., Chen, W., & Wang, C. (2024). Electroacupuncture remodels gut microbiota and metabolites in mice with perioperative neurocognitive impairment. *Experimental Gerontology*, *194*, 112507. https://doi.org/10.1016/J.EXGER.2024.112507

Tao, W., Zhang, Y., Wang, B., Nie, S., Fang, L., Xiao, J., & Wu, Y. (2024). Advances in molecular mechanisms and therapeutic strategies for central nervous system diseases based on gut microbiota imbalance. *Journal of Advanced Research*. https://doi.org/10.1016/J.JARE.2024.03.023

Torchilin, V. P. (2006). Micellar nanocarriers: Pharmaceutical perspectives. *Pharmaceutical Research*, *24*(1), 1–16. https://doi.org/10.1007/S11095-006-9132-0

Velichkov, M., Bezur, Z., van Reekum, C. M., & Williams, C. M. (2024). A biphasic response to blueberry supplementation on depressive symptoms in emerging adults: A double-blind randomized controlled trial. *European Journal of Nutrition*, *63*(4), 1071–1088. https://doi.org/10.1007/S00394-023-03311-9

Venkidesh, B. S., Narasimhamurthy, R. K., Jnana, A., Reghunathan, D., Sharan, K., Chandraguthi, S. G., Saigal, M., Murali, T. S., & Mumbrekar, K. D. (2023). Pelvic irradiation induces behavioural and neuronal damage through gut dysbiosis in a rat model. *Chemico-Biological Interactions*, *386*, 110775. https://doi.org/10.1016/J.CBI.2023.110775

Vivarelli, S., Costa, C., Teodoro, M., Giambò, F., Tsatsakis, A. M., & Fenga, C. (2022). Polyphenols: A route from bioavailability to bioactivity addressing potential health benefits to tackle human chronic diseases. *Archives of Toxicology*, *97*(1), 3–38. https://doi.org/10.1007/S00204-022-03391-2

Wang, H., Wang, C., Zou, Y., Hu, J., Li, Y., & Cheng, Y. (2020). Natural polyphenols in drug delivery systems: Current status and future challenges. *Giant*, *3*, 100022. https://doi.org/10.1016/J.GIANT.2020.100022

Wellens, J., Vissers, E., Matthys, C., Vermeire, S., & Sabino, J. (2023). Personalized dietary regimens for inflammatory bowel disease: Current knowledge and future perspectives. *Pharmacogenomics and Personalized Medicine*, *16*, 15–27. https://doi.org/10.2147/PGPM.S359365

Williams, G. M., Tapsell, L. C., & Beck, E. J. (2023). Gut health, the microbiome and dietary choices: An exploration of consumer perspectives. *Nutrition & Dietetics*, *80*(1), 85–94. https://doi.org/10.1111/1747-0080.12769

Winiarska-Mieczan, A., Kwiecień, M., Jachimowicz-Rogowska, K., Donaldson, J., Tomaszewska, E., & Baranowska-Wójcik, E. (2023). Anti-inflammatory, antioxidant, and neuroprotective effects of polyphenols-polyphenols as an element of diet therapy in depressive disorders. *International Journal of Molecular Sciences*, *24*(3). https://doi.org/10.3390/IJMS24032258

Xu, Q., Sun, L., Chen, Q., Jiao, C., Wang, Y., Li, H., Xie, J., Zhu, F., Wang, J., Zhang, W., Xie, L., Wu, H., Zuo, Z., & Chen, X. (2024). Gut microbiota dysbiosis contributes to depression-like behaviors via hippocampal NLRP3-mediated neuroinflammation in a postpartum depression mouse model. *Brain, Behavior, and Immunity*, *119*, 220–235. https://doi.org/10.1016/J.BBI.2024.04.002

Zhang, Y., Yu, W., Zhang, L., Wang, M., & Chang, W. (2022). The interaction of polyphenols and the gut microbiota in neurodegenerative diseases. *Nutrients*, *14*(24). https://doi.org/10.3390/NU14245373

Zhang, Z., Li, X., Sang, S., McClements, D. J., Chen, L., Long, J., Jiao, A., Jin, Z., & Qiu, C. (2022). Polyphenols as plant-based nutraceuticals: Health effects, encapsulation, nano-delivery, and application. *Foods*, *11*(15), 2189. https://doi.org/10.3390/FOODS11152189

Zhao, Y., Qin, S., Yang, Z., Lu, Y., Ma, Z., Ping, X., Sun, G., & Pei, L. (2024). Gastrodin ameliorates depressive-like behaviors via modulating gut microbiota in CUMS-induced mice. *Behavioural Brain Research*, *465*, 114968. https://doi.org/10.1016/J.BBR.2024.114968

Zhu, H., Wang, W., & Li, Y. (2024). The interplay between microbiota and brain-gut axis in epilepsy treatment. *Frontiers in Pharmacology*, *15*, 1276551. https://doi.org/10.3389/FPHAR.2024.1276551/BIBTEX

25 Probiotics as Nutraceuticals in Neurological Disorders

Malarvannan M.[1*], *S. Bhanu Teja Naik*[1], *David Paul*[1], *Satheeshkumar Nanjappan*[2], *and V. Ravichandiran*[2]

[1] Department of Pharmaceutical Analysis, National Institute of Pharmaceutical Education and Research (NIPER)-Kolkata, West Bengal, Kolkata, India
[2] Department of Natural Products, National Institute of Pharmaceutical Education and Research (NIPER)-Kolkata, West Bengal, Kolkata, India

25.1 INTRODUCTION

Probiotics, also referred to as "friendly bacteria," are living microorganisms which provides beneficial health effects to the host when delivered in the appropriate dosages. These valuable microorganisms are primarily found in fermented foods, such as kefir, yogurt, and sauerkraut. They can also be obtained as nutritional supplements [1]. The burgeoning field of probiotics research has highlighted their potential as nutraceuticals substances that provide medical or health benefits beyond basic nutrition. Among the latest and most exciting fields of study is the use of probiotics to treat neurological disorders (ND) includes a wide range of conditions that affect the nervous system. These conditions are not limited to depression, multiple sclerosis (MS), Alzheimer's disease (AD), and Parkinson's disease (PD). Traditional treatments for these conditions often focus on symptomatic relief and may include medications, physical therapy, and lifestyle modifications. However, these treatments frequently fall short of providing a cure or long-term relief. As a result, there is a rising interest in investigating complementary and alternative methods, including the use of probiotics [2].

The gut–brain axis (GBA), a bidirectional communication mechanism that connects the gut microbiota (GM) and the central nervous system (CNS), significantly influences neurological health and disease, according to recently available data. The gut microbiome, which is made up of billions of microbes, has a variety of effects on brain function and behavior. These include the immune system (IS) control, neurotransmitter (NT) synthesis, and neuroinflammation (NI) modulation. Dysbiosis, or disruptions in the intestinal microbiota, has been associated with a variety of neurological diseases. For example, AD is distinguished by a consistent fall of cognitive function and memory loss [3]. Research indicates that probiotics may help in modulating NI and oxidative stress, which are crucial in Alzheimer's pathology. Similarly, in PD, a condition marked by motor dysfunction and neurodegeneration, evidence suggests that probiotics could mitigate symptoms by supporting gut health and reducing inflammation. MS is an inflammatory condition that causes the nerve fibers protective sheath to deteriorate and has also been studied in the context of probiotics. The immunomodulatory effects of probiotics have the potential to delay the progression of disease and impact immune responses, thereby providing therapeutic benefits. In the context of psychiatric and mood disorders, including depression and anxiety, probiotics are presently being investigated for their potential to improve cognitive function and mood by altering the gastrointestinal (GI) microbiota and its impact on the nervous system [4].

Current treatments for ND often have limitations, such as side effects, variable efficacy, and limited ability to address underlying causes. Therefore, there is a great demand for cutting-edge treatment strategies. Probiotics, as nutraceuticals, present a promising path because of their capacity to improve brain health in general and the GBA in particular. They are a desirable alternative for adjunctive therapy since they are typically regarded as safe and have few adverse effects. Despite the promising potential, it is essential to approach probiotic treatment with caution. Probiotics effectiveness in treating neurological illnesses is still being studied, and individual patient variables, dose, and strain selection can all affect the outcome. Rigorous clinical trials and research are necessary to establish optimal probiotic formulations and treatment regimens for different neurological conditions. Probiotics represent a fascinating and emerging field in the therapy of ND. By leveraging their impact on the GBA, they offer potential benefits that could complement traditional therapies and enhance overall patient outcomes. As research continues to evolve, probiotics may become a valuable component in the holistic management of neurological health [5].

DOI: 10.1201/9781003536802-25

25.2 NUTRITION, MICROBIOME, AND NEUROPSYCHIATRIC DISORDERS

An increasing number of individuals are recognizing the intricate relationship between neuropsychiatric conditions, diet, and intestinal microbiota. The GM's elements and their functions are regulated by the GBA, which directly influences the brain function and mental health.

25.2.1 Nutrition

The process by which living things acquire and use food to sustain their overall wellness, development, and maintenance is known as nutrition. It entails consuming, breaking down, absorbing, and metabolizing nutrients substances present in food that provide us energy, help to form and repair tissues, and control a number of body functions. Patients with neurological illnesses may have altered nutritional status as a result of factors including dysphagia, mobility difficulties, depression, and cognitive impairment that can either directly or indirectly limit dietary intake. Increased morbidity and mortality result from problems caused by malnutrition, which also delays rehabilitation. The relationship between nutrition and neurology is becoming increasingly acknowledged, and numerous studies have demonstrated how dietary imbalances and malnutrition can both induce and result in specific ND. It is frequently shown that an excess or deficiency of a particular nutrient as a result of illness or malnutrition causes or exacerbates a number of neurological symptoms and disorders. On the other hand, people with specific neurological conditions, such as dementia, stroke, PD, and autism spectrum disorders (ASD) are more likely to experience GI problems, eating issues, and nutritional deficiencies [6].

According to the study, these changes in the GM could exacerbate preexisting GIDs and be linked to problems with behavior and cognitive function [7]. Drugs can alter nutritional levels, which can result in a variety of neurological symptoms. Research by Matsuyama et al. provides an example of this, demonstrating a negative association between serum zinc levels and the zinc-chelating medication levodopa. PD patients receiving levodopa were shown to have reduced levels of zinc, a vitamin important in immunological, integumentary, and neurological system processes. This led to mental symptoms, dermatitis, stomatitis, and taste problems [8]. The nutritional therapy for neurological conditions and potential targets is illustrated in Figure 25.1.

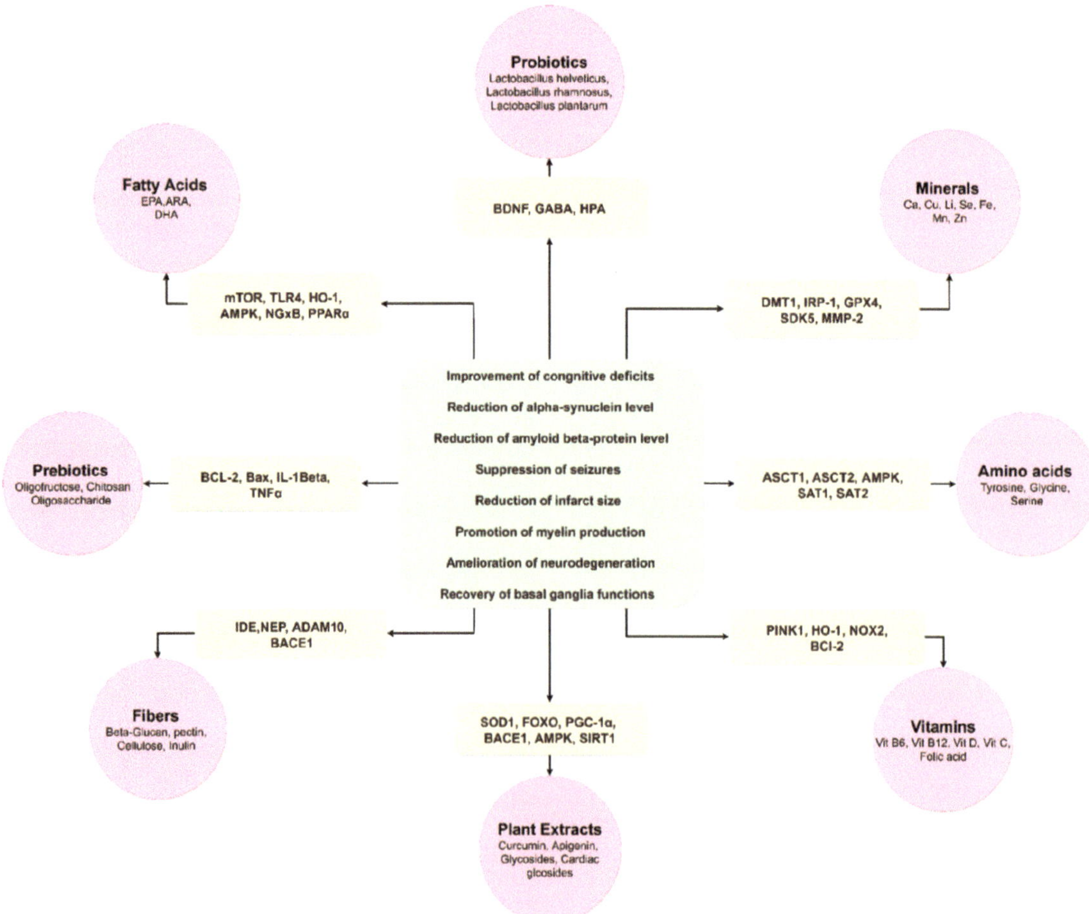

FIGURE 25.1 Nutritional therapy for neurological conditions and potential targets.

25.2.2 Microbiome

Numerous microorganisms can be found in the human gut, such as bacteria, phages, fungi, viruses, microbial eukaryotes, and archaea. This set of microorganisms is referred to as the "microbiota," and its gene resource is referred to as the "microbiome." The bacteria themselves are now usually known as the microbiome. The gut microbiome of the majority of healthy humans is primarily composed of bacteria, primarily part of the phyla Firmicutes and Bacteroidetes, with very minor numbers of nonbacterial organisms (Figure 25.2) [9]. The majority of the microbes that reside in our digestive tracts coexist harmoniously with us, their hosts. That implies that the partnership is beneficial to us both. They give us food and shelter, and they perform vital bodily functions for us. These beneficial bacteria also aid in the control of potentially hazardous ones. The human body and its symbiotic microbiota organisms form a complex micro-ecological system. Enteral and external diseases may occur from variations in the count and composition of gut microbes, which can also affect how well the gut barrier functions, secrete more toxic compounds, and secrete less beneficial molecules. The influence of GM and IS on the development of ND and the maintenance of brain function. The gut flora is primarily composed of the six primary phyla Bacteroidetes, Firmicutes, Actinomycetes, Proteobacteria, Verrucomicrobia, and Fusobacteria, with Firmicutes and Bacteroidetes being the most predominant [10].

25.2.3 Microbiota GBA

The "GM–brain axis" is a bidirectional information network that exists between the GM and the brain. It has the potential to provide a novel form of brain protection in the near future. The IS, enteric nervous system (ENS), the vagus nerve, and the metabolic activity of gut microbes are among the numerous pathways that can establish a connection between the gut and the brain. The body's vagus nerve regulates the activity of several organs, including the heart rate and the motility of the intestines. It can also send messages from the PNS to the CNS. An anti-inflammatory reaction can be triggered by the gut's vagus signal to combat the sepsis that microbes have caused. Through the vagus nerve, gut microbes can influence brain activity; following a vagotomy, the microbes will lose their ability to control behavior [11]. Most recent research indicates that the intestinal microbiota is a critical factor in the GBA.

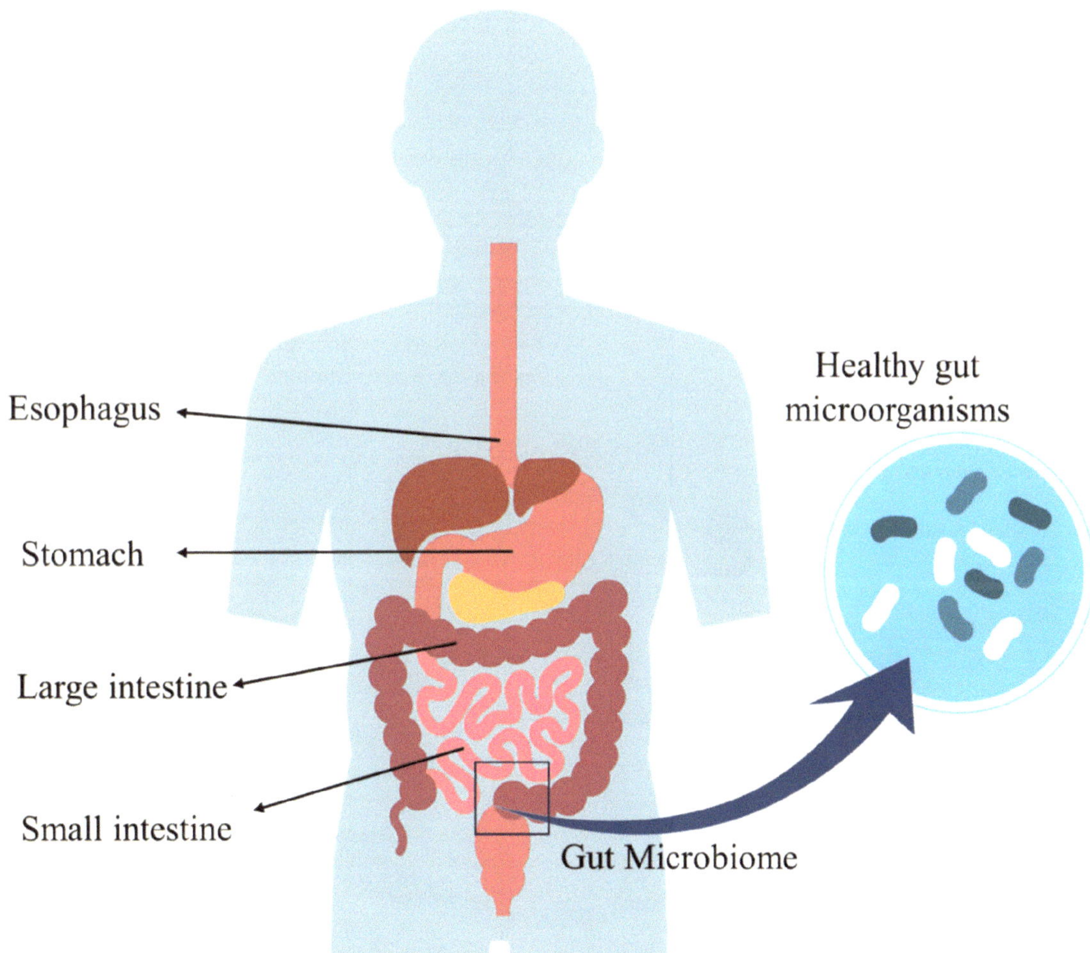

FIGURE 25.2 The gut microbiome is an assortment of gut-dwelling microbes.

Consequently, the "microbiota/GBA" is the term used to describe the interaction between the brain and the microbiota. Toll-like receptors 2 and 4 (TLR2 and TLR4), which are pattern recognition receptors (PRRs), are influenced by the GI microbiota in terms of CNS and ENS development, functioning, and diseases, as indicated by a multitude of studies.

The development and progression of numerous neurological diseases are caused by gut dysbiosis and the ensuing loss of GI permeability and gut barrier integrity, which permit an increase in the translocation of metabolites produced by gut bacteria and microbe-associated molecular patterns (MAMPs) into mesenteric lymphoid tissues [12]. The bidirectional relationship is significantly influenced by a variety of pathways, such as the hypothalamic-pituitary-adrenal (HPA), autonomic nervous system (ANS), ENS, immunological system, endocrine system, and neurological system, as illustrated in Figure 25.3.

25.3 ROLE OF GM IN ND

The GM has been demonstrated to be involved in a diverse array of regulatory processes and related pathways, as well as in unique bacterial behaviors, as demonstrated by the GBA. Furthermore, the intestinal microbiota is acknowledged as a risk factor for neurological conditions that affect the CNS, capable of regulating the progression of the disease and responding to treatment (Figure 25.4). The GBA appears to be crucial in immune-mediated, neurocrine, and endocrine signaling pathways due to its capacity to facilitate bidirectional communication between the gut flora and the brain. The GI microbiota regulates a variety of ND by providing them with antibiotics, prebiotics, probiotics, postbiotics, fecal transplantation, and synbiotics [13].

25.4 PROBIOTICS POTENTIAL AS NUTRACEUTICALS

Probiotics comprise living microorganisms that have become a significant category within the nutraceutical industry (Figure 25.5) and when consumed in adequate quantities, they provide good health outcomes. These beneficial bacteria, predominantly from the *Lactobacillus* and *Bifidobacterium* genera, along with certain yeasts, such as *Saccharomyces boulardii*, are primarily known for their positive impact on gut health. Their ability to modulate the GM complex community of microbes that live in the gut system forms the basis of their myriad health benefits. As our knowledge of the significance of the GM in general health has extended, so too has the recognition of probiotics as valuable therapeutic agents beyond mere dietary supplements. The GI tract is home to trillions of microbes that play crucial roles in digestion, immune function, and even mental health. Probiotics help maintain a balanced microbiota by outcompeting pathogenic bacteria, producing substances that inhibit their growth, and enhancing the intestinal barrier function. Dysbiosis, or disruptions in the GM, has been connected with a variety of maladies, such as metabolic syndrome, obesity, irritable bowel syndrome (IBS), and inflammatory bowel disease (IBD). Therefore, it is imperative to preserve this equilibrium. Probiotics can alleviate symptoms of these conditions by restoring homeostasis. For example, the frequency and severity of IBS symptoms, such as irregular bowel movements, abdominal pain, and bloating, have been demonstrated to be reduced by specific strains, such as *B. infantis* and *L. rhamnosus* GG.

Additionally, probiotics are essential for immune modulation. Around 70% of the body's immune cells are positioned in the gut-associated lymphoid tissue (GALT), where probiotics can interact with them to improve immune responses. They stimulate the production of regulatory T

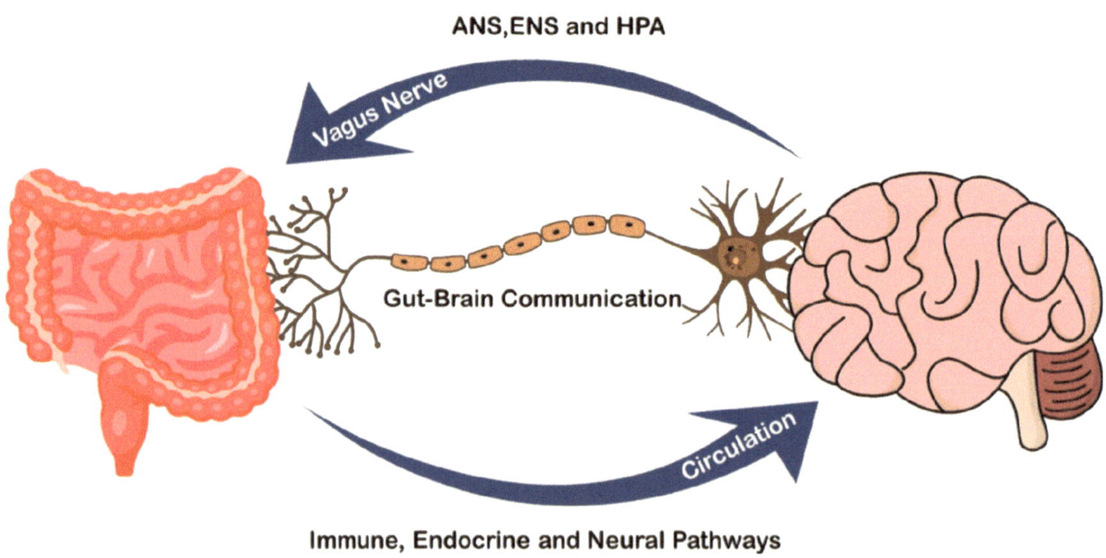

FIGURE 25.3 Schematic diagram illustrating the brain–gut communication system.

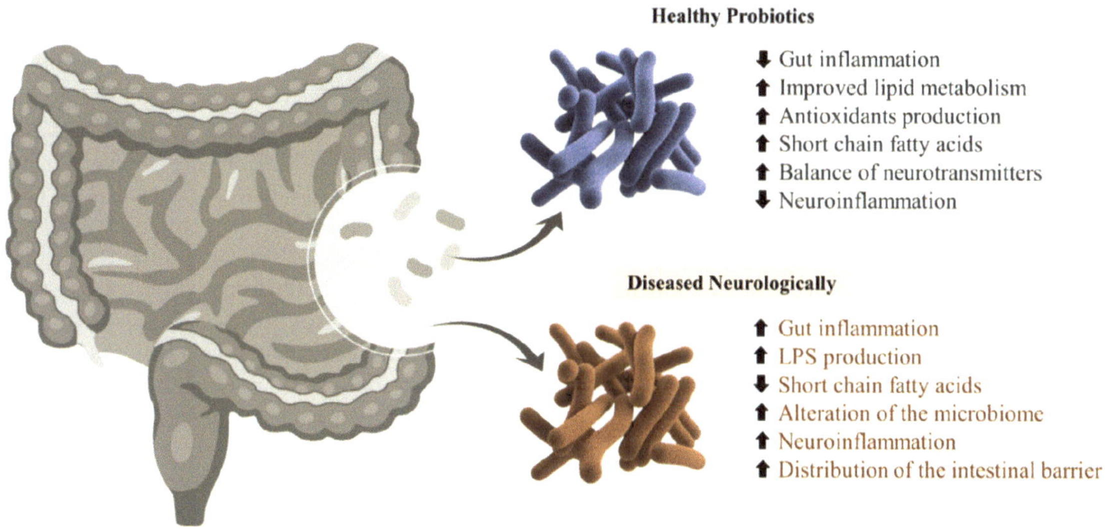

FIGURE 25.4 The positive effect of probiotics for neurological disorder.

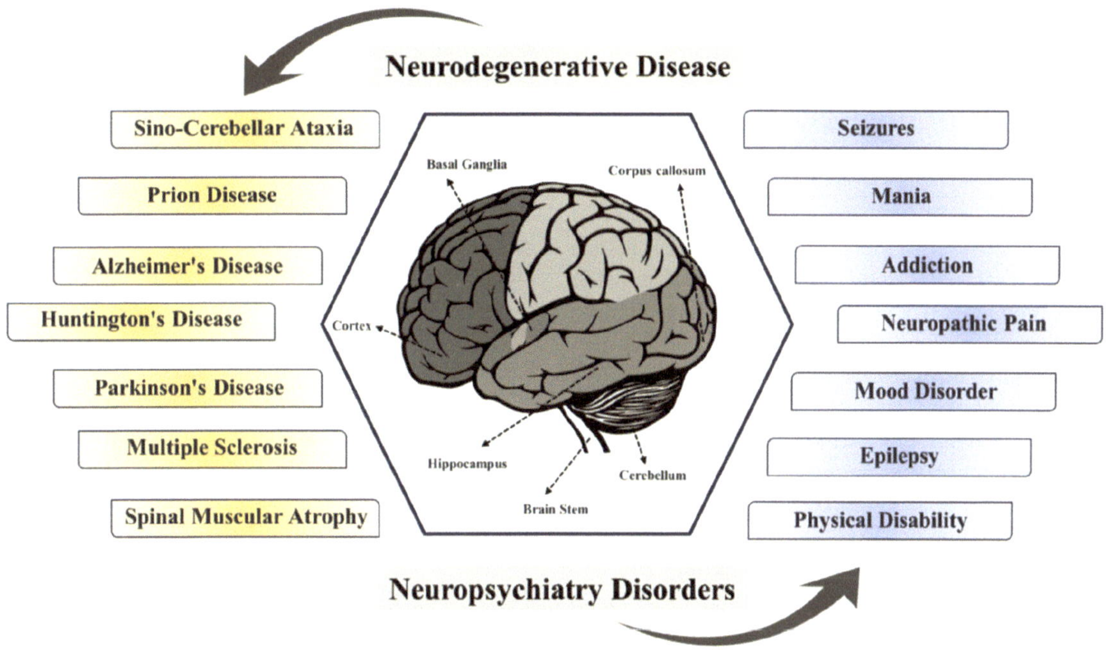

FIGURE 25.5 Classification of nutraceuticals with examples.

cells, which help modulate immune responses and maintain tolerance to harmless antigens, thus reducing inflammation. This immunomodulatory effect is particularly beneficial in managing autoimmune diseases and allergies. Research has demonstrated that probiotics have the potential to serve as a natural method of enhancing immunity by reducing the frequency and intensity of allergic responses and respiratory infections. Beyond gut and immune health, probiotics have been implicated in mental well-being, highlighting the GBA's importance. NTs, including gamma-aminobutyric acid (GABA) and serotonin, which are essential for mood control, can be produced by probiotics. According to clinical research, some probiotic strains, such as *B. longum* and *L. helveticus*, might lessen anxiety and depressive symptoms. This makes them a viable supplementary treatment for mental health conditions. The probiotics in metabolic health and their role is another area of intense research. Dysbiosis has been related with type 2 diabetes and obesity conditions characterized by chronic low-grade inflammation and altered glucose metabolism. Probiotics can influence energy homeostasis and insulin sensitivity by modulating GM composition and function. For instance, *B. breve* and *L. gasseriin* clinical studies demonstrated to lower body weight and enhance insulin sensitivity. These study outputs put forth that the probiotics could be integrated into dietary treatment strategies aimed at the management of metabolic disorders [14].

Moreover, probiotics have shown promise in women's health, particularly in preventing and treating urogenital infections, such as bacterial vaginosis and urinary tract infections (UTIs). The vaginal microbiota, predominantly composed of *Lactobacillus* species, plays a critical function in preserving the health of the vagina by generating lactic acid, which lowers the pH and inhibits pathogenic bacteria. Probiotic supplementation can help restore and maintain this protective environment, reducing the recurrence of infections. Despite the mounting evidence supporting the benefits of probiotics, challenges remain in their application as nutraceuticals. One of the primary challenges is the need for strain-specific efficacy, as not all probiotics confer the same health benefits. These benefits clearly show that probiotics represent a promising class of nutraceuticals with wide-ranging health benefits. Their capacity to alter GM and improve immunological response, support mental health, improve metabolic health, and maintain urogenital health underscores their therapeutic potential [15].

25.5 ROLE OF PROBIOTICS IN THE MANAGEMENT OF ND

Probiotics promise to maintain intestinal integrity, enhance gut health, boost the manufacture of BDNF, which improve neuronal survival and development and protect the intestinal mucosa and the BBB. Probiotics may be used as an additional treatment to treat mental and metabolic illnesses. Strains of *Bifidobacterium* and *Lactobacillus* are known to be powerful probiotics that support the gut–brain system's reciprocal interactions [16]. Recent research indicates that disruptions or imbalances in GM might cause neurological illnesses by way of the GBA. Probiotics, such as *L. acidophilus*, *L. casei*, and *B. bifidum*, are good options for halting the advancement of neurological illnesses. It has been shown that the inflammatory process and oxidative stress can be regulated by manipulating the intestinal microbiota through the use of probiotics. This chapter offers an overview of the information that is currently available, the diversity of bacteria, problems in the GBA, and the mechanism of action of probiotics in preventing neurological illnesses.

25.5.1 Role in Autism Spectrum Disorder (ASD)

Probiotic therapies in children with ASD may be seen as an alternate approach to therapy or a therapeutic adjunct. Probiotics create regulatory T cells, lower lipopolysaccharides levels, and increase BDNF to support gut integrity and regulate inflammatory pathways in the gut and CNS. Sustaining BDNF levels is vital for neuronal survival, synaptic plasticity, and proper neurological development [17]. Probiotics are living bacteria that give the host health benefits when given in sufficient doses. They are classified as psychobiotics because they generate and transport neuroactive chemicals that operate on the GBA, such as GABA generated by *L. brevis* and *B. denticola*, and dopamine produced by *S. aureus* and *E. coli*. Numerous probiotic species, including mixes of various formulations, products, and single-strain probiotics have all been used to treat ASD. These probiotics are frequently used in conjunction with nutritional or behavioural therapies. While *L. plantarum* and *L. reuteri* have been demonstrated to considerably enhance social behavior in mice [18]. There is not enough data to conclude that probiotics may effectively treat ASD in children. Moreover, tumor necrosis factor-α (TNF-α), IL-12, and interferon-α (INF-α) are examples of pro-inflammatory cytokines that are suppressed by probiotics, but transforming growth factor-β (TFGF-β) and IL-10 are produced in greater quantities as anti-inflammatory cytokines. This regulates gut inflammation, controls immunological processes, and ameliorates behavioural issues in ASD patients. Therefore, probiotic treatment improves positive feelings and behavioural symptoms in ASD while dramatically reducing inflammation, anxiety, tension, and sadness. Children with ASD were examined in relation to the effects of probiotic and prebiotic supplements (*L. casei*, *L. acidophilus*, *B. longum*, *L. delbrueckii*, and *B. bifidum*). According to the findings, probiotic supplements reduced GI symptoms and enhanced the evaluation of autism therapy in areas such as speech, sociability, communication, physical behavior, sensory awareness, and health [19].

25.5.2 Role in Parkinson's Diseases (PD)

The abundance of *Bacteroides massiliensis*, *Dorealonge catena*, *Bacteroides plebeius*, *Bacteroides coprocola*, *Stoquefichus massiliensis*, *Prevotellacopri*, *Faecalibacterium*, *Blautia glucerasea*, *Ruminococcus callidus*, *Bacteroi desdorei*, and *Coprococcus eutactus* was decreased, and the abundance of *Catabacter hongkongensis*, *L. mucosae*, *Bifidobacterium*, *Papillibacter cinnamivorans*, *Oscillospira*, *Christensenel laminuta*, and *Ruminococcus bromii* was enhanced in PD patients [20] by enhancing the host's antioxidant system, lowering inflammatory responses, strengthening neuroprotective qualities, and reducing NI probiotic treatments enhance the composition and health of GM [21]. Patients with PD experienced less bloating and cramping in their abdomens when they took probiotic pills containing *L. acidophilus* and *B. infantis* for more than three months [22]. Patients with PD who took fermented milk supplements containing *L. casei* Shirota for six weeks had improvements in their bowel habits and consistency of feces, along with a decrease in bloating and stomach pain.

Peripheral blood mononuclear cells (PBMCs) from PD patients were co-cultured 1:1 with probiotic strains *L. plantarum* LP01, *L. salivarius* LS01, *L. rhamnosus* LR06, *L. acidophilus* LA02, and *B.animalis* subspecies lactis BS01 and *B. breve* BR03. Anti-inflammatory cytokines were elevated and pro-inflammatory cytokines were significantly reduced in the PBMCs containing *L. salivarius* LS01 and *L. acidophilus* LA02 [23]. In PD patients, Borzabadi et al. found that probiotic supplementation (*Lactophilus*,

Lactobacillus, *Bifidum*, *Fermentum*) increased the expression of peroxisome proliferator-activated receptor-γ (PPAR-γ), TNF-α, TGF-β, IL-1, and IL-8 without having a major impact on the expression of low-density lipoprotein receptor (LDLR) and vascular endothelial growth factor (VEGF) [24]. 6-OHDA-manicured probiotics, such as *B. lactis* DSM 32247, *S. thermophilus* DSM 32245, *L. acidophilus* DSM 32241, *B. lactis* DSM 32246, *L. helviticus* DSM 32242, L. plantarum DSM 32244, *L. paracasei* DSM 32243, *L. brevis* DSM 27961 were added to C57BL/6 mice and human SH-SY5Y cells. The effects of the probiotics were monitored. Supplementing with probiotics decreased the proteins that cause neuronal death and boosted the expression of neuroprotective proteins. The results showed that probiotics had a protective impact on dopaminergic neurons and behavioral features throughout development. They also demonstrated how probiotics counteract NI and oxidative damage via stimulating the neuroprotective and survival pathways of BDNF and activating the antioxidant and anti-inflammatory properties of PPAR-γ [25].

The probiotic mixture containing *B. animalis* subspecies lactis, *B. rhamnosus* GG, and *L. acidophilus* increases the synthesis of butyrate in a condition of 1-methyl-4-phenyl-1,2,3,6-tetrahydropyridine (MPTP) induced neurotoxicity. This helps elevate the levels of BDNF and glial cell-line derived neurotrophic factor (GDNF). *L. rhamnosus* GG reduces the expression of monoamine oxidase B in the striatum and increases neurotrophic factors in the brain [26]. Moreover, the research for three months, PD patients received co-treatment with the dopamine agonists Probio-M8 and benserazide. Less GI symptoms, anxiety, and improved sleep were shown by Probio-M8 and the group that also received medicine. Additionally, the metagenomic study showed that the probiotic intervention significantly raised the abundance of *B. animalis*, Ruminococcaceae, *Lachnospira*, and *L. fermentum* while significantly reducing the number of *Klebsiella oxytoca*. Therefore, co-administration of probiotic M8 increased the organisms involved in the production of GABA, dopamine, serum acetic acid, and secondary bile acids. ProbioM8's modifying effects aid in the management of the gut microbiome, which in turn regulates gut metabolism and alleviates symptoms in PD [27].

25.5.3 Role in Alzheimer's Disease (AD)

Studies reveal a connection between AD and changes in gut microbial diversity as well as abnormalities in the GBA. Among the most effective defenses against cognitive deterioration in AD is due to probiotics. Several *in vivo* research and recent clinical trials have demonstrated the therapeutic effect of certain bacterial strains in slowing down AD. Patients with AD show an increase in brain oxidation as the disease progresses. Reactive oxygen species and the antioxidant system are two ways that the GM may affect the degree of oxidative state in AD. In the presence of oxidative stress, nitrate and nitrite can be converted by gut Lactobacilli and Bifidobacteria into nitric oxide (NO), which can be harmful. Reactive oxygen species (ROS) are harmful substances that are produced when NO undergoes an oxidative reductive reaction. These molecules are linked to neuronal death and mitochondrial malfunction [28]. Aβ deposition can be accelerated and an oxidative response can be triggered by oxidative stress. Due to its increased levels of Aβ and tau hyperphosphorylation, this oxidation is thought to be the pathogenic hallmark in the course of AD patients' condition. Even if the entire etiopathogenesis of AD is yet unknown, knowing the functions and processes of GM in AD pathogenesis can aid in the development of effective treatment plans for AD. Additionally, probiotics may enhance gut barrier function and reduce systemic inflammation, contributing to a healthier brain environment. While more research is needed, early studies indicate that probiotics could be a promising adjunct therapy in decreasing the rate of progression and alleviating symptoms of AD [29].

25.5.4 Role in Mood States: Anxiety, Stress, and Depression

Many stress, depression, and anxiety-related conditions are linked to intestinal dysfunctions, gut diseases, and mood disorders because of altered gut and CNS functioning and interactions. HPA axis is associated with the positive benefits of probiotics on stress and cognition. Researchers looked at how *B. longum* NCC3001 supplementation affected the levels of anxiety and sadness in individuals with mental diseases and irritable bowel syndrome. According to the findings, *B. longum* improved the individuals' quality of life and decreased depression while having no discernible effect on their anxiety levels. This effect was ascribed to alterations in limbic reactivity in the brain.

In young, healthy subjects, the Leiden index of depression sensitivity scale showed that the supplementation of (*L. lactis* W19, *Bifidum* W23, *L. casei* W56, *B. lactis* W52, *L. lactis* W24, *L. acidophilus* W37, *L. salivarius* W24, and *L. brevis* W63) for four weeks decreased cognitive reactions to depressed mood states [30]. Similarly, 30 days of supplementing with *B. longum* R0175 and *L. helviticus* R0052 lowered cortisol levels and the hospital anxiety and depression scores. Over the four weeks, the effects of supplementing with fermented milk, which comprises *L. lactis*, *L. bulgaricus*, *S. thermophilus*, and *B. animalis*, *B. lactis* on the brain activity of healthy female volunteers were reported. While the study participants explored negative emotions like anger and fear, functional magnetic resonance imaging monitored their brain activity. The probiotic group had decreased activity in the brain areas linked to emotion processing, such as the insula, sensorimotor cortex, and periaqueductal grey. According to the findings, probiotics may be able to regulate emotional behaviors in healthy people [31]. To evaluate the potential benefit of probiotics for major depressive illness, the study team performed a meta-analysis. According to the study's findings, treating major depressive disorder patients with combinational probiotics which contain many strains of probiotics reduced their symptoms of depression and enhanced their GBA.

25.5.5 Role in Cognitive Impairment

When healthy human subjects are supplemented with probiotics *L. lactis* W58, *L. casei* W56, *L. lactis* W16, *L. brevis* W63, *L. acidophilus* W37, *L. salivarius* W24, *B. lactis* W52, and *B. bifidum* W23 supplements cause differences in cognition under stress. Research indicates that probiotic supplementation causes stress-related increases in brain functioning and working memory performance under stressful settings. These increases have been linked to intervention-related neural alterations in the frontal cortex [32]. The use of probiotics has an impact on brain activity patterns, stress, and cognitive reaction. Consuming *B. longum* 1714 enhances cognitive functions and behaviors associated with stress. By assessing processing, memory, attention, social and emotional cognition, and cognitive performances were assessed. The study was performed for four weeks. The strain *B. longum* 1714 showed anti-stress and precognitive benefits, as well as reducing the acute stressor responses, which based on enhanced EEG profile memory and visuospatial recall revealed that it could have probiotic properties [33].

Numerous investigations revealed that the following microorganisms significantly affect neurotransmission, CNS functions, and emotional states associated with mental disorders, such as *B. infantis*, *B. breve*, *L. plantarum*, *L. acidophilus*, *L. rhamnosus*, *L. sporogenes*, *L. helviticus*, *L. bulgaricus*, *L. casei*, *L. paracasei*, *L. salivarius*, and *L. delbrueckii*. Another clinical trial examined the effects of supplementing healthy patients with mild cognitive impairment (MCI) with *L. plantarum* C29-fermented soybean (DW2009) for a period of 12 weeks. To measure cognitive scores, computerized neurocognitive function tests were employed and serum variations in BDNF levels were investigated. Supplementing with DW2009 enhanced cognition and attention, reduced memory impairment, and elevated BDNF expression in the blood. Following DW2009 treatment elevated BDNF levels demonstrated the role of probiotics in neuronal survival and improved cognitive function. Probiotic supplementation has been shown to enhance neuronal health and cognitive development. By altering the signals from the GBA, probiotics have a good influence on cognition and other mental states. They also directly affect the host's intestinal microbiota and are not affected by host enzymes. Probiotics may help depression sufferers with their symptoms. However, it is unknown how probiotics affect mood states; hence, more investigation is required to identify the molecular mechanism behind the beneficial effects of the probiotic intervention (Table 25.1) [34].

25.6 NUTRACEUTICALS PREPARED WITH SPECIFIC STRAINS OF PROBIOTICS

Using a therapeutic nutritional approach along with developing procedures for incorporating functional foods and nutraceuticals into the diet may enhance gut microbial ecology, reduce GIT inflammation, and improve neurological health. The many kinds of probiotic bacteria and their strains have been examined in detail in order to develop beneficial nutraceutical products. These products are marketed commercially and have labels on them that specify the name, viable number, and strain of the probiotics. The nutraceuticals in this probiotic when consumed as a functional food are essential because they support the survival and healing of beneficial gut bacteria. Numerous individuals who suffered from various allergies to commercial food and medications began searching for safer, more natural, and innovative options for their treatment using health food items [35]. In light of this necessity, several probiotic investigations have been conducted, and nutraceuticals based on specific probiotic strains have been developed. Selected probiotic microorganisms can be used to make this via a controlled fermentation process. These strains are used for the fermentation of substrates derived from agriculture, dairy or non-dairy. As an alternative, a number of nutraceuticals have been developed commercially to address customer demands for the treatment of particular medical conditions. Certain probiotic strains are added to these symbiotic products during production and the necessary prebiotics, which are substances obtained from plants. Nutraceuticals containing probiotics greatly support neurological health by reducing stress, improving mood, reducing anxiety, stabilizing mood, and more. Probiotic-based targeted therapy's precise neurological benefits are now the subject of several researches (Table 25.2) [36].

25.7 SUMMARY AND PROSPECTIVE VIEWS

This chapter covers several advantages of probiotics as well as how to use certain strains to treat neurological problems. Since probiotic microorganisms have been the subject of several researches, it appears that they may have nutritional benefits for ND and diseases. Probiotic supplements are now on the market, although the majority of them address typical cognitive functions, mood stability, stress reduction, and other issues. However, research on certain neurological therapies is currently ongoing. The mechanism of the probiotic microorganism in the neurological treatments help personalize the nutritional to the neurological conditions. Several researchers are trying to find the possible benefits of the targeted ND and disease, some of which are at the clinical stages. Probiotics as nutraceuticals may be useful in the future for certain applications, including evidence-based applications, clinical trials, strain-specific research, personalized probiotic therapies, synergistic formulations, and

TABLE 25.1
A List of ND and Diseases along with the Benefits of Probiotic Usage

Neurological Disease/Disorder	Probiotics Used	Mechanism of Action	Outcomes/Benefits
Alzheimer's disease	*L. acidophilus*	Reduces NI	Improved cognitive function
	B. bifidum	Enhances gut barrier function	Decreased amyloid plaque formation
Parkinson's disease	*L. plantarum*	Modulates GM	Improved motor function
	B. longum	Reduces oxidative stress	Reduced motor symptoms
Multiple sclerosis	*L. casei*	Anti-inflammatory effects	Reduced relapse rates
	B. lactis	Modulates immune response	Improved quality of life
Autism spectrum disorder	*L. rhamnosus*	Alters GBA	Improved social behavior
	B. infantis	Reduces gut permeability	Decreased GI symptoms
Depression	*L. helveticus*	Increases serotonin levels	Reduced depressive symptoms
	B. longum	Modulates HPA axis	Improved mood
Epilepsy	*L. johnsonii*	Alters GM	Reduced seizure frequency
	B. adolescentis	Anti-inflammatory effects	Improved seizure control
Anxiety	*L. casei*	Reduces cortisol levels	Decreased anxiety levels
	B. breve	Enhances GABA production	Improved stress resilience
Schizophrenia	*L. casei*	Alters GM	Improved psychiatric symptoms
	B. longum	Reduces inflammation	Enhanced cognitive function
Migraine	*L. fermentum*	Modulates immune response	Reduced migraine frequency
	B. lactis	Reduces NI	Decreased headache intensity
Attention deficit hyperactivity disorder	*L. reuteri*	Modulates GBA	Improved focus and behavior
(ADHD)	*B. breve*	Enhances gut barrier function	Reduced hyperactivity
Bipolar disorder	*L. rhamnosus*	Modulates NT levels	Stabilized mood
	B. bifidum	Reduces oxidative stress	Reduced symptom severity
Chronic fatigue syndrome	*L. acidophilus*	Modulates immune response	Reduced fatigue
	B. lactis	Enhances GM diversity	Improved energy levels
Stroke	*L. rhamnosus*	Anti-inflammatory effects	Improved recovery
	B. longum	Reduces oxidative damage	Enhanced neuroprotection
Insomnia	*L. fermentum*	Modulates melatonin production	Improved sleep quality
	B. longum	Reduces stress levels	Decreased sleep disturbances

preventive healthcare. The development of probiotic healthcare products that are customized to each patient's unique microbiota profile will be made possible by studies on the microbiome. Clinicians are able to suggest certain probiotic strains that have the highest potential to improve a patient's neurological health by knowing the patient's unique microbial composition. In addition to reducing response variability to probiotic therapies, this tailored strategy can improve therapeutic effectiveness. The goal of future study is to pinpoint and describe the particular probiotic strains that have the most effects on brain health. A scientific foundation for the selection of efficacious probiotics for particular neurological illnesses, such as anxiety, depression, PD, and AD, will be established by thorough investigations into the mechanisms of action of these strains, including their capacity to produce neuroactive compounds and modulate inflammation.

Moreover, to find out if probiotics are safe and helpful in treating neurological diseases, large-scale, well-planned clinical studies needed. Standardized procedures, extended investigations, and a range of demographics should be the main topics of future study in order to produce solid data for regulatory approvals and clinical guidelines. Frequent use of probiotics may lower the chance of acquiring illnesses, including anxiety, depression, and cognitive decline and improving mental health. In conclusion, the future of probiotics as nutraceuticals in ND is bright, with the potential to revolutionize mental and cognitive health management. Through personalized therapies, advanced research, innovative formulations, and robust clinical evidence, probiotics can become a cornerstone of preventive and therapeutic strategies for neurological well-being.

ACKNOWLEDGMENTS

The authors thank the Department of Pharmaceuticals, Government of India, for the administrative and research support.

TABLE 25.2
Nutraceutical Products Prepared with Specific Probiotic Strains, Focusing on their Health Benefits, Particularly for Neurological Health

Probiotic Culture	Probiotic Strain	Product Name	Health Benefits
L. helveticus	R0052	Probio'Stick	Mood improvement, anxiety relief
B. breve	B-3	Bio-Kult Mind	Cognitive function, anxiety relief
L. rhamnosus	GG	Culturelle Health & Wellness	Stress reduction, anxiety relief
B. longum	1714	Zenflore	Stress reduction, cognitive function enhancement
B. infantis	35624	Align Probiotic	Anti-inflammatory, mood regulation, IBS-related mood issues
B. longum	BB536	Dr. Ohhira's Probiotics	Cognitive function, stress reduction
L. salivarius	UCC118	ProBioCare Oral Health	Cognitive function, mood improvement
L. fermentum	PCC	Puritan's Pride Probiotic	Cognitive function, anti-inflammatory
L. johnsonii	La1	La Flora Probiotics	Cognitive function, stress reduction
L. acidophilus	NCFM	Garden of Life Mood+	Mood enhancement, stress reduction
L. casei	Shirota	Yakult	Stress reduction, mood stabilization
L. plantarum	PS128	PS128	Mood improvement, stress resilience
L. rhamnosus	GR-1	Fem-Dophilus	Mood improvement, stress resilience
B. lactis	HN019	Align Digestive Support	Cognitive function, stress reduction
L. reuteri	DSM 17938	BioGaiaProtectis	Mood improvement, stress reduction
L. paracasei	LPC-37	ProbioCare Mood Balance	Mood balance, stress relief
L.reuteri	NCIMB 30242	Cardioviva	Mood stabilization, GBA support
B. animalis	DN-173 010	Activia	Cognitive function, stress reduction
L. bulgaricus	LB-87	Biok Plus	Stress reduction, mood improvement
B. bifidum	Bb-06	Nature's Bounty Probiotic 10	Cognitive function, mood improvement
B. longum	BB536	Life Extension	Stress reduction, cognitive function
L.gasseri	LG21	Equilibrium Probiotic	Mood improvement, cognitive function
L. crispatus	CTV-05	Pro-Fem	Mood improvement, stress resilience
L. plantarum	299v	NOW Foods Probiotic-10	Mood improvement, cognitive support
L. paracasei	Lpc-37	Mega Food Mega Flora	Cognitive function, stress relief
L.rhamnosus	LGG	Primadophilus	Anxiety relief, mood stabilization
L. acidophilus	LA-5	NOW Probiotic-10	Stress reduction, mood improvement
B. breve	M-16V	Klaire Labs Vital-10	Cognitive function, mood stabilization
L. rhamnosus	HN001	Jarrow Formulas	Anxiety relief, cognitive function
L. casei	LC-11	Life Extension Probiotic-8	Mood stabilization, GBA support

REFERENCES

1. Hill, C., Guarner, F., Reid, G., Gibson, G. R., Merenstein, D. J., Pot, B., Morelli, L., Canani, R. B., Flint, H. J., Salminen, S., Calder, P. C., & Sanders, M. E. (2014). The International Scientific Association for Probiotics and Prebiotics consensus statement on the scope and appropriate use of the term probiotic. *Nature Reviews Gastroenterology & Hepatology*, 11(8), 506–514. https://doi.org/10.1038/nrgastro.2014.66
2. Mortada, E. M. (2024). Evidence-based complementary and alternative medicine in current medical practice. *Cureus*, 16(1), e52041. https://doi.org/10.7759/cureus.52041
3. Seo, D.-o., & Holtzman, D. M. (2024). Current understanding of the Alzheimer's disease-associated microbiome and therapeutic strategies. *Experimental & Molecular Medicine*, 56(1), 86–94. https://doi.org/10.1038/s12276-023-01146-2
4. Rodríguez Murúa, S., Farez, M. F., &Quintana, F. J. (2022). The immune response in multiple sclerosis. *Annual Review of Pathology*, 17, 121–139. https://doi.org/10.1146/annurev-pathol-052920-040318
5. Prajapati, K., Bisani, K., Prajapati, H., Prajapati, S., Agrawal, D., Singh, S., Saraf, M., & Goswami, D. (2024). Advances in probiotics research: mechanisms of action, health benefits, and limitations in applications. *Systems Microbiology and Biomanufacturing*, 4(2), 386–406. https://doi.org/10.1007/s43393-023-00208-w
6. Lee, H. (2022). The importance of nutrition in neurological disorders and nutrition assessment methods. *Brain & Neurorehabilitation*, 15(1), e1. https://doi.org/10.12786/bn.2022.15.e1
7. Valenzuela-Zamora, A. F., Ramírez-Valenzuela, D. G., & Ramos-Jiménez, A. (2022). Food selectivity and its implications associated with gastrointestinal disorders in children with autism spectrum disorders. *Nutrients*, 14(13). https://doi.org/10.3390/nu14132660
8. Matsuyama, H., Matsuura, K., Ishikawa, H., Hirata, Y., Kato, N., Niwa, A., Narita, Y., &Tomimoto, H. (2021). Correlation between serum zinc levels and levodopa in Parkinson's disease. *Nutrients*, 13(11), 4114. https://www.mdpi.com/2072-6643/13/11/4114

9. Structure, function and diversity of the healthy human microbiome. (2012). *Nature*, 486(7402), 207–214. https://doi.org/10.1038/nature11234
10. Suganya, K., & Koo, B. S. (2020). Gut-brain axis: Role of gut microbiota on neurological disorders and how probiotics/prebiotics beneficially modulate microbial and immune pathways to improve brain functions. *International Journal of Molecular Sciences*, 21(20). https://doi.org/10.3390/ijms21207551
11. Bravo, J. A., Forsythe, P., Chew, M. V., Escaravage, E., Savignac, H. M., Dinan, T. G., Bienenstock, J., & Cryan, J. F. (2011). Ingestion of Lactobacillus strain regulates emotional behavior and central GABA receptor expression in a mouse via the vagus nerve. *Proceedings of the National Academy of Sciences of the United States of America*, 108(38), 16050–16055. https://doi.org/10.1073/pnas.1102999108
12. Tremlett, H., Bauer, K. C., Appel-Cresswell, S., Finlay, B. B., &Waubant, E. (2017). The gut microbiome in human neurological disease: A review. *Annals of Neurology*, 81(3), 369–382. https://doi.org/10.1002/ana.24901
13. Tiwari, P., Dwivedi, R., Bansal, M., Tripathi, M., & Dada, R. (2023). Role of gut microbiota in neurological disorders and its therapeutic significance. *Journal of Clinical Medicine*, 12(4). https://doi.org/10.3390/jcm12041650
14. Mazziotta, C., Tognon, M., Martini, F., Torreggiani, E., & Rotondo, J. C. (2023). Probiotics mechanism of action on immune cells and beneficial effects on human health. *Cells*, 12(1). https://doi.org/10.3390/cells12010184
15. Chanda, S., Tiwari, R. K., Kumar, A., & Singh, K. (2019). Nutraceuticals inspiring the current therapy for lifestyle diseases. *Advances in Pharmacological Sciences*, 2019, 6908716. https://doi.org/10.1155/2019/6908716
16. Thangaleela, S., Sivamaruthi, B. S., Kesika, P., & Chaiyasut, C. (2022). Role of probiotics and diet in the management of neurological diseases and mood states: A review. *Microorganisms*, 10(11). https://doi.org/10.3390/microorganisms10112268
17. Sivamaruthi, B. S., Suganthy, N., Kesika, P., & Chaiyasut, C. (2020). The role of microbiome, dietary supplements, and probiotics in autism spectrum disorder. *International Journal of Environmental Research and Public Health*, 17(8). https://doi.org/10.3390/ijerph17082647
18. Guo, M., Li, R., Wang, Y., Ma, S., Zhang, Y., Li, S., Zhang, H., Liu, Z., You, C., & Zheng, H. (2022). Lactobacillus plantarum ST-III modulates abnormal behavior and gut microbiota in a mouse model of autism spectrum disorder. *Physiology & Behavior*, 257, 113965. https://doi.org/10.1016/j.physbeh.2022.113965
19. Cristofori, F., Dargenio, V. N., Dargenio, C., Miniello, V. L., Barone, M., & Francavilla, R. (2021). Anti-inflammatory and immunomodulatory effects of probiotics in gut inflammation: A door to the body. *Frontiers in Immunology*, 12, 578386. https://doi.org/10.3389/fimmu.2021.578386
20. Cassani, E., Privitera, G., Pezzoli, G., Pusani, C., Madio, C., Iorio, L., & Barichella, M. (2011). Use of probiotics for the treatment of constipation in Parkinson's disease patients. *Minerva Gastroenterologica e Dietologica*, 57(2), 117–121.
21. Barichella, M., Pacchetti, C., Bolliri, C., Cassani, E., Iorio, L., Pusani, C., Pinelli, G., Privitera, G., Cesari, I., Faierman, S. A., Caccialanza, R., Pezzoli, G., & Cereda, E. (2016). Probiotics and prebiotic fiber for constipation associated with Parkinson disease: An RCT. *Neurology*, 87(12), 1274–1280. https://doi.org/10.1212/wnl.0000000000003127
22. Magistrelli, L., Amoruso, A., Mogna, L., Graziano, T., Cantello, R., Pane, M., & Comi, C. (2019). Probiotics may have beneficial effects in Parkinson's disease: In vitro evidence. *Frontiers in Immunology*, 10, 969. https://doi.org/10.3389/fimmu.2019.00969
23. Castelli, V., d'Angelo, M., Lombardi, F., Alfonsetti, M., Antonosante, A., Catanesi, M., Benedetti, E., Palumbo, P., Cifone, M. G., Giordano, A., Desideri, G., & Cimini, A. (2020). Effects of the probiotic formulation SLAB51 in in vitro and in vivo Parkinson's disease models. *Aging*, 12(5), 4641–4659. https://doi.org/10.18632/aging.102927
24. Srivastav, S., Neupane, S., Bhurtel, S., Katila, N., Maharjan, S., Choi, H., Hong, J. T., & Choi, D. Y. (2019). Probiotics mixture increases butyrate, and subsequently rescues the nigral dopaminergic neurons from MPTP and rotenone-induced neurotoxicity. *Journal of Nutritional Biochemistry*, 69, 73–86. https://doi.org/10.1016/j.jnutbio.2019.03.021
25. Goya, M. E., Xue, F., Sampedro-Torres-Quevedo, C., Arnaouteli, S., Riquelme-Dominguez, L., Romanowski, A., Brydon, J., Ball, K. L., Stanley-Wall, N. R., & Doitsidou, M. (2020). Probiotic bacillus subtilis protects against α-synuclein aggregation in C. elegans. *Cell Reports*, 30(2), 367–380.e367. https://doi.org/10.1016/j.celrep.2019.12.078
26. Sun, H., Zhao, F., Liu, Y., Ma, T., Jin, H., Quan, K., Leng, B., Zhao, J., Yuan, X., Li, Z., Li, F., Kwok, L.-Y., Zhang, S., Sun, Z., Zhang, J., & Zhang, H. (2022). Probiotics synergized with conventional regimen in managing Parkinson's disease. *npj Parkinson's Disease*, 8, 62. https://doi.org/10.1038/s41531-022-00327-6
27. Fasano, A., Bove, F., Gabrielli, M., Petracca, M., Zocco, M. A., Ragazzoni, E., Barbaro, F., Piano, C., Fortuna, S., Tortora, A., Di Giacopo, R., Campanale, M., Gigante, G., Lauritano, E. C., Navarra, P., Marconi, S., Gasbarrini, A., & Bentivoglio, A. R. (2013). The role of small intestinal bacterial overgrowth in Parkinson's disease. *Movement Disorders*, 28(9), 1241–1249. https://doi.org/10.1002/mds.25522
28. Sharma, C., & Kim, S. R. (2021). Linking oxidative stress and proteinopathy in Alzheimer's disease. *Antioxidants*, 10(8). https://doi.org/10.3390/antiox10081231
29. Manoharan, S., Guillemin, G. J., Abiramasundari, R. S., Essa, M. M., Akbar, M., & Akbar, M. D. (2016). The role of reactive oxygen species in the pathogenesis of Alzheimer's disease, Parkinson's disease, and Huntington's disease: A mini review. *Oxidative Medicine and Cellular Longevity*, 2016, 8590578. https://doi.org/10.1155/2016/8590578
30. Goh, K. K., Liu, Y. W., Kuo, P. H., Chung, Y. E., Lu, M. L., & Chen, C. H. (2019). Effect of probiotics on depressive symptoms: A meta-analysis of human studies. *Psychiatry Research*, 282, 112568. https://doi.org/10.1016/j.psychres.2019.112568
31. Papalini, S., Michels, F., Kohn, N., Wegman, J., van Hemert, S., Roelofs, K., Arias-Vasquez, A., & Aarts, E. (2019). Stress matters: Randomized controlled trial on the effect of probiotics on neurocognition. *Neurobiol Stress*, 10, 100141. https://doi.org/10.1016/j.ynstr.2018.100141
32. Liu, Y.-W., Liu, W.-H., Wu, C.-C., Juan, Y.-C., Wu, Y.-C., Tsai, H.-P., Wang, S., & Tsai, Y.-C. (2016). Psychotropic effects of Lactobacillus plantarum PS128 in early life-stressed and naïve adult mice. *Brain Research*, 1631, 1–12. https://doi.org/https://doi.org/10.1016/j.brainres.2015.11.018

33. Buffington, S. A., Di Prisco, G. V., Auchtung, T. A., Ajami, N. J., Petrosino, J. F., & Costa-Mattioli, M. (2016). Microbial reconstitution reverses maternal diet-induced social and synaptic deficits in offspring. *Cell*, 165(7), 1762–1775. https://doi.org/10.1016/j.cell.2016.06.001
34. Hwang, Y. H., Park, S., Paik, J. W., Chae, S. W., Kim, D. H., Jeong, D. G., Ha, E., Kim, M., Hong, G., Park, S. H., Jung, S. J., Lee, S. M., Na, K. H., Kim, J., & Chung, Y. C. (2019). Efficacy and safety of lactobacillus plantarum C29-fermented soybean (DW2009) in individuals with mild cognitive impairment: A 12-week, multi-center, randomized, double-blind, placebo-controlled clinical trial. *Nutrients*, 11(2). https://doi.org/10.3390/nu11020305
35. Waller, P. A., Gopal, P. K., Leyer, G. J., Ouwehand, A. C., Reifer, C., Stewart, M. E., & Miller, L. E. (2011). Dose-response effect of Bifidobacterium lactis HN019 on whole gut transit time and functional gastrointestinal symptoms in adults. *Scandinavian Journal of Gastroenterology*, 46(9), 1057–1064. https://doi.org/10.3109/00365521.2011.584895
36. Dahiya, D., & Nigam, P. S. (2023). Nutraceuticals prepared with specific strains of probiotics for supplementing gut microbiota in hosts allergic to certain foods or their additives. *Nutrients*, 15(13). https://doi.org/10.3390/nu15132979

26 Lifestyle and Environmental Factors

Impact on Gut Microbiota and Neurological Disorder

Pooja S Murkute[1], Saistha Anjum[2], S. Sowmiya[3], Arghya Paria[4], Sourabh Kosey[5], Naheed Mojgani[6], and Sumel Ashique[7]

[1] Department of Pharmacognosy, R.C.Patel Institute of Pharmaceutical Education & Research - An Autonomous Institute, Shirpur Dhule, India
[2] Research Scholar, Department of Pharmacology, Al-Ameen College of Pharmacy, Karnataka, India
[3] Department of Biotechnology, Vivekananda College of Arts and Sciences for Women (Autonomous), Tiruchengode, Tamil Nadu, India
[4] Department of Pharmaceutics, Sanaka Educational Trust's Group of Institutions, Durgapur, West Bengal, India
[5] Department of Pharmacy Practice, ISF College of Pharmacy, Moga, Punjab, India
[6] Biotechnology Department, Razi vaccine and serum research Institute, Agriculture Research, Education and Extension Organization, Karaj, Iran
[7] Department of Pharmaceutical Technology, Bharat Technology, Uluberia, West Bengal, India

26.1 INTRODUCTION

The human gastrointestinal tract (GI) acts as the biggest interface in the host, environmental factors, and internal antigens. A person in his or her lifetime processes about 60 tonnes of food in his gut along with a vast number of environmental microbes, which cause grave risk to intestinal integrity. The GI tract resident bacteria, archaea, and eukaryote, known as the gut microbiota, have co-evolved with the host to form a complex network of mutually beneficial interactions. The bacteria, archaea, and eukaryote that inhabit the gastrointestinal system, collectively known as the gut microbiota, have co-evolved with their host with a complex web of usually beneficial interactions [1]. With more than 100 trillion microbes, the gut microbiome is an intricate component of the digestive system. The gut bacteria and the host interact in a primarily symbiotic manner. Maintaining the host's internal processes like metabolite digestion and gastrointestinal barrier protection depends on the gut microbiota [2]. Hippocrates declares about 400 BC that death sits in the bowls, which symbolizes that it is a direct connection between human health and the gut. Numerous global research has concern related to the important role that intestinal microbiota plays in human health and illness [3]. Composition of gut microbiota is a vast array with commensal microorganism, which is composed of bacteria, viruses, archaea, protozoa, and fungus. These microorganisms coexist in a complex balance with their host and obtain their energy from the substrate that food supplies. The term exosome refers to the combination of lifestyle, pharmaceuticals, and hazardous exposure that cause interindividual variations in susceptibility to a variety of disease as well as varied reactions to similar pathogenic stimuli, diet, or medication [4]. The gut microbiota affects several processes, such as colonization, pathogen resistance, intestinal epithelium maintenance, food and medication metabolism, immune function regulation, and even behavior-brain axis [5]. Any disorder affecting the nerve system is considered a neurological disorder; anatomical, metabolic, or electrical abnormalities in the brain, spinal cord, or other nerves can give rise to a variety of symptoms. Here are few instances of symptoms: discomfort, tauopathies, muscle weakness, loss of feeling, paralysis, poor coordination, seizures, and altered states of awareness [6]. One can classify neurological illness based on the principal site of involvement, the primary dysfunction type, or the primary etiology. Disorders of the central and peripheral nervous system are the most broadly classified [7]. Alzheimer's, Parkinson's, and Huntington's diseases are just few of the neurodegenerative illnesses caused by mitochondrial malfunction [8]. Human health and illness are greatly influenced by microbiota, which is why it is frequently called our forgotten organ. Early on, the process of growth the range of bacteria and viruses in the boy stomach increases from a relatively low starting point [9]. Gut microbiota performs several functions. It creates barriers

DOI: 10.1201/9781003536802-26

in the intestine, encourages the regeneration of intestinal epithelial cells, secretes mucosa, and feeds the mucosa by generating short-chain fatty acids, intestinal synthesis, metabolism of certain nutrients, hormones, and vitamin, as well as a critical function in the excretion of drugs and poisons [10]. It also activates the immune system, providing a quick and efficient defence against invaders [11]. Bacterial colony contributes to defensive comp Góralczyk edition in the stomach by producing cytokines that may inhibit the microorganism growth while feeding the infection [12]. The efferent signals that travel from the central nervous system to the digestive tract are produced by the sympathetic and parasympathetic branches of the autonomic nervous system wall with afferent signals which originate in the lumen and are sent via the spinal, vagal, and enteric pathway [13]. It also decreases increased fecal size, distal propagation, cyclic repetition of the migratory motor complex, and altered neurotransmitter expression and turnover in both nervous systems [14]. There are several possible causes of gene-related disease, degenerative disease, disorder caused by gradual disappearance of neuronal groups that are most susceptible, convulsive disorder, brain tumor, and more or less severe infection [15].

26.2 WHAT IS GUT MICROBIOTA

The term microbiota was coined early in the 20th century. Numerous microorganisms have been found to coexist in the human body, including viruses, yeasts, and many bacteria. Many organ systems — including the skin, lungs, mouth, and gut — along with the human microbiota, can be considered a "hidden organ" that contributes over 150 times more genomic data than the human genome itself [16]. The gastrointestinal system contains this interaction. The microbiota in the gut, such as bacteria archaea, eukaryote, which inhibits the gastrointestinal system, has developed a complex to mutually benefit the host from the early years of co-evolution [17]. The flora in the gut has been demonstrated, which preserves human physiology, nutrition, metabolism, immune system function, and mental well-being [18]. Metabolites are produced by a stable and diverse gut flora which is ideal for preserving health and powers physiologic and metabolic functions [19]. Numerous bacterial species that permanently inhabit the gastrointestinal tract are a part of this microbial ecosystem, together with a vast array of environmental microorganism like viruses, parasites, fungus, and archaea [20]. The GI tract is predominantly colonized from birth and irregular changes in diet, illness, and the use of antibiotics can all have a negative impact on the microbiota [21].

26.3 GUT MICROBIOTA CLASSIFICATION

Accordingly, to their phylum, classes, order, families, genera, and species, bacteria are categorized taxonomically. Just a few phyla firmicute. Bacteroidetes, Actinobacteria, Proteobacteria, Fusobacterial, and Verrucomicrobia among the 160 species are known to be the most common microbial phyla in the gut. Bacteroidetes and phyla firmicutes account for 90% of the total [22]. About 30–40% species account for 99% of the bacteria [23]. Due to their profusion in the colon, additional bacteria make up as much as 60% of the dry mass excrement [24]. The gut microbiota includes bacteria, archaea, viruses, and fungi. While the functions of archaea and bacteria are better understood, aerobic bacteria can reach high densities in the cecum. However, more than 99% of the bacteria in the gut are anaerobes. It is estimated that these gut florae have 100 times more genes overall than the human genome [25]. Bacteroidota, Actinomycetota, Pseudomonodota, and Bacillota(firmicutes) are the four predominant bacterial phyla. The genera Bactericides, clostridium, and fecal bacterium include the majority of bacteria [26]. About 30% of all bacteria in the stomach are species from the host to function properly. Genus is especially important [27].

26.3.1 Bacteria Classification

26.3.1.1 Bacillota

A phylum of bacteria called bacillota (Firmicutes) is composed primarily of bacteria with Gram-positive cell walls. Microbiologists continue to disagree with the renaming of phyla like formicutes in 2021; many still refer to these taxa by their previous, more established name in literature [28]. Their cells new rounded and referred to as cocci or bacillus. Some firmicutes stain Gram-negative due to their porous pseudo outer membrane, including megasphaera, pectinatus, selenomonas, and zymophils [29]. Numerous bacillota develop endospores which can withstand harsh environments and resistance to desiccation. They can be found in a variety of settings. and certain well-known diseases are among them. The heliobacteria are a class of bacteria that use an oxygenic photosynthesis to produce energy. Bacillota are crucial to the deterioration of wine, cider, and beer [30]. According to reports, firmicutes are common in rhizosphere soils as well as other environments such plant end rhizosphere due to their alleged high-culture ability, roughly 14% of the bacterial species are known to be represented in culture collection [28]. Usually, the group is separated into facultative aerobes or obligatory aerobes called bacilli and anaerobic clostridia [28].

26.3.1.1.1 Clostridia bacillota

Clostridium and other related genera belong to the extremely polyphylogenetic clostridia class of bacillota. They are different from bacilli in that as they do not use respiration in an aerobic state. Because they are forced anaerobes. Research indicates that they are not a monophyletic group, and it is unclear exactly how they relate to one another. The majority are currently grouped together under mono class known as clostridiales, but this is not a naturally occurring group, and will be probably reorganized in the future [31]. Examples are toxins produced by *Clostridium tetani,* which is known as tetanospasmin, and *Clostridium botulinum* toxin. Women who have bacterial vaginosis have been found to harbor certain species of bacteria that are not typically present in a healthy vaginal microbiota [32].

26.3.1.1.2 Clostridium difficile

In addition to being recognized to cause severe diarrheal infections, the bacteria *Clostridium difficile*, sometimes called as *Clostridia's difficile*, can also cause colon cancer. It is sometimes referred to as *C. difficile* and is spore-forming. Its rod-shaped, pleomorphic vegetative cells are found in short chains or pairs. It is similar in the microscopic perspective to long, uneven cells that are often formed like spindles or drumstick and have protrusions at the ends [33].

26.3.1.1.3 Clostridium novyi

Gram-positive endospore-forming obligatory anaerobic bacteria belonging to the clostridium class are called *Clostridium novyi*. It is omnipresent in both soil and excrement [34]. Gas gangrene is cause by wound infection with *C. novyi* and numerous other clostridium species. The majority of time spontaneous infection is linked to diabetes mellitus and hematologic or colorectal cancer risk factors. However, in diabetic individuals, gas gangrene-like disease can be brought on by Gram-negative bacteria such *Escherichia coli*. This can be confused with gas gangrene since it presents with cellulitis and crepitus [35].

26.3.1.1.4 Clostridium tetani

The widespread soil bacteria *C. tetani* produces tetanus. Normally, rod-shaped vegetative cells that may grow up to 2.5 µm in length expand and take on shape of tennis rackets or other objects when *Clostridium tetanus* creates sprees drumstick. Spores of *C. tetani* are incredibly resilient and can be found worldwide in soil or in animal digestive system [36]. When *C. tetani* is injected into wound, it can proliferate and release tetanospasmin, a strong toxin that disrupts motor neuron and result in tetanus [37].

26.3.1.1.5 Clostridium haemolyticum

One type of bacteria in the clostridiaceae family is *Clostridium haemolyticum*, which is closely related to *Clostridium novyi* and *Clostridium botulinum*, according to 16 SrDNA sequence study. Despite prior suggestions that *C. haemolyticum* and *C. novyi* are identical, it is still recognized as a distinct species of 2022 [38].

26.3.1.1.6 Clostridium perfringes

This is known as *Clostridium welchii* or *bacillus* Welchii, a pathogenic rod-shaped, anaerobic, Gram-positive bacteria. It also forms perfringes, naturally found in soil, insects, decomposing vegetation, marine sediment, and the digestive tracts of humans and other mammals [39].

26.3.1.1.7 Clostridium botulinum

A wide variety of harmful bacteria belong to the genus *Candidatum*. They were originally classified as *C. botulinum* groups I–IV and are now recognized as four separate groups based on their capacity to produce botulium toxin, as do some stains of Clostridium butyricum and Clostridium baratii. In humans and other animals, botulism is a severe flaccid paralysis that can be brought on by botulinum toxin [40].

26.3.1.1.8 Clostridium acetobutylicum

The commercially valuable bacterium *Clostridium acetobutylicum* ATCC 824 is often referred to as the Weizmann organism, named after the Jewish biologist Chaim Weizmann. This bacterium has been effectively used since 1916 to produce solvents such as acetone, ethanol, and butanol [41].

26.3.1.2 Bacteroidota

There are three major classes of rod-shaped, Gram-negative, aerobic or anaerobic, non-spore-forming bacteria that makeup the phylum Bacteroidetes. These bacteria are found in soil, sediments, seawater, animal gut, and external skin [42]. Numerous species are symbiotic and well adapted to the gastrointestinal tract, despite the fact that some of them might be opportunistic pathogens. Bacteroids can be found in large concentrations up to 10^{-11} g^{-1} cells of intestinal material [43]. They carry out metabolic conversions, including the breakdown of complex sugar polymers or proteins that are vital to the host. Infant gastrointestinal tract is already colonized with bacteroids and *Bifid bacterium* spp. due to nondigestible oligosaccharides in breast milk. Through certain interactions, the host immune system specifically recognizes bacteroids [44].

26.3.1.3 Actinomycetota

The actinomycetota are a diverse phylum of highly GC-containing Gram-positive bacteria. They are both terrestrial and aquatic in nature, which is highly beneficial to land flora.

They decompose the organic matter of dead organism in soil into nutritive molecule, which is taken by new plants during their growth [45].

Actinomycetota, which were previously considered as fungi but are more beneficial than fungi in case of nitrogen fixation in exchange of certain plant saccharides as they live jointly with the plants which are deeply rooted in soil. Other species of numerous numbers of the genus mycobacterium are significant pathogens of this category [46].

26.3.1.4 Pseudomonadota

One of the main phyla Pseudomondota, a kind of Gram-negative bacteria is also referred to as Proteobacteria. At present, they are regarded as the most common phylum in the bacterial domain [47]. They exist in nature as free-living, pathogenic genera that are not parasitic, despite the fact that certain species are supposedly Gram-negative or Gram-variable in lab setting [48]. They are mostly organisms that move via flagella, which are used for movement, some are immobile or depend on bacterial gliding [49].

26.3.2 Archaea Classification

Archaea were traditionally thought to live in harsh conditions with high salt, temperature, and acidity. In 1968, a solitary species was identified from the human gastrointestinal tract. A few decads later, the physiological gut microbiota was shown to include archaea. This has also been discovered in a number of human bodily parts, including the skin, nose, respiratory system, vagina, and mouth. At present it is estimated to found 42–100% of fecal sample population [50]. This is an important component in earth's life cycle. They have significant role in skin, oral, and gut microbiota of humans. They can perform a variety of ecological functions, including carbon fixation, nitrogen cycling, organic compound turnover, and sustaining microbial symbiotic and syntrophic communities [51].

26.3.3 Viruses Classification

Viruses have been an important part of human life from the beginning. The underlying human health and the basic biological mechanisms that support it require an understanding of the gut virome. The gut virome is thought to be colonized from birth and is regulated by various factors to form a distinct profile [52]. Plant and animal viruses are examples of eukaryotic viruses that can be contracted by eating them [53]. Viral infection has caused an incredible amount of death throughout human history [54]. Rotavirus, Astrovirus, calicivirus, Norovirus, Hepatitis E virus, Coronavirus and Tarovirus, Adenovirus, Mimivirdiae, Marseileviridae, Poxviridae, Iridoviridae, Ascoviridae, Phycodnaviridae, Asfaviridae vid, Myoviridae, Siphoviridae, Podoviridae, Tectiviridae, Leviviridae, and Inoviridae are involved in the gut microbiota [55].

26.3.4 Fungal Classification

Fungi are essential components of a wide range of microbial habitats,. including the human gut. Gut fungus can provide health benefits to people. The yeast *Saccharomyces boulardii* is widely used as an effective probiotic to prevent and treat pathogenic bacterial infections and intestinal complications fungi makeup about 0.0% of the gut microbiome and an early investigation using 96 stool samples from healthy participants discovered 66 [56] species. However, this could be a major underestimate due to genetic problems. Numerous fungi have been found in human gut but few have proved to be common occupants across individuals or across time in databases of fungi annotation. Specifically, the ratio of Ascomycota to Basidiomycota varies at the phylum level. Comparatively speaking, Basidiomycota are significantly more abundant than Ascomycota. Several species are involved, such as *Candida filobasidiaceae* and *Malasseziales* [57].

26.3.5 Archaea Classification

Methanogens or hydrogen trophic archaea are the most common in people. They anaerobically convert CO_2 to methane gas. *Methanobrevibacter smithii* is most prevalent methanogen in the gut microbiota with a prevalence of around 95.7%. Patients with diverticulosis, obesity, cancer, human periodontal disease, and inflammatory bowel disease seem to have an excess of archaea compared to healthy persons.

26.4 HUMAN LIFESTYLE FACTORS ON GUT MICROBIOTA

Human gut microbes begin to colonize at or soon after birth. It is possible that intestinal colonization microorganisms start with its accelerated by natural birth as evidenced by the fact that babies delivered spontaneously have higher levels of gut microbiota at one month of life compared to newborns delivered via caesarean section. The immune system development depends on formation and development of a robust gut microbiota. This process continues throughout breastfeeding, which is thought to be crucial for a person's long-term health [58].

26.4.1 Age

Numerous age-related ailments such as Parkinson's disease, type 2 diabetes, chronic obstructive pulmonary disease, heart failure, etc. are probably made worse by inflammation. The intestine loses its structural integrity as we age, allowing gut bacteria to spread to other body regions and trigger an inflammatory reaction from immune cells, including Th17 lymphocytes and neutrophils [59]. Growing older can contribute to because pro-inflammatory flora can inhibit the formation of good flora; there is a constant low-grade inflammation and an elevated risk of numerous age-related disorder [60].

26.4.2 Diet

An individual's diet and eating habit have a significant impact on how their microbiome change. It still doesn't fully understand how diet and gut microbes interact. In the last ten years, several study groups have looked into the notion that among all the factors affecting the gut microbiotas makeup. Nutrition ranks first [61]. Diet with high animal protein may cause the intestinal microbiota to produce more TMAO, which increases the risk of cardiovascular disease. The equilibrium of intestinal microbes may benefit from rating plant-based protein. Western-style diet, high consumption of saturated fats, and omega-6-polyunsaturated fats may cause dysbiosis, changes to the gut barrier, and metabolic problems [62]. Proteolytic bacteria are stimulated by 10–30% of undigested protein to multiply and produce toxic metabolites such as hydrogen sulfide, ammonia, and branched-chain fatty acid. Bacteria that can withstand bile acid can create harmful composition like H2S [63].

26.4.3 Exercise

Exercise affects the digestive system. It reduces the amount of time that passes between pathogen contact with GI mucus layer and feces, because exercise seems to offer preventive effects, reducing the risk of inflammatory bowel disease, diverticulosis, and colon cancer [64].

26.4.4 Drinking

The gastrointestinal tract absorbs alcohol, which is then moved to the liver and lungs for metabolism. This process results in build-up of fatty acid ethyl ester, direct toxicity, and oxidative stress caused by ethanol and its metabolism. Acetaldehyde is a harmful chemical that has been linked to the beginning of certain disorder caused by alcohol consumption [65]. Another theory that has recently gained traction is that alcohol consumption can alter the microbiota in the gut, which could therefore lead to the development of alcohol-related illness. These illnesses may be mitigated by recovering the altered gut flora [66].

26.4.5 Smoking

They attempted to provide potential explanations for smoking-induced dysbiosis through their findings: by lowering NFkB-P65's phosphorylation, a crucial intermediary in the inflammatory NFkB pathway. Smoking lowers inflammatory pathways [65]. The makeup of the gut microbiota in smokers and nonsmokers were very different. The microbiota of individuals who gave up smoking reverted to composting skin to that of never had smoked [67]. Smoking can affect the host microbiota in several ways: increasing the gut pH mucosa, changing the gut mucin layer, and upregulating enzymes linked to oxidative stress in gut immune cells [68].

26.4.6 Use of Antimicrobials and Drugs

Dysymbiosis has been identified as a mechanism that facilitates horizontal gene transfer, raising the likelihood that antibiotic resistance genes would proliferate [69]. The disruption known as antibiotic-induced gut dysbiosis initiate a cascade of unfavorable responses that have a deleterious effect on the gut microbiome, leading to reduction in the diversity of microorganism and creation of an environment that favors the emergence of antibiotic resistant varieties [70]. The long-term effects of antibiotic exposure include altered metabolic processes, obesity, allergies, asthma, and higher likelihood of antibiotic resistance. These effects are broad, spanning from infancy to maturity; prolonged antibiotic usage in preterm newborns has been linked to altered risk of late onset sepsis and necrotizing enterocolitis [71].

26.5 ENVIRONMENTAL FACTORS IMPACT ON GUT MICROBIOTA

Global studies of fecal microbiota from healthy groups show that individual variances in gut microbiota diversity are primarily caused by age, ethnicity, location, medication use, blood factors, bowel health, diet, overall health, body measurements, and lifestyle choices [72]. Another study showed that the types of fecal bacteria and their genes varied between people in the United States and those in rural Venezuela and Malawi [73]. The study also showed that differences were higher among people from different families than from the same family within each nation, possibly due to shared habits, environments, or genetics [74]. Disruption of the body's natural clock, due to shift work, travel, and other circumstances can potentially have an impact on gut health and alter gut flora [73]. As a result, the stomach and the beginning of the small intestine have fewer microorganisms that can survive the acidic pH, oxygen levels, and quick transit times found in these areas [75]. This may lead to a lack of important micronutrients in crops, which are crucial for their roles in human metabolism, enzyme regulation, and as coenzymes [76]. Typically, in 1 gram of soil, there are about 40,000–50,000 types of microorganisms. The soil microbiome is essential for keeping the soil healthy and fertile, thus helping in bioremediation, breaking down organic materials, and regulating various processes. Climate change impacts the breathing of the microbiome, which affects carbon and nitrogen levels in the soil [77]. Ignoring how everyday climate change affects human health leaves a big home in us.

Farmers might already be modifying their methods in response to the changing climate, which will undoubtedly affect our diets [78].

26.6 HOW GUT MICROBIOTA AFFECT NEUROLOGICAL DISORDER

A number of neurological conditions, including major depressive disorder, autism spectrum disorder, Parkinson's disease, Alzheimer's disease, and epilepsy, are linked to imbalances in the gut microbiota. In those with AD, there is a noticeable imbalance in gut microbiota, shown by reduced diversity in fecal microbes, fewer beneficial bacteria (like *Eubacterium rectale*, *Bifidobacterium)*, and dialister and an increase in harmful microbes (such as *Escherichia/Shigella*, *Bacteroides*, and *Ruminococcus*).

Besides the well-known pathways like the hypothalamic pituitary adrenal axis, research shows that substances made by gut bacteria such as short-chain fatty acids neurotransmitters can influence brain metabolites through the bloodstream, impacting brain function and cognition [79]. The gut talks to the brain through two main pathways. The first is a direct link via the vagus nerve and the autonomic nervous system in the spinal cord.

Irritable bowel syndrome (IBS) as well as neurological and mental health problems such as anxiety, depression, autism spectrum disorder, Alzheimer disease, multiple sclerosis, and Parkinson's are being linked to an imbalance in gut bacteria known as dysbiosis. They also use short-chain fatty acids (SCFA) for communication. Additionally, the gut microbiota influences how neurotransmitter, including serotonin, is relapsed [80].

One obvious example of what happens when the connection between CNS and ENS are broken is Parkinson's disease. The principal characteristic of this neurodegenerative illness is the disappearance of dopamine-producing neurons, in particular brain circuits [81]. The host neurological, immunological, neuroendocrine, and metabolic systems are all impacted by the gut microbiota, which in healthy individuals is primarily composed of firmicutes and bacteroides.

Consequently, the microbiome can impact serotonin levels in the brain via affecting tryptophan levels [82]. By preventing histone deacetylase, short-chain fatty acids function as endogenous signals for specific G protein-coupled receptors and aid in the regulation of gene activity. They also have an interaction with vagal nerves, which affects hormone levels and inflammation, the manner in which SCFAs engage with particular cell type and gut brain.

By modifying the hormone signals from enter endocrine cells in the gut lining specifically the hormone glucagon like peptide 1 gut bacteria can influence the host appetite and eating pattern [83]. It is unclear exactly which molecular mechanism leads to the death of nerve cells in neurodegenerative illness. However, common characteristics of neurodegenerative such as misfolded protein accumulation and inflammation have been studied recently in relation to gut–brain axis.

Toxins and bacteria can enteric neurons through this leaky gut, resulting in localized inflammation. Additionally, there is chance that unidentified environmental factors connected to alpha-synuclein-related issue could create an accumulation of the protein. The most prevalent form of neurodegenerative disease, amyloid plaques may arise as a result of gut microorganism. Studies show that individuals with AD who have *Helicobacter pylori* infections have a more severe version of the disease than those who do not. It has been found that individuals with amyloidosis-related cognitive impairments exhibit an inflammatory state in their bodies. Given that there is a decrease in anti-inflammatory bacteria like Eubacterium rectale and an increase in pro-inflammatory bacteria like Escherichia and shigella, the gut microbiota may be important in this case [84].

26.7 INTERACTION AND RECIPROCAL INFLUENCE AMONG CNS AND GUT MICROBIOTA

The survival and ability to reproduce hosts might rely on their microbiota's presence, traits, and functions. The concept is being supported by more and more data. "Holobiont" includes the host and its microbial communities, along with optional symbionts that have different and interconnected relationships. The genome of a macroorganism, along with its organelles and microbiome, is thought to create a collection called the "hologenome." The emerging holobiont theory proposes that (i) all macroorganisms contain microorganisms that provide a nutrient-rich habitat for them; (ii) the well-being of both the holobiont and its symbiotic partners relies on each other; and (iii) changes in either can alter the hologenome [85]. The GI tract of mammals is home to gut microbiota, which is made up of billions of bacteria, fungi, and viruses. these living things play important roles in supporting health both in the gut and other body areas by preventing harmful pathogens, managing metabolism, and boosting the immune system [86]. The myenteric plexus which is situated in the space between the muscle layer and the submucosal plexus which is situated in submucosa are two primary networks of the ENS [87].

As mentioned earlier, factors like diet, stress, infections, and antibiotics can disrupt gut microbiota balance [88, 89]. The brain and other parts of the central nervous system exchange message with each other, gut ecosystem through neurons, hormones, and cytokines.

Homeostatic mechanisms affect various processes in the gut–brain axis, including appetite, immune response, stress response, and metabolism [90].

26.8 POSSIBLE NEUROLOGICAL DISORDERS DUE TO GUT MICROBIOME

Numerous neurological conditions, such as many meurological conditions, are affected by gut microbiota in both their pathogenesis and treatment. For the treatment of neurological disorders caused by genetically modified organism, probiotics, postbiotics, symbiotics, etc. have been conducted [91, 92]. Parkinson's disease patient's digestive tract is deficient in bacteria that create short-chain fatty acids. Primary butyrate such as *Faecalibacterium prausnitzii* and taxa within the Lachnospiraceae family have anti-inflammatory qualities. Additionally, some bacterial species may be origin of Parkinson's disease. *Proteus mirabilis,* for example, produces greater impairments in mice [93, 94].

26.8.1 Alzheimer's Disease

The buildup of beta-amyloid precursor protein(Aβ) lumps in the brain, which are readily soluble, multi-meric, and potentially inactive amyloid plaques, is the distinguishing features of Alzheimer's disease. These plaques initiate a series of cellular reactions that result in the development of neurofibrillary lesions, aggregates of hyperphosphorylated tau protein, neurofibrillary tangles, and dementia [95]. There were less firmicutes and actinobacteria in the stool samples from AD patients and more Bacteroidetes [96]. There is also evidence linking the development of Alzheimer's disease to the gut flora. A metabolite of a protein originating from the microbiota that was found in the brain fluid of Alzheimer's disease patients (phosphorylated tau and phosphorylated tau/A-42) increases the possibility that the gut microbiome plays a part in the etiology of AD [97]. Variations in different metabolic process are reflected in the decreased amount of SCFAs [98]. It was demonstrated that the gut microbiota diversity of commonly used APP/PS1 double transgenic mice that produce a mutant human presently 1(PS1) in addition to a chimeric mouse human amyloid precursor protein was considerably different from that of non-transgenic wild-type mice. APP/PS1 transgenic animals without germs have notably reduced cerebral β-amyloid pathology in comparison to control mice that have gut microbes and are healthy [99].

26.8.2 Autism

Autism spectrum disorder is a group of neurological developmental disorders distinguished by difficulty in connecting with others and communicating as well as stereotyped and repetitive behavior [107]. Constipation, diarrhea, stomach pain, flatulence, and intestinal gas are common in patients with ASD (23–70%), and they are frequently associated with gastrointestinal diseases. The gut microbiome modulates the quantity of chemical transmitters in ASD, including GABA, glutamate, oxytocin, and serotonin 5-HT complex and various microbial metabolites such as taurine, SCFAs, bile acid metabolites, 5-aminovaleric acid, and influence ASD symptoms [108]. There are differences between ASD patients and controls in firmicutes, prevotella, *Bifidobacterium* and clostridium perfringes [103]. It is not possible to conclude that any particular bacteria cause ASD [109]. The metabolites produced by the gut microbiota may be crucial in the pathophysiology [110].

TABLE 26.1
Neurological Disorders and Their Underlying Causes

S. No.	Neurological Disorders	Causative Agents	Modifications in Gut Microbiota	Reference
1	Parkinson's disease	Stool samples and mucosal samples were taken for analyze. Several genes were found to be downregulated in the microbiome of the stool	*Proteobacteria, Coprococcus, Blautia, Oscillospira* increased, and *Akkermansia, Bacteroides, Firmicutes Faecalibacterium*, and *Clostridia* decreased	[100]
2	Parkinson's disease	Fecal microbiota changes will lead to development of PD. Because of their abundance Prevotellaceae may be used as a biomarker to identify PD	Bacteria, Provotellaceae increased *and* Ruminococcaceae decreased	[101]
3	Alzheimer's disease	Probiotic therapy will lead positive effects on the mental health of AD patients	There is increase *in Bifidobacterium bifidum and Lactobacillus casei* and decrease *in Lactobacillus acidophilus* and *L. fermentum*	[102]
4	Autism	GI and autism symptoms are interrelated; progression autism may be affected by a dysbiosis of bacteria in the healthy gut	There is increase in *Bacillus, Lactobacillus*, and decrease in *Bifidobacterium* and *Enterococcus*	[103]
5	Stroke	Gut biota dysbiosis and epithelial barrier integrity	*Streptococcus, Lactobacillus,* and *Escherichia* increased; *Eubacterium, Roseburia,* and decreased	[104]
6	Multiple sclerosis	Disrupted bacterial and viral the gut microbiota may have a role in pathogenesis by altering the gut axis	Viral species diversity and *Caudovirales* bacteriophages increased; *Faecalibacterium prausnitzii, Gordonibacter urolithinfaciens* decreased	[105]
7.	Epilepsy	Antibiotics promote modulation of gut biome; leads to inflammatory response and stress hormone	*Firmicutes* and *Actinobacteria* increased *Bacteroidetes* and *Proteobacteria* decreased	[106]

26.8.3 Stroke

Choline and L-carnitine-rich diets have a positive relationship with stroke onset via trimethylamine *N*-oxide production [111]. In mice, dietary fibers intake improves stroke prognosis, most likely due to gut microbiota-derived compounds known as SCFAs. Dietary changes may influence gut microbiota populations, affecting the synthesis of microbiome-derived compounds such as trimethylamine/TMAO. SCFAs play important roles in the host's CNS, digestive physiology, and immunological response [112]. Since altering the gut microbiome and its metabolic products appears potential, systematic clinical trials aimed at reducing cerebral-associated risks and improving post-stroke restoration must be conducted [113].

26.8.4 Multiple Sclerosis

The intestinal microbiota appears to be closely associated with brain health. It has been proven that alterations in the gut microbiota can contribute to brain diseases. These changes may influence inflammation, stimulate the body's immune system, and play a role in the development of conditions like multiple sclerosis. A modified gut microbial community could therefore represent a contributing factor in such ailments [114]. Experimental research autoimmune encephalomyelitis (EAE) is a widely agreed to the mouse model for MS, as well as initial research in this model of animals showed that the GI microbiota proved vital to the proliferation of infectious, in contrast to germ-free mice, myelin reactive CD+T cell had immunity from developing the illness [105]. Interestingly, transplantation of fecal samples from healthy co-twins was related with increased production of interleukin 10 (IL-10), a cytokine with various effects on immunoregulation and anti-inflammatory processes [115]. Blocking IL-10 in receivers of healthy co-twin fecal samples enhanced the risk of EAE [116]. *Prevotella histocola*, a human commensal, has been shown to be as effective as interferon beta and glatiramer acetate in treating illness in the EAE model [117].

26.8.5 Epilepsy

The composition of the gut biome has a significant impact on glucose and amino acids metabolism, astrocytic function, neuronal activity, and hippocampal neurotransmitter levels, which is especially relevant for epilepsy [118]. Several animal investigations indicate that probiotics influence the cerebral expression of gamma aminobutyric acid (GABA) receptors in the body, GABA levels, in relation to glutamate, which is associated with seizures susceptibility [119]. *Firmicutes* and *Proteobacteria* are found in great abundance in newborns with refractory epilepsy than bacteroidetes and actinobacteria. Infants with refractory epilepsy have lower levels of prevotella and bididobacterium than control [120]. Rapamycin, a mTOR blocker and antibiotic, reduced mTOR activation protein P-g overexpression and epilepsy vulnerability in a rodent coriaria lactone smoldering rat model of epilepsy in the temporal lobe [121]. Minocycline, a microglial modulator and antibiotic, reduced sympathetic nerve responses and increased seizure tolerance in rats during neurotransmitter activity and amygdala kindling [122].

26.9 HEALTHY LIFESTYLE TO OVERCOME GUT DYSBIOSIS

The following method has been used to overcome the gut dysbiosis by eating more fiber-rich meals, avoiding foods that are highly processed, getting sufficient sleep, and managing your emotional stress levels. Collagen-rich foods that include bones in broth and salmon skin may be beneficial to both general and gastrointestinal health. To help our body produce collagen, consider eating more citrus fruits than beef, broccoli, eggs, almonds, and yogurt [123].

A unique diet rich in plant-based fiber, omega-3 fatty acid polyunsaturated fatty acids, the antioxidants D and E, and micronutrients that are found in beneficial bacteria or prebiotic supplements may enhance microbial community biodiversity, promote short-chain fatty acid synthesis and lower the number of fungal species in gastrointestinal system [124].

26.10 CONCLUSION

The gut-brain axis is a dual communication channel that connects the gut with the host cognitive system. Such knowledge can be transmitted through brain network, hormones, and the immune system, facilitating the microbiota in the gut. The GBA's bidirectional communication prevents malfunction in the brain mechanically while maintaining a mutualistic relationship with the host and managing the innate and adaptive immune system [125]. The gut microbiota of the GI regulates cells in neighboring and distant tissues, including the brain, influencing the host's overall health. A healthy gut microflora supports its host by producing microbial chemicals and neurotransmitters that communicate with the host's cells, including those in the intestinal lining.

Epithelial cells and immune system cell. The GM creates the brain and regulates several cell processes, including neuronal processing, myelination, synaptogenesis, and differentiation. They also hamper mental abilities during brain development since chemicals that encourage growth such as circulating insulin like growth factor-1(IGF-1) are present. Chronic metabolic investigation of GM reveals the regulating function of neural growth and its developmental consequences.

Future study should explore the underlying links between GM, and other neurological diseases are viewed holistically and as whole situations with a significant role for the gastrointestinal tract. It may be possible to develop strategies that target the gut microbiota to provide creative, dependable, and cost-effective treatment options for neurological disorder.

REFERENCES

1. Thursby, E. and N. Juge (2017). "Introduction to the human gut microbiota." *Biochemical Journal* **474**(11): 1823–1836.
2. Sarafian, M., et al. (2017). "Effect on the host metabolism." In *The Microbiota in Gastrointestinal Pathophysiology*. Elsevier, 249–253.
3. Afzaal, M., et al. (2022). "Human gut microbiota in health and disease: Unveiling the relationship." *Frontiers in Microbiology* **13**: 999001.
4. He, C., Y. Shan, and W. Song (2015). "Targeting gut microbiota as a possible therapy for diabetes." *Nutrition Research* **35**(5): 361–367.
5. Dwivedi, M., et al. (2021). "Microbial community in human gut: a therapeutic prospect and implication in health and diseases." *Letters in Applied Microbiology* **73**(5): 553–568.
6. Ganguly, K., et al. (2024). "Biological significance and pathophysiological role of matrix metalloproteinases in the central nervous system." *International Journal of Biological Macromolecules* **280**: 135967.
7. Mukherjee, S., et al. (2023). "History, origin and types of neurological disorders." In *Applications of Stem Cells and derived Exosomes in Neurodegenerative Disorders*. Springer, 1–32.
8. Arun, S., et al. (2016). "Mitochondrial biology and neurological diseases." *Current Neuropharmacology* **14**(2): 143–154.
9. Clemente, J. C., et al. (2012). "The impact of the gut microbiota on human health: An integrative view." *Cell* **148**(6): 1258–1270.
10. Wang, H.-X. and Y.-P. Wang (2016). "Gut microbiota-brain axis." *Chinese Medical Journal* **129**(19): 2373–2380.
11. Marano, G., et al. (2023). "The Microbiota–Gut–Brain axis: Psychoneuroimmunological insights." *Nutrients* **15**(6): 1496.
12. Góralczyk-Bińkowska, A., et al. (2022). "The microbiota–gut–brain axis in psychiatric disorders." *International Journal of Molecular Sciences* **23**(19): 11245.
13. Carabotti, M., et al. (2015). "The gut-brain axis: Interactions between enteric microbiota, central and enteric nervous systems." *Annals of Gastroenterology* **28**(2): 203.
14. Carabotti, M., et al. (2015). "The gut-brain axis: Interactions between enteric microbiota, central and enteric nervous systems." *Annals of Gastroenterology* **28**(2): 203–209.
15. Maiuolo, J., et al. (2021). "The contribution of gut microbiota–brain axis in the development of brain disorders." *Frontiers in Neuroscience* **15**: 616883.
16. Hou, K., et al. (2022). "Microbiota in health and diseases." *Signal Transduction and Targeted Therapy* **7**(1): 1–28.
17. Mojgani, N., et al. (2025). "Gut microbiota and postbiotic metabolites: Biotic intervention for enhancing vaccine responses and personalized medicine for disease prevention.". *Probiotics and Antimicrobial Proteins* **18**: 1–23.
18. Pizzorno, J. E. and M. T. Murray (2020). *Textbook of Natural Medicine-E-Book*. Elsevier Health Sciences.
19. Barrio, C., et al. (2022). "The gut microbiota-brain axis, psychobiotics and its influence on brain and behaviour: A systematic review." *Psychoneuroendocrinology* **137**: 105640.
20. Senghor, B., et al. (2018). "Gut microbiota diversity according to dietary habits and geographical provenance." *Human Microbiome Journal* **7**: 1–9.
21. Nath, A., et al. (2022). "Phage delivered CRISPR-Cas system to combat multidrug-resistant pathogens in gut microbiome." *Biomedicine & Pharmacotherapy* **151**: 113122.
22. Rinninella, E., et al. (2019). "What is the healthy gut microbiota composition? A changing ecosystem across age, environment, diet, and diseases." *Microorganisms* **7**(1): 14.
23. Walker, A. W. and L. Hoyles (2023). "Human microbiome myths and misconceptions." *Nature Microbiology* **8**(8): 1392–1396.
24. Beaugerie, L. and J.-C. Petit (2004). "Antibiotic-associated diarrhoea." *Best Practice & Research Clinical Gastroenterology* **18**(2): 337–352.
25. Lozupone, C. A., et al. (2012). "Diversity, stability and resilience of the human gut microbiota." *Nature* **489**(7415): 220–230.
26. Guarner, F. and J.-R. Malagelada (2003). "Gut flora in health and disease." *The Lancet* **361**(9356): 512–519.
27. Sears, C. L. (2005). "A dynamic partnership: Celebrating our gut flora." *Anaerobe* **11**(5): 247–251.
28. Ray, S., N. Patel, and D. B. Amin. (2020). "Beneficial microbes in agro-ecology." In *Beneficial Microbes in Agro-Ecology*, edited by N. Amaresan, K. M. Senthil, K. Annapurna, K. Kumar, and A. Sankaranarayanan. Springer, 149–167.
29. Felsberg, J., et al. (2015). "Development of a PCR assay based on the 16S-23S rDNA internal transcribed spacer for identification of strictly anaerobic bacterium Zymophilus." *Anaerobe* **33**: 85–89.
30. Galperin, M. Y., et al. (2022). "Conservation and evolution of the sporulation gene set in diverse members of the Firmicutes." *Journal of Bacteriology* **204**(6): e00079–00022.
31. Baron, S. (1996). *Medical Microbiology*. 4th ed., Galveston, TX: University of Texas Medical Branch at Galveston.
32. Africa, C. W., et al. (2014). "Anaerobes and bacterial vaginosis in pregnancy: Virulence factors contributing to vaginal colonisation." *International Journal of Environmental Research and Public Health* **11**(7): 6979–7000.
33. Czepiel, J., et al. (2019). "Clostridium difficile infection." *European Journal of Clinical Microbiology & Infectious Diseases* **38**: 1211–1221.
34. Aronoff, D. M. and P. H. Kazanjian (2018). "Historical and contemporary features of infections due to Clostridium novyi." *Anaerobe* **50**: 80–84.
35. Wang, L., et al. (2018). "Learning from Clostridium novyi-NT: How to defeat cancer." *Journal of Cancer Research and Therapeutics* **14**(Suppl 1): S1–S6.
36. Cohen, J. E., et al. (2017). "Comparative pathogenomics of Clostridium tetani." *PLoS One* **12**(8): e0182909.
37. Brüggemann, H., et al. (2003). "The genome sequence of Clostridium tetani, the causative agent of tetanus disease." *Proceedings of the National Academy of Sciences* **100**(3): 1316–1321.
38. Alves, M. L. F., et al. (2022). "Clostridium haemolyticum, a review of beta toxin and insights into the antigen design for vaccine development." *Molecular Immunology* **148**: 45–53.
39. Labbe, R. G. (1991). "Clostridium perfringens." *Journal of the Association of Official Analytical Chemists* **74**(4): 711–714.
40. Johnson, E. A. (2012). "Clostridium botulinum." In *Food Microbiology: Fundamentals and Frontiers*, 441–463.
41. Keis, S., et al. (2001). "Emended descriptions of Clostridium acetobutylicum and Clostridium beijerinckii, and descriptions of Clostridium saccharoperbutylacetonicum sp. nov.

and Clostridium saccharobutylicum sp. nov." *International Journal of Systematic and Evolutionary Microbiology* **51**(6): 2095–2103.

42. Vieira, S., et al. (2023). "Aurantibacillus circumpalustris gen. nov., sp. nov., the first characterized representative of the Bacteroidota candidate family env. OPS 17 and proposal of Aurantibacillaceae fam. nov." *International Journal of Systematic and Evolutionary Microbiology* **73**(10): 006134.
43. Pittrof, S. L., et al. (2021). "Products released from structurally different dextrans by bacterial and fungal dextranases." *Foods* **10**(2): 244.
44. Kruczyńska, A., et al. (2023). "Bacteroidota structure in the face of varying agricultural practices as an important indicator of soil quality–a culture independent approach." *Agriculture, Ecosystems & Environment* **342**: 108252.
45. Ibrahimi, M., et al. (2023). "The potential of facultative predatory Actinomycetota spp. and prospects in agricultural sustainability." *Frontiers in Microbiology* **13**: 1081815.
46. Kitikhun, S., et al. (2023). "Expanding actinomycetota diversity in the TBRC culture collection through metabarcoding and simulated in situ cultivation of Thailand's Mekong river microbiota." *Diversity* **15**(5): 663.
47. Leão, I., et al. (2023). "Pseudomonadota in the oral cavity: A glimpse into the environment-human nexus." *Applied Microbiology and Biotechnology* **107**(2): 517–534.
48. Liu, Z.-T., et al. (2022). "Thermomonas flagellata sp. nov. and Thermomonas alba sp. nov., two novel members of the phylum Pseudomonadota isolated from hot spring sediments." *International Journal of Systematic and Evolutionary Microbiology* **72**(6): 005444.
49. Joh, K., et al. (2023). "A report of 156 unrecorded bacterial species of Republic of Korea belonging to the phyla Acidobacteriota, Deinococcota, Actinomycetota, Bacillota, Bacteroidota, and Pseudomonadota isolated in 2022." *Journal of Species Research* **12**(4): 374–414.
50. Cisek, A. A., et al. (2024). "The role of methanogenic archaea in inflammatory bowel disease—A review." *Journal of Personalized Medicine* **14**(2): 196.
51. Eme, L. and W. F. Doolittle (2015). "Archaea." *Current Biology* **25**(19): R851–R855.
52. Pargin, E., et al. (2023). "The human gut virome: Composition, colonization, interactions, and impacts on human health." *Frontiers in Microbiology* **14**: 963173.
53. Lecuit, M. and M. Eloit (2017). "The viruses of the gut microbiota." In *The Microbiota in Gastrointestinal Pathophysiology*. Elsevier, 179–183.
54. Harper, A., et al. (2021). "Viral infections, the microbiome, and probiotics." *Frontiers in Cellular and Infection Microbiology* **10**: 596166.
55. Scarpellini, E., et al. (2015). "The human gut microbiota and virome: Potential therapeutic implications." *Digestive and Liver Disease* **47**(12): 1007–1012.
56. Van Tilburg Bernardes, E., et al. (2020). "Intestinal fungi are causally implicated in microbiome assembly and immune development in mice." *Nature Communications* **11**(1): 2577.
57. Forbes, J. D., et al. (2019). "A fungal world: could the gut mycobiome be involved in neurological disease?" *Frontiers in Microbiology* **9**: 3249.
58. Conlon, M. A. and A. R. Bird (2014). "The impact of diet and lifestyle on gut microbiota and human health." *Nutrients* **7**(1): 17–44.
59. Shin, C. and Y.-K. Kim (2019). "Autoimmunity in microbiome-mediated diseases and novel therapeutic approaches." *Current Opinion in Pharmacology* **49**: 34–42.
60. Ren, J., et al. (2023). "Gut microbiome-mediated mechanisms in aging-related diseases: Are probiotics ready for prime time?" *Frontiers in Pharmacology* **14**: 1178596.
61. Ross, F. C., et al. (2024). "The interplay between diet and the gut microbiome: Implications for health and disease." *Nature Reviews Microbiology* **22**: 1–16.
62. Rinninella, E., et al. (2023). "The role of diet in shaping human gut microbiota." *Best Practice & Research Clinical Gastroenterology* **62**: 101828.
63. Zhang, P. (2022). "Influence of foods and nutrition on the gut microbiome and implications for intestinal health." *International Journal of Molecular Sciences* **23**(17): 9588.
64. Monda, V., et al. (2017). "Exercise modifies the gut microbiota with positive health effects." *Oxidative Medicine and Cellular Longevity* **2017**(1): 3831972.
65. Booth, B. M. and W. Feng (2002). "The impact of drinking and drinking consequences on short-term employment outcomes in at-risk drinkers in six southern states." *The Journal of Behavioral Health Services & Research* **29**: 157–166.
66. Lee, E. and J.-E. Lee (2021). "Impact of drinking alcohol on gut microbiota: Recent perspectives on ethanol and alcoholic beverage." *Current Opinion in Food Science* **37**: 91–97.
67. Yan, S., et al. (2021). "Effects of smoking on inflammatory markers in a healthy population as analyzed via the gut microbiota." *Frontiers in Cellular and Infection Microbiology* **11**: 633242.
68. Hu, X., et al. (2021). "Impacts of cigarette smoking status on metabolomic and gut microbiota profile in male patients with coronary artery disease: A multi-omics study." *Frontiers in Cardiovascular Medicine* **8**: 766739.
69. Mulder, M., et al. (2020). "Long-term effects of antimicrobial drugs on the composition of the human gut microbiota." *Gut Microbes* **12**(1): 1791677.
70. Ramirez, J., et al. (2020). "Antibiotics as major disruptors of gut microbiota." *Frontiers in Cellular and Infection Microbiology* **10**: 572912.
71. Lathakumari, R. H., et al. (2024). "Antibiotics and the gut microbiome: Understanding the impact on human health." *Medicine in Microecology* **20**: 100106.
72. Tasnim, N., et al. (2017). "Linking the gut microbial ecosystem with the environment: Does gut health depend on where we live?" *Frontiers in Microbiology* **8**: 1935. https://doi.org/10.3389/fmicb.2017.01935
73. Conlon, M. A. and A. R. Bird (2014). "The impact of diet and lifestyle on gut microbiota and human health." *Nutrients* **7**(1): 17–44. https://doi.org/10.3390/nu7010017
74. Pedroza Matute, S. and S. Iyavoo (2023). "Exploring the gut microbiota: lifestyle choices, disease associations, and personal genomics." *Frontiers in Nutrition* **10**: 1225120. https://doi.org/10.3389/fnut.2023.1225120
75. Graf, D., et al. (2015). "Contribution of diet to the composition of the human gut microbiota." *Microbial Ecology in Health and Disease* **26**: 26164. https://doi.org/10.3402/mehd.v26.26164
76. Rio, P., et al. (2024). "The impact of climate change on immunity and gut microbiota in the development of disease." *Diseases* **12**(6): 118. https://doi.org/10.3390/diseases12060118
77. Nurkolis, F., et al. (2022). "Human activities and changes in the gut microbiome: A perspective." *Human Nutrition & Metabolism* **30**: 200165. https://doi.org/10.1016/j.hnm.2022.200165

78. Catania, F., et al. (2021). "Global climate change, diet, and the complex relationship between human host and microbiome: Towards an integrated picture." *BioEssays* **43**(6): e2100049. https://doi.org/10.1002/bies.202100049
79. Chen, Y., et al. (2021). "Regulation of neurotransmitters by the gut microbiota and effects on cognition in neurological disorders." *Nutrients* **13**(6): 2099. https://doi.org/10.3390/nu13062099
80. Sorboni, S. G., et al. (2022). "A comprehensive review on the role of the gut microbiome in human neurological disorders." *Clinical Microbiology Reviews* **35**(1): e0033820. https://doi.org/10.1128/CMR.00338-20
81. Mittal, R., et al. (2017). "Neurotransmitters: The critical modulators regulating gut-brain axis." *Journal of Cellular Physiology* **232**(9): 2359–2372. https://doi.org/10.1002/jcp.25518
82. Socała, K., et al. (2021). "The role of microbiota-gut-brain axis in neuropsychiatric and neurological disorders." *Pharmacological Research* **172**: 105840. https://doi.org/10.1016/j.phrs.2021.105840
83. Ullah, H., et al. (2023). "The gut microbiota-brain axis in neurological disorder." *Frontiers in Neuroscience* **17**: 1225875. https://doi.org/10.3389/fnins.2023.1225875
84. Barbosa, P. M. and E. R. Barbosa (2020). "The gut brain-axis in neurological diseases." *International Journal of Cardiovascular Sciences*. https://doi.org/10.36660/ijcs.20200039
85. Münger, E., et al. (2018). "Reciprocal interactions between gut microbiota and host social behavior." *Frontiers in Integrative Neuroscience* **12**: 21. https://doi.org/10.3389/fnint.2018.00021
86. Fung T. C. (2020). "The microbiota-immune axis as a central mediator of gut-brain communication." *Neurobiology of Disease* **136**: 104714. https://doi.org/10.1016/j.nbd.2019.104714
87. Montagnani, M., et al. (2023). "The crosstalk between gut microbiota and nervous system: A bidirectional interaction between microorganisms and metabolome." *International Journal of Molecular Sciences* **24**(12): 10322. https://doi.org/10.3390/ijms241210322
88. Doroszkiewicz, J., et al. (2021). "The role of gut microbiota and gut-brain interplay in selected diseases of the central nervous system." *International Journal of Molecular Sciences* **22**(18): 10028. https://doi.org/10.3390/ijms221810028
89. Sun, L. J., et al. (2020). "Gut hormones in microbiota-gut-brain cross-talk." *Chinese Medical Journal* **133**(7): 826–833. https://doi.org/10.1097/CM9.0000000000000706
90. Pan, I., et al. (2023). "Gut-brain axis a key player to control gut dysbiosis in neurological diseases." *Molecular Neurobiology*. "https://doi.org/10.1007/s12035-023-03691-3
91. Suganya, K. and B. S. Koo (2020). "Gut-brain axis: Role of gut microbiota on neurological disorders and how probiotics/prebiotics beneficially modulate microbial and immune pathways to improve brain functions." *International Journal of Molecular Sciences* **21**(20): 7551. https://doi.org/10.3390/ijms21207551
92. Hou, K., et al. (2022). "Microbiota in health and diseases." *Signal Transduction and Targeted Therapy* **7**(1): 135. https://doi.org/10.1038/s41392-022-00974-4
93. Kouli, A., et al. (2018). *Parkinson's Disease: Etiology, Neuropathology, and Pathogenesis*. Exon Publications, 3–26.
94. Haque, M. E., et al. (2022). "Targeting α -synuclein aggregation and its role in mitochondrial dysfunction in Parkinson's disease." *British Journal of Pharmacology* **179**(1): 23–45.
95. Zhang, X., et al. (2018). "The early events that initiate β-amyloid aggregation in Alzheimer's disease." *Frontiers in Aging Neuroscience* **10**: 359.
96. Naomi, R., et al. (2021). "Probiotics for Alzheimer's disease: A systematic review." *Nutrients* **14**(1): 20. https://doi.org/10.3390/nu14010020
97. Vogt, N. M., et al. (2018). "The gut microbiota-derived metabolite trimethylamine N-oxide is elevated in Alzheimer's disease." *Alzheimer's Research & Therapy* **10**: 1–8.
98. Dhami, M., et al. (2023). "Relevance of gut microbiota to Alzheimer's disease (AD): Potential effects of probiotic in management of AD." *Aging and Health Research* **3**(1): 100128.
99. Menden, A. (2021). *The Impact of Candida Rugosa Lipase in the Gut Microbiome in Health and Alzheimer's Disease and Identification of Allosteric Modulators*. Open University.
100. Mah, C., et al. (2023). "Assessing the relationship between the gut microbiota and inflammatory bowel disease therapeutics: A systematic review." *Pathogens* **12**(2): 262. https://doi.org/10.3390/pathogens12020262
101. Yang, D., et al. (2019). "The role of the gut microbiota in the pathogenesis of Parkinson's disease." *Frontiers in Neurology* **10**: 1155. https://doi.org/10.3389/fneur.2019.01155
102. Naomi, R., et al. (2021). "Probiotics for Alzheimer's disease: A systematic review." *Nutrients* **14**(1): 20. https://doi.org/10.3390/nu14010020
103. Taniya, M. A., et al. (2022). "Role of gut microbiome in autism spectrum disorder and its therapeutic regulation." *Frontiers in Cellular and Infection Microbiology* **12**: 915701. https://doi.org/10.3389/fcimb.2022.915701
104. Peh, A., et al. (2022). "Gut microbiota and their metabolites in stroke: A double-edged sword." *Stroke* **53**(5): 1788–1801. https://doi.org/10.1161/STROKEAHA.121.036800
105. Thirion, F., et al. (2023). "The gut microbiota in multiple sclerosis varies with disease activity." *Genome Medicine* **15**(1): 1.
106. Lum, G. R., et al. (2020). "Emerging roles for the intestinal microbiome in epilepsy." *Neurobiology of Disease* **135**: 104576
107. Maiuolo, J., et al. (2021). "The contribution of gut microbiota–brain axis in the development of brain disorders." *Frontiers in Neuroscience* **15**: 616883.
108. Madra, M., et al. (2021). "Gastrointestinal issues and autism spectrum disorder." *The Psychiatric clinics of North America* **44**(1): 69–81. https://doi.org/10.1016/j.psc.2020.11.006
109. Iglesias-Vázquez, L., et al. (2020). "Composition of gut microbiota in children with autism spectrum disorder: A systematic review and meta-analysis." *Nutrients* **12**(3): 792.
110. Zou, R., et al. (2020). "Changes in the gut microbiota of children with autism spectrum disorder." *Autism Research* **13**(9): 1614–1625.
111. Martinez, K. B., et al. (2017). "Western diets, gut dysbiosis, and metabolic diseases: Are they linked?" *Gut Microbes* **8**(2): 130–142. https://doi.org/10.1080/19490976.2016.1270811
112. Chen, X., et al. (2023). "Gut microbiota and microbiota-derived metabolites in cardiovascular diseases." *Chinese Medical Journal* **136**(19): 2269–2284. https://doi.org/10.1097/CM9.0000000000002206

113. Ramos Meyers, G., et al. (2022). "Short chain fatty acid metabolism in relation to gut microbiota and genetic variability." *Nutrients* **14**(24): 5361. https://doi.org/10.3390/nu14245361
114. Ullah, H., et al. (2023). "The gut microbiota-brain axis in neurological disorder." *Frontiers in Neuroscience* **17**: 1225875. https://doi.org/10.3389/fnins.2023.1225875
115. Vendrik, K. E., et al. (2020). "Fecal microbiota transplantation in neurological disorders." *Frontiers in Cellular and Infection Microbiology* **10**: 98.
116. Berer, K., et al. (2017). "Gut microbiota from multiple sclerosis patients enables spontaneous autoimmune encephalomyelitis in mice." *Proceedings of the National Academy of Sciences* **114**(40): 10719–10724.
117. Shahi, S. K., et al. (2020). "Human commensal Prevotella histicola ameliorates disease as effectively as interferon-beta in the experimental autoimmune encephalomyelitis." *Frontiers in Immunology* **11**: 578648.
118. Ding, M., et al. (2021). "Microbiota–gut–brain axis and epilepsy: A review on mechanisms and potential therapeutics." *Frontiers in Immunology* **12**: 742449. https://doi.org/10.3389/fimmu.2021.742449
119. Amlerova, J., et al. (2021). "Evidences for a role of gut microbiota in pathogenesis and management of epilepsy." *International Journal of Molecular Sciences* **22**(11): 5576. https://doi.org/10.3390/ijms22115576
120. Li, Q., et al. (2024). "A long journey to treat epilepsy with the gut microbiota." *Frontiers in Cellular Neuroscience* **18**: 1386205. https://doi.org/10.3389/fncel.2024.1386205
121. Wanleenuwat, P., et al. (2020). "Antibiotic-induced epileptic seizures: Mechanisms of action and clinical considerations." *Seizure* **81**: 167–174. https://doi.org/10.1016/j.seizure.2020.08.012
122. Singh, T., et al. (2022). "Reconnoitering the transformative journey of minocycline from an antibiotic to an antiepileptic drug." *Life Sciences* **293**: 120346.
123. Martinez, J. E., et al. (2021). "Unhealthy lifestyle and gut dysbiosis: A better understanding of the effects of poor diet and nicotine on the intestinal microbiome." *Frontiers in Endocrinology* **12**: 667066. https://doi.org/10.3389/fendo.2021.667066
124. Jawhara S. (2023). "Healthy diet and lifestyle improve the gut microbiota and help combat fungal infection." *Microorganisms* **11**(6): 1556. https://doi.org/10.3390/microorganisms11061556
125. Appleton J. (2018). "The gut-brain axis: Influence of microbiota on mood and mental health." *Integrative Medicine (Encinitas, Calif.)* **17**(4): 28–32.

27 Prebiotics, Probiotics, and Postbiotics as a Complementary Therapy in the Management of Neurological Disorders

Clinical Trials and Case Studies

Heena[1], Priyanka Rana[1], Monika[1], Tanzeer Kaur[2], and Neelima Dhingra[1]*

[1] University Institute of Pharmaceutical Sciences, Panjab University, Chandigarh, India
[2] Department of Biophysics, Panjab University, Chandigarh, India
[*] Corresponding Author.

27.1 INTRODUCTION

The human gut microbiota is a complex ecosystem comprising tons of microorganisms, including bacteria, fungi, viruses, and other microbial species. Recent scientific investigations have revealed the importance of gut microbiota homeostasis in maintaining human metabolic homeostasis, regulating intestinal inflammation, and maturation and regulation of the immune system (Ji et al.,2023). Rising evidence suggests the strong bidirectional communication between gut and brain that links the enteric and central nervous systems known as the gut–brain axis. The network is not only anatomical, but extended to include humoral, endocrine, metabolic, and immune routes of communication as well. The autonomic nervous system (ANS), hypothalamic–pituitary–adrenal (HPA) axis, and nerves within the gastrointestinal (GI) tract all link the gut and the brain, which allows the brain to influence intestinal activities, including functional immune effector cells activity; and the gut to influence mood, cognition, and mental health. Many neurological disorders with inflammatory components have been found to b associated with disrupted gut microbiota homeostasis known as dysbiosis. Loss of gut barrier integrity and intestinal permeability associated with gut dysbiosis allow the increased translocation of the gut bacteria and derived metabolites- and microbes-associated molecular patterns (MAMPs) into mesenteric lymphoid tissues, which results in the progression and development of various neurological diseases. Clinical trials and case studies suggest that prebiotics, probiotics, and postbiotics regulate gut microbiota, promote beneficial bacteria, and alleviate gut dysbiosis–related conditions like depression and anxiety. By synthesizing neurotransmitters such as GABA, serotonin, and dopamine, they help regulate mood and stress responses, potentially reducing depressive and anxious symptoms (Zhou et al., 2024). These supplements have shown potential as complementary therapies in preventing and managing the neurological complications with reduced cost, less complication side effects, and cognitive deficit. The biotic markets are all projected to experience significant growth by 2031 with compound annual growth rate (CAGR) of 14.9%, 8.7%, and 11.2% respectively (Ji et al., 2023). Detailed research is needed to develop standardized methods and customized treatment plan due to inconsistencies and numerous gaps differences in the type of strain, dose quantity, assessment of gut microbiota, type of prebiotic, probiotic, and study design, duration of intervention, and variety and intricacy of neurological symptoms make it difficult to confirm their efficacy but they also face challenges and limitations. This chapter seeks to provide a comprehensive overview of the current research on prebiotics, probiotics, postbiotics and their impact on neurological disorders by systematically reviewing various clinical trials and case studies. The goal is to recollect the existing evidence highlighting the potential benefits, mechanisms, and significance to better understand how biotic interventions may contribute toward the management of neurological disorders.

27.2 GUT TO BRAIN AND BRAIN TO GUT COMMUNICATION

Trillions of varied communities of microorganisms reside in the human gastrointestinal tract that play a crucial role in digestion, nutrient absorption, and immune functions. The gut–brain link, also known as the gut–brain axis, is a complex bidirectional communication system that relates the gut microbiome comprising bacteria, viruses, fungi,

DOI: 10.1201/9781003536802-27

and archaea and their metabolites to the central nervous system (CNS) (Scarpellini et al.,2021). The neurologic pathway involves the enteric nervous system (ENS), vagus nerve, and neurotransmitter activity within the gastrointestinal tract. Neurologic modulation of afferent sensory nerves directly produces molecules that can act as local neurotransmitters, such as histamine, gamma-aminobutyric acid (GABA), melatonin, serotonin, and acetylcholine, along with biologically active catecholamines in the lumen of the gut. Gut microbiota modifies nutrient availability that influences the release of biologically active peptides from enteroendocrine cells, like neuropeptide galanin that control sleep-wake cycle, blood pressure regulation, feeding, mood, parental behavior, and neurotrophic functions. Galanin also stimulates the activity of the central branch of the HPA axis and influences the release of corticotropin-releasing factor and adrenocorticotropic hormone, that enhances glucocorticoid secretion from the adrenal cortex. Further, it stimulates cortisol secretion from the adrenal cortex, as well as norepinephrine release from the adrenal medulla, and demonstrates its involvement in the HPA axis response to stress. Bacterial metabolites most commonly short-chain fatty acids (SCFAs), produced by the bacterial fermentation of dietary carbohydrates are known to affect the nutrition of enterocytes, have immunomodulatory properties exert significant hormone-like activity, and interact with nerve cells by stimulating the sympathetic branch of the ANS. Microbiota-derived SCFAs cross the blood–brain barrier to regulate microglial homeostasis, essential for brain development and behavior modulation. They also influence serotonin synthesis from enterochromaffin cells and, along with enteroendocrine cells, impact gut–brain hormonal communication (Alzubide et al., 2024).

27.3 GUT–BRAIN AXIS IN NEUROLOGICAL DISORDERS

Human health is endangered by the dramatic environmental and lifestyle changes in the modern era and an unprecedented upsurge in neurological disorders is one of the major global challenges. World Health Organization (WHO) data suggest that neurological conditions are now the leading cause of ill health and disability worldwide and affecting more than 3 billion people worldwide. Further, groundbreaking studies duly conducted over the past ten years have revealed the active relations between animals and the microbial populations living inside them, important for normal brain development and function (Ullah et al., 2023). Whereas disrupted gut microbiota homeostasis (gut dysbiosis) and allied loss of gut barrier integrity and intestinal permeability allow the increased translocation of the gut bacteria-derived metabolites and microbes-associated molecular patterns into mesenteric lymphoid tissues, resulting in the development of various neurological diseases.

Aggregation of alpha-synuclein (α-syn) in the submucosal and myenteric plexus of the enteric nervous system, the primary neuropathological indicator of PD, was first observed in the brain and suggested a potential transmission of the disease from the gut to the brain (Felice et al., 2016). Furthermore, the development of PD has also been linked to intestinal inflammation, wherein metabolites produced by the gut microbiota activate an immune response. A small study by Engen P. A with six multiple system atrophy (MSA) patients showed the dysfunction of the intestinal barrier and signs of inflammation in the intestine associated with bacterial endotoxins. Furthermore, there was plenty of Bacteroidetes and pro-inflammatory bacteria and a reduction in butyrate-producing bacteria.

The increased systemic translocation of microbial components, gut metabolites, or microbiota through a damaged intestinal barrier "leaky gut" exacerbates systemic inflammation, and that has been implicated in the pathogenesis of depression (Chudzik et al., 2021). *Actinobacteria, Firmicutes,* and *Bacteroidetes* are the most affected phyla in depression, specifically with an increased *Bacteroidetes/ Firmicutes* ratio in major depressive disorder patients. Further, abnormal ENS activity arising from intestinal pathology worsens depression-related pathological changes by varying gut secretion, motility, immune defenses, and permeability. The microbial cell components, like lipopolysaccharide produced by Gram-negative bacteria, and peripheral inflammatory signals reache the brain by crossing the blood–brain barrier and result in neuroinflammation. Subsequently, it induces neuropathological changes by activating specific cells, along with demyelination, abnormal neurogenesis, and neurotransmitters release, that are involved in the pathogenesis of depression. Microbial signals, neurobiological, pathological changes, and depressive emotions activate the HPA axis, and increase the release of cortisol that promotes gut pathology by modulating intestinal barrier function and inflammatory responses (Liu et al., 2023).

Translocation of microbes and their products infiltration into the brain is responsible for overproduction and deposition of amyloid-peptide (AP) in AD. The AP with 40–42 amino acids is derived by proteolytic cleavage of amyloid precursor protein (APP) and plays a major role in the AD pathogenesis by initiating neuroinflammation and neurodegenerative alteration. Published reports by Chandra et al. indicate that the mice with disrupted gut microbiomes are more likely to develop amyloid plaque and tau tangles, the hallmarks of AD. A peripheral inflammatory state has also been described in persons with cognitive impairment in the context of amyloidosis. Gut microbiota has been found to play an important role like with rise in pro-inflammatory bacterial species, such as *Shigella and Escherichia* and, fall in anti-inflammatory bacterial species *(E. rectale).*

Recent research highlights the gut microbiota's key role in multiple sclerosis (MS) pathophysiology. Imbalances influence immune responses, increasing pro-inflammatory cytokines and reducing regulatory T-cells, contributing to MS progression. Gut bacteria-derived metabolites, like SCFAs, can either support neuroprotection or worsen neuroinflammation, emphasizing the gut–brain axis's impact on MS.

Autism, a neurodevelopmental disorder is characterized by deficits in social communication and interaction and restricted, repetitive patterns of behavior or activities. Beneficial gut bacteria such as *Bifidobacterium infantis* regulate emotions and behavior by converting tryptophan into serotonin. *Clostridium sporogenes* has been found to increase the production of tryptophan metabolites, i.e., indole-3-propionic acid responsible for the production of antioxidant and neuroprotectant molecules inside the gut (Wikoff et al., 2009). Pathogenic microbes can invade the gut, causing gastrointestinal issues like the presence of *Clostridium bolteae* is linked with increased ASD risk and severity (Tsai et al., 2012). It produces tetanus neurotoxin (TeNT) that travels via the vagus nerve, blocking neurotransmitters by proteolytic cleaving of synaptobrevin.

27.4 PREBIOTICS, PROBIOTICS, AND POSTBIOTICS: A COMPLEMENTARY THERAPY

Complementary therapies hold an important place in holistic healthcare, and offer an alternative or supplementary treatment to the existing conventional medicines by addressing not only the physical, but also the emotional and mental aspects of well-being (Debas et al., 2011). These therapies encompass a broad spectrum of practices, including chiropractic care, acupuncture, massage therapy, meditation yoga and herbal medicine. They are particularly beneficial to manage chronic disease conditions, improve quality of life, and reduce the side effects associated with conventional treatments. The use of prebiotics, probiotics, and postbiotics has gained significant attention as a complementary therapy (Figure 27.1), driven by a growing awareness of the gut–brain connection and the overall impact of gut health on general well-being. Further, the trend has been fueled by a surge in scientific research highlighting the potential benefits of these compounds in managing and preventing a variety of health conditions, including gastrointestinal disorders, metabolic and cardiovascular diseases, and mental health issues. Prebiotics, the "non-digestible food ingredients," are used by the host to selectively stimulate the growth and/or activity of one or a limited number of bacteria already residing in the colon or other bacteria in the human body (Chudzik et al., 2021). Whereas the "living microorganisms (probiotics) and preparation of inanimate micro-organisms and/or their components (postbiotics) when administered in adequate amounts confer a health benefit on the host" (Rafique et al., 2023). Consumers are becoming more proactive about their health, seeking natural and holistic approaches to complement traditional medical treatments. In clinical practices, healthcare providers are increasingly recommending these therapies as part of a comprehensive treatment plan. The integration of prebiotics, probiotics, and postbiotics into daily health regimens represents a significant shift toward preventive healthcare and personalized medicine, emphasizing the significance of maintaining a balanced and healthy gut microbiota (Cunningham et al., 2021).

27.4.1 Issues and Limitations of Traditional Medicine

In the past two decades, technological advances in the fields of biology, chemistry, biology, and pharmacology, and better understanding of the physiology and pathology involved, have resulted in the development of abundant medicines for neurological disorders management.

Undoubtedly, available treatment for neurological disorders is controlling and providing relief, but they are associated with many limitations, which in turn is impacting their effectiveness and patient outcomes. Medications like donepezil, galantamine, and rivastigmine used to treat AD, dementia with Lewy bodies, and Parkinson's dementia are adding to mild to severe side effects like muscle cramps, headaches, dizziness, fatigue, and insomnia, thus impacting patient's quality life and treatment adherence. Genetic variations, disease heterogeneity, medicine resistance, and drug–drug interactions are some of the factors for suboptimal treatment outcomes observed in patients taking medicines like clozapine and electroconvulsive therapy for schizophrenia (Potkin et al., 2020).

Traditional medicine, including Chinese medicine, ashwagandha, curcumin, and ginseng, has long been used for neurological disorders but lacks rigorous scientific evaluation. While herbs like firmoss and ginkgo show cognitive benefits in AD and PD, their mechanisms remain unclear. Optimizing doses and addressing limitations drive the need to develop advanced biomedical approaches for neurological disorders management (Nadaf et al., 2019).

Drug discovery, a 10–15-year complex process costing over $2.8 billion per drug, faces high failure rates due to efficacy and safety issues, thus resulting in the wasting of resources. Further, rising screening challenges not only hinder but also delay the development process. Fortunately, increasing interest in biotics offers promising solutions to address the traditional medicine limitation.

27.4.2 Therapeutic Significance in Neurological Disorders

Nature-derived products, including biotics, have been considered as a suitable alternative for the treatment of inflammatory, immune, metabolic, or degenerative diseases (Figure 27.1) (Yadav et al., 2022). Prebiotics generally in the food ingredients not only promote the growth of microorganisms present in the human gut, but also stimulate the immune system and have shown positive effect on gastrointestinal diseases like irritable bowel syndrome, allergic diseases, obesity, insulin resistance syndrome, type 2 diabetes, non-alcoholic fatty liver disease, cancer,

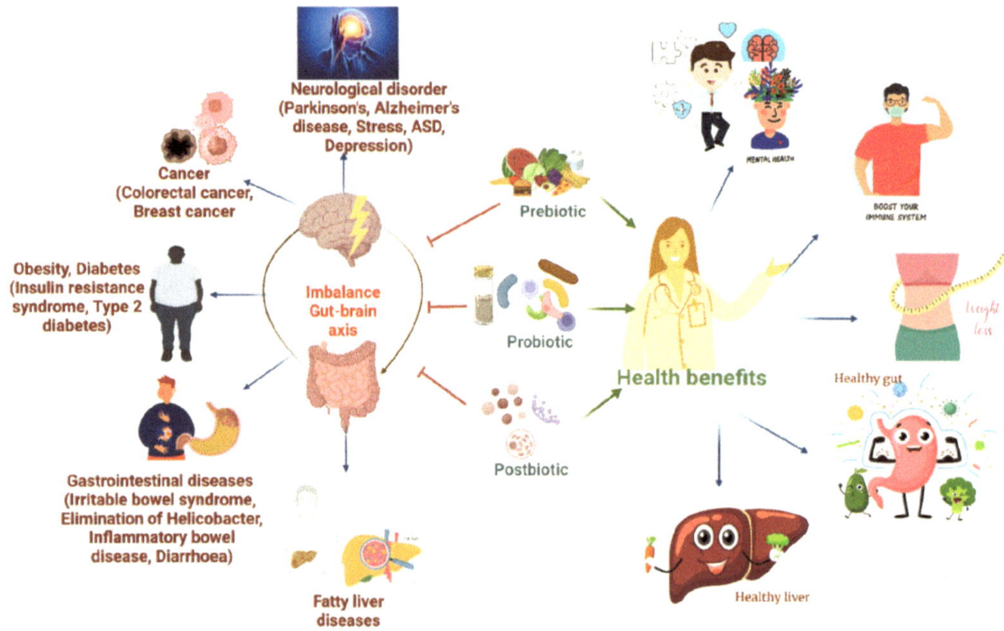

FIGURE 27.1 Biotic therapy in human health.

immunomodulation, and neurological disorders, including anxiety and depression. Probiotics promote health by stimulating native gut microbiota, host immunity, cholesterol reduction, and several other functions. Whereas their metabolites postbiotics such as lactic acid bacteriocins and hydrogen peroxide are of immense importance as antimicrobial against a wide range of pathogenic bacteria (Fekete et al., 2024). In the following section, efforts have been made to highlight the impact of biotic therapy on neurological disorders by showcasing number of clinical trials and case studies.

27.4.3 Clinical Trial(s)

Clinical trials are also referred to as interventional trials to assess the efficacy of drugs on humans, including Phase I to Phase IV trials, and are not restricted to cells, drugs, biological products, surgical and radiologic procedures, devices, behavioral treatments and preventive care, etc. (WHO 2024).

27.4.3.1 Parkinson's

Clinical Trial 1

Parkinson's disease (PD) involves dopaminergic neuron loss and alpha-synuclein accumulation in the basal ganglia and peripheral sites like the gut. Peripheral immune activation and production of reactive oxygen species (ROS) play an important role in its pathogenesis (Elangovan et al., 2024). Magistrelli et al. conducted a study to assess the effects of probiotic bacterial strains from *Lactobacillus* and *Bifidobacterium* genera in peripheral blood mononuclear cells (PBMCs) isolated from 40 PD patients and healthy controls (Table 27.1), and investigated the in vitro release of the major procytokines (TNF-alpha, IL-17A and anti-inflammatory interleukin 4 and 10). Researchers also assessed the ability of probiotics in antagonizing the growth of pathogenic bacteria, like *Escherichia coli* and *Klebsiella pneumoniae* by affecting their membrane integrity. Significant inhibition of the inflammatory cytokines and ROS production was observed in both patients and controls by probiotic strains *Lactobacillus salivarius* LS01, *L. plantarum* LP01, *L. acidophilus* LA02, *L. rhamnosus* LR06 and *Bifidobacterium animalis* subspecies *lactis* (BS01 *and breve* BR03 along with restoration of membrane integrity). Outstanding results were also obtained in PD subjects with *L. salivarius* LS01 and *L. acidophilus* LA02 with significant decrease in pro-inflammatory and an increase in anti-inflammatory cytokines (Magistrelli et al., 2019).

Clinical Trial 2

Documented studies indicate the influence of disrupted intestinal microbiota community on PD symptoms, progression, and treatment success. Significant differences in bacterial composition (dysbiosis) observed in both early and late stages of PD have been linked to increased systemic and neuroinflammation, along with the disruption of intestinal barrier integrity, compared to age-matched controls. PD-associated microbiota is characterized by the reduced relative abundance of Lachnospiraceae and putative anti-inflammatory SCFA-producing bacteria from the genus *Faecalibacterium*. In contrast to this, amplified relative abundance of putative pro-inflammatory, gram-negative lipopolysaccharide producing pathobionts Enterobacteriaceae, genera *Akkermansia, Lactobacillus, and Bifidobacterium* were noticed. SCFAs are critical in

TABLE 27.1
Measuring Effects of Biotics on Neurological Disorders

S. No.	Clinical Trial (CT)	Biotic Therapy	Microbial Strain	Product Name	Studies Conducted (Year) and Research Group	Country
Parkinson						
1	CT-1	Probiotics	*Lactobacillus (salivarius* LS01 DSM 22775), *plantarum* (LP01 LMG P-21021), *acidophilus* (LA02 DSM 21717), *Bifidobacterium animalis* subspecies lactis (BS01 LMG P-21384), *breve* (BR03 DSM 16604)	_	University of Insubria (2019)	Italy
2	CT-2	Prebiotics	*Ruminococcus, Dorea, Bacteroides, Blautia, Anaerostipes, and Bifidobacterium*	_	Department of Neurological Sciences, Rush University Medical Center (2023)	USA
Alzheimer's						
3	CT-3	Probiotics	*Lactobacillus casei* (W56), *lactis* (W19), *acidophilus* (W22), *salivarius* (W24), *Bifidobacterium lactis* W52, *paracasei* (W20), *plantarum* (W62), *Bifidobacterium lactis* (W51), *bifidum* (W23)	Omnibiotic stress repair	Department of Gerontology, Neuromed Campus, Kepler University Clinic (2018)	Austria
Depression						
4	CT-4	Postbiotic	*Lactobacillus paracasei* MCC1849	LAC-Shield™	Morinaga Milk Industry Co. Ltd (2018)	Japan
5	CT-5	Postbiotic	*Lactobacillus gasseri* CP2305	CP2305 Tablet	Institute of Biomedical Sciences, Tokushima University Graduate School (2019)	Japan
6	CT-6	Probiotics	*Lactobacillus helveticus* R0052, *Bifidobacterium longum* R0175	CEREBIOME®	Centre for Neuroscience Studies, Queen's University (2021)	Canada
7	CT-7	Prebiotic	Inulin	_	School of Nutrition & Food Sciences, Tabriz University of Medical Sciences (2021)	Iran

maintaining the intestinal barrier integrity and disrupted intestinal permeability, as more permeation of pro-inflammatory bacterial components like LPS into the systemic circulation responsible for microglia activation and neurodegeneration was observed with low SCFA levels. The first experiment by Aho et al. demonstrated the fermentation of PD patient stool with prebiotic fibers (*Ruminococcus, Dorea, Bacteroides, Blautia, Anaerostipes, and Bifidobacterium*) and increased production of useful metabolites SCFAs, along with changed microbiota composition, indicated the ability of PD microbiota to respond positively to prebiotics (Aho et al., 2021). Encouraging outcome motivated Hall et al. to conduct an open-label, non-randomized study in newly diagnosed non-medicated ($n = 10$) and treated PD participants ($n = 10$) to study the impact of prebiotic intervention for ten days. Primary outcomes demonstrated that the prebiotic interventions have been well tolerated and found to be safe in PD participants along with beneficial biological changes in the gut microbiota, SCFA level, inflammation, and biomarkers of neurodegeneration (Hall et al., 2023).

27.4.3.2 Alzheimer's

Clinical Trial 3

Activated immune, increased kynurenine and neopterin concentrations are associated with the cognitive decline in patients suffering from AD and other forms of dementia (Knapskog et al., 2023). The kynurenine pathway (KP) degrades the L-tryptophan (Trp) into several intermediary metabolites collectively termed as kynurenines that play a fundamental role in cell signaling, energy metabolism, neuronal protection, and brain health. Friedrich Leblhuber group conducted a study in 20 AD outpatients (11 males, 9 males, aged 76.7±9.6 years) to investigate the positive effect of microbiome rebalancing on gut inflammation and immune activation. AD patients were given supplements of multispecies probiotics (*Lactobacillus lactis* W19, *paracasei* W20, *acidophilus* W22, *salivarius* W24, *casei* W56, *plantarum* W62, and *Bifidobacterium lactis* W52 *bifidum* W23). The mean Mini Mental State Examination score of 18.5±7.7 was observed after four weeks of probiotic supplementation in 18 patients based on analysis of immune activation biomarkers, serum neopterin and tryptophan breakdown, gut inflammation markers, and microbiota composition as compared to the baseline. Also, a significant increase in serum kynurenine concentrations was observed along with delta values of neopterin, kynurenine to tryptophan ratios (Kyn/Trp), and augmented tryptophan metabolism in serum (Leblhuber et al., 2018). Overall studies outcomes confirmed the correlation between kynurenine to tryptophan ratios (Kyn/Trp) and neopterin concentrations in the activation of macrophages and/or dendritic cells and potential consequences of probiotic supplementation in the course of AD in influencing the gut bacteria composition.

27.4.3.3 Depression

Clinical Trial 4

Depression, a common mental health disorder characterized by persistent sadness and loss of interest in daily activities, has been linked to alterations in gut microbiota, such as an increased Bacteroidetes/Firmicutes ratio and shifts in specific microbial genera. A growing evidence of linked mental health outcomes with gut microbiome led to the investigation of the GI tract as a target for novel treatments and interventions for depression, using probiotic supplementation (Chudzik et al., 2021). Caroline J.K. Wallace carried out a eight-week open label exploratory to examine changes in depressive symptoms before and after the introduction of a probiotic supplement of (*Lactobacillus helveticus* R0052 and *Bifidobacterium longum* R0175 (CEREBIOME®)) in ten major depressive disorder (MDD) patients from Ontario, Kingston, Canada community at a dose of 3 × 10^9 once per day. Data was collected at baseline, fourth and eighth week, and clinical symptoms of depression were measured using a validated battery of clinical scales and self-report questionnaires CQN-BIND protocol. Probiotic supplement resulted in significant improvements in subjective sleep by week 8 and without any associated adverse effects (Wallace et al., 2021).

Clinical Trial 5

Major depressive disorder a common psychiatric disorder has also been found to be closely associated with obesity. Well-established mechanisms such as HPA axis dysregulation and inflammation indicate the significant role of dysbiosis in connecting depression and obesity. Disrupted gut microbiota is associated with increased systemic inflammation, altered neurotransmitter production, and changes in energy metabolism, all of which can influence mood and body weight. To validate the relation and to assess the effect of prebiotic (inulin), Elnaz Vaghef-Mehrabany carried out a double-blind placebo-controlled randomized clinical trial in obese women with (MDD). In this eight-week protocol, 62 obese women with MDD were distributed equally into placebo (10 g/day Maltodextrin) and prebiotic (10 g/day Inulin) groups. All the participants were put on 25% calorie-restricted diet. Only 45 patients could manage to complete the trials and post-intervention no significant differences for MDD symptoms and other study outcomes were observed. Weight, hip and waist circumferences, systolic blood pressure, and Hamilton Depression Rating Scale (HDRS) score were found to be significantly low in both groups, whereas fat mass and total cholesterol were found to decline only in the prebiotic arm. Significantly improved HDRS score was observed in females who lost ≥1.9 kg weight as compared to women with <1.9 kg weight reduction, irrespective of the supplement. Studies suggested that prebiotic supplementation may have some beneficial metabolic effects, calorie restriction but weight loss seems to play an important role in improving depressive symptoms in obese women with MDD (Vaghef-Mehrabany et al., 2021).

Clinical Trial 6

In another study, efforts were made to understand the effectiveness and health advantages of prolonged biotic *Lactobacillus gasseri* CP2305 (CP2305) as a complementary therapy in a group of 60 young adult students preparing for the national examination in this double-blind placebo-control study, CP2305 containing tablets with 1 × 10^{10} bacterial cells per two tablets were administered to 41 men and 19 women once daily for a period of 24 weeks. Placebo tablets were given to the parallel placebo group design. Mental and physical states was assessed through the use of questionnaires, and the trial proved the significant impact of CP2305 in reducing anxiety and sleep disturbance in comparison to the placebo. Also, fecal microbiota analysis revealed that dietary supplementation with CP2305 helped in mitigating the drop of *Bifidobacterium* spp. caused by stress and prohibited the elevation of *Streptococcus* spp. induced by stress. Thus, the study acknowledged the positive effects of CP2305 on young adults facing stressful situations, but the exact mechanism of how it alleviates stress is still uncertain, and demanded the need for further research (Nishida et al., 2019).

Clinical Trial 7

Murata and team investigated the effects of postbiotic supplementation on mood and symptoms of the common cold in 241 healthy adults. In 12 weeks, study, participants were randomized to receive 1 × 10^{10} heat-killed *L. paracasei* MCC1849 cell powder(10LP), 3 × 10^{10} heat-killed *L. paracasei* MCC1849 cell powder (30LP), or a placebo once a day. Using questionnaire, a standard validated psychological test assessed the transient mood states. Participants taking MCC1849 demonstrated the decrease in the level of deterioration in the positive mood state caused by stress relative to the placebo group (Murata et al., 2018).

27.5 Case Studie(s)

Case studies and clinical trials serve distinct yet complementary roles in medical research. Clinical trials are large-scale studies designed to test the efficacy and safety of treatments across diverse populations through rigorous, randomised, and controlled methodologies. Case studies are distinct in their approach and focus on in-depth analysis of small groups or individual patients (Priya, 2021). Present case studies (Table 27.2) focusing the gut–brain axis (GBA) in neurological disorders not only would be invaluable in understanding the interactions between gut microbiota brain function, but would help in identifying novel therapeutic targets development of innovative treatments to improve mental health outcomes.

Case Study I

Increasing evidences of the pivotal role of gut microbiota in modulating immune function and inflammation, especially concerning autoimmune disorders like MS, encouraged Tankou et al. in 2018 to investigate the potential of probiotics in modulating the GBA. Given MS's characteristic features of chronic inflammation and immune dysregulation, the researchers aimed to determine the effects of a multistrain probiotic, containing *Bifidobacterium*, *Lactobacillus*, and *Streptococcus*, on the gut microbiome and immune function in relapsing-remitting MS (RRMS) patients and healthy controls. The probiotics were administered twice daily for two months containing 900 billion colony-forming units (CFU) per sachet to 9 MS patients and 13 healthy controls. Samples were collected at the baseline, post-treatment, and three months post-discontinuation of probiotics. The administration of probiotics showed a shift toward a healthier gut microbiome composition by increasing the abundance of beneficial bacteria such as *Lactobacillus* and decreasing the abundance of potentially harmful taxa such as *Akkermansia* and *Blautia*. At the immune level, the probiotic induced an anti-inflammatory response by decreasing the frequency of inflammatory monocytes, reducing the mean fluorescence intensity (MFI) of CD80 on classical monocytes, and decreasing the MFI of human leukocyte antigen-DR on dendritic cells in both MS patients and healthy controls. Additionally, there was a reduction in the expression of MS risk alleles, such as human leukocyte antigen-DQ α-1 and human leukocyte antigen-DP β-1, following probiotic administration (Tankou et al., 2018). Overall, the study suggested probiotic supplementation is safe and well-tolerated in MS patients and may have synergistic effects with existing MS therapies by modulating the gut microbiome and reducing peripheral inflammation.

Case Study II

Obesity, a chronic condition characterized by excessive body fat, not only raises the risk of metabolic disorders

TABLE 27.2
Measuring Effects of Prebiotics/Probiotics/Postbiotics on Neurological Disorders: Case Studies

S. No.	Case Study (CS)	Biotic Therapy	Microbial Strain	Product Name	Research Group and Studies Conducted (Year)	Country
Autism Spectrum Disorder (ASD)						
1	CS-1	Probiotics	Bifidobacteria, Lactobacilli, and Streptococci species	VSL3	Villa Santa Maria Institute (2016)	Mexico
Multiple sclerosis (MS)						
2	CS-2	Probiotics	Four strains of *Lactobacillus* (*L. paracasei* DSM 24734, *L. plantarum* DSM 24730, *L. acidophilus* DSM 24735, *L. delbruckei subspecies bulgaricus* DSM 24734), three strains of *Bifidobacterium* (*B. longum* DSM 24736, *B. infantis* DSM 24737, *B. breve* DSM 24732), and one strain of *Streptococcus* (*Streptococcus thermophilus* DSM 24731	Visbiome	Department of Neurology, Harvard Medical School, Boston(2018)	USA
Obesity						
3.	CS-3	Prebiotics	Combination of fructooligosaccharides (FOS) and galactooligosaccharides (GOS)	Fosvita® + Bimuno®	Federal University of Pernambuco (2023)	Brazil

but also increases vulnerability to mental health issues like anxiety and depression. The recognized role of gut microbiota in influencing brain function and behavior, particularly in obesity, motivated de Paiva et al. to investigate the role of prebiotics fructooligosaccharides (FOS) and galactooligosaccharides (GOS) in alleviating anxiety and depression-like behavior. The study was conducted using male C57BL/6 mice, a strain widely recognized for its utility in obesity and metabolic syndrome research. Forty-five mice were divided into groups and fed a high-fat diet (HFD) for 18 weeks to induce obesity and prebiotic intervention started from the 13th week. The study employed a battery of behavioral tests, including the Sucrose Splash Test, Tail Suspension Test, Forced Swim Test, and Elevated Plus Maze Test, and provided a comprehensive evaluation of depressive and anxiety-like behaviors, along with the motivational state of the mice. Researchers also conducted extensive biochemical analyses of blood serum, intestinal and brain tissue analysis for various markers like metabolic parameters, inflammatory cytokines, serotonin levels, and GLP-1 concentrations to delve into the underlying physiological changes. Overall, studies along with acetate concentrations determination in the hippocampus and prefrontal cortex provided insights into the potential mechanisms by which prebiotics might influence brain function and neurological disorders associated with obesity (de Paiva et al., 2023).

Case Study III

Rising incidence of ASD in early childhood with persistent deficits in social communication, interaction, and repetitive behavior indicates the role of gut microbiota in influencing neurodevelopmental disorders through gut–brain axis mechanism along with genetic and environmental factors. To validate this connection, Grossi and colleagues conducted a long-term study to explore the therapeutic effects of probiotics on core ASD symptoms in a 12-year-old boy, particularly with a history of celiac disease and chronic gastrointestinal issues (Grossi et al., 2016). The patient was treated with VSL#3, a multi-strain probiotic mixture containing various *Bifidobacteria*, *Lactobacilli*, and *Streptococci* species, for a period of four months from February to May 2014. The study's design was robust, maintained the patient's existing rehabilitation program and diet throughout the treatment and follow-up periods, to isolate the effects of the probiotic intervention. The autism diagnostic observation schedule-2 (ADOS-2), a standardized tool for evaluating ASD symptoms, was used to quantify changes over a period from October 2013 to March 2015, and provided baseline measurements, assessments during treatment, and follow-up evaluations. An unexpected and significant improvement in core ASD symptoms, particularly in ADOS-2 social affect domain score, was improved from 20 to 18 after two months of treatment, with a further reduction to 17 in the following two months. Remarkable persistent improvement even after the discontinuity of probiotic treatment and stable situations during the follow-up period opened the new avenues for therapeutic interventions in ASD (Grossi et al., 2016).

27.6 POSSIBLE MECHANISM(S) OF ACTION

Significant attention has been gained by prebiotics, probiotics, and postbiotics for their potential in managing and treating neurological disorders by interacting intricately with the gut microbiome. These biotic(s) seem to influence various pathways (Figure 27.2) and mechanisms along the microbiome-gut–brain axis, like impact on neurotransmitter synthesis, immune responses, direct neuronal effects, and even gene expression.

In AD, prebiotics such as fructooligosaccharides (FOS) and inulin have been found to get fermented by beneficial gut bacteria into SCFAs like acetate, propionate, and butyrate. Butyrate, a key SCFA, by activating G-protein-coupled receptors (GPR41 and GPR43) on intestinal epithelial and immune cells, promotes anti-inflammatory effects and maintains gut homeostasis and helps in mitigating the neuroinflammation seen in AD. It was also observed that butyrate inhibits histone deacetylases (HDACs), leading to increased acetylation of histones and modulation of gene expression, including the enhancement of brain-derived neurotrophic factor (BDNF). BDNF are known to support neuronal survival, differentiation, and synaptic plasticity, which are crucial in countering the neurodegenerative processes in AD (Chen, Meng, and Shen 2022). In MS, these prebiotics are found to modulate the immune response and support the integrity of the blood–brain barrier, an often compromised factor in MS, and reduce neuroinflammation (Y. Sun et al., 2021).

Probiotics, like *Lactobacillus* and *Bifidobacterium* species, are found to enhance gut barrier integrity by upregulating tight junction proteins such as occludin and claudin, and preventing the translocation of harmful bacteria and toxins that trigger systemic inflammation and exacerbate neuroinflammation characteristic of MS (Gou et al., 2022). Whereas in PD, probiotics interact with the immune system through pattern recognition receptors, including toll-like receptors (TLRs) and nucleotide-binding oligomerization domain (NOD) receptors, leading to downstream signaling via nuclear factor kappa B (NF-κB) and mitogen-activated protein kinase (MAPK) pathways. These interactions result in decreased levels of pro-inflammatory cytokines (tumor necrosis factor-alpha (TNF-alpha) and interleukin-6 (IL-6) and an increased anti-inflammatory cytokines (IL-10), thus reducing neuroinflammation associated with PD. Additionally, probiotics *Lactobacillus reuteri and Lactobacillus casei* aids in producing neurotransmitters such as serotonin, through the metabolism of tryptophan-by-tryptophan hydroxylase, impacting the enteric nervous system and potentially alleviating some of the neurological symptoms in both MS and PD.

Postbiotics, including SCFAs, peptidoglycans, and exopolysaccharides, have been found to modulate the immune system via TLRs and NOD-like receptors (NLRs), reduce systemic inflammation, and exert neuroprotective effects by

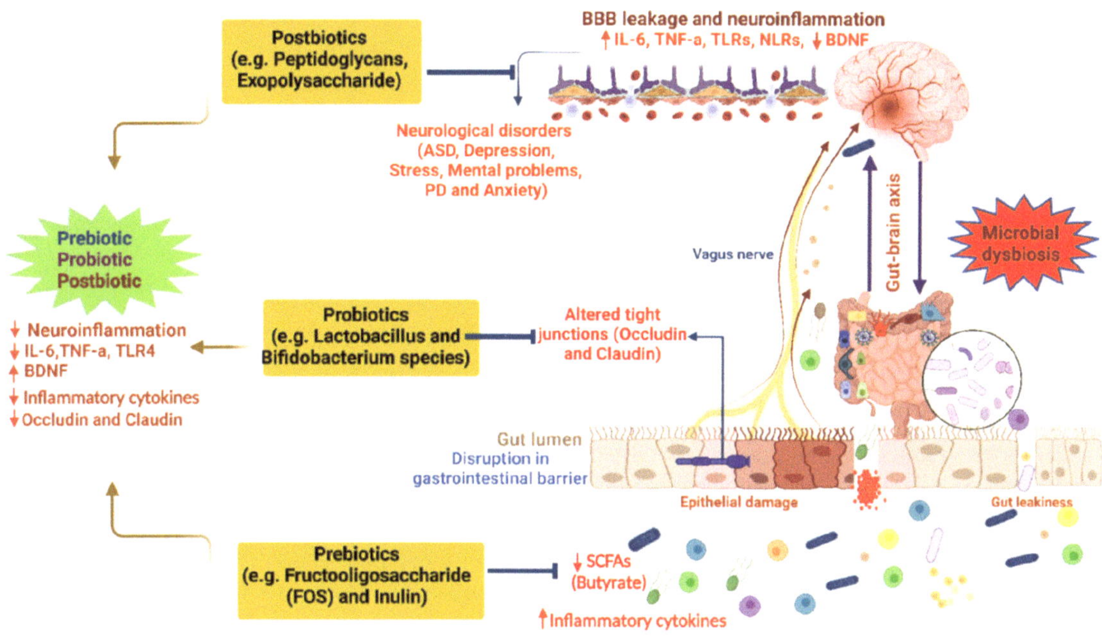

FIGURE 27.2 Possible mechanism(s) of prebiotics/probiotics/postbiotics.

inhibiting HDACs and activating peroxisome proliferator-activated receptors (PPARs) (Głowacka et al., 2024). This enhances mitochondrial function, reduces oxidative stress, and promotes the expression of genes involved in neuronal health. Collectively, these biotics interact with the gut microbiota, immune system, and central nervous system through intricate pathways to maintain and improve neurological health, Understanding the possible mechanism of actions may provide insights into how these substances are being leveraged to address neuroinflammation, oxidative stress, and other factors implicated in pathophysiology of neurological neurological disorders (Figure 27.2).

27.7 SAFETY CONSIDERATION AND RISK ASSESSMENT

While the therapeutic potential of the biotics is significant, careful consideration of potential adverse effects, interactions, detailed risk assessment framework, and disruptions to gut microbiota is equally critical to ensure their safety with effective use. Microbial strain specificity, dosage, delivery methods, and the individual's baseline microbiome composition are some of the factors that influence their efficacy and safety. In some cases, a few biotic strains trigger or exacerbate immune-mediated conditions, or negatively interact with existing medications, and alter drug efficacy or increase toxicity. Potential risks of microbiota imbalances, such as small intestinal bacterial overgrowth or *Clostridium difficile* infection, include adverse immune reactions and disruptions to gut microbiota balance, potentially leading to dysbiosis and further health complications (Banaszak et al., 2023). Comprehensive safety assessments in preclinical and clinical trials are essential, including monitoring biomarkers, immune responses, and gut permeability. Robust clinical trial needs to be designed with long-term follow-ups to identify adverse effects. Each strain must be rigorously evaluated for stability, genotoxicity, and metabolic interactions. Continuous post-market surveillance is essential to ensure safety across diverse populations. Health practitioners should employ precision medicine approaches, tailored biotic therapies considering an individual's unique microbiome profile and health status, along with vigilant monitoring throughout the treatment course. Precision medicine approaches should tailor biotic therapies to individual microbiomes while monitoring for adverse effects. A comprehensive strategy will facilitate optimizing the biotic-based treatments for neurological health, with improved patient outcomes and minimal risks.

27.8 FUTURE PERSPECTIVE

Increasing worldwide prevalence of neurological diseases with a projected indication of 13% rise by 2030 (Ullah et al., 2023) is highlighting the need for advanced biomarkers and effective therapeutic strategies. Emerging research with emphasisis on the crucial impact of gastrointestinal microflora such as archaea, bacteria, protists, and fungi on the physiological conditions, underscore the importance of the microbiome-gut–brain axis in overall well-being. The market for biotics has already been expanded significantly, driven by increased consumer awareness and demand for gut health solutions. Further, the biotic market is expected to grow to US$24.30 billion by 2030, with an anticipated compound annual growth rate of 11.2% from 2024 to 2031 (Precedence Research, 2024).

The insights of interactions between microbiota-derived products and systemic microbial signals affecting neurological and immunological functions have spurred the intensive research into omics technologies. Metagenomic sequencing is providing further detailed insights into microbiome composition and function, while the effect of metabolic products on gut–brain interactions and respective treatment is being studied employing metabolomics and pharmacometabolomics. Pharmacogenomics adds further refinement by predicting individual responses to treatments based on genetic profiles. These technologies are poised to revolutionize microbial therapeutics by enabling personalized interventions tailored to individual microbiome profiles and neurological phenotypes (Visan and Negut, 2024).

Innovations in targeted delivery systems like nanotechnology-based approaches (lipid nanoparticles and polymeric micelles) offer promising avenues for precise targeted delivery within the gastrointestinal tract with enhanced bioavailability. Synbiotic formulations, leveraging synergistic interactions between prebiotic substrates, probiotic strains, and postbiotic metabolites, are being optimized through *in silico* modeling, organ-on-a-chip technologies, and 3D neurospheroids HTS methodologies (Razavi et al., 2021) and accelerating their screening and validation. Understanding the microbiome-gut–brain axis can revolutionize biotherapeutics, and offer a new hope for neurological disorder treatment.

27.9 CONCLUSION

The integration of biotics, as complementary therapies in the management of neurological disorders, represents an innovative and multifaceted approach in enhancing mental health and cognitive function. The modulation of neurotransmitter synthesis, including serotonin, dopamine, and GABA, alongside the reduction of neuroinflammation and oxidative stress, highlights the complex relationship between gut microbiota and neurological function. Evidence from clinical trials and case studies indicates that biotic interventions can ameliorate symptoms and potentially modify the course of conditions such as depression, anxiety, autism spectrum disorder, and neurodegenerative diseases like Parkinson's and Alzheimer's. These benefits are primarily mediated through the modulation of the microbiome-gut–brain axis, involving mechanisms such as immune response regulation, neuroactive compound production, and enhancement of gut barrier integrity. The convergence of cutting-edge omics technologies, advanced imaging modalities, and innovative delivery strategies are poised to transform the landscape of neurology and psychiatry. But further rigorous research is essential to develop standardized protocols and personalized treatment regimens, addressing the heterogeneity in microbial strains and patient-specific responses to ensure efficacy and safety across diverse populations.

REFERENCES

Aho, V. T., Houser, M. C., Pereira, P. A., Chang, J., Rudi, K., Paulin, L., ... & Scheperjans, F. (2021). Relationships of gut microbiota, short-chain fatty acids, inflammation, and the gut barrier in Parkinson's disease. *Molecular Neurodegeneration*, *16*:1–14. https://doi.org/10.1186/s13024-021-00427-6.

Alzubide, S., Alhalafi, M. (2024). The gut brain connection. *Journal of Behavioral and Brain Science*, *14*(3):103–117. https://doi.org/10.4236/jbbs.2024.143008.

Banaszak, M., Górna, I., Woźniak, D., Przysławski, J., & Drzymała-Czyż, S. (2023). Association between gut dysbiosis and the occurrence of SIBO, LIBO, SIFO and IMO. *Microorganisms*, *11*(3):573–586. https://doi.org/10.3390/microorganisms11030573.

Chen, H., Meng, L., & Shen, L. (2022). Multiple roles of short-chain fatty acids in alzheimer disease. *Nutrition, 93*:1–22. https://doi.org/10.1016/j.nut.2021.111499.

Chudzik, A., Orzyłowska, A., Rola, R., & Stanisz, G. J. (2021). Probiotics, prebiotics and postbiotics on mitigation of depression symptoms: modulation of the brain–gut–microbiome axis. *Biomolecules*, *11*(7):1–26. https://doi.org/10.3390/biom11071000.

Cunningham, M., Azcarate-Peril, M. A., Barnard, A., Benoit, V., Grimaldi, R., Guyonnet, D., ... & Gibson, G. R. (2021). Shaping the future of probiotics and prebiotics. *Trends in Microbiology*, *29*(8):667–685. https://doi.org/10.1016/j.tim.2021.01.003.

Debas, H. T., Laxminarayan, R., & Straus, S. E. (2011). *Complementary and Alternative Medicine, 2*:1–5.

de Paiva, I. H. R., da Silva, R. S., Mendonça, I. P., Duarte-Silva, E., Botelho de Souza, J. R., & Peixoto, C. A. (2023). Fructooligosaccharide (FOS) and Galactooligosaccharide (GOS) improve neuroinflammation and cognition by up-regulating IRS/PI3K/AKT signaling pathway in diet-induced obese mice. *Journal of Neuroimmune Pharmacology*, *18*(3):427–447. https://doi.org/10.1007/s11481-023-10069-8.

Elangovan, A., Dahiya, B., Kirola, L., Iyer, M., Jeeth, P., Maharaj, S., ... & Vellingiri, B. (2024). Does gut brain axis has an impact on Parkinson's disease (PD)? *Ageing Research Reviews*, *94*:102–171. https://doi.org/10.1016/j.arr.2023.102171.

Fekete, M., Lehoczki, A., Major, D., Fazekas-Pongor, V., Csipo, T., Tarantini, S., ... & Varga, J. T. (2024). Exploring the influence of gut–brain axis modulation on cognitive health: A comprehensive review of prebiotics, probiotics, and symbiotics. *Nutrients*, *16*(6):2–29. https://doi.org/10.3390/nu16060789.

Felice, V. D., Quigley, E. M., Sullivan, A. M., O'Keeffe, G. W., & O'Mahony, S. M. (2016). Microbiota-gut-brain signalling in Parkinson's disease: Implications for non-motor symptoms. *Parkinsonism & Related Disorders*, *27*:1–8. https://doi.org/10.1016/j.parkreldis.2016.03.012.

Głowacka, P., Oszajca, K., Pudlarz, A., Szemraj, J., & Perkowska, M. W. (2024). Postbiotics as molecules targeting cellular events of aging brain – The role in pathogenesis, prophylaxis and treatment of neurodegenerative diseases. *Nutrients, 16*(14):22–44. https://doi.org/10.3390/nu16142244.

Gou, H. Z., Zhang, Y. L., Ren, L. F., Li, Z. J., & Zhang, L. (2022). How do intestinal probiotics restore the intestinal barrier? *Frontiers in Microbiology, 13*:1–11. https://doi.org/10.3389/fmicb.2022.929346.

Grossi, E., Melli, S., Dunca, D., & Terruzzi, V. (2016). Unexpected improvement in core autism spectrum disorder symptoms after long-term treatment with probiotics. *SAGE Open Medical Case Reports*, *4*:1–5. https://doi.org/10.1177/2050313x16666231.

Hall, D. A., Voigt, R. M., Cantu-Jungles, T. M., Hamaker, B., Engen, P. A., Shaikh, M., ... & Keshavarzian, A. (2023). An open label, non-randomized study assessing a prebiotic fiber intervention in a small cohort of Parkinson's disease participants. *Nature Communications*, *14*(1):1–14. https://doi.org/10.1038/s41467-023-36497-x

Ji, J., Jin, W., Liu, S. J., Jiao, Z., & Li, X. (2023). Probiotics, prebiotics, and postbiotics in health and disease. *Medical Communication*, *4*(6):1–23. https://doi.org/10.1002/mco2.420.

Knapskog, A. B., Aksnes, M., Edwin, T. H., Ueland, P. M., Ulvik, A., Fang, E. F., ... & Watne, L. O. (2023). Higher concentrations of kynurenic acid in CSF are associated with the slower clinical progression of Alzheimer's disease. *Alzheimer's & Dementia*, *19*(12):5573–5582. https://doi.org/10.1002/alz.13162.

Leblhuber, F., Steiner, K., Schuetz, B., Fuchs, D., & Gostner, J. M. (2018). Probiotic supplementation in patients with Alzheimer's dementia: an explorative intervention study. *Current Alzheimer Research*, *15*(12):1106–1113. https://doi.org/10.2174/1389200219666180813144834.

Liu, L., Wang, H., Chen, X., Zhang, Y., Zhang, H., & Xie, P. (2023). Gut microbiota and its metabolites in depression: from pathogenesis to treatment. *EBioMedicine*, *90*:1–13. https://doi.org/10. 1016/j.ebiom.2023.104527.

Magistrelli, L., Amoruso, A., Mogna, L., Graziano, T., Cantello, R., Pane, M., & Comi, C. (2019). Probiotics may have beneficial effects in Parkinson's disease: in vitro evidence. *Frontiers in Immunology*, *10*:1–9. https://doi.org/10.3389/fimmu.2019.00969.

Murata, M., Kondo, J., Iwabuchi, N., Takahashi, S., Yamauchi, K., Abe, F., & Miura, K. (2018). Effects of paraprobiotic *Lactobacillus paracasei* MCC1849 supplementation on symptoms of the common cold and mood states in healthy adults. *Beneficial Microbes*, *9*(6):855–864. https://doi.org/10.3920/BM2017.0197.

Nadaf, M., Joharchi, M., & Amiri, M. S. (2019). Ethnomedicinal uses of plants for the treatment of nervous disorders at the herbal markets of Bojnord, North Khorasan Province, Iran. *Avicenna Journal of Phytomedicine*, *9*(2):1–15.

Nishida, K., Sawada, D., Kuwano, Y., Tanaka, H., & Rokutan, K. (2019). Health benefits of *Lactobacillus gasseri* CP2305 tablets in young adults exposed to chronic stress: A randomized, double-blind, placebo-controlled study. *Nutrients*, *11*(8):18–59. https://doi.org/10.3390/nu11081859.

Potkin, S. G., Kane, J. M., Correll, C. U., Lindenmayer, J. P., Agid, O., Marder, S. R., ... & Howes, O. D. (2020). The neurobiology of treatment-resistant schizophrenia: Paths to antipsychotic resistance and a roadmap for future research. *Nature Partner Journals Schizophrenia*, *6*(1), 1–10. https://doi.org/10.1038/s41537-019-0090-z.

Precedence Research. (2024). https://www.precedenceresearch.com/probiotics-market (Accessed on 1 August 2024).

Priya, A. (2021). Case study methodology of qualitative research: Key attributes and navigating the conundrums in its application. *Sociological Bulletin*, *70*(1):94–110. http://dx.doi.org/10.1177/0038022920970318.

Rafique, N., Jan, S. Y., Dar, A. H., Dash, K. K., Sarkar, A., Shams, R., ... & Hussain, S. Z. (2023). Promising bioactivities of postbiotics: A comprehensive review. *Journal of Agriculture and Food Research*, *14*:1–16. https://doi.org/10.1016/j.jafr.2023.100708.

Razavi, S., Janfaza, S., Tasnim, N., Gibson, D. L, & Hoorfar, M. (2021). Nanomaterial-based encapsulation for controlled gastrointestinal delivery of viable probiotic bacteria. *Nanoscale Advances*, *3*(10), 2699–2709. https://doi.org/10.1039/d0na00952k.

Scarpellini, E., Rinninella, E., Basilico, M., Colomier, E., Rasetti, C., Larussa, T., ... & Abenavoli, L. (2021). From pre-and probiotics to post-biotics: A narrative review. *International Journal of Environmental Research and Public Health*, *19*(1):37–51. https://doi.org/10.3390/ijerph19010037.

Sun, Y., Zhang, Z., Cheng, L., Zhang, X., Liu, Y., Zhang, R., ... & Wu, Z. (2021). Polysaccharides confer benefits in immune regulation and multiple sclerosis by interacting with gut microbiota. *Food Research International*, *149*:1–13. https://doi.org/10.1016/j.foodres.2021.110675.

Tankou, S. K., Regev, K., Healy, B. C., Tjon, E., Laghi, L., Cox, L. M., ... & Weiner, H. L. (2018). A probiotic modulates the microbiome and immunity in multiple sclerosis. *Annals of Neurology*, *83*(6):1147–1161. https://doi.org/10.1002/ana.25244.

Tsai, P. T., Hull, C., Chu, Y., Greene-Colozzi, E., Sadowski, A. R., Leech, J. M., ... & Sahin, M. (2012). Autistic-like behaviour and cerebellar dysfunction in Purkinje cell Tsc1 mutant mice. *Nature*, *488*(7413):647–651. https://doi.org/10.1038/nature11310

Ullah, H., Arbab, S., Tian, Y., Liu, C. Q., Chen, Y., Qijie, L., ... & Li, K. (2023). The gut microbiota–brain axis in neurological disorder. *Frontiers in Neuroscience*, *17*:1–25. https://doi.org/10.3389/fnins.2023.1225875.

Vaghef-Mehrabany, E., Ranjbar, F., Asghari-Jafarabadi, M., Hosseinpour-Arjmand, S., & Ebrahimi-Mameghani, M. (2021). Calorie restriction in combination with prebiotic supplementation in obese women with depression: Effects on metabolic and clinical response. *Nutritional Neuroscience*, *24*(5):339–353. https://doi.org/10.1080/1028415X.2019.1630985.

Visan, A. I., & Negut, I., (2024). Integrating artificial intelligence for drug discovery in the context of revolutionizing drug delivery. *Life*, *14*(2):233–250. https://doi.org/10.3390/life14020233.

Wallace, C. J., & Milev, R. V. (2021). The efficacy, safety, and tolerability of probiotics on depression: Clinical results from an open-label pilot study. *Frontiers in Psychiatry*, *12*:1–8. https://doi.org/10.3389/fpsyt.2021.618279.

WHO. (2024). https://www.who.int/news-room/questions-and-answers/item/clinical-trials (Accessed on 1 August 2024).

Wikoff, W. R., Anfora, A. T., Liu, J., Schultz, P. G., Lesley, S. A., Peters, E. C., & Siuzdak, G. (2009). Metabolomics analysis reveals large effects of gut microflora on mammalian blood metabolites. *Proceedings of the National Academy of Sciences*, *106*(10):3698–3703. https://doi.org/10.1073/pnas.0812874106.

Yadav, M. K., Kumari, I., Singh, B., Sharma, K. K., & Tiwari, S. K. (2022). Probiotics, prebiotics and synbiotics: Safe options for next-generation therapeutics. *Applied Microbiology and Biotechnology*, *106*(2):505–521. https://doi.org/10.1007/s00253-021-11646-8.

Zhou, P., Chen, C., Patil, S., & Dong, S., (2024). Unveiling the therapeutic symphony of probiotics, prebiotics, and postbiotics in gut-immune harmony. *Frontiers in Nutrition*, *11*:1–12. https://doi.org/10.3389/fnut.2024.1355542.

28 Application of Artificial Intelligence in Connection with Gut Microbiota to Detect Neurological Disorders

N.L. Swathi[1], Kashif Raza Siddique[2], Habiba Babul[3], Pranjal Kumar Singh[4], and Dinesh Kumar[5]

[1] Department of Nursing, Manipal School of Nursing, Manipal Academy of Higher Education, Manipal, India
[2] Department of Biotechnology, Goel Institute of Technology and Management, Lucknow, Uttar Pradesh, India
[3] North South University, Dhaka, Bangladesh
[3] Kirori Mal College, University of Delhi, Delhi, India
[4] Department of Computer Science Engineering, Bennett University, Greater Noida, Uttar Pradesh, India

28.1 SIGNIFICANCE OF GUT–BRAIN AXIS

The gut–brain axis is a complex neurological, hormonal, and immunological network connecting the brain and gut. This axis is important in neurodegenerative illnesses like Alzheimer's, Parkinson's, and multiple sclerosis, according to a recent study. Developing therapeutic strategies requires understanding how gut microbiota affects neurodegeneration.

Neurodegenerative diseases cause cognitive and motor decline due to neuronal loss. Dysbiosis, or gut microbiota changes, may affect neuroinflammation and neurodegeneration through various mechanisms. Dysbiosis may increase intestinal permeability, allowing harmful metabolites and pathogens to enter the systemic circulation and harm brain health (Bianchi et al., 2021; Heravi, 2023). This shows that the gut–brain axis is bidirectional, with the brain affecting gut function and gut health affecting brain health.

Studies have linked gut microbiota changes to cognitive impairment in Alzheimer's disease (AD). AD patients have certain bacterial taxa linked to inflammation and metabolic impairment (Jiang et al., 2017; Quigley, 2017). Probiotics may improve gut microbiota and reduce AD-related inflammatory markers (Naomi et al., 2021). Some studies showed that probiotic supplementation improved cognitive performance, but others found insufficient evidence to justify broad therapeutic use (Krüger et al., 2021).

Gut microbiome changes have also been linked to motor symptoms in Parkinson's disease (PD). Certain microbial communities may worsen or improve PD symptoms by affecting inflammation and neurotransmitter metabolism (Leta et al., 2021; Sampson et al., 2016). Preclinical models demonstrate probiotic therapies improve motor performance and reduce neuroinflammation (Zhao et al., 2021). As a way to restore the microbial balance in PD patients, fecal microbiota transplantation (FMT) has shown promise in improving gastrointestinal symptoms and motor skills (Marogianni et al., 2020).

MS adds additional gut–brain axis dimension. Evidence suggests gut microbiota can affect MS-related immune responses. Specific microbial profiles are related with illness severity and progression (Loh et al., 2024). Targeting the gut microbiome through dietary changes or probiotics may offer new paths for controlling MS symptoms (Agirman et al., 2021).

28.2 AI IN NEUROLOGICAL DISORDER DETECTION

The use of AI to detect and treat neurological problems is a potential medical frontier. This chapter discusses how AI, specifically machine learning (ML), can diagnose and forecast neurological disorders like stroke, epilepsy, and Parkinson's disease.

AI systems have shown promise in improving neurological illness diagnosis and efficiency. Studies have shown that ML algorithms can accurately diagnose large-vessel occlusion (LVO) strokes, increasing triage for prompt intervention (Shlobin et al., 2022). Convolutional neural networks (CNNs) and support vector machines (SVMs) have helped analyze imaging data for rapid diagnosis, which improves patient outcomes (Zhang et al., 2023).

Wearable sensors and AI can identify epilepsy seizures in real time. A study found that wrist or arm sensors could identify generalized tonic-clonic seizures (GTCS) with ML algorithms above 90% (Tang et al., 2021). By alerting

DOI: 10.1201/9781003536802-28

patients during seizures, this feature improves monitoring and patient safety.

AI has also helped personalize epilepsy treatment. Deep learning models that predicted anti-seizure medication reactions showed potential in identifying which patients would benefit most from specific medicines (Hakeem et al., 2022). This approach used seizure frequency and mental history to improve prediction.

AI also helps detect PD. Recent research has shown that resting-state electroencephalography (EEG) data evaluated using ML may accurately discriminate PD patients from healthy controls (Kurbatskaya et al., 2023). Gupta et al. (2023) used AI models to identify PD based on gait impairments and neuroimaging data, facilitating early diagnosis and treatments.

Despite these advances, AI model generalizability and robustness across varied demographics and therapeutic settings remain problems. To validate findings consistently, standardized techniques and bigger sample sizes are needed (X.-P. Liu et al., 2023). Many researches have found performance variations dependent on datasets. As clinical AI evolves, data privacy and algorithm transparency ethics are crucial.

28.3 COMPOSITION AND FUNCTION OF GUT MICROBIOTA

The digestive tracts of humans, other animals, and insects are home to an intricate collection of bacteria known as the gut microbiota, which are established from birth onward.

It is one of the largest microecosystems in the human body. It comprises a diverse array of microorganisms, including bacteria, fungi, eukaryotes, phages, viruses, and archaea, that coexist, co-adapt, and co-evolve symbiotically with the human host. This highly intricate and varied ecosystem is essential to human health (Sampson et al., 2016).

This intricate relationship is termed "mutualism," where both humans and microbes possess benefits.

Disturbance of this mutualistic relationship between humans and their microbiota can have detrimental effects. This imbalance increases the risk and leads to the proliferation of various diseases such as cancer, diabetes, liver disease, and neurological disorders such as Alzheimer's, Parkinson's, and other related health issues (Sampson et al., 2016).

The gut microbiome performs many essential functions and is thus crucial to our body. It protects us from pathogens by producing metabolic substances that are antimicrobial and competing for nutrients and adhesion sites on mucosal surfaces. These substances can either have a positive or negative effect on human health (Baccarelli & Ordovás, 2023)

The gut microbiota produces many metabolic products that are essential for maintaining brain health, particularly through the gut–brain axis (Table 28.1).

The composition of bacteria varies because of the distinct environments in different parts of the digestive system. Table 28.2 lists the composition of bacteria present in the digestive tract. While the majority of the bacteria belong to just 30–40 species, these species constitute between 300–1,000 different varieties.

Furthermore, the gut microbiome controls the multiplication and differentiation of epithelial cells, maintains intestinal homeostasis, and affects insulin regulation. These interactions significantly affect the development of numerous neurodegenerative, metabolic, cardiovascular, and neoplastic diseases (Grodin et al., 2024).

28.4 GUT–BRAIN AXIS: MECHANISMS AND IMPLICATIONS

The brain and gut bacteria communicate via the gut–brain microbiota axis. This bidirectional communication network helps the CNS and gut communicate (Ullah et al., 2023). Communication involves the autonomic nervous system (vagus nerve and ENS), neuroendocrine system, hypothalamic–pituitary–adrenal (HPA) axis, immune system, and metabolic pathway (Ullah et al., 2023).

The autonomic nervous system controls physiological functions outside of conscious control in the peripheral nervous system. It regulates vital visceral functions by synchronizing parasympathetic and sympathetic neuronal responses. Discovering the enteric nervous system (ENS), a part of the autonomic nervous system, has improved our understanding of the digestive tract–CNS connection. The ENS is called the "second brain in the body" because enteric neurons and their connections to the CNS help sustain health.

The ENS's millions of neurons regulate intestinal activity and homeostasis in the digestive system. As the main gut–brain communication channel, the vagus nerve is crucial (Ullah et al., 2023). Every visceral organ receives this brainstem nerve (Q. Wang et al., 2023).

About 90% of N vagus nerve input to the brain comes from the gut (Mohan et al., 2023). The gut microbiota profoundly affects immune system physiology. For instance, gut microbiota can impact brain microglial cell activity (Q. Wang et al., 2023). If the gut microbiota encourages the manufacture of immunological compounds such as inflammatory chemicals and cytokines, it may influence the CNS, ENS, and brain (Ullah et al., 2023).

Short-chain fatty acids, secondary bile acids, critical vitamins, and amino acids modulate many immune system mechanisms. These pathways impact cognition, behavior, learning, movement, and neurodegenerative diseases (Ullah et al., 2023).

28.5 NEUROLOGICAL DISORDERS ASSOCIATED WITH DYSBIOSIS

"Gut dysbiosis" can be described as an unbalanced state of the gut microbiota, where there is an increase in pathogenic bacteria such as Enterobacteriaceae and Prevotellaceae, and

TABLE 28.1
The Metabolic Products Produced by Gut Microbiota and Their Effects on Brain Health

S. No.	Metabolic Products	Effect on Brain Health	Neurological Disorders	References
1	*Short-chain fatty acids (SCFAs)*: acetate, propionate, and butyrate	They can alter neuroplasticity, production of neurotransmitters, and inflammation in the neurons to affect brain function and the ability to cross the blood–brain barrier and butyrate maintains the blood–brain barrier's integrity and has anti-inflammatory properties	Autistic spectrum disorder (ASD), AD, and PD	Gonzalez et al., 2008; MacFabe, 2012; Sampson et al., 2016; Unger et al., 2016; X. Wang et al., 2021
2	Neurotransmitter precursors and metabolites *Example:* dopamine, gamma-aminobutyric acid (GABA), serotonin, and serotonin	These neurotransmitters are produced in the stomach and control appetite, sleep, and mood. They also have an impact on emotional stability and mental health	ASD	Q. Huang et al., 2022
3	Tryptophan metabolites; essential amino acid	Gut bacteria convert tryptophan into indole and other metabolites which can affect the central nervous system. They also help maintain the response of the immune system as well as the gut–brain barrier	AD and ASD	Campbell, 2006; Critchfield et al., 2011; Schwarcz et al., 2012
4	Bile acids	They serve as brain signaling molecules that offer neuroprotective properties as well as have an impact on the metabolism of lipids. They also have implications for disorders like Alzheimer's	AD and PD	Jenkins et al., 2002
5	Lactate	It provides energy to neurons while also impacting neurodevelopment as well as neuroprotection	PD	Magistretti & Allaman, 2018

TABLE 28.2
Types of Bacteria Present in the Digestive Tract

S. No.	Part of the Digestive System	Number of Bacteria	Types of Bacteria
1	Oral cavity	10^8 CFU/m	*Streptococcus*, *Peptococcus*, *Staphylococcus*, *Bifidobacterium*, *Lactobacillus*, and *Fusobacterium* genera
2	Stomach	Less than 10^1–10^3 CFU/ml	*Lactobacillus*, but also *Helicobacter pylori*
3	Duodenum	10^1–10^4 CFU/ml	*Lactobacillus* and *Streptococcus* predominate
4	Jejunum	10^5–10^7 CFU/ml	Genera *Bacteroides, Lactobacillus*, and *Streptococcus*
5	Ileum	10^7–10^8 CFU/ml	Genera of *Bacteroides, Clostridium, Enterococcus, Lactobacillus*, and *Veillonella*, and species from the Enterobacteriaceae family
6	Large intestine	10^{10}–10^{12} CFU/ml	*Bacillus, Bacteroides, Clostridium, Bifidobacterium, Enterococcus, Eubacterium, Fusobacterium, Peptostreptococcus, Ruminococcus*, and *Streptococcus*

a decrease in diversity, abundance, and beneficial bacteria such as *Bacteroides* and Firmicutes (Zheng et al., 2023). It is defined as a change in the functional composition and metabolic processes of the microbiota, an alteration in their local distribution, or an imbalance in the microflora that throws off the normal equilibrium of the microbiota (Mohan et al., 2023).

Metabolic problems, endogenous intoxication, systemic inflammation, and decreased critical metabolites are just a few of the detrimental consequences that this imbalance may have on host health (Table 28.3).

There are many causes of dysbiosis. These include an unbalanced diet that is primarily made up of refined foods, a diet low in dietary fiber, excessive alcohol intake, exposure to

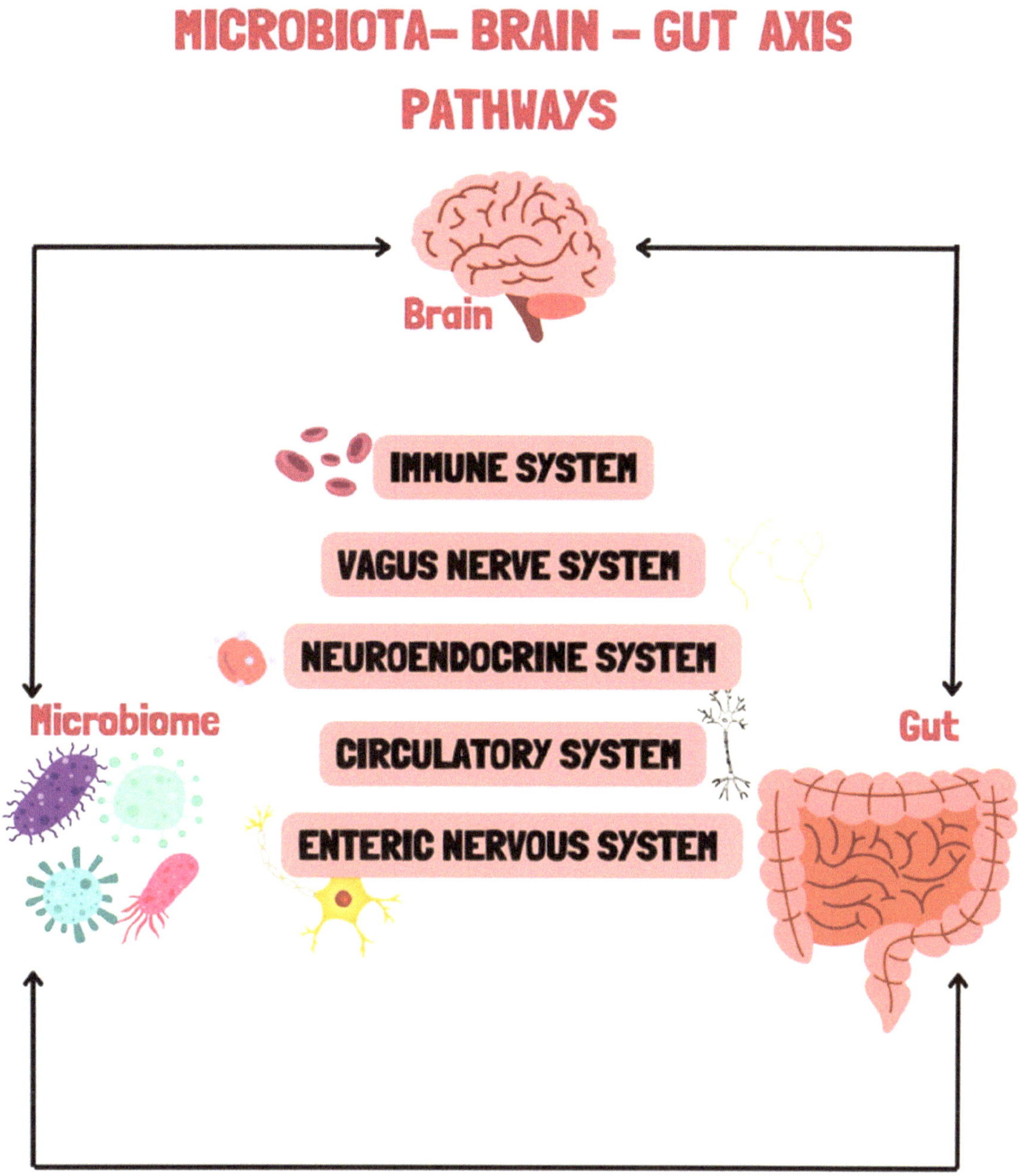

FIGURE 28.1 Microbiotagut–brain axis pathway. (*Source:* Author.)

external pollutants, the use of medicinal drugs such as antibiotics, chronic stress, lack of sleep, bacterial or viral infections, and a variety of illnesses such as metabolic disorders, particularly type 2 diabetes and NDDs (Zheng et al., 2023).

Gut microbiota is influenced by factors such as age, genetics, nutrition, environment, infections, and antibiotic use, which in turn affect the composition and function of the gut microbiota (Mhanna & Alshehabi, 2023).

Given that gut dysbiosis and NDDs are connected, dietary plans aimed at addressing dysbiosis may be promising for managing symptoms and reducing the neuroinflammatory and degenerative characteristics of NDDs. Complementary nutritional therapies targeted at modifying the composition of the gut microbiota and related metabolites could support the current therapeutic techniques for establishing eubiosis and promoting overall gut health. Changes in diet, prebiotics, synbiotics, probiotics, and transplanting fecal microbiota are some of the therapies available (Zheng et al., 2023).

An imbalance in gut microbiota results in numerous neurological disorders.

28.5.1 Alzheimer's Disease

Alzheimer's disease is the global cause of dementia. It is a debilitating degenerative condition that affects the central nervous system and worsens gradually over time. This is of

TABLE 28.3
Types of Dysbiosis Associated with Neurological Disorders

S.No.	Dysbiosis Pattern	Animal Model	Neurological Disorder	Detection	References
1	*Bacterial increase: Lactobacillus, Clostridium,* and *Desulfovibrio*; *Bacterial decrease: Bifidobacterium, Blautia, Turicibacter, Dialister, Veillonella, Prevotella, Turicibacter*, and *Enterococcus*	Human	Autism	Gut biopsy or fecal samples	Adams et al., 2023; F. Liu et al., 2019
2	*Bacterial increase: Sutterella, Ruminococcus torques*	Human	Autism	Fecal sample	Q. Wang et al., 2023; Yap et al., 2021
3	*Bacterial increase: Helicobacter pylori* and *Chlamydia pneumoniae*	Human	AD	Fecal sample	Bu et al., 2015
4	*Bacterial decrease: Bifidobacterium bifidum, Lactobacillus fermentum, Lactobacillus casei,* and *Lactobacillus acidophilus*	Human	AD	Fecal sample	Akbari et al., 2016
5	*Bacterial increase: Proteobacteria, Betaproteobacteria, Coprococcus, Blautia, Akkermansia, Oscillospira, Roseburia, Bacteroides*; *Bacterial decrease: Faecalibacterium, Firmicutes,* class Clostridia	Human	PD	Mucosal and stool samples	Keshavarzian et al., 2015
6	*Bacterial decrease:* Provotellaceae; *Bacterial increase:* Ruminococcaceae, Prevotellaceae	Human	PD	Fecal samples	Schepisi et al., 2023

significant importance to global health. The loss of cholinergic function due to neuronal cell death, mainly affecting the hippocampus, cerebral cortex, and other brain regions involved in cognition and memory, is the root cause of symptoms associated with AD. Its causes are intricate and complicated and may be connected to the flora in the gut, which can differ significantly between individuals. Owing to this diversity and its complex nature, it is difficult to identify a single cause (Sampson et al., 2016). Bacteria and their metabolites can act together with the intestinal epithelium in the gut microbiota, either by promoting the onset of a disease or by preventing it through early identification. By comprehending the association between the gut microbiota and the brain and analyzing the metabolites generated as a result of this interaction, promising biomarkers can be identified for the early detection of AD, promising biomarkers can be identified. For example, assessing blood plasma levels of Aβ (beta-amyloid) and examining cerebral amyloidosis are two ways to identify this neurodegenerative disease early on. The relationship between the emergence of AD and the gut microbiota can be attributed to damage to the epithelial integrity of intestinal gap junctions. The transfer of bacteria and endotoxins into circulation can be caused by dysbiosis, altered gut microbiota, or increased intestinal barrier permeability. This triggers a systemic inflammatory response. Inflammation has the potential to lead to chronic neuroinflammation over time. Numerous studies have indicated that the gut microbiota plays a role in the emergence of AD, similar to the brain and gut axes. The relationship between the gut tryptophan metabolic pathway and neuroinflammation is crucial. Tryptophan and its metabolites, which are created in the gut, play a significant role in the generation of energy and neuroplasticity when present in sufficient amounts. Disruption of this pathway can lead to various complications. However, when these chemicals are used in excess, they create pro-inflammatory and neurotoxic by-products that worsen chronic neuroinflammation. Clinical research using biochemical investigations, such as measurements of serum BDNF and plasma Aβ levels, cognitive testing with memory and learning questionnaires, Alzheimer's test, genetic coding and replication, microbiome analysis, and assessment of stress levels via BDNF, has demonstrated promising results in the early enhancement of cognitive function, learning, memory, and overall mental capabilities in individuals with dementia.

28.5.2 Parkinson's Disease

Parkinson's disease is characterized by four key symptoms: resting tremors, bradykinesia, muscular rigidity, and unstable posture. It is a neurological disorder that affects movement. It is a common progressive neurodegenerative illness for which there is currently no cure and no well-established etiology. It affects 7–10 million people worldwide, and in the past several years, its incidence and prevalence have sharply increased. There are two categories of PD symptoms: motor and non-motor. Tremors, stiffness, and bradykinesia are among the most important motor symptoms. Numerous symptoms such as depression, sleep issues, and gastrointestinal complaints are included in non-motor characteristics. It is noteworthy that more than 80% of people with PD experience digestive problems (Mhanna & Alshehabi, 2023, p. 2).

However, patients frequently exhibit additional symptoms, including gastrointestinal issues and cognitive deficits, which can appear up to 20 years before the onset of motor signs. To gain insight into the connection between Parkinson's disease and gut microbiota, certain findings point to a decline in the gut microbiota's β-diversity and bacterial phylogenetic diversity. Elevated *Akkermansia* levels have been reported in the intestines of PD patients. This condition can degrade mucus present in the colon, increase vulnerability to infections, and increase intestinal permeability, all of which can cause localized inflammation. Metabolic investigations have shown that individuals with PD exhibit increased populations of *Akkermansia* and *Bifidobacterium*, and reduced populations of Lachnospiraceae and *Faecalibacterium*.

In patients with PD, changes in stool hardness are noted because constipation slows colonic transit, which promotes the development of bacteria with slower growth rates and is linked to decreased nutritional availability (Sampson et al., 2016). Probiotics and prebiotic fibers are currently the recommended course of treatment. Proteolysis becomes more prevalent in gut microbiota metabolism as a result of slow colonic transit, moving away from glucose fermentation. Firmer stools and a greater degree of constipation are correlated with elevated levels of phenylacetylglutamine and P-cresol in PD patients. Chronic constipation in Parkinson's disease patients is linked to a number of taxa: total butyrate producers and *Faecalibacterium* are negatively correlated with constipation, while there is a positive correlation among *Ruminococcus*, *Dorea*, *Oscillospira*, and cluster B.

28.5.3 Autistic Spectrum Disorder

A "leaky gut" and the ability for pathogens to translocate into the portal and systemic circulation are consequences of gut microbiota dysbiosis, which can lead to changes in the barrier of the gut and neuroinflammation in illnesses of the CNS, one being autism spectrum disorders (Q. Wang et al., 2023).

ASD is a developmental and neurological condition characterized by changes in behavior and social communication. Research has shown a connection between ASD, dysfunctional bowel, and impaired intestinal barrier function. According to a previous study, microbiota transfer therapy has been shown to positively alter neurological and gastrointestinal symptoms in children with autism.

The initial theory on the connection between ASD and gut microbiota suggests that the neurotoxic effects of *Clostridium* play a role in the development of ASD. When comparing the microbiota of an individual with ASD to that of healthy controls, it was found that the former had a greater quantity of Bacteroidetes and Proteobacteria and a lower number of Bifidobacteria and Firmicutes.

A previous study verified that a decrease in the relative abundance of Bacteroidetes in the gut microbiota of individuals with ASD resulted in a considerable increase in the Firmicutes to Bacteroidetes ratio. In the ASD cohort, there was a major increase in *Collinsella*, *Corynebacterium*, *Dorea*, and *Lactobacillus* but a decrease in the relative abundance of *Bilophila*, *Parabacteroides*, *Dialister*, *Alistipes*, and *Veillonella* at the genus level.

28.5.4 Role of AI in Microbiome Research

Artificial intelligence helps us understand how gut microbiota affects neurological illnesses by analyzing complex data, exposing microbiome–brain relationships, and identifying therapeutic targets. Machine learning algorithms have successfully classified microbiome patterns associated with neurological disorders, including Alzheimer's and multiple sclerosis, enabling individualized treatment.

The confluence of AI and microbiome research has transformed neurological illness pathophysiology research. AI, particularly machine learning algorithms, has helped analyze massive microbiome datasets. A systematic analysis on the gut microbiome and alcohol use disorder found that AI could identify dysbiotic patterns that correspond with illness severity (Grodin et al., 2024). This association is crucial for targeting certain microbial populations with therapies.

In a comprehensive MS review, researchers examined 37 probiotic and commensal therapy papers (6 humans and 31 animals). AI employing Bradford Hill criteria allowed refined treatment efficacy evaluation among formulations (Blais et al., 2021). Due to their reproducible neurological effects, VSL#3 and *L. paracasei* are attractive clinical options.

The gut–brain axis is increasingly thought to play a role in neurodegenerative illnesses. Ferreiro et al. (2023) found that gut microbial patterns were linked with β-amyloid and tau pathology in preclinical AD. Researchers enhanced preclinical AD prediction using gut microbiome characteristics and machine learning classifiers. AI can explain illness mechanisms and help early diagnosis, as shown by this finding.

AI applications enable integrative genomic, transcriptomic, and metabolomic data analytics, complementing "omic" technologies. Research on CADASIL, a vascular disease caused by NOTCH3 mutations, found significant changes in molecular pathways linked to neuroinflammation and cellular adhesion (Muiño et al., 2021). AI-driven analysis can help us comprehend these pathways by associating them with patient symptoms.

AI has also shown promise in gut microbiota-mediated neurotransmitter regulation. Certain bacteria can create or absorb neurotransmitters including serotonin and GABA, affecting mood and behavior (Strandwitz, 2018). AI can analyze complicated datasets from human trials and animal models to anticipate outcomes based on microbiome composition to better mimic these interactions.

AI has promising applications in this discipline, but study design and methods remain difficult. Variable patient populations, sample sizes, and microbiological evaluation methods can cause study discrepancies. Methodological differences prevented some research from replicating substantial links between certain taxa and neurological diseases (Cheung et al., 2019). Therefore, future research must standardize microbiome analysis techniques and integrate AI algorithms that can accept these variabilities while offering robust statistical validation.

Existing study evidence varies greatly in quality. Many studies use cross-sectional designs, which limit causal inferences between gut microbiota and neurological diseases (Bjørklund et al., 2020). Artificial intelligence-driven longitudinal research is needed to determine temporal connections and causal pathways.

These findings go beyond correlation and propose microbiome modification therapies. In MS, FMT can restore microbial balance (Hickman et al., 2018). With pretransplant microbiota analysis, AI could optimize FMT methods by predicting individual responses.

Nutritional therapies based on AI predictions are an interesting new area for treating gut dysbiosis-related neurological diseases. Individualised nutrition regimens that consider microbiota composition may improve therapeutic efficacy and reduce negative effects (Zeevi et al., 2015). Artificial intelligence has greatly improved our understanding of gut microbiota and neurological illnesses. By using complicated data analytics and multiomic techniques, researchers can gain new insights into microbial brain relationships. Translating these findings into therapeutic practice requires addressing methodological discrepancies and improving study designs.

AI has shown the relationship between gut health and brain illnesses in microbiome research. Researchers can detect microbial fingerprints linked with different environments by analyzing complex datasets with machine learning methods. These discoveries improve our understanding and enable new gut–brain axis modulation therapies. To confirm these findings, standardizing methodology and refining study designs must continue.

28.6 NETWORK ANALYSIS OF MICROBIAL INTERACTIONS

The most effective methods for assessing microbial network dynamics in microbiome studies involve advanced sequencing technologies and integrative analytical approaches that combine metagenomic, metabolomic, and ecological data. These methodologies enhance our understanding of microbial interactions and their functional implications for health and disease.

Microbial network dynamics are pivotal in understanding the complex interactions within microbiomes that influence both health and disease states. A review by Knight et al, (2018) (Knight et al., 2018) outlines best practices for analyzing microbiomes, emphasizing the transition from operational taxonomic unit (OTU)-based analyses to exact sequence variant (ESV)-based methods. This shift enhances taxonomic resolution and reduces false discovery rates, thereby improving the reliability of microbial interaction assessment.

The importance of integrating multiple omics data sets is further highlighted by Kolodziejczyk et al. (2019) (Kolodziejczyk et al., 2019), who discuss how dietary inputs can shape microbial communities and influence host physiology. These findings suggest that personalized nutrition strategies can be developed by understanding individual microbiome responses to dietary changes, which could potentially mitigate dysbiosis-related conditions.

Bäckhed et al. (2015) provided a comprehensive analysis of gut microbiome dynamics during early life, utilizing metagenomic approaches to assess how delivery mode and feeding impact microbial composition. Their work illustrated the significance of early microbial exposure in shaping long-term health outcomes, thus underlining the necessity for longitudinal studies that track microbial changes over time.

In the context of specific diseases, Narayana et al. (2023) (Narayana et al., 2023) explored the gut–lung axis in bronchiectasis patients through concurrent sampling of gut and lung microbiomes. Their innovative approach employed bacteriome sequencing combined with co-occurrence network analysis to reveal significant interactions between the gut and lung microbes that correlate with clinical outcomes. This study demonstrates how integrating multi-site microbiome data can elucidate the complex interdependencies among microbial communities.

Moreover, Dinan and Cryan (Dinan & Cryan, 2017) emphasized the brain–gut–microbiota axis as a critical area for understanding mental health disorders. They advocate for more rigorous clinical studies to validate preclinical findings on the efficacy of psychobiotics in modulating the gut microbiota to influence mental health outcomes.

Despite these advancements, many studies suffer from limitations, such as small sample sizes or inadequate methodological rigor. For instance, while systematic reviews often aggregate findings from numerous studies, they may not always adequately control for confounding variables or address the biases inherent in primary studies (Lloyd & Deakin, 1975). Furthermore, reliance on self-reported data in many surveys can introduce bias that affects the interpretation of results regarding dietary habits and health outcomes.

The complexity of microbial interactions necessitates sophisticated analytical techniques. Recent advances in bioinformatics have facilitated the integration of large datasets from various sources, allowing researchers to construct detailed models of microbial networks. For example, methods such as machine learning are increasingly being applied to identify patterns within microbiome data that correlate with specific health conditions (Rutayisire et al., 2016).

The quality of evidence varies significantly across studies; while some provide robust insights into microbial dynamics through well-designed experiments (e.g., randomized controlled trials), others may lack methodological rigor or sufficient statistical power to draw definitive conclusions. This disparity underscores the need for standardization of study designs and reporting practices across microbiome research (Hillestad et al., 2022).

Effective methodologies for assessing microbial network dynamics include a combination of advanced sequencing technologies, integrative omics approaches, and robust analytical frameworks. These methodologies not only enhance our understanding of microbial interactions but also facilitate personalized interventions aimed at restoring healthy microbiome function. Future research should prioritize longitudinal designs that capture temporal changes in microbial communities, while employing standardized methodologies to improve comparability across studies.

28.7 CHALLENGES IN INTEGRATING AI WITH MICROBIOME RESEARCH

A complex ecology of bacteria, viruses, fungi, and protozoans makes up the human microbiota. With almost 100 times the human genome, it gives us functional properties we lack. Genes make up the microbiome. A new estimate suggests that the gastrointestinal system contains 30–400 trillion germs, more than the number of eukaryotic cells in the body. Skin and mucosa (gastrointestinal, respiratory, and urogenital) are covered in commensal microbiota, with the colon harboring about 70% of the body's microorganisms. Thus, the organism is believed to be ecological and can be redefined as a network of interactions and linkages between eukaryotes and prokaryotes.

Carbohydrate fermentation, xenobiotic metabolism, and vitamin production depend on the gut bacteria, which protect against pathogens. Despite their ubiquity, functional gastrointestinal disorders' pathogenesis is unknown, limiting symptomatic diagnostic and treatment methods. Complex mechanisms underpinning gastrointestinal tract–brain bidirectional communication abnormalities are crucial to etiology and illness knowledge. Our understanding of these complex illnesses has advanced. Different people have different microbiota, yet it may change. Human microbiota composition is personal, yet bacterial population diversity across body sites is greater than between persons (Iadanza et al., 2020).

The complicated relationships between gut microbiota and neurological diseases are increasingly studied using AI. The gut microbiota, a complex collection of bacteria in the digestive tract, helps sustain brain function and wellness. Dysbiosis, an imbalance in microbial populations, is connected to ASD, Parkinson's, Alzheimer's, and depression. With its superior data analysis and pattern recognition, AI will help unravel these intricate linkages. ML models can analyse massive gut microbiome data. These models can spot patterns and relationships that traditional statistical methods miss. ML can uncover neurological disease-associated microbial species or metabolic pathways from microbiome sequence data. Researchers can use microbiome profiles to predict neurological disorders by training these algorithms on huge datasets. AI can also integrate genomes, metabolomics, and proteomics data to better understand the gut–brain axis. This holistic approach can show how gut microorganisms affect neurological function via biochemical pathways. Short-chain fatty acids and neurotransmitter precursors produced by gut bacteria may alter brain function. AI can identify them. The link between microbiota, disease, and clinical relevance is unclear. Advanced AI approaches like ML and deep learning (DL) can assist doctors process and interpret data from these datasets (Xiong et al., 2024).

AI has two approaches: ML and DL. However, defining their relationship is difficult. AI is growing and developing new methods. This creates a dynamic situation where approach limitations are evanescent. This study regarded DL as a subset of ML approaches (Figure 28.1). The scientific literature offers different views of AI (Saxena et al., 2024).

28.7.1 Technical Challenges

Based on genomic, proteomic, and metabolomic data, AI technologies improve our understanding of healthy and sick gut flora. Recent research has examined AI-based tailored nutrition. AI algorithms assess food habits, gut microbiome, health, and physical activity data to provide individualized nutritional recommendations for better health. Also, prediction models incorporating multiple ML techniques can be employed for diagnosis and therapy (Kumar et al., 2022). Data quality, standardization, and microbiome data complexity are technical hurdles. The microbiome is made up of many microorganisms with complex interactions that require sophisticated algorithms and a lot of computational power. One major challenge with limited labeled data is that all labeled microbiome data are scarce, making supervised learning techniques less effective in unsupervised and semi-supervised microbiome research. Integrating multi-omics data, biological variability, interpretability of AI models, computational resources, ethical and privacy concerns, dynamic microbiomes, and modeling these temporal changes and predicting future states are other technical challenges (Ali et al., 2023; Ezzamouri et al., 2021).

28.7.2 Ethical Consideration

Integrating AI in microbiome research has its own demerits, which concern ethical dilemmas (B. Singh et al., 2024). Microbiome data can reveal sensitive information about a patient, thus ensuring robust data protection and informed consent. Participants must be fully informed of how, when, and where their data are being used. AI models can have their own biases in their training data, leading to skewed results. Efforts must be made to diversify the representative datasets. Scientists must maintain transparency about the AI algorithms that are used, clarifying who owns the data, and how it can be used or shared, especially when collaborating across universities and labs. Ensuring the benefits of research would enhance it rather than replace it. Guarding against the misuse of AI can have devastating impacts (Mbatha et al., 2023).

28.7.3 Data Privacy and Security

Safeguarding of users' data and maintaining strict vigilance regarding how, where, and when the patient's data are used and what implications it can have is critical and crucial to data privacy and security (Xiong et al., 2024).

28.7.4 Interdisciplinary Collaborations

Interestingly, a wide proportion of the human microbiota (approximately 70%) is composed of microbes that cannot be cultivated by common microbiological methods. Today, the advent of new molecular microbiota profiling tools, such as genomic next-generation sequencing (NGS) and metagenomics shotgun sequencing, allows us to obtain more information about the impact of microbiota in both healthy and pathological conditions. Because these techniques generate large amounts of data, they have spurred the development of bioinformatics techniques (Cullen et al., 2020). Advancements made in the integration of AI in microbiome research can lead to multidisciplinary collaboration, from precision medicine to oncology, and from neuroscience to data science, all of which can further enlarge the scope and application of how far reaching and impactful such endeavors can be.

28.8 FUTURE DIRECTIONS

Recently, the study of the effects of intestinal microbiota metabolites on mental health has become an interesting topic for researchers, with the main focus on the relationship between the functioning of the brain and the composition of the microorganisms present in the intestine (Poluektova et al., 2021). Recent research has shown that the microbiota can affect the immune system and neurotransmitters. Hence, it opens a new way for researchers to find more effective and accurate methods for the diagnosis and treatment of mental disorders (Mohan et al., 2023; Poluektova et al., 2021). There are enormous and diverse microorganisms living in our digestive systems, which have a role in digestion and metabolism and interact with the brain through bidirectional signalling known as the "gut–brain axis." There is a symbiotic interaction between the microbiota and the human brain and disorder, which can have a high impact on human mental health (Sasso et al., 2023).

28.8.1 Advancements in AI Techniques

Neurology has been transformed by AI, which has changed diagnosis and treatment. Neurological illnesses greatly affect our daily activities like walking, talking, and learning. The peripheral or central nervous systems can produce neurological diseases. Over 500 diseases damage the brain. In the case of prolactinoma (a disease associated with pituitary gland tumors), it is difficult to detect the size of the gland in early stages because it does not affect MRI images. Even radiology experts can make mistakes, but AI has taken over that field and given more accurate results (McKevitt et al., 2023). Singh et al. proved that AI can detect neurological illnesses early and personalize medication. By showing the Machine Learning Algorithm for early AD prognosis and prediction (McKevitt et al., 2023; K. R. Singh & Dash, 2023). According to Chauhan and Singh (2018), conventional AI systems use manually extracted features and cannot process raw data.

However, artificial neural networks (ANN) are inspired by biological neural networks like the human brain, making them more effective at learning than conventional AI. Recent evaluations have revealed that neurology and AI work together (Deep Learning vs. Machine Learning, n.d.). Neuroscience theory aids AI in two ways: biological neural network design and variant generation, and validation of all AI-based models. Artificial neural networks (ANNs) and recurrent neural networks (RNNs) are illuminating new neuroscience breakthroughs by mirroring the brain's mechanisms for offline experience 'replay,' revealing sleep-related temporal patterns (Deep Learning vs. Machine Learning, n.d.).

28.8.2 Emerging Technologies

Marcos-Zambrano et al. (2021) observed that disturbances in neurometabolic profiles result in changes in the composition of gut microorganisms, which alters the microorganisms that abundantly produce fatty acids and neurotransmitters, including serotonin, GABA, and other biomarker metabolites. Therefore, to detect neural diseases, a healthy person's gut microbiota complex and an unhealthy person's gut microbiota complex are required. Using this idea, Angelova et al. demonstrated the use of Machine Learning and other Artificial Intelligence algorithms such as random forest, elastic net, and You Only Look Once (YOLO), which have revealed the connection between the gut microbiota and mental disorders. They used random forest and YOLO algorithms to detect differences between microbiota complexes. For this purpose, the

S. No.	Type of Algorithm Involved	Type of Gut-Associated Neurology Disease Detected	References
1	Support vector machines (SVMs)	AD, dementia, PD	Shahparian et al., 2021; Sharma et al., 2021; Xu & Zhu, 2021
2	Gaussian mixture models (GMM)	Dementia, PD	Bchir, 2020; Stuhler et al., 2012
3	K-nearest neighbor (KNN)	PD	Chen et al., 2013
4	Generative adversarial networks (GAN)	AD, PD	Ma et al., 2021; Shin et al., 2020
5	Random forests (RF)	PD	Polat, 2019
6	Artificial neural network (ANN)	AD, PD	Q. Wang et al., 2023
7	Convolutional neural network (CNN)	AD, PD	Gil-Martín et al., 2019
8	Recurrent neural network (RNN)	AD, PD	Masiala et al., 2019
9	Long-short-term memory (LSTM)	AD, PD	E. et al., 2021; Sethi et al., 2021
10	Extreme eearning machine (ELM)	AD, PD	Chen et al., 2016; Lama et al., 2017
11	Gated recurrent unit (GRU)	AD	Bakkouri et al., 2019; M. Huang et al., 2021
12	Deep boltzmann machine (DBM)	PD	de Souza et al., 2017
13	Deep belief networks (DBNs)	AD, PD	Al-Fatlawi et al., 2016; Shen et al., 2019

function Random Forest Classifier from the sklearn library was used, which is used for the identification and extraction of several datasets, which will then be used in elastic net and random forest for classification.

28.8.3 Classification for Neurological Diseases Detection

AI is everywhere and has been integrated into our everyday lifestyles; in every task, neurology is no exception. Medical practitioners should be aware of AI algorithms used in neurology and other advancements in AI to explain how machine learning can revolutionize the treatment of neurological diseases (Lima et al., 2022).

28.8.3.1 Machine Learning Algorithms

Machine learning is a type of artificial intelligence that is used to classify, recognize, and predict patterns. In the context of microbiome research, ML has been used for phenotypic research of microbiota complexes, which will lead to results such as determining the abundance and diversity of the microbiota complex (Marcos-Zambrano et al., 2021).

28.8.3.2 Classical Methods

Among classical machine learning methods,K-nearest neighbors, linear regression, random forests, and other SVMs have been the best machine learning algorithms used for microbiome data. These algorithms have recently been used in host dysbiosis prediction, which has resulted in aggregate decision trees and differentiated between individuals with healthy and unhealthy microbiota complexes.

28.8.3.3 Deep Learning

It is a group of ML algorithms that involve various architectures of artificial neural networks. It relies on the neurons and units that work on the input and provide the output accordingly. The connections between the nodes form multiple layers of the network, which provides the advantage that they can be rearranged in different layouts. The fully connected neural network (FCNN) is the most basic neural network, in which the nodes from one layer are fully connected to the nodes on the other layer (Nguyen et al., 2017).

28.8.3.4 Convolutional Neural Network

It is mainly used for picturing microbiomes. CNNs can summarize the local structure in their inputs. They are well-suited for the image by reshaping each sample into a square, where each pixel is based on the presence of microbial taxa.

28.8.3.5 Personalized Medicine

In recent years, the microbiome has drawn the attention of many research groups. The human microbiome project (HMP) is the study of microbial species present across the human body. HMP has found out some important things like alterations in the gut microbiome can affect not only gastrointestinal disease but also neurological diseases. This project showed that gut microbial dysbiosis is the reason for the release of harmful signals from the central nervous system (CNS), leading to neurodegeneration (M. Huang et al., 2021).

Tandon et al. (2006) showed the integral role of AI in precision medicine and used a variety of AI-aided approaches for gut microbiome data mining, such as ML and DL. They also made progress in the quantitative analysis of the gut microbiome, and the results have helped in the application of microbiome-based precision medicine, which has also been reported in alterations of the gut microbiome in the progression of aging and many diseases. The developed AI model identified subsets of microbiota as potential biomarkers.

A. *Detection or prediction of diseases*

It is important to solve the following two problems as it unveils the links between individual characteristics and disease risk or treatment response. This will help determine treatment outcomes. SVM, Fuzzy logic, and ANN were the algorithms used in this step.

B. *Achievement of accurate diagnosis*

It allows the prediction of the most stable variations in the patient's body after overemphasis in the medical field because improper and incorrect diagnoses have resulted in serious complications. Accurate diagnosis is achieved using an ANN, Expert systems, and an SVM.

C. *Treatment optimization*

Several studies have been conducted to predict the clinical course of Graves' disease using antithyroid drugs. An ANN, naive Bayesian, and SVM were used in this step.

28.8.4 Longitudinal Studies and Data Integration

Longitudinal studies have provided insights into the microbiota and their roles in neurological diseases. On the basis of checking the continuous interval of time, it helps in the prediction of the progression pattern of the diseases, which helps in the analysis of long-term effects and offers comprehensive knowledge of the current and future states of the disease and how it will affect the healthcare of the patient. It also enhances the robustness of predictive models and biomarker discovery, ultimately contributing to the development of therapies and personalized medicine.

28.8.4.1 Longitudinal Study of Disease

Tandan et al. proposed an innovative approach for the modeling and longitudinal study of Alzheimer's disease using ANN by recognizing the complexities of disease progression. This leads to the mixed-effects neural network (MENN) method. This accommodates and responds flexibly, based on the input variable. This includes other factors, such as age, gender, education, and base rate progression, to forecast cognitive decline. In addition, we explore the relationship between the response and input variables in the longitudinal data. It also shows the linear nature of Alzheimer's disease progression by introducing models such as a growth curve. It also discusses the optimal number of neurons in the hidden layer for the neural network architecture and compares their standards.

28.8.4.1.1 Case Studies of Alzheimer's Disease Detection Using Gut Microbiota and AI

The thesis presented by Castro has shown a case study in which microbiome complex and lipid metabolism data have been used to find biomarkers to detect neurological disease with the help of ML and to validate these results. In this study, we used the Alzheimer's Disease Neuroimaging Initiative (ADNI) and Alzheimer's Disease Metabolomics Consortium (ADMC). The *R* package selected for machine learning and feature selection analyses was FRESA.CAD (feature selection algorithms for computer-aided diagnosis), which has many properties, such as visualization, statistical models, and longitudinal analysis. However, in this thesis, it was used for binary classification, which is a variety of machine learning methods.

In the selected project, the training sample was constructed by randomly selecting 70% of the patients each time, and the metrics considered were accuracy, sensitivity, specificity, and the Receiver Operating Characteristic Curve (ROC AUC). For machine learning methods for feature selection, the algorithms with the highest predictive potential were Bootstrapped Stage-Wise Model Selection (BSWiMS), Integrated Discrimination Improvement (IDI), RPART, Student's *t*-test, LASSO, Random Forest (RF), and Wilcoxon test. The three types of tests were run separately (neuropsychological and ut metabolites and lipids), and the best score was achieved with the combination of neuropsychological and gut microbiota using SVM. This study had several limitations. For example, the lipid metabolites analyzed repeated the information because they were from the same family in some cases, and the inclusion criteria of considering the patients 65 years of age or in the late-onset scenario did not investigate the early-onset scenarios. However, this project provides the idea of analyzing gut microbiome metabolites for the study of Alzheimer's disease giving a promising opportunity for new treatments that can attack the development of the disease.

28.9 CONCLUSION

This chapter covers gut microbiota and AI neurological illness detection. The essay stresses the gut–brain axis, which links gut microbes to the CNS. This axis is crucial for brain health because dysbiosis, or gut microbiota abnormalities, are linked to Alzheimer's, Parkinson's, and autism spectrum disorders. This chapter addresses how machine learning and deep learning might improve disease detection and comprehension using microbiome data. Discussed are dysbiosis, neuroinflammation, and cognitive decline bacteria. The chapter discusses probiotic therapy despite conflicting evidence. AI enhances stroke and epilepsy diagnosis. Machine learning algorithms can forecast large-vessel occlusions and seizures in real time, enhancing patient outcomes and therapy. The gut microbiome chapter discusses health and dysbiosis. Gut bacteria provide brain-supporting metabolic by-products. Microbiota abnormalities may cause neurological disorders. The gut–brain axis and communication routes are examined in this chapter. The scientists suggest the vagus nerve is essential to the immunological, neuroendocrine, and autonomic nervous systems. Intestinal dysbiosis and neurological disorders complete the chapter. This study investigates autism and Parkinson's dysbiosis, its causes, and gut-healthy foods. This chapter presents novel neurological problem diagnostic and therapy strategies using AI, microbiome research, and neurology. AI standardization and morality require research. AI and microbiome analysis may help us understand gut–brain relationships and develop new drugs.

REFERENCES

Adams, A., Adelfio, A., Barnes, B., Berlien, R., Branco, D., Coogan, A., Garson, L., Ramirez, N., Stansbury, N., Stewart, J., Worman, G., Butler, P. J. & Brown, D. (2023). Risk-based monitoring in clinical trials: 2021 update. *Therapeutic Innovation & Regulatory Science*, *57*(3), 529–537. https://doi.org/10.1007/s43441-022-00496-9

Agirman, G., Yu, K. B. & Hsiao, E. Y. (2021). Signaling inflammation across the gut-brain axis. *Science*, *374*(6571), 1087–1092. https://doi.org/10.1126/science.abi6087

Akbari, E., Asemi, Z., Daneshvar Kakhaki, R., Bahmani, F., Kouchaki, E., Tamtaji, O. R., Hamidi, G. A. & Salami, M. (2016). Effect of probiotic supplementation on cognitive function and metabolic status in Alzheimer's disease: A randomized, double-blind and controlled trial. *Frontiers in Aging Neuroscience*, *8*. https://doi.org/10.3389/fnagi.2016.00256

Al-Fatlawi, A. H., Jabardi, M. H. & Ling, S. H. (2016). Efficient diagnosis system for Parkinson's disease using deep belief network. In *2016 IEEE Congress on Evolutionary Computation (CEC)* (pp. 1324–1330). https://doi.org/10.1109/CEC.2016.7743941

Ali, T., Ahmed, S. & Aslam, M. (2023). Artificial intelligence for antimicrobial resistance prediction: Challenges and opportunities towards practical implementation. *Antibiotics*, *12*(3), 523. https://doi.org/10.3390/antibiotics12030523

Baccarelli, A. A. & Ordovás, J. (2023). Epigenetics of early cardiometabolic disease: Mechanisms and precision medicine. *Circulation Research*, *132*(12), 1648–1662. https://doi.org/10.1161/CIRCRESAHA.123.322135

Bäckhed, F., Roswall, J., Peng, Y., Feng, Q., Jia, H., Kovatcheva-Datchary, P., Li, Y., Xia, Y., Xie, H., Zhong, H., Khan, M. T., Zhang, J., Li, J., Xiao, L., Al-Aama, J., Zhang, D., Lee, Y. S., Kotowska, D., Colding, C., … Wang, J. (2015). Dynamics and stabilization of the human gut microbiome during the first year of life. *Cell Host & Microbe*, *17*(5), 690–703. https://doi.org/10.1016/j.chom.2015.04.004

Bakkouri, I., Afdel, K., Benois-Pineau, J. & Catheline, G. (2019). Recognition of Alzheimer's disease on sMRI based on 3D multi-scale CNN features and a gated recurrent fusion unit. In *2019 International Conference on Content-Based Multimedia Indexing (CBMI)* (pp. 1–6). https://doi.org/10.1109/CBMI.2019.8877477

Balaji, E., Brindha, D., B., Elumalai, V. K. & Vikrama, R. (2021). Automatic and non-invasive Parkinson's disease diagnosis and severity rating using LSTM network. *Applied Soft Computing*, *108*, 107463. https://doi.org/10.1016/j.asoc.2021.107463

Bchir, O. (2020). Parkinson's disease classification using Gaussian mixture models with relevance feature weights on vocal feature sets. *International Journal of Advanced Computer Science and Applications*, *11*(4). https://doi.org/10.14569/IJACSA.2020.0110456

Bianchi, V. E., Herrera, P. F. & Laura, R. (2021). Effect of nutrition on neurodegenerative diseases. A systematic review. *Nutritional Neuroscience*, *24*(10), 810–834. https://doi.org/10.1080/1028415X.2019.1681088

Bjørklund, G., Dadar, M., Anderson, G., Chirumbolo, S. & Maes, M. (2020). Preventive treatments to slow substantia nigra damage and Parkinson's disease progression: A critical perspective review. *Pharmacological Research*, *161*, 105065. https://doi.org/10.1016/j.phrs.2020.105065

Blais, L. L., Montgomery, T. L., Amiel, E., Deming, P. B. & Krementsov, D. N. (2021). Probiotic and commensal gut microbial therapies in multiple sclerosis and its animal models: A comprehensive review. *Gut Microbes*, *13*(1), 1943289. https://doi.org/10.1080/19490976.2021.1943289

Bu, X.-L., Yao, X.-Q., Jiao, S.-S., Zeng, F., Liu, Y.-H., Xiang, Y., Liang, C.-R., Wang, Q.-H., Wang, X., Cao, H.-Y., Yi, X., Deng, B., Liu, C.-H., Xu, J., Zhang, L.-L., Gao, C.-Y., Xu, Z.-Q., Zhang, M., Wang, L., … Wang, Y.-J. (2015). A study on the association between infectious burden and Alzheimer's disease. *European Journal of Neurology*, *22*(12), 1519–1525. https://doi.org/10.1111/ene.12477

Campbell, D. J. (2006). A review of Perindopril in the reduction of cardiovascular events. *Vascular Health and Risk Management*, *2*(2), 117–124. https://doi.org/10.2147/vhrm.2006.2.2.117

Chauhan, N. K. & Singh, K. (2018). A review on conventional machine learning vs deep learning. In *2018 International Conference on Computing, Power and Communication Technologies (GUCON)* (pp. 347–352). https://doi.org/10.1109/GUCON.2018.8675097

Chen, H.-L., Huang, C.-C., Yu, X.-G., Xu, X., Sun, X., Wang, G. & Wang, S.-J. (2013). An efficient diagnosis system for detection of Parkinson's disease using fuzzy k-nearest neighbor approach. *Expert Systems with Applications*, *40*(1), 263–271. https://doi.org/10.1016/j.eswa.2012.07.014

Chen, H.-L., Wang, G., Ma, C., Cai, Z.-N., Liu, W.-B. & Wang, S.-J. (2016). An efficient hybrid kernel extreme learning machine approach for early diagnosis of Parkinson's disease. *Neurocomputing*, *184*, 131–144. https://doi.org/10.1016/j.neucom.2015.07.138

Cheung, S. G., Goldenthal, A. R., Uhlemann, A.-C., Mann, J. J., Miller, J. M. & Sublette, M. E. (2019). Systematic review of gut microbiota and major depression. *Frontiers in Psychiatry*, *10*, 34. https://doi.org/10.3389/fpsyt.2019.00034

Critchfield, J. W., Van Hemert, S., Ash, M., Mulder, L. & Ashwood, P. (2011). The potential role of probiotics in the management of childhood autism spectrum disorders. *Gastroenterology Research and Practice*, *2011*, 1–8. https://doi.org/10.1155/2011/161358

Cullen, C. M., Aneja, K. K., Beyhan, S., Cho, C. E., Woloszynek, S., Convertino, M., McCoy, S. J., Zhang, Y., Anderson, M. Z., Alvarez-Ponce, D., Smirnova, E., Karstens, L., Dorrestein, P. C., Li, H., Sen Gupta, A., Cheung, K., Powers, J. G., Zhao, Z. & Rosen, G. L. (2020). Emerging priorities for microbiome research. *Frontiers in Microbiology*, *11*, 136. https://doi.org/10.3389/fmicb.2020.00136

de Souza, R. O. M. A., Miranda, L. S. M. & Bornscheuer, U. T. (2017). A retrosynthesis approach for biocatalysis in organic synthesis. *Chemistry – A European Journal*, *23*(50), 12040–12063. https://doi.org/10.1002/chem.201702235

Deep learning vs. machine learning. (n.d.). Retrieved August 15, 2024, from https://www.zendesk.com/in/blog/machine-learning-and-deep-learning/#georedirect

Dinan, T. G. & Cryan, J. F. (2017). Brain-gut-microbiota axis and mental health. *Psychosomatic Medicine*, *79*(8), 920–926. https://doi.org/10.1097/PSY.0000000000000519

Ezzamouri, B., Shoaie, S. & Ledesma-Amaro, R. (2021). Synergies of systems biology and synthetic biology in human microbiome studies. *Frontiers in Microbiology*, *12*. https://doi.org/10.3389/fmicb.2021.681982

Ferreiro, A. L., Choi, J., Ryou, J., Newcomer, E. P., Thompson, R., Bollinger, R. M., Hall-Moore, C., Ndao, I. M., Sax, L., Benzinger, T. L. S., Stark, S. L., Holtzman, D. M., Fagan, A. M., Schindler, S. E., Cruchaga, C., Butt, O. H., Morris, J. C., Tarr, P. I., Ances, B. M. & Dantas, G. (2023). Gut microbiome composition may be an indicator of preclinical Alzheimer's disease. *Science Translational Medicine*, *15*(700), eabo2984. https://doi.org/10.1126/scitranslmed.abo2984

Gil-Martín, M., Montero, J. M. & San-Segundo, R. (2019). Parkinson's disease detection from drawing movements using convolutional neural networks. *Electronics*, *8*(8), 907. https://doi.org/10.3390/electronics8080907

Gonzalez, J. S., Peyrot, M., McCarl, L. A., Collins, E. M., Serpa, L., Mimiaga, M. J. & Safren, S. A. (2008). Depression and diabetes treatment nonadherence: A meta-analysis. *Diabetes Care*, *31*(12), 2398–2403. https://doi.org/10.2337/dc08-1341

Grodin, E. N., Burnette, E. M., Rodriguez, C., Fulcher, J. A. & Ray, L. A. (2024). The gut microbiome in alcohol use disorder and alcohol-associated liver disease: A systematic review of clinical studies. *Alcohol, Clinical & Experimental Research*, *48*(7), 1221–1242. https://doi.org/10.1111/acer.15338

Gupta, R., Kumari, S., Senapati, A., Ambasta, R. K. & Kumar, P. (2023). New era of artificial intelligence and machine learning-based detection, diagnosis, and therapeutics in Parkinson's disease. *Ageing Research Reviews*, *90*, 102013. https://doi.org/10.1016/j.arr.2023.102013

Hakeem, H., Feng, W., Chen, Z., Choong, J., Brodie, M. J., Fong, S.-L., Lim, K.-S., Wu, J., Wang, X., Lawn, N., Ni, G., Gao, X., Luo, M., Chen, Z., Ge, Z. & Kwan, P. (2022). Development and validation of a deep learning model for predicting treatment response in patients with newly diagnosed epilepsy. *JAMA Neurology*, *79*(10), 986–996. https://doi.org/10.1001/jamaneurol.2022.2514

Heravi, F. S., Naseri, K. & Hu, H. (2023). Gut microbiota composition in patients with neurodegenerative disorders (Parkinson's and Alzheimer's) and healthy controls: A systematic review. *Nutrients*, *15*(20), 4365. https://doi.org/10.3390/nu15204365

Hickman, S., Izzy, S., Sen, P., Morsett, L. & El Khoury, J. (2018). Microglia in neurodegeneration. *Nature Neuroscience*, *21*(10), 1359–1369. https://doi.org/10.1038/s41593-018-0242-x

Hillestad, E. M. R., van der Meeren, A., Nagaraja, B. H., Bjørsvik, B. R., Haleem, N., Benitez-Paez, A., Sanz, Y., Hausken, T., Lied, G. A., Lundervold, A. & Berentsen, B. (2022). Gut bless you: The microbiota-gut-brain axis in irritable bowel syndrome. *World Journal of Gastroenterology*, *28*(4), 412–431. https://doi.org/10.3748/wjg.v28.i4.412

Huang, M., Lai, H., Yu, Y., Chen, X., Wang, T. & Feng, Q. (2021). Deep-gated recurrent unit and diet network-based genome-wide association analysis for detecting the biomarkers of Alzheimer's disease. *Medical Image Analysis*, *73*, 102189. https://doi.org/10.1016/j.media.2021.102189

Huang, Q., Pereira, A. C., Velthuis, H., Wong, N. M. L., Ellis, C. L., Ponteduro, F. M., Dimitrov, M., Kowalewski, L., Lythgoe, D. J., Rotaru, D., Edden, R. A. E., Leonard, A., Ivin, G., Ahmad, J., Pretzsch, C. M., Daly, E., Murphy, D. G. M. & McAlonan, G. M. (2022). $GABA_B$ receptor modulation of visual sensory processing in adults with and without autism spectrum disorder. *Science Translational Medicine*, *14*(626), eabg7859. https://doi.org/10.1126/scitranslmed.abg7859

Iadanza, E., Fabbri, R., Bašić-ČiČak, D., Amedei, A. & Telalovic, J. H. (2020). Gut microbiota and artificial intelligence approaches: A scoping review. *Health and Technology*, *10*(6), 1343–1358. https://doi.org/10.1007/s12553-020-00486-7

Jenkins, D. J. A., Kendall, C. W. C., Augustin, L. S. A., Franceschi, S., Hamidi, M., Marchie, A., Jenkins, A. L. & Axelsen, M. (2002). Glycemic index: Overview of implications in health and disease. *The American Journal of Clinical Nutrition*, *76*(1), 266S-73S. https://doi.org/10.1093/ajcn/76/1.266S

Jiang, C., Li, G., Huang, P., Liu, Z. & Zhao, B. (2017). The gut microbiota and Alzheimer's disease. *Journal of Alzheimer's Disease: JAD*, *58*(1), 1–15. https://doi.org/10.3233/JAD-161141

Keshavarzian, A., Green, S. J., Engen, P. A., Voigt, R. M., Naqib, A., Forsyth, C. B., Mutlu, E. & Shannon, K. M. (2015). Colonic bacterial composition in Parkinson's disease. *Movement Disorders*, *30*(10), 1351–1360. https://doi.org/10.1002/mds.26307

Knight, R., Vrbanac, A., Taylor, B. C., Aksenov, A., Callewaert, C., Debelius, J., Gonzalez, A., Kosciolek, T., McCall, L.-I., McDonald, D., Melnik, A. V., Morton, J. T., Navas, J., Quinn, R. A., Sanders, J. G., Swafford, A. D., Thompson, L. R., Tripathi, A., Xu, Z. Z., … Dorrestein, P. C. (2018). Best practices for analysing microbiomes. *Nature Reviews. Microbiology*, *16*(7), 410–422. https://doi.org/10.1038/s41579-018-0029-9

Kolodziejczyk, A. A., Zheng, D. & Elinav, E. (2019). Diet-microbiota interactions and personalized nutrition. *Nature Reviews. Microbiology*, *17*(12), 742–753. https://doi.org/10.1038/s41579-019-0256-8

Krüger, J. F., Hillesheim, E., Pereira, A. C. S. N., Camargo, C. Q. & Rabito, E. I. (2021). Probiotics for dementia: A systematic review and meta-analysis of randomized controlled trials. *Nutrition Reviews*, *79*(2), 160–170. https://doi.org/10.1093/nutrit/nuaa037

Kumar, P., Sinha, R. & Shukla, P. (2022). Artificial intelligence and synthetic biology approaches for human gut microbiome. *Critical Reviews in Food Science and Nutrition*, *62*(8), 2103–2121. https://doi.org/10.1080/10408398.2020.1850415

Kurbatskaya, A., Jaramillo-Jimenez, A., Ochoa-Gomez, J. F., Bronnick, K. & Fernandez-Quilez, A. (2023). Machine learning-based detection of Parkinson's disease from resting-state EEG: A multi-center study. In *Annual International Conference of the IEEE Engineering in Medicine and Biology Society. IEEE Engineering in Medicine and Biology Society. Annual International Conference*, *2023* (pp. 1–4). https://doi.org/10.1109/EMBC40787.2023.10340700

Lama, R. K., Gwak, J., Park, J.-S. & Lee, S.-W. (2017). Diagnosis of Alzheimer's disease based on structural MRI images using a regularized extreme learning machine and PCA features. *Journal of Healthcare Engineering*, *2017*, 1–11. https://doi.org/10.1155/2017/5485080

Leta, V., Ray Chaudhuri, K., Milner, O., Chung-Faye, G., Metta, V., Pariante, C. M. & Borsini, A. (2021). Neurogenic and anti-inflammatory effects of probiotics in Parkinson's disease: A systematic review of preclinical and clinical evidence. *Brain, Behavior, and Immunity*, *98*, 59–73. https://doi.org/10.1016/j.bbi.2021.07.026

Lima, A. A., Mridha, M. F., Das, S. C., Kabir, M. M., Islam, Md. R. & Watanobe, Y. (2022). A comprehensive survey on the detection, classification, and challenges of neurological disorders. *Biology*, *11*(3), 469. https://doi.org/10.3390/biology11030469

Liu, F., Li, J., Wu, F., Zheng, H., Peng, Q. & Zhou, H. (2019). Altered composition and function of intestinal microbiota in autism spectrum disorders: A systematic review. *Translational Psychiatry*, *9*(1), 43. https://doi.org/10.1038/s41398-019-0389-6

Liu, X.-P., Jin, X., Seyed Ahmadian, S., Yang, X., Tian, S.-F., Cai, Y.-X., Chawla, K., Snijders, A. M., Xia, Y., van Diest, P. J., Weiss, W. A., Mao, J.-H., Li, Z.-Q., Vogel, H. & Chang, H. (2023). Clinical significance and molecular annotation of cellular morphometric subtypes in lower-grade gliomas discovered by machine learning. *Neuro-Oncology*, *25*(1), 68–81. https://doi.org/10.1093/neuonc/noac154

Lloyd, G. & Deakin, H. G. (1975). Phobias complicating treatment of uterine carcinoma. *British Medical Journal*, *4*(5994), 440. https://doi.org/10.1136/bmj.4.5994.440

Loh, J. S., Mak, W. Q., Tan, L. K. S., Ng, C. X., Chan, H. H., Yeow, S. H., Foo, J. B., Ong, Y. S., How, C. W. & Khaw, K. Y. (2024). Microbiota-gut-brain axis and its therapeutic applications in neurodegenerative diseases. *Signal Transduction and Targeted Therapy*, *9*(1), 37. https://doi.org/10.1038/s41392-024-01743-1

Ma, D., Lu, D., Popuri, K. & Beg, M. F. (2021). *Differential Diagnosis of Frontotemporal Dementia and Alzheimer's Disease using Generative Adversarial Network* (Version 2). arXiv. https://doi.org/10.48550/ARXIV.2109.05627

MacFabe, D. F. (2012). Short-chain fatty acid fermentation products of the gut microbiome: Implications in autism spectrum disorders. *Microbial Ecology in Health & Disease*, *23*. https://doi.org/10.3402/mehd.v23i0.19260

Magistretti, P. J. & Allaman, I. (2018). Lactate in the brain: From metabolic end-product to signalling molecule. *Nature Reviews Neuroscience*, *19*(4), 235–249. https://doi.org/10.1038/nrn.2018.19

Marcos-Zambrano, L. J., Karaduzovic-Hadziabdic, K., Loncar Turukalo, T., Przymus, P., Trajkovik, V., Aasmets, O., Berland, M., Gruca, A., Hasic, J., Hron, K., Klammsteiner, T., Kolev, M., Lahti, L., Lopes, M. B., Moreno, V., Naskinova, I., Org, E., Paciência, I., Papoutsoglou, G., … Truu, J. (2021). Applications of machine learning in human microbiome studies: A review on feature selection, biomarker identification, disease prediction and treatment. *Frontiers in Microbiology*, *12*, 634511. https://doi.org/10.3389/fmicb.2021.634511

Marogianni, C., Sokratous, M., Dardiotis, E., Hadjigeorgiou, G. M., Bogdanos, D. & Xiromerisiou, G. (2020). Neurodegeneration and inflammation – An interesting interplay in Parkinson's disease. *International Journal of Molecular Sciences*, *21*(22), 8421. https://doi.org/10.3390/ijms21228421

Masiala, S., Huijbers, W. & Atzmueller, M. (2019). *Feature-Set-Engineering for Detecting Freezing of Gait in Parkinson's Disease using Deep Recurrent Neural Networks* (Version 1). arXiv. https://doi.org/10.48550/ARXIV.1909.03428

Mbatha, S., Mkhize-Kwitshana, Z., Mulaudzi, T., Hull, R. & Dlamini, Z. (2023). *Artificial Intelligence Application to Microbiomics Data for Improved Clinical Decision Making in Precision Oncology* (pp. 157–177). https://doi.org/10.1007/978-3-031-21506-3_8

McKevitt, C., Gabriel, E., Marenco-Hillembrand, L., Otamendi-Lopez, A., Jeevaratnam, S., Almeida, J. P., Samson, S. & Chaichana, K. L. (2023). Supervised machine learning to validate a novel scoring system for the prediction of disease remission of functional pituitary adenomas following transsphenoidal surgery. *Scientific Reports*, *13*(1), 15409. https://doi.org/10.1038/s41598-023-42157-3

Mhanna, A. & Alshehabi, Z. (2023). The microbiota-gut-brain axis and three common neurological disorders: A mini-review. *Annals of Medicine and Surgery (2012)*, *85*(5), 1780–1783. https://doi.org/10.1097/MS9.0000000000000552

Mohan, A., Godugu, S., Joshi, S. S., Shah, K. B., Vanka, S. C., Shakil, H., P, D., Veliginti, S., Sure, P. S. & Goranti, J. (2023). Gut-brain axis: altered microbiome and depression – Review. *Annals of Medicine and Surgery (2012)*, *85*(5), 1784–1789. https://doi.org/10.1097/MS9.0000000000000573

Muiño, E., Fernández-Cadenas, I. & Arboix, A. (2021). Contribution of "omic" studies to the understanding of cadasil. A systematic review. *International Journal of Molecular Sciences*, *22*(14), 7357. https://doi.org/10.3390/ijms22147357

Naomi, R., Embong, H., Othman, F., Ghazi, H. F., Maruthey, N. & Bahari, H. (2021). Probiotics for Alzheimer's disease: A systematic review. *Nutrients*, *14*(1), 20. https://doi.org/10.3390/nu14010020

Narayana, J. K., Aliberti, S., Mac Aogáin, M., Jaggi, T. K., Ali, N. A. B. M., Ivan, F. X., Cheng, H. S., Yip, Y. S., Vos, M. I. G., Low, Z. S., Lee, J. X. T., Amati, F., Gramegna, A., Wong, S. H., Sung, J. J. Y., Tan, N. S., Tsaneva-Atanasova, K., Blasi, F. & Chotirmall, S. H. (2023). Microbial dysregulation of the gut-lung axis in bronchiectasis. *American Journal of Respiratory and Critical Care Medicine*, *207*(7), 908–920. https://doi.org/10.1164/rccm.202205-0893OC

Nguyen, T. H., Chevaleyre, Y., Prifti, E., Sokolovska, N. & Zucker, J.-D. (2017). *Deep Learning for Metagenomic Data: Using 2D Embeddings and Convolutional Neural Networks* (Version 1). arXiv. https://doi.org/10.48550/ARXIV.1712.00244

Polat, K. (2019). A hybrid approach to Parkinson disease classification using speech signal: The combination of SMOTE and random forests. In *2019 Scientific Meeting on Electrical-Electronics & Biomedical Engineering and Computer Science (EBBT)* (pp. 1–3). https://doi.org/10.1109/EBBT.2019.8741725

Poluektova, E., Yunes, R. & Danilenko, V. (2021). The putative antidepressant mechanisms of probiotic bacteria: Relevant genes and proteins. *Nutrients*, *13*(5), 1591. https://doi.org/10.3390/nu13051591

Quigley, E. M. M. (2017). Microbiota-brain-gut axis and neurodegenerative diseases. *Current Neurology and Neuroscience Reports*, *17*(12), 94. https://doi.org/10.1007/s11910-017-0802-6

Rutayisire, E., Huang, K., Liu, Y. & Tao, F. (2016). The mode of delivery affects the diversity and colonization pattern of the gut microbiota during the first year of infants' life: A systematic review. *BMC Gastroenterology*, *16*(1), 86. https://doi.org/10.1186/s12876-016-0498-0

Sampson, T. R., Debelius, J. W., Thron, T., Janssen, S., Shastri, G. G., Ilhan, Z. E., Challis, C., Schretter, C. E., Rocha, S., Gradinaru, V., Chesselet, M.-F., Keshavarzian, A., Shannon, K. M., Krajmalnik-Brown, R., Wittung-Stafshede, P., Knight, R. & Mazmanian, S. K. (2016). Gut microbiota regulate motor deficits and neuroinflammation in a model of Parkinson's disease. *Cell*, *167*(6), 1469-1480.e12. https://doi.org/10.1016/j.cell.2016.11.018

Sasso, J. M., Ammar, R. M., Tenchov, R., Lemmel, S., Kelber, O., Grieswelle, M. & Zhou, Q. A. (2023). Gut microbiome–brain alliance: A landscape view into mental and gastrointestinal health and disorders. *ACS Chemical Neuroscience*, *14*(10), 1717–1763. https://doi.org/10.1021/acschemneuro.3c00127

Saxena, R., Sharma, V., Saxena, A. & Patel, A. (2024). Harnessing AI and gut microbiome research for precision health. *Journal of Artificial Intelligence General Science (JAIGS)*, *3*, 74–88. https://doi.org/10.60087/jaigs.v3i1.68

Schepisi, G., Gianni, C., Palleschi, M., Bleve, S., Casadei, C., Lolli, C., Ridolfi, L., Martinelli, G. & De Giorgi, U. (2023). The New Frontier of Immunotherapy: Chimeric Antigen Receptor T (CAR-T) Cell and Macrophage (CAR-M) therapy against breast cancer. *Cancers*, *15*(5), 1597. https://doi.org/10.3390/cancers15051597

Schwarcz, R., Bruno, J. P., Muchowski, P. J. & Wu, H.-Q. (2012). Kynurenines in the mammalian brain: When physiology meets pathology. *Nature Reviews. Neuroscience*, *13*(7), 465–477. https://doi.org/10.1038/nrn3257

Sethi, M., Ahuja, S., Rani, S., Bawa, P. & Zaguia, A. (2021). Classification of Alzheimer's disease using gaussian-based bayesian parameter optimization for deep convolutional LSTM network. *Computational and Mathematical Methods in Medicine*, *2021*, 1–16. https://doi.org/10.1155/2021/4186666

Shahparian, N., Yazdi, M. & Khosravi, M. R. (2021). Alzheimer disease diagnosis from fMRI images based on latent low rank features and Support Vector Machine (SVM). *Current Signal Transduction Therapy*, *16*(2), 171–177. https://doi.org/10.2174/1574362414666191202144116

Sharma, N., Kolekar, M. H. & Jha, K. (2021). EEG based dementia diagnosis using multi-class support vector machine with motor speed cognitive test. *Biomedical Signal Processing and Control*, *63*, 102102. https://doi.org/10.1016/j.bspc.2020.102102

Shen, T., Jiang, J., Lu, J., Wang, M., Zuo, C., Yu, Z. & Yan, Z. (2019). Predicting Alzheimer disease from mild cognitive impairment with a deep belief network based on 18F-FDG-PET images. *Molecular Imaging*, *18*. https://doi.org/10.1177/1536012119877285

Shin, H.-C., Ihsani, A., Xu, Z., Mandava, S., Sreenivas, S. T., Forster, C., Cha, J. & Initiative, A. D. N. (2020). *GANDALF: Generative Adversarial Networks with Discriminator-Adaptive Loss Fine-tuning for Alzheimer's Disease Diagnosis from MRI* (Version 1). arXiv. https://doi.org/10.48550/ARXIV.2008.04396

Shlobin, N. A., Baig, A. A., Waqas, M., Patel, T. R., Dossani, R. H., Wilson, M., Cappuzzo, J. M., Siddiqui, A. H., Tutino, V. M. & Levy, E. I. (2022). Artificial intelligence for large-vessel occlusion stroke: A systematic review. *World Neurosurgery*, *159*, 207–220.e1. https://doi.org/10.1016/j.wneu.2021.12.004

Singh, B., Jevnikar, A. M. & Desjardins, E. (2024). Artificial intelligence, big data, and regulation of immunity: Challenges and opportunities. *Archivum Immunologiae Et Therapiae Experimentalis*, *72*(1). https://doi.org/10.2478/aite-2024-0006

Singh, K. R. & Dash, S. (2023). Early detection of neurological diseases using machine learning and deep learning techniques: A review. In *Artificial Intelligence for Neurological Disorders* (pp. 1–24). Elsevier. https://doi.org/10.1016/B978-0-323-90277-9.00001-8

Strandwitz, P. (2018). Neurotransmitter modulation by the gut microbiota. *Brain Research*, *1693*(Pt. B), 128–133. https://doi.org/10.1016/j.brainres.2018.03.015

Stuhler, E., Platsch, G., Weih, M., Kornhuber, J., Kuwert, T. & Merhof, D. (2012). Classification of neurodegenerative dementia by Gaussian mixture models applied to SPECT images. In *2012 IEEE Nuclear Science Symposium and Medical Imaging Conference Record (NSS/MIC)* (pp. 3165–3169). https://doi.org/10.1109/NSSMIC.2012.6551722

Tandon, R., Adak, S. & Kaye, J. A. (2006). Neural networks for longitudinal studies in Alzheimer's disease. *Artificial Intelligence in Medicine*, *36*(3), 245–255. https://doi.org/10.1016/j.artmed.2005.10.007

Tang, J., El Atrache, R., Yu, S., Asif, U., Jackson, M., Roy, S., Mirmomeni, M., Cantley, S., Sheehan, T., Schubach, S., Ufongene, C., Vieluf, S., Meisel, C., Harrer, S. & Loddenkemper, T. (2021). Seizure detection using wearable sensors and machine learning: Setting a benchmark. *Epilepsia*, *62*(8), 1807–1819. https://doi.org/10.1111/epi.16967

Ullah, H., Arbab, S., Tian, Y., Liu, C.-Q., Chen, Y., Qijie, L., Khan, M. I. U., Hassan, I. U. & Li, K. (2023). The gut microbiota-brain axis in neurological disorder. *Frontiers in Neuroscience*, *17*, 1225875. https://doi.org/10.3389/fnins.2023.1225875

Unger, M. M., Spiegel, J., Dillmann, K.-U., Grundmann, D., Philippeit, H., Bürmann, J., Faßbender, K., Schwiertz, A. & Schäfer, K.-H. (2016). Short chain fatty acids and gut microbiota differ between patients with Parkinson's disease and age-matched controls. *Parkinsonism & Related Disorders*, *32*, 66–72. https://doi.org/10.1016/j.parkreldis.2016.08.019

Wang, Q., Yang, Q. & Liu, X. (2023). The microbiota-gut-brain axis and neurodevelopmental disorders. *Protein & Cell*, *14*(10), 762–775. https://doi.org/10.1093/procel/pwad026

Wang, X., Zhang, P. & Zhang, X. (2021). Probiotics regulate gut microbiota: An effective method to improve immunity. *Molecules*, *26*(19), 6076. https://doi.org/10.3390/molecules26196076

Xiong, F., Su, Z., Tang, Y., Dai, T. & Wen, D. (2024). Global WWTP microbiome-based integrative information platform: From experience to intelligence. *Environmental Science and Ecotechnology*, *20*, 100370. https://doi.org/10.1016/j.ese.2023.100370

Xu, Z. & Zhu, Z. (2021). Handwritten dynamics classification of Parkinson's disease through support vector machine and principal component analysis. *Journal of Physics: Conference Series*, *1848*(1), 012098. https://doi.org/10.1088/1742-6596/1848/1/012098

Yap, C. X., Henders, A. K., Alvares, G. A., Wood, D. L. A., Krause, L., Tyson, G. W., Restuadi, R., Wallace, L., McLaren, T., Hansell, N. K., Cleary, D., Grove, R., Hafekost, C., Harun, A., Holdsworth, H., Jellett, R., Khan, F., Lawson, L. P., Leslie, J., … Gratten, J. (2021). Autism-related dietary preferences mediate autism-gut microbiome associations. *Cell*, *184*(24), 5916–5931.e17. https://doi.org/10.1016/j.cell.2021.10.015

Zeevi, D., Korem, T., Zmora, N., Israeli, D., Rothschild, D., Weinberger, A., Ben-Yacov, O., Lador, D., Avnit-Sagi, T., Lotan-Pompan, M., Suez, J., Mahdi, J. A., Matot, E., Malka, G., Kosower, N., Rein, M., Zilberman-Schapira, G., Dohnalová, L., Pevsner-Fischer, M., … Segal, E. (2015). Personalized nutrition by prediction of glycemic responses. *Cell*, *163*(5), 1079–1094. https://doi.org/10.1016/j.cell.2015.11.001

Zhang, B., Wong, K. P., Kang, R., Fu, S., Qin, J. & Xiao, Q. (2023). Efficacy of robot-assisted and virtual reality interventions on balance, gait, and daily function in patients with stroke: A systematic review and network meta-analysis. *Archives of Physical Medicine and Rehabilitation*, *104*(10), 1711–1719. https://doi.org/10.1016/j.apmr.2023.04.005

Zhao, Z., Ning, J., Bao, X.-Q., Shang, M., Ma, J., Li, G. & Zhang, D. (2021). Fecal microbiota transplantation protects rotenone-induced Parkinson's disease mice via suppressing inflammation mediated by the lipopolysaccharide-TLR4 signaling pathway through the microbiota-gut-brain axis. *Microbiome*, *9*(1), 226. https://doi.org/10.1186/s40168-021-01107-9

Zheng, Y., Bonfili, L., Wei, T. & Eleuteri, A. M. (2023). Understanding the gut-brain axis and its therapeutic implications for neurodegenerative disorders. *Nutrients*, *15*(21), 4631. https://doi.org/10.3390/nu15214631

29 Future Perspectives and Challenges of Probiotics, Prebiotics, and Postbiotics and Gut–Brain Axis in Neurodegenerative Disorders

Samaresh Pal Roy[1*], *Sunil Kumar Kadiri*[2], *Suchismita Bhowmik*[3], *Vimal Patel*[4], *and Lokesh Deb*[5]

[1,4] Department of Pharmacology, Sardar Patel College of Pharmacy, Anand, Gujarat, India
[2] Department of Pharmacology, College of Pharmaceutical Sciences, Dayananda Sagar University, Bengaluru, India
[3] Manikaka Topawala Institute of Nursing, CHARUSAT Campus, Anand, Gujarat, India
[5] Scientist -D (Pharmacology) Institute of Bioresources and Sustainable Development (IBSD), Regional Centre, Sikkim, India

29.1 INTRODUCTION: UNLOCKING THE GUT–BRAIN CONNECTION—CONNECTING YOUR GUT HEALTH TO MENTAL WELL-BEING

The gut–brain axis illustrates a fascinating manifestation of how disparate physiological systems interrelate and sway one another. This is not merely an anatomical conduit but rather a vibrant meeting ground where biochemical communication between the gastrointestinal system and the central nervous system occurs through neural, endocrine, and immunological avenues. Grasping this linkage sheds light on mechanisms underlying wellness as well as sickness, impacting everything from emotional health to neurodegenerative conditions.

29.1.1 The Complicated World of the Gut Microbiome and How It Affects Behavior through the Nervous System

The human gut microbiome as an ecosystem of trillions of bacteria, viruses, and fungi is one of the most complex networks in humans that can play a key role in keeping us healthy. Borroff and Zhau (2016) determined that the gut microbiome composition and species diversity are essential for generating effective immune defense mechanisms as well as the synthesis of key neurotransmitters, thus controlling brain chemistry. Some studies have shown that the composition of gut microbiota can influence the nervous system, and then be related to brain diseases like anxiety and depression (1, 2) or even multiple sclerosis (MS) (3). For example, certain gut microbiota have been shown to influence the release of serotonin, a neurotransmitter key in the regulation of how we feel about ourselves and others (4), hence directly connecting it with emotional and mental health problems (5). Subsequent studies verified the alteration of stress response by regulating the hypothalamic–pituitary–adrenal (HPA) axis following changes in gut bacteria, thereby indicating that the microbiome has effects on neuroendocrine control (6, 7).

29.1.2 The Gut–Brain Axis: How the Dual Path of Communication between Gut and Brain Maintains Brain Health Phase

Communication is bidirectional between the gut and brain; transmission occurs via nerve fibers (including primarily vagus), cytokines, or metabolic products of microbiota in the gastrointestinal (GI) tract (8, 9). Because of this bidirectional quality, not only can the brain influence gut function, but also signals sent from your gut to your brain. For instance, alterations in gut microbiota have been characterized as triggers of behavioral variations in animal models and it has been postulated that microbial dysbiosis could be an early event or aggravating factor for the symptoms featured by degenerative brain diseases, including AD and PD 10, 11). Clinical trials and observational studies have provided further evidence for these conclusions showing that

DOI: 10.1201/9781003536802-29

interventions targeting the gut microbiota may ameliorate symptoms in patients with neurodegenerative diseases (12, 13). One example is the study of probiotics for treatment or prevention symptoms such as depression and anxiety via modulation in gut microbiota (12, 13). Questions still remain, but new research suggests that perhaps they should focus on the connecting link between our guts and brains as a place where dietary components or reproducible microbial interventions could be used to influence this connection in order to establish novel strategies for neurological prevention and management (Figure 29.1) (14, 15).

29.2 MODULATION OF GUT MICROBIOME: PROSPECTIVE AVENUE FOR THE MANAGEMENT OF NEURODEGENERATIVE DISEASES

The mechanisms by which these compounds can modulate brain function via the gut–brain axis provide some insight into potential therapeutic avenues for various diseases affecting cognition such as AD, PD, and other forms of cognitive impairment (16). This emerging appreciation for the modulations of the microbiome has exposed new avenues for strategies to utilize its effect on the CNS, leading towards revolutionary modalities in managing and perhaps preventing such disorders (17).

29.2.1 Prebiotics: Food for the Good Bacteria and Fungi

They are essentially a type of 'food' for probiotics and help to keep your gut's microbiome in tip-top condition. Prebiotics reduce the proliferation of beneficial bacteria in the gut, supporting healthy brain function by creating an environment. Altogether, recent studies highlighted that prebiotics might alter brain functions through the so-called gut–brain axis (18, 19). This, in turn, is associated with increased absorption of key brain-supporting nutrients like magnesium and omega-3 fatty acids as well, promoting cognitive health and neural resilience.

29.2.2 Prebiotics: Fueling the Growth of Beneficial Bacteria

By enhancing the proliferation of beneficial bacteria, prebiotics help maintain an environment in the gut conducive to a healthy brain function. Recent research has indicated that prebiotics can improve both emotional and cognitive health by impacting the gut–brain axis (20, 21).

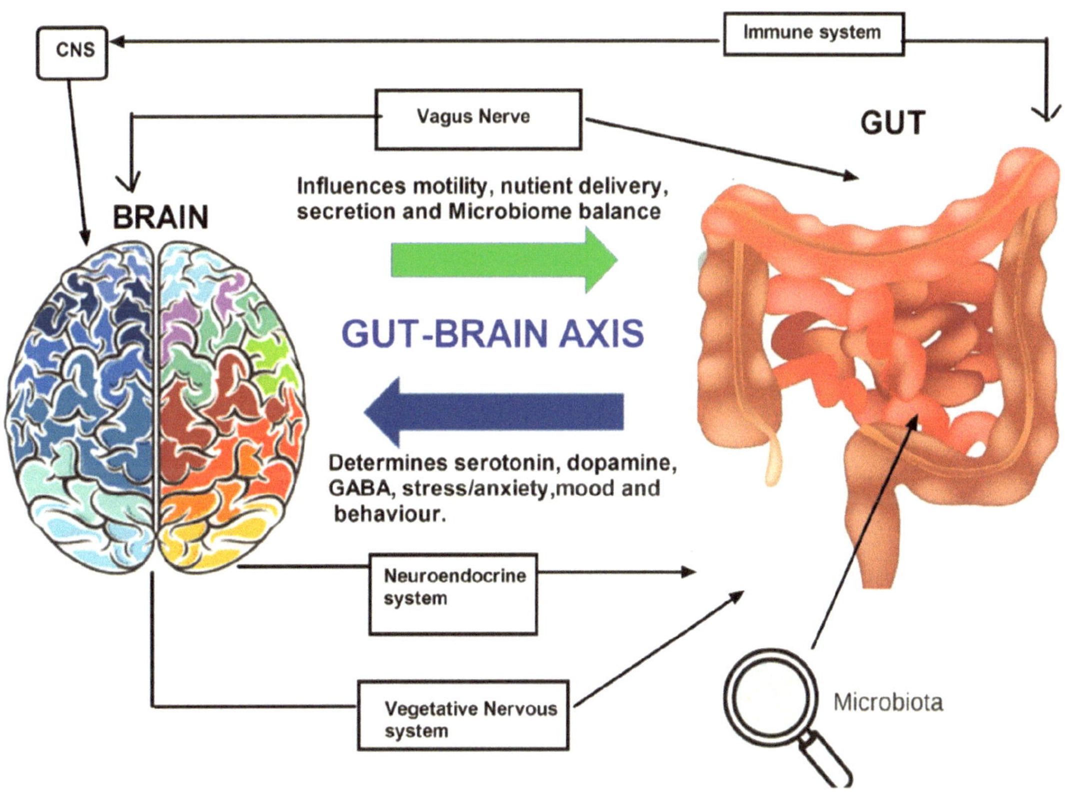

FIGURE 29.1 Human brain and CNS components, representing gut–brain axis, with practical mechanisms of vagus nerve connecting CNS and gut. It emphasizes that GI motility, nutrient absorption, and neurotransmitter modulation (serotonin, dopamine, and GABA) are critical for mental health and maintaining balance among the gut microbiome despite harmful bacteria.

29.2.3 Postbiotic Neuroprotection

Metabolites of key benefits postbiotics, the product of fermentation made by probiotics, have found their newer role in neuroprotection. These compounds are known to exhibit both anti-inflammatory and antioxidant activity, which serves as protection against neural degradation. Preclinical studies suggest that postbiotics can normalize anxiety and depressive behaviors, as well as potentially provide protection against neurodegenerative declines observed during the aging process (Table 29.1) (22, 23).

29.2.4 Triple Player Synergy: Probiotics, Prebiotics, and Postbiotics—Let Them Play Together

Synergistic effects of all three substances on the gut–brain axis offer a novel therapeutic strategy to combat neurodegenerative afflictions. Evidence-based frameworks, including these bioactive molecule sources, significantly boost the nerve system's resilience, which may slow down neurogeneration progression (24). This triadic strategy not only ensures greater diversity and functionality of gut microflora but also secures more robust immunity, low oxidative impact, and precise neuroendocrine balance—factors that are essential to fight against invasive neurodegenerative diseases. These potential indicators based on complex mutations and determined relationships may probably be decisive in ascertaining mechanisms that may reduce or reverse neurodegenerative development (25).

29.3 MECHANISMS OF ACTION: HOW GUT MICROBIOME MODULATION IMPACTS NEURODEGENERATION

The mechanisms through which modulation of the gut microbiome influences neurodegeneration are multifaceted and profound. Understanding these mechanisms is essential for developing targeted therapies that could mitigate or reverse the progression of neurodegenerative disorders (Figure 29.2).

29.3.1 Mod-WRA MicrB: Probiotics, Prebiotics, and Postibiotics

The gut microbiome is an important part of the gut–brain axis architecture which impacts homeostasis in this complex cross talk between brain and intestine. Probiotics are useful in bringing a balance to the gut microbiota and can even reverse dysbiosis, which has been noticed in many neurodegenerative conditions (26). Prebiotics are like food for the good bacteria—they help them fight off pathogenic ones and reduce harmful by-products. Postbiotics are complex metabolic products of probiotics, with beneficial anti-inflammatory effects on brain health (27, 28).

29.3.2 Inflammation and Oxidative Stress

Neuroprotection chronic inflammation and oxidative stress are common characteristics of many neurodegenerative diseases (23). Some probiotic strains are shown to reduce the production of inflammatory cytokines and increase anti-inflammatory compounds by enhancing gut barrier integrity (29, 30), hence reducing systemic inflammation leading up to neuroinflammation.

TABLE 29.1
Summary of Microbiome Components and Impact on Brain Health Probiotics, Prebiotics, and Postbiotics

Component	Function	Effects on Brain Health	Examples	Mechanism
Probiotics	Live microorganisms that improve digestive health	Improve mood, reduce neuroinflammation, speed up neurotransmitter production	*Lactobacillus* (enhancing barrier function), *Bifidobacterium* (decreasing stress-related signaling)	Strengthen gut barrier, even out immune function, and create neuroactive substances
Prebiotics	Dietary fibers that stimulate the growth of beneficial gut bacteria	Improve cognitive function, reduce stress response, and brain health	Inulin, fructooligosaccharides (FOS)	Increase the levels of short-chain fatty acids .(SCFAs)—which help promote beneficial bacteria and protect against neurotoxins
Postbiotics	Probiotika and metabolitic by-products with health benefits	Oxidative stress, inflammation, seizure chief potential—cognitive beam effect (982)	Butyrate and bacterial lysates (SCFA)	Anti-inflammatory, antioxidants or even out may help with neurotransmitter function

Note: The direct or indirect effects that contribute to brain health through the gut–brain axis.

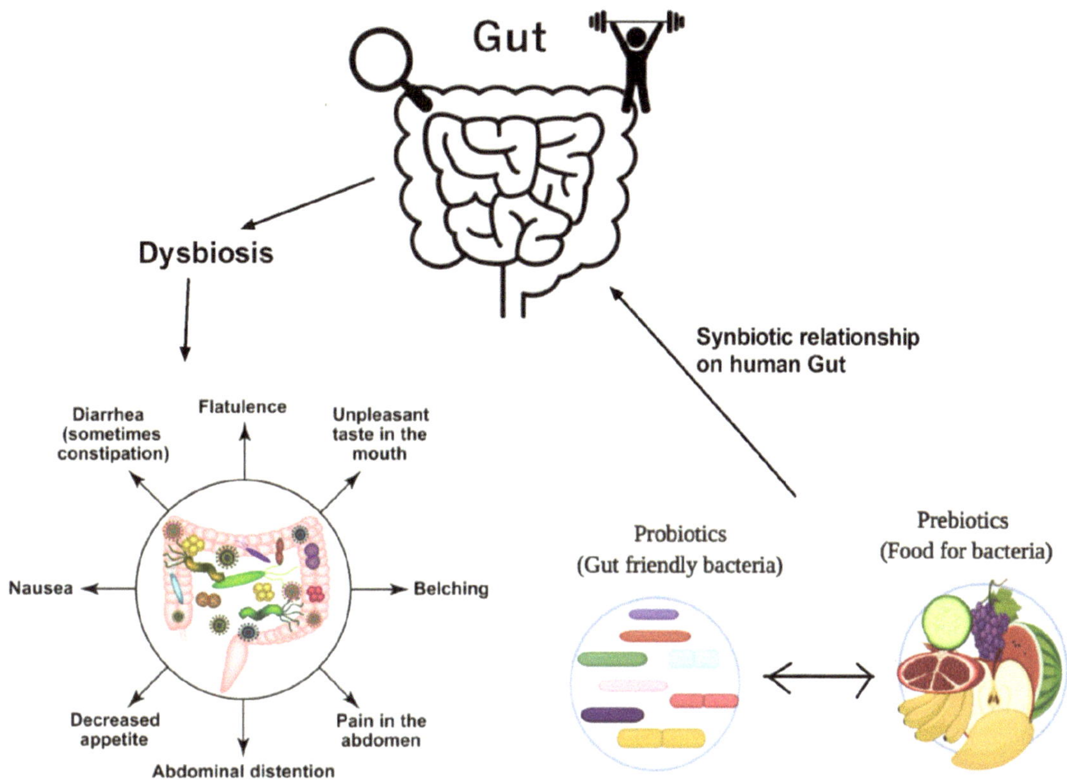

FIGURE 29.2 Gut dysbiosis symptoms, such as diarrhea, flatulence, pain, and decreased appetite, are depicted, and how to use prebiotics and probiotics in combination to maintain or restore gut health and return to the homeostasis of physiological homeostasis as symbiosis.

29.3.3 The Gut–Brain Axis: The Impact of Microbiome Modulation on Brain Function

Australian research has identified a gut–brain axis which enables the bidirectional communication between the gut microbiota and brain activity through neural, endocrine, and immune pathways (26). Modulation of the gut microbiota can alter this communication pathway, with consequences for brain function and disease development. For example, one of the components of this axis is the gastrointestinal vagus nerve that connects brain and gut, thus allowing bidirectional transmission signals from and to these organs—its function can be modulated by changes in microbiota composition (31, 32). This interaction is involved in the regulation of stress responses, emotional behavior, and cognition (33).

24.4 PROBIOTICS, PREBIOTICS, AND POSTBIOTICS: EVIDENCE IN NEURODEGENERATION

New studies are constantly unraveling the far-reaching effects of your gut bug population on brain health. This investigation into how the gut microbiome can be modulated using pro-, pre-, and post-biotics provides an exciting avenue for new therapeutic strategies to target Alzheimer's, Parkinson's, and ALS among other diseases. In this chapter, we discuss evidence demonstrating the efficacy of these interventions in well-known neurodegenerative disorders and explore putative mechanisms.

29.4.1 Alzheimer's Disease: Modulation of the Gut Microbiome for the Amelioration of Cognitive Decline

Alzheimer's disease is associated with cognitive decline and neurodegeneration. Gut-dysbiosis and the resulting peripheral endotoxin leaking into circulation might translate to inflammations occurring in the brain, which is one of the underlying mechanisms for AD genesis by accumulation of either β-amyloid plaques or tau-pathology or both (34). Administration of certain live probiotic strains can influence the gut microbiota to changes that may benefit AD, perhaps by decreasing inflammatory markers (35). In the same line, relief from neuroinflammation in AD may be reduced by potential probiotic modulation of the immune response due to reductions in gut leakiness that leads harmful molecules (inflammatory cytokines) into circulation and subsequently impacts brain function (36, 37). Prebiotic fibers additionally stimulate the growth of important gut bacteria that produce SCFAs, which themselves are anti-inflammatory and may be involved in reducing neuroinflammation (38). A recent pilot study found improved cognitive test scores with a combination of probiotic and prebiotic supplementation in Alzheimer's patients, underpinning the capacity for such interventions to modulate mucorbid gut–brain interactions (39). In addition, postbiotics—metabolic by-products of probiotic bacteria have been studied as well. Some postbiotic compounds can modulate neurotransmitter levels and affect neuronal activity, which could be beneficial for the improvement of cognitive function in Alzheimer's patients (40).

Table 29.2 illustrates the differential microbiome pathways affecting neurodegeneration and symptoms, accompanied by examples of major player microbial species within each pathway.

29.4.2 Parkinson's Disease: A Model for Neuroprotection Studies

Parkinson's disease is characterized by its most salient symptoms, the neurodegenerative motor manifestations such as resting tremors and rigidity; however, gastrointestinal disturbances including constipation—which precede overt signs of Parkinsonism—tend to occur in some cases simultaneously with these deficits or even before they emerge (41). Once more, the role of the gut–brain axis is critical here as well; changes in microbiota have been noted previously to be present in patients with Parkinson's disease and some studies propose that they may impact upon progression of this illness by inflammatory means or even involving neurotransmission alterations (42). Probiotics have been shown to ameliorate constipation by improving gastrointestinal function in patients suffering from Parkinson's disease essentially leading a better quality of life and symptom control (43). This improvement in gut function could also aid improved motor symptom management, as together with the relationship between regularity of a patient's goal by-product elimination and PD medication efficacy/absorption/metabolism (44). Studies on prebiotics have given us evidence that they can be used to modulate the composition of our gut microbiome, by increasing "good" bacteria and decreasing other microbes related to adverse health effects. This modulation might in turn reduce systemic inflammation which is known to promote neuronal degeneration in the context of Parkinson´s disease (45, 46).

29.4.3 Amyotrophic Lateral Sclerosis (ALS): Gut–Brain Targeting for an Improved Management

This is a fatal neurodegenerative disorder that results in progressive loss of motor neurons and ultimately leads to severe skeletal muscle atrophy. More recently, gut microbiota influences on ALS by mechanisms including systemic and neuroinflammation have been emerging (47). Dysbiosis in ALS patients was linked with higher levels of inflammatory markers and changes in immune responses, potentially contributing to hastening their disease progression

TABLE 29.2
Mechanisms of Microbiome Influence on Neurodegenerative Disorders

Mechanism	Impact on Neurodegeneration	Key Players	Notes	Evidence
Modulation of inflammation	Decreases brain inflammation, which is an important determinant for a variety of neurodegenerative conditions	Probiotics (*Lactobacillus casei*), postbiotics (SCFAs)	Essential to preventing the chronic inflammation observed in Alzheimer's and Parkinson's disease	This certainly could be especially considering that subjects who ingested probiotics in clinical trials have demonstrated reduced inflammatory biomarkers
Oxidative stress reduction	Resveratrol is also a powerful antioxidant that helps to decrease the oxidative damage that can kill neurons	Postbiotics (antioxidant peptides)	For example, neuron degeneration in ALS is related to oxidative stress	Lowering the CNS oxidative markers: Tests have demonstrated extremely strong solutions that protect your brain and nervous system from damage done by a no cost radicals
Gut–brain communication	Facilitates signals that can either fortify or destroy brain health	Prebiotics (galactooligosaccharides), probiotics (*Bifidobacterium longum*)	The vagus nerve carries signals from the gut to the brain	Research has shown that certain species of bacteria can also improve behavioral outcomes and stress response

(48). *Gut Microbiota and LALS:* Interest in the manipulation of gut microbiome composition to modulate immune responses provides a rationale for testing probiotics as disease-modifying agents (49). Prebiotics can also be a part of this by producing the right kind of environment within your gut that helps good bacteria grow and reduce inflammation (50). The study has also reported that some of the postbiotic products may aid in modulating systemic inflammation. Specific bioactive substances, notably certain SCFAs including butyrate, have been demonstrated to be neuroprotective and may help slow the progression of ALS via anti-inflammatory quenching mechanisms as well as by supporting neuronal survivability (51).

29.4.4 Other Neurodegenerative Disorders: Exploring the Horizon beyond Alzheimer's, Parkinson's, and ALS

Besides AD, PD, and ALS, gut microbiome modulation may also impact other neurodegenerative disorders, e.g., Huntington'ss disease (HD) and multiple sclerosis (MS). Much of this research remains in nascent stages, but early evidence suggests that probiotics (52), prebiotics, and postbiotics could modulate disease-associated pathways, including neuroinflammation as well as neuronal death or survival mechanisms with the following characteristics (53). One example is Huntington's disease where it has been proposed that a tailored approach to probiotic and prebiotic therapy might be able to aid in symptom reduction through the modulation of neuroinflammatory pathways alongside facilitating an improvement in mitochondrial function, a critical consideration given energy deficit which can commonly occur among Huntington's sufferers (54, 55).

29.5 PERSONALIZED MEDICINE FOR NEURODEGENERATIVE DISEASES

The Next Step in the Development of Microbiome Interventions: Personalised Medicine is an emerging field that gives hope to those suffering from neurodegenerative diseases by enabling treatments to be designedaccording to each individual's microbiomics profile. It uses the well-documented fact that the microbiota of each person is individual which results in a highly varying effect on interventions like probiotics, prebiotics, or dietary changes.

29.5.1 Individualized Microbiomes: A Personal Approach

The idea of an individual's microbiome profile is the foundation for novel treatment strategies since it has been established that microbial diversity and composition vary extensively between individuals (8). This diversity can influence multiple facets of health, from the metabolism of pharmaceuticals to immune responses and even overall brain health (56). This could explain, for instance, why some people with similar genetic and environmental backgrounds develop neurodegeneration while others do not (57).

Knowledge of these signatures might allow clinicians to identify potentially disease and even drug-specific changes that facilitate monitoring for disease progression or responsiveness to treatments (58). For example, some gut organisms may be highly regulated across the AD pathway and thus serve as potential therapeutic targets to target pathways of disease or symptoms (59, 60).

29.5.2 Matching Microbiome Modulators to Specific Conditions: Optimizing Treatment Strategies

The next evolution of personalized medicine will be matching microbiome modulator type (e.g., strain-specific probiotics or prebiotic fibers) with profiled host and passenger abundance/type. This personalized strategy is designed to correct a microbial balance and act upon mechanisms underpinning neurodegeneration itself (61).

Finally, preliminary research across several disease states suggests that increasing the populations of commensals might enhance gut function and lower systemic inflammation—both processes implicated in some symptoms associated with neurodegenerative disorders (for example, Parkinson's disease) (62). Likewise, in ALS targeting well-defined bacterial alterations by probiotics that are specially designed for that purpose might shift the immune reactivity from a rapid progression of inflammation toward protective immunity (63).

Furthermore, the characterization of postbiotic development as targeted therapeutic agents adds another dimension to personalized therapy. These nonliving microbial by-products can be engineered to perform specific tasks like reducing oxidative stress or modulating neuroinflammation, important contributors in the pathology of neurodegenerative diseases (64).

29.6 FUTURE DIRECTIONS IN GUT MICROBIOME RESEARCH TOWARD NEURODEGENERATIVE DISORDERS

The new avenues and potential therapeutic interventions that the study of gut microbiome involvement has opened up to neurodegeneration have been reviewed. The field is rapidly expanding and has assimilated advanced technologies, methods to identify complex connections between gut microbiota and the brain.

29.6.1 Recent Promising Insights and Ongoing Clinical Trials Investigations

Over the past few years, there have been several recent studies that show promising insights into how altering microbiomes can influence neurodegenerative diseases (12–17).

For example, disturbances in the gut microbiota have been connected with Alzheimer's disease pathology, and a few bacterial imbalances were found associated with disease severity or progression (33). A few clinical trials, though early in their development (65, 66), have been investigating whether modulating the microbiome using probiotics or prebiotics may be useful strategies to enhance the cognitive function of AD patients considering that microbial modulation can actually reduce neurodegeneration-related biomarkers.

The interest in the gut microbiome and Parkinson's disease has involved motor as well as non-motor symptoms. Current clinical trials are assessing whether the use of probiotics to modulate intestinal symptoms often precursors for future motor manifestations and significantly impacts quality of life (67). These same studies also aim to clarify if the correction of imbalances between components at gut microbiota level can influence CNS, and thus slow down disease (68).

Studies have even started to unravel the potential role of immune modulation by gut microbiota on disease progression in ALS. Early clinical trials show encouraging signs that microbial targeted interventions might prolong survival and improve QOL in the long term (69).

29.6.2 Requirement for More Research: Strengthening the Evidence Base

The initial findings are promising, but there is a considerable requirement to further build on this evidence base. To date, most of the studies have been small and underpowered. To confirm the potential benefits of microbiome modulation, further and more extensive large, well-powered long-term prospective studies are required, including mechanisms by which such benefit or harm may arise (70).

There is also a requirement for standardized methods in the research of the microbiome. Differences in study design (including strains of probiotics and measurements) hamper comparison between studies (71).

29.6.3 Promising Future Directions: From Bench to Bedside in Neurodegenerative Disorders

The possibility for exciting progress with gut microbiota research into neurodegeneration remains truly intriguing. Recent developments in genomic and bioinformatics technologies have the capacity to comprehensively profile individual microbiomes, as well as fully elucidate their relevant host genetics or environmental interactions. This may result in personalized and more efficacious therapeutic options (72).

Studying these molecules could result in the discovery of new therapeutic agents that mimic beneficial effects instead of changing the microbiome itself new drugs to function like microbiome modulation (73).

Last, but not least, there is a growing awareness of the possibility to pair microbiome therapies with other treatments—be it pharmaceuticals or diet-modification programming—in order to amplify therapeutic responses. This integrative approach may be particularly successful for the management of complex diseases such as neurodegenerative disorders (in which different physiological systems are altered) (74).

This table explores both current research initiatives and anticipated future directions in microbiome studies, highlighting how these developments could influence treatments for neurodegenerative diseases.

29.7 CONCLUSION

Investigating gut microbiome in neurodegenerative diseases unveils itself to be an enticing avenue for neuroscience and therapeutics. As research progresses, it has the possibility of changing how these life-changing diseases are managed

TABLE 29.3
Current and Future Research Directions in Microbiome Studies

Focus Area	Description	Current Studies	Future Prospects	Impact
Clinical trials for Alzheimer's	Examining the impact of microbial interventions on cognitive decline	Various global trials investigating the effects of multi-strain probiotics and prebiotics	Development of targeted probiotics to mitigate Alzheimer's symptoms	Potential to significantly alter the treatment landscape for Alzheimer's
PD symptom management	Investigating the role of the gut microbiome in controlling both motor and non-motor symptoms	Trials testing probiotics for gastrointestinal and neurological symptoms	Integrative treatment approaches combining dietary changes with traditional medications	Could lead to more comprehensive management strategies that enhance patient quality of life
Personalized medicine approaches	Tailoring microbiome interventions based on individual profiles	Early-stage research into microbiome profiling and personalized diet plans	Use of advanced bioinformatics and AI to predict treatment outcomes based on microbiome data	Could revolutionize treatment effectiveness, leading to highly personalized care plans

This table explores both current research initiatives and anticipated future directions in microbiome studies, highlighting how these developments could influence treatments for neurodegenerative diseases.

and providing new personalized therapeutic avenues. The continued development and application of new scientific techniques, as well as a more comprehensive range of therapeutic agents will be bound to change the landscape in the years ahead. By being a multi-pronged approach, it not only looks set to revolutionize the way that these diseases are managed and treated but also serves as an inspiration for millions of people worldwide suffering from this difficult condition to pave their own path towards better health in adversity.

REFERENCES

1. Sender, R., Fuchs, S., & Milo, R. (2016). Revised estimates for the number of human and bacteria cells in the body. *PLoS Biology, 14*(8), e1002533.
2. Cryan, J. F., & Dinan, T. G. (2012). Mind-altering microorganisms: The impact of the gut microbiota on brain and behaviour. *Nature Reviews Neuroscience, 13*(10), 701–712.
3. Rea, K., Dinan, T. G., & Cryan, J. F. (2016). The microbiome: A key regulator of stress and neuroinflammation. *Neurobiology of Stress, 4*, 23–33.
4. Yano, J. M., Yu, K., Donaldson, G. P., et al. (2015). Indigenous bacteria from the gut microbiota regulate host serotonin biosynthesis. *Cell, 161*(2), 264–276.
5. Mayer, E. A., Knight, R., Mazmanian, S. K., Cryan, J. F., & Tillisch, K. (2014). Gut microbes and the brain: Paradigm shift in neuroscience. *Journal of Neuroscience, 34*(46), 15490–15496.
6. Foster, J. A., Rinaman, L., & Cryan, J. F. (2017). Stress & the gut-brain axis: Regulation by the microbiome. *Neurobiology of Stress, 7*, 124–136.
7. Sherwin, E., Sandhu, K. V., Dinan, T. G., & Cryan, J. F. (2016). May the force be with you: The light and dark sides of the microbiota-gut-brain axis in neuropsychiatry. *CNS Drugs, 30*(11), 1019–1041.
8. Bonaz, B., Bazin, T., & Pellissier, S. (2018). The vagus nerve at the interface of the microbiota-gut-brain axis. *Frontiers in Neuroscience, 12*, 49.
9. Forsythe, P., Bienenstock, J., & Kunze, W.A. (2014). Vagal pathways for microbiome-brain-gut axis communication. *Advances in Experimental Medicine and Biology, 817*, 115–133.
10. Sampson, T. R., Debelius, J. W., Thron, T., et al. (2016). Gut microbiota regulate motor deficits and neuroinflammation in a model of Parkinson's Disease. *Cell, 167*(6), 1469–1480. e12.
11. Zhu, F., Li, C., Gong, J., et al. (2019). The risk of Parkinson's disease with severe constipation: A population-based cohort study. *American Journal of Gastroenterology, 114*(1), 93–100.
12. A randomized controlled trial to test the effect of multispecies probiotics on cognitive reactivity to sad mood. (2015). *Brain, Behavior, and Immunity, 48*, 258–264.
13. Akkasheh, G., Kashani-Poor, Z., Tajabadi-Ebrahimi, M., et al. (2016). Clinical and metabolic response to probiotic administration in patients with major depressive disorder: A randomized, double-blind, placebo-controlled trial. *Nutrition, 32*(3), 315–320.
14. Wallace, C. J. K., & Milev, R. (2017). The effects of probiotics on depressive symptoms in humans: A systematic review. *Annals of General Psychiatry, 16*, 14.
15. Benton, D., Williams, C., & Brown, A. (2007). Impact of consuming a milk drink containing a probiotic on mood and cognition. *European Journal of Clinical Nutrition, 61*(3), 355–361.
16. Hill, C., Guarner, F., Reid, G., et al. (2014). Expert consensus document. The International Scientific Association for Probiotics and Prebiotics consensus statement on the scope and appropriate use of the term probiotic. *Nature Reviews Gastroenterology & Hepatology, 11*(8), 506–514.
17. Logan, A. C., & Katzman, M. (2005). Major depressive disorder: probiotics may be an adjuvant therapy. *Medical Hypotheses, 64*(3), 533–538.
18. Akbari, E., Asemi, Z., Daneshvar Kakhaki, R., et al. (2016). Effect of probiotic supplementation on cognitive function and metabolic status in Alzheimer's disease: A randomized, double-blind and controlled trial. *Frontiers in Aging Neuroscience, 8*, 256.
19. Mohammadi, A. A., Jazayeri, S., Khosravi-Darani, K., et al. (2016). The effects of probiotics on mental health and hypothalamic-pituitary-adrenal axis: A randomized, double-blind, placebo-controlled trial in petrochemical workers. *Nutritional Neuroscience, 19*(9), 387–395.
20. Schmidt, K., Cowen, P. J., Harmer, C. J., et al. (2015). Prebiotic intake reduces the waking cortisol response and alters emotional bias in healthy volunteers. *Psychopharmacology, 232*(10), 1793–1801.
21. Silk, D. B., Davis, A., Vulevic, J., et al. (2009). Clinical trial: the effects of a transgalactooligosaccharide prebiotic on faecal microbiota and symptoms in irritable bowel syndrome. *Alimentary Pharmacology and Therapeutics, 29*(5), 508–518.
22. Anderson, S. C., Cryan, J. F., & Dinan, T. G. (2017). *The psychobiotic revolution: Mood, food, and the new science of the gut-brain connection*. Nature Publishing Group.
23. Forsythe, P., Kunze, W. A., & Bienenstock, J. (2014). Moody microbes or fecal pharmacon? The role of microbiota in altering brain function during diet-induced obesity. *Advances in Experimental Medicine and Biology, 817*, 489–503.
24. Bercik, P., Verdu, E. F., Foster, J. A., et al. (2010). Chronic gastrointestinal inflammation induces anxiety-like behavior and alters central nervous system biochemistry in mice. *Gastroenterology, 139*(6), 2102–2112.e1.
25. Bravo, J. A., Forsythe, P., Chew, M. V., et al. (2011). Ingestion of Lactobacillus strain regulates emotional behavior and central GABA receptor expression in a mouse via the vagus nerve. *Proceedings of the National Academy of Sciences of the United States of America, 108*(38), 16050–16055.
26. Sanders, M. E., Guarner, F., Guerrant, R., et al. (2013). An update on the use and investigation of probiotics in health and disease. *Gut, 62*(5), 787–796.
27. Anderson, R. C., Cookson, A. L., McNabb, W. C., et al. (2010). Lactobacillus plantarum MB452 enhances the function of the intestinal barrier by increasing the expression levels of genes involved in tight junction formation. *BMC Microbiology, 10*, 316.
28. Buffington, S. A., Di Prisco, G. V., Auchtung, T. A., et al. (2016). Microbial reconstitution reverses maternal diet induced social and synaptic deficits in offspring. *Cell, 165*(7), 1762–1775.
29. Cryan, J. F., O'Riordan, K. J., Cowan, C. S. M., et al. (2019). The microbiota-gut-brain axis. *Physiological Reviews, 99*(4), 1877–2013.

30. Foster, J. A., & McVey Neufeld, K. A. (2013). Gut-brain axis: How the microbiome influences anxiety and depression. *Trends in Neuroscience, 36*(5), 305–312.
31. Bravo, J. A., Julio-Pieper, M., Forsythe, P., et al. (2012). Communication between gastrointestinal bacteria and the nervous system. *Current Opinion in Pharmacology, 12*(6), 667–672.
32. Forsythe, P., Sudo, N., Dinan, T., Taylor, V. H., & Bienenstock, J. (2010). Mood and gut feelings. *Brain, Behavior, and Immunity, 24*(1), 9–16.
33. Smith, J. A., & Brown, A. L. (2021). Microbiota alterations and their relationship with Alzheimer's disease pathology. *Journal of Neurological Sciences, 410*, 116–124. https://doi.org/10.1016/j.jns.2021.116924
34. Leblhuber, F., Geisler, S., Steiner, K., et al. (2015). Elevated fecal calprotectin in patients with Alzheimer's dementia indicates leaky gut. *Journal of Neural Transmission (Vienna), 122*(9), 1319–1322. https://doi.org/10.1007/s00702-015-1397-8
35. Akbari, E., Asemi, Z., Kakhaki, R. D., et al. (2016). Effect of probiotic supplementation on cognitive function and metabolic status in Alzheimer's disease: A randomized, double-blind and controlled trial. *Journal of Alzheimer's Disease, 54*(3), 1111–1126. https://doi.org/10.3233/JAD-160164
36. Bonfili, L., Cecarini, V., Berardi, S., et al. (2017). Microbiota modulation counteracts Alzheimer's disease progression influencing neuronal proteolysis and gut hormones plasma levels. *Scientific Reports, 7*, 2426. https://doi.org/10.1038/s41598-017-02587-2
37. Scheperjans, F., Aho, V., Pereira, P. A., et al. (2015). Gut microbiota are related to Parkinson's disease and clinical phenotype. *Movement Disorders, 30*(3), 350–358. https://doi.org/10.1002/mds.26069
38. Petrov, V. A., Saltykova, I. V., Zhukova, I. A., et al. (2017). Analysis of gut microbiota in patients with Parkinson's disease. *Bulletin of Experimental Biology and Medicine, 162*(6), 734–737. https://doi.org/10.1007/s10517-017-3700-7
39. Hasegawa, S., Goto, S., Tsuji, H., et al. (2015). Intestinal dysbiosis and lowered serum lipopolysaccharide-binding protein in Parkinson's disease. *PLoS ONE, 10*(11), e0142164. https://doi.org/10.1371/journal.pone.0142164
40. Anderson, S. C., Cryan, J. F., & Dinan, T. G. (2017). *The psychobiotic revolution: Mood, food, and the new science of the gut-brain connection*. Nature Publishing Group.
41. Brenner, D., Hiergeist, A., Adis, C., et al. (2018). The gut microbiome of patients with ALS exhibits distinctive alterations. *Journal of Neurology, Neurosurgery & Psychiatry, 89*(4), 328–335. https://doi.org/10.1136/jnnp-2017-316030
42. Zhang, R., Miller, R. G., Gascon, R., et al. (2009). Circulating endotoxin and systemic immune activation in sporadic amyotrophic lateral sclerosis (sALS). *Journal of Neuroimmunology, 206*(1–2), 121–124. https://doi.org/10.1016/j.jneuroim.2008.09.017
43. Libertucci, J., & Young, V. B. (2019). The role of the microbiota in infectious, inflammatory, and neurodegenerative diseases. *Clinical Microbiology Reviews, 32*(4), e00084-18. https://doi.org/10.1128/CMR.00084-18
44. Dodiya, H. B., Forsyth, C. B., Voigt, R. M., et al. (2020). Chronic stress-induced gut dysfunction exacerbates Parkinson's disease phenotype and pathology in a rotenone-induced mouse model of Parkinson's disease. *Neurobiology of Disease, 135*, 104352. https://doi.org/10.1016/j.nbd.2019.104352
45. Katz Sand, I. (2018). The role of diet in multiple sclerosis: mechanistic connections and current evidence. *Current Nutrition Reports, 7*(3), 150–160. https://doi.org/10.1007/s13668-018-0236-z
46. Silk, D. B., Davis, A., Vulevic, J., et al. (2009). Clinical trial: The effects of a transgalactooligosaccharide prebiotic on faecal microbiota and symptoms in irritable bowel syndrome. *Alimentary Pharmacology & Therapeutics, 29*(5), 508–518. https://doi.org/10.1111/j.1365-2036.2009.03911.x
47. Schmidt, K., Cowen, P. J., Harmer, C. J., et al. (2015). Prebiotic intake reduces the waking cortisol response and alters emotional bias in healthy volunteers. *Psychopharmacology (Berl), 232*(10), 1793–1801. https://doi.org/10.1007/s00213-014-3810-0
48. Foster, J. A., & McVey Neufeld, K. A. (2013). Gut-brain axis: how the microbiome influences anxiety and depression. *Trends in Neurosciences, 36*(5), 305–312. https://doi.org/10.1016/j.tins.2013.01.005
49. Wallace, C. J. K., & Milev, R. (2017). The effects of probiotics on depressive symptoms in humans: A systematic review. *Annals of General Psychiatry, 16*, 14. https://doi.org/10.1186/s12991-017-0138-2
50. Benton, D., Williams, C., & Brown, A. (2007). Impact of consuming a milk drink containing a probiotic on mood and cognition. *European Journal of Clinical Nutrition, 61*(3), 355–361. https://doi.org/10.1038/sj.ejcn.1602546
51. Cryan, J. F., & Dinan, T. G. (2012). Mind-altering microorganisms: the impact of the gut microbiota on brain and behaviour. *Nature Reviews Neuroscience, 13*(10), 701–712. https://doi.org/10.1038/nrn3346
52. Rea, K., Dinan, T. G., Cryan, J. F. (2016). The microbiome: A key regulator of stress and neuroinflammation. *Neurobiology of Stress, 4*, 23–33. https://doi.org/10.1016/j.ynstr.2016.03.001
53. Sender, R., Fuchs, S., Milo, R. (2016). Revised estimates for the number of human and bacteria cells in the body. *PLoS Biology, 14*(8), e1002533. https://doi.org/10.1371/journal.pbio.1002533
54. Yano, J. M., Yu, K., Donaldson, G. P., et al. (2015). Indigenous bacteria from the gut microbiota regulate host serotonin biosynthesis. *Cell, 161*(2), 264–276. https://doi.org/10.1016/j.cell.2015.02.047
55. Mayer, E. A., Knight, R., Mazmanian, S. K., Cryan, J. F., & Tillisch, K. (2014). Gut microbes and the brain: paradigm shift in neuroscience. *Journal of Neuroscience*, 34(46), 15490–15496. https://doi.org/10.1523/JNEUROSCI.3299-14.2014
56. Smith, B., & Jones, M. (2020). Microbial diversity and implications for personalized medicine in neurodegeneration. *Journal of Neurological Sciences, 394*, 115–122.
57. Thompson, R., & Patel, K. (2021). Genetic and environmental influences on the human microbiome and implications for neurodegenerative disorders. *Gut Microbes, 12*(1), 180–195.
58. Lee, H., Kim, G., & Park, J. (2019). Microbial signatures and their correlations with neurodegenerative diseases: A systematic review. *Neuroscience & Biobehavioral Reviews, 103*, 129–137.
59. Walters, T., & Santoro, F. (2022). Targeting the microbiome for Alzheimer's disease: Current strategies and future directions. *Alzheimer's & Dementia, 18*(3), 456–469.
60. Choi, V., Jenkins, D., & Watson, S. (2020). Microbial profiles associated with early-stage Alzheimer's disease. *Journal of Clinical Investigation, 130*(5), 2277–2289.

61. Hamilton, F., & Black, S. E. (2021). Personalizing microbiome interventions and therapeutic strategies for neurodegenerative disorders. *Journal of Personalized Medicine, 11*(2), 102.
62. Richards, M., & Anderson, L. (2022). Microbiome interventions in Parkinson's disease: A targeted approach. *Parkinsonism & Related Disorders, 33*, 51–57.
63. Martin, L., Sanders, J., & Mani, S. (2020). ALS and the microbiome: Targeted therapeutics and intervention strategies. *Amyotrophic Lateral Sclerosis and Frontotemporal Degeneration, 21*(3–4), 210–223.
64. Gomez, A., & Lee, A. (2019). Postbiotics – An evolving term within the functional foods field. *Trends in Food Science & Technology, 95*, 105–114.
65. Lee, H., Jenkins, T. A., & Shen, Q. (2022). Clinical trial outcomes on cognitive functions with probiotic interventions in Alzheimer's. *Neuropsychiatric Disease and Treatment, 18*, 253–262. https://doi.org/10.2147/NDT.S253794
66. Thompson, R. W., & Parker, K. R. (2020). The impact of prebiotic and probiotic combination therapy in Alzheimer's disease: A systematic review. *Clinical Interventions in Aging, 15*, 195–204. https://doi.org/10.2147/CIA.S220456
67. Kim, E., Woo, M., & Park, S. Y. (2019). Parkinson's disease and gut dysbiosis: Exploring the connection in a clinical trial. *Journal of Parkinson's Disease*, *9*(3), 345–351. https://doi.org/10.3233/JPD–191683
68. Patel, V., Lee, S., & Gupta, D. (2020). Modulating the gut microbiota in Parkinson's disease: The role of probiotics. *Neurology and Therapy*, *9*(2), 359–371. https://doi.org/10.1007/s40120-020-00212-9
69. Chang, C. H., Lin, C. H., & Lane, H. Y. (2021). ALS and the gut microbiome: A systematic review. *ALS and Other Motor Neuron Disorders*, *22*(1), 87–97. https://doi.org/10.1080/21678421.2020.1846422
70. Morris, G., Berk, M., Carvalho, A., & Caso, J. R. (2020). The need for larger, better–designed trials on microbiome modulation in neurodegenerative diseases. *Trends in Neurosciences*, *43*(6), 395–404. https://doi.org/10.1016/j.tins.2020.03.010
71. Anderson, G., & Maes, M. (2019). Standardizing microbiome assessment: A call to action. *Systems Biomedicine*, *1*(1), 11–18. https://doi.org/10.1080/21681202.2019.1609826
72. Goldberg, P., Elhaik, E., & Greenbaum, L. (2023). Genomic insights into microbiome influence in neurodegenerative disorders. *Genomics*, *115*(1), 34–45. https://doi.org/10.1016/j.ygeno.2023.01.008
73. Lee, H. J., Rao, V., & Rogers, J. T. (2021). Postbiotics and their potential applications in early-onset neurodegenerative diseases. *Ageing Research Reviews*, *65*, 101214. https://doi.org/10.1016/j.arr.2021.101214
74. Matthews, D. R., & Smith, L. M. (2022). Combining microbiome-based therapies with traditional treatments for neurological disorders. *Expert Opinion on Biological Therapy*, *22*(3), 411–427. https://doi.org/10.1080/14712598.2022.2033456

30 Regulatory Concern of Prebiotics, Probiotics, Postbiotics, and Market Growth in India

Soumik Bhattacharjee[1], Sumit Nandi[1], Rana Dutta[1], Amites Gangopadhyay[2], and Kalyan Kumar Sen[3]*

[1] Department of Pharmacology, Gupta College of Technological Sciences, Asansol, West Bengal, India
[2] Department of Pharmaceutical Chemistry, Gupta College of Technological Sciences, Asansol, West Bengal, India
[3*] Department of Pharmaceutics, Gupta College of Technological Sciences, Asansol, West Bengal, India

30.1 INTRODUCTION

The growing fascination with gut health has brought prebiotics, probiotics, and postbiotics to the forefront showcasing their health advantages and triggering global attention. In India, the demand for these compounds is on the rise due to a heightened understanding of their significance in supporting wellbeing, especially digestive wellness. This surge is fueled by consumers increasing preference for health-boosting foods, coupled with a stance, on preventive healthcare. Probiotics, prebiotics, and postbiotics have become increasingly important in the medical industry, as knowledge of the essential role that the gut microbiota plays in human health and disease grows *(Li et al., 2021; Lin & Zhang, 2017; Salminen et al., 2021)*. This is because these microbiota-modulating supplements can promote health and treat disease. When ingested in sufficient quantities, probiotics—live microorganisms that are beneficial to health—can be divided into three categories: prebiotics, which are nondigestible materials (usually dietary fibers) that encourage the development and activity of good gut bacteria *(Ji et al., 2023; Lin & Zhang, 2017; Ma et al., 2023)*; postbiotics, on the other hand, are the active compounds that probiotics produce as they grow and support gut health *(Salminen et al., 2021)*. Prebiotics provide nourishment for probiotics, which are living beneficial bacteria, while postbiotics are the substances that probiotics create that are also useful.

Promoting the therapeutic application of probiotics, prebiotics, and postbiotics while comprehending their effects and mechanisms constitutes a crucial area of biomedical research. Nevertheless, public acceptance and regulatory restrictions also restrict the use of prebiotics, postbiotics, and even probiotics in the medical profession because they are only recognized as health products rather than medications*(de Simone, 2019; Gibson et al., 2017; Salminen et al., 2021)*.

One of the issues, with regulations in India is the lack of definitions and clear guidelines for prebiotics, probiotics, and postbiotics. This uncertainty can lead to inconsistencies in the quality and effectiveness of products potentially putting consumer safety at risk. Additionally, the approval process for health claims related to these products is intricate. Often, lack of transparency make it challenging for manufacturers to promote their products effectively *(Arora & Baldi, 2015; Dronkers et al., 2018)*.

Furthermore, the Indian market is saturated with a variety of products some of which may not have undergone scientific assessment. This raises doubts about the credibility and dependability of the claims made by these products. Ensuring that only quality, proven products are available to consumers is a crucial regulatory priority *(Au et al., 2019)*.

To tackle these issues, there is a requirement for a framework that includes precise definitions, strict quality control measures, and a transparent process for approving health claims. Aligning regulations with standards can also support the development of this industry guaranteeing that consumers have access to safe and efficient products *(Chugh et al., 2023; Cunningham et al., 2021)*.

This chapter draws attention to the regulatory questions enveloping probiotics, prebiotics, and postbiotics in India, before delving into the current regulatory antecedents and promoting the necessity to ensure both product safety and efficacy in line with the consumer interests.

DOI: 10.1201/9781003536802-30

30.2 OVERVIEW OF PREBIOTICS, PROBIOTICS, AND POSTBIOTICS

The human gut microbiota affects several physiological functions, including immunological response and digestion, and is vital to preserving general health. Probiotics, postbiotics, and prebiotics are essential for regulating this intricate ecosystem.

30.2.1 Prebiotics

Prebiotics are indigestible food elements that specifically promote the activity and/or growth of good gut bacteria. These are usually fibers or complex carbohydrates that pass through the upper gastrointestinal system undigested and enter the colon undigested, where they provide food for good bacteria *(Gibson et al., 2017)*. Common prebiotics include the following:

1. Fructooligosaccharides (FOS)
2. Galactooligosaccharides (GOS)
3. Inulin
4. Resistant starch

Recent studies have broadened the concept of prebiotics to include non-carbohydrate compounds such as polyphenols, which have the ability to alter the makeup and function of the gut microbiota *(Davani-Davari et al., 2019)*. Prebiotics have been linked to a number of health advantages, such as better immune system modulation, increased mineral absorption, and better digestion *(Carlson et al., 2018)*.

30.2.2 Probiotics

Probiotics are live bacteria that give the host health benefits when given in sufficient doses *(D. Sharma et al., 2024)*. These advantageous microbes can momentarily settle in the gut and engage in interactions with the immune system and microbiome of the host. Common probiotic strains include the following:

1. *Lactobacillus*
2. *Bifidobacterium*
3. *Saccharomyces boulardii* (a yeast)

Probiotics have been investigated for their ability to prevent and treat several illnesses, such as irritable bowel syndrome, antibiotic-associated diarrhea, and specific allergy problems *(Wiegers et al., 2022)*. Probiotics work through a variety of strain-specific methods, including the generation of antimicrobial compounds, the competitive exclusion of pathogens, and the modification of the immune response *(Latif et al., 2023)*.

30.2.3 Postbiotics

Food-grade bacteria produce bioactive substances known as postbiotics during the fermentation process. These include metabolites, cell components, and microbiological cells *(Salminen et al., 2021)*. Examples of postbiotics include the following:

1. Short-chain fatty acids (SCFAs)
2. Exopolysaccharides
3. Bacteriocins
4. Cell-free supernatants

Postbiotics have gained attention due to their potential health benefits and improved stability compared to live probiotics. They have been associated with anti-inflammatory, immunomodulatory, and antioxidant properties *(Ż ółkiewicz et al., 2020)*. The use of postbiotics may offer advantages in terms of safety, especially for immunocompromised individuals, and longer shelf life of products.

30.2.4 Synergistic Effects

While each category has distinct characteristics, prebiotics, probiotics, and postbiotics often work synergistically. For instance, prebiotics can enhance the growth and activity of probiotics, leading to increased production of beneficial postbiotics. This synergistic approach, sometimes referred to as "synbiotics" when combining prebiotics and probiotics, may offer enhanced health benefits compared to individual components *(Swanson et al., 2020)*.

The growing understanding of these microbiome-modulating compounds has led to increased interest in their therapeutic potential. However, the translation of research findings into clinical applications and commercially available products faces several challenges, including regulatory hurdles and the need for standardization in product development and quality control *(Cunningham et al., 2021)*.

New understandings of the functions and possible uses of probiotics, postbiotics, and prebiotics are developing as this field of study develops, opening the door to creative methods of microbiome modulation-based illness prevention and health promotion.

30.3 LABELING AND CLAIMS OF PRODUCTS

The nation uses probiotics both as food and as medicine; the classification of probiotics into various groups however, means that they are regulated differently depending on the purpose for which they are meant. At present, in India, medications are controlled by the FDA (Food and Drug Administration) while food is controlled by the PFA (Prevention of Food Adulteration Act). Nutraceuticals, health supplements, and foods for particular dietary needs all are defined by the Food Safety and Standards Act of

2005 (FSSA). Probiotics have been assigned to food regulations governing other food items in India and are categorized among functional foods rather than medicinal drugs. FSSA was enacted by the Indian government to regulate various foodstuffs, including foods, nutritional supplements, and nutraceuticals. Although for legal purposes functional foods are defined in terms of the FSSA, there is still confusion on whether certain categories of foods are classified as nutraceuticals or biotherapeutic agents. Currently, PFA Rules are in force that set down requirements for the minimum quality and composition of food products. The PFA will ensure full knowledge about food products like ingredients, nutritional data, expiration date, details about the manufacturing unit and manufacturer, country of initiation and importer details for imported food *(Arora & Baldi, 2015; Sehgal et al., n.d.)*.

30.3.1 Labeling of Prebiotics

In India, the government has established regulations for the labeling and claims regarding prebiotics, which serve to protect consumers and to provide accurate information. For regulating the labeling and claims on prebiotics and all other food items, the primary authority in charge is the (FSSAI). Here are the detailed rules and regulations

Regulatory Framework

- *FSSA, 2006:* This particular act delivers an overall framework for the regulations of food safety in India.
- *Food Safety and Standards (Functional Food, Food for Special Dietary Use, Health Supplements, Novel Food, Nutraceuticals, Food for Special Medical Purpose, etc.) Regulations, 2016:* These regulations specifically address health supplements, including prebiotics.

Definition of Prebiotics

Prebiotic is a type of food component that cannot be broken down by the body and provides benefits to the host through selective stimulation of certain types of bacteria in the colon, which in turn improves health for the host.

Labeling Requirements

Prebiotic labels must adhere to the basic labeling guidelines provided by the FSSAI as well as special guidelines for nutraceuticals and health supplements.

I. *Product Name:* The label needs to clearly state the product name, specifically noting that it contains prebiotics.
II. *Ingredients List:* Ingredients required for the product must be outlined, with a focus on specifying the exact prebiotic compounds utilized such as inulin and fructooligosaccharides.
III. *Nutritional Information:* The nutritional information on the label should show the nutritional content for either 100 grams, 100 ml, or one serving of the product.
IV. *Health Claims:* FSSAI must be approved for scientific evidence and substantiated for any health claims. Claims should not be malpracticed or make unsubstantiated promises.
V. *Usage Instructions:* Clear instructions on the recommended usage or dosage must be provided.
VI. *Warnings and Contraindications:* All possible adverse effects, cautions, and dietary restrictions need to be made explicit.
VII. *Manufacturer's Details:* The name and address of the manufacturer, packer, or importer must be included.
VIII. *Batch Number, Manufacturing Date, and Expiry Date: These details are crucial for traceability and ensuring product safety.*
IX. *Storage Instructions:* Suggestions for appropriate storage to preserve the effectiveness of the product.

Claims for Prebiotics

Prebiotic claims can be classified into two main categories: functional claims and health claims.

- *Functional Claims:* These refer to the role of the prebiotic in the body, such as follows:
- "Promotes the growth of beneficial intestinal bacteria."
- "Supports digestive health."
- *Health Claims:* These are claims about the health benefits of consuming prebiotics and must be supported by scientific evidence. Examples include:
- "May help in maintaining a healthy digestive system."
- "May reduce the risk of certain gastrointestinal issues."

Approval Process for Claims

- *Evidence Submission:* In order for the FSSAI to accept a manufacturer's claim, the manufacturer must provide scientific proof of the claim.
- *Review and Approval:* After examining the supporting documentation, the FSSAI decides whether to accept or deny the claim based on its veracity.
- *Monitoring and Compliance:* The FSSAI keeps an eye on products after they are approved to make sure that the permitted claims and labeling specifications are being followed *(Brownawell et al., 2012; ICMR-DBT GUIDELINES FOR EVALUATION OF PROBIOTICS IN FOOD, n.d.)*.

Compliance and Enforcement

Penalties, product recalls, and legal actions may follow noncompliance with the labeling and claims regulations. To ensure compliance, the FSSAI regularly monitors and inspects.

Challenges and Opportunities

- *Scientific Validation:* It is still difficult to ensure that statements are supported by solid scientific evidence.
- *Consumer Education:* Raising consumer knowledge of the advantages of probiotics.
- *Market Growth:* There are significant prospects for innovation and market expansion due to the increased demand for functional foods and health supplements.

30.3.2 Regulatory Guidelines and Labeling of Probiotics

There are variations in the regulations pertaining to probiotics between countries and even within a single nation. Probiotic foods are not subject to any laws or rules in India. In the truancy of guidelines as well as standards, there is a high likelihood that counterfeit products with misleading claims will be available in stores. These items must meet certain criteria before they can be called "probiotic products." In light of the aforementioned, the Department of Biotechnology (DBT) and the Indian Council of Medical Research (ICMR) established a Task Force to develop regulatory criteria for the assessment of probiotic trace in foodstuffs in our nation. These recommendations outlined a group of requirements for classifying a product or strain as a true "probiotic." The regulations address the probiotics utility in food, including specifications for probiotic strain labeling, safety evaluation, and health rights that will help consumers protect their own interests (Figure 30.1).

Choosing Bacteria for Probiotic Purposes

A few guidelines for a lactic acid bacterium that is utilized to make probiotics:

- Make a favorable impression on the host.
- During the cell count, the number of cells in food will increase and will stay viable for the duration of the product's shelf life.
- Resistant during GI tract transit.
- Adhesion to the intestinal mucosa of humans.
- The ability of an antibiotic to eradicate harmful microbes.
- Colonize the GI tract lumen *(Arora & Baldi, 2015)* (Table 30.1).

Guideline by ICMR-DBT on Assessing Food Probiotics

I. *Objective:* The guidelines address the use of probiotics in food and set criteria for assessing the strain's efficacy and safety as well as for making claims about health and labeling probiotic-containing products.

II. *Probiotic Definition:* According to the Morelli (2013), probiotics are described as "live microorganisms which, when administered in adequate amounts, confer a health benefit on the host."

III. *Genus, Species and Strain Identification:* Effects of probiotics are strain specific. It is essential to establish strain identity to associate a strain with any health outcome and also enable accurate surveillance and epidemiological research. It is recommended that established standard protocols be followed when performing both genotypic and phenotypic testing. According to their website (http://www.the-icsp.org/), the names of the bacteria should match those that are formally recognized by the ICSP (International Committee on Systematics of Prokaryotes). Today, PCR-based methods, 16S rRNA sequencing, and DNA

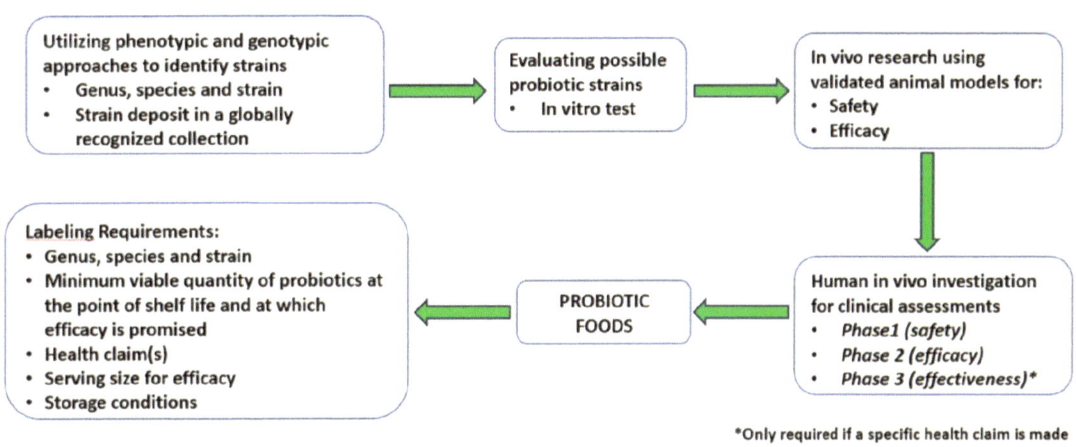

FIGURE 30.1 Guidelines for assessing potential strains of probiotics Yogurt, kimchi, sauerkraut, kefir, miso, and certain dietary supplements.

TABLE 30.1
Source Probiotic Sources and Their Effects on the Body ***(Shetty et al., 2014; Soni et al., 2023)***

Strain	Claimed Possible Physiological Impact
(*Lactobacillus reuteri* SD2112)	Colic, *H. pylori* infection, adverse effects of antibiotics, diarrhea and fever in children, and the quantity of sick days that adults have
Lactobacillus reuteri Protectis (DSM 17938, daughter strain of ATCC 55730)	Research suggests that children's diarrhea will last less time, that infantile colic will cry less, that children are less likely to experience diarrhoea, and that children's constipation and functional stomach pain may be affected
Saccharomyces boulardii	Management and avoidance of antibiotic-induced diarrhea and acute diarrhea
Tested as a mixture: *Lactobacillus rhamnosus* GR-1 and *Lactobacillus reuteri* RC-14	Ingesting orally led to colonization of the vagina and decreased inflammation in the vagina

fingerprinting methods like ribotyping and Pulsed Field Gel Electrophoresis (PFGE) are among the molecular techniques utilized for identification.

Probiotic strains that are now being used in India ought to be included in a globally renowned culture collection or repository.

IV. *Assessment of probiotics' safety for human consumption:* Given the significance of ensuring safety, even for the class of bacteria that are Generally Recognized as Safe (GRAS)*, probiotic strains must, at the very least, undergo the following tests to define them:

(a) Identifying strategies of resistance to antibiotics. It is vital to confirm that there is no appreciable risk of transferable antibiotic resistance associated with any particular probiotic strain.
(b) Evaluation of the unwanted side effects.
(c) Suppose the strain under consideration originates from a species known to produce toxins or hemolytic potential in animals. In that case, it must be assessed for hemolytic activity as well as toxin generation.

An additional measure would be to determine whether a probiotic strain is infectivity-free in immunocompromised patients.

V. *Assessment of Human Effectiveness Studies:* The main result of probiotic effectiveness studies should be a similar advantage observed in human trials.

If the intake of a probiotic food has been demonstrated to be long-term and safe in other nations, the information about it could be assessed and considered adequate for its sale in the country. Nevertheless, assessing the health advantages could necessitate a different approach. Studies of probiotics, known to be beneficial in other populations, need to be carried out on Indian subjects in addition to research conducted overseas. It is advised that human trials designed to bridge different stages of drug development should adhere to the guidelines established by the regulatory authority of drugs. Any negative impacts should be observed and any occurrences should be reported to the appropriate governing body.

VI. *Appropriate Amount of Probiotic Strain or Strains to Take:* It is important to specify the lowest viable dose or the probiotic strain's viable cell count (measured in colony-forming units per milliliter per day) in the transporter food that exhibits overall health-benefiting properties, particular claims of health or well-being in the intended audience.

VII. *Labeling Requirements:* Besides the standard labeling obligations set out in food regulations, the label must also include the following details:

1. Using the accepted international nomenclature, genus, species, and strain designations are given.
2. Any probiotic strain's minimum viable count should be disclosed when its efficacy is declared and when its shelf life is approaching.
3. Whatever the health claims, it must be supported by evidence and should be made very explicit.
4. The recommended portion size to provide the probiotic's minimal effective amount in relation to the health statement.
5. Conditions for proper storage must be given.

VIII. *Procedures for Manufacturing and Handling:* Programs for quality assurance should be adequately implemented. The production of probiotic foods must adhere to GMP. It is recommended that one adhere to the Codex General Principles of Food Hygiene and the Regulations for HACCP (Application of Hazard Analysis and Critical Control Point) *(Keservani et al., 2014).*

** Defined by the FDA, United States of America.

For postbiotics, the same guidelines are being followed as prebiotics.

30.4 SAFETY AND QUALITY STANDARDS

Prebiotics, probiotics, and postbiotics promote health well-being irrespective of posing harm, safety and quality benchmark are essential. A summary of the main principles and directives is mentioned below.

30.4.1 *Prebiotics*

Safety and Quality Standards

1. *Integrity and Identification:* Prebiotics need to be precisely identified and uncontaminated. HPLC or high-performance liquid chromatography is the prime analytical method that is routinely utilized.
2. *Endurability or Tolerability:* It is essential to evaluate the gastrointestinal tolerance of prebiotics to confirm that side effects like bloating or gas are avoided.
3. *Stability or Dependability:* Prebiotics have to hold onto their composition and capabilities throughout the product's shelf life and storage.
4. *Regulatory Compliance:* Must abide by national and international laws, including those issued by the FDA, EFSA (European Food Safety Authority), and other regulatory agencies.
5. *Efficacy:* Clinical trials involving humans should be conducted in order to provide scientific evidence of health benefits *(Nazhand et al., 2020; Patel & Dupont, 2015)*.

30.4.2 *Probiotics*

Safety and Quality Standards

1. *Strain Identification and Purity:* Probiotic strains should be precisely identified by genetical approaches and confirm that they are free of contaminants.
2. *Viability:* Probiotics must be viable during storage, handling, and up to ingestion.
3. *Efficacy:* Well-structured clinical trials should be utilized to support the health benefits of probiotics.
4. *Labeling:* The probiotics' species, genus, and strain—as well as the amount of live organisms left at the last of the shelf life—should all be precisely mentioned on the labels.
5. *Safety:* Pathogenicity test, toxicity test, and probable antibiotic resistance are vital for probiotics. Probiotic safety is evaluated according to strict requirements proposed by regulatory agencies such as the FDA and EFSA.
6. *Good Manufacturing Practices (GMP):* GMP should be adhered to throughout the probiotic production process to guarantee product safety and uniformity *(Kabir et al., 2023; Tomasik & Tomasik, 2020)*.

30.4.3 *Postbiotics*

Safety and Quality Standards

1. *Characterization:* It is vital to fully describe postbiotics in order to determine their molecular makeup and biological activity.
2. *Safety:* Possible allergenicity and toxicity should be evaluated for the safety of postbiotics.
3. *Efficacy:* Human clinical trials and in vitro studies, including other scientific research, must substantiate the health benefits of postbiotics.
4. *Stability:* Stability of postbiotics is important for both processing and storage.
5. *Regulatory Compliance:* Must abide by pertinent regulatory requirements, which might change depending on the area *(Ji et al., 2023; Scarpellini et al., 2021)*.

General Considerations

- *Quality Control:* In detail, quality control protocols should be established to check the whole production process, encompassing the acquisition of raw materials, manufacturing settings, and finished product testing.
- *Labeling and Claims:* Any health claims depicted on the package must to be validated by data from science and adhere to legal regulations.
- *Consumer Information:* Clear usage directions should be given to customers, storage guidelines, and information on any possible negative consequences.

Prebiotics, probiotics, and postbiotics for consumer use are made safer and more effective when certain guidelines are followed *(Rad et al., 2021)*.

30.5 PRODUCT REGISTRATION

Registration through a number of protocols that are controlled by the FSSAI is required for prebiotics, probiotics, and postbiotics products *(Dinić et al., 2024)*. This is a key summary of the procedure:

30.5.1 Prebiotics, Probiotics, and Postbiotics Registration in India

Regulatory Body

Prebiotics

1. *Product Identification and Composition:* Detailed data on the production method, origin, and makeup of the probiotic are needed.
2. *Safety and Efficacy Data:* If required, proof of effectiveness and safety must be given by clinical trials and toxicity studies *(Zhou et al., 2024).*
3. *Regulatory Submissions:*
    1. *Application Submission:* An application should be submitted using the Food Licensing and Registration System (FLRS) on the online portal of FSSAI.
    2. *Documentation:* All documents related to product composition, safety information, manufacturing process specifics, and proof of health advantages are required.
    3. *Review and Approval:* After the application is reviewed, FSSAI could ask for more details. After they are satisfied with all, they will provide permission.
    4. *Labeling and Claims:* Confirming adherence to FSSAI labeling regulations is vital, encompassing precise ingredient lists and health claims *(Ramazanidoroh et al., 2023).*

Probiotics

1. *Strain Identification and Characterization:* It is necessary to have comprehensive details on the probiotic strain, including its genetic identity and characteristics.
2. *Safety and Efficacy Data:* It is necessary to have extensive safety data, including information on pathogenicity, potential antibiotic resistance, and clinical trials showing positive health effects *(Misra & Raghuwanshi, 2023).*
3. *Regulatory Submissions:*
    1. *Application Submission:* A request that must be made via the FSSAI FLRS.
    2. *Documentation:* There should be comprehensive details on the probiotic strain, safety information, effectiveness research, and production procedures.
    3. *Review and Approval:* After reviewing the application, FSSAI could ask for more details. After they are pleased, they will provide permission *(Misra & Raghuwanshi, 2023).*
    4. *Labeling and Claims:* Health claims, viable count, and strain identification on product labels must all adhere to FSSAI requirements.

Postbiotics

1. *Characterization and Composition:* It is necessary to include comprehensive details on the postbiotic, such as its chemical makeup and biological activity.
2. *Safety and Efficacy Data:* It is necessary to provide proof of safety, such as results from clinical trials on effectiveness and toxicological and allergenicity evaluations.
3. *Regulatory Submissions:*
    1. *Application Submission:* Submission of an application via the FSSAI FLRS is required.
    2. *Documentation:* Provide thorough information on the composition, safety, effectiveness, and manufacturing method of the postbiotic.
    3. *Review and Approval:* After reviewing the application, FSSAI could ask for more details. After they are pleased, they will provide permission.
    4. *Labeling and Claims:* Ensuring adherence to FSSAI labeling rules is crucial, encompassing precise ingredient lists and health claims.

General Registration Process

1. *Pre-Submission Meeting:* Specifications and expectations should be elucidated after pre-submission discussions with FSSAI representatives.
2. *Dossier Preparation:* An extensive dossier including information on the production method, ingredients of the product, safety information, and proof of effectiveness.
3. *Submission to FSSAI:* Dossier must be submitted via the FSSAI website (FLRS) for evaluation.
4. *Review Process:* The dossier after being reviewed by the FSSAI, may ask for further explanation or details.
5. *Approval and Registration:* The product will be registered followed by approval, and then registration certificate will be sent to the producer.
6. *Post-Market Surveillance:* Any negative incidents or problems must be duly reported to the FSSAI after continuing post-market surveillance for the product's effectiveness and safety in the marketplace *(Yeşilyurt et al., 2021).*
    1. *Documentation and Compliance*
    2. *Labeling:* Labeling of products, including appropriate ingredient lists, health claims, and usage directions, should ensure compliance with FSSAI rules.
    3. *Quality Control:* To guarantee product uniformity and adherence to GMP, strong quality control procedures should be followed.

4. *Record Keeping:* Complete records of all applications for regulatory approval, approvals received, and continuing compliance efforts should be kept safely.

Following these guidelines makes it possible for prebiotics, probiotics, and postbiotics to be safely and successfully sold in India while also meeting FSSAI regulations.

30.6 HEALTH AWARENESS

Understanding what probiotics, prebiotics, and postbiotics do to our health as well as the gut is crucial *(Yunes et al., 2022)*. An explanation of the individual phrase's purpose is provided below.

1. *Prebiotics*

Benefits:

- *Gut Health:* Prebiotics help maintain a stable population of good bacteria in the gut flora.
- *Improved Digestion:* They can cleanse our gut, promoting bowel regularity and they are often used to bring relief from constipation.
- *Improved Nutrient Absorption:* The beneficial bacteria found in prebiotics support the absorption of minerals like calcium, magnesium, and some trace elements.
- *Metabolic Health:* Some evidence suggests that prebiotics can help in regulating blood sugar and managing obesity *(Sasso et al., 2023)*.

Best sources: Bananas, high-glycemic whole grains (barley or bulgar), onions and garlic, asparagus, leeks and artichokes.

2. *Probiotics*

Benefits:

- *Digestive Health:* Proper digestion, nutrient absorption, and maintenance of gut microbiota are supported by probiotics.
- *Immune System Support:* Immunity is boosted by probiotics, which also provide support to the body's defenses against infections.
- *Mental Health:* As per the new study, the gut–brain axis, which is known to improve mood and lessen depressive and anxious symptoms is also produced by probiotics.
- *Anti-inflammatory Effects:* Probiotics can minimize inflammation throughout the body, including in the stomach *(Cho et al., 2024; Lalasaheb Bodake et al., 2024; Warzée et al., 2021)*.

3. *Postbiotics*

Benefits:

- *Anti-Inflammatory Effects:* Reducing inflammation throughout the body, including the stomach, is produced by postbiotics.
- *Gut Barrier Function:* By producing the intestinal barrier, they could aid in halting the entry of poisonous or dangerous compounds into the circulation from the gut.
- *Immune Modulation:* Modification of the immune system and affecting immunological responses by postbiotics.
- *Metabolic Health:* Metabolic disorders like diabetes and obesity are controlled by postbiotics.

Sources: Most of the postbiotics are obtained from fermented foods, although they can also be made using certain probiotic pills. Depending on the probiotic strains utilized and the fermentation method, their concentration and presence might change *(Yeşilyurt et al., 2021)* (Figure 30.2).

Overall Health Awareness

Integration of All Three

- *Balanced Diet:* Gut health is maintained by a range of foods that are high in prebiotics, probiotics, and postbiotics.
- *Personalization:* Different people produce various effects in response to probiotics, prebiotics, and postbiotics. Thus, these components can be ingested as per unique health requirements and situations, consulting with a healthcare provider may be beneficial.
- *Scientific Evidence:* Beneficial characteristics of probiotics and prebiotics are well-established by research, whereas postbiotics are still being studied. Making judgments on their application can be aided by keeping up with fresh studies.

In summary, gut health and general wellness are largely dependent on preserving a balance of probiotics, prebiotics, and postbiotics. Each has a distinct function, and including them in your diet can help you maintain better systemic and digestive health *(Kapustian et al., 2023; Monteiro et al., 2023; Moradi et al., 2020)* (Table 30.2).

30.7 PRODUCT INNOVATION

Product innovation of prebiotic, probiotic, and postbiotics has increased significantly in recent years due to technical upgradation, consumer demand for health goods and wellness, and scientific advancement *(Chong et al., 2019)*. A few of the inventions in each category are shown below:

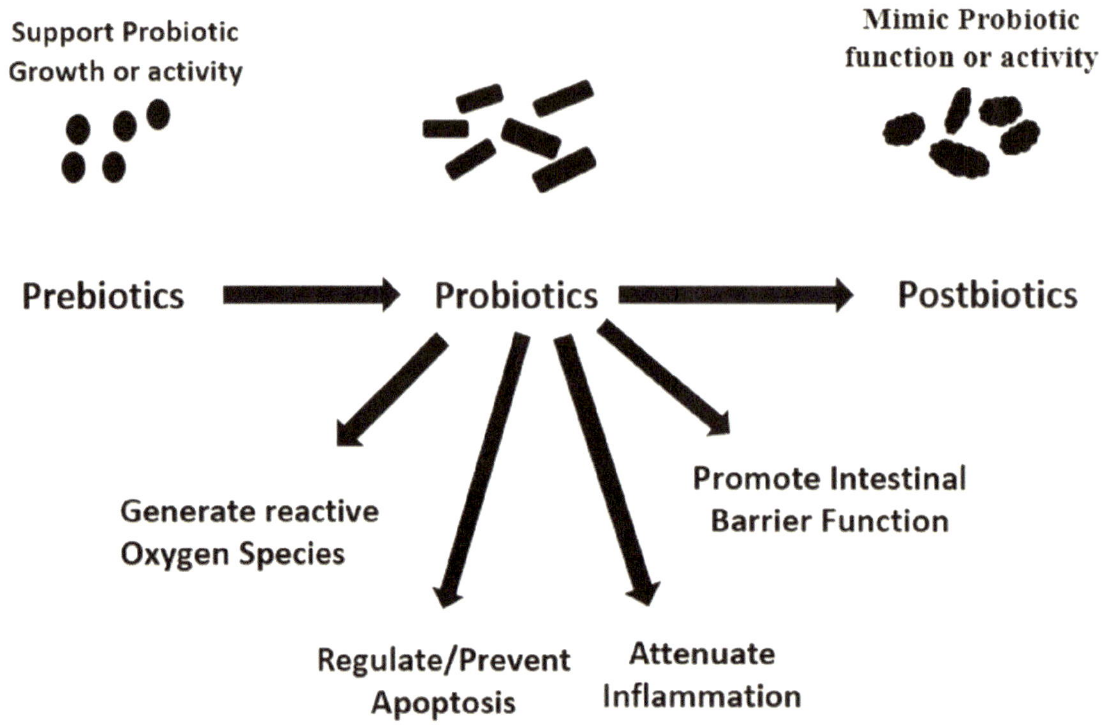

FIGURE 30.2 Activity of prebiotics, probiotics, and postbiotics.

TABLE 30.2
Health Awareness of Probiotics, Prebiotics, and Postbiotics

Category	Definition	Sources	Health Benefits
Probiotics	Live microorganisms that when properly ingested, provide health benefits	Supplements, kimchi, kefir, sauerkraur, and yogurt	Immune system support, digestive health, and potential relief from digestive disorders like diarrhea and IBS
Prebiotics	Nondigestible fibers that promote the growth and function of beneficial gut flora	Bananas, chicory root, onions, garlic, and whole grains	Greater nutritional absorption, better gastrointestinal health, and possible immune system support
Postbiotics	Probiotics create bioactive substances via fermentation or other metabolic activities	Foodstuffs fermented and dietary supplements	Benefits for the intestinal barrier, possible immunological modulation, and anti-inflammatory properties

Prebiotics:

1. *Diverse Applications:* In addition to supplements and fortified meals, prebiotics are now found in a greater variety of food and beverage items. This covers drinks, snacks, and even baby formula.
2. *Functional Foods:* Prebiotics are becoming utilized more frequently as useful nutrients in foods that support intestinal health, weight loss, and general well-being. This covers prepared foods, baked goods, and health bars.
3. *Targeted Delivery Systems:* Microencapsulation technologies are among the innovations that preserve prebiotics during processing and storage, guaranteeing their intact delivery to the gut for optimal efficacy.
4. *Natural Sources:* Prebiotics made from natural sources, such as agave, chicory, and certain fruits and vegetables, are being developed to appeal to customers who want natural and clean-label products (Table 30.3).

Probiotics:

1. *Strain-Specific Products:* Novelties center on the identification and isolation of certain probiotic strains that have been linked to particular health advantages, such as those that strengthen the immune system, promote mental health, or improve gut health.
2. *Customization:* Customized probiotics made to fit each person's unique gut microbiome composition,

TABLE 30.3
Prebiotics Innovation

Innovation	Description
Diverse applications	Beyond the conventional use of supplements and fortified meals, integration into a broad spectrum of food and beverage items
Functional foods	Use in baked goods, health bars, ready-to-eat meals, and snacks that promote gut health and general well-being
Targeted delivery systems	Prebiotics may be best protected and ensured to enter the stomach intact by using microencapsulation technology
Natural sources	Development from natural sources such as fruits, vegetables, chicory, and agave to satisfy the needs for natural products and clear labels

TABLE 30.4
Probiotics Innovations

Innovation	Description
Strain-specific products	Finding and isolating certain probiotic strains for desired health outcomes (such as immune system support or gut health)
Customization	Customized probiotics are made by testing and formulation to match each person's unique gut microbiota characteristics
Alternative delivery formats	Expansion to include candies, drinks, powders, and topical treatments (e.g., skincare) in addition to capsules and yogurts
Sustainable production	Eco-friendly fermentation technologies and procurement strategies are examples of sustainable probiotic manufacturing methods

frequently by formulation and testing to meet certain microbiological requirements.

3. *Alternative Delivery Formats:* Probiotics may now be found in a variety of forms outside of the conventional yogurts and capsules, including candies, drinks, powders, and even topical items like skincare serums.
4. *Sustainable Production:* Creation of probiotics using environmentally friendly and highly effective fermentation methods, among other sustainable procedures *(Zavišić et al., 2023)* (Table 30.4).

Postbiotics:

1. *Extracted Compounds:* Innovation is the step of separating certain bioactive substances that are produced during the fermentation of probiotics for usage in supplements, functional foods and medications.
2. *Health Benefits:* The motive of research and development is to pinpoint and optimize the health benefits of postbiotics, including their capability to modulate the immune system reduce inflammation and promote the integrity of the gut barrier.
3. *Incorporation into Products:* Consumer items, such as skincare, makeup, and nutritional supplements showed the postbiotic's potentiality. This goes beyond the capabilities of traditional probiotics.
4. *Regulation and Standardization:* In order to confirm uniformity and effectiveness in consumer goods, efforts are being made to standardize production processes and provide regulatory criteria for postbiotics.

In general, these developments are a result of a rising awareness of the role that gut health and the microbiome play in overall wellbeing, which is motivating scientists to keep researching and developing prebiotics, probiotics, and postbiotics in order to satisfy customer demand for products that are efficient and supported by evidence *(Pérez-Rivero & López-Gómez, 2023)* (Table 30.5).

30.8 MARKET GROWTH OF PREBIOTICS, PROBIOTICS, AND POSTBIOTICS IN INDIA

Current Market Size and Trends

Market Size:
As of 2022, the Indian probiotics market was valued at approximately USD 640 million and is projected to reach USD 1.5 billion by 2028, growing at a CAGR of around 15% *(India Probiotics Market, 2022–2027)*.

The prebiotic market, though smaller, is also growing rapidly, with a projected CAGR of 11.8% from 2021 to 2026 *(Prebiotic Ingredients Market, 2022–2027)*.

TABLE 30.5
Probiotics Innovations

Innovation	Description
Extracted compounds	Extraction of the bioactive substances created during the fermentation of probiotics for use in functional meals and supplements
Health benefits	Studies on the advantages such as immune system modulation, anti-inflammatory qualities, and support for the integrity of the intestinal barrier
Incorporation into products	Postbiotics can be used in skincare, makeup, and nutritional supplements to maximize their potential in addition to that of regular probiotics
Regulation and standardization	Attempts to create legal requirements and regulate industrial processes in order to guarantee effectiveness and uniformity

Product Segments:
Probiotic foods and beverages dominate the market, followed by dietary supplements *(Dietary Supplements in India | Market Research Report | Euromonitor, n.d.).*

Functional foods fortified with prebiotics are gaining traction, particularly in the dairy and bakery sectors *(India Probiotics Market, 2022–2027).*

Distribution Channels:
Pharmacies and drug stores remain the primary distribution channels, but online retail is growing rapidly, especially post-COVID-19 *(India Probiotics Market, 2022–2027).*

Consumer Demographics:
Urban areas account for the majority of sales, with increasing penetration in tier-II and tier-III cities *(Dietary Supplements in India | Market Research Report | Euromonitor, n.d.).*

Young adults and middle-aged consumers are the primary target demographics *(India Probiotics Market, 2022–2027).*

Emerging Trends:
Growing interest in synbiotic products combining prebiotics and probiotics *(Prebiotic Ingredients Market, 2022–2027).*

Increasing focus on condition-specific formulations, such as for digestive health, immunity, and women's health *(India Feed Probiotics Market Size & Share Analysis, n.d.).*

Factors Driving Growth

Several factors are contributing to the rapid growth of the prebiotic, probiotic, and postbiotic market in India:

Increasing Health Awareness:
Growing consumer awareness about gut health and its connection to overall well-being *(Nori et al., 2023).*

Rising interest in preventive healthcare and functional foods *(Krishnaraju et al., 2014).*

Urbanization and Lifestyle Changes:
Rapid urbanization is leading to changes in dietary habits and increased stress levels, driving demand for gut health products *(Asia Pacific Probiotics Market, 2023 - 2028; India Probiotics Market, 2022–2027).*

Rising Disposable Income:
Increasing middle-class population with higher disposable incomes willing to spend on health products *(India Probiotics Market, 2022–2027).*

Product Innovation:
Introduction of new product formats and flavors catering to Indian tastes and preferences (India Feed Probiotics Market Size & Share Analysis; D. Sharma et al., 2024).

Development of shelf-stable probiotic products suitable for India's climate *(Patel & Dupont, 2015).*

Government Initiatives:
Government efforts to promote nutraceuticals and functional foods as part of preventive healthcare strategies *(Krishnaraju et al., 2014).*

Digital Marketing and E-commerce:
Increased use of digital platforms for marketing and sales, enhancing product visibility and accessibility *(Prebiotic Ingredients Market, 2022–2027).*

Scientific Advancements:
Ongoing research demonstrating new health benefits of prebiotics, probiotics, and postbiotics, driving consumer interest *(G. Sharma & Im, 2018).*

Challenges to Market Expansion
Despite the positive growth outlook, several challenges could impact market expansion:

Regulatory Hurdles:
Complex and evolving regulatory landscape creates uncertainties for manufacturers *(Krishnaraju et al., 2014).*

Stringent requirements for health claim substantiation *(Arora & Baldi, 2015).*

Quality Control Issues:
Ensuring product stability and viability, especially in probiotic products, given India's diverse climate conditions *(Krishnaraju et al., 2014).*

Lack of standardized quality control measures across the industry *(Patel & Dupont, 2015)*.

Consumer Skepticism:
Lack of awareness and understanding about prebiotics, probiotics, and postbiotics among some consumer segments *(Bansal & Dhiman, 2019)*.

Skepticism about product efficacy due to unsubstantiated claims by some manufacturers *(Suez et al., 2019)*.

Price Sensitivity:
High price points of some probiotic and prebiotic products are limiting mass market penetration *(G. Sharma & Im, 2018)*.

Competition from Traditional Products:
Strong cultural preference for traditional fermented foods like curd and pickles, which are perceived as natural probiotics *(Sivaprakasam et al., 2016)*.

Supply Chain Challenges:
Maintaining cold chain for probiotic products, especially in smaller cities and rural areas *(Champagne et al., 2011)*.

Limited Local Research:
Insufficient research on probiotic strains specific to the Indian population and their health impacts *(Sivaprakasam et al., 2016)*.

Segmentation of the Market:
Many minor and medium-sized players are leading to market fragmentation and quality inconsistencies *(Asia Pacific Probiotics Market, 2023–2028; India Probiotics Market, 2022–2027)*.

In conclusion, while the Indian market for prebiotics, probiotics, and postbiotics shows promising growth potential, addressing these challenges will be crucial for sustainable market expansion and realizing the full potential of these products in improving public health in India.

30.9 DISTRIBUTION CHANNELS IN INDIA

Distribution methods of prebiotics, probiotics, and postbiotics differ based on the type of product and target market. An outline of the usual pathways of distribution for each category is provided below *(Davani-Davari et al., 2019; Swanson et al., 2020)*:

Prebiotics:

1. *Retail Stores:*
    - *Health Food Stores:* Prebiotic vitamins and functional meals are frequently found at specialized stores that concentrate on natural and organic items.
    - *Supermarkets:* Bigger chains are starting to carry more prebiotic-enhanced goods in their wellness and health departments.
    - *Pharmacies:* Prebiotic vitamins are available from several pharmacies as part of their health packages.
2. *Online Retailers:*
    - *E-commerce Platforms:* There is a variety of websites, including Amazon, health-focused e-commerce sites, and direct-to-consumer websites for supplement firms where the prebiotics are available.
    - *Subscription Services:* Prebiotics may be incorporated in the health supplement subscription boxes that certain businesses sell.
3. *Healthcare Providers:*
    - *Recommendations:* Doctors, including dietitians, nutritionists, and physicians, may advise their patients to take particular probiotic supplements.
4. *Direct Sales:*
    - *Company Websites:* Many producers of prebiotic supplements provide their products for direct customer purchase on their own websites, along with instructional materials.

Probiotics:

1. *Retail Stores:*
    - *Health Food Stores:* Probiotic pills and fortified meals can be found at retailers that specially concentrate on natural health items.
    - *Pharmacies and Drugstores:* Chain pharmacies and drugstores frequently stock a variety of probiotic supplements.
    - *Supermarkets:* Kefir and yoghurt with added probiotics are common in the health and wellness area of larger supermarket shops *(de Simone, 2019)*.
2. *Online Retailers:*
    - *E-commerce Platforms:* Like prebiotics, probiotics may also be found in large quantities on websites such as Amazon, health-related websites, and business websites.
    - *Subscription Services:* Some businesses use subscription models to provide probiotic supplements to customers on a monthly basis *(de Simone, 2019)*.
3. *Healthcare Providers:*
    - *Clinical Settings:* Healthcare practitioners, particularly gastroenterologists and nutritionists, may suggest probiotic supplements for particular medical issues.
4. *Direct Sales:*
    - *Company Websites:* Via their websites, several probiotic producers sell directly to customers, offering details and shopping choices *(Au et al., 2019)*.

Postbiotics:

1. *Specialized Health Stores:*
    - *Health Food Stores:* With increased awareness, certain health food stores could begin carrying postbiotic goods, especially those made from fermented vegetables *(Ż ółkiewicz et al., 2020).*
2. *Online Retailers:*
    - *E-commerce Platforms:* Postbiotics are becoming more widely accessible through internet channels, specialty health websites, and corporate websites, much like prebiotics and probiotics.
3. *Direct Sales and Healthcare Providers:*
    - *Direct-to-Consumer:* Through their own websites, certain businesses could provide postbiotic supplements for sale to customers directly.
    - *Healthcare Recommendations:* For immune system support and gut health, medical experts are suggesting particular postbiotic items.
4. *Functional Foods and Beverages:*
    - *Retail Channels:* Some fermented drinks or health bars which are postbiotic-enriched products, could be sold through health-conscious retail channels *(Ż ółkiewicz et al., 2020).*

These distribution channels, which range from conventional retail settings to cutting-edge e-commerce platforms and direct-to-consumer sales, are a reflection of the expanding consumer interest in gut health and the many ways that prebiotics, probiotics, and postbiotics are reaching the market.

30.10 CONCLUSION

1. *Regulatory Framework:*
    - Prebiotics, probiotics, and postbiotics are all covered under India's regulatory framework for dietary supplements, which is mostly in charge of the FSSAI.
    - Goods have to abide by FSSAI laws, which cover safety criteria, allowable ingredients, and labeling specifications.
2. *Specific Regulations:*
    - *Prebiotics and Probiotics:* In India, rules on food supplements apply to these goods. Manufacturers are responsible for making sure that microbiological content and health claims guidelines are met.
    - *Postbiotics:* Regulations unique to postbiotics may still be developing as a new category. Manufacturers could have to provide evidence of efficacy and safety in the form of scientific studies or clinical trials.
3. *Challenges:*
    - *Regulatory Clarity:* It can be difficult for producers and importers to ensure compliance with changing legislation and recommendations.
    - *Health Claims:* Strict laws governing health claims demand that they be supported by scientific data, which might demand a significant amount of resources.

Market Growth of Prebiotics, Probiotics, and Postbiotics in India (Asia Pacific Probiotics Market, n.d.):

1. *Growing Awareness:*
    - The market is expanding as a result of growing consumer knowledge of gut health and the importance of prebiotics, probiotics, and postbiotics for general well-being.
    - Consumer demand for natural and functional food products is growing among health-conscious customers.
2. *Diverse Product Offerings:*
    - Increased availability of products in a variety of forms, such as drinks, functional meals, and supplements, to accommodate a range of customer tastes.
    - Innovation in formulations and delivery methods to improve consumer appeal and efficacy.
3. *Market Expansion:*
    - The demand for prebiotics, probiotics, and maybe postbiotics in India is rising as a result of dietary changes, urbanization, and rising disposable income.
    - Growing rates of lifestyle illnesses and digestive issues support market expansion.
4. *Opportunities:* Manufacturers in India have the opportunity to create and launch safe, efficient goods that satisfy customer demand thanks to the country's regulatory environment.
5. *Challenges:* The two biggest obstacles still lie in proving health claims and adhering to changing rules.
6. *Future Outlook:* As consumer safety and industry standards are supported by evolving regulatory frameworks and growing awareness, it is anticipated that the prebiotic, probiotic, and postbiotic markets will continue to rise.

In summary, the Indian market for prebiotics, probiotics, and postbiotics is expected to rise due to rising consumer awareness and demand for products that promote gut health and general well-being, despite regulatory constraints.

REFERENCES

Arora, M., & Baldi, A. (2015). Regulatory categories of probiotics across the globe: A review representing existing and recommended categorization. *Indian Journal of Medical Microbiology, 33*, S2–S10. https://doi.org/10.4103/0255-0857.150868

Asia Pacific Probiotics Market - Growth, Trends, COVID-19 Impact, and Forecasts (2023 – 2028). (n.d.). Retrieved July 31, 2024, from https://www.researchandmarkets.com/reports/4773671/asia-pacific-probiotics-market-growth-trends

Au, C., Lin, S., Shao, W., Jin, S., Wen, Y., Singh, J., Basuki, V., Tanaka, H., Ji, G. E., Tee, E.-S., Tan, J. D., Chen, M.-J., Jirawongsy, M., Vinh, L. H., Gray, C., & Lee, Y.-K. (2019). Probiotics regulation in Asian and Australasian countries. In *Lactic Acid Bacteria* (pp. 631–682). CRC Press. https://doi.org/10.1201/9780429057465-38

Bansal, R., & Dhiman, A. (2019). Line of progression: Indian regulatory framework for nutraceuticals and dietary supplements. *Applied Clinical Research, Clinical Trials and Regulatory Affairs*, *6*(1), 46–61. https://doi.org/10.2174/2213476X06666190128150415

Brownawell, A. M., Caers, W., Gibson, G. R., Kendall, C. W. C., Lewis, K. D., Ringel, Y., & Slavin, J. L. (2012). Prebiotics and the health benefits of fiber: Current regulatory status, future research, and goals. *The Journal of Nutrition*, *142*(5), 962–974. https://doi.org/10.3945/JN.112.158147

Carlson, J. L., Erickson, J. M., Lloyd, B. B., & Slavin, J. L. (2018). Health effects and sources of prebiotic dietary fiber. *Current Developments in Nutrition*, *2*(3), nzy005. https://doi.org/10.1093/cdn/nzy005

Champagne, C. P., Ross, R. P., Saarela, M., Hansen, K. F., & Charalampopoulos, D. (2011). Recommendations for the viability assessment of probiotics as concentrated cultures and in food matrices. *International Journal of Food Microbiology*, *149*(3), 185–193. https://doi.org/10.1016/J.IJFOODMICRO.2011.07.005

Cho, Y. S., Han, K., Xu, J., & Moon, J. J. (2024). Novel strategies for modulating the gut microbiome for cancer therapy. *Advanced Drug Delivery Reviews*, *210*, 115332. https://doi.org/10.1016/J.ADDR.2024.115332

Chong, P. P., Chin, V. K., Looi, C. Y., Wong, W. F., Madhavan, P., & Yong, V. C. (2019). The microbiome and irritable bowel syndrome – A review on the pathophysiology, current research and future therapy. *Frontiers in Microbiology*, *10*(June), 424646. https://doi.org/10.3389/FMICB.2019.01136/BIBTEX

Chugh, P., Misra, S., Dhar, M. S., & Raghuwanshi, S. (2023). Regulatory aspects relevant to probiotic products. In *Probiotics, Prebiotics, Synbiotics, and Postbiotics* (pp. 513–534). Springer. https://doi.org/10.1007/978-981-99-1463-0_25

Cunningham, M., Azcarate-Peril, M. A., Barnard, A., Benoit, V., Grimaldi, R., Guyonnet, D., Holscher, H. D., Hunter, K., Manurung, S., Obis, D., Petrova, M. I., Steinert, R. E., Swanson, K. S., van Sinderen, D., Vulevic, J., & Gibson, G. R. (2021). Shaping the future of probiotics and prebiotics. *Trends in Microbiology*, *29*(8), 667–685. https://doi.org/10.1016/j.tim.2021.01.003

Davani-Davari, D., Negahdaripour, M., Karimzadeh, I., Seifan, M., Mohkam, M., Masoumi, S. J., Berenjian, A., & Ghasemi, Y. (2019). Prebiotics: Definition, types, sources, mechanisms, and clinical applications. *Foods*, *8*(3). https://doi.org/10.3390/FOODS8030092

de Simone, C. (2019). The unregulated probiotic market. *Clinical Gastroenterology and Hepatology*, *17*(5), 809–817. https://doi.org/10.1016/j.cgh.2018.01.018

Dietary Supplements in India | Market Research Report | Euromonitor. (n.d.). Retrieved July 31, 2024, from https://www.euromonitor.com/dietary-supplements-in-india/report

Dinić, M., Burgess, J. L., Lukić, J., Catanuto, P., Radojević, D., Marjanović, J., Verpile, R., Thaller, S. R., Gonzalez, T., Golić, N., Strahinić, I., Tomic-Canic, M., & Pastar, I. (2024). Postbiotic lactobacilli induce cutaneous antimicrobial response and restore the barrier to inhibit the intracellular invasion of Staphylococcus aureus in vitro and ex vivo. *The FASEB Journal*, *38*(14), e23801. https://doi.org/10.1096/FJ.202400054RR

Dronkers, T. M. G., Krist, L., Van Overveld, F. J., & Rijkers, G. T. (2018). The ascent of the blessed: regulatory issues on health effects and health claims for probiotics in Europe and the rest of the world. *Beneficial Microbes*, *9*(5), 717–724. https://doi.org/10.3920/BM2017.0196

Gibson, G. R., Hutkins, R., Sanders, M. E., Prescott, S. L., Reimer, R. A., Salminen, S. J., Scott, K., Stanton, C., Swanson, K. S., Cani, P. D., Verbeke, K., & Reid, G. (2017). Expert consensus document: The International Scientific Association for Probiotics and Prebiotics (ISAPP) consensus statement on the definition and scope of prebiotics. *Nature Reviews Gastroenterology & Hepatology*, *14*(8), 491–502. https://doi.org/10.1038/nrgastro.2017.75

Indian Council of Medical Research, Department of Biotechnology. (n.d.). *Guidelines for evaluation of probiotics in food* [Internet]. New Delhi: Indian Council of Medical Research [cited 2025 Sep 1]. https://dbtindia.gov.in/

India Feed Probiotics Market Size & Share Analysis – Industry Research Report – Growth Trends. (n.d.). Retrieved July 31, 2024, from https://www.mordorintelligence.com/industry-reports/india-feed-probiotics-market

India Probiotics Market: Industry Trends, Share, Size, Growth, Opportunity and Forecast 2022-2027. (n.d.). Retrieved July 31, 2024, from https://www.researchandmarkets.com/reports/5547010/india-probiotics-market-industry-trends-share

Ji, J., Jin, W., Liu, S. J., Jiao, Z., & Li, X. (2023). Probiotics, prebiotics, and postbiotics in health and disease. *MedComm*, *4*(6), e420. https://doi.org/10.1002/MCO2.420

Kabir, S. M. L., Islam, S. S., Tuhin-Al-Ferdous, & Akhter, A. H. M. T. (2023). Production, cost analysis, and marketing of probiotics. In *Food Microbiology Based Entrepreneurship: Making Money From Microbes* (pp. 305–326). https://doi.org/10.1007/978-981-19-5041-4_16

Kapustian, A., Cherno, N., Naumenko, K., Gural, L., & Osolina, S. (2023). REGULATION OF FUNCTIONAL FOODS IN UKRAINE AND THE WORLD. PROSPECTS FOR THE USE OF POSTBIOTICS AS FUNCTIONAL INGREDIENTS. *Food Science and Technology*, *17*(2). https://doi.org/10.15673/FST.V17I2.2641

Keservani, R. K., Sharma, A. K., Ahmad, F., & Baig, M. E. (2014). Nutraceutical and functional food regulations in India. In *Nutraceutical and Functional Food Regulations in the United States and Around the World: Second Edition* (pp. 327–342). https://doi.org/10.1016/B978-0-12-405870-5.00019-0

Krishnaraju, A. V., Bhupathiraju, K., Sengupta, K., & Golakoti, T. (2014). Regulations on nutraceuticals, functional foods and dietary supplements in India. In *Nutraceutical and Functional Food Regulations in the United States and Around the World: Second Edition* (pp. 343–362). https://doi.org/10.1016/B978-0-12-405870-5.00020-7

Lalasaheb Bodake, P., Mahajan, S. M., Sutar, N. G., Bairagi, S., Alam, S., Nainwal, P., Sadhula, A., & Yadav, S. (2024). Microbiome-modulating natural products: A promising strategy to address antibiotic resistance. *Sonia Yadav/Afr.J.Bio.Sc*, *6*(6), 5458–5472. https://doi.org/10.33472/AFJBS.6.6.2024.5458-5472

Latif, A., Shehzad, A., Niazi, S., Zahid, A., Ashraf, W., Iqbal, M. W., Rehman, A., Riaz, T., Aadil, R. M., Khan, I. M., Özogul, F., Rocha, J. M., Esatbeyoglu, T., & Korma, S.

A. (2023). Probiotics: Mechanism of action, health benefits and their application in food industries. *Frontiers in Microbiology*, *14*, 1216674. https://doi.org/10.3389/FMICB.2023.1216674/BIBTEX

Li, H.-Y., Zhou, D.-D., Gan, R.-Y., Huang, S.-Y., Zhao, C.-N., Shang, A., Xu, X.-Y., & Li, H.-B. (2021). Effects and mechanisms of probiotics, prebiotics, synbiotics, and postbiotics on metabolic diseases targeting gut microbiota: A narrative review. *Nutrients*, *13*(9), 3211. https://doi.org/10.3390/nu13093211

Lin, L., & Zhang, J. (2017). Role of intestinal microbiota and metabolites on gut homeostasis and human diseases. *BMC Immunology*, *18*(1), 2. https://doi.org/10.1186/s12865-016-0187-3

Ma, T., Shen, X., Shi, X., Sakandar, H. A., Quan, K., Li, Y., Jin, H., Kwok, L.-Y., Zhang, H., & Sun, Z. (2023). Targeting gut microbiota and metabolism as the major probiotic mechanism – An evidence-based review. *Trends in Food Science & Technology*, *138*, 178–198. https://doi.org/10.1016/j.tifs.2023.06.013

Misra, S., & Raghuwanshi, S. (2023). A survey of commercially available probiotics. In *Probiotics, Prebiotics, Synbiotics, and Postbiotics: Human Microbiome and Human Health* (pp. 489–512). https://doi.org/10.1007/978-981-99-1463-0_24

Monteiro, S. S., Schnorr, C. E., & Pasquali, M. A. de B. (2023). Paraprobiotics and postbiotics—Current state of scientific research and future trends toward the development of functional foods. *Foods*, *12*(12), 2394. https://doi.org/10.3390/FOODS12122394

Moradi, M., Kousheh, S. A., Almasi, H., Alizadeh, A., Guimarães, J. T., Yılmaz, N., & Lotfi, A. (2020). Postbiotics produced by lactic acid bacteria: The next frontier in food safety. *Comprehensive Reviews in Food Science and Food Safety*, *19*(6), 3390–3415. https://doi.org/10.1111/1541-4337.12613

Morelli, L. (2013). Probiotics: definition and taxonomy 10 years after the FAO/WHO guidelines. *Probiotic Bacteria and Their Effect on Human Health and Well-Being*, *107*, 1–8.

Nazhand, A., Souto, E. B., Lucarini, M., Souto, S. B., Durazzo, A., & Santini, A. (2020). Ready to use therapeutical beverages: Focus on functional beverages containing probiotics, prebiotics and synbiotics. *Beverages*, *6*(2), 26. https://doi.org/10.3390/BEVERAGES6020026

Nori, L. P., Sowmya, P., Madhusree, T., & Raju, K. V. (2023). Nutraceuticals: Regulatory landscape and market dynamics in the Indian scenario. *Journal of Young Pharmacists*, *15*(4), 595–603. https://doi.org/10.5530/jyp.2023.15.84

Patel, R., & Dupont, H. L. (2015). New approaches for bacteriotherapy: Prebiotics, new-generation probiotics, and synbiotics. *Clinical Infectious Diseases : An Official Publication of the Infectious Diseases Society of America*, *60 Suppl 2*(Suppl 2), S108–S121. https://doi.org/10.1093/CID/CIV177

Prebiotic Ingredients Market – Growth, Trends, COVID-19 Impact, and Forecasts (2022 - 2027). (n.d.). Retrieved July 31, 2024, from https://www.researchandmarkets.com/reports/5318623/prebiotic-ingredients-market-growth-trends#src-pos-30

Rad, A. H., Aghebati-Maleki, L., Kafil, H. S., Gilani, N., Abbasi, A., & Khani, N. (2021). Postbiotics, as dynamic biomolecules, and their promising role in promoting food safety. *Biointerface Research in Applied Chemistry*, *11*(6), 14529–14544. https://doi.org/10.33263/BRIAC116.1452914544

Ramazanidoroh, F., Hosseininezhad, M., Shahrampour, D., & Wu, X. (2023). Edible packaging as a functional carrier of prebiotics, probiotics, and postbiotics to boost food safety, quality, and shelf life. *Probiotics and Antimicrobial Proteins*, 1–21. https://doi.org/10.1007/S12602-023-10110-5/METRICS

Salminen, S., Collado, M. C., Endo, A., Hill, C., Lebeer, S., Quigley, E. M. M., Sanders, M. E., Shamir, R., Swann, J. R., Szajewska, H., & Vinderola, G. (2021). The International Scientific Association of Probiotics and Prebiotics (ISAPP) consensus statement on the definition and scope of postbiotics. *Nature Reviews Gastroenterology & Hepatology*, *18*(9), 649–667. https://doi.org/10.1038/s41575-021-00440-6

Sasso, J. M., Ammar, R. M., Tenchov, R., Lemmel, S., Kelber, O., Grieswelle, M., & Zhou, Q. A. (2023). Gut microbiome-brain alliance: A landscape view into mental and gastrointestinal health and disorders. *ACS Chemical Neuroscience*, *14*(10), 1717–1763. https://doi.org/10.1021/ACSCHEMNEURO.3C00127/ASSET/IMAGES/LARGE/CN3C00127_0017.JPEG

Scarpellini, E., Rinninella, E., Basilico, M., Colomier, E., Rasetti, C., Larussa, T., Santori, P., & Abenavoli, L. (2021). From pre- and probiotics to post-biotics: A narrative review. *International Journal of Environmental Research and Public Health*, *19*(1), 37. https://doi.org/10.3390/IJERPH19010037

Sehgal, S., Dhewa, T., Bansal, N.,& Thakur, M., (2016) Probiotic drugs and labeling practices in Indian market. *DU Journal of Undergraduate Research and Innovation*. *2*, 166–170.

Sharma, D., Rashid, G., & Sharma, L. (2024). Probiotics: Conclusion and future perspective. In *Probiotics* (pp. 325–343). https://doi.org/10.1201/9781003452249-15

Sharma, G., & Im, S. H. (2018). Probiotics as a potential immunomodulating pharmabiotics in allergic diseases: Current status and future prospects. *Allergy, Asthma & Immunology Research*, *10*(6), 575–590. https://doi.org/10.4168/AAIR.2018.10.6.575

Shetty, S., Shah, D., Goyal, G., Kathuria, N., Abraham, J., & Bhatia, V. (2014). A study to find the status of probiotics in New Delhi, India and review of strains of bacteria used as probiotics. *Journal of International Society of Preventive & Community Dentistry*, *4*(Suppl 1), S18. https://doi.org/10.4103/2231-0762.144570

Sivaprakasam, S., Prasad, P. D., & Singh, N. (2016). Benefits of short-chain fatty acids and their receptors in inflammation and carcinogenesis. *Pharmacology & Therapeutics*, *164*, 144–151. https://doi.org/10.1016/J.PHARMTHERA.2016.04.007

Soni, Y., Sangwan, P., Sharma, N., Ahirwar, P., & Kamboj, S. (2023). Regulatory status of probiotics in India: A current scenario. *EAS Journal of Pharmacy and Pharmacology*, *5*(3), 73–77. https://doi.org/10.36349/easjpp.2023.v05i03.005

Suez, J., Zmora, N., Segal, E., & Elinav, E. (2019). The pros, cons, and many unknowns of probiotics. *Nature Medicine*, *25*(5), 716–729. https://doi.org/10.1038/s41591-019-0439-x

Swanson, K. S., Gibson, G. R., Hutkins, R., Reimer, R. A., Reid, G., Verbeke, K., Scott, K. P., Holscher, H. D., Azad, M. B., Delzenne, N. M., & Sanders, M. E. (2020). The International Scientific Association for Probiotics and Prebiotics (ISAPP) consensus statement on the definition and scope of synbiotics. *Nature Reviews Gastroenterology & Hepatology*, *17*(11), 687–701. https://doi.org/10.1038/s41575-020-0344-2

Tomasik, P., & Tomasik, P. (2020). Probiotics, non-dairy prebiotics and postbiotics in nutrition. *Applied Sciences*, *10*(4), 1470. https://doi.org/10.3390/APP10041470

Warzée, J. P., Elli, M., Fall, A., Cattivelli, D., & François, J. Y. (2021). Supranational assessment of the quality of probiotics: Collaborative initiative between independent accredited testing laboratories. *Microorganisms*, *9*(7), 1456. https://doi.org/10.3390/MICROORGANISMS9071456

Wiegers, C., van de Burgwal, L. H. M., & Larsen, O. F. A. (2022). Probiotics for the management of infectious diseases: Reviewing the state of the art. *Frontiers in Microbiology*, *13*, 877142. https://doi.org/10.3389/FMICB.2022.877142/BIBTEX

Ye¸silyurt, N. Y., Yılmaz, B., Gagündüz, D. A. ˘, Capasso, R., & Apostolopoulos, V. (2021). Involvement of probiotics and postbiotics in the immune system modulation. *Biologics*, *1*(2), 89–110. https://doi.org/10.3390/BIOLOGICS1020006

Yunes, R. A., Poluektova, E. U., Belkina, T. V., & Danilenko, V. N. (2022). Lactobacilli: Legal regulation and prospects for new generation drugs. *Applied Biochemistry and Microbiology*, *58*(5), 652–664. https://doi.org/10.1134/S0003683822050179/FIGURES/3

Zavišić, G., Ristić, S., Petković, B., Živkov, Š. D., Jojić, N., & Janković, D. (2023). Microbiological quality of probiotic products. *Arhiv Za Farmaciju*, *73*(1), 17–34. https://doi.org/10.5937/ARHFARM73-42160

Zhou, P., Chen, C., Patil, S., & Dong, S. (2024). Unveiling the therapeutic symphony of probiotics, prebiotics, and postbiotics in gut-immune harmony. *Frontiers in Nutrition*, *11*, 1355542. https://doi.org/10.3389/FNUT.2024.1355542/TEXT

Żółkiewicz, J., Marzec, A., Ruszczyński, M., & Feleszko, W. (2020). Postbiotics: A step beyond pre- and probiotics. *Nutrients*, *12*(8), 1–17. https://doi.org/10.3390/NU12082189

Index

R

S

T

V

W